LINEAR MODEL THEORY

LINEAR MODEL THEORY

Univariate, Multivariate, and Mixed Models

Keith E. Muller
University of North Carolina
Department of Biostatistics
Chapel Hill, NC

Paul W. Stewart
University of North Carolina
Department of Biostatistics
Chapel Hill, NC

WILEY-INTERSCIENCE

A JOHN WILEY & SONS, INC., PUBLICATION

Published by John Wiley & Sons, Inc., Hoboken, New Jersey.
Published simultaneously in Canada.

For general information on our other products and services or for technical support, please contact our Customer Care Department within the United States at (800) 762-2974, outside the United States at (317) 572-3993 or fax (317) 572-4002.

Wiley also publishes its books in a variety of electronic formats. Some content that appears in print may not be available in electronic format. For information about Wiley products, visit our web site at www.wiley.com.

Library of Congress Cataloging-in-Publication Data:

Muller, Keith E.
Linear model theory : univariate, multivariate, and mixed models / Keith E. Muller, Paul W. Stewart.
p. cm.
Includes bibliographical references and index.
ISBN-10 0-471-21488-4
ISBN-13 978-0-471-21488-5
1. Linear models (Statistics). I. Stewart, Paul Wilder. II. Title.

QA279.M86 2006
519.5—dc22 2006044266

Contents

Preface **xiii**

1 Matrix Algebra for Linear Models **1**

1.1 Notation
1.2 Some Operators and Special Types of Matrices
1.3 Five Kinds of Multiplication
1.4 The Direct Sum
1.5 Rules of Operations
1.6 Other Special Types of Matrices
1.7 Quadratic and Bilinear Forms
1.8 Vector Spaces and Rank
1.9 Finding Rank
1.10 Determinants
1.11 The Inverse and Generalized Inverse
1.12 Eigenanalysis (Spectral Decomposition)
1.13 Some Factors of Symmetric Matrices
1.14 Singular Value Decomposition
1.15 Projections and Other Functions of a Design Matrix
1.16 Special Properties of Patterned Matrices
1.17 Function Optimization and Matrix Derivatives
1.18 Statistical Notation Involving Matrices
1.19 Statistical Formulas
1.20 Principal Components
1.21 Special Covariance Patterns

2 The General Linear Univariate Model **39**

2.1 Motivation
2.2 Model Concepts
2.3 The General Linear Univariate Linear Model
2.4 The Univariate General Linear Hypothesis
2.5 Tests about Variances
2.6 The Role of the Intercept
2.7 Population Correlation and Strength of Relationship
2.8 Statistical Estimates
2.9 Testing the General Linear Hypothesis
2.10 Confidence Regions for $\boldsymbol{\theta}$
2.11 Sufficient Statistics for the Univariate Model
Exercises

3 The General Linear Multivariate Model **55**

3.1 Motivation
3.2 Definition of the Multivariate Model
3.3 The Multivariate General Linear Hypothesis
3.4 Tests About Covariance Matrices
3.5 Population Correlation
3.6 Statistical Estimates
3.7 Overview of Testing Multivariate Hypotheses
3.8 Computing MULTIREP Tests
3.9 Computing UNIREP Tests
3.10 Confidence Regions for $\boldsymbol{\Theta}$
3.11 Sufficient Statistics for the Multivariate Model
3.12 Allowing Missing Data in the Multivariate Model
Exercises

4 Generalizations of the Multivariate Linear Model **79**

4.1 Motivation
4.2 The Generalized General Linear Univariate Model: Exact and Approximate Weighted Least Squares
4.3 Doubly Multivariate Models
4.4 Seemingly Unrelated Regressions
4.5 Growth Curve Models (GMANOVA)
4.6 The Relationship of the GCM to the Multivariate Model
4.7 Mixed, Hierarchical, and Related Models

5 The Linear Mixed Model **91**

5.1 Motivation
5.2 Definition of the Mixed Model
5.3 Distribution-Free and Noniterative Estimates
5.4 Gaussian Likelihood and Iterative Estimates
5.5 Tests about $\boldsymbol{\beta}$ (Means, Fixed Effects)
5.6 Tests of Covariance Parameters, $\boldsymbol{\tau}$ (Random Effects)
Exercises

6 Choosing the Form of a Linear Model for Analysis **101**

6.1 The Importance of Understanding Dependence
6.2 How Many Variables per Independent Sampling Unit?
6.3 What Types of Variables Play a Role?
6.4 What Repeated Sampling Scheme Was Used?
6.5 Analysis Strategies for Multivariate Data
6.6 Cautions and Recommendations
6.7 Review of Linear Model Notation

7 General Theory of Multivariate Distributions **115**

7.1 Motivation
7.2 Notation and Concepts
7.3 Families of Distributions
7.4 Cumulative Distribution Function
7.5 Probability Density Function
7.6 Formulas for Probabilities and Moments
7.7 Characteristic Function
7.8 Moment Generating Function
7.9 Cumulant Generating Function
7.10 Transforming Random Variables
7.11 Marginal Distributions
7.12 Independence of Random Vectors
7.13 Conditional Distributions
7.14 (Joint) Moments of Multivariate Distributions
7.15 Conditional Moments of Distributions
7.16 Special Considerations for Random Matrices

8 Scalar, Vector, and Matrix Gaussian Distributions **139**

8.1 Motivation
8.2 The Scalar Gaussian Distribution
8.3 The Vector ("Multivariate") Gaussian Distribution
8.4 Marginal Distributions
8.5 Independence
8.6 Conditional Distributions
8.7 Asymptotic Properties
8.8 The Matrix Gaussian Distribution
8.9 Assessing Multivariate Gaussian Distribution
8.10 Tests for Gaussian Distribution
Exercises

9 Univariate Quadratic Forms **169**

9.1 Motivation
9.2 Chi-Square Distributions
9.3 General Properties of Quadratic Forms
9.4 Properties of Quadratic Forms in Gaussian Vectors
9.5 Independence among Linear and Quadratic Forms
9.6 The ANOVA Theorem
9.7 Ratios Involving Quadratic Forms
Exercises

10 Multivariate Quadratic Forms **193**

10.1 The Wishart Distribution
10.2 The Characteristic Function of the Wishart
10.3 Properties of the Wishart
10.4 The Inverse Wishart
10.5 Related Distributions
Exercises

11 Estimation for Univariate and Weighted Linear Models **209**

11.1 Motivation
11.2 Statement of the Problem
11.3 (Unrestricted) Linearly Equivalent Linear Models
11.4 Estimability and Criteria for Checking It
11.5 Coding Schemes and the Essence Matrix
11.6 Unrestricted Maximum Likelihood Estimation of $\boldsymbol{\beta}$
11.7 Unrestricted BLUE Estimation of $\boldsymbol{\beta}$
11.8 Unrestricted Least Squares Estimation of $\boldsymbol{\beta}$

11.9 Unrestricted Maximum Likelihood Estimation of $\boldsymbol{\theta}$
11.10 Unrestricted BLUE Estimation of $\boldsymbol{\theta}$
11.11 Related Distributions
11.12 Formulations of Explicit Restrictions of $\boldsymbol{\beta}$ and $\boldsymbol{\theta}$
11.13 Restricted Estimation Via Equivalent Models
11.14 Fitting Piecewise Polynomial Models Via Splines
11.15 Estimation for the GGLM: Weighted Least Squares
Exercises

12 Estimation for Multivariate Linear Models **243**

12.1 Alternate Formulations of the Model
12.2 Estimability in the Multivariate GLM
12.3 Unrestricted Likelihood Estimation
12.4 Estimation of Secondary Parameters
12.5 Estimation with Multivariate Restrictions
12.6 Unrestricted Estimation With Compound Symmetry: the "Univariate" Approach to Repeated Measures
Exercises

13 Estimation for Generalizations of Multivariate Models **263**

13.1 Motivation
13.2 Criteria and Algorithms
13.3 Weighted Estimation of $\boldsymbol{B}$ and $\boldsymbol{\Sigma}$
13.4 Transformations among Growth Curve Designs
13.5 Within-Individual Design Matrices
13.6 Estimation Methods
13.7 Relationships to the Univariate and Mixed Models
Exercises

14 Estimation for Linear Mixed Models **279**

14.1 Motivation
14.2 Statement of the General Linear Mixed Model
14.3 Estimation and Estimability
14.4 Some Special Types of Models
14.5 ML Estimation
14.6 REML Estimation
14.7 Small-Sample Properties of Estimators
14.8 Large-Sample Properties of Variance Estimators
14.9 Conditional Estimation of $\boldsymbol{d}_i$ and BLUP Prediction
Exercises

15 Tests for Univariate Linear Models **289**

15.1 Motivation
15.2 Testability of Univariate Hypotheses
15.3 Tests of a Priori Hypotheses
15.4 Related Distributions
15.5 Transformations and Invariance Properties
15.6 Confidence Regions for $\boldsymbol{\theta}$
Exercises

16 Tests for Multivariate Linear Models **311**

16.1 Motivation
16.2 Testability of Multivariate Hypotheses
16.3 Tests of a Priori Hypotheses
16.4 Linear Invariance
16.5 Four Multivariate Test Statistics
16.6 Which Multivariate Test Is Best?
16.7 Univariate Approach to Repeated Measures: UNIREP
16.8 More on Invariance Properties
16.9 Tests of Hypotheses about $\boldsymbol{\Sigma}$
16.10 Confidence Regions for $\boldsymbol{\Theta}$
Exercises

17 Tests for Generalizations of Multivariate Linear Models **337**

17.1 Motivation
17.2 Doubly Multivariate Models
17.3 Missing Responses in Multivariate Linear Models
17.4 Exact and Approximate Weighted Least Squares
17.5 Seemingly Unrelated Regressions
17.6 Growth Curve Models (GMANOVA)
17.7 Testing Hypotheses in the GCM
17.8 Confidence Bands for Growth Curves

18 Tests for Linear Mixed Models **341**

18.1 Overview
18.2 Estimability of $\boldsymbol{\theta} = \boldsymbol{C\beta}$
18.3 Likelihood Ratio Tests of $\boldsymbol{C\beta}$
18.4 Likelihood Ratio Tests Involving $\boldsymbol{\tau}$
18.5 Test Size of Wald-Type Tests of $\boldsymbol{\beta}$ Using REML
18.6 Using Wald-Type Tests of $\boldsymbol{\beta}$ with REML
18.7 Using Wald-Type Tests of $\{\boldsymbol{\beta}, \boldsymbol{\tau}\}$ with REML

19 A Review of Multivariate and Univariate Linear Models **349**

19.1 Matrix Gaussian and Wishart Properties
19.2 Design Matrix Properties
19.3 Model Components
19.4 Primary Parameter and Related Estimators
19.5 Secondary Parameter Estimators
19.6 Added-Last and Added-in-Order Tests

20 Sample Size for Univariate Linear Models **361**

20.1 Sample Size Consulting: Before You Begin
20.2 The Machinery of a Power Analysis
20.3 Independent t Example
20.4 Paired t Example
20.5 The Impact of Using $\widehat{\sigma}^2$ or $\widehat{\beta}$ in Power Analysis
20.6 Random Predictors
20.7 Internal Pilot Designs
20.8 Other Criteria for Choosing a Sample Size
Exercises

21 Sample Size for Multivariate Linear Models **371**

21.1 The Machinery of a Power Analysis
21.2 Paired t Example
21.3 Time by Treatment Example
21.4 Comparing between and within Designs
21.5 Some Invariance Properties
21.6 Random Predictors
21.7 Internal Pilot Designs
Exercises

22 Sample Size for Generalizations of Multivariate Models **383**

22.1 Motivation
22.2 Sample Size Methods for Growth Curve Models

23 Sample Size for Linear Mixed Models **385**

23.1 Motivation
23.2 Methods
23.3 Internal Pilot Designs

Appendix: Computing Resources 387

References 393

Index 405

Preface

MOTIVATION FOR THE BOOK

Statisticians often use linear models for data analysis and for developing new statistical methods. Success in either endeavor requires a solid understanding of the underlying theory. Historically, univariate, multivariate, and mixed linear models have been discussed separately. In contrast, we give a *unified* treatment in order to make clear the distinctions among the three classes of models. No single model class proves uniformly best. Therefore choosing the best approach requires a detailed knowledge of the differences and similarities.

A student needs to acquire four sets of skills. (1) Using all three classes of linear models correctly in practice requires knowing enough theory, especially a deep understanding of assumptions and their possible violations. However, we leave detailed discussion of diagnostics to others. (2) Correct use also requires knowing when to choose one type over another. (3) Understanding the theory helps guide when *not* to use any of the models. (4) Finally, developing new methods requires a detailed knowledge of known work.

TOPICS COVERED

We focus on linear models of interval scale responses with finite second moments, especially models with correlated observations and Gaussian errors. Such correlations always create additional complexity. In contrast to most "multivariate" books, most classical techniques, including cluster analysis, factor analysis, discriminant analysis, and canonical correlation, receive little attention.

Meeting our goals in a book appropriate for a one-semester course required omitting many worthwhile topics. Even so, the book includes more material than usually covered entirely in a four-credit class. On the other hand, it may seem wasteful to include separate and overlapping treatments of univariate, multivariate, and mixed models. However, our students have adamantly preferred the current organization. We accept some duplication for the sake of clarity.

NOTATION

We sought a precise but accessible presentation of the theory underlying practice, illustrated with examples. Fairness to our students and readers required creating compatible univariate, multivariate, and mixed model notation. Many authors describe a univariate model in terms of a single observation. Others describe univariate and multivariate models in matrix notation for all observations and all independent sampling units. In contrast, mixed model discussions typically describe the observations for only one independent sampling unit. Table 6.5 serves as a Rosetta stone by allowing translations between models and publications. Discussing all three forms of all models clarifies differences and helps avoid many sources of confusion.

PREREQUISITES

We developed the book for a four-hour class required of our doctoral students. Most have the equivalent of a Master's Degree in Biostatistics, including two semesters of probability and inference, at the level of Wackerly, Mendenhall, and Scheaffer (1996). Students also need a solid background in applied univariate linear models from a matrix perspective, as in Muller and Fetterman (2002).

Most explanations and proofs use matrix algebra. Rank, basis space, eigenanalysis, the singular value decomposition, and generalized inverses play central roles. Although we summarize key matrix results in Chapter 1, the reader without sufficient background should invest time in studying one of the many good matrix theory books, such as Schott (2005). Section 1.15 merits special and repeated attention in understanding and proving many results in estimation and inference. Prerequisite material in sections 1.11-1.14 also may require study.

We use the most basic properties of complex arithmetic to allow simplifying some proofs by using characteristic functions. Measure theoretic and contour integration methods are mentioned only occasionally, and not required.

ACKNOWLEDGMENTS

We owe a great debt to the authors of many earlier books about linear models. We give explicit citations for particular results and also for conceptual approaches. We wrote the book while the first author was a Professor in the Department of Biostatistics at the University of North Carolina Hill.

Many colleagues helped us along the way. We especially thank Ronald W. Helms, Lloyd J. Edwards, and Christopher S. Coffey for their guidance and friendship. Teaching assistants Stacey Major, William K. Pan, Hae-Young Kim, and J. (Chris) Slaughter, as well as a number of graders, helped shape the organization and choice of topics by providing a student's perspective. Our hard-working and enthusiastic students motivated us to start the book and kept us going to the finish. As in all such endeavors, the book would not exist without the love, support, and forbearance of our families, especially our wives, Sally and Dawn.

CHAPTER 1

Matrix Algebra for Linear Models

1.1 NOTATION

Graybill (1969), Searle (1982), Harville (1996), and Schott (2005) provided thorough introductions to matrix algebra for statistics. We summarize only key results here. Substantial omissions include the deletion of nearly all proofs, as well as consideration of more general forms not commonly used in statistics. Also, many issues of numerical accuracy have been ignored. Some of the formulas described here, although very useful for understanding concepts, prove numerically unstable with typical computer precision.

Braces, { }, indicate sets and *brackets*, [ ], indicate matrices or vectors (arrays). In a distinct use of the same symbols, mathematical expressions will be grouped by using the nesting sequence {[()]}, which may be iterated as {[({[()]})]}.

> **Definition 1.1 (a)** A *matrix* is a rectangular, two-dimensional array of *elements*. Writing $\boldsymbol{A} = \{a_{ij}\}$ says $\boldsymbol{A}$ is the matrix with element a_{ij} at row i and column j. Here i is the row index, while j is the column index, which are always written in row-column order.
>
> **(b)** A *vector* is any matrix with exactly one column, such as $\boldsymbol{v} = \begin{bmatrix} 7 \\ 8 \end{bmatrix}$.
>
> **(c)** A *scalar*, such as $s = 6$, can be expressed as a vector with one row or as a matrix with one row and one column, written $s = \boldsymbol{s} = \boldsymbol{S}$.

We restrict attention to real numbers and finite dimensions. With r rows, c columns, $\boldsymbol{A}$ is $r \times c$ (r by c):

$$\boldsymbol{A} = \begin{bmatrix} a_{11} & a_{12} & \cdots & a_{1c} \\ a_{21} & & & \vdots \\ \vdots & & & \vdots \\ a_{r1} & \cdots & \cdots & a_{rc} \end{bmatrix}. \tag{1.1}$$

In turn, $\langle \boldsymbol{A} \rangle_{ij} = a_{ij}$ indicates element i, j has been extracted from $\boldsymbol{A}$.

Although not all authors do so, we are scrupulous about the distinction between a matrix of one row, such as $\boldsymbol{A} = [\, a_1 \;\; a_2 \,]$, and a vector, $\boldsymbol{b} = \begin{bmatrix} a_1 \\ a_1 \end{bmatrix}$. The vector $\boldsymbol{b}$

can also be written in terms of a transpose (defined in the next section), $\boldsymbol{b} = \boldsymbol{A}'$. Doing so not only avoids notational ambiguity, but also builds in many consistency checks by requiring dimensions and symbol types ($\boldsymbol{a}$ or $\boldsymbol{A}$) to align.

As an aid to working with matrices, we always use bold typeface in word processing software, as in the present book. Most, but not all, statistical journals require the convention. With the convention, a represents a scalar, $\boldsymbol{a}$ represents a vector, and $\boldsymbol{A}$ represents a matrix. When handwriting expressions, we highly recommend always putting a tilde or dash under any matrix or vector to indicate boldface. It is *extremely* helpful to write the dimensions of each matrix in an equation underneath the equation. Using the transpose, matrix multiplication, and inverse operators (introduced later in the chapter) illustrates the idea, with

$$\widehat{\boldsymbol{\beta}} = (\boldsymbol{X}'\boldsymbol{X})^{-1}\boldsymbol{X}'\boldsymbol{y} \tag{1.2}$$

being much less informative than

$$\underset{p\times 1}{\widehat{\boldsymbol{\beta}}} = \underset{[(p\times n)(n\times p)](p\times n)(n\times 1)}{(\boldsymbol{X}'\boldsymbol{X})^{-1}\boldsymbol{X}'\boldsymbol{y}}\,. \tag{1.3}$$

The practice saves a great deal of time (otherwise spent being confused). More bluntly, if one does not know the dimensions, one cannot understand the equation.

We reserve superscripts for operators and use subscripts for descriptors, such as in x, x^2, x_2. Often, functional notation, such as $x(c, \alpha)$, provides a better alternative than a long and elaborate subscript descriptor.

1.2 SOME OPERATORS AND SPECIAL TYPES OF MATRICES

Definition 1.2 (a) A *square* matrix has the same number of rows as columns.
(b) For $\boldsymbol{A}$ $(r \times c)$, the (main) *diagonal* of $\boldsymbol{A}$ is $\{a_{11}, a_{22}, \ldots, a_{cc}\}$.
(c) A square matrix is *diagonal* if all elements off the main diagonal are zero; if $i \neq j$, then $a_{ij} = 0$, while a_{ii} can be anything.

Definition 1.3 Writing $\mathrm{Dg}(\boldsymbol{v}) = \mathrm{Dg}(\{v_j\})$ indicates creating a square diagonal matrix from a vector or elements of a set, as in

$$\mathrm{Dg}\left(\begin{bmatrix}1\\2\\3\end{bmatrix}\right) = \begin{bmatrix}1&0&0\\0&2&0\\0&0&3\end{bmatrix}. \tag{1.4}$$

Definition 1.4 A prime indicates *transpose*. If $\boldsymbol{A} = \{a_{ij}\}$ is $r \times c$, then $\boldsymbol{A}' = \{a_{ji}\}$ is $c \times r$. The transpose operation causes rows to become columns and columns to become rows.

Definition 1.5 A *symmetric* matrix is a square matrix such that $a_{ij} = a_{ji}$. Equivalently, $\boldsymbol{A}' = \boldsymbol{A}$.

As mentioned earlier, we use the term "vector" only for $n \times 1$ arrays and never for $1 \times n$ arrays. A $1 \times n$ array will always be written as a matrix, such as $\boldsymbol{A} = [1\ 2\ 3]$, or as a transposed vector, $\boldsymbol{b}' = [1\ 2\ 3]$. Here $\boldsymbol{A} = \boldsymbol{b}'$ and $\boldsymbol{A}' = \boldsymbol{b}$.

Definition 1.6 An *identity* matrix, $\boldsymbol{I}$ or $\boldsymbol{I}_n$, is a square matrix with all 1's on the main diagonal, and all 0's off-diagonal. Equivalently, $a_{ij} = 0$ if $i \neq j$ and $a_{ij} = 1$ if $i = j$.

Definition 1.7 A *zero* matrix, $\mathbf{0}$, has $a_{ij} \equiv 0$ and may be written $\mathbf{0}_r$ to indicate a vector or $\mathbf{0}_{r \times c}$ to indicate a matrix for clarity.

Similarly an $n \times 1$ vector with all elements 1 is written $\mathbf{1}$ or $\mathbf{1}_n = [1\ 1\ \cdots\ 1]'$. Also, $\mathbf{1}_n \mathbf{1}_n'$ is an $n \times n$ matrix of all 1's.

Definition 1.8 An *upper triangular* matrix has $a_{ij} = 0$ for $i > j$, such as

$$\boldsymbol{U} = \begin{bmatrix} 1 & 5 & 6 \\ 0 & 2 & 4 \\ 0 & 0 & 3 \end{bmatrix}. \tag{1.5}$$

A *lower triangular* matrix has $a_{ij} = 0$ for $i < j$:

$$\boldsymbol{L} = \begin{bmatrix} 1 & 0 & 0 \\ 7 & 2 & 0 \\ 8 & 9 & 3 \end{bmatrix}. \tag{1.6}$$

Definition 1.9 A *partitioned* matrix (supermatrix) has elements grouped meaningfully by combinations of vertical and horizontal slicing, indicated $\boldsymbol{A} = \{\boldsymbol{A}_{jk}\}$. Necessarily $\boldsymbol{A}_{jk}$ and $\boldsymbol{A}_{jk'}$ have the same number of rows, while $\boldsymbol{A}_{jk}$ and $\boldsymbol{A}_{j'k}$ have the same number of columns.

Definition 1.10 A *block diagonal* matrix is a partitioned matrix with all partitions zero except possibly $\{\boldsymbol{A}_{jj}\}$.

Two examples are the block diagonal matrix

$$\boldsymbol{A} = \left[\begin{array}{cc:cc} 1 & 2 & 0 & 0 \\ 3 & 4 & 0 & 0 \\ \hdashline 0 & 0 & 5 & 6 \\ 0 & 0 & 7 & 8 \end{array}\right] \tag{1.7}$$

and the general partitioned matrix

$$B = \begin{bmatrix} 1 & \vdots & 0 & 0 \\ 1 & \vdots & 0 & 0 \\ 2 & \vdots & 0 & 0 \\ 3 & \vdots & 0 & 0 \end{bmatrix}. \tag{1.8}$$

Conformation of shapes requires consistent partitioning. If

$$A = \begin{bmatrix} B & C & D \\ E & F & G \end{bmatrix} \tag{1.9}$$

then B, C, and D have the same number of rows, while B and E have the same number of columns, etc. However, complete uniformity of dimensions is not required (the number of rows of B need not equal the number of rows of E). It would be hard to overemphasize the value of partitioned matrices in deriving algebraic and statistical properties for linear models. Expressions can often be greatly simplified by taking advantage of special properties of partitioned matrices for basic operations (matrix summation, multiplication, etc.) and more complicated operations (determinants, inverses, etc.).

Definition 1.11 For $r \times c$ A, writing $\text{col}_k(A) = a_k$ indicates extracting $r \times 1$ column j from A. Writing $A_j = \text{row}_j(A)$ indicates extracting a particular $1 \times c$ row.

As an important example of partitioning, $r \times c$ A can be expressed in terms of its c column vectors, $\{a_j\}$, with a_j of dimension $r \times 1$, or its r rows, $\{A_k\}$, with A_k of dimension $c \times 1$. In summary,

$$\begin{aligned} A &= [\, a_1 \; a_2 \; \cdots \; a_c \,] \\ &= \begin{bmatrix} A_1 \\ A_2 \\ \vdots \\ A_r \end{bmatrix}. \end{aligned} \tag{1.10}$$

Definition 1.12 (a) Writing $A = [\, a_1 \; a_2 \cdots a_c \,]$, which requires $\{a_j\}$ to be $r \times 1$, indicates $\{a_j\}$ have been *horizontally concatenated.*

(b) Writing $A = \begin{bmatrix} A_1 \\ \vdots \\ A_r \end{bmatrix}$, which requires $\{A_k\}$ to be $c \times 1$, indicates $\{A_k\}$ have been *vertically concatenated.*

Definition 1.13 Writing vec( ) indicates all elements of a matrix have been stacked *by column*, as in $b_1 = \text{vec}(A)$, because it creates an $(rc) \times 1$ vector from an $r \times c$ matrix. Equivalently, the columns have been vertically concatenated.

If $\boldsymbol{A} = [\,\boldsymbol{a}_1 \;\; \boldsymbol{a}_2 \;\; \cdots \;\; \boldsymbol{a}_c\,]$ then

$$\text{vec}(\boldsymbol{A}) = \begin{bmatrix} \boldsymbol{a}_1 \\ \boldsymbol{a}_2 \\ \vdots \\ \boldsymbol{a}_c \end{bmatrix}. \tag{1.11}$$

In turn, $\boldsymbol{b}_2 = \text{vec}(\boldsymbol{A}')$ also creates an $(rc) \times 1$ vector which differs from $\boldsymbol{b}_1$ only by permutation of the rows. Creating $\text{vec}(\boldsymbol{A}')$ stacks the matrix by rows:

$$\text{vec}(\boldsymbol{A}') = \begin{bmatrix} \boldsymbol{A}'_1 \\ \boldsymbol{A}'_2 \\ \vdots \\ \boldsymbol{A}'_r \end{bmatrix}. \tag{1.12}$$

If $r = c$ and $\boldsymbol{A} = \boldsymbol{A}'$ then only $r(r+1)/2$ elements are distinct. The r^2 elements are not functionally independent. The vech( ) operator extracts the distinct elements into a vector:

$$\boldsymbol{z} = \text{vech}\left(\begin{bmatrix} a & b & c \\ b & d & e \\ c & e & f \end{bmatrix}\right) = \begin{bmatrix} a \\ b \\ c \\ d \\ e \\ f \end{bmatrix}. \tag{1.13}$$

Definition 1.14 The *trace* of an $n \times n$ (square) matrix is $\text{tr}(\boldsymbol{A}) = \sum_{i=1}^{n} a_{ii}$.

Definition 1.15 Matrices *conform* for an operation if their sizes allow the result of the operation to exist. Matrices do not conform for an operation if their dimensions do not allow the desired operation.

Definition 1.16 *Matrix addition* yields $\boldsymbol{A} + \boldsymbol{B} = \{a_{ij} + b_{ij}\}$ while *matrix subtraction* yields $\boldsymbol{A} - \boldsymbol{B} = \{a_{ij} - b_{ij}\}$. Either result exists only if $\boldsymbol{A}$ and $\boldsymbol{B}$ are the same size (and thereby conform for the operation).

1.3 FIVE KINDS OF MULTIPLICATION

Definition 1.17 *Scalar* multiplication of a matrix gives $\boldsymbol{A}b = b\boldsymbol{A} = \{ba_{ij}\}$.

Definition 1.18 If $\boldsymbol{A}$ and $\boldsymbol{B}$ are both $r \times c$, then *elementwise multiplication* gives $\boldsymbol{A}\#\boldsymbol{B} = \{a_{ij}b_{ij}\} = \boldsymbol{C}$, with $\boldsymbol{C}$ also $r \times c$.

Definition 1.19 (a) *Matrix multiplication* of $\boldsymbol{A}$ $(r \times c)$ and $\boldsymbol{B}$ $(c \times d)$ gives $\boldsymbol{AB} = \boldsymbol{C} = \{c_{jk}\}$ for $c_{jk} = \sum_{m=1}^{c} a_{km} b_{mk}$, with $\boldsymbol{C}$ $r \times d$.
(b) If $\boldsymbol{A}$ is $r \times r$, the *matrix power* of $\boldsymbol{A}$ is $\boldsymbol{A}^k = \boldsymbol{A}_1 \boldsymbol{A}_2 \cdots \boldsymbol{A}_k$, with $\boldsymbol{A}_j = \boldsymbol{A}$.

Definition 1.20 (a) The *cross product* of $r \times 1$ $\boldsymbol{a}$ and $\boldsymbol{b}$ is $\boldsymbol{a}'\boldsymbol{b} = \sum_{k=1}^{r} a_k b_k$.
(b) The *dot product* of $r \times 1$ $\boldsymbol{a}$ and $\boldsymbol{b}$ is $\boldsymbol{a} \bullet \boldsymbol{b} = \cos(\theta)\sqrt{\boldsymbol{a}'\boldsymbol{a}\boldsymbol{b}'\boldsymbol{b}}$, with θ the angle between the two vectors. Necessarily $\boldsymbol{a}'\boldsymbol{b} = 0$ if and only if $\theta = 90°$.
(c) If $\boldsymbol{a}$ is $r \times 1$, then $\boldsymbol{a}'\boldsymbol{a} = \sum_{k=1}^{r} a_k^2$ is the *inner product* of the vector.
(d) If $\boldsymbol{a}$ is $r \times 1$, then $\boldsymbol{a}\boldsymbol{a}' = \{a_i a_j\}$ is the *outer product* of the vector.

Matrix multiplication can be expressed as a collection of cross products. Multiplying row j of $\boldsymbol{A}$ with column j of $\boldsymbol{B}$ yields $c_{jk} = \{\text{row}_j(\boldsymbol{A})\text{col}_k(\boldsymbol{B})\}$:

$$\overset{\text{row}}{\longrightarrow}\qquad \begin{bmatrix} \rightarrow \rightarrow \rightarrow \\ \\ \\ \end{bmatrix} \begin{bmatrix} \downarrow & \\ \downarrow & \\ \downarrow & \end{bmatrix} \downarrow \text{column}\,.$$

Lemma 1.1 (a) Premultiplying by a (square) diagonal matrix scales the rows, and postmultiplying by a (square) diagonal matrix scales the columns.
(b) The result generalizes to partitioned matrices with conforming partitions.

In particular, for conforming matrices $\boldsymbol{IA} \equiv \boldsymbol{A}$ and $\boldsymbol{AI} \equiv \boldsymbol{A}$. For a 2×3 matrix, $\boldsymbol{A}$, multiplication by diagonal matrices gives

$$\boldsymbol{GA} = \boldsymbol{GAI} = \begin{bmatrix} g_1 & 0 \\ 0 & g_2 \end{bmatrix} \begin{bmatrix} a & b & c \\ d & e & f \end{bmatrix} \begin{bmatrix} 1 & 0 & 0 \\ 0 & 1 & 0 \\ 0 & 0 & 1 \end{bmatrix} = \begin{bmatrix} g_1 a & g_1 b & g_1 c \\ g_2 d & g_2 e & g_2 f \end{bmatrix} \tag{1.14}$$

$$\boldsymbol{AH} = \boldsymbol{IAH} = \begin{bmatrix} 1 & 0 \\ 0 & 1 \end{bmatrix} \begin{bmatrix} a & b & c \\ d & e & f \end{bmatrix} \begin{bmatrix} h_1 & 0 & 0 \\ 0 & h_2 & 0 \\ 0 & 0 & h_3 \end{bmatrix} = \begin{bmatrix} ah_1 & bh_2 & ch_3 \\ dh_1 & eh_2 & fh_3 \end{bmatrix}. \tag{1.15}$$

Definition 1.21 The *horizontal direct product* creates a new matrix by elementwise multiplication of pairs of columns from two matrices with the same number of rows:

$$\boldsymbol{A} \odot \boldsymbol{B} = \begin{bmatrix} a & b \\ c & d \\ e & f \end{bmatrix} \odot \begin{bmatrix} r & s & t \\ u & v & w \\ x & y & z \end{bmatrix}$$
$$= \begin{bmatrix} ar & as & at & br & bs & bt \\ cu & cv & cw & du & dv & dw \\ ex & ey & ez & fx & fy & fz \end{bmatrix}. \qquad (1.16)$$

To operate on rows (1) transpose each operand, (2) compute the product, and (3) transpose the result, as with $(\boldsymbol{A}' \odot \boldsymbol{B}')'$. The operator could be called the vertical or column direct product.

Definition 1.22 The *direct* (or Kronecker) product is

$$\boldsymbol{A} \otimes \boldsymbol{B} = \{a_{ij}\boldsymbol{B}\}$$
$$= \begin{bmatrix} a_{11}\boldsymbol{B} & \cdots & a_{1c}\boldsymbol{B} \\ a_{21}\boldsymbol{B} & \ddots & \vdots \\ \vdots & & \vdots \\ a_{r1}\boldsymbol{B} & \cdots & a_{rc}\boldsymbol{B} \end{bmatrix}. \qquad (1.17)$$

With $r \times c$ $\boldsymbol{A}$ and $s \times d$ $\boldsymbol{B}$, the result has dimension $(rs) \times (cd)$ = (rows × columns). Some authors choose to define $\{\boldsymbol{A}b_{ij}\}$ as the direct product, which produces a different matrix.

1.4 THE DIRECT SUM

Definition 1.23 The *direct sum* operator creates a block diagonal matrix from any set of square matrices:

$$\bigoplus_{j=1}^{J} \boldsymbol{A}_j = \boldsymbol{A}_1 \oplus \boldsymbol{A}_2 \oplus \cdots \oplus \boldsymbol{A}_J$$
$$= \begin{bmatrix} \boldsymbol{A}_1 & \mathbf{0} & \cdots & \mathbf{0} \\ \mathbf{0} & \boldsymbol{A}_2 & \mathbf{0} & \vdots \\ \vdots & \mathbf{0} & \ddots & \mathbf{0} \\ \mathbf{0} & \cdots & \mathbf{0} & \boldsymbol{A}_J \end{bmatrix}. \qquad (1.18)$$

Lemma 1.2 In general $\mathrm{tr}(\boldsymbol{A}_1 \oplus \boldsymbol{A}_2) = \mathrm{tr}(\boldsymbol{A}_1) + \mathrm{tr}(\boldsymbol{A}_2)$. If $\boldsymbol{A}_j$ and $\boldsymbol{B}_j$ are both $n_j \times n_j$, then

$$\left(\bigoplus_{j=1}^{J} \boldsymbol{A}_j\right) + \left(\bigoplus_{j=1}^{J} \boldsymbol{B}_j\right) = \bigoplus_{j=1}^{J} (\boldsymbol{A}_j + \boldsymbol{B}_j) \tag{1.19}$$

$$\left(\bigoplus_{j=1}^{J} \boldsymbol{A}_j\right)\left(\bigoplus_{j=1}^{J} \boldsymbol{B}_j\right) = \bigoplus_{j=1}^{J} (\boldsymbol{A}_j \boldsymbol{B}_j) . \tag{1.20}$$

Any direct product of the form $\boldsymbol{I} \otimes \boldsymbol{B}$ is a special case of the direct sum:

$$\boldsymbol{I}_J \otimes \boldsymbol{B} = \bigoplus_{j=1}^{J} \boldsymbol{B} . \tag{1.21}$$

Direct products including an identity matrix, $\boldsymbol{A} \otimes \boldsymbol{I}$ or $\boldsymbol{I} \otimes \boldsymbol{B}$, occur often in expressions for covariance matrices of data in clusters of fixed size. A common form occurs in describing the covariance matrix of data observed in N clusters of constant size, with homogeneity of covariance between clusters:

$$\begin{aligned} \Xi &= \boldsymbol{I}_N \otimes \Sigma \\ &= \begin{bmatrix} \Sigma & \mathbf{0} & \cdots & \mathbf{0} \\ \mathbf{0} & \Sigma & \mathbf{0} & \vdots \\ \vdots & \mathbf{0} & \ddots & \mathbf{0} \\ \mathbf{0} & \cdots & \mathbf{0} & \Sigma \end{bmatrix} . \end{aligned} \tag{1.22}$$

If the dimension or elements of Σ_i vary with i, then Ξ cannot be written as a direct product. The direct sum allows writing

$$\begin{aligned} \Xi &= \bigoplus_{i=1}^{N} \Sigma_i \\ &= \begin{bmatrix} \Sigma_1 & \mathbf{0} & \cdots & \mathbf{0} \\ \mathbf{0} & \Sigma_2 & \mathbf{0} & \vdots \\ \vdots & \mathbf{0} & \ddots & \mathbf{0} \\ \mathbf{0} & \cdots & \mathbf{0} & \Sigma_N \end{bmatrix} . \end{aligned} \tag{1.23}$$

1.5 RULES OF OPERATIONS

Unless otherwise specified, we assume $\boldsymbol{A}$ and $\boldsymbol{B}$ conform for the operations in question. Without additional knowledge of the matrices involved, the following are true.

Theorem 1.1 Some operations obey *commutative laws*:

$$\begin{aligned} \boldsymbol{A}+\boldsymbol{B} &= \boldsymbol{B}+\boldsymbol{A} && (1.24)\\ \boldsymbol{A}-\boldsymbol{B} &= \boldsymbol{B}-\boldsymbol{A} && (1.25)\\ \boldsymbol{A}\,\#\,\boldsymbol{B} &= \boldsymbol{B}\,\#\boldsymbol{A} && (1.26)\\ a\boldsymbol{B} &= \boldsymbol{B}a\,. && (1.27) \end{aligned}$$

It is important to recognize that $\boldsymbol{AB} \neq \boldsymbol{BA}$ and $\boldsymbol{A} \otimes \boldsymbol{B} \neq \boldsymbol{B} \otimes \boldsymbol{A}$, except in special cases.

Theorem 1.2 Some operations obey *associative laws*:

$$\begin{aligned} (\boldsymbol{A}+\boldsymbol{B})+\boldsymbol{C} &= \boldsymbol{A}+(\boldsymbol{B}+\boldsymbol{C}) && (1.28)\\ (\boldsymbol{AB})\boldsymbol{C} &= \boldsymbol{A}(\boldsymbol{BC}) && (1.29)\\ (\boldsymbol{A}\otimes\boldsymbol{B})\otimes\boldsymbol{C} &= \boldsymbol{A}\otimes(\boldsymbol{B}\otimes\boldsymbol{C}) && (1.30)\\ (a\boldsymbol{B})\otimes(c\boldsymbol{D}) &= ac(\boldsymbol{B}\otimes\boldsymbol{D}) && (1.31)\\ a\otimes\boldsymbol{B} &= \boldsymbol{B}\otimes a = a\boldsymbol{B}\,. && (1.32) \end{aligned}$$

Theorem 1.3 Some operations obey *distributive laws*:

$$\begin{aligned} \boldsymbol{A}(\boldsymbol{B}+\boldsymbol{C}) &= \boldsymbol{AB}+\boldsymbol{AC} && (1.33)\\ \boldsymbol{A}(\boldsymbol{B}-\boldsymbol{C}) &= \boldsymbol{AB}-\boldsymbol{AC} && (1.34)\\ (\boldsymbol{B}+\boldsymbol{C})\boldsymbol{D} &= \boldsymbol{BD}+\boldsymbol{CD} && (1.35)\\ (\boldsymbol{B}-\boldsymbol{C})\boldsymbol{D} &= \boldsymbol{BD}-\boldsymbol{CD} && (1.36)\\ a(\boldsymbol{B}+\boldsymbol{C}) &= a\boldsymbol{B}+a\boldsymbol{C} = (\boldsymbol{B}+\boldsymbol{C})a && (1.37)\\ (\boldsymbol{B}-\boldsymbol{C}) &= a\boldsymbol{B}-a\boldsymbol{C} = (\boldsymbol{B}-\boldsymbol{C})a && (1.38)\\ (\boldsymbol{A}+\boldsymbol{B})' &= \boldsymbol{A}'+\boldsymbol{B}' && (1.39)\\ (\boldsymbol{A}-\boldsymbol{B})' &= \boldsymbol{A}'-\boldsymbol{B}'\,. && (1.40) \end{aligned}$$

Theorem 1.4 The transpose has some special operational properties:

$$\begin{aligned} (a\boldsymbol{B})' &= a\boldsymbol{B}' = \boldsymbol{B}'a && (1.41)\\ (\boldsymbol{ABC}\cdots)' &= \cdots\boldsymbol{C}'\boldsymbol{B}'\boldsymbol{A}' && (1.42)\\ (\boldsymbol{A}\otimes\boldsymbol{B}\otimes\boldsymbol{C}\otimes\cdots)' &= \boldsymbol{A}'\otimes\boldsymbol{B}'\otimes\boldsymbol{C}'\otimes\cdots && (1.43)\\ \boldsymbol{ab}' &= \boldsymbol{a}\otimes\boldsymbol{b}' = \boldsymbol{b}'\otimes\boldsymbol{a}\,. && (1.44) \end{aligned}$$

Theorem 1.5 For conforming matrices,

$$\begin{aligned} \mathrm{vec}(\boldsymbol{AB}) &= (\boldsymbol{I}\otimes\boldsymbol{A})\mathrm{vec}(\boldsymbol{B}) = (\boldsymbol{B}'\otimes\boldsymbol{I})\mathrm{vec}(\boldsymbol{A}) && (1.45)\\ \mathrm{vec}(\boldsymbol{ABC}) &= (\boldsymbol{C}'\otimes\boldsymbol{A})\mathrm{vec}(\boldsymbol{B}) && (1.46)\\ \mathrm{vec}(\boldsymbol{ABCD}) &= (\boldsymbol{I}\otimes\boldsymbol{A})(\boldsymbol{I}\otimes\boldsymbol{B})(\boldsymbol{I}\otimes\boldsymbol{C})\mathrm{vec}(\boldsymbol{D})\,. && (1.47) \end{aligned}$$

Theorem 1.6 **(a)** For any matrix pair, $\text{tr}(\boldsymbol{A} \otimes \boldsymbol{B}) = \text{tr}(\boldsymbol{A})\text{tr}(\boldsymbol{B})$.
(b) For conforming matrices, $\text{tr}(\boldsymbol{AB}) = \text{tr}(\boldsymbol{BA})$.

1.6 OTHER SPECIAL TYPES OF MATRICES

Definition 1.24 A matrix of the form $\boldsymbol{A}'\boldsymbol{A}$ is an *inner product* and $\boldsymbol{AA}'$ is an *outer product.*

If $\boldsymbol{A} = [\boldsymbol{a}_1 \cdots \boldsymbol{a}_c]$ is $r \times c$, then $\boldsymbol{AA}' = \sum_{j=1}^{c} \boldsymbol{a}_j\boldsymbol{a}_j'$ is $r \times r$ and equals the sum of the c outer products of the c columns of $\boldsymbol{A}$. If $\boldsymbol{A} = \begin{bmatrix} \boldsymbol{A}_1 \\ \vdots \\ \boldsymbol{A}_r \end{bmatrix} = \begin{bmatrix} \boldsymbol{b}_1' \\ \vdots \\ \boldsymbol{b}_r' \end{bmatrix} = \boldsymbol{B}'$ is $r \times c$, then $\boldsymbol{A}'\boldsymbol{A} = \sum_{k=1}^{r} \boldsymbol{b}_k\boldsymbol{b}_k'$ is the sum of r outer products of the columns of $\boldsymbol{B}$, which are the rows of $\boldsymbol{A}$. Both inner and outer products are always symmetric. Using concepts introduced later in the chapter, inner and outer products are always either positive definite (all eigenvalues real and positive) or positive semidefinite (all real eigenvalues, with some positive and some zero). Inner and outer products always have the same rank, which equals the rank of $\boldsymbol{A}$. They also have the same eigenvalues, except for some zeros if $\boldsymbol{A}$ is not square.

Definition 1.25 A matrix is (columnwise) *orthogonal* if $\boldsymbol{A}'\boldsymbol{A}$ (the inner product) is diagonal, and a matrix is (rowwise) *orthogonal* if $\boldsymbol{AA}'$ (the outer product) is diagonal. A matrix is (columnwise) *orthonormal* if $\boldsymbol{A}'\boldsymbol{A} = \boldsymbol{I}$, and a matrix is (rowwise) *orthonormal* if $\boldsymbol{AA}' = \boldsymbol{I}$. Two matrices are *biorthogonal* if $\boldsymbol{AB} = \boldsymbol{0}$.

In the preceding definition, neither $\boldsymbol{A}$ nor $\boldsymbol{B}$ need be square.

Definition 1.26 Any square matrix is described as *idempotent* if $\boldsymbol{A} = \boldsymbol{A}^2$.

Lemma 1.3 If $\boldsymbol{A}$ is idempotent, then $\boldsymbol{I} - \boldsymbol{A}$ is also idempotent and $\boldsymbol{A}(\boldsymbol{I} - \boldsymbol{A}) = \boldsymbol{0}$.

Idempotent matrices play important roles in discovering properties of quadratic forms, especially independence.

1.7 QUADRATIC AND BILINEAR FORMS

Definition 1.27 **(a)** For square $\boldsymbol{A} = \boldsymbol{A}'$ and conforming $\boldsymbol{x}$, the expression $q = \boldsymbol{x}'\boldsymbol{Ax}$ is a *quadratic form* in $\boldsymbol{x}$.
(b) The expression $b = \boldsymbol{x}_1'\boldsymbol{Bx}_2$, for $\boldsymbol{B}$ not necessarily symmetric or square is a *bilinear form* in conforming vectors $\boldsymbol{x}_1$ and $\boldsymbol{x}_2$,

If $\boldsymbol{x}$ (2×1) is free to vary and $q_0 > 0$ is constant, then $q_0 = \boldsymbol{x}'\boldsymbol{A}\boldsymbol{x}$ is the equation of an ellipse. The result generalizes to higher dimensions. A quadratic form lies at the heart of the density of a vector Gaussian and consequently leads to ellipsoidal probability contours.

Lemma 1.4 If $q = \boldsymbol{x}'\boldsymbol{A}\boldsymbol{x}$, then without loss of generality $\boldsymbol{A}$ may be assumed to be symmetric.

Proof. If $\boldsymbol{C} = (\boldsymbol{A} + \boldsymbol{A}')/2$ then $\boldsymbol{C} = \boldsymbol{C}'$ and

$$\begin{aligned}\boldsymbol{x}'\boldsymbol{C}\boldsymbol{x} &= \boldsymbol{x}'[(\boldsymbol{A} + \boldsymbol{A}')/2]\boldsymbol{x} \\ &= [\boldsymbol{x}'\boldsymbol{A}\boldsymbol{x} + (\boldsymbol{x}'\boldsymbol{A}\boldsymbol{x})']/2 \\ &= \boldsymbol{x}'\boldsymbol{A}\boldsymbol{x}\,. \end{aligned} \tag{1.48}$$ □

Lemma 1.5 Without loss of generality, any bilinear form, $\boldsymbol{x}_1\boldsymbol{B}\boldsymbol{x}_2$, may be expressed as a quadratic form, $\boldsymbol{y}'\boldsymbol{D}\boldsymbol{y}$, with $\boldsymbol{y}' = [\,\boldsymbol{x}_1'\ \boldsymbol{x}_2'\,]$ and $\boldsymbol{D} = \begin{bmatrix} \boldsymbol{0} & \boldsymbol{B} \\ \boldsymbol{B}' & \boldsymbol{0} \end{bmatrix}/2$.

Proof. Here $\boldsymbol{D} = \boldsymbol{D}'$ and

$$\begin{aligned}\boldsymbol{y}'\boldsymbol{D}\boldsymbol{y} &= \begin{bmatrix} \boldsymbol{x}_1 \\ \boldsymbol{x}_2 \end{bmatrix}' \left(\frac{1}{2}\begin{bmatrix} \boldsymbol{0} & \boldsymbol{B} \\ \boldsymbol{B}' & \boldsymbol{0} \end{bmatrix}\right) \begin{bmatrix} \boldsymbol{x}_1 \\ \boldsymbol{x}_2 \end{bmatrix} \\ &= [\,\boldsymbol{x}_1'\ \boldsymbol{x}_2'\,] \begin{bmatrix} \boldsymbol{B}\boldsymbol{x}_2 \\ \boldsymbol{B}'\boldsymbol{x}_1 \end{bmatrix} /2 \\ &= (\boldsymbol{x}_1'\boldsymbol{B}\boldsymbol{x}_2 + \boldsymbol{x}_2'\boldsymbol{B}'\boldsymbol{x}_1)/2 \\ &= \boldsymbol{x}_1'\boldsymbol{B}\boldsymbol{x}_2\,. \end{aligned} \tag{1.49}$$ □

1.8 VECTOR SPACES AND RANK

Definition 1.28 A set of $n \times 1$ vectors $\{\boldsymbol{x}_1, \ldots, \boldsymbol{x}_p\}$ is *linearly dependent* if a set of scalar coefficients $\{a_1, \ldots, a_p\}$, not all zero, exist such that

$$\sum_{i=1}^{p} a_i\boldsymbol{x}_i = a_1\boldsymbol{x}_1 + a_2\boldsymbol{x}_2 + \cdots + a_p\boldsymbol{x}_p = \boldsymbol{0}\,. \tag{1.50}$$

If no such set of a_i exists then the set of $\boldsymbol{x}_i$ is *linearly independent*. The single equation about $n \times 1$ vectors defines a set of n scalar equations.

Definition 1.29 (a) Any finite set of $n \times 1$ vectors generates a *vector space*, namely the (usually infinite) collection of all possible vectors created by any combination of multiplications of one vector by a constant, or the addition of two vectors.

(b) Any set of vectors which generate the particular set of vectors *spans* the vector space, and provides a *basis* for the space.

Definition 1.30 (a) The *rank* of the vector space equals the smallest possible number of linearly independent vectors which span the space. The rank of a set equals zero if and only if the only member of the set is $\boldsymbol{x}_i \equiv \mathbf{0}$. The rank of a set of p vectors, necessarily an integer, ranges from zero to p.
(b) A set with rank p is *full rank*, while a set with rank strictly less than p is *less than full rank*.

Any two distinct vectors $\boldsymbol{x}_1$ and $\boldsymbol{x}_2$ are orthogonal if and only if $\boldsymbol{x}_1'\boldsymbol{x}_2 = 0$. An orthogonal basis provides the most convenient form and has $\boldsymbol{x}_j'\boldsymbol{x}_{j'} = 0$ if $j \neq j'$. Spectral (eigenvalue) decomposition provides an orthonormal basis for any square and symmetric matrix, and some nonsymmetric square matrices. The singular value decomposition provides a convenient way for any matrix, symmetric or not, square or not. Both are discussed later in the chapter.

An $r \times c$ matrix, $\boldsymbol{A}$, can be thought of as a collection of c vectors, the columns, each $r \times 1$. Alternately, considering the columns of $\boldsymbol{A}'$ allows describing the matrix as a collection of r vectors, the transposed rows, each $c \times 1$. The rank of a matrix may be found by decomposing it into its columns and treating them as a set of vectors:

$$\begin{aligned} \boldsymbol{A} &= \begin{bmatrix} 1 & 4 & 7 \\ 2 & 5 & 8 \\ 3 & 6 & 9 \\ 4 & 9 & 2 \end{bmatrix} \\ &= \begin{bmatrix} \boldsymbol{a}_1 & \boldsymbol{a}_2 & \boldsymbol{a}_3 \end{bmatrix} \qquad (1.51) \\ &\Leftrightarrow \\ \{\boldsymbol{a}_1, \boldsymbol{a}_2, \boldsymbol{a}_3\} &= \left\{ \begin{bmatrix} 1 \\ 2 \\ 3 \\ 4 \end{bmatrix}, \begin{bmatrix} 4 \\ 5 \\ 6 \\ 9 \end{bmatrix}, \begin{bmatrix} 7 \\ 8 \\ 9 \\ 2 \end{bmatrix} \right\}. \qquad (1.52) \end{aligned}$$

Transposing the matrix allows applying the same process to the rows. The resulting row rank always equals the column rank, which leads to the following.

Definition 1.31 The *rank of a matrix* equals the maximum number of linearly independent rows or columns, indicated rank$(\boldsymbol{A})$. An $r \times c$ matrix is *full rank* if rank$(\boldsymbol{A}) = \min(r, c)$ and *less than full rank* otherwise. The only matrix of rank zero is a matrix of all zeros, $\mathbf{0}_{n \times m}$.

It would be hard to overemphasize the importance of the concepts of vector space, span, basis, and orthogonal basis in the study of linear models. The concepts lie at the heart of many theorems, proofs and computational methods.

The same ideas provide crucial tools in understanding the logic of hypothesis tests, constrained models, and equivalence between models and parameterizations.

1.9 FINDING RANK

One method to find rank uses only elementary row operations or only elementary column operations to produce a canonical form, which is necessarily of equivalent rank. In particular, using (only) elementary row operations allows transforming a matrix to a triangular one (triangularize the matrix). The three elementary row (column) operations are: (1) multiplying a row (column) by a nonzero constant, (2) adding one row (column) to another, and (3) exchanging two rows (columns).

A second approach to finding rank uses the spectral or singular value decompositions described later in the chapter. The rank of any square symmetric matrix equals the number of nonzero eigenvalues (although the relationship may not hold for square but not symmetric matrices). However, *any* matrix has a singular value decomposition, with the number of nonzero singular values equal to the rank of the matrix.

The concepts of similarity and congruence (defined in Section 1.12) can be used to simplify the task. The following lemmas also prove useful.

Lemma 1.6 In general

$$0 \leq \text{rank}(\boldsymbol{A}) \leq \min(r, c) \tag{1.53}$$
$$\text{rank}(\boldsymbol{AB}) \leq \min[\text{rank}(\boldsymbol{A}), \text{rank}(\boldsymbol{B})] \tag{1.54}$$
$$\text{rank}(\boldsymbol{A}+\boldsymbol{B}) \leq \text{rank}(\boldsymbol{A}) + \text{rank}(\boldsymbol{B})\,. \tag{1.55}$$

Lemma 1.7 (a) The rank of a diagonal matrix equals the number of nonzero diagonal elements.
(b) If $b \neq 0$, then

$$\text{rank}(\boldsymbol{A}) = \text{rank}(b\boldsymbol{A})\,. \tag{1.56}$$

The case of $b = -1$ provides an example: $\text{rank}(\boldsymbol{A}) = \text{rank}(-\boldsymbol{A})$.
(c) For any matrix

$$\text{rank}(\boldsymbol{A}) = \text{rank}(\boldsymbol{A}') = \text{rank}(\boldsymbol{A}'\boldsymbol{A}) = \text{rank}(\boldsymbol{A}\boldsymbol{A}')\,. \tag{1.57}$$

Lemma 1.8 Multiplying by a square full-rank matrix does not change rank. If $\boldsymbol{A}$ is $r \times r$ of rank r, $\boldsymbol{B}$ is $r \times c$ with $0 \leq \text{rank}(\boldsymbol{B}) \leq \min(r, c)$, and $\boldsymbol{C}$ is $c \times c$ of rank c, then

$$\text{rank}(\boldsymbol{ABC}) = \text{rank}(\boldsymbol{AB}) = \text{rank}(\boldsymbol{BC}) = \text{rank}(\boldsymbol{B})\,. \tag{1.58}$$

Lemma 1.9 If $\boldsymbol{A}$ is $r \times c$ of rank c and $\boldsymbol{B}$ is $c \times d$ with $\text{rank}(\boldsymbol{B}) = c$, then

$$\text{rank}(\boldsymbol{AB}) = \text{rank}(\boldsymbol{A}). \tag{1.59}$$

Lemma 1.10 For any $\boldsymbol{A}$ and $\boldsymbol{B}$

$$\text{rank}(\boldsymbol{A} \otimes \boldsymbol{B}) = \text{rank}(\boldsymbol{A})\text{rank}(\boldsymbol{B}) \tag{1.60}$$

and for any square $\boldsymbol{A}$ and any square $\boldsymbol{B}$

$$\text{rank}(\boldsymbol{A} \oplus \boldsymbol{B}) = \text{rank}(\boldsymbol{A}) + \text{rank}(\boldsymbol{B}). \tag{1.61}$$

1.10 DETERMINANTS

Definition 1.32 The *determinant*, a scalar, is indicated $|\boldsymbol{A}|$ and is defined only for a square matrix. Also $|\boldsymbol{A}'| = |\boldsymbol{A}|$.

For any 2×2 matrix

$$\left| \begin{bmatrix} a & b \\ c & d \end{bmatrix} \right| = ad - bc. \tag{1.62}$$

The determinant of a diagonal or triangular matrix equals the product of the diagonal values. For 3×3 matrices

$$\left| \begin{bmatrix} a & 0 & 0 \\ 0 & b & 0 \\ 0 & 0 & c \end{bmatrix} \right| = \left| \begin{bmatrix} a & 0 & 0 \\ x & b & 0 \\ y & z & c \end{bmatrix} \right| = \left| \begin{bmatrix} a & x & y \\ 0 & b & z \\ 0 & 0 & c \end{bmatrix} \right| = abc. \tag{1.63}$$

For any general 3×3 matrix

$$\left| \begin{bmatrix} a & b & c \\ d & e & f \\ g & h & i \end{bmatrix} \right| = aei + bfg + chd - ceg - afh - bdi. \tag{1.64}$$

In general the determinant equals the sum of products with alternating sign of elements of the matrix. The determinant equals the product of the eigenvalues (discussed later in the present chapter), which provides the most useful interpretation of the determinant for statistical analysis. Determinants have many useful properties related to rank.

Lemma 1.11 For $\boldsymbol{A}$ $(n \times n)$,

$$\begin{aligned} |\boldsymbol{A}| = 0 &\Leftrightarrow \text{rank}(\boldsymbol{A}) < n \\ &\Leftrightarrow \boldsymbol{A} \text{ less than full rank} \\ &\Leftrightarrow \boldsymbol{A}^{-1} \text{ does } \textit{not} \text{ exist,} \end{aligned}$$

while

$$\begin{aligned}|\boldsymbol{A}| \neq 0 &\Leftrightarrow \text{rank}(\boldsymbol{A}) = n \\ &\Leftrightarrow \boldsymbol{A} \text{ full rank} \\ &\Leftrightarrow \boldsymbol{A}^{-1} \text{ exists,} \\ &\Leftrightarrow \text{all columns (rows) of } \boldsymbol{A} \text{ linearly independent.}\end{aligned}$$

Definition 1.33 A less-than-full-rank square matrix may be described as *singular* and a full-rank square matrix as *nonsingular.*

Lemma 1.12 (a) $|\boldsymbol{A}| = |\boldsymbol{A}'|$.
(b) If $\boldsymbol{A}$ is $m \times m$ and $\boldsymbol{B}$ is $n \times n$ then $|\boldsymbol{A} \oplus \boldsymbol{B}| = |\boldsymbol{A}||\boldsymbol{B}|$ and
(c) $|\boldsymbol{A} \otimes \boldsymbol{B}| = |\boldsymbol{A}|^m |\boldsymbol{B}|^n$.
(d) If $m = n$, then $|\boldsymbol{AB}| = |\boldsymbol{A}||\boldsymbol{B}| = |\boldsymbol{BA}|$.

1.11 THE INVERSE AND GENERALIZED INVERSE

Theorem 1.7 (a) An $n \times n$ matrix $\boldsymbol{A}$ of full rank has $\text{rank}(\boldsymbol{A}) = n$, and there exists a unique matrix $\boldsymbol{A}^{-1}$, called the *inverse* of $\boldsymbol{A}$, such that

$$\boldsymbol{A}^{-1}\boldsymbol{A} = \boldsymbol{A}\boldsymbol{A}^{-1} = \boldsymbol{I}. \tag{1.65}$$

(b) If $n = 1$, then $\boldsymbol{A}$ is a scalar, and $\boldsymbol{A}^{-1} = 1/a$ exists if and only if $a \neq 0$.
(c) The inverse of a full-rank diagonal matrix equals the diagonal matrix of reciprocals of the diagonal elements.

Lemma 1.13 For square, conforming (same-size) and full-rank matrices

$$(\boldsymbol{AB})^{-1} = \boldsymbol{B}^{-1}\boldsymbol{A}^{-1}. \tag{1.66}$$

More generally, for any finite set of full-rank (and same size) matrices

$$(\boldsymbol{ABCD}\cdots)^{-1} = \cdots \boldsymbol{D}^{-1}\boldsymbol{C}^{-1}\boldsymbol{B}^{-1}\boldsymbol{A}^{-1}. \tag{1.67}$$

The inverse has some symmetries. Not surprisingly, a symmetric matrix has a symmetric inverse. When they exist, the inverse of the transpose equals the transpose of the inverse and hence may be written unequivocally as $\boldsymbol{A}^{-t}$:

$$\boldsymbol{A}^{-t} = (\boldsymbol{A}')^{-1} = (\boldsymbol{A}^{-1})'. \tag{1.68}$$

Furthermore, if $\boldsymbol{A}^{-1}$ exists, then $|\boldsymbol{A}^{-1}| = (|\boldsymbol{A}|)^{-1}$.

Lemma 1.14 (a) If $\boldsymbol{V} = \boldsymbol{R} + \boldsymbol{ZGZ}'$ and $\boldsymbol{V}$, $\boldsymbol{R}$, and $\boldsymbol{G}$ are all square and full rank, then $\boldsymbol{V}^{-1} = \boldsymbol{R}^{-1}(\boldsymbol{R} - \boldsymbol{ZBZ}')\boldsymbol{R}^{-1}$, with $\boldsymbol{B} = (\boldsymbol{G}^{-1} + \boldsymbol{Z}'\boldsymbol{R}^{-1}\boldsymbol{Z})^{-1} = \boldsymbol{G}(\boldsymbol{G}^{-1} - \boldsymbol{Z}'\boldsymbol{V}^{-1}\boldsymbol{Z})\boldsymbol{G}$.
(b) If $\boldsymbol{V} = \boldsymbol{R} + g\boldsymbol{ZZ}'$ and $\boldsymbol{V}$, $\boldsymbol{R}$, and g are all square and full rank, then

$(\boldsymbol{R}+g\boldsymbol{Z}\boldsymbol{Z}')^{-1} = \boldsymbol{R}^{-1}(\boldsymbol{R}-b\boldsymbol{Z}\boldsymbol{Z}')\boldsymbol{R}^{-1}$, with $b = (g^{-1}+\boldsymbol{Z}'\boldsymbol{R}^{-1}\boldsymbol{Z})^{-1}$.
(c) For $\boldsymbol{A}$ $(r \times c)$ and $\boldsymbol{B}$ $(c \times r)$, if $\text{rank}(\boldsymbol{I}+\boldsymbol{A}\boldsymbol{B}) = r$, then $(\boldsymbol{I}+\boldsymbol{A}\boldsymbol{B})^{-1} = \boldsymbol{I}-\boldsymbol{A}(\boldsymbol{I}+\boldsymbol{B}\boldsymbol{A})^{-1}\boldsymbol{B}$.
(d) Applying (c) recursively gives $(\boldsymbol{I}_r - \boldsymbol{C})^{-1} = \boldsymbol{I} + \sum_{j=1}^{\infty} \boldsymbol{C}^j$ (when it exists).

Theorem 1.8 For any matrix $\boldsymbol{A}$, square or not, a nonunique *generalized inverse* $\boldsymbol{A}^-$ always exists such that

$$\boldsymbol{A}\boldsymbol{A}^-\boldsymbol{A} = \boldsymbol{A}. \tag{1.69}$$

One particularly inconvenient property of the one-condition generalized inverse is that $\boldsymbol{A}^-$ need not be symmetric even though $\boldsymbol{A}$ is symmetric.

Theorem 1.9 For any matrix $\boldsymbol{A}$, square or not, the unique *Moore-Penrose (generalized) inverse* $\boldsymbol{A}^+$ always exists and satisfies four conditions:

$$1.\quad \boldsymbol{A}\boldsymbol{A}^+\boldsymbol{A} = \boldsymbol{A} \tag{1.70}$$
$$2.\quad \boldsymbol{A}^+\boldsymbol{A}\boldsymbol{A}^+ = \boldsymbol{A}^+ \tag{1.71}$$
$$3.\quad (\boldsymbol{A}^+\boldsymbol{A})' = \boldsymbol{A}^+\boldsymbol{A} \tag{1.72}$$
$$4.\quad (\boldsymbol{A}\boldsymbol{A}^+)' = \boldsymbol{A}\boldsymbol{A}^+. \tag{1.73}$$

If $\boldsymbol{A}$ is symmetric, then $\boldsymbol{A}^+$ is always symmetric. Matrices meeting only subsets of conditions, such as 1 and 2, (a two-condition inverse), or conditions 1, 2, and 3 (a three-condition inverse) have also been studied. Although a few proofs in linear models require a two- or three-condition inverse, we nearly always restrict attention to either one- or four-condition inverses.

Both the Moore-Penrose (four-condition) and any one-condition generalized inverse always coincide with the regular inverse for a square and full-rank matrix. Furthermore any one-condition inverse always coincides with the four-condition inverses for a nonsquare and full-rank matrix.

For a wide variety of linear models with a less-than-full-rank design matrix $\boldsymbol{X}$, the form $(\boldsymbol{X}'\boldsymbol{X})^-$ occurs often in expressions for estimators and distribution parameters. Very conveniently, Theorem 1.15 in the next section eliminates the need to distinguish between $(\boldsymbol{X}'\boldsymbol{X})^-$ and $(\boldsymbol{X}'\boldsymbol{X})^+$ in a wide range of linear model applications.

Theorem 1.10 (a) The Moore-Penrose inverse of a less-than-full-rank diagonal matrix is the diagonal matrix with reciprocals of the nonzero elements on the diagonal in the same locations as the nonzero elements and zero elsewhere.
(b) The Moore-Penrose inverse of the transpose equals the transpose of the generalized inverse and hence may be written unequivocally as $\boldsymbol{A}^{+t}$:

$$\boldsymbol{A}^{+t} = (\boldsymbol{A}')^+ = (\boldsymbol{A}^+)'. \tag{1.74}$$

(c) If $r \times c$ $\boldsymbol{A}$ has rank$(\boldsymbol{A}) = r \le c$, then $\boldsymbol{A}^+$ has dimensions $c \times r$ and

$$\boldsymbol{A}^+ = \boldsymbol{A}'(\boldsymbol{A}\boldsymbol{A}')^{-1}. \tag{1.75}$$

(d) If $r \times c$ $\boldsymbol{A}$ has $r \ge c =$ rank$(\boldsymbol{A})$, then $\boldsymbol{A}^+$ has dimensions $c \times r$ and

$$\boldsymbol{A}^+ = (\boldsymbol{A}'\boldsymbol{A})^{-1}\boldsymbol{A}'. \tag{1.76}$$

In contrast to the regular inverse, $(\boldsymbol{AB})^+$ may or may not equal $\boldsymbol{B}^+\boldsymbol{A}^+$. The next lemma lists two side conditions which suffice to ensure that the result holds.

Lemma 1.15 (a) If $\boldsymbol{A}$ is $r \times c$, $\boldsymbol{P}$ is $r \times r$, and $\boldsymbol{Q}$ is $c \times c$, with $\boldsymbol{P}'\boldsymbol{P} = \boldsymbol{I}_r$ and $\boldsymbol{Q}'\boldsymbol{Q} = \boldsymbol{I}_c$, then

$$(\boldsymbol{PAQ}')^+ = \boldsymbol{PA}^+\boldsymbol{Q}'. \tag{1.77}$$

(b) If $\boldsymbol{A}$ is $r \times c$ of rank c and $\boldsymbol{B}$ is $c \times d$ of rank c, then

$$(\boldsymbol{AB})^+ = \boldsymbol{B}^+\boldsymbol{A}^+. \tag{1.78}$$

Definition 1.34 For known $\boldsymbol{A}$ ($r \times c$) and known $\boldsymbol{b}$ ($r \times 1$), system $\boldsymbol{Ax} = \boldsymbol{b}$ is *consistent* whenever any linear relationships existing among the rows of $\boldsymbol{A}$ also exist among the corresponding rows of $\boldsymbol{b}$ ($r \times 1$). Equivalently $\boldsymbol{c}'\boldsymbol{A} = \boldsymbol{0}$ $\Rightarrow$ $\boldsymbol{c}'\boldsymbol{b} = 0$, $\Leftrightarrow$ the system has one or more solutions, $\boldsymbol{x}_0$.

By the definition, a system of equations is consistent if at least one solution set exists. More than one or even infinitely many solutions may exist. With $\boldsymbol{A}$ and $\boldsymbol{b}$ known constants and $\boldsymbol{x}$ unknown, the equation $\boldsymbol{Ax} = \boldsymbol{b}$ defines a system of equations (one per row of $\boldsymbol{A}$ and $\boldsymbol{b}$).

Lemma 1.16 For known $\boldsymbol{A}$ ($r \times c$) and known $\boldsymbol{b}$ ($r \times 1$), if the system $\boldsymbol{Ax} = \boldsymbol{b}$ is consistent, then the following all hold.
(a) If $r = c$ and rank$(\boldsymbol{A}) = r$, then the system has a unique solution given by $\boldsymbol{x} = \boldsymbol{A}^{-1}\boldsymbol{b}$.
(b) If $r \ne c$ or rank$(\boldsymbol{A}) \ne r$ or both and $\boldsymbol{A}^-$ is any generalized inverse of $\boldsymbol{A}$ (i.e., $\boldsymbol{AA}^-\boldsymbol{A} = \boldsymbol{A}$), then $\boldsymbol{x} = \boldsymbol{A}^-\boldsymbol{b}$ is one solution of infinitely many.

Proof of (b). $\boldsymbol{AA}^-\boldsymbol{A} = \boldsymbol{A}$ $\Rightarrow (\boldsymbol{AA}^-\boldsymbol{A})\boldsymbol{x} = \boldsymbol{b}$ $\Rightarrow$ $\boldsymbol{AA}^-(\boldsymbol{Ax}) = \boldsymbol{b}$ $\Rightarrow \boldsymbol{AA}^-\boldsymbol{b} = \boldsymbol{b}$. □

Lemma 1.17 If $\boldsymbol{A}$ ($r \times c$) has $(\boldsymbol{A}^-)_1$ as a particular one-condition generalized inverse and $\boldsymbol{B}$ is $c \times r$, then

$$(\boldsymbol{A}^-)_2 = (\boldsymbol{A}^-)_1 + \boldsymbol{B} - (\boldsymbol{A}^-)_1\boldsymbol{ABA}(\boldsymbol{A}^-)_1 \tag{1.79}$$

is also a one-condition generalized inverse for $\boldsymbol{A}$.

Corollary 1.17 For $\boldsymbol{A}$ $(r \times c)$ and any $\boldsymbol{A}^-$ $(c \times r)$, $\boldsymbol{B}$ and $\boldsymbol{C}$ exist such that

$$\boldsymbol{A}^- = \boldsymbol{A}^+ + \boldsymbol{B} - \boldsymbol{A}^+\boldsymbol{A}\boldsymbol{B}\boldsymbol{A}\boldsymbol{A}^+ \tag{1.80}$$

$$\boldsymbol{A}^+ = \boldsymbol{A}^- + \boldsymbol{C} - \boldsymbol{A}^-\boldsymbol{A}\boldsymbol{C}\boldsymbol{A}\boldsymbol{A}^- . \tag{1.81}$$

Lemma 1.18 If $\boldsymbol{A}^{-1}$ exists, then

$$\operatorname{rank}\left(\boldsymbol{A}^{-1}\right) = \operatorname{rank}(\boldsymbol{A}) . \tag{1.82}$$

For $\boldsymbol{A}$ of any dimension and any rank

$$\operatorname{rank}(\boldsymbol{A}^+) = \operatorname{rank}(\boldsymbol{A}\boldsymbol{A}^+) = \operatorname{rank}(\boldsymbol{A}^+\boldsymbol{A}) = \operatorname{rank}(\boldsymbol{A}) \tag{1.83}$$

and

$$\operatorname{rank}(\boldsymbol{A}^-) \geq \operatorname{rank}(\boldsymbol{A}\boldsymbol{A}^-) = \operatorname{rank}(\boldsymbol{A}^-\boldsymbol{A}) = \operatorname{rank}(\boldsymbol{A}) . \tag{1.84}$$

Lemma 1.19 If $\boldsymbol{A}$ is $r \times c$, then $\boldsymbol{A}^-\boldsymbol{A}$ is $c \times c$, $\boldsymbol{A}\boldsymbol{A}^-$ is $r \times r$, and both are idempotent. The same properties hold if $\boldsymbol{A}^+$ replaces $\boldsymbol{A}^-$.

Proof. $\boldsymbol{A}^-\boldsymbol{A} = \boldsymbol{A}^-\boldsymbol{A}\boldsymbol{A}^-\boldsymbol{A} = (\boldsymbol{A}^-\boldsymbol{A})^2$ and $\boldsymbol{A}\boldsymbol{A}^- = \boldsymbol{A}\boldsymbol{A}^-\boldsymbol{A}\boldsymbol{A}^- = (\boldsymbol{A}\boldsymbol{A}^-)^2$. □

Lemma 1.20 For any $\boldsymbol{A}$ and $\boldsymbol{B}$

$$(\boldsymbol{A} \otimes \boldsymbol{B})^- = \boldsymbol{A}^- \otimes \boldsymbol{B}^- \tag{1.85}$$

$$(\boldsymbol{A} \otimes \boldsymbol{B})^+ = \boldsymbol{A}^+ \otimes \boldsymbol{B}^+ , \tag{1.86}$$

and for any square $\boldsymbol{A}$ and square $\boldsymbol{B}$

$$(\boldsymbol{A} \oplus \boldsymbol{B})^- = \boldsymbol{A}^- \oplus \boldsymbol{B}^- \tag{1.87}$$

$$(\boldsymbol{A} \oplus \boldsymbol{B})^+ = \boldsymbol{A}^+ \oplus \boldsymbol{B}^+ . \tag{1.88}$$

If square $\boldsymbol{A}$ and square $\boldsymbol{B}$ are both full rank, then

$$(\boldsymbol{A} \oplus \boldsymbol{B})^{-1} = \boldsymbol{A}^{-1} \oplus \boldsymbol{B}^{-1} \tag{1.89}$$

$$(\boldsymbol{A} \otimes \boldsymbol{B})^{-1} = \boldsymbol{A}^{-1} \otimes \boldsymbol{B}^{-1} . \tag{1.90}$$

1.12 EIGENANALYSIS (SPECTRAL DECOMPOSITION)

Eigenanalysis is only defined for square $\boldsymbol{A}$. Nearly all interest in decomposing matrices in statistics lies with symmetric matrices. The symmetry (in statistical applications) arises because the matrices of interest are inner products or outer products. However, for the moment, we consider any square matrix, $n \times n$, symmetric or not.

Definition 1.35 A *right eigenvector* of $\boldsymbol{A}$ is an $n \times 1$ vector, $\boldsymbol{x} \neq \boldsymbol{0}$, such that

$$\boldsymbol{A}\boldsymbol{x} = \lambda \boldsymbol{x}, \tag{1.91}$$

with λ the *eigenvalue* corresponding to $\boldsymbol{x}$. A *left eigenvector* is an $n \times 1$ vector $\boldsymbol{y}$ such that $\boldsymbol{y}'\boldsymbol{A} = \lambda \boldsymbol{y}'$. A left eigenvector of $\boldsymbol{A}$ is a right eigenvector of $\boldsymbol{A}'$: $\boldsymbol{A}'\boldsymbol{y} = \lambda \boldsymbol{y}$. The set of eigenvalues is sometimes referred to as the *spectrum* of the matrix.

Eigen means "characteristic" in German, which leads to the alternative descriptions as *characteristic values* and *characteristic vectors*. The eigenvector $\boldsymbol{x}$ has the special property of projecting $\boldsymbol{A}$ back into itself $(\boldsymbol{x})$ times a constant. The self-replicating feature leads to thinking of eigenvectors as matrix DNA.

Statisticians most often apply eigenanalysis to symmetric matrices. Right and left eigenvectors coincide for a symmetric matrix but usually do not for a nonsymmetric matrix and may not even exist. One important exception occurs in multivariate linear models for which the test statistics and associated contrasts correspond to eigenvalues and eigenvectors of a nonsymmetric matrix. For computational purposes, although not for scientific interpretation, the task can be expressed in terms of a symmetric matrix.

Theorem 1.11 (a) The roots of the *characteristic equation*, $|\boldsymbol{A} - \lambda \boldsymbol{I}| = 0$, equal the eigenvalues (characteristic values).
(b) Also

$$\boldsymbol{A}\boldsymbol{x} = \lambda \boldsymbol{x} \tag{1.92}$$

$$\Leftrightarrow$$

$$(\boldsymbol{A} - \lambda \boldsymbol{I})\boldsymbol{x} = \boldsymbol{0} \tag{1.93}$$

$$\Leftrightarrow$$

$$|\boldsymbol{A} - \lambda \boldsymbol{I}| = 0\,. \tag{1.94}$$

(c) The characteristic equation of an $n \times n$ matrix equals a polynomial in λ of order n.

For a 2×2

$$|\boldsymbol{A} - \lambda \boldsymbol{I}| = \left| \begin{bmatrix} a & b \\ c & d \end{bmatrix} - \lambda \begin{bmatrix} 1 & 0 \\ 0 & 1 \end{bmatrix} \right| \tag{1.95}$$

implies

$$\begin{aligned} 0 = \left| \begin{bmatrix} a & b \\ c & d \end{bmatrix} - \begin{bmatrix} \lambda & 0 \\ 0 & \lambda \end{bmatrix} \right| &= \begin{vmatrix} a - \lambda & b \\ c & d - \lambda \end{vmatrix} \\ &= (a - \lambda)(d - \lambda) - bc \\ &= \lambda^2 - \lambda(a + d) + ad - bc\,. \end{aligned} \tag{1.96}$$

Of course the quadratic formula allows solving for λ here and gives two values.

The properties of eigenvalues and eigenvectors vary substantially, with many awkward possibilities. As for the roots of any polynomial in real numbers, the eigenvalues of a real matrix may be imaginary. Furthermore, even though an $n \times n$ matrix always has n eigenvalues, not all $n \times n$ matrices have a complete set of n distinct eigenvectors. The number of distinct eigenvalues ranges from 1 to n.

Although distinct eigenvalues are unique, eigenvectors (when they exist) are not completely unique. If $\boldsymbol{x}$ is an eigenvector of $\boldsymbol{A}$, then $c\boldsymbol{x}$ is also, for $c \neq 0$, because $\boldsymbol{A}\boldsymbol{x} = \lambda\boldsymbol{x}$ implies $\boldsymbol{A}\boldsymbol{x}c = \lambda\boldsymbol{x}c$. By strong convention, $\boldsymbol{x}$ is usually scaled such that $\boldsymbol{x}'\boldsymbol{x} = 1$, described as *normalized to unit length*. However, a normalized eigenvector is still not unique because both $\boldsymbol{x}$ and $-\boldsymbol{x}$ are normalized eigenvectors for $\boldsymbol{A}$. Geometrically, reversing the sign of the vector corresponds to a reflection about an axis. Although the sign ambiguity affects some computations, the subtlety can usually be ignored in discussing "the" (normalized) eigenvectors.

Definition 1.36 (a) *Algebraic multiplicity* is the number of times a particular distinct eigenvalue, λ_j, occurs as a root of the characteristic equation.
(b) The *geometric multiplicity* of λ_j equals the number of distinct eigenvectors associated with λ_j.
(c) A *simple* matrix has geometric multiplicity equal to algebraic multiplicity.

It follows from the definitions that geometric multiplicity must be less than or equal to algebraic multiplicity. The definitions have the important implication that any simple matrix (necessarily square but not necessarily symmetric) has a decomposition in terms of the diagonal matrix of eigenvalues.

Theorem 1.12 (a) All symmetric matrices are simple.
(b) A real and symmetric matrix has real eigenvalues and eigenvectors.
(c) Any real and symmetric matrix ($n \times n$) has a *spectral* decomposition,

$$\boldsymbol{A} = \boldsymbol{V}\text{Dg}(\boldsymbol{\lambda})\boldsymbol{V}', \tag{1.97}$$

with $\boldsymbol{V}$ a set of orthogonal eigenvectors and $\boldsymbol{\lambda}$ the corresponding eigenvalues in the same order (usually sorted from largest to smallest).
(d) By convention, and without loss of generality, $\boldsymbol{V}'\boldsymbol{V} = \boldsymbol{V}\boldsymbol{V}' = \boldsymbol{I}_n$, giving orthonormal $\boldsymbol{V}$ and eigenvectors ($\boldsymbol{v}_j'\boldsymbol{v}_j = 1$, $\boldsymbol{v}_j'\boldsymbol{v}_k = 0$ for $j \neq k$).
(e) If $\boldsymbol{D}$ ($n \times n$) is a diagonal matrix with diagonal elements freely chosen from $\{+1, -1\}$, without loss of generality $\boldsymbol{V}$ may be taken to be $\boldsymbol{V}\boldsymbol{D}$.

The columns of $\boldsymbol{V} = [\,\boldsymbol{v}_1\, \boldsymbol{v}_2 \cdots \boldsymbol{v}_n\,]$ are "the" normalized eigenvectors of $\boldsymbol{A}$, corresponding to the eigenvalues. The fact that $\boldsymbol{D}\boldsymbol{D} = \boldsymbol{I}$ gives $\boldsymbol{D}\text{Dg}(\boldsymbol{\lambda})\boldsymbol{D} = \text{Dg}(\boldsymbol{\lambda})$ and $\boldsymbol{A} = (\boldsymbol{V}\boldsymbol{D})\text{Dg}(\boldsymbol{\lambda})(\boldsymbol{V}\boldsymbol{D})'$. Eigenvalues and eigenvectors *must* be in a corresponding order. Also $\boldsymbol{V}$ always has full rank and $\boldsymbol{V}^{-1} = \boldsymbol{V}'$ (true for square, orthonormal matrices). Without loss of generality, but for convenience and by strong convention, $\boldsymbol{V}$ is scaled to unit length ($\boldsymbol{v}_j'\boldsymbol{v}_j = 1$). A simple example is

$$\begin{aligned} \boldsymbol{A} &= \begin{bmatrix} 1.5 & -0.5 \\ -0.5 & 1.5 \end{bmatrix} \\ &= \boldsymbol{V}\mathrm{Dg}(\boldsymbol{\lambda})\boldsymbol{V}' \\ &= \left(\begin{bmatrix} 1 & -1 \\ 1 & 1 \end{bmatrix}/\sqrt{2}\right)\begin{bmatrix} 2 & 0 \\ 0 & 1 \end{bmatrix}\left(\begin{bmatrix} 1 & -1 \\ 1 & 1 \end{bmatrix}/\sqrt{2}\right), \end{aligned} \tag{1.98}$$

with $\boldsymbol{\lambda} = [2\ 1]'$, $\boldsymbol{V} = [\boldsymbol{v}_1\ \boldsymbol{v}_2]$.

Despite the many awkward possibilities, statistical applications of eigenanalysis nearly always involve well behaved matrices such as $\boldsymbol{X}'\boldsymbol{X}$ or $\boldsymbol{X}\boldsymbol{X}'$. Such inner and outer products are always symmetric and therefore simple with real eigenvalues. Furthermore, they never have negative eigenvalues (all eigenvalues are positive or zero; nonnegative definite). Sums of squares, covariance, and correlation matrices can all be expressed as inner or outer products. In the study of linear models, the few computations requiring eigenanalysis of nonsymmetric matrices can be expressed in terms of closely related symmetric matrices.

Definition 1.37 (a) If $\boldsymbol{A}$, $\boldsymbol{B}$, and $\boldsymbol{T}$ are $n \times n$ with $\boldsymbol{T}$ full rank, then $\boldsymbol{A}$ and $\boldsymbol{B}$ are said to be *similar* if and only if $\boldsymbol{B} = \boldsymbol{T}\boldsymbol{A}\boldsymbol{T}^{-1}$.
(b) Matrices $\boldsymbol{A}$ and $\boldsymbol{B}$ are *congruent* if and only if $\boldsymbol{B} = \boldsymbol{T}\boldsymbol{A}\boldsymbol{T}'$.

Lemma 1.21 (a) Similar matrices have all the same eigenvalues.
(b) Any matrix similar to a diagonal matrix has rank equal to the number of nonzero eigenvalues.
(c) Every symmetric matrix is similar to a diagonal matrix, namely a diagonal matrix of the eigenvalues of $\boldsymbol{A}$, and always has rank equal to the number of nonzero eigenvalues.
(d) Geometrically, the eigenvectors corresponding to the nonzero eigenvalues of a simple matrix (including all symmetric matrices) span and provide an orthogonal basis for the full rank subspace spanned by $\boldsymbol{A}$.
(e) Sylvester's law of inertia (Lancaster, 1969, p 90) guarantees that congruent matrices have the same number of positive, negative and zero eigenvalues (but not necessarily the same values).

In contrast to the nice properties in the last lemma, the rank of a square and nonsymmetric matrix may not equal the number of nonzero eigenvalues. Although the matrix $\boldsymbol{A} = \begin{bmatrix} 0 & 1 \\ 0 & 0 \end{bmatrix}$ has rank 1, both eigenvalues are zero.

Lemma 1.22 For any square matrix, the trace and determinant equal functions of the eigenvalues:

$$\mathrm{tr}(\boldsymbol{A}) = \sum_{j=1}^{n} \lambda_j \tag{1.99}$$

and

$$|\boldsymbol{A}| = \prod_{j=1}^{n} \lambda_j . \tag{1.100}$$

Lemma 1.23 (a) $|\boldsymbol{A}| = 0 \Leftrightarrow$ at least one eigenvalue equals zero $\Leftrightarrow$ $\boldsymbol{A}$ is less than full rank; $|\boldsymbol{A}| \neq 0 \Leftrightarrow$ no eigenvalue equals zero $\Leftrightarrow$ $\boldsymbol{A}$ is full rank.
(b) The eigenvalues of a (square) diagonal or triangular matrix are the diagonal elements.
(c) For square and full-rank $\boldsymbol{B}$ and square $\boldsymbol{A}$ of the same dimension, the eigenvalues of $\boldsymbol{BAB}^{-1}$ are the eigenvalues of $\boldsymbol{A}$.
(d) The eigenvalues of a (square) orthonormal matrix are ± 1.

Definition 1.38 (a) Square $\boldsymbol{A}$ is *positive definite* if $\boldsymbol{x}'\boldsymbol{A}\boldsymbol{x} > 0$ for any conforming finite $\boldsymbol{x}$.
(b) Square $\boldsymbol{A}$ is *positive semidefinite* if $\boldsymbol{x}'\boldsymbol{A}\boldsymbol{x} \geq 0$ and $\boldsymbol{x}'\boldsymbol{A}\boldsymbol{x} = 0$ for at least one $\boldsymbol{x} \neq \boldsymbol{0}$.
(c) Square $\boldsymbol{A}$ is *negative definite* if $\boldsymbol{x}'\boldsymbol{A}\boldsymbol{x} < 0$.
(d) Square $\boldsymbol{A}$ is *negative semidefinite* if $\boldsymbol{x}'\boldsymbol{A}\boldsymbol{x} \leq 0$ and $\boldsymbol{x}'\boldsymbol{A}\boldsymbol{x} = 0$ for at least one $\boldsymbol{x} \neq \boldsymbol{0}$.
(e) A *nonnegative definite* matrix is either positive definite or positive semidefinite.
(f) A *nonpositive definite* matrix is either negative definite or negative semidefinite.

A symmetric matrix $\boldsymbol{A}$ with eigenvalues $\{\lambda_j\}$ and $\min(\lambda_j) > 0$ is always *positive definite*. If $\min(\lambda_j) = 0$, then $\boldsymbol{A}$ is *positive semidefinite*. Similarly, $\boldsymbol{A}$ is *negative definite* if $\max(\lambda_j) < 0 \Leftrightarrow \boldsymbol{x}'\boldsymbol{A}\boldsymbol{x} < 0$ and *negative semidefinite* if $\max(\lambda_j) = 0$. Positive definite and negative definite matrices are full rank.

Inner and outer products, such as $\boldsymbol{X}'\boldsymbol{X}$ and $\boldsymbol{X}\boldsymbol{X}'$, are symmetric and therefore simple with a spectral decomposition. They are also necessarily positive definite or positive semidefinite (nonnegative definite).

Lemma 1.24 (a) For symmetric $\boldsymbol{A} = \boldsymbol{V}\text{Dg}(\boldsymbol{\lambda})\boldsymbol{V}'$ of full rank

$$\boldsymbol{A}^{-1} = \boldsymbol{V}\text{Dg}(\boldsymbol{\lambda})^{-1}\boldsymbol{V}' . \tag{1.101}$$

(b) For symmetric $\boldsymbol{A}$ of any rank

$$\boldsymbol{A}^{+} = \boldsymbol{V}[\text{Dg}(\boldsymbol{\lambda})]^{+}\boldsymbol{V}' . \tag{1.102}$$

(c) For symmetric $\boldsymbol{A} = \boldsymbol{V}\text{Dg}(\boldsymbol{\lambda})\boldsymbol{V}'$ of any rank and finite integer $p > 0$

$$\boldsymbol{A}^{p} = \boldsymbol{V}[\text{Dg}(\boldsymbol{\lambda})]^{p}\boldsymbol{V}' . \tag{1.103}$$

(d) For symmetric $\boldsymbol{A} = \boldsymbol{V}\text{Dg}(\boldsymbol{\lambda})\boldsymbol{V}'$ of full rank and finite integer $p > 0$

$$A^{-p} = (A^{-1})^p = (A^p)^{-1} = V[\mathrm{Dg}(\boldsymbol{\lambda})]^{-p}V' \,. \tag{1.104}$$

(e) For symmetric $A = V\mathrm{Dg}(\boldsymbol{\lambda})V'$ of any rank and finite integer $p > 0$

$$(A^+)^p = (A^p)^+ = V\{[\mathrm{Dg}(\boldsymbol{\lambda})]^p\}^+V' \,. \tag{1.105}$$

The lemma illustrates the extent to which the spectral decomposition characterizes the matrix. Merely computing the reciprocals of the eigenvalues allows computing the inverse or generalized inverse of a symmetric matrix. Although we do not pursue the idea here, the concept of a matrix function of a symmetric matrix is well defined. If a function $f(\lambda_j)$ has a valid Taylor series expansion for the eigenvalues of $A = A'$, namely $\{\lambda_j\}$, then $f(A) = V[\mathrm{Dg}(\{f(\lambda_j)\})]V'$ is a well-defined matrix.

Lemma 1.25 (a) For $V = [\,v_j \cdots v_n\,]$ with $V'V = VV' = I_n$, the $n \times n$ and symmetric matrix A with $\mathrm{rank}(A) = n_1 \leq n$ has n_1 nonzero eigenvalues and spectral decomposition $A = V\mathrm{Dg}(\boldsymbol{\lambda})V'$.

(b) The corresponding *constituent matrix decomposition* is

$$A = \sum_{j=1}^{n} \lambda_j v_j v_j' = \sum_{j=1}^{n_1} \lambda_j v_j v_j \,. \tag{1.106}$$

(c) Constituent matrix $G_j = v_j v_j'$ is symmetric and idempotent, which gives $\mathrm{rank}(v_j v_j') = \mathrm{tr}(v_j v_j') = \mathrm{tr}(v_j' v_j) = 1$.

(d) Aggregating the eigenvalues and corresponding eigenvectors into two mutually exclusive and together exhaustive groups of sizes n_1 and n_0, with $n = n_1 + n_0$, allows writing

$$\begin{aligned} A &= [\,V_1 \;\; V_0\,]\begin{bmatrix} \mathrm{Dg}(\boldsymbol{\lambda}_1) & \mathbf{0} \\ 0 & \mathrm{Dg}(\boldsymbol{\lambda}_0) \end{bmatrix}\begin{bmatrix} V_1' \\ V_0' \end{bmatrix} \\ &= V_1\mathrm{Dg}(\boldsymbol{\lambda}_1)V_1' + V_0\mathrm{Dg}(\boldsymbol{\lambda}_0)V_0' \,. \end{aligned} \tag{1.107}$$

(e) If $\boldsymbol{\lambda}_1$ contains the n_1 nonzero eigenvalues then $\boldsymbol{\lambda}_0 = \mathbf{0}$ and $A = V_1\mathrm{Dg}(\boldsymbol{\lambda}_1)V_1'$, with V_1 $n \times n_1$ ($V_1'V_1 = I_{n_1}$), while $\mathrm{Dg}(\boldsymbol{\lambda}_1)$ is $n_1 \times n_1$ of full rank.

(f) The decomposition in part (d), in terms of two groups of eigenvalues and vectors, generalizes to three or more groups.

Lemma 1.26 Any $n \times n$ symmetric and idempotent matrix A of rank $n_1 \leq n$ has **(a)** n_1 eigenvalues of 1, **(b)** all remaining $n - n_1$ eigenvalues of 0, and **(c)** $\mathrm{rank}(A) = \mathrm{tr}(A)$. **(d)** The eigenvectors $V = [\,V_1\; V_0\,]$ may be arranged and scaled such that $V_1'V_1 = I_{n_1}$, $V_0'V_0 = I_{n-n_1}$, $V_1'V_0 = \mathbf{0}$, and

$$A = [\,V_1\; V_0\,]\begin{bmatrix} I_{n_1} & \mathbf{0} \\ \mathbf{0} & \mathbf{0}_{n-n_1} \end{bmatrix}\begin{bmatrix} V_1' \\ V_0' \end{bmatrix} = V_1V_1' \,. \tag{1.108}$$

Lemma 1.27 (a) If $m \times 1$ $\boldsymbol{a}$ contains the m eigenvalues of $m \times m$ $\boldsymbol{A}$, while $n \times 1$ $\boldsymbol{b}$ contains the n eigenvalues of $n \times n$ $\boldsymbol{B}$, then the eigenvalues of $\boldsymbol{C} = \boldsymbol{A} \oplus \boldsymbol{B}$ are the $m+n$ elements of $\boldsymbol{c} = [\,\boldsymbol{a}' \quad \boldsymbol{b}'\,]'$.
(b) The eigenvalues of $\boldsymbol{D} = \boldsymbol{A} \otimes \boldsymbol{B}$ are the mn elements of $\boldsymbol{d} = \boldsymbol{a} \otimes \boldsymbol{b}$.
Neither $\boldsymbol{c}$ nor $\boldsymbol{d}$ is necessarily sorted by size.

Many regression diagnostics methods implicitly focus on finding and avoiding small eigenvalues of $\boldsymbol{X}'\boldsymbol{X}$. The spectral decomposition of the sums of squares and cross products matrix (SSCP, Definition 1.44) is

$$\boldsymbol{X}'\boldsymbol{X} = \boldsymbol{V}\mathrm{Dg}(\boldsymbol{d})\boldsymbol{V}'. \tag{1.109}$$

If $\boldsymbol{X}$ has full rank, then $(\boldsymbol{X}'\boldsymbol{X})^{-1} = \boldsymbol{V}[\mathrm{Dg}(\boldsymbol{d})]^{-1}\boldsymbol{V}'$ exists and

$$\left|(\boldsymbol{X}'\boldsymbol{X})^{-1}\right| = (|(\boldsymbol{X}'\boldsymbol{X})|)^{-1} = \left(\prod_{k=1}^{q} d_k\right)^{-1}. \tag{1.110}$$

As $\min(d_k) \downarrow 0$ the determinant of the inverse $\uparrow \infty$. If $\boldsymbol{X}$ is less than full rank, then $(\boldsymbol{X}'\boldsymbol{X})^{+} = \boldsymbol{V}[\mathrm{Dg}(\boldsymbol{d})]^{+}\boldsymbol{V}'$ and $\boldsymbol{d}$ includes some zeros. Equivalently $(\boldsymbol{X}'\boldsymbol{X})^{+} = \boldsymbol{V}\mathrm{Dg}(\boldsymbol{d}^{-1}, \boldsymbol{0})\boldsymbol{V}'$. The condition of the numerical properties of the matrix may be judged in terms of condition values, such as $\sqrt{\max(\boldsymbol{d})/\min(\boldsymbol{d})}$.

1.13 SOME FACTORS OF SYMMETRIC MATRICES

For a real and nonnegative number, such as $a = 25$, one can find its square root, $\sqrt{a}$, such that $(\sqrt{a})(\sqrt{a}) = a$. A diagonal matrix of nonnegative numbers $\boldsymbol{D} = \mathrm{Dg}(\{d_1, \ldots, d_p\})$ allows defining $\boldsymbol{D}^{1/2} = \mathrm{Dg}(\{d_1^{1/2}, \ldots, d_p^{1/2}\})$, with $\boldsymbol{D}^{1/2}\boldsymbol{D}^{1/2} = \boldsymbol{D}$. The concept extends to square and symmetric matrices in a variety of ways.

Definition 1.39 For $p \times p$ and symmetric $\boldsymbol{A}$, a $p_1 \times p$ *factor* with $p_1 \leq p$ is any $\boldsymbol{F}$ such that $\boldsymbol{A} = \boldsymbol{F}\boldsymbol{F}'$.

The definition does not suffice to guarantee a unique factor, even with full rank $\boldsymbol{A}$. Depending on the side conditions desired, more than one factoring exists. In some analytic or computational settings any choice will serve, while other settings demand a particular choice. Full rank $\boldsymbol{A}$ does always imply any factor is full rank, necessarily $p \times p$, and $\boldsymbol{A}^{-1} = \boldsymbol{F}^{-t}\boldsymbol{F}^{-1}$. More generally, $\mathrm{rank}(\boldsymbol{F}) = \mathrm{rank}(\boldsymbol{A})$, but it may not have the same dimensions when $\boldsymbol{A}$ is not full rank.

Theorem 1.13 Any $p \times p$ and symmetric $\boldsymbol{A}$ can be factored in three ways.
(a) The *square root*, or *Cholesky*, factor adds the side condition of lower triangular factor:

$$A = LL', \tag{1.111}$$

with L $p \times p$, lower triangular, with nonnegative diagonal elements. If $\text{rank}(A) = p$, then L is also full rank with strictly positive diagonal elements.
(b) The spectral decomposition, $A = V\text{Dg}(\lambda)V'$, implies a distinct factoring. If

$$F = V\text{Dg}(\lambda)^{1/2}, \tag{1.112}$$

then $A = FF'$, with $\text{Dg}(\lambda)^{1/2} = \text{Dg}(\{\lambda_j^{1/2}\})$ and $VV' = V'V = I_p$ (without loss of generality). Like A, the factor F is $p \times p$ and $\text{rank}(F) = \text{rank}(A)$.
(c) If $\text{rank}(A) = p_1 \leq p$ and λ_1 indicates the vector of p_1 nonzero eigenvalues, then without loss of generality we may choose $F = [\, V_+ \; V_0 \,]\text{Dg}(\lambda_1, \mathbf{0})^{1/2}$, which is $p \times p$ and rank p_1. In contrast, the matrix

$$F_1 = V_1\text{Dg}(\lambda_1)^{1/2} \tag{1.113}$$

is $p \times p_1$ and rank p_1. In turn

$$\begin{aligned} A &= F_1F_1' \\ &= V_1\text{Dg}(\lambda_1)V_1'. \end{aligned} \tag{1.114}$$

(d) Except in special cases, L, F and F_1 are not symmetric. In some situations a symmetric factor may be preferred. If so, then choosing

$$F_s = V\text{Dg}(\lambda)^{1/2}V' \tag{1.115}$$

ensures $F_s = F_s'$ and $F_sF_s' = F_s'F_s = F_sF_s = A$, with F_s $p \times p$ and $\text{rank}(F_s) = p_1$.
(e) An alternative symmetric factor is given by

$$F_{s+} = V_1\text{Dg}(\lambda_1)^{1/2}V_1', \tag{1.116}$$

with F_{s+} $p \times p$ and $\text{rank}(F_{s+}) = p_1$.
(f) If A has any negative eigenvalues, complex variables will occur in the factor.

We use the theorem most often to factor a covariance matrix. The spectral factor becomes particularly convenient in proofs with less-than-full-rank covariance. The nice properties of diagonal and orthonormal matrices lead to simple expressions for useful generalized inverses, including $(F_1)^+ = \text{Dg}(\lambda_1)^{-1/2}V_1'$. In contrast, for computational purposes the Cholesky factor and related QR decomposition should always be used.

1.14 SINGULAR VALUE DECOMPOSITION

Theorem 1.14 (a) Any $m \times n$ matrix A has a *singular value decomposition,* $A = USV'$, with orthonormal $m \times m$ U of rank m, orthonormal $n \times n$ V of rank n, and $s_{jk} = 0$ except for the main diagonal elements, $\{s_{jj}\}$, which are nonnegative. If $m \geq n \geq n_1 = \text{rank}(A)$, then s has n_1 strictly positive values.

If $m \geq n \geq n_1$, then

$$A = USV' \tag{1.117}$$
$$U'U = UU' = I_m \tag{1.118}$$
$$V'V = VV' = I_n \tag{1.119}$$
$$S = \begin{bmatrix} \mathrm{Dg}(s) \\ 0_{(m-n)\times n} \end{bmatrix}, \tag{1.120}$$

If $m < n$, then a similar construction applies to A'.

(b) The vector s contains the *singular values*, some or all of which may be zero. Necessarily $\mathrm{Dg}(s)^2 = \mathrm{Dg}(\{s_i^2\})$. Singular values equal the positive square roots of the eigenvalues of $n \times n$ $A'A$, which coincide, except perhaps for some zeros, with the eigenvalues of $m \times m$ AA'.

(c) Without loss of generality, elements of s may be assumed to be sorted from largest to smallest. If so, and $m \geq n \geq n_1 = \mathrm{rank}(A)$, then $s' = [\, s_1' \quad 0_{n-n_1}' \,]$ with s_1 the $n_1 \times 1$ vector of strictly positive elements. In turn $U = [\, U_1 \quad U_0 \,]$ with U_1 $m \times n_1$, $U_1'U_1 = I_{n_1}$, $V = [\, V_1 \quad V_0 \,]$ with V_1 $n \times n_1$, $V_1'V_1 = I_{n_1}$ and

$$\begin{aligned} A &= [\, U_1 \; U_0 \,] \begin{bmatrix} \mathrm{Dg}(s_1) & 0 \\ 0 & 0 \end{bmatrix} \begin{bmatrix} V_1' \\ V_0' \end{bmatrix} \\ &= [\, U_1\mathrm{Dg}(s_1) \;\; 0_{m\times(n-n_1)} \,] \begin{bmatrix} V_1' \\ V_0' \end{bmatrix} \\ &= U_1\mathrm{Dg}(s_1)V_1' + 0_{m\times n} \\ &= U_1\mathrm{Dg}(s_1)V_1' \, . \end{aligned} \tag{1.121}$$

(d) Without loss of generality, V and U may be chosen such that

$$A'A = V\mathrm{Dg}(s)^2V' = V_1\mathrm{Dg}(s_1)^2V_1' \tag{1.122}$$

and

$$AA' = U\mathrm{Dg}(s, 0_{m-n})^2U' = U_1\mathrm{Dg}(s_1)^2U_1' \, . \tag{1.123}$$

In practice, the particular choices for U and V are intertwined due to the fact that an eigenvector remains an eigenvector if it is multiplied by -1. In order to account for the signs, having chosen V_1 requires choosing $U_1 = AV_1\mathrm{Dg}(s_1)^{-1}$. Alternately, having chosen U_1 requires choosing $V_1' = \mathrm{Dg}(s_1)^{-1}U_1'A$. If needed, the definitions of U and V may be completed by adding the eigenvectors which correspond to zero eigenvalues of AA' and $A'A$.

Lemma 1.28 For any $m \times n$ matrix A, the $n \times m$ Moore-Penrose (four-condition, unique) generalized inverse is

$$
\begin{aligned}
\boldsymbol{A}^{+} &= \boldsymbol{V}\boldsymbol{S}^{+}\boldsymbol{U}' \\
&= \boldsymbol{V}\big[\, [\mathrm{Dg}(\boldsymbol{s})]^{+} \;\; \mathbf{0}_{n\times(m-n)} \,\big]\boldsymbol{U}' \\
&= \boldsymbol{V}_1[\mathrm{Dg}(\boldsymbol{s}_1)]^{-1}\boldsymbol{U}_1' \,.
\end{aligned} \tag{1.124}
$$

Of course $[\mathrm{Dg}(\boldsymbol{s})]^{+} = \mathrm{Dg}(\{s_1^{-1},\ldots,s_r^{-1},0,\ldots,0\})$.

The singular value and spectral decomposition can be chosen to coincide for square, symmetric, positive definite and semidefinite matrices (such as inner and outer product matrices).

1.15 PROJECTIONS AND OTHER FUNCTIONS OF A DESIGN MATRIX

Design matrices play a central role in the study of linear models. A design matrix, $\boldsymbol{X}$, for N observations and q variables is of dimension $N \times q$ with $N \geq q$ and $\mathrm{rank}(\boldsymbol{X}) = r \leq q$. Many times throughout the book we shall take advantage of various properties of the singular value decomposition (SVD) of $\boldsymbol{X}$, as well as the closely related spectral decompositions of $\boldsymbol{X}'\boldsymbol{X}$ and $\boldsymbol{X}\boldsymbol{X}'$. The following lemma summarizes the properties in an integrated presentation.

Lemma 1.29 (a) For $N \times q$ matrix $\boldsymbol{X}$, with $N \geq q$ and $\mathrm{rank}(\boldsymbol{X}) = r \leq q$, the SVD gives

$$
\begin{aligned}
\boldsymbol{X} &= \boldsymbol{L}\begin{bmatrix}\mathrm{Dg}(\boldsymbol{s}) \\ \mathbf{0}\end{bmatrix}\boldsymbol{R}' \\
&= [\,\boldsymbol{L}_1 \;\; \boldsymbol{L}_0\,]\begin{bmatrix}\mathrm{Dg}(\boldsymbol{s}_1) & \mathbf{0} \\ \mathbf{0} & \mathbf{0}\end{bmatrix}\begin{bmatrix}\boldsymbol{R}_1' \\ \boldsymbol{R}_0'\end{bmatrix} \\
&= \boldsymbol{L}_1\mathrm{Dg}(\boldsymbol{s}_1)\boldsymbol{R}_1' \,.
\end{aligned} \tag{1.125}
$$

The first product has dimensions $(N \times N)(N \times q)(q \times q)$, and the last product has dimensions $(N \times r)(r \times r)(r \times q)$. Without loss of generality, singular values are sorted from largest to smallest, $\boldsymbol{s} = [\,\boldsymbol{s}_1' \; \mathbf{0}_{n-n_1}\,]$, $\boldsymbol{s}_1$ is $n_1 \times 1$ with all strictly positive values. Also $\boldsymbol{R} = [\,\boldsymbol{R}_1 \; \boldsymbol{R}_0\,]$, $\boldsymbol{R}'\boldsymbol{R} = \boldsymbol{R}\boldsymbol{R}' = \boldsymbol{I}_q$ with $\boldsymbol{R}_1$ $q \times n_1$. Similarly $\boldsymbol{L} = [\,\boldsymbol{L}_1 \; \boldsymbol{L}_0\,]$ with $\boldsymbol{L}_1$ $n \times n_1$ and $\boldsymbol{L}'\boldsymbol{L} = \boldsymbol{L}\boldsymbol{L}' = \boldsymbol{I}_N$.

(b) Here $\boldsymbol{R}_1$ and $\boldsymbol{R}_0$ are sets of orthonormal eigenvectors (unique, up to reflections) corresponding to nonzero ($\boldsymbol{R}_1$) and zero ($\boldsymbol{R}_0$) eigenvalues of $\boldsymbol{X}'\boldsymbol{X}$. Specifically, without loss of generality, SVD properties correspond to assuming

$$
\begin{aligned}
\boldsymbol{X}'\boldsymbol{X} &= \boldsymbol{R}\mathrm{Dg}(\boldsymbol{s})^2\boldsymbol{R}' \\
&= \boldsymbol{R}_1\mathrm{Dg}(\boldsymbol{s}_1)^2\boldsymbol{R}_1' \,.
\end{aligned} \tag{1.126}
$$

(c) Similarly, $\boldsymbol{L}_1$ and $\boldsymbol{L}_0$ are (nonunique) sets of orthonormal eigenvectors corresponding to nonzero ($\boldsymbol{L}_1$) and zero ($\boldsymbol{L}_0$) eigenvalues of $\boldsymbol{X}\boldsymbol{X}'$. Specifically, without loss of generality, SVD properties correspond to assuming

$$\begin{aligned} XX' &= L\begin{bmatrix} \mathrm{Dg}(s_1, 0)^2 & 0 \\ 0 & 0 \end{bmatrix} L' \\ &= L_1 \mathrm{Dg}(s_1)^2 L_1' . \end{aligned} \tag{1.127}$$

(d) *Given* R_1 and s_1, necessarily

$$L_1 = XR_1\mathrm{Dg}(s_1)^{-1} . \tag{1.128}$$

Alternately, *given* R_1 and s_1, necessarily

$$R_1 = \mathrm{Dg}(s_1)^{-1} L_1' X . \tag{1.129}$$

However, only one of L_1 and R_1 can be chosen freely, due to the need to account for the sign ambiguity of eigenvectors (as detailed in Theorem 1.12).

(e) Furthermore

$$\begin{aligned} (X'X)^+ &= R[\mathrm{Dg}(s)^2]^+ R' \\ &= R_1 \mathrm{Dg}(s_1)^{-2} R_1' . \end{aligned} \tag{1.130}$$

Proof. The results follow from the SVD definition, based on properties of orthonormal matrices, inverses, and eigenanalysis of symmetric matrices.

Many functions of design matrices, especially ones involving $(X'X)^-$, must be understood to develop the theory of linear models. The value of the results lies in guaranteeing uniqueness, symmetry, and the ability to simplify expressions in estimators. Even though $(X'X)$ is always symmetric, $(X'X)^-$ need not be symmetric. Furthermore a less-than-full-rank X leads to infinitely many choices for $(X'X)^-$, while $(X'X)^+$ is unique. The theorem gives many equalities useful for simplification and many invariance properties guaranteeing uniqueness. The corollary summarizes many special properties associated with the matrix $H = X(X'X)^- X'$, which plays a key role in linear models estimation and distribution theory.

Definition 1.40 (a) The matrix $H = X(X'X)^- X'$ is a *projection matrix*.
(b) Matrix $X(X'X)^- X'Y$ is the *projection of* Y into the space spanned by the columns of X.

Theorem 1.15 For any matrix X, the following all hold.
(a) $[(X'X)^-]'$ is also a generalized inverse of $(X'X)$.
(b) $X(X'X)^- X'X = X$; hence $(X'X)^- X'$ is a generalized inverse of X.
(c) $X(X'X)^- X'$ is invariant to $(X'X)^-$.
(d) $X(X'X)^- X'$ is always symmetric, even if $(X'X)^-$ is not symmetric.
(e) $X[(X'X)^-]'X'X = X$.
(f) $X'X(X'X)^- X' = X'$.
(g) $X'X[(X'X)^-]'X' = X'$.

(h) $X[(X'X)^{-}]'X' = X(X'X)^{-}X'$.
(i) $X[(X'X)^{-}]'X'$ is always symmetric even if $[(X'X)^{-}]'$ is not symmetric.
(j) $H = X(X'X)^{-}X' = X(X'X)^{+}X'$.
(k) *All* of the matrices mentioned in the theorem have rank $r = \text{rank}(X)$.

Proof. Searle (1971) stated and proved parts (a)–(d) as his Theorem 7 and parts (e)–(i) as a corollary. Part (j) follows from uniqueness of the four-condition inverse. Part (k) follows from four- and one- condition inverse properties, as well as Lemma 1.29. □

Corollary 1.15 For any $N \times q$ matrix X with $\text{rank}(X) = r \leq q \leq N$ the projection matrix, $H = X(X'X)^{-}X'$, is **(a)** unique, **(b)** symmetric, **(c)** idempotent and **(d)** rank r, with **(e)** r eigenvalues of one and $N - r$ of zero. Furthermore, $I - H$ is **(f)** unique, **(g)** symmetric, **(h)** idempotent and **(i)** rank $N - r$, with **(j)** $N - r$ eigenvalues of one and r of zero. **(k)** Also, $H(I - H) = 0$. **(l)** With exactly the notation of Lemma 1.29,

$$\begin{aligned} H &= L\begin{bmatrix} I_r & 0 \\ 0 & 0_{N-r} \end{bmatrix} L' \\ &= [\,L_1\ L_0\,]\begin{bmatrix} I_r & 0 \\ 0 & 0_{N-r} \end{bmatrix}\begin{bmatrix} L_1' \\ L_0' \end{bmatrix} \\ &= L_1 L_1' \end{aligned} \tag{1.131}$$

and

$$\begin{aligned} I - H &= L\begin{bmatrix} I_r & 0 \\ 0 & 0_{N-r} \end{bmatrix} L' \\ &= [\,L_1\ L_0\,]\begin{bmatrix} I_r & 0 \\ 0 & 0_{N-r} \end{bmatrix}\begin{bmatrix} L_1' \\ L_0' \end{bmatrix} \\ &= L_0 L_0'\,. \end{aligned} \tag{1.132}$$

Here $L'L = LL' = I_N$, while $L_1'L_1 = I_r$ and $L_0'L_0 = I_{N-r}$. **(m)** Finally, L contains orthonormal eigenvectors corresponding to nonzero (L_1) and zero (L_0) eigenvalues of XX':

$$\begin{aligned} XX' &= L\begin{bmatrix} \text{Dg}(s_1, 0)^2 & 0 \\ 0 & 0 \end{bmatrix} L' \\ &= L_1 \text{Dg}(s_1)^2 L_1'\,. \end{aligned} \tag{1.133}$$

Proof. Results follow from combining forms in the theorem and Lemma 1.29. Writing $H^2 = (L_1L_1')(L_1L_1') = L_1L_1'$ demonstrates that H is idempotent. □

The matrix $H = X(X'X)^{-}X' = X(X'X)^{+}X'$ earned the nickname "hat matrix" in the linear model $y = X\beta + e$ because $\hat{y} = Hy$. Uniqueness of H

implies uniqueness of many functions of $\boldsymbol{H}$, including $\widehat{\boldsymbol{y}}$, as well as $\widehat{\boldsymbol{e}} = (\boldsymbol{I} - \boldsymbol{H})\boldsymbol{y}$ and $\widehat{\sigma}^2 = \widehat{\boldsymbol{e}}'\widehat{\boldsymbol{e}}/(N-r)$.

Lemma 1.30 If $\boldsymbol{A}$ $(n \times n)$ is a constant matrix and $\boldsymbol{F}$ $(n \times n)$ is nonsingular with $\boldsymbol{B} = \boldsymbol{F}\boldsymbol{F}'$, then $\boldsymbol{F}'\boldsymbol{A}\boldsymbol{F}$ is idempotent $\Leftrightarrow$ $\boldsymbol{A}\boldsymbol{B}$ is idempotent.

Proof. $\boldsymbol{F}\boldsymbol{F}' = \boldsymbol{B}$ implies $\boldsymbol{A}\boldsymbol{F}\boldsymbol{F}' = \boldsymbol{A}\boldsymbol{B}$. Premultiplying by $\boldsymbol{F}'$ and postmultiplying by $\boldsymbol{F}^{-t}$ gives $\boldsymbol{F}'\boldsymbol{A}\boldsymbol{F} = \boldsymbol{F}'\boldsymbol{A}\boldsymbol{B}\boldsymbol{F}^{-t}$. Hence $(\boldsymbol{F}'\boldsymbol{A}\boldsymbol{F})^2 = \boldsymbol{F}'\boldsymbol{A}\boldsymbol{B}\boldsymbol{A}\boldsymbol{B}\boldsymbol{F}^{-t}$. If $(\boldsymbol{F}'\boldsymbol{A}\boldsymbol{F})^2 = \boldsymbol{F}'\boldsymbol{A}\boldsymbol{F}$ then $(\boldsymbol{A}\boldsymbol{B})^2 = \boldsymbol{A}\boldsymbol{B}$, which implies $(\boldsymbol{F}'\boldsymbol{A}\boldsymbol{F})^2 = (\boldsymbol{F}'\boldsymbol{A}\boldsymbol{F})$. □

Definition 1.41 Matrix $\boldsymbol{P}$ $(N \times N)$ is a *permutation matrix* if it is obtained by permuting the rows (only) of $\boldsymbol{I}_N$.

Given the definition, premultiplying by a conforming permutation matrix exchanges rows. Postmultiplying by a conforming transposed permutation matrix exchanges columns. A permutation matrix is always orthonormal and full rank, which implies $\boldsymbol{P}^{-1} = \boldsymbol{P}'$. A permutation matrix may always be found which, when multiplied times a conforming matrix, permutes the rows (or columns) to any new order desired.

1.16 SPECIAL PROPERTIES OF PATTERNED MATRICES

A partitioned matrix may be thought of as a supermatrix, a matrix containing matrices. Most importantly, if the partitions conform for addition and matrix multiplication then the results can be expressed in terms of the partitions, without considering particular elements. *It is crucial to verify conformation of each pair of partitions as well as the total matrices.* When the partitions do conform, the rules of matrix multiplication apply to create what might be thought of as "super" or "meta" multiplication or addition. Examples include the following (when the matrices conform for the operations):

$$\boldsymbol{A}[\boldsymbol{B}_1 \;\; \boldsymbol{B}_2] = [\boldsymbol{A}\boldsymbol{B}_1 \;\; \boldsymbol{A}\boldsymbol{B}_2] \tag{1.134}$$

$$[\boldsymbol{B}_1 \;\; \boldsymbol{B}_2]\begin{bmatrix} \boldsymbol{C}_1 \\ \boldsymbol{C}_2 \end{bmatrix} = \boldsymbol{B}_1\boldsymbol{C}_1 + \boldsymbol{B}_2\boldsymbol{C}_2 \tag{1.135}$$

$$[\boldsymbol{B}_1 \;\; \boldsymbol{B}_2][\boldsymbol{B}_1 \;\; \boldsymbol{B}_2]' = \boldsymbol{B}_1\boldsymbol{B}_1' + \boldsymbol{B}_2\boldsymbol{B}_2' \tag{1.136}$$

$$[\boldsymbol{B}_1 \;\; \boldsymbol{B}_2]'[\boldsymbol{B}_1 \;\; \boldsymbol{B}_2] = \begin{bmatrix} \boldsymbol{B}_1'\boldsymbol{B}_1 & \boldsymbol{B}_1'\boldsymbol{B}_2 \\ \boldsymbol{B}_2'\boldsymbol{B}_1 & \boldsymbol{B}_2'\boldsymbol{B}_2 \end{bmatrix}. \tag{1.137}$$

Not only multiplication and addition, but also determinants and inverses often can be expressed compactly and conveniently in terms of partitions.

Theorem 1.16 If $p \times p$ $\boldsymbol{A}$ is positive definite and symmetric, $\boldsymbol{B} = \boldsymbol{A}^{-1}$, and $\boldsymbol{A}$ and $\boldsymbol{B}$ are partitioned with corresponding submatrices of the same sizes,

$$A = \begin{bmatrix} A_{11} & A_{12} \\ A_{21} & A_{22} \end{bmatrix} \tag{1.138}$$

$$B = \begin{bmatrix} B_{11} & B_{12} \\ B_{21} & B_{22} \end{bmatrix}, \tag{1.139}$$

then (given the indicated operations are valid)

$$B_{11} = (A_{11} - A_{12}A_{22}^{-1}A_{21})^{-1} = A_{11}^{-1} + A_{11}^{-1}A_{12}B_{22}A_{21}A_{11}^{-1} \tag{1.140}$$

$$B_{12} = -A_{11}^{-1}A_{12}B_{22} \tag{1.141}$$

$$B_{21} = -A_{22}^{-1}A_{21}B_{11} \tag{1.142}$$

$$B_{22} = (A_{22} - A_{21}A_{11}^{-1}A_{12})^{-1} = A_{22}^{-1} + A_{22}^{-1}A_{21}B_{11}A_{12}A_{22}^{-1}. \tag{1.143}$$

The subscript pattern gives a mnemonic device for remembering the formulas.

Theorem 1.17 For any $p \times p$ A partitioned as

$$A = \begin{bmatrix} A_{11} & A_{12} \\ A_{21} & A_{22} \end{bmatrix}, \tag{1.144}$$

(a) if either $A_{12} = 0$ or $A_{21} = 0$, then $|A| = |A_{11}||A_{22}|$.
(b) For any (conforming) $\{A_{11}, A_{12}, A_{21}\}$, if A_{22} is full rank, then $|A| = |A_{22}||A_{11} - A_{12}A_{22}^{-1}A_{21}|$.
(c) For any A_{12} or A_{21}, if A_{11} is full rank, then $|A| = |A_{11}||A_{22} - A_{21}A_{11}^{-1}A_{12}|$.

Theorem 1.18 (a) If A and B are the same size then

$$(A+B) \otimes C = (A \otimes C) + (B \otimes C) \tag{1.145}$$

and

$$C \otimes (A+B) = (C \otimes A) + (C \otimes B). \tag{1.146}$$

(b) If A, B, C, and D are $m \times h$, $p \times k$, $h \times n$ and $k \times q$, respectively, then

$$(A \otimes B)(C \otimes D) = (AC) \otimes (BD). \tag{1.147}$$

1.17 FUNCTION OPTIMIZATION AND MATRIX DERIVATIVES

Deriving properties of linear models often leads to the need to maximize or minimize a smooth function. In addition, side conditions may be desired which impose constraints on the optimization. The most convenient approach usually involves creating a system of equations to be solved by determining derivatives of the function. In turn, the introduction of Lagrangian multipliers often satisfies the need to impose side conditions.

Given the focus of the book, we assume the reader has substantial knowledge of the theory and practical use of derivatives for the analysis of scalar functions. In the few instances in the book when we need to know specific vector and matrix forms of derivatives we present the required results without comment. A defensible description of even the basic rules of matrix derivatives requires attention to concepts in real analysis rather far from the topic at hand. We recommend the reader seeking the motivation for the derivatives used here consult the excellent book by Magnus and Neudecker (1988) for a definitive treatment. Schott (2005) provided a more brief but carefully constructed introduction.

1.18 STATISTICAL NOTATION INVOLVING MATRICES

A common notation for the study of probability uses Greek letters for parameters, uppercase Roman letters for random variables, and lowercase Roman letters for realizations of random variables (particular sample values). The convention conflicts with matrix notation distinctions between scalars, vectors, and matrices. The importance of matrix expressions throughout the book means that matrix notation must dominate. Consequently the reader must often distinguish fixed from random, known from unknown, observed from unobserved, and observable from unobservable via the context of the discussion. When in doubt about a particular item, simply search backwards in the text to discover where the variable was introduced.

Expressions involving random variables use the portion of conventional notation that does not conflict with the matrix notation introduced in Section 1.1. Roman letters towards the beginning of the alphabet, such as $\{c, \boldsymbol{c}, \boldsymbol{C}\}$, will usually represent constants (either known or unspecified). In contrast, Roman letters towards the end of the alphabet, such as $\{y, \boldsymbol{y}, \boldsymbol{Y}\}$, will usually represent random variables (which may take on infinitely many values). In turn, $\{y_*, \boldsymbol{y}_*, \boldsymbol{Y}_*\}$ will represent an arbitrary possible value, such as a variable of integration, while $\{y_0, \boldsymbol{y}_0, \boldsymbol{Y}_0\}$ will represent a single but unspecified particular value. For two jointly absolutely continuous random variables, $\{x, y\}$ the rules lead to the expression $\mathrm{E}(x|y = y_0) = \int x_* f_{x,y}(x_*, y_0)dx_*$. Greek letters, such as $\{\beta, \boldsymbol{\beta}, \boldsymbol{B}\}$, will represent parameters (fixed and unknown population properties). Corresponding (random) estimators will be indicated $\{\widehat{\beta}, \widehat{\boldsymbol{\beta}}, \widehat{\boldsymbol{B}}\}$ or $\{\widetilde{\beta}, \widetilde{\boldsymbol{\beta}}, \widetilde{\boldsymbol{B}}\}$.

1.19 STATISTICAL FORMULAS

With $N \gg p$, matrix $\boldsymbol{Y} = [\boldsymbol{y}_1\ \boldsymbol{y}_2 \cdots \boldsymbol{y}_p]$ contains a collection of scores on p variables. Rows are independent sampling units and columns are variables. Although discussed in more detail in Chapter 7, the concepts of mean and variance for a vector are useful here.

Definition 1.42 (a) For $\boldsymbol{y} = \{y_j\}$ a vector of random variables with a well-defined joint distribution $\mathrm{E}(\boldsymbol{y}) = \{\mathrm{E}(y_j)\}$ defines the *population mean*, when $\mathrm{E}(y_j)$ exists $\forall j$.
(b) Similarly, $\mathrm{E}(\boldsymbol{Y}) = \{\mathrm{E}(y_{jk})\}$ when it exists.
(c) For $\boldsymbol{y} = \{y_j\}$ a vector of random variables with a well-defined joint distribution and $\mathrm{E}(\boldsymbol{y}) = \boldsymbol{\mu}$, $\mathcal{V}(\boldsymbol{y}) = \{\mathrm{E}[(y_j - \mu_j)(y_k - \mu_k)]\}$ defines the *population covariance* matrix, when $\mathrm{E}[(y_j - \mu_j)(y_k - \mu_k)]$ exists $\forall (j, k)$.

The concept of $\mathcal{V}(\boldsymbol{Y})$ is not well defined. It is customary and fully satisfactory to consider either $\mathcal{V}[\mathrm{vec}(\boldsymbol{Y})]$ or $\mathcal{V}[\mathrm{vec}(\boldsymbol{Y}')]$, which are well defined. They differ only by a permutation of rows, which can be achieved by multiplication with a permutation matrix. The following definition and lemma are repeated and discussed in Chapter 7. They are presented here to allow a precise description of principal components analysis of random data.

Definition 1.43 (a) A random vector $\boldsymbol{y}$ $(p \times 1)$ with finite second moments has an associated *covariance matrix*

$$\mathcal{V}(\boldsymbol{y}) = \{\mathrm{E}(y_j y_k)\} - \{\mathrm{E}(y_j)\mathrm{E}(y_k)\} = \boldsymbol{\Sigma}. \tag{1.148}$$

(b) If all $\sigma_{jj} = \langle \boldsymbol{\Sigma} \rangle_{jj}$ are such that $0 < \sigma_{jj} < \infty$, then $\boldsymbol{y}$ has *correlation matrix*

$$\begin{aligned} \boldsymbol{P} &= \mathrm{Dg}(\{\sigma_{11}, \ldots, \sigma_{pp}\})^{-1/2} \boldsymbol{\Sigma} \mathrm{Dg}(\{\sigma_{11}, \ldots, \sigma_{pp}\})^{-1/2} \\ &= \{\rho_{jk}\}. \end{aligned} \tag{1.149}$$

Lemma 1.31 Finite second moments for a random vector $\boldsymbol{y}$ $(p \times 1)$ guarantee the existence of finite covariance and correlation matrices for the population.
(a) The population covariance matrix can be expressed as

$$\begin{aligned} \mathcal{V}(\boldsymbol{y}) &= \{\mathrm{E}(y_j y_k)\} - \{\mathrm{E}(y_j)\mathrm{E}(y_k)\} \\ &= \{\mathrm{E}(y_j y_k)\} - \{\mu_j \mu_k\} \\ &= \mathrm{E}(\boldsymbol{y}\boldsymbol{y}') - \boldsymbol{\mu}\boldsymbol{\mu}' \\ &= \mathrm{E}(\boldsymbol{y}\boldsymbol{y}') - \mathrm{E}(\boldsymbol{y})[\mathrm{E}(\boldsymbol{y})]' \\ &= \boldsymbol{\Sigma}. \end{aligned} \tag{1.150}$$

It is symmetric and nonnegative definite (positive definite or positive semidefinite). The covariance matrix contains centered, average cross products, with diagonal element σ_{jj} the variance of y_j.
(b) If $\sigma_{jj} > 0$ $\forall j$, then the population correlation matrix $\boldsymbol{P}$ is well defined and nonnegative definite, with $\rho_{jk} = \sigma_{jk} / \sqrt{\sigma_{jj}\sigma_{kk}}$. The correlation matrix contains centered and scaled average cross products.

Definition 1.44 (a) A set of N observations of a $p \times 1$ random vector arranged in an $N \times p$ matrix, with each observation forming a row, provides a *data matrix*, such as $\boldsymbol{Y}$.
(b) With $N \times 1$ vector $\mathbf{1}_N = [1\,1 \cdots 1]'$, the $p \times 1$ *sample mean* vector is

$$\overline{\boldsymbol{y}} = \boldsymbol{Y}'\mathbf{1}_N/N. \tag{1.151}$$

(c) The $p \times p$ *sample SSCP matrix* $\boldsymbol{Y}'\boldsymbol{Y}$ contains sums of squares and cross products.
(d) The $p \times p$ *sample covariance matrix* is

$$\begin{aligned}\boldsymbol{S} &= (\boldsymbol{Y} - \mathbf{1}_N\overline{\boldsymbol{y}}')'(\boldsymbol{Y} - \mathbf{1}_N\overline{\boldsymbol{y}}')/N \\ &= \boldsymbol{Y}'[\boldsymbol{I}_N - \mathbf{1}(\mathbf{1}'\mathbf{1})^{-1}\mathbf{1}']\boldsymbol{Y}/N \\ &= \boldsymbol{Y}'\boldsymbol{Y}/N - \overline{\boldsymbol{y}}\,\overline{\boldsymbol{y}}'.\end{aligned} \tag{1.152}$$

(e) If $s_{jj} > 0$, then the $p \times p$ *sample correlation matrix* is

$$\boldsymbol{R} = \mathrm{Dg}(\{s_{11}, \ldots, s_{pp}\})^{-1/2}\boldsymbol{S}\mathrm{Dg}(\{s_{11}, \ldots, s_{pp}\})^{-1/2}, \tag{1.153}$$

with $r_{jk} = s_{jk}/\sqrt{s_{jj}s_{kk}}$.

Lemma 1.32 (a) If data matrix $\boldsymbol{Y}$ has independent rows with $\mathrm{E}[\mathrm{row}_i(\boldsymbol{Y})] \equiv \boldsymbol{\mu}'$ and $\mathcal{V}[\mathrm{row}_i(\boldsymbol{Y})] = \boldsymbol{\Sigma}$, then

$$\widehat{\boldsymbol{\Sigma}} = \boldsymbol{S}N/(N-1) \tag{1.154}$$

and $\mathrm{E}(\widehat{\boldsymbol{\Sigma}}) = \boldsymbol{\Sigma}$. More generally, in fitting multivariate linear models, $\widehat{\boldsymbol{\Sigma}} = \boldsymbol{S}N/(N-r)$, in which r equals the rank of the design matrix, provides an unbiased estimator when $N > r$.
(b) If the sample correlation matrix estimates the population matrix, $\boldsymbol{R} = \widehat{\boldsymbol{P}}$, then $\mathrm{E}(\widehat{\boldsymbol{P}}) \neq \boldsymbol{P}$, except when $\boldsymbol{P} = \boldsymbol{I}_p$ (which implies $\rho_{jk} = 0$ if $j \neq k$). If $\boldsymbol{P} \neq \boldsymbol{I}_p$, then no constant c can be found to make $c\widehat{\boldsymbol{P}}$ unbiased.

If r_{-jj} indicates diagonal element j of $\boldsymbol{R}^{-1}$, then $R_j^2 = (r_{-jj} - 1)/\,r_{-jj}$ is the squared multiple correlation between variable j and the remaining $p-1$ variables.

Partitioning the variables into two sets, with $\boldsymbol{Y} = [\,\boldsymbol{Y}_1\ \boldsymbol{Y}_2\,]$, gives p_1 variable in $\boldsymbol{Y}_1$ and p_2 variables in $\boldsymbol{Y}_2$. A corresponding partitioning of the covariance matrix gives

$$\boldsymbol{S} = \begin{bmatrix}\boldsymbol{S}_{11} & \boldsymbol{S}_{12} \\ \boldsymbol{S}_{21} & \boldsymbol{S}_{22}\end{bmatrix}. \tag{1.155}$$

The sample covariance for $\boldsymbol{Y}_1|\boldsymbol{Y}_2$ equals

$$\boldsymbol{S}_{Y_1|Y_2} = \boldsymbol{S}_{11} - \boldsymbol{S}_{12}\boldsymbol{S}_{22}^{-1}\boldsymbol{S}_{21}, \tag{1.156}$$

with corresponding unbiased estimator $\boldsymbol{S}_{Y_1|Y_2}N/(N - p_2 - 1)$. If $\boldsymbol{D}_{Y_1|Y_2}$ indicates the diagonal matrix containing the diagonal elements of $\boldsymbol{S}_{Y_1|Y_2}$, the corresponding correlation matrix equals

$$\boldsymbol{R}_{Y_1|Y_2} = \boldsymbol{D}_{Y_1|Y_2}^{-1/2} \boldsymbol{S}_{Y_1|Y_2} \boldsymbol{D}_{Y_1|Y_2}^{-1/2} . \tag{1.157}$$

Elements of $\boldsymbol{R}_{Y_1|Y_2}$ equal full partial correlations among the variables in $\boldsymbol{Y}_1$ because any two variables in $\boldsymbol{Y}_1$ have both been adjusted for variables in $\boldsymbol{Y}_2$. Muller and Fetterman (2002, Chapter 6) gave a brief overview of various sorts of partial correlations in the context of univariate linear models regression.

1.20 PRINCIPAL COMPONENTS

Given a set of observations on a group of variables, it may be helpful to describe a simple model for the structure of the corresponding cross products, covariance, or correlation matrix. Three distinct applications may motivate the process: (1) analytic decomposition of population variables in a proof, (2) numerical analysis of observed values on a set of variables, and (3) data analysis for exploratory or confirmatory purposes. In the present section and the remainder of the book, we focus on the first application. Muller and Fetterman (2002, Chapter 8) provided a brief introduction to using principal components analysis for regression diagnostics. Timm (2002, Chapter 8) detailed the basic methods of principal components in the context of factor analysis models. Both texts also include useful additional references. Jackson (1991), and Jolliffe (2002) provided entire books about component analysis, while Basilevsky (1994) discussed component analysis within the more general context of factor analysis.

We strongly prefer factor analysis methods and related covariance structure models over principal components analysis for building and evaluating covariance models. One important reason arises from the concept of robustness to overfitting. A principal components analysis (PCA) model defines a special case of a factor analysis (FA) model. If the PCA model holds but the data analyst uses the FA model, then no harm should result. An adequate sample and analysis strategy should lead to reducing the model appropriately. In contrast, fitting a PCA model when the FA model holds always leads to bias and an invalid model, no matter how large the sample size. A parallel conclusion holds in seeking the best model for the response mean in a multiple regression. Overfitting only costs a bit of sensitivity, while underfitting creates bias and invalid models. Widaman (2004; also 1993) provided an extensive discussion of the question.

The discussion of principal components analysis will be cast primarily in terms of decomposing a population covariance matrix $\boldsymbol{\Sigma}$. The underlying data are $N \times p$ $\boldsymbol{Y} = [\,\boldsymbol{y}_1\, \boldsymbol{y}_2 \cdots \boldsymbol{y}_p\,]$, with $N \gg p$, a collection of scores on p variables. Here rows of $\boldsymbol{Y}$ are independent, with common covariance $\mathcal{V}[\text{row}_i(\boldsymbol{Y})] = \boldsymbol{\Sigma}$. Most often, analysis involves $\boldsymbol{Y}$ or $\boldsymbol{Y}_d = [\boldsymbol{I}_N - \boldsymbol{1}(\boldsymbol{1}'\boldsymbol{1})^{-1}\boldsymbol{1}']\boldsymbol{Y} = \boldsymbol{Y} - \boldsymbol{1}\overline{\boldsymbol{y}}' = \{y_{ij} - \overline{y}_j\}$, the mean-centered transformation of $\boldsymbol{Y}$. With the impact of the means (first moments) suppressed, the focus lies on modeling the second moments, which are variances and covariances.

Definition 1.45 (a) The first *principal component* equals the linear combination of a set of variables which has maximum variance, among all such combinations with unit-length coefficient vectors.
(b) Such a linear combination of a set of variables defines a new variable, which will be called a *variate*.

Any covariance matrix has many special properties. A population or sample covariance matrix is always symmetric and can be expressed as an inner product. It always has a spectral decomposition with only positive or zero eigenvalues, which are the variances of the principal component variables. Each corresponding eigenvector holds a set of regression weights for the original variables which define the principal variables. The principal component variables are uncorrelated, with successively maximum variances (the eigenvalues).

The principal components have corresponding principal component scores, $\boldsymbol{Y}_c = [\,\boldsymbol{y}_{c,1}\;\boldsymbol{y}_{c,2}\cdots\boldsymbol{y}_{c,p}\,]$, an $N\times p$ matrix. The scores have many special properties. (1) Each column of $\boldsymbol{Y}_c$ equals a linear combination of $\boldsymbol{Y}_d$: $\boldsymbol{Y}_c = \boldsymbol{Y}_d\boldsymbol{\Upsilon}$. (2) $\mathcal{V}[\text{row}_i(\boldsymbol{Y}_c)] = \text{Dg}(\boldsymbol{\lambda})$. (3) The first set of N scores, $\boldsymbol{y}_{c,1}$, has maximum variance among all linear combinations (subject to the unit-length constraint $\boldsymbol{v}_1'\boldsymbol{v}_1 = 1$), the second set of scores, $\boldsymbol{y}_{c,2}$, has maximum variance among all linear combinations given $\boldsymbol{y}_{c,1}$, etc. (4) The weight matrix $\boldsymbol{\Upsilon} = [\,\boldsymbol{v}_1\;\boldsymbol{v}_2\cdots\boldsymbol{v}_p\,]$ must be the (orthonormal and full-rank) eigenvectors of the covariance matrix, $\boldsymbol{\Sigma} = \boldsymbol{\Upsilon}\text{Dg}(\boldsymbol{\lambda})\boldsymbol{\Upsilon}'$ with $\boldsymbol{\Upsilon}'\boldsymbol{\Upsilon} = \boldsymbol{\Upsilon}\boldsymbol{\Upsilon}' = \boldsymbol{I}_p$. (5) For $k \neq k'$, the covariance and correlation between $\boldsymbol{y}_{c,k}$ and $\boldsymbol{y}_{c,k'}$ are zero.

Spectral decomposition of a sample estimator gives $\widehat{\boldsymbol{\Sigma}} = \widehat{\boldsymbol{\Upsilon}}\text{Dg}(\widehat{\boldsymbol{\lambda}})\widehat{\boldsymbol{\Upsilon}}'$. For the particular set of sample values in hand, $\boldsymbol{Y}_0$, which corresponds to $\widehat{\boldsymbol{\Sigma}}_0 = \widehat{\boldsymbol{\Upsilon}}_0\text{Dg}(\widehat{\boldsymbol{\lambda}}_0)\widehat{\boldsymbol{\Upsilon}}_0'$, sample estimates attain optimal numerical properties in parallel to population properties. The component score estimator is $\widehat{\boldsymbol{Y}}_c = \boldsymbol{Y}_d\widehat{\boldsymbol{\Upsilon}}$, with corresponding estimate $\widehat{\boldsymbol{Y}}_{0c} = \boldsymbol{Y}_{0d}\widehat{\boldsymbol{\Upsilon}}_0$. Each $(p\times 1)$ column of $\widehat{\boldsymbol{\Upsilon}}_0$ serves as the estimate of a set of p regression coefficients.

Here $\boldsymbol{Y}_{0c}$ $(N\times p)$ is a matrix with each column a set of component score estimates. Also $\overline{\boldsymbol{y}}_{0d} = \boldsymbol{0}$ implies $\overline{\boldsymbol{y}}_{0c} = \widehat{\boldsymbol{\Upsilon}}_0\overline{\boldsymbol{y}}_{0d} = \boldsymbol{0}$. The covariance matrix observed among the component score estimates is

$$\begin{aligned}
\boldsymbol{S}_{0c} &= (\boldsymbol{Y}_{0c}'\boldsymbol{Y}_{0c}/N - \overline{\boldsymbol{y}}_{0c}\overline{\boldsymbol{y}}_{0c}')N/(N-1)\\
&= \boldsymbol{Y}_{0c}'\boldsymbol{Y}_{0c}/(N-1)\\
&= \widehat{\boldsymbol{\Upsilon}}_0'\boldsymbol{Y}_{0d}'\boldsymbol{Y}_{0d}\widehat{\boldsymbol{\Upsilon}}_0/(N-1)\\
&= \widehat{\boldsymbol{\Upsilon}}_0'\widehat{\boldsymbol{\Sigma}}_0\widehat{\boldsymbol{\Upsilon}}_0\\
&= \text{Dg}(\widehat{\boldsymbol{\lambda}}_0).
\end{aligned} \tag{1.158}$$

Consequently the estimated component scores $\boldsymbol{Y}_{0c} = \boldsymbol{Y}_{0d}\widehat{\boldsymbol{\Upsilon}}_0$ are uncorrelated.

Also, the variance of the first component is $\widehat{\lambda}_{0,1}$, the largest eigenvalue of $\widehat{\Sigma}_0$, the variance of the second component is $\widehat{\lambda}_{0,2}$, the second largest eigenvalue of $\widehat{\Sigma}_0$, etc. The $p \times p$ factor (matrix) for $\widehat{\Sigma}$, based on a spectral decomposition, equals $\widehat{\boldsymbol{\Phi}}_0 = \widehat{\boldsymbol{\Upsilon}}_0 \mathrm{Dg}(\widehat{\boldsymbol{\lambda}}_0)^{1/2}$, with $\widehat{\Sigma}_0 = \widehat{\boldsymbol{\Phi}}_0 \widehat{\boldsymbol{\Phi}}_0'$. The elements of $\widehat{\boldsymbol{\Phi}}_0 = \{\widehat{\phi}_{0.jk}\}$ do *not* equal covariances between variable i and component j. The factor matrix is distinct from the eigenvectors (coefficients) and from the component scores.

1.21 SPECIAL COVARIANCE PATTERNS

Any $p \times p$ symmetric and nonnegative definite matrix provides a valid covariance matrix with up to $p(p+1)$ distinct elements. Some special sampling schemes generate patterned covariance matrices, with elements expressible as a function of a small number of parameters. In the simplest case, complete independence (and finite second moments) gives $\Sigma = \sigma^2 \boldsymbol{I}_p$. The study of time series leads to considering models such as autoregressive and moving average covariance patterns (Box, Jenkins, and Reinsel, 1994). The study of spatial statistics has also led to the development of a large range of covariance models (Cressie, 1991). The following definitions will be used throughout the book.

Definition 1.46 A set of jointly Gaussian variables with $\Sigma = \sigma^2 \boldsymbol{I}_p$ have a *spherical* distribution in that equal-probability regions centered at the mean vector are circles for $p = 2$, spheres for $p = 3$, and hyperspheres for $p > 3$. In some contexts, *sphericity* is present when the weaker condition $\Sigma = \boldsymbol{V}(\sigma^2 \boldsymbol{I}_p)\boldsymbol{V}'$ holds.

Definition 1.47 Any square, symmetric and nonnegative definite matrix with finite elements describes an *unstructured covariance matrix.*

Definition 1.48 A $p \times p$ *compound symmetric covariance matrix* may be expressed as $\Sigma = \sigma^2[\mathbf{1}_p\mathbf{1}_p'\rho + \boldsymbol{I}_p(1-\rho)]$, with $0 < \sigma^2 < \infty$ and $-1/(p-1) < \rho < 1$.

Lemma 1.33 A $p \times p$ compound symmetric covariance matrix may be written

$$\begin{aligned}\Sigma &= \sigma^2[\mathbf{1}_p\mathbf{1}_p'\rho + \boldsymbol{I}_p(1-\rho)] \\ &= \boldsymbol{V}\mathrm{Dg}(\boldsymbol{\lambda})\boldsymbol{V}' \\ &= \begin{bmatrix}\boldsymbol{v}_0 & \boldsymbol{V}_t\end{bmatrix}\begin{bmatrix}\lambda_1 & \mathbf{0} \\ \mathbf{0} & \lambda_2 \boldsymbol{I}_{p-1}\end{bmatrix}\begin{bmatrix}\boldsymbol{v}_0' \\ \boldsymbol{V}_t'\end{bmatrix}.\end{aligned} \tag{1.159}$$

The $p \times p$ matrix $\mathrm{Dg}(\boldsymbol{\lambda}) = \mathrm{Dg}(\lambda_1, \lambda_2 \mathbf{1}_{p-1})$ has

$$\lambda_1 = \sigma^2[1 + (p-1)\rho] \tag{1.160}$$

$$\lambda_2 = \sigma^2(1-\rho). \tag{1.161}$$

The restrictions

$$-1/(p-1) < \rho < 1 \tag{1.162}$$
$$0 < \sigma^2 < \infty \tag{1.163}$$

guarantee $0 < \lambda_j < \infty$ and therefore $\boldsymbol{\Sigma}$ will be positive definite. The first eigenvector, corresponding to λ_1, is $\boldsymbol{v}_0 = \mathbf{1}_p/p^{1/2}$ (normalized to unit length). The first eigenvector may be described as spanning (1) the sum when scaled $\mathbf{1}_p$, (2) the average when scaled $\mathbf{1}_p/p$, or (3) the zero-order polynomial trend when scaled $\mathbf{1}_p/p^{1/2}$. The remaining $p-1$ eigenvectors correspond to λ_2, which has multiplicity $p-1$. The $p \times (p-1)$ matrix $\boldsymbol{V}_t$ may be taken to contain first-(linear) through $(p-1)$-order orthogonal polynomial trend coefficients (scaled to unit length and hence orthonormal). Although $\boldsymbol{v}_0$ is unique (up to scaling), $\boldsymbol{V}_t$ is not. Any other $p \times (p-1)$ orthonormal matrix which is orthogonal to $\boldsymbol{v}_0$ will also suffice. In any case, the eigenvectors can always be expressed as a known and constant matrix, no matter what the unknown parameters (ρ, σ^2).

Proof. Left as an exercise. Hints: verify directly that the eigenvectors reproduce themselves; eigenvalues must be positive, finite and nonzero.

A compound symmetric covariance matrix may be described as having one eigenvalue, λ_1, of multiplicity 1 with corresponding normalized eigenvector $\boldsymbol{v}_0$, and a second eigenvalue, λ_2, of multiplicity $p-1$ with corresponding normalized eigenvectors $\boldsymbol{V}_t$. Any $p \times (p-1)$ and columnwise orthonormal matrix also orthogonal to $\boldsymbol{u}_0$ could be chosen in lieu of the trends.

Definition 1.49 An AR(1), *autoregressive, order 1*, covariance matrix is $\sigma^2\boldsymbol{P}$, with $\boldsymbol{P} = \{\rho_{jk}\}$, with $\rho_{jk} = \rho^{|j-k|}$, for $0 \leq \rho < 1$ and $0 < \sigma^2 < \infty$.

CHAPTER 2

The General Linear Univariate Model

2.1 MOTIVATION

Chapter 2 centers on providing a careful statement of assumptions most often used with the general linear univariate model. A number of specific examples illustrate the basic ideas. Chapters 3, 4, and 5 have the same structure for the multivariate model, multivariate generalizations, and mixed models. Together, applications and properties of the models in Chapters 2–5 will illustrate the need for and uses of the theory in the remainder of the book. Later chapters contain explicit proofs of nearly all basic results in Chapters 2–5. Most others can be deduced with only modest effort from the ones provided.

2.2 MODEL CONCEPTS

Both in the title of and purpose for the book, the univariate, multivariate, and mixed linear models share equal billing. However, the univariate model likely has more pages devoted to it than the multivariate model, which likely has more pages devoted to it than the mixed model. The unevenness reflects the fact that a solid understanding of univariate models allows quickly generalizing many results to multivariate and mixed models. In turn, multivariate theory helps develop mixed model results. The principle holds most often for estimation, while hypothesis testing and inference usually differ in basic ways.

Figure 2.1 illustrates the four aspects of any model: scientific *meaning* of the model, *estimation* of parameters, *inference* about parameters, and *numerical* methods. In Chapters 2–5 we look at the practical interpretations of linear models and focus on scientific meaning. Subsequent groups of chapters center on distribution theory, estimation, inference, and sample size. Although rarely discussed here, accurate numerical methods must be used to ensure the validity of any data analysis.

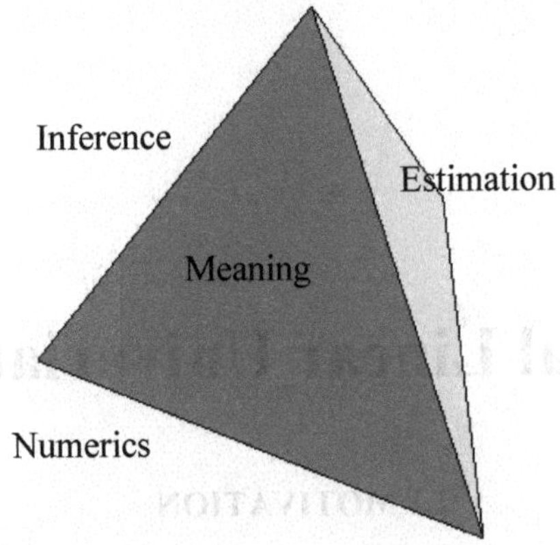

Figure 2.1 Four aspects of a statistical model.

2.3 THE GENERAL LINEAR UNIVARIATE LINEAR MODEL

Definition 2.1 A *general linear univariate model* will be indicated by $\text{GLM}_{N,q}(y_i; \boldsymbol{X}_i\boldsymbol{\beta}, \sigma^2)$, with *primary parameters* $\{\boldsymbol{\beta}, \sigma^2\}$, and includes the following assumptions.

1. The elements of the $N \times 1$ random vector $\boldsymbol{y} = \{y_i\}$ are mutually independent.
2. With $\boldsymbol{X}_i = \text{row}_i(\boldsymbol{X})$, the $N \times q$ *design matrix*, $\boldsymbol{X}$ has $\text{rank}(\boldsymbol{X}) = r \leq q \leq N$, and is fixed and known without appreciable error, conditional on knowing the sampling units, for data analysis. Power analysis requires knowing the predictor distribution in the population.
3. Elements of $\boldsymbol{\beta}$ $(q \times 1)$ are fixed and unknown and typically regression coefficients or means.
4. The mean of $\boldsymbol{y}$ is $\text{E}(\boldsymbol{y}) = \boldsymbol{X\beta}$.
5. Response y_i has finite variance $\sigma^2 \geq 0$, which is fixed and unknown.

Writing $\text{GLM}_{N,q}(y_i; \boldsymbol{X}_i\boldsymbol{\beta} | \boldsymbol{R\beta} = \boldsymbol{a}, \sigma^2)$ specifies *explicit restrictions* on parameters in $\boldsymbol{\beta}$ through the fixed and known constants $\boldsymbol{R}$ and $\boldsymbol{a}$.

The model is described as *full rank* (FR) if $r = \text{rank}([\boldsymbol{X}' \; \boldsymbol{R}']') = q$ and otherwise as *less than full rank* (LTFR) if $r < q$. Clarity may require writing $\text{GLM}_{N,q}\text{FR}()$ or $\text{GLM}_{N,q}\text{LTFR}()$.

The definition of a GLM describes the "least squares" assumptions because they guarantee estimates for the primary parameters $\boldsymbol{\beta}$ and σ^2 exist which satisfy the least squares criterion. It is very important to recognize that no particular distribution has been specified for any random variable. The rather modest requirement of finite second (and implicitly first) moments is made. Much more importantly, the requirements of independent and homogeneous observations place strong restrictions on the range of models. The model definition specifies three components: the response for the independent sampling unit, the mean of the response, and the variance of the response.

A number of implications of the GLM definition (least squares assumptions) may be deduced easily. First, response vector $\boldsymbol{y}$ has finite covariance $\mathcal{V}(\boldsymbol{y}) = \sigma^2 \boldsymbol{I}$. Independence makes the off-diagonal elements zero (independence implies zero covariance and correlation), while the homogeneity assumption provides the diagonal elements. Second, the response vector may be separated into purely fixed and purely random vectors by centering the responses to define

$$\boldsymbol{e} = \boldsymbol{y} - \mathrm{E}(\boldsymbol{y}) = \boldsymbol{y} - \boldsymbol{X}\boldsymbol{\beta}, \tag{2.1}$$

with $\mathrm{E}(\boldsymbol{e}) = \boldsymbol{0}$ and $\mathcal{V}(\boldsymbol{e}) = \sigma^2 \boldsymbol{I}$. Third,

$$\boldsymbol{y} = \boldsymbol{X}\boldsymbol{\beta} + \boldsymbol{e}. \tag{2.2}$$

The $N \times 1$ constant vector $\boldsymbol{X}\boldsymbol{\beta}$ describes (models) the first moment of the responses, the mean vector. The $N \times 1$ random vector $\boldsymbol{e}$ describes (models) the second and higher moments.

Choosing estimators for the primary parameters, $\boldsymbol{\beta}$ and σ^2, which satisfy a variety of optimal properties which are exact, even in small samples, does not require any particular choice of distribution function for the responses. In contrast, the desire to test hypotheses leads to describing distributions of test statistics. Finding exact distributions for small samples usually requires explicit and particular specification of the distribution of the data.

Data analysts often assume the data have a Gaussian distribution. As detailed in Chapter 8, $\boldsymbol{y}_1 \sim \mathcal{N}_n(\boldsymbol{\mu}_1, \boldsymbol{\Sigma}_1)$ indicates the vector $\boldsymbol{y}_1$ $(n \times 1)$ has a Gaussian distribution, with finite mean $\boldsymbol{\mu}_1$, finite and positive definite covariance matrix $\boldsymbol{\Sigma}_1$, and density $f_1(\boldsymbol{y}_1) = (2\pi)^{-n/2}|\boldsymbol{\Sigma}_1|^{-1/2}\exp\left[-(\boldsymbol{y}_1-\boldsymbol{\mu}_1)'\boldsymbol{\Sigma}_1^{-1}(\boldsymbol{y}_1-\boldsymbol{\mu}_1)/2\right]$. In turn, constant $\boldsymbol{T}$, $m \times n$ with $m > n$, makes $\boldsymbol{y}_2 = \boldsymbol{T}\boldsymbol{y}_1 \sim \mathcal{SN}_m(\boldsymbol{\mu}_2, \boldsymbol{\Sigma}_2)$, singular Gaussian, for $\boldsymbol{\mu}_2 = \boldsymbol{T}\boldsymbol{\mu}_1$ and $\boldsymbol{\Sigma}_2 = \boldsymbol{T}\boldsymbol{\Sigma}_1\boldsymbol{T}'$. The deficient rank of $\boldsymbol{\Sigma}_2$ disallows the existence of a density (the distribution function remains well defined). Writing $(\mathcal{S})\mathcal{N}_m(\boldsymbol{\mu}_2, \boldsymbol{\Sigma}_2)$ indicates the distribution may or may not be singular.

Definition 2.2 Writing $\mathrm{GLM}_{N,q}(y_i; \boldsymbol{X}_i\boldsymbol{\beta}, \sigma^2)$ *with Gaussian errors* indicates $y_i \sim \mathcal{N}_1[\mathrm{row}_i(\boldsymbol{X})\boldsymbol{\beta}, \sigma^2]$.

Following Kleinbaum, Kupper, Muller, and Nizam (1998) and Muller and Fetterman (2002) the GLM assumptions may be summarized with the mnemonic *HILE Gauss*: homogeneity $[\mathcal{V}(y_i) = \mathcal{V}(y_{i'}) = \sigma^2]$, independence ($y_i \perp\!\!\!\perp y_{i'}$ if $i \neq i'$), linearity $[\mathrm{E}(\boldsymbol{y}) = \boldsymbol{X}\boldsymbol{\beta}]$, existence of finite second moments, and, *optionally*, Gaussian observations. The mnemonic groups the least squares assumptions together and separates the distribution assumption.

Adding the assumption of Gaussian errors allows deducing additional properties. With HILE Gauss, essentially all of the assumptions are captured by the statement $\boldsymbol{y} \sim \mathcal{N}_N(\boldsymbol{X\beta}, \sigma^2\boldsymbol{I})$. Obviously the parameters of the Gaussian distribution coincide with the parameters of the GLM. First and second moments fully characterize both. Also, $\boldsymbol{e} \sim \mathcal{N}_N(\boldsymbol{0}, \sigma^2\boldsymbol{I})$.

The theory of the GLM applies for two apparently disparate classes of applications: regression models and Analysis-of-Variance (ANOVA) models. ANOVA models were developed to test the effects of one or more categorical predictors on a Gaussian response with independent and homogenous errors. Regression models were developed to express a continuous response as a function of one or more continuous predictors. Models with both categorical and continuous predictors fall in between and are best thought of simply as linear models. The underlying theory, for data analysis if not always for power analysis, coincides for all of them.

Example 2.1 Benignus, Muller, Smith, Pieper, and Prah (1990) exposed 74 participants to one of five profiles of carbon monoxide (CO) in the air breathed during the study: Air, Low, Medium, High, or Slow. Accuracy of eye-hand coordination was measured before exposure and four times during exposure. Experience in a series of similar studies, coupled with knowledge of the underlying physical process and measurements, convinced the investigators that the logarithm of an accuracy measure follows a Gaussian distribution. The primary planned analysis compared the five groups on mean change from baseline to time 4.

A GLM for the study may be written in many ways. Muller and Fetterman (2002, Chapter 12) reviewed coding schemes for design matrices. Group sizes were $\{14, 15, 15, 15, 15\}$ for Air, Low, Medium, High, and Slow. A classical LTFR ANOVA coding scheme for the 74 responses, $\{d_i\}$, with $d_i = y_{i4} - y_{i0}$ (y_{i4} is the hour 4 response, y_{i0} is the hour 0 response, the baseline), may be written, with all elements of the design (super-) matrix being conforming vectors,

$$\boldsymbol{d} = \begin{bmatrix} \mathbf{1}_{14} & 1 & 0 & 0 & 0 & 0 \\ \mathbf{1}_{15} & 0 & 1 & 0 & 0 & 0 \\ \mathbf{1}_{15} & 0 & 0 & 1 & 0 & 0 \\ \mathbf{1}_{15} & 0 & 0 & 0 & 1 & 0 \\ \mathbf{1}_{15} & 0 & 0 & 0 & 0 & 1 \end{bmatrix} \begin{bmatrix} \mu \\ \alpha_{\mathrm{A}} \\ \alpha_{\mathrm{L}} \\ \alpha_{\mathrm{M}} \\ \alpha_{\mathrm{H}} \\ \alpha_{\mathrm{S}} \end{bmatrix} + \boldsymbol{e}. \tag{2.3}$$

The parameters are not well defined without adding a constraint. Traditionally, it is assumed, with $\alpha_g \in \{\alpha_{\mathrm{A}}, \alpha_{\mathrm{S}}, \alpha_{\mathrm{M}}, \alpha_{\mathrm{H}}, \alpha_{\mathrm{S}}\}$, that $\sum \alpha_g = 0$. Choosing $\boldsymbol{R} = [0 \;\; \mathbf{1}_5']$ and $a = 0$ specifies the constraints in the context of a restricted model, with restrictions $\boldsymbol{R\beta} = \boldsymbol{a}$.

Example 2.2 Deleting the first column of the design matrix in equation (2.3) creates a cell mean coding and a full rank-design matrix. No constraints are needed.

Example 2.3 Deleting any one of columns 2–6 in the design matrix for the first example creates a reference cell coding, also full rank, with the deleted column indicating the reference cell. Chapters 12 and 13 in Muller and Fetterman (2002) describe coding schemes and practical aspects of one-way ANOVA.

Example 2.4 An alternative analysis would use a more general model, the full model in every cell. With $y_{0,j}$ indicating the vector of time 0 (baseline) responses and y_4 the 74×1 vector of responses at time 4, the model is

$$y_4 = \begin{bmatrix} \mathbf{1}_{14} & \mathbf{0} & \mathbf{0} & \mathbf{0} & \mathbf{0} & y_{0,\mathrm{A}} & \mathbf{0} & \mathbf{0} & \mathbf{0} & \mathbf{0} \\ \mathbf{0} & \mathbf{1}_{15} & \mathbf{0} & \mathbf{0} & \mathbf{0} & \mathbf{0} & y_{0,\mathrm{L}} & \mathbf{0} & \mathbf{0} & \mathbf{0} \\ \mathbf{0} & \mathbf{0} & \mathbf{1}_{15} & \mathbf{0} & \mathbf{0} & \mathbf{0} & \mathbf{0} & y_{0,\mathrm{M}} & \mathbf{0} & \mathbf{0} \\ \mathbf{0} & \mathbf{0} & \mathbf{0} & \mathbf{1}_{15} & \mathbf{0} & \mathbf{0} & \mathbf{0} & \mathbf{0} & y_{0,\mathrm{H}} & \mathbf{0} \\ \mathbf{0} & \mathbf{0} & \mathbf{0} & \mathbf{0} & \mathbf{1}_{15} & \mathbf{0} & \mathbf{0} & \mathbf{0} & \mathbf{0} & y_{0,\mathrm{S}} \end{bmatrix} \begin{bmatrix} \beta_{0\mathrm{A}} \\ \beta_{0\mathrm{L}} \\ \beta_{0\mathrm{M}} \\ \beta_{0\mathrm{H}} \\ \beta_{0\mathrm{S}} \\ \beta_{1\mathrm{A}} \\ \beta_{1\mathrm{L}} \\ \beta_{1\mathrm{M}} \\ \beta_{1\mathrm{H}} \\ \beta_{1\mathrm{S}} \end{bmatrix} + e. \tag{2.4}$$

Chapter 16 in Muller and Fetterman (2002) provides details of practical aspects of coding schemes, estimation, and testing for the full model in every cell. Special cases include ANCOVA and difference scores.

Example 2.5 An Analysis of Covariance (ANCOVA) model may be coded

$$y_4 = \begin{bmatrix} \mathbf{1}_{14} & \mathbf{0} & \mathbf{0} & \mathbf{0} & \mathbf{0} & y_{0,\mathrm{A}} \\ \mathbf{0} & \mathbf{1}_{15} & \mathbf{0} & \mathbf{0} & \mathbf{0} & y_{0,\mathrm{L}} \\ \mathbf{0} & \mathbf{0} & \mathbf{1}_{15} & \mathbf{0} & \mathbf{0} & y_{0,\mathrm{M}} \\ \mathbf{0} & \mathbf{0} & \mathbf{0} & \mathbf{1}_{15} & \mathbf{0} & y_{0,\mathrm{H}} \\ \mathbf{0} & \mathbf{0} & \mathbf{0} & \mathbf{0} & \mathbf{1}_{15} & y_{0,\mathrm{S}} \end{bmatrix} \begin{bmatrix} \beta_{0\mathrm{A}} \\ \beta_{0\mathrm{L}} \\ \beta_{0\mathrm{M}} \\ \beta_{0\mathrm{H}} \\ \beta_{0\mathrm{S}} \\ \beta_{1} \end{bmatrix} + e. \tag{2.5}$$

The model assumes equal slopes and is a special case of the full model in every cell. In turn, analysis of difference scores (as discussed earlier) is a special case of an ANCOVA model because it assumes the common slope is 1.0.

2.4 THE UNIVARIATE GENERAL LINEAR HYPOTHESIS

Definition 2.3 In the $\mathrm{GLM}_{N,q}(y_i; \boldsymbol{X}_i\boldsymbol{\beta}, \sigma^2)$, functions of the primary parameters are *secondary parameters*.

Three important examples deserve specific mention because they lie at the heart of the general linear hypothesis (GLH). For $\boldsymbol{C}$ an $a \times q$ matrix of known constants and $\boldsymbol{\theta}_0$ an $a \times 1$ vector of known constants, $a \times 1$ vectors $\boldsymbol{\theta} = \boldsymbol{C}\boldsymbol{\beta}$ and $\boldsymbol{\theta} - \boldsymbol{\theta}_0$ are secondary parameters. Furthermore, the covariance matrix of the estimators, namely $\mathcal{V}(\widehat{\boldsymbol{\theta}}) = \mathcal{V}(\widehat{\boldsymbol{\theta}} - \boldsymbol{\theta}_0)$, is also a secondary parameter.

The general linear (null) hypothesis may be stated in two equivalent ways:

$$\begin{aligned} H_0 &: \boldsymbol{C}\boldsymbol{\beta} = \boldsymbol{\theta}_0 \\ H_0 &: \boldsymbol{\theta} = \boldsymbol{\theta}_0 \,. \end{aligned} \tag{2.6}$$

If $a > 1$, then the alternative hypothesis necessarily is $H_A : \boldsymbol{\theta} \neq \boldsymbol{\theta}_0$. If $a = 1$ (and θ is a scalar, such as a mean difference), then a one-sided alternative may occasionally be preferred. The notation of a Boolean algebra (as in the following definition) greatly simplifies discussions of inference about hypotheses.

Definition 2.4 (a) A *Boolean algebra* (Weisstein, 2003) is the partial order on subsets (of a collection of sets) which is closed under finite union (OR, $\vee$), intersection (AND, $\wedge$) and complementation (NOT, $\neg$).
(b) Each element defines a *Boolean function*, $\mathbb{B}(\{\})$.
(c) By convention, a two-valued Boolean algebra has values TRUE or FALSE, with $\mathbb{B}(\{\}) = 1$ (TRUE) or $\mathbb{B}(\{\}) = 0$ (FALSE).

Definition 2.5 (a) The *null hypothesis* may be written $H_0 = \mathbb{B}(\boldsymbol{\theta} = \boldsymbol{\theta}_0)$.
(b) The *alternative hypothesis* may be written $H_A = \mathbb{B}(\boldsymbol{\theta} \neq \boldsymbol{\theta}_0)$.

For linear models, the secondary parameter *noncentrality* characterizes the changes in distributions due to changing hypotheses. As discussed later in the chapter, the test statistic is essentially a noncentrality estimator.

Definition 2.6 (a) For a $\text{GLM}_{N,q}(y_i; \boldsymbol{X}_i\boldsymbol{\beta}, \sigma^2)$ with $\boldsymbol{X}$ fixed and known, the *shift parameter* is

$$\begin{aligned} \delta &= (\boldsymbol{\theta} - \boldsymbol{\theta}_0)'\boldsymbol{M}^{-1}(\boldsymbol{\theta} - \boldsymbol{\theta}_0) \\ &= (\boldsymbol{\theta} - \boldsymbol{\theta}_0)'\left[\boldsymbol{C}(\boldsymbol{X}'\boldsymbol{X})^{-}\boldsymbol{C}'\right]^{-1}(\boldsymbol{\theta} - \boldsymbol{\theta}_0) \,. \end{aligned} \tag{2.7}$$

(b) The *noncentrality parameter* is

$$\omega = \delta/\sigma^2 = (\boldsymbol{\theta} - \boldsymbol{\theta}_0)'\left[\boldsymbol{C}(\boldsymbol{X}'\boldsymbol{X})^{-}\boldsymbol{C}'\right]^{-1}(\boldsymbol{\theta} - \boldsymbol{\theta}_0)/\sigma^2 \,. \tag{2.8}$$

The parameters ω and δ play central roles in the distribution theory of hypothesis tests. The parameter ω is scale free in the sense that multiplying $\boldsymbol{y}$ by a nonzero constant does not change it. Neither δ nor ω vary under a full-rank transformation of the rows of both $\boldsymbol{C}$ and $\boldsymbol{\theta}_0$, of the form $\boldsymbol{C}_T = \boldsymbol{T}\boldsymbol{C}$ and $\boldsymbol{\theta}_{0T} = \boldsymbol{T}\boldsymbol{\theta}_0$, with $\boldsymbol{T}$ $(a \times a)$ of rank a. The null hypothesis is $\boldsymbol{\theta} = \boldsymbol{\theta}_0$, which implies $\delta = (\boldsymbol{0})'\boldsymbol{M}^{-1}(\boldsymbol{0}) = 0$ and therefore $\omega = 0$. The alternative hypothesis has

$\boldsymbol{\theta} \neq \boldsymbol{\theta}_0$, which guarantees $\delta > 0$ and therefore $\omega > 0$. The reader should be cautioned that some authors include a factor of $1/2$ in the definition of ω. The presence or absence of the factor must always be checked in any discussion.

Example 2.6 An independent groups t test with equal sample sizes may be conducted with the following cell mean coding model:

$$\begin{aligned} \boldsymbol{y} &= \quad \boldsymbol{X\beta} \quad + \boldsymbol{e} \\ &= \begin{bmatrix} \mathbf{1}_{N/2} & \mathbf{0} \\ \mathbf{0} & \mathbf{1}_{N/2} \end{bmatrix} \begin{bmatrix} \mu_1 \\ \mu_2 \end{bmatrix} + \boldsymbol{e}. \end{aligned} \tag{2.9}$$

Testing $H_0 : \mu_1 = \mu_2$ leads to using $\boldsymbol{C} = [1 \quad -1]$ and $\boldsymbol{\theta}_0 = 0$. In turn,

$$\begin{aligned} \boldsymbol{M} &= \boldsymbol{C}(\boldsymbol{X}'\boldsymbol{X})^{-}\boldsymbol{C}' \\ &= [1 \quad -1][(N/2)\boldsymbol{I}_2]^{-1}[1 \quad -1]' \\ &= 4/N, \end{aligned} \tag{2.10}$$

$$\begin{aligned} \boldsymbol{\theta} - \boldsymbol{\theta}_0 &= \boldsymbol{C\beta} - \boldsymbol{\theta}_0 \\ &= [1 \, -1] \begin{bmatrix} \mu_1 \\ \mu_2 \end{bmatrix} - 0 \\ &= \mu_1 - \mu_2, \end{aligned} \tag{2.11}$$

$$\begin{aligned} \delta &= (\boldsymbol{\theta} - \boldsymbol{\theta}_0)'\boldsymbol{M}^{-1}(\boldsymbol{\theta} - \boldsymbol{\theta}_0) \\ &= (\mu_1 - \mu_2)'(4/N)^{-1}(\mu_1 - \mu_2) \\ &= (\mu_1 - \mu_2)^2(N/4), \end{aligned} \tag{2.12}$$

and

$$\omega = (N/4)(\mu_1 - \mu_2)^2/\sigma^2. \tag{2.13}$$

The final form illustrates the principle that noncentrality in the linear model with fixed predictors depends only on sample size, mean differences, and error variance.

A GLH describes a set of linear constraints on $\boldsymbol{\beta}$, namely $\boldsymbol{C\beta} = \boldsymbol{\theta}_0$, through the fixed and known constants $\boldsymbol{C}$ and $\boldsymbol{\theta}_0$. Relative to the unconstrained model, $\text{GLM}_{N,q}(y_i; \boldsymbol{X}_i\boldsymbol{\beta}, \sigma^2)$, writing $\text{GLM}_{N,q}(y_i; \boldsymbol{X}_i\boldsymbol{\beta}|\boldsymbol{C\beta} = \boldsymbol{\theta}_0, \sigma_0^2)$ specifies a constrained model. *Hence every GLH corresponds to comparing a full and constrained model.* Any $\boldsymbol{C}$ matrix with all rows selected from a q-dimensional identity matrix leads to an easily understood constrained model. In such a case, the constrained model may be produced from the full model simply by deleting appropriate columns of $\boldsymbol{X}$ and corresponding rows of $\boldsymbol{\beta}$. Also, the hypothesis compares the original (full) and the reduced models. The parameter δ has a simple form in the context of comparing the full and constrained models,

$$\delta = N\sigma_0^2 - N\sigma^2. \tag{2.14}$$

2.5 TESTS ABOUT VARIANCES

The general linear hypothesis makes a statement about expected value parameters, elements of $\boldsymbol{\beta}$. It is less common, although perfectly reasonable, to consider hypotheses about variances, such as σ^2. Exact forms for a confidence interval for σ^2 and exact tests of $H_0 : \sigma^2 = \sigma_0^2$ as well as $H_0 : \sigma_1^2 = \sigma_2^2 = \cdots = \sigma_q^2$ are known (with fully independent and Gaussian data).

Tests of variance equality should *not* be used to check the assumption of homogeneity in a linear model. A number of authors have provided analytic and simulation results to support the claim. Instead, robust tests, such as the Satterthwaite approach for the t test, should be used. Similar conclusions apply to testing hypotheses about variances. O'Brien (1979) reviewed available methods and made helpful recommendations.

2.6 THE ROLE OF THE INTERCEPT

The design matrix for the great majority, but not all, linear models either directly includes a column with all 1's, or has columns that span a column of 1's. The following definition formalizes the concept.

Definition 2.7 (a) If constant $\boldsymbol{t}_0$ $(q \times 1)$ exists such that $\boldsymbol{X}\boldsymbol{t}_0 = \mathbf{1}_N$ for $N \times q$ design matrix $\boldsymbol{X}$, then the design matrix, and also the associated linear model, *spans an intercept.*
(b) If such a $\boldsymbol{t}_0$ exists and $\boldsymbol{C}\boldsymbol{t}_0 = \mathbf{0}$, then the hypothesis $H_0 : \boldsymbol{C}\boldsymbol{\beta} = \boldsymbol{\theta}_0$ *excludes the intercept.*

In the great majority of cases, scientific considerations alone dictate that the model should include an intercept. Most often, but not always, the mean response contains an arbitrary (location) constant and therefore requires the model to span an intercept. Temperature measured in degrees centigrade uses the arbitrary zero point of the temperature at which water freezes. A clinical trial comparing two treatments to reduce fever could compare mean body temperature, with

$$H_0 : \mu_1 = \mu_2 \Leftrightarrow$$
$$H_0 : (\mu_1 - \mu_2) = 0 .$$

The parameter $\theta = (\mu_1 - \mu_2)$ does not vary due to adding a constant to the data. Including an intercept allows conducting a location-invariant test. Just as most models span an intercept, most hypotheses exclude the intercept, for the same reasons of scientific irrelevance and lack of meaning.

In some cases, it makes sense to compare the model with and without an intercept. The data for an astronomical study of the temperature of a comet orbiting far from the sun may use values recorded in degrees Kelvin, for which the value zero has meaning. If so, then either excluding an intercept or including one and testing whether it equals zero may be reasonable. Alternately, Casella (1983) argued that comparing the models with and without an intercept provides a natural way to characterize and understand the statistical value of the intercept.

The preceding brief comments have a number of implications for discussions of the theory of linear models. (1) Most applications will involve models that span an intercept and tests that exclude the intercept. Such tests have location invariance. (2) Completely general results must allow for tests involving the intercept. (3) Completely general results must allow for models (and tests) which do not span an intercept, which corresponds to assuming the intercept equals zero. We urge the reader to always remember the complexities that can occur. Muller and Fetterman (2002, Chapters 4–6) included extensive discussion of the role of the intercept in univariate linear models, tests, and correlation.

2.7 POPULATION CORRELATION AND STRENGTH OF RELATIONSHIP

Data analysts often wish to consider the entire collection of predictors in order to evaluate how well the model fits. Most often, such tests exclude the intercept (for scientific reasons). In the following, β_0 represents the intercept of a GLM, and the intercept may be constrained to be zero, which corresponds to excluding the intercept. In such a model, the special case of a test of all parameters other than the intercept equal to zero compares the original model, $\text{GLM}_{N,q}(y_i; \boldsymbol{X}_i\boldsymbol{\beta}, \sigma^2)$, to $\text{GLM}_{N,1}(y_i; 1 \cdot \beta_0, \sigma_0^2)$ or to $\text{GLM}_{N,1}(y_i; 0, \sigma_0^2)$.

Definition 2.8 For the $\text{GLM}_{N,q}(y_i; \boldsymbol{X}_i\boldsymbol{\beta}, \sigma^2)$, the population value of the *coefficient of determination*, the proportion of variance accounted for by the model equals

$$\rho_u^2 = \frac{N\sigma_0^2 - N\sigma^2}{(N\sigma_0^2 - N\sigma^2) + N\sigma^2} = \frac{\sigma_0^2 - \sigma^2}{\sigma_0^2}. \tag{2.15}$$

For a model spanning an intercept,

$$\rho^2 = \frac{\mathcal{V}(y_i) - \mathcal{V}(y_i|\boldsymbol{X}_i)}{\mathcal{V}(y_i)} = 1 - \frac{\mathcal{V}(y_i|\boldsymbol{X}_i)}{\mathcal{V}(y_i)}, \tag{2.16}$$

with ρ the multiple correlation coefficient. In contrast, without an intercept, the general form reduces to

$$\rho_u^2 = \frac{\mathrm{E}(y_i^2) - \mathrm{E}\left[(y_i - \boldsymbol{X}_i\boldsymbol{\beta})^2\right]}{\mathrm{E}(y_i^2)}, \tag{2.17}$$

and ρ_u equals an "uncorrected" (for the intercept) or "generalized" multiple correlation. The parameters ρ^2 and ρ_u^2 may be interpreted as the proportion of response variance controlled by the predictors (predictable by the model). Both have scale invariance (do not change if the response or any predictors are multiplied by nonzero constants) and ρ^2 also has location invariance (does not change if constants are added to the response or predictors). Kvålseth (1985), Willett and Singer (1988), and Scott and Wild (1991) provided additional useful guidance.

More generally, interest may center on the strength of the relationship between the response variable and some linear combination of predictors, such as a subset, corresponding to a general linear hypothesis, $H_0 : \boldsymbol{C\beta} = \boldsymbol{\theta}_0$. The resulting parameter equals a squared semipartial multiple correlation. It measures the strength of the relationship between the response and the variables included in the hypothesis, with the predictors in the hypothesis adjusted for predictors not included in the hypothesis. Muller and Fetterman (2002, Chapter 6) discussed correlations in univariate linear models in detail. In the population, the proportion of the response variance controlled by the predictors underlying the general linear hypothesis $H_0 : \boldsymbol{C\beta} = \boldsymbol{\theta}_0$ may be expressed as

$$\rho_\delta^2 = \frac{\delta}{\delta + N\sigma^2} = \frac{\omega}{\omega + N} . \tag{2.18}$$

Furthermore

$$\omega = \frac{\rho_\delta^2}{(1-\rho_\delta^2)/N} = N\frac{\rho_\delta^2}{1-\rho_\delta^2} . \tag{2.19}$$

Consequently, testing the general linear hypothesis $H_0 : \boldsymbol{C\beta} = \boldsymbol{\theta}_0$ is equivalent to testing $H_0 : \rho_\delta^2 = 0$, or $H_0 : \rho_\delta = 0$, or $H_0 : \omega = 0$, with $\omega = \delta/\sigma^2$. In general, the coefficient of determination and the noncentrality for the hypothesis are one-to-one functions of each other. The coefficient of determination population value does not vary with sample size (while the estimator does). The value falls in the unit interval $0 \le \rho_\delta^2 \le 1$, with $\rho_\delta^2 = 0$ [which implies $\mathcal{V}(y_i) = \mathcal{V}(e_i) = \sigma^2 > 0$] occurring only under the null hypothesis and $\rho_\delta^2 = 1$ indicating perfect predictability [and $\mathcal{V}(y_i) > \mathcal{V}(e_i) = \sigma^2 = 0$]. The squared multiple correlation for the entire model arises as a special case by choosing an appropriate $\boldsymbol{C}$ matrix.

2.8 STATISTICAL ESTIMATES

A less-than-full-rank model has $r = \text{rank}(\boldsymbol{X}) < q$, and estimation relies on a nonunique generalized inverse $(\boldsymbol{X}'\boldsymbol{X})^-$. The unique inverse $(\boldsymbol{X}'\boldsymbol{X})^{-1}$ exists only when $\boldsymbol{X}$ has full rank, $\text{rank}(\boldsymbol{X}) = q$. With less than full rank,

$$\tilde{\boldsymbol{\beta}} = (\boldsymbol{X}'\boldsymbol{X})^- \boldsymbol{X}'\boldsymbol{y} \tag{2.20}$$

provides a proper, although biased, estimator of $\boldsymbol{\beta}$, which varies with the choice of generalized inverse. In contrast, a full-rank model defines $\boldsymbol{\beta}$ *differently* and thereby creates a unique and unbiased estimator, namely

$$\widetilde{\boldsymbol{\beta}} = (\boldsymbol{X}'\boldsymbol{X})^{-1}\boldsymbol{X}'\boldsymbol{y}\,. \tag{2.21}$$

Hence a full-rank design defines $\boldsymbol{\beta}$ in such a way as to guarantee it is *estimable*, in the sense that a known linear function of the data provides an unbiased estimator. The loose definition of "estimable" will be refined in Chapter 11. The obvious estimator of $\boldsymbol{\theta}$ may be written

$$\widehat{\boldsymbol{\theta}} = \boldsymbol{C}\widetilde{\boldsymbol{\beta}}\,. \tag{2.22}$$

We delay the discussion as to how to determine whether $\widehat{\boldsymbol{\theta}}$ is a good estimator, along with the proofs of all estimation properties, to Chapter 11.

Definition 2.9 For the $\text{GLM}_{N,q}(y_i; \boldsymbol{X}_i\boldsymbol{\beta}, \sigma^2)$, **(a)** *predicted values* are

$$\begin{aligned}\widehat{\boldsymbol{y}} &= \boldsymbol{X}\widetilde{\boldsymbol{\beta}} \\ &= \boldsymbol{X}(\boldsymbol{X}'\boldsymbol{X})^{-}\boldsymbol{X}'\boldsymbol{y} = \boldsymbol{H}\boldsymbol{y}\,.\end{aligned} \tag{2.23}$$

(b) In turn, the estimated errors, the *residuals*, are

$$\begin{aligned}\widehat{\boldsymbol{e}} &= \boldsymbol{y} - \boldsymbol{X}\widetilde{\boldsymbol{\beta}} \\ &= [\boldsymbol{I} - \boldsymbol{X}(\boldsymbol{X}'\boldsymbol{X})^{-}\boldsymbol{X}']\boldsymbol{y} = (\boldsymbol{I} - \boldsymbol{H})\boldsymbol{y}\,.\end{aligned} \tag{2.24}$$

The projection matrix, $\boldsymbol{H} = \boldsymbol{X}(\boldsymbol{X}'\boldsymbol{X})^{-}\boldsymbol{X}'$, earned the nickname "hat matrix" because $\widehat{\boldsymbol{y}} = \boldsymbol{H}\boldsymbol{y}$. Very conveniently, $\boldsymbol{H}$ is unique (Corollary 1.15) even though $(\boldsymbol{X}'\boldsymbol{X})^{-}$ is not, which ensures the predicted values and residuals also are unique.

Definition 2.10 For the $\text{GLM}_{N,q}(y_i; \boldsymbol{X}_i\boldsymbol{\beta}, \sigma^2)$ and the general linear hypothesis $H_0 : \boldsymbol{C}\boldsymbol{\beta} = \boldsymbol{\theta}_0$, the *error sum of squares* is $SSE = \widehat{\boldsymbol{e}}'\widehat{\boldsymbol{e}}$ and the *hypothesis sum of squares* is $SSH = \widehat{\delta} = (\widehat{\boldsymbol{\theta}} - \boldsymbol{\theta}_0)'\boldsymbol{M}^{-1}(\widehat{\boldsymbol{\theta}} - \boldsymbol{\theta}_0)$.

With $r = \text{rank}(\boldsymbol{X})$ in a univariate model,

$$\begin{aligned}\widehat{\sigma}^2 &= \widehat{\boldsymbol{e}}'\widehat{\boldsymbol{e}}/(N - r) \\ &= \boldsymbol{y}'[\boldsymbol{I} - \boldsymbol{X}(\boldsymbol{X}'\boldsymbol{X})^{-}\boldsymbol{X}']\boldsymbol{y}/(N - r) \\ &= \boldsymbol{y}'(\boldsymbol{I} - \boldsymbol{H})\boldsymbol{y}/(N - r)\end{aligned} \tag{2.25}$$

is unique and also can be proven to be an unbiased estimator of σ^2.

The obvious estimator of noncentrality is

$$\begin{aligned}\widehat{\omega} &= \widehat{\delta}/\widehat{\sigma}^2 \\ &= \frac{(\widehat{\boldsymbol{\theta}} - \boldsymbol{\theta}_0)'\boldsymbol{M}^{-1}(\widehat{\boldsymbol{\theta}} - \boldsymbol{\theta}_0)}{\widehat{\sigma}^2} \\ &= \frac{SSH}{SSE/(N - r)}\,.\end{aligned} \tag{2.26}$$

In turn, the usual F statistic is simply $F = \widehat{\omega}/a$, which can be characterized as the average amount of noncentrality per degree of freedom.

The various correlation parameter estimators may be computed in various ways. A general approach may be inferred from

$$\widehat{\rho}_{\delta}^{2} = \frac{SSH}{SSH + SSE} \,. \tag{2.27}$$

The special case of the usual squared correlation for a model including an intercept (a special case of spanning an intercept) can be computed in the terms of a test of all slopes equal zero. Alternately, for any model spanning an intercept,

$$\widehat{\rho}^{2} = 1 - \frac{\boldsymbol{y}'[\boldsymbol{I} - \boldsymbol{X}(\boldsymbol{X}'\boldsymbol{X})^{-}\boldsymbol{X}']\boldsymbol{y}}{\boldsymbol{y}'\left[\boldsymbol{I} - \mathbf{1}(\mathbf{1}'\mathbf{1})^{-1}\mathbf{1}'\right]\boldsymbol{y}} \,. \tag{2.28}$$

The "uncorrected" (for the intercept) or "generalized" multiple correlation estimator assumes $\boldsymbol{X}$ does not include or span $\mathbf{1}_N$. It takes the form

$$\widehat{\rho}_{u}^{2} = 1 - \frac{\boldsymbol{y}'[\boldsymbol{I} - \boldsymbol{X}(\boldsymbol{X}'\boldsymbol{X})^{-}\boldsymbol{X}']\boldsymbol{y}}{\boldsymbol{y}'\boldsymbol{y}} \,. \tag{2.29}$$

2.9 TESTING THE GENERAL LINEAR HYPOTHESIS

Throughout, we assume $F \sim F(\nu_1, \nu_2, \omega)$ indicates the random variable F follows a noncentral F distribution with ν_1 numerator degrees of freedom, ν_2 denominator degrees of freedom, and noncentrality ω, as detailed in Chapter 9. A central case has $\omega = 0$ and is indicated $F \sim F(\nu_1, \nu_2)$.

As detailed in Chapter 15, well-defined tests of *testable* hypotheses require, with $\boldsymbol{M} = \boldsymbol{C}(\boldsymbol{X}'\boldsymbol{X})^{-}\boldsymbol{C}'$ of dimensions $a \times a$,

$$\text{rank}(\boldsymbol{M}) = a \,, \tag{2.30}$$

which means $\boldsymbol{M}$ is full rank and invertible, as well as

$$\boldsymbol{C} = \boldsymbol{C}(\boldsymbol{X}'\boldsymbol{X})^{-}(\boldsymbol{X}'\boldsymbol{X}) \,. \tag{2.31}$$

Only less-than-full-rank models require checking the second condition, which can be replaced by a variety of equivalent conditions (which are usually less computationally convenient). Here we restrict attention to such testable hypotheses. The first condition implicitly requires $\boldsymbol{C}$ has full (row) rank of a, which provides a sufficient condition to ensure testability for full-rank models but not for less-than-full-rank models.

Definition 2.11 (a) The *test* of hypothesis H_0 is a Boolean function, and also a random variable, $\phi(\boldsymbol{y}) = \mathbb{B}(\boldsymbol{y} \in RR)$, and is defined by
(b) *rejection region* (*critical region*) $RR = \{\boldsymbol{y} : t > t_0\}$ for
(c) *test statistic* t, a function of the data, $\boldsymbol{y}$, and an
(d) appropriate *critical value*, t_0.
(e) Here $\phi(\boldsymbol{y}) = 1$ if $t > t_0$ and $\phi(\boldsymbol{y}) = 0$ if $t \leq t_0$.
(f) The complement of RR is the *acceptance region*, AR.
(g) The *size* of the test is $\alpha = \Pr\{\boldsymbol{y} \in RR | H_0 = 1\}$ (the null is true).
(h) *Power* is $1 - \beta = \Pr\{\boldsymbol{y} \in RR | H_A = 1\}$ (the alternative is true).
More generally, the *power function* is $\Pr\{\boldsymbol{y} \in RR | H\}$.
(i) The probability of making a *type I error* (false positive) is α.
(j) The probability of making a *type II error* (false negative) is β.

The generality of the power function allows disccussing both null and alternative hypotheses. If the size of the test (size of RR) is α, then $\Pr\{\boldsymbol{y} \in RR | H = H_0\} = \alpha$. Furthermore, $\Pr\{\boldsymbol{y} \in RR | H = H_A\} = 1 - \beta$.

A general linear hypothesis may be tested with the statistic

$$\begin{aligned} F &= \frac{(\widehat{\boldsymbol{\theta}} - \boldsymbol{\theta}_0)' \boldsymbol{M}^{-1} (\widehat{\boldsymbol{\theta}} - \boldsymbol{\theta}_0)/a}{\widehat{\sigma}^2} \\ &= \frac{SSH/a}{SSE/(N-r)} = \widehat{\omega}/a \\ &\sim F(a, N-r, \omega) \,. \end{aligned} \tag{2.32}$$

Under the null, $\omega = 0 \Leftrightarrow \rho_\delta^2 = 0 \Leftrightarrow \delta = 0 \Leftrightarrow \boldsymbol{\theta} = \boldsymbol{\theta}_0$. Except for the simple constant factor of a, the hypothesis degrees of freedom, the usual test statistic merely estimates the noncentrality parameter, ω. The population value (ω) may be interpreted as a times the F statistic which would occur if $\widehat{\boldsymbol{\theta}} = \boldsymbol{\theta}$ and $\widehat{\sigma}^2 = \sigma^2$.

The test has many optimal properties. It provides the likelihood ratio test and the union-intersection principle test. It always has scale invariance and also has location invariance if the model spans an intercept while the test excludes the intercept. The test has uniformly most power among all similarly invariant tests of size α. Finally, it always has exact size α.

Estimators from $\text{GLM}_{N,q}(y_i; \boldsymbol{X}_i\boldsymbol{\beta}, \sigma^2)$, and $\text{GLM}_{N,q}(y_i; \boldsymbol{X}_i\boldsymbol{\beta} | \boldsymbol{C}\boldsymbol{\beta} = \boldsymbol{\theta}_0, \sigma_0^2)$ allow writing an alternate form

$$\widehat{\delta} = (N - r + a)\widehat{\sigma}_0^2 - (N - r)\widehat{\sigma}^2 \,. \tag{2.33}$$

In turn,

$$\begin{aligned} \widehat{\rho}_\delta^2 &= \frac{\widehat{\delta}}{\widehat{\delta} + (N-r)\widehat{\sigma}^2} = \frac{\widehat{\omega}}{\widehat{\omega} + (N-r)} \\ &= \frac{aF}{aF + (N-r)} = \frac{SSH}{SSH + SSE} \,, \end{aligned} \tag{2.34}$$

and

$$\widehat{\omega} = \frac{\widehat{\rho}_\delta^2}{(1-\widehat{\rho}_\delta^2)/(N-r)} = (N-r)\frac{\widehat{\rho}_\delta^2}{1-\widehat{\rho}_\delta^2} \,. \tag{2.35}$$

Also, the test statistic may be written as

$$F = \frac{\widehat{\rho}_\delta^2/a}{(1-\widehat{\rho}_\delta^2)/(N-r)} \,. \tag{2.36}$$

The forms for parameters and estimators just given prove useful in both developing and understanding discussions of scale and location invariance properties.

2.10 CONFIDENCE REGIONS FOR θ

Confidence intervals, and more general confidence regions, convey useful information about the location of a parameter by combining precision and location properties of parameter estimators. Hypothesis tests also combine information about location and precision. Although tests provide apparently distinct information to that provided by confidence regions, an invertible one-to-one relationship exists between confidence regions and hypothesis tests. Confidence regions can be obtained by inverting hypothesis tests, and a confidence region can be inverted to yield a hypothesis test. In Section 15.6 we explain how the inversions are performed and prove that confidence regions exist only for parameters that are estimable and testable.

Definition 2.12 (a) If data vector $\boldsymbol{y}$ depends on a primary or secondary parameter vector $\boldsymbol{\theta}$ $(a \times 1)$ with an unknown true value in parameter space S, then $R(\boldsymbol{y}) \in S$ with $\Pr\{\boldsymbol{\theta} \in R(\boldsymbol{y})\} = c(\alpha) \in [0,1]$ and boundaries defined by a vector-valued function $\boldsymbol{g}(\boldsymbol{y};\alpha)$ is a *confidence region* for $\boldsymbol{\theta}$.
(b) Here $c(\alpha)$ is a *confidence coefficient*.
(c) An exact confidence region has $c(\alpha) = 1-\alpha$.
(d) An approximate confidence region has $c(\alpha) \approx 1-\alpha$.
(e) A conservative confidence region has $c(\alpha) \geq 1-\alpha$.

At least in linear models, $g(\boldsymbol{y},\alpha)$ usually has a closed-form representation. Consequently, computing confidence intervals for parameters of linears models typically proves quite simple.

2.11 SUFFICIENT STATISTICS FOR THE UNIVARIATE MODEL

If the univariate linear model $\text{GLM}_{N,q}(y_i; \boldsymbol{X}_i\boldsymbol{\beta}, \sigma^2)$ with Gaussian distribution is correct, then the SSCP matrix

$$\boldsymbol{S} = \begin{bmatrix} \boldsymbol{X}'\boldsymbol{X} & \boldsymbol{X}'\boldsymbol{y} \\ \boldsymbol{y}'\boldsymbol{X} & \boldsymbol{y}'\boldsymbol{y} \end{bmatrix} = \begin{bmatrix} \boldsymbol{X}' \\ \boldsymbol{y}' \end{bmatrix} [\boldsymbol{X}\ \boldsymbol{y}] \tag{2.37}$$

contains all of the complete sufficient statistics. If $\boldsymbol{X}$ contains an intercept, then, without loss of generality, $\boldsymbol{X}$ may be arranged with the intercept in column 1, with $\boldsymbol{X} = [\,\boldsymbol{1}_N\ \boldsymbol{X}_1\,]$, for $\boldsymbol{X}_1$ of dimension $N \times (q-1)$. In turn, the SSCP matrix

$$\boldsymbol{S} = \begin{bmatrix} \boldsymbol{1}'\boldsymbol{1} & \boldsymbol{1}'\boldsymbol{X}_1 & \boldsymbol{1}'\boldsymbol{y} \\ \boldsymbol{X}_1'\boldsymbol{1} & \boldsymbol{X}_1'\boldsymbol{X}_1 & \boldsymbol{X}'\boldsymbol{y} \\ \boldsymbol{y}'\boldsymbol{1} & \boldsymbol{y}'\boldsymbol{X}_1 & \boldsymbol{y}'\boldsymbol{y} \end{bmatrix} = \begin{bmatrix} \boldsymbol{1}' \\ \boldsymbol{X}_1' \\ \boldsymbol{y}' \end{bmatrix} [\boldsymbol{1}\ \boldsymbol{X}_1\ \boldsymbol{y}] \tag{2.38}$$

contains all of the complete sufficient statistics for estimation of all parameters of the regression model identified by the relationship

$$\text{E}(\boldsymbol{y}|\boldsymbol{X})\,. \tag{2.39}$$

Here $[\boldsymbol{1}\ \boldsymbol{X}_1\ \boldsymbol{y}]$ is $N \times (q+1)$ so $\boldsymbol{S}$ is $(q+1) \times (q+1)$. All parameter estimators and general linear hypothesis tests depend on the data only through the elements of $\boldsymbol{S}$. Very conveniently, the raw data are not needed for parameter estimation or testing the general linear hypothesis.

EXERCISES

2.1 Provide an explicit interpretation of each parameter in Example 2.5.

2.2 Provide a reference-cell-style coding design matrix for Example 2.5. Include clear specifications of all dimensions. Provide an explicit interpretation of each parameter.

2.3 Provide an effect style coding design matrix for Example 2.6. Include clear specifications of all dimensions. Provide an explicit interpretation of each parameter.

2.4 Explicitly describe the vector of constants $\boldsymbol{t}$ which demonstrates that the design matrix for Example 2.2 spans an intercept.

2.5 Explicitly describe the vector of constants $\boldsymbol{t}$ which demonstrates that the design matrix for Example 2.4 spans an intercept.

2.6 For a GLM, briefly explain why multiplying the response values by a nonzero constant automatically implies that $\boldsymbol{\beta}$ and $\boldsymbol{e}$ have also been multiplied by the same nonzero constant for the model to remain valid.

2.7 Prove explicitly that multiplying the model equation by a nonzero constant does not change ω.

2.8 Prove explicitly that multiplying the observed response values by a nonzero constant does not change $\widehat{\omega}$ or the observed statistic.

2.9 Give four words or short phrases which highlight the four aspects of a statistical model.

2.10 By default in an ANOVA model, SAS version 8 PROC GLM with the CLASS statement creates a less-than-full-rank design matrix. It creates the design matrix in the following steps: (1) include a column of 1's for the intercept; (2) if a factor has G levels, generate G columns in the design matrix, with each column an indicator variable for one of the levels of the factor. The algorithm (a sweep method; Goodnight, 1979) leads to a "bottom right" reference cell coding scheme, based on the sort order of the formatted values in the class variables (*not* the sort order of the data in the file being analyzed). If factor A has levels $\{1, 2, 3\}$ and factor B has levels $\{x, y, z\}$, the reference cell for a complete two-way design would be $3, z$. Additional detail is provided in SAS documentation.

Assume that factor C has four levels $\{1, 2, 3, 4\}$, and it is the only factor in the design (and the class statement). Also let n_i, $i \in \{1, 2, 3, 4\}$, be the number of participants at factor level i and $N = \sum_{i=1}^{4} n_i$ be the total number of observations.
2.10.1 Explicitly describe the design matrix originally created by SAS GLM and an associated parameter matrix (give all dimensions and provide brief interpretations). Allow for an unbalanced design (but no missing cells).
2.10.2 Explicitly specify the design matrix and associated parameters implicitly used after the sweep algorithm has been applied.
2.10.3 Explicitly specify constraint matrices and a constrained version of the original model which corresponds to choosing parameters in the reference cell model created by the sweep algorithm.

2.11 Give an example of a nonlinear model which is inherently linear.

2.12 Give an example of a nonlinear model which is not inherently linear.

CHAPTER 3

The General Linear Multivariate Model

3.1 MOTIVATION

The multivariate linear model generalizes the univariate linear model by allowing two or more responses to be measured on each independent sampling unit. Implicitly the model requires that the same design matrix apply to every response and every independent sampling unit have the same set of responses variables. The material in the present chapter summarizes some basic theory needed to fit such models and test hypotheses about predictors and responses.

The multivariate model allows generalizing univariate results for estimation to the multivariate model with very little complication. In contrast, measures of association and hypothesis testing become far more complicated to derive and discuss. The complexity arises from the fact that various criteria for tests lead to a total of nine (yes, nine) commonly used different test methods. Unavoidably, responsible analysis of a multivariate model requires an a priori choice of test method to avoid bias introduced by post hoc p value shopping.

Not surprisingly, the relative appeal of the many tests varies with the nature of the data and the scientific goals of the analysis. Therefore the chapter begins with some suggestions for characterizing dependent responses.

Example 3.1 Example 1.1 contained a classical (LTFR) ANOVA coding for data from Benignus, Muller, Smith, Pieper, and Prah (1990). They exposed 74 human participants to one of five profiles of carbon monoxide (CO) in the air breathed during the study: Air, Low, Medium, High, or Slow. The planned primary analysis compared the five group mean changes from baseline to time 4.

An alternate analysis used four responses, the hour 1–4 differences from baseline. Group sizes were $\{14, 15, 15, 15, 15\}$ for Air, Low, Medium, High, and Slow. The design matrix remains the same as for the univariate model. A multivariate general linear model requires the same design matrix for all responses. LTFR ANOVA coding for the 74 sets of 4 responses, $\{d_{i1}, d_{i1}, d_{i1}, d_{i4}\}$, may be written, with all elements of the design (super-) matrix being conforming vectors,

$$[\boldsymbol{d}_1\ \boldsymbol{d}_2\ \boldsymbol{d}_3\ \boldsymbol{d}_4] = \begin{bmatrix} \mathbf{1}_{14} & 1 & 0 & 0 & 0 & 0 \\ \mathbf{1}_{15} & 0 & 1 & 0 & 0 & 0 \\ \mathbf{1}_{15} & 0 & 0 & 1 & 0 & 0 \\ \mathbf{1}_{15} & 0 & 0 & 0 & 1 & 0 \\ \mathbf{1}_{15} & 0 & 0 & 0 & 0 & 1 \end{bmatrix} \begin{bmatrix} \mu_1 & \mu_2 & \mu_3 & \mu_4 \\ \alpha_{\mathrm{A},1} & \alpha_{\mathrm{A},2} & \alpha_{\mathrm{A},3} & \alpha_{\mathrm{A},4} \\ \alpha_{\mathrm{L},1} & \alpha_{\mathrm{L},2} & \alpha_{\mathrm{L},3} & \alpha_{\mathrm{L},4} \\ \alpha_{\mathrm{M},1} & \alpha_{\mathrm{M},2} & \alpha_{\mathrm{M},3} & \alpha_{\mathrm{M},4} \\ \alpha_{\mathrm{H},1} & \alpha_{\mathrm{H},2} & \alpha_{\mathrm{H},3} & \alpha_{\mathrm{H},4} \\ \alpha_{\mathrm{S},1} & \alpha_{\mathrm{S},2} & \alpha_{\mathrm{S},3} & \alpha_{\mathrm{S},4} \end{bmatrix} + \boldsymbol{E}. \tag{3.1}$$

Here $[\boldsymbol{d}_1\ \boldsymbol{d}_2\ \boldsymbol{d}_3\ \boldsymbol{d}_4]$ is 74×4, with columns corresponding to hour and rows to participants. The error matrix has the same dimensions and pattern. Each column in $\boldsymbol{B}$ contains parameters for a particular hour. The parameters are not well-defined without adding a constraint. For $\alpha_{g,t} \in \{\alpha_{\mathrm{A}}, \alpha_{\mathrm{S}}, \alpha_{\mathrm{M}}, \alpha_{\mathrm{H}}, \alpha_{\mathrm{S}}\}$, it was traditionally assumed $\sum_g \alpha_{g,t} = 0$ holds separately for each value of $t \in \{1,2,3,4\}$. Choosing $\boldsymbol{R}_x = [0\ \mathbf{1}_5']$, $\boldsymbol{R}_y = \boldsymbol{I}_4$, and $\boldsymbol{A} = \mathbf{0}_{1\times 4}$ implements the constraints in a restricted model with restrictions $\boldsymbol{R}_x\boldsymbol{B}\boldsymbol{R}_y = \boldsymbol{A}$.

3.2 DEFINITION OF THE MULTIVARIATE MODEL

Definition 3.1 A *general linear multivariate model* will be indicated by $\mathrm{GLM}_{N,p,q}(\boldsymbol{Y}_i; \boldsymbol{X}_i\boldsymbol{B}, \boldsymbol{\Sigma})$, with *primary parameters* $\{\boldsymbol{B}, \boldsymbol{\Sigma}\}$, and includes the following assumptions.

1. The rows of the $N \times p$ random matrix $\boldsymbol{Y}$ are mutually independent, with $\boldsymbol{Y}_i = \mathrm{row}_i(\boldsymbol{Y}) = [y_{i1}\ y_{i2}\ \cdots\ y_{ip}]$.
2. With $\boldsymbol{X}_i = \mathrm{row}_i(\boldsymbol{X})$, the $N \times q$ *design matrix* $\boldsymbol{X}$ has $\mathrm{rank}(\boldsymbol{X}) = r \leq q \leq N$, and is fixed and known without appreciable error, conditional on knowing the sampling units, for data analysis. Power analysis requires knowing the predictor distribution in the population.
3. Elements of $\boldsymbol{B}$ ($q \times p$) are fixed and unknown and often regression coefficients or means.
4. The mean of $\boldsymbol{Y}$ is $\mathrm{E}(\boldsymbol{Y}) = \boldsymbol{X}\boldsymbol{B}$.
5. Response matrix $\boldsymbol{Y}_i$ has finite covariance matrix $\boldsymbol{\Sigma} = \boldsymbol{\Sigma}'$, which is fixed, unknown, and positive definite or positive semidefinite.

Writing $\mathrm{GLM}_{N,p,q}(\boldsymbol{Y}_i; \boldsymbol{X}_i\boldsymbol{B} | \boldsymbol{R}_x\boldsymbol{B}\boldsymbol{R}_y = \boldsymbol{A}, \boldsymbol{\Sigma})$ specifies *explicit restrictions* on parameters in $\boldsymbol{B}$ through the fixed and known constants $\boldsymbol{R}_x$, $\boldsymbol{R}_y$, and $\boldsymbol{A}$.

The model is described as *full rank* (FR) if $r = \mathrm{rank}([\boldsymbol{X}'\ \boldsymbol{R}_x{}']') = q$ and otherwise as *less than full rank* (LTFR) if $r < q$. Clarity may require writing $\mathrm{GLM}_{N,p,q}\mathrm{FR}()$ or $\mathrm{GLM}_{N,p,q}\mathrm{LTFR}()$.

The definition of a GLM describes the "least squares" assumptions because they guarantee estimates for the primary parameters $\boldsymbol{B}$ and $\boldsymbol{\Sigma}$ exist which satisfy the least squares criterion. It is very important to recognize that no particular distribution has been specified for any random variable. The rather modest requirement of finite second (and implicitly first) moments is made. Much more importantly, the requirements of independent rows of $\boldsymbol{Y}$ with common covariance and a common design matrix for every column place strong restrictions on the range of models. The model definition specifies three components: the response matrix for the independent sampling unit, the mean of the response matrix, and the covariance of the response matrix.

A number of implications of the multivariate GLM definition (least squares assumptions) may be deduced easily. The first implication is that elements of response matrix $\boldsymbol{Y}$ have a finite covariance matrix following a simple pattern. The result may be expressed in terms of either vec($\boldsymbol{Y}'$) or vec($\boldsymbol{Y}$), which are both $Np \times 1$ vectors and are merely permutations of each other. The responses may be decomposed by row or column:

$$\boldsymbol{Y} = \begin{bmatrix} \boldsymbol{Y}_1 \\ \boldsymbol{Y}_2 \\ \vdots \\ \boldsymbol{Y}_N \end{bmatrix} = [\boldsymbol{y}_1 \quad \boldsymbol{y}_2 \quad \cdots \quad \boldsymbol{y}_p]. \tag{3.2}$$

Each row of $\boldsymbol{Y}$ contains the p responses for a single independent sampling unit, while each column contains the N responses for a single variable.

With $p \times N$ $\boldsymbol{Y}'$, stacking the N transposed rows (each $p \times 1$) creates an $Np \times 1$ vector,

$$\text{vec}(\boldsymbol{Y}') = \begin{bmatrix} \boldsymbol{Y}_1' \\ \boldsymbol{Y}_2' \\ \vdots \\ \boldsymbol{Y}_N' \end{bmatrix}, \tag{3.3}$$

which has $Np \times Np$ covariance matrix

$$\mathcal{V}[\text{vec}(\boldsymbol{Y}')] = \boldsymbol{I}_N \otimes \boldsymbol{\Sigma} = \begin{bmatrix} \boldsymbol{\Sigma} & \boldsymbol{0} & \cdots & \boldsymbol{0} \\ \boldsymbol{0} & \boldsymbol{\Sigma} & \cdots & \boldsymbol{0} \\ \vdots & \vdots & \ddots & \vdots \\ \boldsymbol{0} & \boldsymbol{0} & \cdots & \boldsymbol{\Sigma} \end{bmatrix}. \tag{3.4}$$

The last matrix contains N^2 submatrices, each of them $p \times p$, with a value of either $\boldsymbol{\Sigma}$ or $\boldsymbol{0}$. Independence between the N rows of $\boldsymbol{Y}$ makes the off diagonal matrices $\boldsymbol{0}$ (independence implies zero covariance and correlation). The homogeneity assumption provides the diagonal matrices.

In contrast, stacking the columns presents exactly the same values in a permuted form. The $Np \times 1$ vector

$$\begin{bmatrix} \boldsymbol{y}_1 \\ \boldsymbol{y}_2 \\ \vdots \\ \boldsymbol{y}_p \end{bmatrix} = \text{vec}(\boldsymbol{Y}) \tag{3.5}$$

has $Np \times Np$ covariance matrix

$$\begin{aligned} \mathcal{V}[\text{vec}(\boldsymbol{Y})] &= \boldsymbol{\Sigma} \otimes \boldsymbol{I}_N \\ &= \begin{bmatrix} \sigma_{11}\boldsymbol{I} & \sigma_{12}\boldsymbol{I} & \cdots & \sigma_{1p}\boldsymbol{I} \\ \sigma_{21}\boldsymbol{I} & \sigma_{22}\boldsymbol{I} & \cdots & \sigma_{2p}\boldsymbol{I} \\ \vdots & \vdots & \ddots & \vdots \\ \sigma_{p1}\boldsymbol{I} & \sigma_{p2}\boldsymbol{I} & \cdots & \sigma_{pp}\boldsymbol{I} \end{bmatrix}. \end{aligned} \tag{3.6}$$

In the last equation there are p^2 submatrices, each of them $N \times N$, with the i, j submatrix having the value $\sigma_{i,j}\boldsymbol{I}_N$. We allow heterogeneity and correlation across columns, but not rows of $\boldsymbol{Y}$.

A second implication is that the response matrix may be separated into purely fixed and purely random matrices by centering the responses to define

$$\begin{aligned} \boldsymbol{E} &= \boldsymbol{Y} - \text{E}(\boldsymbol{Y}) \\ &= \boldsymbol{Y} - \boldsymbol{XB}. \end{aligned} \tag{3.7}$$

with $\text{E}(\boldsymbol{E}) = \boldsymbol{0}$ and $\mathcal{V}[\text{vec}(\boldsymbol{E}')] = \mathcal{V}[\text{vec}(\boldsymbol{Y}')] = \boldsymbol{I}_N \otimes \boldsymbol{\Sigma}$. A closely related third implication of the assumptions is that

$$\boldsymbol{Y} = \boldsymbol{XB} + \boldsymbol{E}. \tag{3.8}$$

The $N \times p$ constant matrix $\boldsymbol{XB}$ describes (models) the first moment of the responses, the mean matrix. The $N \times p$ random matrix $\boldsymbol{E}$ describes (models) the second and higher moments.

Finding estimators for $\boldsymbol{B}$ and $\boldsymbol{\Sigma}$ which satisfy a variety of optimal properties in small samples does not require any particular distribution function for the data. In contrast, the desire to test hypotheses leads to describing distributions of test statistics as a function of explicit specification of the data distribution.

Data analysts often assume the errors follow a Gaussian distribution. As detailed in Chapter 8, writing $\boldsymbol{Y} \sim \mathcal{N}_{n,m}(\boldsymbol{M}, \Xi, \Sigma)$ indicates Ξ and Σ are symmetric and positive definite or positive semidefinite, and $\text{vec}(\boldsymbol{Y}') \sim \mathcal{N}_{n,m}[\text{vec}(\boldsymbol{M}'), \Xi \otimes \Sigma]$, a direct product matrix Gaussian. A singular distribution may be specified by writing $\mathcal{SN}_{n,m}(\boldsymbol{M}, \Xi, \Sigma)$, while writing $(\mathcal{S})\mathcal{N}_{n,m}(\boldsymbol{M}, \Xi, \Sigma)$ indicates the distribution may or may not be singular. If either Ξ or Σ is singular, then so is $\Xi \otimes \Sigma$ because the eigenvalues of the direct product are products of the eigenvalues of the original matrices. Also $\text{rank}(\Xi \otimes \Sigma) = \text{rank}(\Xi)\text{rank}(\Sigma)$.

> **Definition 3.2** Writing $\text{GLM}_{N,p,q}(\boldsymbol{Y}_i; \boldsymbol{X}_i\boldsymbol{B}|\boldsymbol{R}_x\boldsymbol{B}\boldsymbol{R}_y = \boldsymbol{A}, \boldsymbol{\Sigma})$ *with Gaussian errors* indicates $\boldsymbol{Y}_i' \sim \mathcal{N}_p\{[\text{row}_i(\boldsymbol{X})\boldsymbol{B}]', \boldsymbol{\Sigma}\}$.
> Equivalently, $\boldsymbol{Y} \sim \mathcal{N}_{N,p}(\boldsymbol{XB}, \boldsymbol{I}_N, \boldsymbol{\Sigma})$.

Following Kleinbaum, Kupper, Muller and Nizam (1998), and Muller and Fetterman (2002), the GLM assumptions may be summarized with the mnemonic *HILE Gauss*: Homogeneity ($\mathcal{V}(\boldsymbol{Y}_i') = \mathcal{V}(\boldsymbol{Y}_{i'}') = \boldsymbol{\Sigma}$), Independence ($\boldsymbol{Y}_i \perp\!\!\!\perp \boldsymbol{Y}_{i'}$ if $i \neq i'$), Linearity ($\text{E}(\boldsymbol{Y}) = \boldsymbol{XB}$), existence of finite second moments, and *optionally*, Gaussian observations. The mnemonic groups the least squares assumptions together and separates the distribution assumption.

Adding the assumption of Gaussian errors allows deducing additional properties. With HILE Gauss, essentially all of the assumptions are captured by the statement $\boldsymbol{Y} \sim \mathcal{N}_{N,p}(\boldsymbol{XB}, \boldsymbol{I}_N, \boldsymbol{\Sigma})$. Obviously the parameters of the Gaussian distribution coincide with the parameters of the GLM. First and second moments fully characterize both. Also, $\boldsymbol{E} \sim \mathcal{N}_{N,p}(\boldsymbol{0}, \boldsymbol{I}_N, \boldsymbol{\Sigma})$.

As in univariate models, the theory of the multivariate GLM applies for two apparently disparate classes of models: linear regression and ANOVA models. Both are special cases of the multivariate GLM. ANOVA models were developed to test the effects of one or more categorical predictors on two or more Gaussian responses with independent and homogenous errors. Multivariate regression models express a set of continuous responses as a function of one or more continuous predictors. Models with both categorical and continuous predictors fall in between and are best thought of simply as multivariate linear models. The underlying theory, for data analysis, if not always for power analysis, coincides for all of them.

3.3 THE MULTIVARIATE GENERAL LINEAR HYPOTHESIS

> **Definition 3.3** In the $\text{GLM}_{N,p,q}(\boldsymbol{Y}_i; \boldsymbol{X}_i\boldsymbol{B}, \boldsymbol{\Sigma})$, functions of the primary parameters are *secondary parameters*.

Four important examples deserve specific mention because they lie at the heart of the multivariate general linear hypothesis (GLH). With $\boldsymbol{C}$ an $a \times q$ matrix of known constants, $\boldsymbol{U}$ a $p \times b$ matrix of known constants, and $\boldsymbol{\Theta}_0$ an $a \times b$ vector of known constants, $a \times b$ matrices $\boldsymbol{\Theta} = \boldsymbol{CBU}$ and $\boldsymbol{\Theta} - \boldsymbol{\Theta}_0$ are secondary parameters. Furthermore, the covariance matrix of the estimators, namely $\mathcal{V}[\text{vec}(\widehat{\boldsymbol{\Theta}})] = \mathcal{V}[\text{vec}(\widehat{\boldsymbol{\Theta}} - \boldsymbol{\Theta}_0)]$, as well as $\boldsymbol{\Sigma}_* = \boldsymbol{U}'\boldsymbol{\Sigma}\boldsymbol{U}$, are secondary parameters.

The general linear (null) hypothesis may be stated in two equivalent ways:

$$H_0 : \boldsymbol{CBU} = \boldsymbol{\Theta}_0$$
$$H_0 : \boldsymbol{\Theta} = \boldsymbol{\Theta}_0 \, . \tag{3.9}$$

If $\max(a, b) > 1$, then the alternative hypothesis is necessarily $H_A : \boldsymbol{\Theta} \neq \boldsymbol{\Theta}_0$. A one-sided alternative may occasionally be preferred when $\max(a, b) = 1$ (and $\boldsymbol{\Theta}$ reduces to θ, a scalar, such as a mean difference).

The $\boldsymbol{C}$ matrix $(a \times q)$ defines contrasts *between* groups or levels of predictors by computing linear combinations of coefficients of predictor variables, such as means. The $\boldsymbol{C}$ matrix implicitly computes and explicitly allows testing linear combinations of columns of $\boldsymbol{X}$, the predictor variables.

The $\boldsymbol{U}$ matrix $(p \times b)$ defines contrasts *within* an independent sampling unit (such as person) or level of response character (such as time) by computing linear combinations of coefficients of response variables, such as means. The $\boldsymbol{U}$ matrix implicitly computes and explicitly allows testing linear combinations of columns of $\boldsymbol{Y}$, the response variables in the model $\boldsymbol{Y} = \boldsymbol{XB} + \boldsymbol{E}$. The parameter $\boldsymbol{\Sigma}$ provides the $p \times p$ covariance matrix among response variables. The parameter $\boldsymbol{\Sigma}_* = \boldsymbol{U}'\boldsymbol{\Sigma U}$ provides the $b \times b$ covariance matrix among transformed (hypothesis) variables in the model $\boldsymbol{YU} = \boldsymbol{XBU} + \boldsymbol{EU}$. With repeated measures in $\boldsymbol{Y}$, often $\boldsymbol{U}$ contains orthogonal or orthonormal polynomial trend contrasts (linear, quadratic, etc., through order $p - 1$). The zero-order trend corresponds to the mean across the times.

Definition 3.4 For a $\text{GLM}_{N,p,q}(\boldsymbol{Y}_i; \boldsymbol{X}_i\boldsymbol{B}, \boldsymbol{\Sigma})$ with $\boldsymbol{X}$ fixed and full-rank $\boldsymbol{M} = \boldsymbol{C}(\boldsymbol{X}'\boldsymbol{X})^{-}\boldsymbol{C}'$, **(a)** the *shift parameter* is the $b \times b$ matrix

$$\boldsymbol{\Delta} = (\boldsymbol{\Theta} - \boldsymbol{\Theta}_0)'\boldsymbol{M}^{-1}(\boldsymbol{\Theta} - \boldsymbol{\Theta}_0) \, . \tag{3.10}$$

(b) For full-rank $\boldsymbol{\Sigma}_* = \boldsymbol{U}'\boldsymbol{\Sigma U}$ $(b \times b)$, the covariance matrix of transformed (hypothesis) variables, the *noncentrality parameter* is the $b \times b$ matrix

$$\boldsymbol{\Omega} = \boldsymbol{\Delta \Sigma}_*^{-1} = (\boldsymbol{\Theta} - \boldsymbol{\Theta}_0)'\boldsymbol{M}^{-1}(\boldsymbol{\Theta} - \boldsymbol{\Theta}_0)\boldsymbol{\Sigma}_*^{-1} \, . \tag{3.11}$$

Parameters $\boldsymbol{\Delta}$ and $\boldsymbol{\Omega}$ play central roles in the distribution theory of multivariate hypothesis tests. The noncentrality parameter $\boldsymbol{\Omega}$ is scale free in the sense of corresponding to hypothesis variables standardized to have covariance $\boldsymbol{I}_b$.

As in the univariate case, both $\boldsymbol{\Delta}$ and $\boldsymbol{\Omega}$ do not vary under any full-rank transformation of the rows of both $\boldsymbol{C}$ and $\boldsymbol{\Theta}_0$, of the form $\boldsymbol{C}_T = \boldsymbol{T}_\text{B}\boldsymbol{C}$ and $\boldsymbol{\Theta}_{0T} = \boldsymbol{T}_\text{B}\boldsymbol{\Theta}$, for $\boldsymbol{T}_\text{B}$ $(a \times a)$ of rank a. On the other hand, both $\boldsymbol{\Delta}$ and $\boldsymbol{\Omega}$ *do* vary under simultaneous full-rank transformation of the columns of $\boldsymbol{U}$ and $\boldsymbol{\Theta}_0$, of the form $\boldsymbol{U}_T = \boldsymbol{UT}_\text{W}$ and $\boldsymbol{\Theta}_{0T} = \boldsymbol{\Theta}_0\boldsymbol{T}_\text{W}$, for $\boldsymbol{T}_\text{W}$ $(b \times b)$ of rank b. However, the eigenvalues of $\boldsymbol{\Omega}$ do not vary under such a transformation. As discussed briefly below and in more detail in Chapter 16, the eigenvalues of $\boldsymbol{\Omega}$, in addition to the dimensions of the problem, suffice to fully determine the distribution of the multivariate test statistics. The multivariate test statistics are functions only of the eigenvalues of $\widehat{\boldsymbol{\Omega}}$ and the dimensions. The statistics do not vary under any full-rank transformation of C or $\boldsymbol{U}$ (and $\boldsymbol{\Theta}_0$) as just discussed.

A wide variety of apparently distinct methods correspond to special cases of the multivariate test statistics, include multivariate ANOVA (MANOVA) and the multivariate approach to repeated measures (MULTIREP). However, tests arising from the univariate approach to repeated measures (UNIREP) differ fundamentally in both origin and many properties. The present chapter contains a brief overview, while Chapter 16 has full details. Both the multivariate (MULTIREP) and UNIREP tests arise from the same estimation theory.

With the spectral decomposition $\boldsymbol{\Sigma}_* = \boldsymbol{\Upsilon}\mathrm{Dg}(\boldsymbol{\lambda})\boldsymbol{\Upsilon}'$, and $\boldsymbol{\Delta}_\Upsilon = \boldsymbol{\Upsilon}'\boldsymbol{\Delta}\boldsymbol{\Upsilon}$, the distributions of the tests in the UNIREP tests depend on the b eigenvalues of $\boldsymbol{\Sigma}_*$, $\boldsymbol{\lambda}$, and the diagonal elements (not the eigenvalues) of $\boldsymbol{\Omega}_\Upsilon = \boldsymbol{\Delta}_\Upsilon\mathrm{Dg}(\boldsymbol{\lambda})^{-1} = \boldsymbol{\Upsilon}'\boldsymbol{\Delta}\boldsymbol{\Sigma}_*^{-1}\boldsymbol{\Upsilon} = \boldsymbol{\Upsilon}'\boldsymbol{\Omega}\boldsymbol{\Upsilon}$. As functions of the eigenvalues of both $\widehat{\boldsymbol{\Delta}}$ and $\widehat{\boldsymbol{\Sigma}}_*$ separately, the corresponding test statistics *do* vary under full rank transformation of the columns of $\boldsymbol{U}$ and $\boldsymbol{\Theta}_0$, of the form $\boldsymbol{U}_T = \boldsymbol{U}\boldsymbol{T}_\mathrm{W}$ and $\boldsymbol{\Theta}_{0T} = \boldsymbol{\Theta}_0\boldsymbol{T}_\mathrm{W}$, with $\boldsymbol{T}$ $(b \times b)$ of rank b.

Example 3.2 A multivariate independent groups t test is sometimes described as a two sample Hotelling T^2 test. The setting also corresponds to the special case of a MANOVA test for two groups, which implies $\boldsymbol{U} = \boldsymbol{I}_p$ for p responses. With N_g observations in group $g \in \{\mathrm{A}, \mathrm{B}\}$, $\boldsymbol{Y}$ and $\boldsymbol{E}$ are $(N_\mathrm{A} + N_\mathrm{B}) \times p$. A reference cell coding, with $\alpha_j = \mu_{\mathrm{B},j} - \mu_{\mathrm{A},j}$ gives $\boldsymbol{X}$ of dimension $(N_\mathrm{A} + N_\mathrm{B}) \times 2$ and

$$\boldsymbol{Y} = \begin{bmatrix} \mathbf{1}_{N_\mathrm{A}} & \mathbf{0} \\ \mathbf{1}_{N_\mathrm{B}} & \mathbf{1}_{N_\mathrm{B}} \end{bmatrix} \begin{bmatrix} \mu_{\mathrm{A},1} & \mu_{\mathrm{A},2} & \cdots & \mu_{\mathrm{A},p} \\ \alpha_1 & \alpha_2 & \cdots & \alpha_p \end{bmatrix} + \boldsymbol{E}. \tag{3.12}$$

Therefore, with $\boldsymbol{C} = [0 \quad 1]$ and $\boldsymbol{U} = \boldsymbol{I}_p$,

$$\begin{aligned} \boldsymbol{M} &= [0\,1]\begin{bmatrix} N_\mathrm{A} + N_\mathrm{B} & N_\mathrm{B} \\ N_\mathrm{B} & N_\mathrm{B} \end{bmatrix}^{-1} [0\,1]' \\ &= [0\,1]\begin{bmatrix} 1/N_\mathrm{A} & -1/N_\mathrm{A} \\ -1/N_\mathrm{A} & 1/N_\mathrm{B} + 1/N_\mathrm{A} \end{bmatrix}[0\,1]' \\ &= (N_\mathrm{A} + N_\mathrm{B})/(N_\mathrm{A} N_\mathrm{B}), \end{aligned} \tag{3.13}$$

$$\boldsymbol{\Sigma}_* = \boldsymbol{I}_p'\boldsymbol{\Sigma}\boldsymbol{I}_p = \boldsymbol{\Sigma} \tag{3.14}$$

$$\begin{aligned} \boldsymbol{\Theta} - \boldsymbol{\Theta}_0 &= [0\,1]\begin{bmatrix} \mu_{\mathrm{A},1} & \mu_{\mathrm{A},2} & \cdots & \mu_{\mathrm{A},p} \\ \alpha_1 & \alpha_2 & \cdots & \alpha_p \end{bmatrix}\boldsymbol{I}_p - \mathbf{0}_{1\times p} \\ &= [\alpha_1\,\alpha_2 \cdots \alpha_p]. \end{aligned} \tag{3.15}$$

If $\boldsymbol{A} = \{\alpha_j \alpha_{j'}\}$ then

$$\begin{aligned}\boldsymbol{\Omega} &= [\,\alpha_1\,\alpha_2\cdots\alpha_p\,]'[(N_{\mathrm{A}}+N_{\mathrm{B}})/(N_{\mathrm{A}}N_{\mathrm{B}})]^{-1}[\,\alpha_1\,\alpha_2\cdots\alpha_p\,]\boldsymbol{\Sigma}^{-1} \\ &= (N_{\mathrm{A}}N_{\mathrm{B}})(N_{\mathrm{A}}+N_{\mathrm{B}})^{-1}\boldsymbol{A}\boldsymbol{\Sigma}^{-1}\,. \end{aligned} \tag{3.16}$$

As in the univariate case, noncentrality depends only on sample size, mean differences, and error variance.

The rank of $\boldsymbol{\Omega}$ plays an important role in the distribution theory of multivariate linear models. In general, with $s_* = \mathrm{rank}(\boldsymbol{\Omega})$, it follows that $0 \le s_* \le \min(a,b) = s$, with $s_* = 0$ only when $\boldsymbol{\Theta} = \boldsymbol{\Theta}_0$.

Explaining the origin of the bounds on s_* describes many of the relationships among the dimensions and parameters. The matrices $\boldsymbol{\Delta}$ and $\boldsymbol{\Sigma}_*^{-1}$ are symmetric and $b \times b$. A well-defined test requires full (row) rank of a for $\boldsymbol{C}$. It also requires full (column) rank of b for $\boldsymbol{U}$, which ensures full rank of b for $\boldsymbol{\Sigma}_* = \boldsymbol{U}'\boldsymbol{\Sigma}\boldsymbol{U}$ (given a full-rank $\boldsymbol{\Sigma}$, as nearly always assumed) and also $\boldsymbol{\Sigma}_*^{-1}$. The $b \times b$ noncentrality matrix, $\boldsymbol{\Omega} = \boldsymbol{\Delta}\boldsymbol{\Sigma}_*^{-1}$, will not be symmetric, except for special cases. Furthermore, $\mathrm{rank}(\boldsymbol{\Omega}) = \mathrm{rank}(\boldsymbol{\Delta})$, with $\boldsymbol{\Delta} = (\boldsymbol{\Theta} - \boldsymbol{\Theta}_0)'\boldsymbol{M}^{-1}(\boldsymbol{\Theta} - \boldsymbol{\Theta}_0)$. The $a \times a$ $\boldsymbol{M}$ matrix must be full rank for a well-defined test, which ensures the rank of $\boldsymbol{\Delta}$ $(b \times b)$ equals the rank of $(\boldsymbol{\Theta} - \boldsymbol{\Theta}_0)$ $(a \times b)$. Obviously $0 \le \mathrm{rank}(\boldsymbol{\Theta} - \boldsymbol{\Theta}_0) \le \min(a,b)$. Here $\mathrm{rank}(\boldsymbol{\Theta} - \boldsymbol{\Theta}_0) = 0$ only when $\boldsymbol{\Theta} = \boldsymbol{\Theta}_0$.

Example 3.3 The matrix $\boldsymbol{\Omega} = [(N_{\mathrm{A}}N_{\mathrm{B}})/(N_{\mathrm{A}}+N_{\mathrm{B}})]\boldsymbol{A}\boldsymbol{\Sigma}^{-1}$ from the last example has rank 1, which can be proven as follows. Square and full rank $\boldsymbol{\Sigma}^{-1}$ ensures $\mathrm{rank}(\boldsymbol{\Omega}) = \mathrm{rank}(\boldsymbol{A})$. If $\boldsymbol{\alpha} = [\,\alpha_1\,\alpha_2\cdots\alpha_p\,]'$, then $\boldsymbol{A} = \boldsymbol{\alpha}\boldsymbol{\alpha}'$. Hence $\mathrm{rank}(\boldsymbol{A}) = \mathrm{rank}(\boldsymbol{\alpha}\boldsymbol{\alpha}') = \mathrm{rank}(\boldsymbol{\alpha}'\boldsymbol{\alpha}) = 1$.

Alternately, if $\boldsymbol{\alpha}_1 = \boldsymbol{\alpha}/\sqrt{\boldsymbol{\alpha}'\boldsymbol{\alpha}}$, then there exists orthonormal $\boldsymbol{A}_0$ of dimension $p \times (p-1)$ with $\boldsymbol{A}_0'\boldsymbol{\alpha}_1 = \boldsymbol{0}$. The spectral decomposition of $\boldsymbol{A}$ is

$$\begin{aligned}\boldsymbol{A} = \boldsymbol{\alpha}\boldsymbol{\alpha}' &= \frac{1}{\sqrt{\boldsymbol{\alpha}'\boldsymbol{\alpha}}}\boldsymbol{\alpha}(\boldsymbol{\alpha}'\boldsymbol{\alpha})\boldsymbol{\alpha}'\frac{1}{\sqrt{\boldsymbol{\alpha}'\boldsymbol{\alpha}}} \\ &= \boldsymbol{\alpha}_1(\boldsymbol{\alpha}'\boldsymbol{\alpha})\boldsymbol{\alpha}_1' \\ &= [\,\boldsymbol{\alpha}_1\ \ \boldsymbol{A}_0\,]\begin{bmatrix}(\boldsymbol{\alpha}'\boldsymbol{\alpha}) & 0 \\ 0 & 0\end{bmatrix}\begin{bmatrix}\boldsymbol{\alpha}_1' \\ \boldsymbol{A}_0'\end{bmatrix}. \end{aligned} \tag{3.17}$$

The eigenvalues of $\boldsymbol{\Omega} = (N_{\mathrm{A}}N_{\mathrm{B}})(N_{\mathrm{A}}+N_{\mathrm{B}})^{-1}\boldsymbol{\alpha}\boldsymbol{\alpha}'\boldsymbol{\Sigma}^{-1}$ and $\boldsymbol{\Omega}_1 = (N_{\mathrm{A}}N_{\mathrm{B}})(N_{\mathrm{A}}+N_{\mathrm{B}})^{-1}\boldsymbol{\alpha}'\boldsymbol{\Sigma}^{-1}\boldsymbol{\alpha}$ coincide, but not the eigenvectors. Writing $\omega_1 = (N_{\mathrm{A}}N_{\mathrm{B}})(N_{\mathrm{A}}+N_{\mathrm{B}})^{-1}\boldsymbol{\alpha}'\boldsymbol{\Sigma}^{-1}\boldsymbol{\alpha}$ is more precise because ω_1 is 1×1, a scalar. Any scalar has a single eigenvalue, the scalar itself. The expression may be written

$$\omega_1 = \frac{N_{\mathrm{A}}N_{\mathrm{B}}}{N_{\mathrm{A}}+N_{\mathrm{B}}}[\,\mu_{\mathrm{B},1}-\mu_{\mathrm{A},1}\cdots\mu_{\mathrm{B},p}-\mu_{\mathrm{A},p}\,]\boldsymbol{\Sigma}^{-1}\begin{bmatrix}\mu_{\mathrm{B},1}-\mu_{\mathrm{A},1} \\ \vdots \\ \mu_{\mathrm{B},p}-\mu_{\mathrm{A},p}\end{bmatrix}. \tag{3.18}$$

The parameter ω_1 is the Mahalanobis distance from the origin of the vector of group differences. If the responses are uncorrelated, then $\boldsymbol{\Sigma}$ is diagonal and

$$\begin{aligned}\omega_1 &= \frac{N_A N_B}{N_A + N_B}\boldsymbol{\alpha}'\left[\mathrm{Dg}(\{\sigma_j^2\})\right]^{-1}\boldsymbol{\alpha} \\ &= \frac{N_A N_B}{N_A + N_B}\sum_{j=1}^{p}\frac{(\mu_{B,j} - \mu_{A,j})^2}{\sigma_j^2} . \end{aligned} \tag{3.19}$$

If $p = 1$ and $N_A = N_B$, then $(N_A N_B)(N_A + N_B)^{-1} = N/4$ and ω_1 reduces to the t-test result in Chapter 2.

Writing $\mathrm{GLM}_{N,p,q}(\boldsymbol{Y}_i; \boldsymbol{X}_i\boldsymbol{B}, \boldsymbol{\Sigma})$ specifies an unconstrained model. A GLH (general linear hypothesis) describes a set of linear constraints on $\boldsymbol{B}$, namely $\boldsymbol{CBU} = \boldsymbol{\Theta}_0$, through the fixed and known constants $\boldsymbol{C}$, $\boldsymbol{U}$, and $\boldsymbol{\Theta}_0$. Writing $\mathrm{GLM}_{N,p,q}(\boldsymbol{Y}_i; \boldsymbol{X}_i\boldsymbol{B}|\boldsymbol{CBU} = \boldsymbol{\Theta}_0, \boldsymbol{\Sigma})$ specifies a constrained model. *Hence every GLH corresponds to comparing a full and constrained model.* Choosing $\boldsymbol{U} = \boldsymbol{I}_p$ has the same effect as not using any $\boldsymbol{U}$ matrix. With $\boldsymbol{U} = \boldsymbol{I}_p$, any $\boldsymbol{C}$ with all rows selected from $\boldsymbol{I}_q$ leads to an easily understood constrained model. In such a case, the constrained model may be produced from the full model simply by deleting appropriate columns of $\boldsymbol{X}$ and corresponding rows of $\boldsymbol{B}$. Similarly,with $\boldsymbol{C} = \boldsymbol{I}_q$, any $\boldsymbol{U}$ matrix with all columns selected from $\boldsymbol{I}_p$ leads to an easily understood constrained model. In such a case, the constrained model may be produced from the full model simply by deleting appropriate columns of $\boldsymbol{Y}$ and corresponding columns of $\boldsymbol{B}$. More generally, $\boldsymbol{U}$ defines a priori linear transformations of the response variables, such as an average or set of difference scores. The hypothesis compares the original (full) and the reduced models.

3.4 TESTS ABOUT COVARIANCE MATRICES

The general linear hypothesis makes a statement about expected value parameters, elements of $\boldsymbol{\beta}$. It is less common, although perfectly reasonable, to consider hypotheses about covariance matrices, such as $\boldsymbol{\Sigma}$ or $\boldsymbol{\Sigma}_*$. Some exact tests have been developed. Morrison (1990), Anderson (2004), and Timm (2002, Section 3.8) have additional details.

Tests of covariance pattern should *not* be used to check the assumption of homogeneity in a linear model. A number of authors have provided analytic and simulation results to support the proposition. Chapter 16 includes discussion of tests about covariance matrices. The tests do perform appropriately when used as they were intended.

3.5 POPULATION CORRELATION

The univariate correlation applies to one response and one predictor. Recalling $\widehat{y}_i = \beta_0 + \beta_1 x_i$ implies that the squared correlation of the response and predictor is the same as that between the response and the predicted value, $\rho^2(y_i, x_i) = \rho^2(y_i, \widehat{y}_i)$. Allowing many predictors gives the squared multiple correlation $\rho^2(y_i, \{x_{i1}, \ldots, x_{iq-1}\}) = \rho^2(y_i, \widehat{y}_i)$ with $\widehat{y}_i = \widehat{\beta}_0 + \sum_{j=1}^{q-1}\widehat{\beta}_j x_{ij}$ for a model with an intercept [more generally, $\widehat{y}_i = \boldsymbol{X}_i(\boldsymbol{X}'\boldsymbol{X})^{-}\boldsymbol{X}'\boldsymbol{y}$]. Generalizing to p responses and q predictors (which span an intercept) leads to considering a set of $s = \min(p, q-1)$ squared canonical correlations. Muller (1982) described the model underlying the traditional approach to canonical correlation as a measure of association between two sets of variables. Canonical correlation also lie at the heart of the multivariate linear model.

Definition 3.5 For the $\text{GLM}_{N,p,q}(\boldsymbol{Y}_i; \boldsymbol{X}_i\boldsymbol{B}, \boldsymbol{\Sigma})$ and associated general linear hypothesis $H_0 : \boldsymbol{CBU} = \boldsymbol{\Theta}_0$, the *(generalized) squared canonical correlations* $\{\rho_{*k}^2\}$ are the eigenvalues of the $b \times b$ matrix $(\boldsymbol{\Delta} + N\boldsymbol{\Sigma}_*)(N\boldsymbol{\Sigma}_*)^{-1}$.

The eigenvalues of $\boldsymbol{\Omega} = \boldsymbol{\Delta}\boldsymbol{\Sigma}_*^{-1}$ are $\{\omega_k\}$, with $\omega_k = N\rho_{*k}^2/(1-\rho_{*k}^2)$. With $\boldsymbol{\Sigma}_* = \boldsymbol{\Phi}\boldsymbol{\Phi}'$, the eigenvalues of $\boldsymbol{\Omega}$ coincide with the eigenvalues of the symmetric and positive definite or positive semidefinite and $b \times b$ matrix $\boldsymbol{\Omega}_{\Phi} = \boldsymbol{\Phi}^{-1}\boldsymbol{\Delta}\boldsymbol{\Phi}^{-t}$. In all cases s_* nonzero values of ρ_{*k}^2 occur, with $s_* = \text{rank}(\boldsymbol{\Omega}_{\Phi}) = \text{rank}(\boldsymbol{\Omega}) = \text{rank}(\boldsymbol{\Delta}) = \text{rank}(\boldsymbol{\Theta} - \boldsymbol{\Theta}_0)$. Also, $0 \leq s_* \leq s = \min(a, b)$. Only under the null hypothesis does $s_* = 0$. In general,

$$\omega_k = N\frac{\rho_{*k}^2}{1-\rho_{*k}^2} \tag{3.20}$$

and

$$\rho_{*k}^2 = \frac{\omega_k}{\omega_k + N}\,. \tag{3.21}$$

Naturally the multivariate formulas reduce to the univariate formulas if $p = 1$. In turn, $b = 1$ implies $s = 1$, $s_* = 1$ under the alternative, and $s_* = 0$ under the null.

As in the univariate case, $\{\omega_k\}$ and $\{\rho_{*k}^2\}$ always have scale invariance. If the model spans an intercept and the hypothesis excludes it, the hypothesis also has location invariance. If so, the generalized canonical correlations are appropriately described simply as canonical correlations.

3.6 STATISTICAL ESTIMATES

In practice, generalizing formulas for estimates from univariate to multivariate models simply requires replacing the $N \times 1$ vectors $\boldsymbol{y}$ and $\boldsymbol{e}$ by the $N \times p$ matrices $\boldsymbol{Y}$ and $\boldsymbol{E}$, with the additional columns corresponding to the p responses. Similarly, $N \times 1$ $\widehat{\boldsymbol{y}}$ and $\widehat{\boldsymbol{e}}$ become $N \times p$ $\widehat{\boldsymbol{Y}}$ and $\widehat{\boldsymbol{E}}$, while $q \times 1$ $\boldsymbol{\beta}$ and $\widehat{\boldsymbol{\beta}}$ become $q \times p$ $\boldsymbol{B}$ and $\widehat{\boldsymbol{B}}$. The reason lies hidden in three implicit assumptions: (1) the same design matrix applies to all response variables, which correspond to columns of $\boldsymbol{Y}$; (2) there are no missing data; and (3) each value within a variable (column of $\boldsymbol{Y}$) was measured in a consistent way (no appreciable mistiming is allowed).

Adding responses adds columns in $\boldsymbol{Y}$ but does not change the properties of the design matrix. In a LTFR model $[r = \text{rank}(\boldsymbol{X}) < q]$ estimation relies on a nonunique generalized inverse $(\boldsymbol{X}'\boldsymbol{X})^{-}$. In contrast, $(\boldsymbol{X}'\boldsymbol{X})^{-1}$ exists only when $\boldsymbol{X}$ has full rank $[\text{rank}(\boldsymbol{X}) = q]$. With less than full rank, the matrix

$$\widetilde{\boldsymbol{B}} = (\boldsymbol{X}'\boldsymbol{X})^{-}\boldsymbol{X}'\boldsymbol{Y} \tag{3.22}$$

is a proper estimator of $\boldsymbol{B}$. However, it is biased and varies with the choice of generalized inverse. In contrast, a full-rank model defines $\boldsymbol{B}$ differently and thereby makes possible a unique and unbiased estimator, namely

$$\widehat{\boldsymbol{B}} = (\boldsymbol{X}'\boldsymbol{X})^{-1}\boldsymbol{X}'\boldsymbol{Y}. \tag{3.23}$$

A full rank design defines $\boldsymbol{B}$ so that it is *estimable*, which means that a known function of the data is an unbiased estimator. The definition will be refined later.

Definition 3.6 For the $\text{GLM}_{N,p,q}(\boldsymbol{Y}_i; \boldsymbol{X}_i\boldsymbol{B}, \boldsymbol{\Sigma})$,

(a) *predicted values* may be expressed as

$$\begin{aligned}\widehat{\boldsymbol{Y}} &= \boldsymbol{X}\widetilde{\boldsymbol{B}}\\ &= \boldsymbol{X}(\boldsymbol{X}'\boldsymbol{X})^{-}\boldsymbol{X}'\boldsymbol{Y}\\ &= \boldsymbol{H}\boldsymbol{Y}.\end{aligned} \tag{3.24}$$

(b) In turn, the estimated errors, the *residuals*, are

$$\begin{aligned}\widehat{\boldsymbol{E}} &= \boldsymbol{Y} - \boldsymbol{X}\widetilde{\boldsymbol{B}}\\ &= [\boldsymbol{I} - \boldsymbol{X}(\boldsymbol{X}'\boldsymbol{X})^{-}\boldsymbol{X}']\boldsymbol{Y}\\ &= (\boldsymbol{I} - \boldsymbol{H})\boldsymbol{Y}.\end{aligned} \tag{3.25}$$

The "hat matrix" $\boldsymbol{H}$ is unique even though $(\boldsymbol{X}'\boldsymbol{X})^{-}$ is not, which also makes the predicted values and residuals unique.

With $\boldsymbol{C}$ an $a \times q$ matrix of known constants and $\boldsymbol{U}$ a $p \times b$ matrix of known constants, $\boldsymbol{\Theta} = \boldsymbol{C}\boldsymbol{B}\boldsymbol{U}$ defines an $a \times b$ matrix of secondary parameters, with corresponding estimator $\widehat{\boldsymbol{\Theta}} = \boldsymbol{C}\widehat{\boldsymbol{B}}\boldsymbol{U}$ or $\boldsymbol{C}\widetilde{\boldsymbol{B}}\boldsymbol{U}$. The covariance matrix of the estimator may also be described as a secondary parameter. We delay the discussion as to how to determine whether $\widehat{\boldsymbol{\Theta}}$ is a good estimator, along with the proofs of all estimation properties, to Chapter 12.

Definition 3.7 For the $\text{GLM}_{N,p,q}(\boldsymbol{Y}_i; \boldsymbol{X}_i\boldsymbol{B}, \boldsymbol{\Sigma})$ and associated general linear hypothesis $H_0 : \boldsymbol{C}\boldsymbol{B}\boldsymbol{U} = \boldsymbol{\Theta}_0$, the *error sum of squares* is $\boldsymbol{S}_e = \boldsymbol{U}'\widehat{\boldsymbol{E}}'\widehat{\boldsymbol{E}}\boldsymbol{U}$, and the *hypothesis sum of squares* is $\boldsymbol{S}_h = \widehat{\boldsymbol{\Delta}} = (\widehat{\boldsymbol{\Theta}} - \boldsymbol{\Theta}_0)'\boldsymbol{M}^{-1}(\widehat{\boldsymbol{\Theta}} - \boldsymbol{\Theta}_0)$, which are both $b \times b$ matrices. Choosing $\boldsymbol{U} = \boldsymbol{I}_p$ gives $\boldsymbol{S}_e = \widehat{\boldsymbol{E}}'\widehat{\boldsymbol{E}}$, the error sums squares for the model.

The sum of squared errors $SSE = \widehat{e}'\widehat{e}$ generalizes from a scalar to a $b \times b$ matrix $\boldsymbol{U}'\widehat{\boldsymbol{E}}'\widehat{\boldsymbol{E}}\boldsymbol{U}$, and scalar $\widehat{\sigma}^2$ becomes the $b \times b$ matrix $\widehat{\boldsymbol{\Sigma}}_* = \boldsymbol{U}'\widehat{\boldsymbol{\Sigma}}\boldsymbol{U}$. Row and column labels of $\boldsymbol{U}'\widehat{\boldsymbol{E}}'\widehat{\boldsymbol{E}}\boldsymbol{U}$ and $\widehat{\boldsymbol{\Sigma}}_*$ are the names of transformed response variables (columns of $\boldsymbol{YU}$, $\boldsymbol{EU}$, $\widehat{\boldsymbol{Y}}\boldsymbol{U}$, and $\widehat{\boldsymbol{E}}\boldsymbol{U}$). As with $p = 1$ (the univariate model), $r = \text{rank}(\boldsymbol{X})$ and

$$\begin{aligned}\widehat{\boldsymbol{\Sigma}} &= \widehat{\boldsymbol{E}}'\widehat{\boldsymbol{E}}/(N-r) \\ &= \boldsymbol{Y}'[\boldsymbol{I} - \boldsymbol{X}(\boldsymbol{X}'\boldsymbol{X})^{-}\boldsymbol{X}']\boldsymbol{Y}/(N-r) \\ &= \boldsymbol{Y}'(\boldsymbol{I} - \boldsymbol{H})\boldsymbol{Y}/(N-r) \end{aligned} \tag{3.26}$$

is unique and also can be proven to be an unbiased estimator of $\boldsymbol{\Sigma}$. Furthermore

$$\widehat{\boldsymbol{\Sigma}}_* = \boldsymbol{U}'\boldsymbol{Y}'(\boldsymbol{I} - \boldsymbol{H})\boldsymbol{Y}\boldsymbol{U}/(N-r) \tag{3.27}$$

is also unique and unbiased. With $p = 1$, the univariate model, necessarily $\boldsymbol{U} = [1] = \boldsymbol{I}_1$. Multivariate models allow for $\boldsymbol{U} \neq \boldsymbol{I}_p$, which leads to interest in $\boldsymbol{\Sigma}_* = \boldsymbol{U}'\boldsymbol{\Sigma}\boldsymbol{U}$.

The obvious estimators are $\widehat{\boldsymbol{\Delta}} = (\widehat{\boldsymbol{\Theta}} - \boldsymbol{\Theta}_0)'\boldsymbol{M}^{-1}(\widehat{\boldsymbol{\Theta}} - \boldsymbol{\Theta}_0) = \boldsymbol{S}_h$ and

$$\widehat{\boldsymbol{\Omega}} = \widehat{\boldsymbol{\Delta}}\widehat{\boldsymbol{\Sigma}}_*^{-1} = (\widehat{\boldsymbol{\Theta}} - \boldsymbol{\Theta}_0)'\boldsymbol{M}^{-1}(\widehat{\boldsymbol{\Theta}} - \boldsymbol{\Theta}_0)\widehat{\boldsymbol{\Sigma}}_*^{-1}. \tag{3.28}$$

3.7 OVERVIEW OF TESTING MULTIVARIATE HYPOTHESES

As detailed in Chapter 16, well-defined multivariate tests of *testable* hypotheses require three properties. With $\boldsymbol{M} = \boldsymbol{C}(\boldsymbol{X}'\boldsymbol{X})^{-}\boldsymbol{C}'$ of dimensions $a \times a$,

$$\text{rank}(\boldsymbol{M}) = a\,, \tag{3.29}$$

which means $\boldsymbol{M}$ is full rank and invertible. Testable hypotheses require estimable parameters, which is guaranteed when

$$\boldsymbol{C} = \boldsymbol{C}(\boldsymbol{X}'\boldsymbol{X})^{-}(\boldsymbol{X}'\boldsymbol{X})\,. \tag{3.30}$$

Ensuring a testable multivariate hypothesis imposes an additional requirement, namely full (column) rank of the $p \times b$ matrix $\boldsymbol{U}$:

$$\text{rank}(\boldsymbol{U}) = b\,, \tag{3.31}$$

A testable hypothesis necessarily has $b \leq p$. Only less-than-full-rank models require checking the estimability condition, which can be replaced by a variety of equivalent conditions (which are usually less computationally convenient). The first condition implicitly requires $\boldsymbol{C}$, of dimension $a \times q$, have full (row) rank a, which provides a sufficient condition to ensure testability for full-rank models, but not for less-than-full-rank models. In any case, a testable hypothesis necessarily has $a \leq q$.

A multivariate set of response variables may be tested with any one of nine different tests, which fall into three groups, as summarized in Table 3.1. In contrast to the univariate setting, no single test satisfies the various measures of goodness for the multivariate general linear hypothesis $H_0 : \boldsymbol{CBU} = \boldsymbol{\Theta}_0$ for all conditions. Although the Wilks statistic (WLK) provides the likelihood ratio test, Roy's largest root (RLR) provides the union-intersection principle test. Equally important, the relative powers of the various tests vary with the pattern of noncentralities, which implies each may be preferred in some settings. We shall provide more detail after providing explicit forms for the tests.

Table 3.1 Tests for Multivariate Hypotheses

Approach	Test	$\boldsymbol{\Sigma}_*$ Eigenvalues	Test Size	Best Power?
Bonferroni		Any	$\leq \alpha$	$\boldsymbol{\Sigma} = \mathrm{Dg}(\{\sigma_j^2\})$?
MULTIREP	Hotelling-Lawley (HLT)	Any	$= \alpha$	$s_* > 1$
(MANOVA)	Pillai-Bartlett (PBT)	Any	$= \alpha$	$s_* > 1$
	Wilks Likelihood (WLK)	Any	$= \alpha$	$s_* > 1$
	Roy's Largest Root (RLR)	Any	$= \alpha$	$s_* = 1$
UNIREP	Box Conservative	Any	$\lll \alpha$	$\epsilon = 1/b$
	Geisser-Greenhouse (GG)	Any	$\lesssim \alpha$	ϵ near 1
	Huynh-Feldt (HF)	Any	$\approx \alpha$	ϵ near 1
	Uncorrected (UN)	Any	$\gg \alpha$	does not apply
		$\epsilon = 1$	$= \alpha$	UMP[1]

[1]Uniformly most powerful given assumptions

With p variables, the simplest approach uses a Bonferroni correction with α/p test size in separate univariate analyses of each response variable. The approach does not completely test the multivariate general linear hypothesis $H_0 : \boldsymbol{CBU} = \boldsymbol{\Theta}_0$. Instead, for each $\boldsymbol{C}$ (between subject, a group contrast), a set of p distinct $\boldsymbol{U}$ matrices are created, with each a distinct column of $\boldsymbol{I}_p$. The approach has two particular strengths. First, it tolerates missing data and different design matrices for each response. Second, it allows allocating more test size to the important variables, which may better meet the scientific goals and desires of the investigators. The approach obviously inherits the invariance properties of the univariate model.

The second approach, labeled MULTIREP, groups four tests together, traditionally described as the multivariate tests because they were specifically developed for the multivariate general linear hypothesis $H_0 : \boldsymbol{CBU} = \boldsymbol{\Theta}_0$. They share the important property of invariance to the value of $\boldsymbol{\Sigma}$ and therefore to $\boldsymbol{\Sigma}_*$. The tests may be employed with the "multivariate" approach to repeated measures, in contrast to the "univariate" approach to repeated measures (UNIREP). Both approaches to repeated measures decompose the p dimensional response space into the $p-1$ dimensions within- plus 1 dimension between-"subject" (independent sampling unit) subspaces. Tests involving contrasts between subject use $\boldsymbol{u}_{\rm B} = \mathbf{1}_p/p$, of dimension $p \times 1$, which computes the average response across repeated measures. Tests involving contrasts within subject's use $\boldsymbol{U}_{\rm W}$ of dimension $p \times (p-1)$, with full column rank of $p-1$, such that $\boldsymbol{I}_p = [\,\boldsymbol{u}_{\rm B}\;\boldsymbol{U}_{\rm W}\,]\boldsymbol{T}$, with $\boldsymbol{T}$ full rank and $p \times p$. Two common choices include $\boldsymbol{U}_{\rm W} = [\,\mathbf{1}_{p-1}\; -\boldsymbol{I}_{p-1}\,]'$, which gives pairwise contrasts, and the $p \times (p-1)$ matrix of orthonormal polynomial trend coefficients, with columns corresponding to linear, quadratic, cubic, etc. Values of the continuous variable labeling the columns of $\boldsymbol{Y}$, such as time, provide the points needed to generate the coefficients.

The four MULTIREP tests also provide an appropriate test of a MANOVA hypothesis, which always uses $\boldsymbol{U} = \boldsymbol{I}_p$. The fact that $\boldsymbol{I}_p = [\,\boldsymbol{u}_{\rm B}\;\boldsymbol{U}_{\rm W}\,]\boldsymbol{T}$, with $\boldsymbol{T}$ full rank and $p \times p$ allows concluding that the MANOVA hypothesis spans both the between and within hypothesis contrasts. It asks whether *any* linear combination of the responses included in the hypothesis (by the choice of $\boldsymbol{C}$) is related to the predictors.

The uncorrected UNIREP test was developed long before any of the other UNIREP tests. Validity of the uncorrected test requires $\boldsymbol{\Sigma}_* = \boldsymbol{U}'\boldsymbol{\Sigma}\boldsymbol{U} = \boldsymbol{I}_b\sigma_*^2$, namely sphericity of the b transformed responses corresponding to the general linear hypothesis. With sphericity the uncorrected test provides an exact size-α test with uniformly most power among all similarly invariant tests. Without sphericity the uncorrected test can have greatly inflated test size.

The covariance matrix of the original responses achieves compound symmetry when all p response have the same variance, σ^2, and all $p(p-1)/2$ pairs of distinct responses have the same correlation, ρ (Lemma 1.33 summarizes properties). Choosing $\boldsymbol{U}$ to be either $\boldsymbol{u}_0 = \mathbf{1}_p/p^{1/2}$ or any $\boldsymbol{U}_t$ with (1) $\boldsymbol{U}_t'\mathbf{1}_p = \mathbf{0}$ *and* (2) $\boldsymbol{U}_t'\boldsymbol{U}_t = \boldsymbol{I}_b$ combines with compound symmetry to provide a sufficient (but not necessary) set of conditions to guarantee sphericity. If $\boldsymbol{U} = \boldsymbol{u}_0$, then $b = 1$ and $\boldsymbol{U}'\boldsymbol{\Sigma}_{\rm CS}\boldsymbol{U} = \lambda_1 = \sigma^2[1 + (p-1)\rho]$. With repeated measures, the data may be arranged, without loss of generality, such that all observations in column j were collected at time t_j, with $t_j \in \{t_1, t_2, \ldots, t_p\}$ and $t_j < t_{j+1}$. Here and throughout the book, time may be thought of as a metameter representing any interval- or ratio-scale dimension along which the observations vary within a subject (independent sampling unit, ISU). The orthonormal polynomial trends provide one convenient choice for $\boldsymbol{U}_t$.

Example 3.4 If $p = 3$ and the times are equally spaced, the orthonormal trends may be taken to be

$$\boldsymbol{U}_t = \begin{bmatrix} -1 & 1 \\ 0 & -2 \\ 1 & 1 \end{bmatrix} \begin{bmatrix} 1/\sqrt{2} & 0 \\ 0 & 1/\sqrt{6} \end{bmatrix}. \tag{3.32}$$

The first column provides the linear trend and the second column provides the quadratic trend. In turn, $\boldsymbol{U}_t'\boldsymbol{\Sigma}_{\rm CS}\boldsymbol{U}_t = \lambda_2\boldsymbol{I}_{p-1} = [\sigma^2(1-\rho)]\boldsymbol{I}_{p-1}$. Combining the forms gives the spectral decomposition

$$\boldsymbol{\Sigma}_{\text{CS}} = [\,\boldsymbol{u}_0\ \boldsymbol{U}_t\,]\begin{bmatrix}\lambda_1 & \mathbf{0}\\ \mathbf{0} & \lambda_2\boldsymbol{I}_{p-1}\end{bmatrix}\begin{bmatrix}\boldsymbol{u}_0'\\ \boldsymbol{U}_t'\end{bmatrix}. \tag{3.33}$$

Without sphericity and with λ_k indicating one of the b possibly distinct eigenvalues of $\boldsymbol{\Sigma}_*$, Box (1954a, b) proposed quantifying the deviation from sphericity with the parameter

$$\begin{aligned}\epsilon &= \frac{\text{tr}^2(\boldsymbol{\Sigma}_*)}{b\text{tr}(\boldsymbol{\Sigma}_*^2)}\\ &= \frac{\left(\sum_{k=1}^{b}\lambda_j/b\right)^2}{\left(\sum_{k=1}^{b}\lambda_j^2\right)/b} = \frac{\left(\overline{\lambda}\right)^2}{\overline{\lambda^2}}.\end{aligned} \tag{3.34}$$

Under sphericity $\epsilon = 1$, while in general $1/b \leq \epsilon \leq 1$. Test size inflation for $\epsilon < 1$ led to the development of the Box conservative, Geisser-Greenhouse, and Huynh-Feldt tests. All use the same test statistic as the uncorrected test but use distinct and more stringent critical values.

Allowing for any covariance pattern and for $s = \min(a, b) > 1$, no single test provides the uniformly most powerful test (among size α and similarly invariant tests). The special case $b = 1$ implies $s = \min(a, b) = 1$ and causes all MULTIREP and UNIREP tests to become equivalent by providing exactly the same F statistic and p value. Furthermore, if $b = 1$ the single test provides the usual exact size-α and uniformly most powerful test (among the class of unbiased and scale invariant tests). Furthermore, $b > 1$ and $a = 1$ imply $s = \min(a, b) = 1$. If $a = 1$, then the MULTIREP tests become equivalent to each other by providing exactly the same F statistic and p value. Similarly, if $a = 1$, then the (single) MULTIREP test provides an exact size-α and uniformly most powerful test among the class of unbiased and scale invariant tests. However, if $a = 1$ and $b > 1$ the UNIREP tests are not equivalent to each other or to the MULTIREP test. The differences arise from the fact that the UNIREP tests are only invariant to an orthonormal transformation, rather than fully scale invariant. Both UNIREP and MULTIREP tests correspond to transforming the model and hypothesis as follows:

$$\begin{aligned}\boldsymbol{YU} &= \boldsymbol{XBU} + \boldsymbol{EU}\\ \boldsymbol{Y}_u &= \boldsymbol{XB}_u + \boldsymbol{E}_u\,,\end{aligned} \tag{3.35}$$

$\boldsymbol{E}_u$ has $b \times b$ covariance $\boldsymbol{\Sigma}_* = \boldsymbol{U}'\boldsymbol{\Sigma}\boldsymbol{U}$,

$$\begin{aligned}H_0 &: \boldsymbol{CBU} = \boldsymbol{\Theta}_0\\ H_0 &: \boldsymbol{CB}_u\boldsymbol{I}_b = \boldsymbol{\Theta}_0\,.\end{aligned} \tag{3.36}$$

With spectral decomposition $\boldsymbol{\Sigma}_* = \boldsymbol{\Upsilon}\text{Dg}(\boldsymbol{\lambda})\boldsymbol{\Upsilon}'$ and $\boldsymbol{\Upsilon}'\boldsymbol{\Upsilon} = \boldsymbol{I}_b$ the MULTIREP and UNIREP tests do not vary under orthonormal transformation of the model and $\boldsymbol{\Theta}_0$. In particular

$$\begin{aligned} \boldsymbol{Y}_u \boldsymbol{\Upsilon} &= \boldsymbol{X}\boldsymbol{B}_u \boldsymbol{\Upsilon} + \boldsymbol{E}_u \boldsymbol{\Upsilon} \\ \boldsymbol{Y}_{\Upsilon} &= \boldsymbol{X}\boldsymbol{B}_{\Upsilon} + \boldsymbol{E}_{\Upsilon} \end{aligned} \tag{3.37}$$

and $\boldsymbol{\Theta}_{0\Upsilon} = \boldsymbol{\Theta}_0 \boldsymbol{\Upsilon}$. Here $\boldsymbol{E}_{\Upsilon}$ has $b \times b$ covariance $\boldsymbol{\Sigma}_{\Upsilon} = \text{Dg}(\boldsymbol{\lambda})$. If $\boldsymbol{\Phi} = \boldsymbol{\Upsilon}\text{Dg}(\boldsymbol{\lambda})^{1/2}$, then $\boldsymbol{\Phi}^{-t} = \boldsymbol{\Upsilon}\text{Dg}(\boldsymbol{\lambda})^{-1/2}$. The eigenvector transformation eliminates all correlations. With Gaussian data, the process creates independent but typically heterogeneous responses. In contrast to the situation for orthonormal invariance, in general the MULTIREP tests do *not* vary while the UNIREP tests *do* vary under the (particular scale) transformation

$$\begin{aligned} \boldsymbol{Y}_{\Upsilon}\text{Dg}(\boldsymbol{\lambda})^{-1/2} &= \boldsymbol{X}\boldsymbol{B}_{\Upsilon}\text{Dg}(\boldsymbol{\lambda})^{-1/2} + \boldsymbol{E}_{\Upsilon}\text{Dg}(\boldsymbol{\lambda})^{-1/2} \\ \boldsymbol{Y}_u \boldsymbol{\Phi}^{-t} &= \boldsymbol{X}\boldsymbol{B}_u \boldsymbol{\Phi}^{-t} + \boldsymbol{E}_u \boldsymbol{\Phi}^{-t} \\ \boldsymbol{Y}_{\boldsymbol{\Phi}} &= \boldsymbol{X}\boldsymbol{B}_{\boldsymbol{\Phi}} + \boldsymbol{E}_{\boldsymbol{\Phi}} . \end{aligned} \tag{3.38}$$

Here $\boldsymbol{E}_{\boldsymbol{\Phi}}$ has $b \times b$ covariance $\boldsymbol{\Sigma}_{\boldsymbol{\Phi}} = \boldsymbol{I}_b$. The only exception occurs when $\boldsymbol{\Sigma}_* = \sigma_*^2 \boldsymbol{I}_b$, namely sphericity, with $\epsilon = 1$, and then the uncorrected test achieves size α and uniformly more power than the MULTIREP test. If $a = 1$, $b > 1$, and $\epsilon < 1$, examples can be found in which either the MULTIREP test (exact size α) or a corrected UNIREP test (at least approximately size α) can be more powerful than the other for a particular $\boldsymbol{\Sigma}_*$. The same statement holds for $a > 1$, $b > 1$ [$s = \min(a, b) > 1$], and $\epsilon < 1$.

3.8 COMPUTING MULTIREP TESTS

The four MULTIREP test statistics are all simple functions of the $s = \min(a, b)$ nonzero estimators of the generalized squared canonical correlations $\{\widehat{\rho}_{*k}^2\}$. The correlations are generalized in the sense that they may or may not be adjusted for an intercept. The $\{\widehat{\rho}_{*k}^2\}$ are the nonzero eigenvalues of the $b \times b$ matrix $\widehat{\boldsymbol{\Delta}}[\widehat{\boldsymbol{\Delta}} + (N - r)\widehat{\boldsymbol{\Sigma}}_*]^{-1}$, with $r = \text{rank}(\boldsymbol{X})$.

Definition 3.8 A measure of *multivariate association* generalizes the concept of a squared multiple correlation, the proportion of variance controlled by the hypothesis. Each of the multivariate test statistics leads to a different measure of association. For test statistic m, $0 \leq \eta_m \leq 1$, with $\eta_m = 0$ corresponding to no relationship and $\eta_m = 1$ to a perfect relationship.

Table 3.2 gives expressions for each of the MULTIREP test statistics. For compactness and to emphasize the sums-of-squares nature of the matrices, the alternate notations $\nu_e = N - \text{rank}(\boldsymbol{X})$, $\boldsymbol{S}_h = \widehat{\boldsymbol{\Delta}}$, and $\boldsymbol{S}_e = (N - r)\widehat{\boldsymbol{\Sigma}}_*$ are used. The column labeled $\widehat{\eta}_m$ defines a measure of the strength of multivariate association for test m which corresponds to the form of the test statistic. Cramer and Nicewander (1979) reviewed measures of multivariate association in the context of two models, $\boldsymbol{Y} = \boldsymbol{X}\boldsymbol{B}_{Y|X} + \boldsymbol{E}_{Y|X}$ and $\boldsymbol{X} = \boldsymbol{Y}\boldsymbol{B}_{X|Y} + \boldsymbol{E}_{X|Y}$. Such measures of correlation all share the property of symmetry with respect to the two sets of variables $\boldsymbol{X}$ and $\boldsymbol{Y}$ in the GLM: $\eta_m(\boldsymbol{Y},\boldsymbol{X}) = \eta_m(\boldsymbol{X},\boldsymbol{Y})$. In contrast, regression coefficients and other properties of regression models are generally not symmetric in the roles of the variables: $\boldsymbol{B}_{\boldsymbol{Y}|\boldsymbol{X}} \neq \boldsymbol{B}_{\boldsymbol{X}|\boldsymbol{Y}}$.

Table 3.2 Tests for Multivariate Hypotheses
$\widehat{\eta}_m$ = Strength of Multivariate Association of Test m
$\boldsymbol{S}_h = \widehat{\boldsymbol{\Delta}}$ and $\boldsymbol{S}_e = (N-r)\widehat{\boldsymbol{\Sigma}}_*$

Name	Statistic	Principle	$\widehat{\eta}_m$	Univariate Case
HLT	$\sum_{k=1}^{s} \frac{\widehat{\rho}_k^2}{(1-\widehat{\rho}_k^2)} = \mathrm{tr}(\boldsymbol{S}_h \boldsymbol{S}_e^{-1})$	ANOVA analog	$\frac{\mathrm{HLT}/s}{1+\mathrm{HLT}/s}$	$\frac{\widehat{\rho}^2}{(1-\widehat{\rho}^2)} = \frac{SSH}{SSE}$
PBT	$\sum_{k=1}^{s} \widehat{\rho}_k^2 = \mathrm{tr}[\boldsymbol{S}_h(\boldsymbol{S}_h+\boldsymbol{S}_e)^{-1}]$	Substitution	$\frac{\mathrm{PBT}}{s}$	$\widehat{\rho}^2 = \frac{SSH}{SSH+SSE}$
WLK	$\prod_{k=1}^{s}(1-\widehat{\rho}_k^2) = \lvert \boldsymbol{S}_e(\boldsymbol{S}_h+\boldsymbol{S}_e)^{-1} \rvert$	Likelihood ratio	$1-\mathrm{WLK}^{1/g}$	$1-\widehat{\rho}^2 = \frac{SSE}{SSH+SSE}$
RLR	$\max_k \widehat{\rho}_k^2 =$ max eigenvalue $\boldsymbol{S}_h(\boldsymbol{S}_h+\boldsymbol{S}_e)^{-1}$	Union-intersection	$\widehat{\rho}_1^2$	$\widehat{\rho}^2 = \frac{SSH}{SSH+SSE}$
UNIREP	$\frac{\mathrm{tr}(\boldsymbol{S}_h)}{\mathrm{tr}(\boldsymbol{S}_e)}$	Best with sphericity	$\frac{\mathrm{tr}(\boldsymbol{S}_h)}{\mathrm{tr}(\boldsymbol{S}_h+\boldsymbol{S}_e)}$	$\widehat{\rho}^2 = \frac{SSH}{SSH+SSE}$

For consistency with the approximate distribution for WLK discussed in the remainder of the section,

$$g = \begin{cases} 1 & a^2b^2 \leq 4 \\ [(a^2b^2-4)/(a^2+b^2-5)]^{1/2} & \text{otherwise}. \end{cases} \tag{3.39}$$

However, choosing $g = s$ leads to a simpler interpretation of $\widehat{\eta}_{\mathrm{WLK}}$, the measure of multivariate association, in terms of a geometric mean of (canonical) error variances. In fact, if $s \leq 2$, then $g = s$.

The $b \times b$ matrix $\widehat{\boldsymbol{\Delta}}$ contains the sums of squares for the hypothesis and reduces to the scalar sum-of-squares hypothesis, *SSH*, whenever $b = 1$. The $b \times b$ matrix $(N-r)\widehat{\boldsymbol{\Sigma}}_*$ contains the sums of squares for error and reduces to the scalar sum of squares *SSE* whenever $b = 1$. Also, $\widehat{\boldsymbol{\Delta}}[\widehat{\boldsymbol{\Delta}} + (N-r)\widehat{\boldsymbol{\Sigma}}_*]^{-1}$ reduces to the scalar $SSH/(SSH+SSE) = \widehat{\rho}^2$, the estimated squared multiple correlation (which may or may not be adjusted for an intercept). Of course, any 1×1 matrix has only one eigenvalue, the scalar itself. Under the null, with $s = \min(a, b) = 1$, each of the four MULTIREP statistics can be expressed exactly as a one-to-one function of each other, and of an F random variable with numerator degrees of freedom $\nu_1 = a$ and denominator degrees of freedom $\nu_2 = N - r = \nu_e$.

Under the null, with $s > 1$, the MULTIREP statistics are not one-to-one functions of each other, and exact distributions are known only for special cases.

However, approximations based on an F random variable which match two moments are available for three of the four statistics, as detailed in Table 3.3. When $s > 1$ and $m \in \{\text{HLT, PBT, WLK}\}$, an approximate p value may be computed as follows. Denominator degrees of freedom (df) $\nu_2(m)$ are in Table 3.3. With $\nu_e = N - r$, the numerator degrees of freedom $\nu_1(m)$ are $\nu_1(\text{HLT}) = ab$, $\nu_1(\text{WLK}) = ab$, and

$$\nu_1(\text{PBT}) = ab \frac{1}{s(\nu_e + a)} \left[\frac{s(\nu_e + s - b)(\nu_e + a + 2)(\nu_e + a - 1)}{\nu_e(\nu_e + a - b)} - 2 \right]. \quad (3.40)$$

Table 3.3 Denominator df for F Approximations

Test	$\nu_2(m)$	Author
HLT	$\dfrac{[\nu_e^2 - \nu_e(2b+3) + b(b+3)](ab+2)}{\nu_e(a+b+1) - (a+2b+b^2-1)} + 4$	McKeon (1974)
PBT	$\dfrac{\nu_e+s-b}{\nu_e+a} \left[\dfrac{s(\nu_e+s-b)(\nu_e+a+2)(\nu_e+a-1)}{\nu_e(\nu_e+a-b)} - 2 \right]$	Muller (1998)
WLK	$g[\nu_e - (b - a + 1)/2] - (ab - 2)/2$	Rao (1951)

Computing

$$f_{\text{obs}}(m) = \frac{\widehat{\eta}_m / \nu_1(m)}{(1 - \widehat{\eta}_m)/\nu_2(m)} \quad (3.41)$$

leads to the associated approximate p value

$$p(m) = 1 - F_F[f_{\text{obs}}(m); \nu_1(m), \nu_2(m)]. \quad (3.42)$$

Harris (1975, Appendix B) provided a useful method for directly approximating tail probabilities of RLR.

Eigenvalue k of $\boldsymbol{S}_h \boldsymbol{S}_e^{-1} = \widehat{\boldsymbol{\Omega}}/\nu_e$ is a one-to-one function of eigenvalue k of $\boldsymbol{S}_h(\boldsymbol{S}_h + \boldsymbol{S}_e)^{-1}$ and also of eigenvalue k of $\boldsymbol{S}_e(\boldsymbol{S}_h + \boldsymbol{S}_e)^{-1}$. Computational accuracy considerations lead to preferring to compute the eigenvalues of $\boldsymbol{S}_h \boldsymbol{S}_e^{-1}$, namely $\{\widehat{\rho}_k^2/(1 - \widehat{\rho}_k^2)\}$. A standard and simple approach allows converting the nonsymmetric matrix to a symmetric matrix which has the same eigenvalues (and different eigenvectors which allow computing the eigenvectors of the original matrix). The Cholesky method, among other methods, allows finding $\widehat{\boldsymbol{\Phi}}$ such that $\widehat{\boldsymbol{\Sigma}}_* = \widehat{\boldsymbol{\Phi}} \widehat{\boldsymbol{\Phi}}'$, which implies $\widehat{\boldsymbol{\Sigma}}_*^{-1} = \widehat{\boldsymbol{\Phi}}^{-t} \widehat{\boldsymbol{\Phi}}^{-1}$. In turn, it is straightforward to prove that the eigenvalues of $\widehat{\boldsymbol{\Omega}}$ coincide with the eigenvalues of the symmetric matrix

$$\widehat{\boldsymbol{\Omega}}_\Phi = \widehat{\boldsymbol{\Phi}}^{-1} \widehat{\boldsymbol{\Delta}} \widehat{\boldsymbol{\Phi}}^{-t}. \quad (3.43)$$

Similarly, if $\boldsymbol{S}_e = \nu_e \widehat{\boldsymbol{\Sigma}}_* = \boldsymbol{F}_e \boldsymbol{F}_e'$, then the eigenvalues of $\boldsymbol{S}_h \boldsymbol{S}_e^{-1}$, namely $\{b_k\} = \{\widehat{\rho}_k^2/(1 - \widehat{\rho}_k^2)\}$, coincide with the eigenvalues of

$$\widehat{\boldsymbol{\Omega}}_{\Phi}/\nu_e = \boldsymbol{F}_e^{-1}\boldsymbol{S}_h\boldsymbol{F}_e^{-t}\,. \tag{3.44}$$

Obviously $\widehat{\rho}_k^2 = b_k/(1+b_k)$ gives the squared canonical correlations, from which all of the test statistics may be computed, as detailed above.

3.9 COMPUTING UNIREP TESTS

The four UNIREP tests are *not* functions of the eigenvalues of $\boldsymbol{S}_h\boldsymbol{S}_e^{-1} = \widehat{\boldsymbol{\Delta}}(\widehat{\boldsymbol{\Delta}} + \nu_e\widehat{\boldsymbol{\Sigma}}_*)^{-1}$. They all use the same test statistic,

$$f_{\text{obs}}(\text{U}) = \frac{\text{tr}(\boldsymbol{S}_h)/(ab)}{\text{tr}(\boldsymbol{S}_e)/(b\nu_e)} = \frac{\text{tr}(\widehat{\boldsymbol{\Delta}})/a}{\text{tr}(\widehat{\boldsymbol{\Sigma}}_*)}\,, \tag{3.45}$$

and corresponding measure of multivariate association,

$$\widehat{\eta}_{\text{U}} = \frac{\text{tr}(\boldsymbol{S}_h)/\text{tr}(\boldsymbol{S}_e)}{1+\text{tr}(\boldsymbol{S}_h)/\text{tr}(\boldsymbol{S}_e)} = \frac{\text{tr}(\boldsymbol{S}_h)}{\text{tr}(\boldsymbol{S}_h+\boldsymbol{S}_e)}\,. \tag{3.46}$$

The uncorrected test uses the p value

$$p(\text{Un}) = 1 - F_F[f_{\text{obs}}(\text{U}); ab, b\nu_e]\,. \tag{3.47}$$

The Geisser-Greenhouse test reduces degrees of freedom by the maximum likelihood estimator of ϵ, namely $\widehat{\epsilon} = b^{-1}\text{tr}^2(\widehat{\boldsymbol{\Sigma}}_*)/\text{tr}(\widehat{\boldsymbol{\Sigma}}_*^2)$:

$$p(\text{GG}) = 1 - F_F[f_{\text{obs}}(\text{U}); ab\widehat{\epsilon}, b\nu_e\widehat{\epsilon}]\,. \tag{3.48}$$

In seeking an approximately unbiased estimator, the Huynh-Feldt test uses $\widetilde{\epsilon} = (Nb\widehat{\epsilon} - 2)/[b(\nu_e - b\widehat{\epsilon})]$:

$$p(\text{HF}) = 1 - F_F[f_{\text{obs}}(\text{U}); ab\widetilde{\epsilon}, b\nu_e\widetilde{\epsilon}]\,. \tag{3.49}$$

The fact that $\widetilde{\epsilon}$ may exceed 1.0 leads to using $\widetilde{\epsilon}_t = \min(\widetilde{\epsilon}, 1)$. The Box conservative test uses the lower bound for ϵ, namely $1/b$:

$$p(\text{Box}) = 1 - F_F[f_{\text{obs}}(\text{U}); a, \nu_e]\,. \tag{3.50}$$

For data analysis, UNIREP tests differ only due to the degrees of freedom multipliers, which are always in the same order: Box, GG, HF, and uncorrected, with values $1/b \leq \widehat{\epsilon} \leq \widetilde{\epsilon}_t \leq 1$. Furthermore, the p values will always be in the reverse order.

If all $\lambda_k = \lambda_1$, then $\epsilon = 1$ and $\text{Dg}(\boldsymbol{\lambda}) = \lambda_1\boldsymbol{I}_b$, which corresponds to a spherical Gaussian distribution. Under sphericity $f_{\text{obs}}(\text{U}) \sim F\{ab, b\nu_e, \text{tr}(\boldsymbol{\Omega})\}$ (exactly), the test is exactly size α and uniformly most powerful (among similarly invariant tests). Box (1954a, b) observed that $1/b \leq \epsilon \leq 1$ and that $\epsilon < 1$ implies, under the null, $f_{\text{obs}}(\text{U}) \stackrel{.}{\sim} F(ab\epsilon, b\nu_e\epsilon)$, an *approximate* result.

Muller, LaVange, Ramey, and Ramey (1992) reviewed power approximation for both UNIREP and MULTIREP tests. We leave the topic until Chapter 21.

3.10 CONFIDENCE REGIONS FOR Θ

As noted in Section 2.10, confidence regions can be obtained by inverting hypothesis tests, and a confidence region can be inverted to yield a hypothesis test. The definiton in Section 2.10 can be extended easily from vector data $\boldsymbol{y}$ and vector parameter $\boldsymbol{\theta}$ (or σ^2) to matrix data, $\boldsymbol{Y}$, and matrix parameter $\boldsymbol{\Theta}$ (or $\boldsymbol{\Sigma}$). Doing so merely requires replacing $\boldsymbol{y}$ by vec($\boldsymbol{Y}$) and $\boldsymbol{\theta}$ by vec($\boldsymbol{\Theta}$) [or σ^2 by vec($\boldsymbol{\Sigma}$)] in Definition 2.12. In Section 15.6 we prove a variety of results for confidence intervals in univariate models. In Section 16.10 we describe extensions to multivariate models.

3.11 SUFFICIENT STATISTICS FOR THE MULTIVARIATE MODEL

If the multivariate linear model $\text{GLM}_{N,p,q}(\boldsymbol{Y}_i; \boldsymbol{X}_i\boldsymbol{B}, \boldsymbol{\Sigma})$ with Gaussian distribution is correct, then the matrix

$$\begin{aligned} \boldsymbol{S} &= \begin{bmatrix} \boldsymbol{X}'\boldsymbol{X} & \boldsymbol{X}'\boldsymbol{Y} \\ \boldsymbol{Y}'\boldsymbol{X} & \boldsymbol{Y}'\boldsymbol{Y} \end{bmatrix} \\ &= \begin{bmatrix} \boldsymbol{X}' \\ \boldsymbol{Y}' \end{bmatrix} [\boldsymbol{X}\ \boldsymbol{Y}] \end{aligned} \tag{3.51}$$

contains all of the complete sufficient statistics. If $\boldsymbol{X}$ contains an intercept, then, without loss of generality, $\boldsymbol{X}$ may be arranged with the intercept in column 1, with $\boldsymbol{X} = [\,\boldsymbol{1}_N\ \boldsymbol{X}_1\,]$, for $\boldsymbol{X}_1$ of dimension $N \times (q-1)$. In turn,

$$\begin{aligned} \boldsymbol{S} &= \begin{bmatrix} \boldsymbol{1}'\boldsymbol{1} & \boldsymbol{1}'\boldsymbol{X}_1 & \boldsymbol{1}'\boldsymbol{Y} \\ \boldsymbol{X}_1'\boldsymbol{1} & \boldsymbol{X}_1'\boldsymbol{X}_1 & \boldsymbol{X}'\boldsymbol{Y} \\ \boldsymbol{Y}'\boldsymbol{1} & \boldsymbol{Y}'\boldsymbol{X}_1 & \boldsymbol{Y}'\boldsymbol{Y} \end{bmatrix} \\ &= \begin{bmatrix} \boldsymbol{1}' \\ \boldsymbol{X}_1' \\ \boldsymbol{Y}' \end{bmatrix} [\,\boldsymbol{1}\ \boldsymbol{X}_1\ \boldsymbol{Y}\,] \end{aligned} \tag{3.52}$$

contains all of the complete sufficient statistics for estimation of all parameters of the regression models (one for each column of $\boldsymbol{Y}$) identified by the relationship

$$\text{E}(\boldsymbol{Y}|\boldsymbol{X})\,. \tag{3.53}$$

Here $[\,\boldsymbol{1}\ \boldsymbol{X}_1\ \boldsymbol{Y}\,]$ is $N \times (q+p)$ so $\boldsymbol{S}$ is $(q+p) \times (q+p)$. All parameter estimators and general linear hypothesis tests (both MULTIREP and UNIREP) depend on the data only through the elements of $\boldsymbol{S}$. Conveniently, the raw data are not needed for parameter estimation or testing the general linear hypothesis.

3.12 ALLOWING MISSING DATA IN THE MULTIVARIATE MODEL

Definition 3.9 (a) If $(1 \times p_i)$ random matrix $\boldsymbol{Y}_i = \{y_{ij}\}$ represents the potential response of ISU i, then an element y_{ij} is said to be *missing* if a realized value for y_{ij} is not included in the statistical analysis.
(b) If only two patterns of data occur, namely either all elements of $\boldsymbol{Y}_i$ are present or all elements are missing, the observations are described as having only *casewise missing* observations.

Examples of missing values include outlier values intentionally omitted by the analyst, recorded values that were lost during data entry, values that were never recorded because the ISU (individual participant) was not available for evaluation, and interval-censored values treated as unknown by the analyst. The definition applies to the multivariate model ($p_i \equiv p \, \forall i$) as well as to clustered data in general.

The standard methods described in the present chapter conveniently allow casewise missing observations in the multivariate (and univariate) linear model. If the mechanism causing data to be missing does not lead to selection biases, then the approach gives optimal estimators and exact tests. Partially missing $\boldsymbol{X}_i$ or $\boldsymbol{Y}_i$ do not have such nice properties. In the multivariate linear model, nearly all research on the topic has focused on having complete $\boldsymbol{X}$ and partially missing $\boldsymbol{Y}_i$. Such patterns occur naturally in an experiment with random assignment to treatment, as in a typical clinical trial. Little (1992) reviewed methods for regression with incomplete $\boldsymbol{X}$. Little and Rubin (2002) reviewed methods for missing data, with particular emphasis on estimation from a likelihood perspective.

Using the notation of Stewart (2000) and others, a binary random variable $r_{ij} = 1$ indicates y_{ij} is not missing while $r_{ij} = 0$ indicates that it is missing. With $(1 \times p_i)$ matrix $\boldsymbol{R}_i$ containing all r_{ij} for ISU i, specifying the entire response for ISU i requires knowing $\{\boldsymbol{Y}_i, \boldsymbol{R}_i\}$. Vertical concatenation of $\{\boldsymbol{Y}_i'\}$ and similar concatenation of $\{\boldsymbol{R}_i'\}$ yield the pair of random vectors $\{\boldsymbol{y}, \boldsymbol{r}\}$. It is assumed that an appropriate parametric statistical model can be formulated for $\boldsymbol{y}$. If $\boldsymbol{y}$ has a density function, then the model can be represented as $\{\boldsymbol{y}, f_y(\boldsymbol{y}_* | \boldsymbol{X}; \boldsymbol{\theta}), \boldsymbol{\theta} \in \boldsymbol{\Theta}\}$.

Information for estimation and inferences about $\boldsymbol{\theta}$ is available only via $\{\boldsymbol{y}_{\text{obs}}, \boldsymbol{R}\}$, in which $\boldsymbol{y}_{\text{obs}}$ is the vector of nonmissing values. The missing values are denoted $\boldsymbol{y}_{\text{mis}}$. Usually, most or all of this information about $\boldsymbol{\theta}$ is anticipated to come from $\boldsymbol{y}_{\text{obs}}$. If the observed pattern of missing values $\boldsymbol{R}_*$ contains no additional information about $\boldsymbol{\theta}$, then the underlying missing-data mechanism is said to be ignorable as defined by Rubin (1976). We then say the missing values are ignorably missing.

In general, the likelihood function (any function proportional to the joint density function of $\{\boldsymbol{y}_{\text{obs}}, \boldsymbol{r}\}$) is obtained by an integration over possible values of $\boldsymbol{y}_{\text{mis}}$

(Little and Rubin, 2002). The process creates a marginal density:

$$f_y(\boldsymbol{y}_{*\text{obs}}, \boldsymbol{r}_* | \boldsymbol{X}; \boldsymbol{\theta}, \boldsymbol{\psi}) = \int f_y(\boldsymbol{y}_{*\text{obs}}, \boldsymbol{y}_{*\text{mis}} | \boldsymbol{X}; \boldsymbol{\theta}) f_r(\boldsymbol{r}_* | \boldsymbol{y}_{*\text{obs}}, \boldsymbol{X}; \boldsymbol{\psi}, \boldsymbol{\theta})\, d\boldsymbol{y}_{*\text{mis}} . \quad (3.54)$$

Definition 3.10 The following definitions follow Rubin (1976).
(a) If $\boldsymbol{r}$ and $\boldsymbol{y}$ are statistically independent, given $\boldsymbol{X}$, then the joint density is

$$f_y(\boldsymbol{y}_{*\text{obs}}, \boldsymbol{r}_* | \boldsymbol{X}; \boldsymbol{\theta}, \boldsymbol{\psi}) = f_y(\boldsymbol{y}_{*\text{obs}} | \boldsymbol{X}; \boldsymbol{\theta}) f_r(\boldsymbol{r}_* | \boldsymbol{X}; \boldsymbol{\psi}, \boldsymbol{\theta}) , \quad (3.55)$$

and the missing values are said to be *missing completely at random* (MCAR).
(b) The missing-data mechanism is said to be *ignorable* if the following factorization holds true:

$$f_y(\boldsymbol{y}_{*\text{obs}}, \boldsymbol{r}_* | \boldsymbol{X}; \boldsymbol{\theta}, \boldsymbol{\psi}) = f_y(\boldsymbol{y}_{*\text{obs}} | \boldsymbol{X}; \boldsymbol{\theta}) f_r(\boldsymbol{r}_* | \boldsymbol{y}_{*\text{obs}}, \boldsymbol{X}; \boldsymbol{\psi}) . \quad (3.56)$$

(c) Invariance of $f_r(\boldsymbol{r}_* | \boldsymbol{y}_{*\text{obs}}, \boldsymbol{y}_{*\text{mis}}, \boldsymbol{X}; \boldsymbol{\psi}, \boldsymbol{\theta})$ to the possible values of $\boldsymbol{y}_{*\text{mis}}$ gives data described as *missing at random* (MAR).

The definitions lead to the following observations. If the conditional distribution of $\boldsymbol{r}$ in equation (3.55) does not depend on $\boldsymbol{\theta}$, then no information about $\boldsymbol{\theta}$ is neglected if $\boldsymbol{r}$ is ignored. Rubin (1976) proved that (3.56) holds if and only if $f_r(\boldsymbol{r}_* | \boldsymbol{y}_{*\text{obs}}, \boldsymbol{y}_{*\text{mis}}, \boldsymbol{X}; \boldsymbol{\psi}, \boldsymbol{\theta})$ is invariant to the possible values of $\{\boldsymbol{y}_{*\text{mis}}, \boldsymbol{\theta}\}$ when evaluated at $\{\boldsymbol{r}_*, \boldsymbol{y}_{*\text{obs}}\}$. The invariance with respect to $\boldsymbol{\theta}$ is a condition Rubin (1976) referred to as the absence of a priori ties between the parameters of the two densities in the factorization.

If factorization (3.55) or (3.56) holds, then maximizing the $\{\boldsymbol{y}_{\text{obs}}, \boldsymbol{r}\}$ likelihood with respect to $\boldsymbol{\theta}$ is equivalent to maximizing $f_y(\boldsymbol{y}_{*\text{obs}} | \boldsymbol{X}; \boldsymbol{\theta})$. It is in this sense that $\boldsymbol{r}$ can be ignored. It is important to realize, however, that maximization of a likelihood can yield parameter estimates and likelihood ratios without providing standard errors for the parameter estimators. Verbeke and Molenberghs (2000, Chapter 21) gave examples of bias stemming from using the expected information matrix for approximating standard errors in the MAR case. Heitjan (1994) gave a clear statement of the problem. Diggle and Kenward (1994), Stewart (2000), and Lipsitz et al. (2002) discussed illustrative MAR examples.

With all assumptions of the GLM$_{N,p,q}(\boldsymbol{Y}_i; \boldsymbol{X}_i\boldsymbol{B}, \Sigma)$ with Gaussian errors met, having MAR or MCAR data allows computing maximum likelihood estimates of $\{\boldsymbol{B}, \Sigma\}$, although usually through iterative methods. One convenient approach uses the estimates in complete data formulas for confidence intervals and testing hypotheses. Unfortunately, Barton and Cramer (1989) demonstrated that the naive approach leads to optimistic (biased) estimates of precision with small to moderate sample sizes and MCAR data. The same authors, as well as Catellier and Muller (2000), described very simple adjustments to the degrees of freedom for the MULTIREP and UNIREP tests in the MCAR case. In simulations, the adjusted tests completely or nearly control test size, even in very small samples ($N = 12$

and $p = 6$). In contrast, standard linear mixed model tests greatly inflated test size (up to 0.40 with a nominal value of $\alpha = 0.05$). The Appendix (Section A.2) contains descriptions of free SAS/IML® code, and where to find it on the Web, which implements the methods.

Informatively missing data present one of the most vexing problems in statistical modeling. Almost unavoidably, quite specific assumptions must be made which derive from a particular scientific setting and analysis goal. As noted earlier, Little and Rubin (2002) provided the best starting point for further reading. General techniques have not been developed, and new developments continue. An analyst seeking the best method available for a specific analysis would be wise to carefully review the statistical literature. Using the Current Index to Statistics and other electronic databases greatly eases the pain of the search.

EXERCISES

All exercises refer to the breast cancer example described in the Appendix (Section A.1). Use P0104.SD2 for any data analysis. When necessary, consider "Benign" as the reference cell, which corresponds to choosing "MALIGN" as a predictor. Use a nominal test size of 0.025 (chosen due to conducting two planned analyses in the original study).

When needing to understand a test or contrast, one can often ignore the multivariate nature of the design and apply the logic of univariate ANOVA design and interpretation. (The approach fails for derivations of distribution theory.) To specify a within-subject contrast matrix, it may help to first specify a contrast matrix for between-subject effects based on cell mean coding (with the correct dimensions and factor pattern) and then transpose it.

3.1 One focus of the study was testing the hypothesis of no difference in the variables DLOGROI1–DLOGROI3 DLOG_P_1–DLOG_P_3 between patients with malignant and benign pathology. Briefly specify an appropriate multivariate linear model $\mathrm{GLM}_{N,q,p}(\boldsymbol{Y}_i; \boldsymbol{X}_i\boldsymbol{B}, \boldsymbol{\Sigma})$ with Gaussian errors. Use reference cell coding. Include values of all dimensions for the specific data in hand as well as brief definitions of parameter matrix elements.

3.2.1 For the model chosen in exercise 3.1, explicitly specify all contrast matrices needed to test a "MANOVA" hypothesis of no differences between benign and malignant. Include all dimensions for the specific data in hand and brief definitions of parameter matrix elements.

3.2.2 Choose a test statistic for 3.2.1 and briefly justify your choice. This must be done a priori (before looking at *any* data).

3.2.3 Assuming that the overall MANOVA test just discussed is significant, explicitly specify a modest number of scientifically interesting and appropriate stepdown tests and associated contrast matrices. It is acceptable to use the simplest approach to control multiple testing bias, namely a Bonferroni correction. Include all dimensions for the specific data in hand and brief definitions of parameter matrix elements.

3.3 Treating tissue type [(region of interest (ROI), versus parenchyma] and time (with three levels) as factors in a factorial design with benign versus malignant may have more appeal than a MANOVA analysis.
3.3.1 Use cell mean style coding in this exercise. Specify an appropriate multivariate model.
3.3.2 Assuming a factorial approach describe an appropriate "source table" by listing the sources to be tested, and associated hypothesis degrees of freedom.
3.3.3 Explicitly specify all contrast matrices for each source listed in 3.1. Include all dimensions for the specific data in hand and brief definitions of parameter matrix elements. Use individual rows (or columns) that test *differences in means* (which are the secondary parameters).

3.4 Choose a test statistic for 3.1 and 3.2 and briefly justify your choice. This must be done a priori (before looking at *any* data).

3.5 Explicitly specify all contrast matrices for each source listed in 3.1. Include all dimensions for the specific data in hand and brief definitions of parameter matrix elements. Use individual rows (or columns) that test *polynomial trends* which are the secondary parameters, even though not all factors have levels defined in terms of a continuous variable.

3.6 Using PROC GLM and the data supplied, implement the two-way model described in 3.1, 3.3, and 3.4. *Hint*: Use a particular statement type available with GLM which does most of the coding and testing work associated with repeated measures automatically.
3.6.1 Provide sufficient source code and a compact numerical version of a source table. Include numerator degrees of freedom, a test statistic p value, and an appropriate measure of multivariate association.
3.6.2 Provide a brief scientific interpretation of the results.

3.7 Construct three new difference variables at times 1, 2, and 3: DLOGROI1–DLOG_P_1, DLOGROI2–DLOG_P_2, and DLOGROI3–DLOG_P_3.
3.7.1 Fit a one-way model, again using the same statement approach in PROC GLM, with Time as a factor. Report and interpret an appropriate test of no difference in the constructed variables between patients with malignant and benign pathology.
3.7.2 How does the model relate to the two-way model in exercise 3.4?

3.8 Examine the maximum likelihood estimate of ϵ for the UNIREP tests. Compare the p values of the four UNIREP tests and also the four MULTIREP tests. Make a recommendation for a choice of statistic in future studies of the same sort.

3.9 (*optional, noncredit*) Use LINMOD to repeat the analysis. Select appropriate options to enrich the output and help understanding. If you used LINMOD in the first place, then use PROC GLM or a procedure in another computer language. You should be able to reproduce almost exactly nearly all values [except for some multivariate p values if $s = \min(a, b) > 1$].

The Appendix (Section A.2) contains a brief description of the free software LINMOD and where to find it on the Web.

CHAPTER 4

Generalizations of the Multivariate Linear Model

4.1 MOTIVATION

As mentioned in the previous chapter, the multivariate general linear model has a number of limitations: The multivariate model does not directly tolerate incomplete or mistimed data; the multivariate model does not allow the design matrix to vary across responses; the multivariate model does not explicitly allow modeling the covariance structure. A number of generalizations of the multivariate model have been developed to avoid the limitations.

In the present chapter we briefly survey some generalizations of the multivariate linear model and its special case, the univariate linear model. Many of them stand somewhere in between the mixed and multivariate linear models, in terms of both theory and applications. As a broad generality, all provide well-behaved estimation but may have difficulty providing completely accurate inference in small samples. However, the alternative use of a mixed model may provide even less accuracy, depending on the situation. Lacking exact and perfect methods, good statistical practice, as always, centers on using the best available approximation.

In contrast, the general linear mixed model has none of the limitations. Unfortunately, the generality of the mixed model may come at a steep price. Even with a correctly specified model, the approach can lead to extremely inaccurate inference (optimistically small confidence intervals and inflated test size), especially with small to moderate sample sizes. The inaccuracy arises from what Littel (2003) described as "approximations piled on approximations." Furthermore, limitations of current methods make it difficult to check the validity of the model, especially the covariance component. Simulation results make it clear that misspecification of the covariance model may introduce (additional) substantial inaccuracy in inference (Park, Park, and Davis, 2001; Muller, Edwards, Simpson, and Taylor, 2006).

Although linear in the expected-value parameters, the likelihood varies nonlinearly as a function of the covariance parameters. As a consequence, computing estimates for a linear mixed model requires iterative solution of a system of simultaneous nonlinear equations. Collinearity arising from less than careful attention to data scaling or location and less-than-full-rank coding schemes easily and often degrades the performance of the algorithms needed. The resulting difficulties in computing estimates can lead the unsophisticated user to propose a valid model which fails to converge. Simplifying the covariance model greatly increases the chances of achieving convergence. However, the simplification may also misspecify the covariance model and thereby introduce severe bias.

A simple example illustrates the concern. Clinical trials of a pharmaceutical agent which involve repeated measures frequently lead to some missing and mistimed data, which make computing estimates for linear mixed models more difficult. In many such cases, analysts have assumed compound symmetry of covariance. Although extremely convenient (due to helping convergence), the assumption often seems implausible for a sequence of (time) ordered responses.

4.2 THE GENERALIZED GENERAL LINEAR UNIVARIATE MODEL: EXACT AND APPROXIMATE WEIGHTED LEAST SQUARES

The $\text{GLM}_{N,q}(y_i; \boldsymbol{X}_i\boldsymbol{\beta}, \sigma^2)$ can be generalized in many ways. The motivation lies in the need to allow for patterns of dependence, rather than complete independence, among response values. The simplest way to allow such dependence is to assume $\mathcal{V}(\boldsymbol{y}) = \mathcal{V}(\boldsymbol{e}) = \boldsymbol{\Upsilon}$, with $\boldsymbol{\Upsilon}$ of dimension $N \times N$, symmetric, and positive definite or positive semidefinite, (which allows any covariance matrix). In contrast, however, the approach allows describing only extremely limited results of little practical value. The limitations arise from the fact that $\boldsymbol{\Upsilon}$ has $N(N+1)/2$ distinct parameters, which exceeds N, the number of observations. Increasing sample size only worsens the problem because the number of parameters increases more rapidly than N. Even assuming complete independence while allowing complete heterogeneity does not solve the problem. In that case $\boldsymbol{\Upsilon} = \text{Dg}(\boldsymbol{v})$ has N parameters, which implies increasing sample size never allows the number of observations to exceed the number of parameters to be estimated. Avoiding the problem requires adding assumptions which impose structure on $\boldsymbol{\Upsilon}$. Necessarily the number of parameters must grow more slowly than the sample size if reasonable estimators are to exist. The multivariate and mixed linear models both generalize the univariate linear in the same fashion, although in very different directions. Subsequent chapters are devoted to properties of two approaches.

Following McCullagh and Nelder (1989), the term "generalized linear model" refers to a model with expected values linear or loglinear in the parameters and the response distribution any member of the exponential family, not just the Gaussian. To avoid confusion, we introduce the following definitions.

Definition 4.1 A *generalized GLM* is indicated by the notation $\text{GGLM}_{N,q}(\boldsymbol{y}; \boldsymbol{X\beta}|\boldsymbol{R\beta} = \boldsymbol{a}, \boldsymbol{\Upsilon})$, which describes all observations, not just a single observation for an independent sampling unit. The assumptions differ from a $\text{GLM}_{N,q}()$ in only one way, which is important. The assumption "elements of the $N \times 1$ random vector $\boldsymbol{y}$ are mutually independent." is replaced by the assumption "elements of the $N \times 1$ random vector $\boldsymbol{y}$ have (finite) $N \times N$, constant, covariance matrix $\boldsymbol{\Upsilon}$, which may be at least partially known."

The $\text{GGLM}_{N,q}()$ notation describes all N observations at once because they may not be independent, while all other model notations used here describe the observations for a single independent sampling unit. In fact, if all off-diagonal elements of a full-rank $\boldsymbol{\Upsilon}$ are nonzero, then a GGLM has only one independent sampling unit. In the special case of a grand mean model, which has $\boldsymbol{X\beta} = \boldsymbol{1}_N\mu$, the $\text{GGLM}_{N,q}(\boldsymbol{y}; \boldsymbol{1}_N\mu, \boldsymbol{\Upsilon})$ in many ways corresponds to a multivariate model with one observation, $\text{GLM}_{1,N,1}(\boldsymbol{y}'; \boldsymbol{1}'_N\mu, \boldsymbol{\Upsilon})$. With Gaussian data, the special case has $\boldsymbol{y} \sim \mathcal{N}_N(\boldsymbol{1}_N, \boldsymbol{\Upsilon})$.

Definition 4.2 A GGLM with Gaussian errors refers to a setting in which $\boldsymbol{y} \sim \mathcal{N}_N(\boldsymbol{X\beta}|\boldsymbol{R\beta} = \boldsymbol{a}, \boldsymbol{\Upsilon})$, which is an assumption of joint ("multivariate") Gaussian distribution, not merely marginally Gaussian $\{y_i\}$. As for a univariate GLM, a GGLM may be described as either FR or LTFR and restricted or unrestricted, depending on $\boldsymbol{X}$, $\boldsymbol{R}$, and $\boldsymbol{a}$.

Lemma 4.1 For $\text{GGLM}_{N,q}(\boldsymbol{y}; \boldsymbol{X\beta}|\boldsymbol{R\beta} = \boldsymbol{a}, \sigma^2\boldsymbol{I}_N)$, elements of $\boldsymbol{\Upsilon} = \mathcal{V}(\boldsymbol{y}) = \mathcal{V}(\boldsymbol{e})$ control many properties of the model.

(a) If any off-diagonal element of $\boldsymbol{\Upsilon}$ is not equal to zero, the model does not have independent observations.

(b) If $\boldsymbol{\Upsilon}$ has two or more distinct diagonal elements, the model does not have homogeneity of variance.

(c) If $\boldsymbol{\Upsilon} = \sigma^2\boldsymbol{I}_N$, then the model meets the assumptions of the univariate $\text{GLM}_{N,q}(y_i; \boldsymbol{X}_i\boldsymbol{\beta}|\boldsymbol{R\beta} = \boldsymbol{a}, \sigma^2)$, which means the univariate $\text{GLM}_{N,q}()$ is a special case, namely $\text{GGLM}_{N,q}(\boldsymbol{y}; \boldsymbol{X\beta}|\boldsymbol{R\beta} = \boldsymbol{a}, \sigma^2\boldsymbol{I}_N)$.

(d) More generally, the stringent condition $\boldsymbol{\Upsilon} = \sigma^2\boldsymbol{D}$ with $\boldsymbol{D}$ symmetric, positive definite or semidefinite of rank $N_1 \leq N$, *known*, and not needing to be estimated allows converting a $\text{GGLM}_{N,q}()$ with Gaussian errors to a univariate $\text{GLM}_{N_1,q}()$ with Gaussian errors.

Proof. Parts **(a)**, **(b)**, and **(c)** are true due to properties of second moments. Part **(d)** has $0 < \text{rank}(\boldsymbol{D}) = N_1 \leq N$. With $\boldsymbol{V}_1'\boldsymbol{V}_1 = \boldsymbol{I}_{N_1}$, spectral decomposition gives $\boldsymbol{D} = \boldsymbol{V}_1\text{Dg}(\boldsymbol{d}_1)\boldsymbol{V}_1'$. If $\boldsymbol{F}' = \text{Dg}(\boldsymbol{d}_1)^{-1/2}\boldsymbol{V}_1'$, then $\boldsymbol{D}^+ = \boldsymbol{V}_1\text{Dg}(\boldsymbol{d}_1)^{-1}\boldsymbol{V}_1' = \boldsymbol{FF}'$. The original data satisfy the equation $\boldsymbol{y} = \boldsymbol{X\beta} + \boldsymbol{e}$, with $\boldsymbol{e} \sim (\mathcal{S})\mathcal{N}_N(\boldsymbol{0}, \sigma^2\boldsymbol{D})$. In turn, knowing $\boldsymbol{D}$ allows transforming the model:

$$\boldsymbol{F}'\boldsymbol{y} = \boldsymbol{F}'\boldsymbol{X\beta} + \boldsymbol{F}'\boldsymbol{e} \tag{4.1}$$

$$\boldsymbol{y}_F = \boldsymbol{X}_F\boldsymbol{\beta} + \boldsymbol{e}_F\,. \tag{4.2}$$

The relationship $\boldsymbol{F}'\boldsymbol{DF} = \boldsymbol{I}_{N_1}$ allows concluding that $\boldsymbol{e}_F \sim \mathcal{N}_{N_1}(\boldsymbol{0}, \sigma^2\boldsymbol{I}_{N_1})$. Any jointly Gaussian variables with zero covariance are statistically independent.

Therefore the $\{y_{Fi}\}$ meet the assumptions of $\text{GLM}_{N,q}(y_{Fi}; \boldsymbol{X}_{Fi}\boldsymbol{\beta}|\boldsymbol{R\beta} = \boldsymbol{a}, \sigma^2)$ with Gaussian errors. □

Definition 4.3 (a) Part (d) of the last lemma and the associated proof defines a model and data analysis process often described as *exact weighted least squares* or just *weighted least squares* (*WLS*).
(b) Using observed data to estimate any property of $\boldsymbol{D}$ gives *approximate weighted least squares* (*AWLS*) analysis.
(c) *Iterated approximate least squares* (*ITAWLS*) centers on alternately updating the estimates of mean and covariance parameters.

Such methods apply whenever all assumptions of a $\text{GLM}_{N,q}(y_i; \boldsymbol{X}_i\boldsymbol{\beta}, \sigma^2)$ hold except homogeneity of variance, and $\mathcal{V}(y_i|\boldsymbol{X}_i\boldsymbol{\beta}) = \sigma_i^2 = \sigma^2 w_i$, with $\{w_i\}$ known and σ^2 unknown. As discussed in later chapters, the model can be transformed exactly to a GLM with all assumptions holding, including homogeneity. In such cases the transformation leads to optimal estimators and exact (and optimal) tests with respect to the parameters of the original model.

To be precise, weighted least squares may be referred to as exact weighted least squares in order to distinguish it from approximate weighted least squares (AWLS), which estimates some features of $\boldsymbol{D}$ in estimating $\boldsymbol{\Upsilon} = \sigma^2\boldsymbol{D}$. Iterated approximate least squares (ITAWLS) relies on an alternating updating of the estimates of mean and covariance parameters. With Gaussian errors, in many settings the approach leads to maximum likelihood estimation. Not surprisingly, without appropriate adjustments, inference may be inaccurate in small to moderate sample sizes.

Many popular statistical methods, including most mixed model analyses, implicitly include some variation of ITAWLS or AWLS and then operate as though the estimated covariance structure was the population structure. In small to moderate sample sizes the approach can lead to substantial optimistic bias in confidence intervals and hypothesis tests. We urge the reader to always distinguish between the two approaches in reading and evaluating the work of others and in reporting analyses using either method. Including some indication of the expected impact of estimating the weights seems necessary for nonstatisticians to appreciate the amount of uncertainty introduced by an approximate analysis.

Example 4.1 Most methods for confidence intervals and hypothesis tests in current mixed model software implicitly depend on a large-sample assumption: Using covariance estimates in weighted least squares forms introduces no bias. Later chapters centered on mixed models include the details.

4.3 DOUBLY MULTIVARIATE MODELS

The standard multivariate linear applies to the doubly multivariate setting and provides exact size-α tests. However, with p response variables, each measured at t times, the approach implies estimating a $(pt) \times (pt)$ fully unstructured covariance matrix with $pt(pt+1)/2$ distinct elements. The structure of the data makes it very appealing to assume the covariance matrix equals $\boldsymbol{\Sigma}_1 \otimes \boldsymbol{\Sigma}_2$, with $\boldsymbol{\Sigma}_1$ the $p \times p$ covariance among responses and $\boldsymbol{\Sigma}_2$ the $t \times t$ covariance among times. The direct-product form has far fewer covariance parameters, $p(p+1)/2 + t(t+1)/2$. Even in the simplest case with $p = t = 2$, the unstructured model has 10 covariance parameters, while the direct-product form has 6. In turn, $p = t = 5$ gives 325 versus 111 covariance parameters. Timm (2002) reviewed doubly multivariate models based on the direct-product covariance assumption. The work of Boik deserves special attention.

4.4 SEEMINGLY UNRELATED REGRESSIONS

Definition 4.4 A *seemingly unrelated regression model* (sometimes called a multiple design matrix model) corresponds to a $\text{GLM}_{N,p,q}(\boldsymbol{Y}_i; \boldsymbol{X}_i\boldsymbol{B}, \boldsymbol{\Sigma})$ generalized to allow the design matrix and associated parameters to vary across columns of $\boldsymbol{Y}_i$.

Much of the work on seemingly unrelated regression was motivated by econometric applications. Srivastava and Giles (1987) provided a book-length treatment.

Repeated-measures settings may naturally lead to the desire to vary the design matrix across response values. A clinical trial of a drug in the elderly may need to use the dose of a second drug as a covariate. The dose of the second drug may represent a time-varying covariate (or repeated covariate). Two natural variations occur. In the first, only the contemporaneous dose of the second drug matters, while in the second the contemporaneous and all preceding doses matter. If cast as a $\text{GLM}_{N,p,q}(\boldsymbol{Y}_i; \boldsymbol{X}_i\boldsymbol{B}, \boldsymbol{\Sigma})$, the first setting implies the desire for a block of the regression coefficients matrix $\boldsymbol{B}$ to be diagonal. The second setting implies the desire for a block of $\boldsymbol{B}$ to be triangular. Either condition requires imposing nonlinear constraints on $\boldsymbol{B}$ [which correspond to linear constraints on vec($\boldsymbol{B}$)].

4.5 GROWTH CURVE MODELS (GMANOVA)

When growth is observed over time by repeated measurement of a characteristic, the recorded longitudinal pattern can be plotted in two dimensions as a response-versus-time growth curve. The ordered responses of interest might be childrens' linearly increasing weights recorded at ages 1, 2.5, 3, and 4.5 months.

Alternately, the ordered responses might be longitudinal measurements of a drug's exponentially decreasing serum concentration in healthy volunteers. By definition, growth curve model (GCM) analysis focuses on investigating the functional relationship among ordered responses. Conventional GCM methods apply to growth data (indexed by *time* or *age*) and to other analogs such as dose-response data (indexed by *dose*), location-response data (indexed by *distance*), or response-surface data (indexed by two or more variables such as *latitude* and *longitude*), for example. Growth data may exhibit either positive or negative growth, as in the case of the rise or decline of bacterial colonies grown in laboratory dishes. Although most applications of GCM methods center on longitudinal observations on a one-dimensional characteristic (e.g., weight of children), the methods can also apply to multidimensional characteristics such as {weight, height}. The GCM discussed in the present chapter is the classical model considered by Potthoff and Roy (1964), Grizzle and Allen (1969) and Rao (1973). It is also known as a GMANOVA model. Kshirsagar and Smith (1995) provided the best single source.

The scope of our discussion will focus on any such one-dimensional collection of ordered responses with *consistently timed observations*; that is, all the ISUs studied have been observed (or not) on the same occasions (*ages* or *times* or *doses*, etc.). In such cases the observational design must have specified recording the response of interest at p different *times*, $\{t_1, t_2, \ldots, t_p\}$, *doses*, $\{d_1, d_2, \ldots, d_p\}$, etc. The observation process creates a matrix of responses, $\boldsymbol{Y}$ $(N \times p)$. If $N = 1$ and the response of interest is a child's weight, then plotting weight at several ages indicates a temporal pattern of growth. A univariate linear model for weight given age could be fitted with a design matrix $\boldsymbol{T}$ $(p \times m)$ expressing the child's central tendency as a linear or curvilinear function of age. Here $\boldsymbol{T}$ is an example of a *within-subject* design matrix. If $N > 1$, then a separate curve could be fitted for each child to obtain a separate $(m \times 1)$ matrix of regression parameter estimators for each ISU, $\{\widehat{\boldsymbol{B}}_i = \boldsymbol{Y}_i\boldsymbol{T}(\boldsymbol{T}'\boldsymbol{T})^{-1} : i \in \{1, 2, \ldots, N\}\}$. A simple average of the N fitted curves is a proper (if not efficient) estimator of the population growth curve: $\widehat{\boldsymbol{B}} = (\widehat{\boldsymbol{B}}_1 + \widehat{\boldsymbol{B}}_2 + \cdots + \widehat{\boldsymbol{B}}_N)/N$. In the following, $\boldsymbol{X}$ $(N \times q)$ represents a *between-subject* design matrix, which contains intersubject explanatory variables such as gender. The $(q \times m)$ efficient estimator has the form

$$\widehat{\boldsymbol{B}} = [(\boldsymbol{X}'\boldsymbol{X})^{-1}\boldsymbol{X}']\boldsymbol{Y}[\boldsymbol{T}(\boldsymbol{T}'\boldsymbol{T})^{-1}]. \tag{4.3}$$

If the subjects are a homogeneous group, then $\boldsymbol{X} = \boldsymbol{1}$ $(N \times 1)$ is the appropriate choice for computing $\widehat{\boldsymbol{B}}$. The choice of $\boldsymbol{T}$ defines the functional form of the population growth curve by describing a functional relationship between weight and age. It also defines functional dependencies among the age-specific mean heights. If mean height is linear in age, then the mean at 3.5 years is constrained by linearity to lie halfway between the means at 2.5 and 4.5 years. Necessarily the means are collinear. The GCM explicitly addresses the dependencies via the within-subject design matrix $\boldsymbol{T}$ and simultaneously addresses intersubject factors via the between-subject design matrix $\boldsymbol{X}$.

The GCM can be fitted using methods of either a restricted multivariate linear model for $\boldsymbol{Y}$ $(N \times p)$ or a linear mixed model for vec$(\boldsymbol{Y})$. The GCM thus provides common ground for our discussion of connections between multivariate linear models and linear mixed models.

For $\boldsymbol{Y}_i$ $(1 \times p)$ representing growth or dose-response in individual i, a preliminary multivariate model incorporating only a between-subject design matrix $\boldsymbol{X}$ is $\text{GLM}_{N,p,q}\left(\boldsymbol{Y}_i; \boldsymbol{X}_i \boldsymbol{B}_{Y|X}, \boldsymbol{\Sigma}_{Y|X}\right)$. The key feature of the general model is that the $(1 \times p)$ matrix of means for individual i, $\boldsymbol{B}_i = \boldsymbol{\mu}_i' = \text{E}(\boldsymbol{Y}_i) = \boldsymbol{X}_i \boldsymbol{B}_{Y|X}$, is constrained to be a linear combination of the columns of $\boldsymbol{X}_i$ $(1 \times r)$. In turn, the GCM is only of interest when the variation of the central tendency within individual i as a function of time (dose) satisfies a linear model, $\text{E}(\boldsymbol{Y}_i') = \boldsymbol{T}' \boldsymbol{B}_{Y|T}$. It follows that we also wish to constrain $\boldsymbol{\mu}_i'$ to be a linear combination of the rows of $\boldsymbol{T}$ $(m \times p)$, $\boldsymbol{\mu}_i' = \text{E}(\boldsymbol{Y}_i) = \boldsymbol{B}_{Y|T} \boldsymbol{T}$. The constraint can be imposed by modifying the preliminary multivariate model with the restrictions $m \leq p$ and

$$\underset{q \times p}{\boldsymbol{B}_{\boldsymbol{Y}|\boldsymbol{X}}} = \underset{(q \times m)(m \times p)}{\boldsymbol{B}\boldsymbol{T}} \tag{4.4}$$

With $\boldsymbol{T}$ treated as a given constant, the constrained model may be represented as

$$\text{GLM}_{N,p,q}\left(\boldsymbol{Y}_i; \boldsymbol{X}_i \boldsymbol{B}_{Y|X} | \boldsymbol{B}_{Y|X} = \boldsymbol{B}\boldsymbol{T}, \boldsymbol{\Sigma}_{Y|X}\right). \tag{4.5}$$

We assign a special notation to the corresponding restricted multivariate GLM in the following definition. Hopefully $\left|\boldsymbol{\Sigma}_{Y|X,T}\right|$ will be noticeably smaller than $\left|\boldsymbol{\Sigma}_{Y|X}\right|$ because conditioning on both $\boldsymbol{X}$ and $\boldsymbol{T}$ should partially account for both inter-individual variance and intra-individual variance.

Definition 4.5 A *growth curve model* will be indicated by $\text{GCM}_{N,p,q,m}(\boldsymbol{Y}_i; \boldsymbol{X}_i \boldsymbol{B}\boldsymbol{T}, \boldsymbol{\Sigma})$ and includes the following assumptions.

1. The rows of the $N \times p$ random matrix $\boldsymbol{Y}$ are mutually independent. With $\boldsymbol{Y}_i = \text{row}_i(\boldsymbol{Y}) = [\, y_{i1} \; y_{i2} \; \cdots \; y_{ip} \,]$, columns correspond to p ordered responses for p doses, times, etc., arrayed in vector $\boldsymbol{d} = [\, d_1 \; d_2 \; \cdots \; d_p \,]'$.
2. *Within-subject design matrix*, $\boldsymbol{T}$ $(m \times p)$, has $\text{rank}(\boldsymbol{T}) = m \leq p$ and is a fixed, known function of $\boldsymbol{d}$, and known without appreciable error. Consistent timing ensures $\boldsymbol{d}$ and $\boldsymbol{T}$ are constant $\forall i$.
3. With $\boldsymbol{X}_i = \text{row}_i(\boldsymbol{X})$, the $N \times q$ *between-subjects design matrix* $\boldsymbol{X}$ has $\text{rank}(\boldsymbol{X}) = r \leq q \leq N$, and is fixed and known without appreciable error, conditional on knowing the sampling units, for data analysis.
4. Elements of $\boldsymbol{B}$ $(q \times m)$ are fixed and unknown regression coefficients.
5. The mean of $\boldsymbol{Y}_i$ is $\text{E}(\boldsymbol{Y}_i | \boldsymbol{X}_i, \boldsymbol{T}) = \boldsymbol{X}_i \boldsymbol{B}\boldsymbol{T}$ $(1 \times q)(q \times m)(m \times p)$.
6. Response $\boldsymbol{Y}_i$ $(1 \times p)$ has finite covariance matrix, $\boldsymbol{\Sigma}$ $(p \times p)$, which is fixed, unknown, and positive definite or positive semidefinite. Also $\mathcal{V}[\text{vec}(\boldsymbol{Y}') | \boldsymbol{X}, \boldsymbol{T}] = \boldsymbol{I} \otimes \boldsymbol{\Sigma}$.

Writing $\text{GCM}_{N,p,q}(\boldsymbol{Y}_i; \boldsymbol{X}_i \boldsymbol{B}\boldsymbol{T} | \boldsymbol{R}_x \boldsymbol{B} \boldsymbol{R}_y = \boldsymbol{A}, \boldsymbol{\Sigma})$ specifies additional *explicit restrictions* on parameters in $\boldsymbol{B}$ through the fixed and known constants $\boldsymbol{R}_x$, $\boldsymbol{R}_y$, and $\boldsymbol{A}$.

The model is described as *full rank* (FR) if $r = \text{rank}([\boldsymbol{X}' \ \ \boldsymbol{R}_x{}']') = q$ and otherwise as *less than full rank* (LTFR) if $r < q$.

Optionally, $\boldsymbol{Y}_i$ may be assumed to follow a jointly Gaussian distribution.

In the present section we assume $\boldsymbol{\Sigma}$ is unstructured. In later sections we will consider the case in which the $p(p+1)/2$ unique elements of $\boldsymbol{\Sigma}$ are functions of a smaller number of parameters $\boldsymbol{\tau}$ $(k \times 1)$, with $k < p(p+1)/2$.

In the following, $\boldsymbol{t}_j = \text{col}_j(\boldsymbol{T})$. The bilinear form, $\text{E}[\boldsymbol{y}_{ij}|\text{row}(\boldsymbol{X}_i), \boldsymbol{t}_j] = \boldsymbol{\mu}_{ij} = \boldsymbol{X}_i\boldsymbol{B}\boldsymbol{t}_j$, has two interpretations. The first interpretation is that the mean is a linear function of $\boldsymbol{t}_j$ with regression coefficients $(\boldsymbol{\theta}_i)$ that are themselves functions of the characteristics of the participants (e.g., gender), $\mu_{ij} = \boldsymbol{\theta}_i\boldsymbol{t}_j$ with $\boldsymbol{\theta}_i = \boldsymbol{X}_i\boldsymbol{B}$. The second interpretation is that the mean is a linear function of $\boldsymbol{X}_i$ with regression coefficients $(\boldsymbol{\psi}_j)$ which are themselves functions of dose or time, etc., $\mu_{ij} = \boldsymbol{X}_i\boldsymbol{\psi}_j$ with $\boldsymbol{\psi}_j = \boldsymbol{B}\boldsymbol{t}_j$. The concept of statistical interaction unifies the two interpretations. In particular, the magnitudes of the main effects and slopes for the variables represented in $\boldsymbol{X}$ (or $\boldsymbol{T}$) depend on values in $\boldsymbol{T}$ (or $\boldsymbol{X}$). The next lemma allows concluding that the GCM represents mean response as a linear function of the qm cross products of the q explanatory variables represented in row $\boldsymbol{X}_i$ and the m explanatory variables represented in column $\boldsymbol{t}_j$. If $\boldsymbol{X}_i = [\, x_{i1} \ x_{i2} \ x_{i3} \,]$ and $\boldsymbol{t}_j = [\, 1 \ d_j \ d_j^2 \,]'$, then the expected value for participant i on occasion j contains nine cross products, $\{x_{i1}, x_{i1}d_j, x_{i1}d_j^2, x_{i2}, x_{i2}d_j, x_{i2}d_j^2, x_{i3}x_{i3}d_j, x_{i3}d_j^2\}$.

Lemma 4.2 The GCM assumptions have the following implications. For participant i on occasion j the mean response is

$$\text{E}(\boldsymbol{y}_{ij}|\boldsymbol{X},\boldsymbol{T}) = \text{vec}(\boldsymbol{X}_i\boldsymbol{B}\boldsymbol{t}_j) = (\boldsymbol{X}_i \otimes \boldsymbol{t}_j')\text{vec}(\boldsymbol{B}') \,. \tag{4.6}$$

For participant i the mean response vector for all times together is

$$\text{E}(\boldsymbol{Y}_i'|\boldsymbol{X},\boldsymbol{T}) = \text{vec}[(\boldsymbol{X}_i\boldsymbol{B}\boldsymbol{T})'] = (\boldsymbol{X}_i \otimes \boldsymbol{T}')\text{vec}(\boldsymbol{B}') \,. \tag{4.7}$$

For all observations simultaneously

$$\text{E}[\text{vec}(\boldsymbol{Y}')|\boldsymbol{X},\boldsymbol{T}] = \text{vec}[(\boldsymbol{X}\boldsymbol{B}\boldsymbol{T})'] = (\boldsymbol{X} \otimes \boldsymbol{T}')\text{vec}(\boldsymbol{B}') \,, \tag{4.8}$$

which is of the form $\boldsymbol{X}_*\boldsymbol{\beta}_*$ $(Np \times qm)(qm \times 1)$.

Data analysts often assume the errors follow a Gaussian distribution. As detailed in Chapter 8, writing $\boldsymbol{Y} \sim \mathcal{N}_{n,m}(\boldsymbol{M}, \boldsymbol{\Xi}, \boldsymbol{\Sigma})$ indicates $\boldsymbol{Y}$ follows a direct-product matrix Gaussian distribution. By definition, $\boldsymbol{\Xi}$ and $\boldsymbol{\Sigma}$ are symmetric and positive definite or positive semidefinite, and $\text{vec}(\boldsymbol{Y}') \sim \mathcal{N}_{n \cdot m}[\text{vec}(\boldsymbol{M}'), \boldsymbol{\Xi} \otimes \boldsymbol{\Sigma}]$.

Definition 4.6 Writing $\text{GCM}_{N,p,q,m}(\boldsymbol{Y}_i; \boldsymbol{X}_i\boldsymbol{BT}, \boldsymbol{\Sigma})$ *with Gaussian errors* indicates $\boldsymbol{Y}_i' \sim \mathcal{N}_p\{[\text{row}_i(\boldsymbol{X})\boldsymbol{B}]', \boldsymbol{\Sigma}\}$. Equivalently, $\boldsymbol{Y} \sim \mathcal{N}_{N,p}(\boldsymbol{XB}, \boldsymbol{I}_N, \boldsymbol{\Sigma})$.

Example 4.2 Ordered responses of interest are repeated measures of height in centimeters recorded at ages 2, 3, 4, and 6 years for n boys and n girls. It may be plausible to suppose mean height increases linearly with age in years 2–6. In the context of $\text{E}(\boldsymbol{Y}|\boldsymbol{X},\boldsymbol{T}) = \boldsymbol{XBT}$ $(N \times q)(q \times m)(m \times p)$ there are $n = N/2$ participants per group, with $q = 2$ groups. The $p = 4$ measurements per participant are indexed by age, $\boldsymbol{d}' = [\,2\;3\;4\;6\,]$, and

$$\begin{aligned}\boldsymbol{XBT} &= \begin{bmatrix} \mathbf{1}_n & \mathbf{0} \\ \mathbf{0} & \mathbf{1}_n \end{bmatrix} \begin{bmatrix} \beta_{11} & \beta_{12} \\ \beta_{21} & \beta_{22} \end{bmatrix} \begin{bmatrix} \mathbf{1}_p' \\ \boldsymbol{d}' \end{bmatrix} . \\ &= (N \times q)(q \times m)(m \times p) \end{aligned} \tag{4.9}$$

Here $\text{row}_i(\boldsymbol{X}) = \boldsymbol{X}_i = [\,x_{i1}\; x_{i2}\,]$ is $[\,1\;0\,]$ for girls and $[\,0\;1\,]$ for boys. Parameters β_{11} and β_{12} are the intercept and slope, respectively, for girls and parameters β_{21} and β_{22} are the intercept and slope, respectively, for boys. In terms of $\boldsymbol{X}_i = \text{row}_i(\boldsymbol{X})$ and $\boldsymbol{t}_j = \text{col}_j(\boldsymbol{T})$, the mean for participant i on occasion j is $\mu_{ij} = \boldsymbol{X}_i\boldsymbol{B}\boldsymbol{t}_j$. By the lemma,

$$\mu_{ij} = [\,x_{i1} \quad x_{i1}d_j \quad x_{i2} \quad x_{i2}d_j\,]\text{vec}(\boldsymbol{B}') \,, \tag{4.10}$$

and the mean of the vertical concatenation of all the rows of $\boldsymbol{Y}$ is

$$\begin{aligned}\text{E}[\text{vec}(\boldsymbol{Y}')|\boldsymbol{X},\boldsymbol{T}] &= \left(\begin{bmatrix} \mathbf{1}_n & \mathbf{0} \\ \mathbf{0} & \mathbf{1}_n \end{bmatrix} \otimes [\,\mathbf{1}_p \;\; \boldsymbol{d}\,] \right) \text{vec}(\boldsymbol{B}') \\ &= \begin{bmatrix} \mathbf{1}_{n \cdot p} & \boldsymbol{d}_* & \mathbf{0} & \mathbf{0} \\ \mathbf{0} & \mathbf{0} & \mathbf{1}_{n \cdot p} & \boldsymbol{d}_* \end{bmatrix} \begin{bmatrix} \beta_{11} \\ \beta_{12} \\ \beta_{21} \\ \beta_{22} \end{bmatrix} , \end{aligned} \tag{4.11}$$

in which $\boldsymbol{d}_* = (\mathbf{1}_n \otimes \boldsymbol{d})$ is a column vector. The difference between slopes, $(\beta_{12} - \beta_{22})$, is an example of gender-by-age interaction. The mean for girls, $\mu_{ij} = \beta_{11} + \beta_{12}d_j$, and the mean for boys, $\mu_{ij} = \beta_{21} + \beta_{22}d_j$, are necessarily of the same functional form because the within-subject design matrix is assumed to be common to all subjects.

4.6 THE RELATIONSHIP OF THE GCM TO THE MULTIVARIATE MODEL

The $\text{GCM}_{N,p,s,q}(\boldsymbol{Y}_i\boldsymbol{T}; \boldsymbol{X}_i\boldsymbol{BT}, \boldsymbol{T}'\boldsymbol{\Sigma}\boldsymbol{T})$ can be interpreted as a transformed multivariate linear model. The corresponding model for all of the data is

$$\boldsymbol{YT} = \boldsymbol{XBT} + \boldsymbol{ET} \,. \tag{4.12}$$

For a single independent sampling unit

$$\boldsymbol{Y}_i\boldsymbol{T} = \boldsymbol{X}_i\boldsymbol{B}\boldsymbol{T} + \boldsymbol{E}_i\boldsymbol{T} \tag{4.13}$$

$$\boldsymbol{Y}_{iT} = \boldsymbol{X}_i\boldsymbol{B}_{iT} + \boldsymbol{E}_{iT}\,. \tag{4.14}$$

To avoid over parameterization the design matrix $\boldsymbol{T}$ $(m \times p)$ must have $m \leq p$. Both sides of the equation

$$\underset{N\times p}{\mathrm{E}(\boldsymbol{Y}|\boldsymbol{X},\boldsymbol{T})} = \underset{(N\times q)(q\times m)(m\times p)}{\boldsymbol{X}\boldsymbol{B}\boldsymbol{T}} \tag{4.15}$$

can be postmultiplied by a generalized inverse such as $\boldsymbol{T}^+ = \boldsymbol{T}'(\boldsymbol{T}\boldsymbol{T}')^{-1}$ $(p \times m)$ or $\boldsymbol{T}^- = \boldsymbol{V}^{-1}\boldsymbol{T}'(\boldsymbol{T}\boldsymbol{V}^{-1}\boldsymbol{T}')^{-1}$, in which $\boldsymbol{V}^{-1}$ is arbitrary, $p \times p$, and nonsingular. Doing so gives

$$\mathrm{E}[\boldsymbol{Y}_i\boldsymbol{T}'(\boldsymbol{T}\boldsymbol{T}')^{-1}|\boldsymbol{X},\boldsymbol{T}] = \boldsymbol{X}_i\boldsymbol{B} \tag{4.16}$$

and

$$\mathrm{E}\left[\boldsymbol{Y}_i\boldsymbol{V}^{-1}\boldsymbol{T}'(\boldsymbol{T}\boldsymbol{V}^{-1}\boldsymbol{T}')^{-1}|\boldsymbol{X},\boldsymbol{T}\right] = \boldsymbol{X}_i\boldsymbol{B}\,. \tag{4.17}$$

The right-hand side, $\boldsymbol{\theta}_i = \boldsymbol{X}_i\boldsymbol{B}$, is invariant to the choice of generalized inverse. The vector on the left, $\widehat{\boldsymbol{\theta}}_i' = (\boldsymbol{T}\boldsymbol{V}^{-1}\boldsymbol{T}')^{-1}\boldsymbol{T}\boldsymbol{V}^{-1}\boldsymbol{Y}_i'$, is easily recognized as being a weighted least squares estimator (or unweighted if $\boldsymbol{V} = \boldsymbol{I}$) for the subject-specific model $\mathrm{E}(\boldsymbol{Y}_i'|\boldsymbol{T}) = \boldsymbol{\theta}_i'\boldsymbol{T}$. If $m = p$, then $\boldsymbol{T}^- = \boldsymbol{T}^{-1}$ $(p \times p)$ and

$$\underset{1\times p}{\mathrm{E}\left(\boldsymbol{Y}_i\boldsymbol{T}^{-1}|\boldsymbol{X},\boldsymbol{T}\right)} = \underset{(1\times q)(q\times p)}{\boldsymbol{X}_i\boldsymbol{B}}\,. \tag{4.18}$$

In the simple example of heights measured among boys and girls, measurements were made on four occasions and growth was assumed to be a linear function of age. It is possible to include additional terms in the within-subject regression equation, such as a quadratic term. However, it was assumed that higher order terms are not needed. The assumption constrains the boys' mean, $\mu_{ij} = \beta_{11} + \beta_{12}d_j + \beta_{13}d_j^2 + \beta_{14}d_j^3$, with $\beta_{13} = 0$ and $\beta_{14} = 0$. The same constraint applies for the girls. The general notation for the constrained representation is

$$\begin{aligned}\mathrm{E}(\boldsymbol{Y}|\boldsymbol{X},\boldsymbol{T}) &= \boldsymbol{X}\boldsymbol{B}\boldsymbol{T}\\ &= \boldsymbol{X}[\boldsymbol{B}_1\ \ \boldsymbol{B}_2]\begin{bmatrix}\boldsymbol{T}_1\\ \boldsymbol{T}_2\end{bmatrix}\\ &= \begin{bmatrix}\mathbf{1} & \mathbf{0}\\ \mathbf{0} & \mathbf{1}\end{bmatrix}\begin{bmatrix}\beta_{11} & \beta_{12} & \beta_{13} & \beta_{14}\\ \beta_{21} & \beta_{22} & \beta_{23} & \beta_{24}\end{bmatrix}\begin{bmatrix}1 & 1 & 1 & 1\\ 2 & 3 & 4 & 6\\ 2^2 & 3^2 & 4^2 & 6^2\\ 2^3 & 3^3 & 4^3 & 6^3\end{bmatrix},\end{aligned} \tag{4.19}$$

constrained by $\boldsymbol{B}_2 = \mathbf{0}$ with dimensions $\boldsymbol{B}_1$ $(q \times m)$ and $\boldsymbol{B}_2$ $q \times (p-m)$. Here $m = 2$ and $\boldsymbol{T}$ is square and full rank. Also

$$\begin{aligned}\mathrm{E}(\boldsymbol{Y}\boldsymbol{T}^{-1}|\boldsymbol{X},\boldsymbol{T}) &= \boldsymbol{X}\boldsymbol{B} \\ &= \boldsymbol{X}[\,\boldsymbol{B}_1\ \boldsymbol{B}_2\,], \end{aligned} \tag{4.20}$$

constrained by $\boldsymbol{B}_2 = \boldsymbol{0}$. The form suggests partitioning the left-hand side as $\boldsymbol{Y}\boldsymbol{T}^{-1} = [\,\boldsymbol{Y}_{*1}\ \boldsymbol{Y}_{*2}\,]$, with $\mathrm{E}(\boldsymbol{Y}_{*1}|\boldsymbol{X},\boldsymbol{T}) = \boldsymbol{X}\boldsymbol{B}_1$ while $\mathrm{E}(\boldsymbol{Y}_{*2}|\boldsymbol{X},\boldsymbol{T}) = \boldsymbol{0}$. Rao (1965) first proposed the reduction of the GCM to an ordinary multivariate GLM.

The GCM is a multivariate GLM constrained by $p - m$ linear restrictions on the regression parameter matrix $\boldsymbol{B}$. The unconstrained model is denoted $\mathrm{GLM}_{N,p,q}(\boldsymbol{Y}_i; \boldsymbol{X}_i\boldsymbol{B}_{\boldsymbol{Y}|\boldsymbol{X}}, \boldsymbol{\Sigma}_{\boldsymbol{Y}|\boldsymbol{X}})$ with $\boldsymbol{B}_{\boldsymbol{Y}|\boldsymbol{X}}$ $(q \times p)$. The assumption of a full rank within-subject design matrix $\boldsymbol{T}$ $(q \times m)$ with $m \leq p$ defines the linear constraints $\boldsymbol{B}_{\boldsymbol{Y}|\boldsymbol{X}} = \boldsymbol{B}\boldsymbol{T}$. If $m = p$, then $\boldsymbol{T}$ is a square nonsingular matrix and the number of linear restrictions is zero.

Example 4.3 An example is given by

$$\begin{aligned}\boldsymbol{T} &= \{t_{ij} : t_{ij} = d_j^{i-1}\} \\ &= \begin{bmatrix} 1 & 1 & 1 & 1 \\ d_1 & d_2 & d_3 & d_4 \\ d_1^2 & d_2^2 & d_3^2 & d_4^2 \\ d_1^3 & d_2^3 & d_3^3 & d_4^3 \end{bmatrix} = \begin{bmatrix} 1 & 1 & 1 & 1 \\ 2 & 3 & 4 & 6 \\ 2^2 & 3^2 & 4^2 & 6^2 \\ 2^3 & 3^3 & 4^3 & 6^3 \end{bmatrix}. \end{aligned} \tag{4.21}$$

In turn,

$$\mathrm{E}(\boldsymbol{Y}|\boldsymbol{X},\boldsymbol{T}) = \begin{bmatrix} 1\,0 \\ 0\,1 \end{bmatrix} \begin{bmatrix} \beta_{11}\ \beta_{12}\ \beta_{13}\ \beta_{14} \\ \beta_{21}\ \beta_{22}\ \beta_{23}\ \beta_{24} \end{bmatrix} \begin{bmatrix} 1 & 1 & 1 & 1 \\ 2 & 3 & 4 & 6 \\ 2^2 & 3^2 & 4^2 & 6^2 \\ 2^3 & 3^3 & 4^3 & 6^3 \end{bmatrix}. \tag{4.22}$$

Usually some columns of $\boldsymbol{B}$ will be assumed to be zero. If $\boldsymbol{T}$ represents polynomial regression of y_{ij} on d_j, the last few columns of $\boldsymbol{B}$ correspond to the highest order polynomial terms. The constraint on $\boldsymbol{B}$ induces a partitioning of both $\boldsymbol{B}$ and $\boldsymbol{T}$, with $\boldsymbol{B}_2 = \boldsymbol{0}$ $[(q \times (p-m)]$ and $\boldsymbol{B}_1$ $(q \times m)$ is the set of nonzero columns. The constraint $\boldsymbol{B}_2 = \boldsymbol{0}$ corresponds to omitting some of the rows of $\boldsymbol{T}$ from the model, which is conceptually no different from deciding to omit some columns of $\boldsymbol{X}$. Usually $\boldsymbol{B}_1$ is the first few columns of $\boldsymbol{B}$, but any columns of $\boldsymbol{B}$ may be required to be zero. Since the rows of $\boldsymbol{T}$ and the columns of $\boldsymbol{B}$ can be permuted, there is no loss in generality in using the notation given above. When some columns of $\boldsymbol{B}$ are required to be zero, then only the q nonzero columns, $\boldsymbol{B}_1$, must be estimated $(2 \leq m \leq p)$. The model can be represented by $\mathrm{GCM}_{N,p,q,m}(\boldsymbol{Y}_i; \boldsymbol{X}_i\boldsymbol{B}_1\boldsymbol{T}_1, \boldsymbol{\Sigma})$ with $\boldsymbol{B}_1$ $(q \times m)$ being a subset of the columns of $\boldsymbol{B}$ $(q \times p)$.

Definition 4.7 In the GCM, the mean response as a function of dose d or time is of the form $\mu(d; \boldsymbol{C}) = \boldsymbol{C}\boldsymbol{B}\boldsymbol{u}_d$ $(1 \times q \times m \times 1)$ in which

$\boldsymbol{u}_d = [1 \; d \; \cdots \; d^{m-1}]'$ $(m \times 1)$. In the special case of $d = d_j$ and $\boldsymbol{C} = \boldsymbol{X}_i$ we have $\mu(d_j; \boldsymbol{X}_i) = \mathrm{E}(y_{ij}|\boldsymbol{X}_i, \boldsymbol{t}_j) = \boldsymbol{X}_i \boldsymbol{B} \boldsymbol{t}_j$. A set of coordinates $\{[d, \mu(d; \boldsymbol{C})] : d \in [d_1, d_p]\}$ defines a *growth curve*.

4.7 MIXED, HIERARCHICAL, AND RELATED MODELS

The general linear mixed model encompasses the most general range of models considered in detail in the present book. In terms of expressions for population parameters, many other models mentioned can be thought of as special cases. The special relationships have led to the widespread misconception that similar special case relationships hold for estimates and tests. However, appropriate tests occur as special cases for only a very limited range of models. The mixed model has mostly approximate results and very few exact results for inference. The exact results for estimates and tests for a (univariate) $\mathrm{GLM}_{N,q}(y_i; \boldsymbol{X}_i\boldsymbol{\beta}, \sigma^2)$ with Gaussian errors do occur automatically as special cases of mixed model approximations. In contrast, for the $\mathrm{GLM}_{N,p,q}(\boldsymbol{Y}_i; \boldsymbol{X}_i\boldsymbol{B}, \Sigma)$ with Gaussian errors, only the maximum likelihood estimates occur automatically as special cases of mixed model results. None of the commonly used tests in mixed models correspond to multivariate model tests, except asymptotically or in the special case of a univariate model.

The approximate theory of mixed models also applies to many models described as "hierarchical," which allow additional freedom in specifying the random components (in contrast to general linear mixed models). Raudenbush and Bryk (2002) provided a book-length treatment. Mixed model theory also appears to encompass and apply to a wide range of "state-space" models (Billio and Monfort, 1998, provided an example).

CHAPTER 5

The Linear Mixed Model

5.1 MOTIVATION

As discussed in earlier chapters, the general linear mixed model allows missing or mistimed data as well as repeated covariates. The approach also allows specifying covariance structures as a function of a small number of parameters. In most uses, the model implicitly assumes commensurate data (all measured in the same units). Most, but certainly not all, applications involve repeated measures.

Many different classes of models have been described as "mixed models." The term dates back to early developments in ANOVA. The simplest ANOVA design involves one factor, a categorical predictor with G levels defining G groups, and $n_g = N/G$ independent sampling units in each group (cell). Classical less-than-full-rank coding led to writing the model as a scalar equation, with $i \in \{1, 2, \ldots, n_g\}$ and $g \in \{1, 2, \ldots, G\}$:

$$y_{ig} = \mu + \alpha_g + e_{ig} \,. \tag{5.1}$$

Here μ and $\{\alpha_g\}$ are fixed and unknown finite constants (parameters) which characterize the means and $e_{ig} \sim \mathcal{N}(0, \sigma_e^2)$, with σ_e^2 a fixed and unknown finite constant, the variance. The nature of $\{\alpha_g\}$ led to describing the predictor variable as a "fixed effect" and the model as a fixed effect model. In contrast, a "random effect" model assumes $\{\alpha_g\}$ are randomly selected from an infinite population, with independent and identically distributed $\alpha_g \sim \mathcal{N}(0, \sigma_g^2)$ independent of $\{e_{ig}\}$. Here μ alone represents the mean, while $\alpha_g + e_{ig}$ represents total variance in terms of two components. More generally, a model with two or more fixed effects was described as a fixed effects model, while a model with two or more random effects was described as a random effects model. In turn, a model with one or more fixed effects and one or more random effects was referred to as a mixed effects model.

The terms random effect, fixed effect, and mixed effect do not always clearly convey the underlying simple structure of a mixed model to readers not intimately familiar with the theory. We prefer to discuss the parameters of a mixed model in terms of two separate components: the model for the means and the model for the covariance. With Gaussian distributions, specifying the first two moments fully determines the distributions and therefore implicitly all derived properties. The approach allows couching most discussions about the mixed model in terms of the simple concepts of means, variances and correlations.

Mixed models are a useful tool for longitudinal data exhibiting missing values, inconsistently timed observations, or mistimed observations. Incomplete data put the analyst at a disadvantage because critically important information is missing. Analysis cannot proceed unless the missing information is replaced by assumptions. The mixed model approach presents intuitively appealing assumptions and has some procedures for inference that work well at least with large and some moderate sample sizes.

The approach involves building a model for the expected values of the data and also building a model for the covariances of the data. The linear model for the mean allows the flexibility of using polynomials, trigonometric functions, and regression splines, among others. Both linear and nonlinear models for the covariance structure are useful.

We begin by considering an extremely general class of linear models. As stated, the model allows specifying a particular covariance model in a variety of ways. The generality allows ambiguity without adding further constraints. However, it serves well as the basis for special cases of interest.

The interested reader seeking additional details may wish to consult any of a number of excellent book-length treatments centered on mixed models, including Vonesh and Chinchilli (1997), Khuri, Mathew, and Sinha (1998), Verbeke and Molenberghs (2000), and Demidenko (2004). Timm (2002) discussed the mixed model as an alternative to a wide variety of multivariate methods.

5.2 DEFINITION OF THE MIXED MODEL

Definition 5.1 A *general linear mixed model* will be indicated by $\text{LMM}_{N,p_i,q,m}[\boldsymbol{y}_i; \boldsymbol{X}_i\boldsymbol{\beta}, \boldsymbol{Z}_i\Sigma_{di}(\boldsymbol{\tau}_d)\boldsymbol{Z}_i' + \Sigma_{ei}(\boldsymbol{\tau}_e)]$ and includes the following assumptions. When no clarity will be lost, the model may be abbreviated $\text{LMM}_{N,p_i,q,m}(\boldsymbol{y}_i; \boldsymbol{X}_i\boldsymbol{\beta}, \boldsymbol{Z}_i\Sigma_{di}\boldsymbol{Z}_i' + \Sigma_{ei})$.

1. For $i \in \{1, 2, \ldots, N\}$,
(a) the N random vectors, $\{\boldsymbol{e}_i\}$, are $p_i \times 1$ and mutually independent,
(b) the N random vectors, $\{\boldsymbol{d}_i\}$ are $m \times 1$ and mutually independent, and
(c) the $\{\boldsymbol{e}_i\}$ and $\{\boldsymbol{d}_i\}$ are all mutually independent.
2. Each $\boldsymbol{X}_i$, the $p_i \times q$ *expected value design matrix* for independent sampling unit i, is fixed and known without appreciable error, conditional on knowing the sampling units, for data analysis.
3. Elements of $\boldsymbol{\beta}$ $(q \times 1)$ are fixed and unknown and often regression coefficients or means.
4. Each $\boldsymbol{Z}_i$, the $p_i \times m$ *covariance design matrix* for independent sampling unit i, is fixed and known without appreciable error, conditional on knowing the sampling units, for data analysis.

5. For $i \in \{1, 2, \ldots, N\}$, (a) $\mathrm{E}(\boldsymbol{e}_i) = \mathbf{0}$ and (b) $\mathrm{E}(\boldsymbol{d}_i) = \mathbf{0}$.
6. With $\boldsymbol{\tau}' = [\,\boldsymbol{\tau}_d' \; \boldsymbol{\tau}_e'\,]$ fixed and unknown, $\boldsymbol{d}_i$ and $\boldsymbol{e}_i$ have a finite, fixed, and unknown (joint) covariance matrix which is either positive definite or positive semidefinite:

$$\mathcal{V}\left(\begin{bmatrix} \boldsymbol{d}_i \\ \boldsymbol{e}_i \end{bmatrix}\right) = \begin{bmatrix} \boldsymbol{\Sigma}_{di}(\boldsymbol{\tau}_d) & \mathbf{0} \\ \mathbf{0} & \boldsymbol{\Sigma}_{ei}(\boldsymbol{\tau}_e) \end{bmatrix}. \tag{5.2}$$

Elements of $\boldsymbol{\Sigma}_{di}(\boldsymbol{\tau}_d)$ are twice differentiable functions of $\boldsymbol{\tau}_d$, a vector of no more than $m(m+1)/2$ covariance parameters. Elements of $\boldsymbol{\Sigma}_{ei}(\boldsymbol{\tau}_e)$ are twice differentiable functions of $\boldsymbol{\tau}_e$, a vector of no more than $\max_i[p_i(p_i+1)/2]$ covariance parameters.
7. The $p_i \times 1$ response vector $\boldsymbol{y}_i$ is expressed as $\boldsymbol{y}_i = \boldsymbol{X}_i\boldsymbol{\beta} + \boldsymbol{Z}_i\boldsymbol{d}_i + \boldsymbol{e}_i$ with

$$\mathrm{E}(\boldsymbol{y}_i) = \boldsymbol{X}_i\boldsymbol{\beta} \tag{5.3}$$

and fixed, unknown, and positive definite or positive semidefinite covariance

$$\mathcal{V}(\boldsymbol{y}_i) = \boldsymbol{\Sigma}_i(\boldsymbol{\tau}) = \boldsymbol{Z}_i\boldsymbol{\Sigma}_{di}(\boldsymbol{\tau}_d)\boldsymbol{Z}_i' + \boldsymbol{\Sigma}_{ei}(\boldsymbol{\tau}_e)\,. \tag{5.4}$$

When no clarity will be lost, the covariance model may be abbreviated $\boldsymbol{\Sigma}_i$. Writing $\mathrm{LMM}_{N,p_i,q,m}(\boldsymbol{y}_i; \boldsymbol{X}_i\boldsymbol{\beta}|\boldsymbol{R}_x\boldsymbol{\beta}{=}\boldsymbol{a}, \boldsymbol{Z}_i\boldsymbol{\Sigma}_{di}\boldsymbol{Z}_i'{+}\boldsymbol{\Sigma}_{ei}|\boldsymbol{R}_d\boldsymbol{d}_i{=}\mathbf{0})$ specifies explicit restrictions on parameters in $\boldsymbol{\beta}$ through the fixed and known constants $\boldsymbol{R}_x$ and $\boldsymbol{a}$ and explicit restrictions on $\boldsymbol{d}_i$ through the fixed and known constant $\boldsymbol{R}_d$.

The definition contains what may be described as the "approximate generalized least squares" assumptions. In combination with mild restrictions on dimensions and ranks of $\{\boldsymbol{X}_i, \boldsymbol{\Sigma}_i\}$, they guarantee the existence of estimators for $\boldsymbol{\beta}$ and $\boldsymbol{\tau}$ which satisfy an approximate generalized least squares criterion. It is very important to recognize that no particular distribution has been specified for any random variable. Only the rather modest requirement of finite second (and implicitly first) moments is made.

The model definition specifies three components: the response vector for the independent sampling unit, the mean of the response vector, and the covariance of the response vector. In turn, the covariance of the response consists of two components, corresponding to the two unobservable random vectors $\boldsymbol{d}_i$ and $\boldsymbol{e}_i$.

The data may be "stacked" to represent a combined model. With $n = \sum_{i=1}^{N} p_i$, it is often convenient to write $\boldsymbol{y}_s' = [\,\boldsymbol{y}_1' \; \boldsymbol{y}_2' \cdots \boldsymbol{y}_N'\,]$, which implies $\boldsymbol{y}_s$ is $n \times 1$. Similarly, $\boldsymbol{d}_s' = [\,\boldsymbol{d}_1' \; \boldsymbol{d}_2' \cdots \boldsymbol{d}_N'\,]$ implies $\boldsymbol{d}_s$ is $Nm \times 1$ and $\boldsymbol{e}_s' = [\,\boldsymbol{e}_1' \; \boldsymbol{e}_2' \cdots \boldsymbol{e}_N'\,]$ implies $\boldsymbol{e}_s$ is $n \times 1$. In turn, concatenation of the predictor matrices gives

$$\boldsymbol{X}_s = \begin{bmatrix} \boldsymbol{X}_1 \\ \boldsymbol{X}_2 \\ \vdots \\ \boldsymbol{X}_N \end{bmatrix}, \tag{5.5}$$

an $n \times q$ matrix. In contrast,

$$Z_s = \begin{bmatrix} Z_1 & 0 & \cdots & 0 \\ 0 & Z_2 & & \vdots \\ \vdots & & \ddots & 0 \\ 0 & \cdots & 0 & Z_N \end{bmatrix} \tag{5.6}$$

is an $n \times (Nm)$ block diagonal matrix. Necessarily the rows of $\boldsymbol{y}_s$, $\boldsymbol{d}_s$, $\boldsymbol{e}_s$, $\boldsymbol{X}_s$, and $\boldsymbol{Z}_s$ must be sorted in the same order (both within and between independent sampling units) for the following equation, the stacked data model, to be valid:

$$\boldsymbol{y}_s = \boldsymbol{X}_s\boldsymbol{\beta} + \boldsymbol{Z}_s\boldsymbol{d}_s + \boldsymbol{e}_s\,. \tag{5.7}$$

Consequently

$$\mathrm{E}(\boldsymbol{y}_s) = \boldsymbol{X}_s\boldsymbol{\beta} \tag{5.8}$$

and, with $\boldsymbol{\Sigma}_{ds} = \bigoplus_{i=1}^{N}\boldsymbol{\Sigma}_{di}$,

$$\begin{aligned} \mathcal{V}(\boldsymbol{y}_s) = \boldsymbol{\Sigma}_s &= \bigoplus_{i=1}^{N}(\boldsymbol{Z}_i\boldsymbol{\Sigma}_{di}\boldsymbol{Z}_i' + \boldsymbol{\Sigma}_{ei}) \\ &= \bigoplus_{i=1}^{N}(\boldsymbol{Z}_i\boldsymbol{\Sigma}_{di}\boldsymbol{Z}_i') + \bigoplus_{i=1}^{N}\boldsymbol{\Sigma}_{ei} \\ &= \boldsymbol{Z}_s\boldsymbol{\Sigma}_{ds}\boldsymbol{Z}_s' + \boldsymbol{\Sigma}_{es}\,. \end{aligned} \tag{5.9}$$

The model $\boldsymbol{y}_s = \boldsymbol{X}_s\boldsymbol{\beta} + \boldsymbol{Z}_s\boldsymbol{d}_s + \boldsymbol{e}_s$ expresses the observations as a function of three terms. The term $\boldsymbol{X}_s\boldsymbol{\beta}$ describes the fixed contribution of the population, conditional on the predictor values (which often define subpopulations). Each row of $\boldsymbol{Z}_s\boldsymbol{d}_s$ describes a random deviation from the population value due to observing a particular person (ISU). Each row of $\boldsymbol{e}_s$ describes an additional and distinct random deviation due to observing a particular person on a particular occasion. In summary, a mixed model expresses an observation as a *subpopulation mean* plus a *random deviation due to person* plus an additional *random deviation due to occasion*.

Alternately, the response vector may be expressed in terms of one purely fixed and one purely random vector. If $\boldsymbol{e}_{+s} = \boldsymbol{Z}_s\boldsymbol{d}_s + \boldsymbol{e}_s$, then

$$\begin{aligned} \boldsymbol{y}_s &= \boldsymbol{X}_s\boldsymbol{\beta} + \boldsymbol{Z}_s\boldsymbol{d}_s + \boldsymbol{e}_s \\ &= \boldsymbol{X}_s\boldsymbol{\beta} + \quad \boldsymbol{e}_{+s} \\ &= \text{fixed} + \text{random} \\ & \quad\;\, \uparrow \qquad\quad \uparrow \\ & \;\mathrm{E}(\boldsymbol{y}_s) \quad \mathcal{V}(\boldsymbol{y}_s) \\ & \;\text{model} \quad\; \text{model}\,. \end{aligned} \tag{5.10}$$

The fixed component $\boldsymbol{X}_s\boldsymbol{\beta}$ completely determines the mean, the first moment, of $\boldsymbol{y}_s$ and has no effect on any variance or covariance. The random component $\boldsymbol{e}_{+s}$ completely determines all variances and covariances, the second moments, and has no effect on the mean. The model for a particular independent sampling unit naturally follows the same pattern:

$$\begin{aligned} \boldsymbol{y}_i &= \boldsymbol{X}_i\boldsymbol{\beta} + \boldsymbol{Z}_i\boldsymbol{d}_i + \boldsymbol{e}_i \\ &= \boldsymbol{X}_i\boldsymbol{\beta} + \quad \boldsymbol{e}_{+i} \\ &= \text{fixed} + \text{random}. \\ &\quad\ \ \uparrow \qquad\qquad \uparrow \\ &\quad \mathrm{E}(\boldsymbol{y}_i) \qquad \mathcal{V}(\boldsymbol{y}_i) \\ &\quad \text{model} \qquad \text{model} \end{aligned} \tag{5.11}$$

As in univariate and multivariate linear models, a less-than-full-rank design matrix $\boldsymbol{X}_s$ prevents unique identification and unbiased estimation of $\boldsymbol{\beta}$. Side conditions must be imposed to avoid the problem. We shall usually refer to the side conditions as "restrictions" or "constraints." As the model is defined in Definition 1.1, side conditions also must be imposed on the components of $\boldsymbol{\Sigma}_i$ to uniquely define the covariance model parameters and estimates. Given appropriate side conditions, modern computing tools often make it straightforward to find estimates for $\boldsymbol{\beta}$ and $\boldsymbol{\tau}$ which satisfy an approximate least squares criterion or an iterated approximate least squares criterion. Except for special cases or in large samples, the estimators have few optimal properties. Kackar and Harville (1984) proved that $\widehat{\boldsymbol{\beta}}$ from iterated approximate least squares (sometimes called estimated generalized least squares, among other names) is unbiased for a Gaussian (or any other symmetric) distribution. However, covariance parameter estimators, at least in small samples, typically have substantial bias. Littel (2003) provided an excellent overview.

With additional restrictions on the covariance parameters $\boldsymbol{\tau} = [\boldsymbol{\tau}_d' \ \boldsymbol{\tau}_e']'$ the model defined in Definition 1.1 becomes sufficiently well-defined to allow computing parameter estimators with reasonable properties. Typically data analysts greatly simplify or completely eliminate either $\boldsymbol{\tau}_d$ or $\boldsymbol{\tau}_e$. As an example, assuming response vector i has a first-order autocorrelation covariance pattern requires the j, k element to be $\{\boldsymbol{\Sigma}_i(\boldsymbol{\tau})\}_{j,k} = \sigma^2\rho^{|t_j - t_k|}$. Choosing $\boldsymbol{Z}_i = \boldsymbol{0}$, $\boldsymbol{\Sigma}_{ei}(\boldsymbol{\tau}_e) = \{\sigma^2\rho^{|t_j-t_k|}\}$, and $\boldsymbol{\tau}_e = [\sigma^2 \ \rho]'$ achieves the desired pattern. Doing so expresses the variances and covariances as a nonlinear function of the two parameters σ^2 and ρ. Implicitly, the observations for sampling unit i follow a stationary time series. For nonstationary processes an inherently linear model for the covariances is often assumed. In particular, specifying $\{\boldsymbol{G}_{ik}\}$ as known constant matrices allows writing

$$\boldsymbol{\Sigma}_i(\boldsymbol{\tau}) = \sum_{k=1}^{t} \tau_k(\boldsymbol{Z}_i\boldsymbol{G}_{ik}\boldsymbol{Z}_i') + \sigma^2\boldsymbol{I}_{p_i}\,. \tag{5.12}$$

With such a structure $\boldsymbol{\Sigma}_i(\boldsymbol{\tau})$ is a linear function of $t+1$ unknown parameters.

Example 5.1 The ambiguity in representing the covariance model may be illustrated quite easily. The response vector (and the purely random part of the mixed model) will have a compound symmetric covariance structure if $\mathcal{V}(\boldsymbol{e}_i) = \sigma^2[\boldsymbol{1}_{p_i}\boldsymbol{1}_{p_i}'\rho + \boldsymbol{I}_{p_i}(1-\rho)]$ and $\boldsymbol{Z}_i = \boldsymbol{0}$ [$\boldsymbol{Z}_i = \boldsymbol{0}$ has exactly the same effect as assuming $\mathcal{V}(\boldsymbol{d}_i) = \mathbf{0}$]. In either case $\mathcal{V}(\boldsymbol{y}_i) = \mathcal{V}(\boldsymbol{e}_i)$. Alternately, choosing $\boldsymbol{Z}_i = \mathbf{1}_{p_i}$, $\mathcal{V}(\boldsymbol{d}_i) = \sigma^2\rho$, and $\mathcal{V}(\boldsymbol{e}_i) = \sigma^2(1-\rho)\boldsymbol{I}_{p_i}$ gives exactly the same $\mathcal{V}(\boldsymbol{y}_i)$, but with a different $\mathcal{V}(\boldsymbol{e}_i)$.

Most presentations of the general linear mixed model include the assumption that the random variables follow a Gaussian distribution. However, in contrast to the univariate and multivariate linear models, the assumption does not lead to closed formed expressions for estimates.

Definition 5.2 Writing
$\text{LMM}_{N,p_i,q,m}(\boldsymbol{y}_i; \boldsymbol{X}_i\boldsymbol{\beta}|\boldsymbol{R}_x\boldsymbol{\beta}=\boldsymbol{a}, \boldsymbol{Z}_i\boldsymbol{\Sigma}_{di}(\boldsymbol{\tau}_d)\boldsymbol{Z}_i'+\boldsymbol{\Sigma}_{ei}(\boldsymbol{\tau}_e)|\boldsymbol{R}_d\boldsymbol{d}_i=\mathbf{0})$
with Gaussian errors indicates

$$\begin{bmatrix}\boldsymbol{d}_i\\ \boldsymbol{e}_i\end{bmatrix} \sim \mathcal{N}_{m+p_i}\left\{\begin{bmatrix}\mathbf{0}\\ \mathbf{0}\end{bmatrix}, \begin{bmatrix}\boldsymbol{\Sigma}_{di}(\boldsymbol{\tau}_d) & \mathbf{0}\\ \mathbf{0} & \boldsymbol{\Sigma}_{ei}(\boldsymbol{\tau}_e)\end{bmatrix}\right\}, \tag{5.13}$$

which is equivalent to

$$\begin{bmatrix}\boldsymbol{d}_s\\ \boldsymbol{e}_s\end{bmatrix} \sim \mathcal{N}_{2n}\left\{\begin{bmatrix}\mathbf{0}\\ \mathbf{0}\end{bmatrix}, \begin{bmatrix}\bigoplus_{i=1}^{N}\boldsymbol{\Sigma}_{di}(\boldsymbol{\tau}_d) & \mathbf{0}\\ \mathbf{0} & \bigoplus_{i=1}^{N}\boldsymbol{\Sigma}_{ei}(\boldsymbol{\tau}_e)\end{bmatrix}\right\}. \tag{5.14}$$

Recalling $\boldsymbol{\Sigma}_i(\boldsymbol{\tau}) = \boldsymbol{Z}_i\boldsymbol{\Sigma}_{di}(\boldsymbol{\tau}_d)\boldsymbol{Z}_i' + \boldsymbol{\Sigma}_{ei}(\boldsymbol{\tau}_e)$, it follows that

$$\boldsymbol{y}_i \sim \mathcal{N}_{p_i}[\boldsymbol{X}_i\boldsymbol{\beta}, \boldsymbol{\Sigma}_i(\boldsymbol{\tau})]\,. \tag{5.15}$$

Equivalently,

$$\boldsymbol{y}_s \sim \mathcal{N}_n\left[\boldsymbol{X}_s\boldsymbol{\beta}, \bigoplus_{i=1}^{N}\boldsymbol{\Sigma}_i(\boldsymbol{\tau})\right]. \tag{5.16}$$

Some mixed models, members of a class often referred to as components of variance models, assume compound symmetric covariance among *all* observations, which implies having only one independent sampling unit. The special properties of compound symmetry allow using exact weighted least squares methods to transform the model to one with completely independent observations (and some heterogeneity).

The mnemonic HILE Gauss must be interpreted carefully in the mixed model setting because allowing p_i to vary causes $\boldsymbol{\Sigma}_i(\boldsymbol{\tau})$ to vary. *Independence* of sampling units remains the cornerstone (even given the special handling required for some components of variance models). In turn, *linearity* of the response expected value (mean) as a function of the parameters also holds, as does the assumption of finite second moments (*existence*). However, describing the model as having *homogeneity* of second moments must be interpreted to indicate that a single covariance *model* holds for all sampling units, while $\boldsymbol{\Sigma}_i(\boldsymbol{\tau}) = \boldsymbol{\Sigma}_j(\boldsymbol{\tau})$ only if $\boldsymbol{Z}_i = \boldsymbol{Z}_j$ and $p_i = p_j$.

5.3 DISTRIBUTION-FREE AND NONITERATIVE ESTIMATES

As always in the history of mixed models, computing difficulties often dominate theory and practice. Although current computing hardware speed and associated software have virtually eliminated some problems, we still face many issues considered by Henderson (1953). Method-of-moments and MINQUE (minimum variance quadratic unbiased estimation) provide noniterative estimates (Searle, Casella, and McCulloch, 1992). Although the methods have many good properties, they are currently much less popular than likelihood methods.

5.4 GAUSSIAN LIKELIHOOD AND ITERATIVE ESTIMATES

The nearly ubiquitous use of the Gaussian assumption leads most data analysts to seek either maximum likelihood (ML) estimates or *restricted maximum likelihood* (REML) estimates (Chapter 14 has some details). The joint log likelihood is, with $n = \sum_{i=1}^{N} p_i$, $\boldsymbol{e}'_{+i} = \boldsymbol{y}_i - \boldsymbol{X}_i\boldsymbol{\beta}$ and $\boldsymbol{\Sigma}_i(\boldsymbol{\tau})$ abbreviated as $\boldsymbol{\Sigma}_i$,

$$-2\log L(\boldsymbol{\beta}, \boldsymbol{\tau}) = n\log(2\pi) - \frac{1}{2}\sum_{i=1}^{N}\left[\log|\boldsymbol{\Sigma}_i| + (\boldsymbol{y}_i - \boldsymbol{X}_i\boldsymbol{\beta})'\boldsymbol{\Sigma}_i^{-1}(\boldsymbol{y}_i - \boldsymbol{X}_i\boldsymbol{\beta})\right]. \quad (5.17)$$

Iterative methods must be used to solve the system of equations, which are nonlinear in the parameters $\{\boldsymbol{\beta}, \boldsymbol{\tau}\}$. Without the Gaussian assumption, the resulting estimates satisfy the iterated approximate weighted least squares criterion (ITAWLS). Either with or without the Gaussian assumption, the resulting estimates are biased in small samples. As mentioned earlier, the estimate of $\boldsymbol{\beta}$ is unbiased, while the estimate of $\boldsymbol{\tau}$ (and $\{\boldsymbol{\Sigma}_i\}$) typically has substantial bias in small samples.

REML estimates of $\boldsymbol{\tau}$ (and $\{\boldsymbol{\Sigma}_i\}$) have less bias than ML estimates. REML estimates arise from maximizing the reduced profile log-likelihood equation, based on $\widehat{\boldsymbol{e}}'_{+i} = \boldsymbol{y}_i - \boldsymbol{X}_i\widehat{\boldsymbol{\beta}}$:

$$-2\log\widehat{L}_{\text{REML}}(\boldsymbol{\tau}) = (n-q)\log(2\pi) + \sum_{i=1}^{N}\left(\log|\boldsymbol{\Sigma}_i| + \widehat{\boldsymbol{e}}'_{+i}\boldsymbol{\Sigma}_i^{-1}\widehat{\boldsymbol{e}}_{+i}\right) + \log\left|\sum_{i=1}^{N}\boldsymbol{X}'_i\boldsymbol{\Sigma}_i^{-1}\boldsymbol{X}_i\right|. \quad (5.18)$$

Although we leave the details to Chapter 14, it is worth observing the following. In the univariate linear model, a special case of the mixed model, the maximum likelihood estimate of the error variance, $\widetilde{\sigma}^2 = \boldsymbol{y}'[\boldsymbol{I}_N - \boldsymbol{X}(\boldsymbol{X}'\boldsymbol{X})^{-}\boldsymbol{X}']\boldsymbol{y}/N$, has an expected value smaller than σ^2. In contrast, $\widehat{\sigma}^2 = \widetilde{\sigma}^2 N/[N - \text{rank}(\boldsymbol{X})]$ is the unbiased REML estimate of σ^2.

Estimates of both the primary parameters and a variety of secondary parameters are usually desired. Interest may center on either estimates of population location parameters (elements of $\boldsymbol{\theta} = \boldsymbol{C\beta}$) which apply to all sampling units or estimates of properties particular to a specific independent sampling unit. We leave further discussion to the more detailed treatment of estimation and testing in later chapters.

5.5 TESTS ABOUT β (MEANS, FIXED EFFECTS)

In the context of the mixed model, data analysts often wish to test a hypothesis of the form

$$\begin{aligned} H_0 &: \boldsymbol{C\beta} = \boldsymbol{\theta}_0 \\ H_0 &: \boldsymbol{\theta} = \boldsymbol{\theta}_0 \,. \end{aligned} \tag{5.19}$$

Only low order approximate tests have been described. Not surprisingly, the tests often perform very poorly in small samples.

Given the assumption of Gaussian data and the strong parallels of the forms to the univariate linear model, it is straightforward to define the form of the likelihood ratio test statistic. Computing the statistic requires iterative calculations to successfully fit two models, the full model and the constrained model, at least one of which must be false. In practice, the false model is less likely to converge. If both models converge, then the test is well defined. The difficulty with the likelihood ratio test lies in finding an adequate approximation to the distribution of the test statistic. The approximation $-2\log[L(\widehat{\boldsymbol{\theta}}|\boldsymbol{C\beta} = \boldsymbol{\theta}_0)/L(\widehat{\boldsymbol{\theta}})] \mathrel{\dot\sim} \chi^2(a)$ with $\boldsymbol{\theta}$ of dimension $a \times 1$ relies on the log likelihood being approximately quadratic in shape. From the perspective of a Taylor series expansion for the test statistic, the approximation is less than first order (it matches the first moment only asymptotically). The inaccuracy arises from ignoring the variability due to replacing $\boldsymbol{\Sigma}_i$ by $\widehat{\boldsymbol{\Sigma}}_i$.

It is also possible to create tests based on model comparisons with estimates based on REML estimates. Again, little is known about the resulting distributions.

A variety of alternative tests have been suggested based on assuming the distribution of the test statistic may be approximated by an F distribution (sometimes referred to as a Wald type test). Such tests can be based on ML or REML estimates. A key advantage lies in only needing to fit a single model (which ideally provides an essentially correct model). With estimates $\widehat{\boldsymbol{\theta}} = \boldsymbol{C}\widehat{\boldsymbol{\beta}}$ and

$$\widehat{\boldsymbol{\Sigma}}_s = \bigoplus_{i=1}^{N} \widehat{\boldsymbol{\Sigma}}_i \tag{5.20}$$

and $\boldsymbol{C}$ of dimension $a \times q$ and rank a,

$$\mathcal{V}(\widehat{\boldsymbol{\theta}}) = \mathcal{V}(\widehat{\boldsymbol{\theta}} - \boldsymbol{\theta}_0) = \boldsymbol{C}(\boldsymbol{X}_s'\widehat{\boldsymbol{\Sigma}}_s^{-1}\boldsymbol{X}_s)^{-1}\boldsymbol{C}' \tag{5.21}$$

and

$$F_m = (\widehat{\boldsymbol{\theta}} - \boldsymbol{\theta}_0)'[\boldsymbol{C}(\boldsymbol{X}_s'\widehat{\boldsymbol{\Sigma}}_s^{-1}\boldsymbol{X}_s)^{-1}\boldsymbol{C}']^{-1}(\widehat{\boldsymbol{\theta}} - \boldsymbol{\theta}_0)/a\,. \tag{5.22}$$

Among the widely available methods, the Kenward and Roger (1997) method appears to provide the most accurate test size. However, room for substantial improvement remains (Schaalje, McBride, and Fellingham, 2003). The Kenward Roger method starts with REML estimates and then creates an improved estimate of $\boldsymbol{\Sigma}_s$. Sample values are used to estimate a scale parameter λ and a degrees-of-freedom parameter ν with

$$F_m\widehat{\lambda} \mathrel{\dot\sim} F(a, \nu)\,. \tag{5.23}$$

5.6 TESTS OF COVARIANCE PARAMETERS, $\boldsymbol{\tau}$ (RANDOM EFFECTS)

As in the univariate and multivariate linear models, a data analyst may wish to test hypotheses about variance or covariance parameters. The likelihood ratio test provides a reasonable approach. A variety of approximate F approaches also have been proposed, based on extending consideration to

$$\begin{bmatrix} \boldsymbol{\theta}_\beta \\ \boldsymbol{\theta}_\tau \end{bmatrix} = [\boldsymbol{C}_\beta \quad \boldsymbol{C}_\tau]\begin{bmatrix} \boldsymbol{\beta} \\ \boldsymbol{\tau} \end{bmatrix}. \tag{5.24}$$

Such tests have not received much attention. Some work of a similar nature has been done in univariate and multivariate linear models.

EXERCISES

5.1 Clearly specify values for every dimension and parameter which reduce a $\text{LMM}_{N,p_i,q,m}[\boldsymbol{y}_i; \boldsymbol{X}_i\boldsymbol{\beta}, \boldsymbol{Z}_i\boldsymbol{\Sigma}_{di}(\boldsymbol{\tau}_d)\boldsymbol{Z}_i' + \boldsymbol{\Sigma}_{ei}(\boldsymbol{\tau}_e)]$ with Gaussian errors to a $\text{GLM}_{N,q}(y_i; \boldsymbol{X}_i\boldsymbol{\beta}, \sigma^2)$ with Gaussian errors. Some choices will not be unique.

5.2 For a $\text{GLM}_{N,p,q}(\boldsymbol{Y}_i; \boldsymbol{X}_i\boldsymbol{B}, \boldsymbol{\Sigma})$ with Gaussian errors, the model may be written as $\boldsymbol{Y} = \boldsymbol{X}\boldsymbol{B} + \boldsymbol{E}$. Assume $\boldsymbol{u}_s = \text{vec}(\boldsymbol{Y}')$, and $\boldsymbol{u}_i = [\text{row}_i(\boldsymbol{Y})]'$.
5.2.1 Clearly specify values for every dimension and parameter which specify a $\text{LMM}_{N,p_i,q,m}[\boldsymbol{u}_i; \boldsymbol{X}_{iu}\boldsymbol{\beta}_u, \boldsymbol{Z}_i\boldsymbol{\Sigma}_{di}(\boldsymbol{\tau}_d)\boldsymbol{Z}_i' + \boldsymbol{\Sigma}_{ei}(\boldsymbol{\tau}_e)]$ with Gaussian errors. Some choices will not be unique.
5.2.2 Clearly specify (and simplify the expressions when possible) all matrices (including dimensions and pattern of elements) in the stacked-data form $\boldsymbol{u}_s = \boldsymbol{X}_s\boldsymbol{\beta}_u + \boldsymbol{e}_u$ corresponding to the choices you made in 5.2.1.

5.3 Consider the magnetic resonance imaging (MRI) data used for the Chapter 3 exercises. Ignoring all data for parenchyma and fat gives a multivariate model with three columns for the three region-of-interest (ROI) measurements.

5.3.1 Briefly specify an appropriate multivariate model, $\text{GLM}_{N,p,q}(\boldsymbol{Y}_i; \boldsymbol{X}_i\boldsymbol{B}, \boldsymbol{\Sigma})$ with Gaussian errors. Use reference cell coding. Specify all dimensions for the particular data in hand, as well as brief definitions of parameter matrix elements.
5.3.2 Assume $\boldsymbol{Z}_i \equiv \mathbf{0}$ and an unstructured covariance pattern across time. Clearly specify values for every dimension and parameter which specify a $\text{LMM}_{N,p_i,q,m}[\boldsymbol{y}_i; \boldsymbol{X}_i\boldsymbol{\beta}, \boldsymbol{Z}_i\boldsymbol{\Sigma}_{di}(\boldsymbol{\tau}_d)\boldsymbol{Z}_i' + \boldsymbol{\Sigma}_{ei}(\boldsymbol{\tau}_e)]$ with Gaussian errors appropriate for the data which gives estimates of the same parameters as in the GLM() model.
5.3.3 Clearly specify (and simplify the expressions where possible) all matrices (including dimensions and pattern of elements) in the stacked data form (for *all* observations on *all* participants) corresponding to the choices you made in 5.3.2.

5.4 In the MRI analysis, the multivariate model applies and therefore should always be used in preference to the mixed model. However, to improve understanding of linear models, the data may be analyzed with a mixed model. For the following questions, we are only interested in analyzing the ROI tissue type (ignore all data for parenchyma and fat).
5.4.1 Briefly explain *why* a multivariate model is preferred to a mixed model here.
5.4.2 Use a "stacked" version of the data (in file P0105) to fit the mixed model from exercise 5.3 using SAS PROC MIXED. You will need to use the CLASS and REPEATED statements in PROC MIXED and specify a compound symmetric covariance matrix. Provide tests for the Time and Diagnosis main effects as well as the Time x Diagnosis interaction (malignant versus benign) which correspond to the test that a multivariate model would provide.
5.4.3 Starting with the P0104 file, use PROC GLM with the REPEATED statement to compute MULTIREP and UNIREP tests of Time, Diagnosis, and Time by Diagnosis.
5.4.4 Compare and discuss the degrees of freedom, F statistics, and p values from exercises 5.4.2 and 5.4.3.
5.4.5 Implement a new version of the PROC GLM analysis and use reference cell coding in lieu of the CLASS statement for coding benign versus malignant. What changes result?
5.4.6 (*Optional, noncredit*) Use LINMOD to compute MULTIREP and UNIREP tests of the same three hypotheses.

5.5 (*Optional, noncredit*) Considering both the ROI and Parenchyma tissue type, we will now consider a factorial analysis that includes Time (three levels), Diagnosis (two levels), and Tissue (two levels). Again, the multivariate model is appropriate and should be used, but conducting a mixed model analysis will illustrate many issues.
5.5.1 (*Optional, noncredit*) Starting with the P0105.sd2 file, create a file of stacked data which includes all covariates and the ROI and Parenchyma response variables. Using SAS PROC MIXED with the CLASS and REPEATED statements, provide tests for the Time, Diagnosis, and Tissue main effects as well as the three two-way and one three-way interaction.
5.5.2 (*Optional, noncredit*) Compare and discuss the degrees of freedom, F statistics, and p values from exercise 5.5.1, with exercise 3.4 from Chapter 3.

CHAPTER 6

Choosing the Form of a Linear Model for Analysis

6.1 THE IMPORTANCE OF UNDERSTANDING DEPENDENCE

The pattern of dependence among observations can have more effect on the validity and quality of a statistical analysis than any other feature. Therefore choosing an analysis must begin with determining the logical properties of the sampling scheme and thereby characterizing the patterns of independence and correlation among observations.

In the simplest case, as in the general linear univariate model, all response values, all observations, are statistically independent. Relatively simple and well-behaved methods are nearly always available for completely independent observations. In the most complicated case, all observations are correlated with each other in idiosyncratic ways. Few methods with desirable properties can be found for such general problems. The following definitions help characterize fundamental properties of sampling schemes central to the choice of an analysis.

Definition 6.1 (a) *Independent* observations have values which are statistically independent.
(b) An *independent sampling unit* (ISU) provides one or more observations such that observations from one unit are statistically independent from any other distinct unit while observations from the same unit may be correlated.
(c) The *observational unit* distinguishes one correlated observation from another within the ISU.
(d) Observing the same variable in two or more instances across time, space, or other dimension within an ISU creates *repeated measures*.
(e) *Commensurate* observations share the same measurement scale and units.
(f) *Multivariate* outcomes arise from a single ISU and therefore are not independent and need not be commensurate.
(g) *Doubly multivariate* outcomes include repeated measures of two or more noncommensurate variables.

Developers of multivariate statistics have progressed by considering a relatively small number of limited and systematic patterns of dependence among observations. A simple taxonomy useful for describing any type of data and pattern of dependence may be developed by considering three diagnostic questions. We consider each separately.

6.2 HOW MANY VARIABLES PER INDEPENDENT SAMPLING UNIT?

Table 6.1 summarizes some dimensions for describing models. In a clinical trial of a new pharmaceutical with random assignment to treatment, usually the individual participants in the trial are the independent sampling units. Each person may have more than one response measured, such as red blood cell count and blood cholesterol level, which leads to two observations per ISU. For the example, a single value of a blood assay represents the observation unit. As a second example, an educator who randomly assigns all children in a classroom to a particular curriculum must treat scores from children within a classroom as correlated.

Table 6.1 How Many Variables?

Number of Responses	Number of Predictors	Model Description
1	1	Univariate
1	Many	Multivariable
Many	1 or many	Multivariate
Many	Many	Multivariate
Repeated	1 or many	Repeated measures

In the latter setting, classroom becomes the ISU and child the observational unit. Recording performance on the same test for every child leads to repeated measures occurring within the classroom (ISU). Alternately, measuring only one child from each classroom once per month for three months also leads to repeated measures. All repeated measures have commensurate observations (all measured in the same units, on the same scale). However, not all sets of commensurate data are analyzed appropriately with repeated-measures models, which usually imply interest in contrasts corresponding to polynomial trends across a repeated-measure dimension. A simple example arises in measuring drug or toxicant levels in a variety of organs in the body of an animal. Although levels in the brain, muscle, and kidney would all be reported in the same units (the data are commensurate), trends across organs have no scientific appeal.

The multivariate profile of drug or toxicant levels for a variety of organs does pique scientific interest. Statisticians usually describe any setting with two or more response variables as multivariate. From the perspective of mathematical and statistical theory, repeated measures merely represent a special case of multivariate responses. Modeling and testing differences *between* independent sampling units often allow using simple univariate theory, while modeling and testing differences *within* independent sampling units usually require more complex multivariate theory, as do differences involving both within and between dimensions.

Some research involves collecting two or more distinct response variables on two or more occasions. Many longitudinal studies of human development and clinical trials of treatments for chronic diseases have such data, which may be described as doubly multivariate.

In summary, completely independent observations may be contrasted with many patterns of nonindependent observations. The presence of a nonzero correlation provides the simplest way to identify nonindependent observations, because any correlated variables are necessarily dependent (*not* independent). Although uncommon, examples of uncorrelated and nonindependent variables do exist. Hence a lack of correlation does not guarantee independence. However, in the special case of jointly Gaussian variables, the converse does hold: Uncorrelated and jointly Gaussian random variables are necessarily independent.

The emphasis on recognizing the many forms of nonindependence reflects the crucial importance that the pattern of dependence plays in determining the underlying distribution theory and the choices of parameter estimates and tests. Improper analysis can severely bias results for estimation and inference. Furthermore, in sharp contrast to most other assumptions in linear models, access to large samples will not always overcome the problem.

The term "large" is ambiguous in the presence of multivariate data. If a sample contains p observations on each of N independent sampling units, consideration must be given to increasing N alone (the most common meaning), p alone, N and p in a fixed ratio, or N and p in a varying ratio (with many variations).

Fortunately, the theory for multivariate and repeated-measures data coincides for the multivariate general linear model. Only the particulars of the scientific context and goals can indicate which analysis method to choose for a specific application. The practical use and interpretation of multivariate theory varies greatly across applications, in contrast to the theory itself.

6.3 WHAT TYPES OF VARIABLES PLAY A ROLE?

Definition 6.2 (a) *Nominal* scales only define categories or groups of observations.

(b) *Ordinal* scales provide numeric values sufficient only to rank observations.

(c) *Interval* scales provide numeric values with all differences of the same size being equivalent.

(d) *Ratio* scales give numeric values for which ratios of the same size are equivalent.

(e) *Continuous* data may include any sort of interval- and ratio-scale variables.

The type of measurement scale, especially for random components such as error terms, typically plays a major role in determining the choice and validity of a particular analysis method. Stevens (1946, 1951) distinguished among four levels of measurement. Values on a nominal (categorical) scale only label objects. As an example, the chemical name of a compound only distinguishes it from other compounds. Values on an ordinal scale rank objects (and name them) but carry no other information. A person's finishing position in a 100-meter dash provides only ordinal information. Values on an interval scale provide information about differences between objects (and rank them and name them), such as acidity of a solution measured on the pH scale or temperature in degrees centigrade. Values on a ratio scale provide the additional information of relative size, such as the mass of an object in kilograms.

The value of recognizing the scale of a response variable lies in the guidance it provides in choosing a data analysis. However, we agree with Velleman and Wilkinson's (1993) cautions about not being too rigid in using scale to choose a data analysis. Nominal or ordered categorical data typically lead to using categorical methods for data analysis. In turn, special "distribution-free" statistical methods have been developed for ordinal data (with few ties). Both interval and ratio data tend to be considered together (at some peril in guiding a choice of valid analysis) as continuous data. Typically we will classify such data as either Gaussian or not Gaussian.

Ratio scale data are necessarily nonnegative. In such scales, a value of zero indicates the absence of the property and negative values have no meaning. In practice, such data often are positively skewed and have variance increasing with the mean, especially for a sufficient wide range of conditions. Concentration of a pollutant in a river, concentration of a drug in a person's bloodstream, and volume of lava emitted by a volcano in a month might be expected to have such a variance pattern. The data often appear Gaussian after a logarithmic or similar power transformation, including square or cube root. Muller and Fetterman (Chapter 7, 2002) described the practical use of power (Box-Cox) type transformations.

6.4 WHAT REPEATED SAMPLING SCHEME WAS USED?

The distinction between the independent sampling unit (ISU) and the observational unit plays a key role in describing the sample scheme. Table 6.2 summarizes some repeated sampling schemes. The table defines rough categories, with blurry distinctions between neighbors. Although various terms in the table are used interchangeably in the literature, the definitions and distinctions made here reflect the spirit and practice in the biological and behavioral sciences. The column for "timing" might correspond to a wide variety of dimensions, other than time, such as distance from a town, location on the surface of the earth, or amount of a treatment (measured on an interval or ratio scale).

Table 6.2 Some Simple Repeated Sampling Schemes (i Indicates a Particular ISU)

Design	Number of Times	Number of ISUs	Typical Timing
Cross sectional	1	N	None
Repeated measures	$p > 1$	N	Consistent
Crossover	$p > 1$	N	Consistent
Longitudinal	$p_i > 1$	N	Inconsistent
Time series	N	1	Regular

Some terms in the table deserve clarification. Consistent timing simply requires all ISUs to be evaluated at the same times, such as Monday, Tuesday, and Thursday. In contrast, inconsistent timing allows one participant to appear on Monday, Tuesday, and Thursday, while another appears on Monday, Tuesday, and Friday. Inconsistent timing may arise due to an inability to fully control the timing of data collection. Such mistimed data often arise in human clinical trials. Regular spacing requires a constant distance between times, such as measuring air temperature at a weather station once per week for a set of N consecutive weeks. The study of univariate ANOVA models (for cross-sectional designs) led to the definition of certain terms that generalize to settings with repeated observations.

Definition 6.3 (a) A *complete* design has at least one observation per treatment combination (cell).
(b) *Balanced* designs have an equal number of observations in each cell.
(c) *Exchangeable* observations may be correlated but have identical distributions and identical relationships to other observations with which they may be exchanged.

Losing one or more independent sampling units and all associated observations usually creates unbalanced or incomplete designs. Both univariate and multivariate linear models can tolerate such deviations (in "between-subject" design) and retain excellent properties. In contrast, losing only a fraction of the observations for one or more ISUs greatly complicates the task of finding accurate estimates. Furthermore, creating accurate inferences (tests and confidence intervals) usually becomes extremely difficult in the presence of missing data, especially in small to moderate samples. The size of a sample has many possible variations in the presence of repeated measures. Most often, large-sample results refer to a setting with a fixed (or finitely bounded) number of repeated observations and an arbitrarily large number of independent sampling units.

Unfortunately, many data analysts refer to any sort of repeated sampling scheme as involving repeated measures. The failure to recognize special cases has often had the regrettable effect of more general and less accurate statistical methods being used when more accurate and easier to use methods were available.

An important special case of repeated-measures designs arises when the observations for each ISU are exchangeable; the order is arbitrary and the variable distinguishing observations within an ISU only labels them, thus providing only a nominal (categorical) scale. The term split-plot design reflects the origin of the term in agricultural statistics. A study aimed at finding the best amount of fertilizer to apply to corn might use such a design. A set of N fields from different farms represent the ISUs. With three levels of fertilizer of interest, the great variability between fields (plots of land) militates toward splitting each plot into three subplots and randomly assigning one of the three fertilizer levels to each. Local differences in such things as the quality of soil, rainfall, and cultivation equipment all contribute to between-plot variation but not within. Assuming equal variability and correlations (compound symmetry of the covariance matrix) of yields across subplots (the repeated measures) seems completely reasonable. Compound symmetry arises naturally with exchangeable observations, while nonexchangeable repeated measures rarely achieve compound symmetry.

Time as the repeated-measure dimension provides the most common example of nonexchangeable repeated observations that seem extremely unlikely to have compound symmetry. Concern about the assumption led to documentation of dramatic inflation of type I error rates under violation of the assumption of compound symmetry (Box, 1954a, b). Subsequently, statisticians developed tests robust to violation of the assumption, which implicitly allow for an arbitrary covariance structure (Geisser and Greenhouse, 1958; Greenhouse and Geisser, 1959; Huynh and Feldt, 1976) within the traditional split-plot or "univariate" approach to repeated measures. The "multivariate" test statistics, developed between roughly 1930 and 1970, avoid compound symmetry and begin with an assumption of unstructured covariance matrix.

Every analysis of repeated measures makes an implicit or explicit choice of model for the covariance pattern within a participant (ISU) across repeated measures. With Gaussian data, the validity of a choice among a univariate, multivariate, or mixed linear model depends almost entirely on the actual covariance pattern among observations. Covariance models discussed here are usually one of four types: complete independence, compound symmetry, partially structured, and completely unstructured. The four are ordered from simplest to most complex.

6.5 ANALYSIS STRATEGIES FOR MULTIVARIATE DATA

Data suitable for analysis with general linear multivariate models may be grouped into four classes, depending on scientific considerations: pure multivariate, commensurate multivariate, repeated measures (also commensurate), and doubly multivariate. The scientific goals drive the choice of analysis strategy for any particular set of data. We focus on the following possible strategies: (1) a collection of univariate analyses with a Bonferroni correction, (2) multivariate analysis of Variance (MANOVA), (3) a collection of multivariate analyses with a Bonferroni correction, (4) repeated measures with a "multivariate" approach (MULTIREP), (5) repeated measures with a "univariate" approach (UNIREP), and (6) repeated measures with a "mixed" model approach. (7) Special purpose methods include growth curves, seemingly unrelated regression (SUR), a doubly-multivariate model (DMM), and missing data methods. Given a particular analysis strategy, more than one approach to inference may apply. Table 6.3 summarizes a number of cases.

Table 6.3 Linear Model Analysis Strategies for Gaussian Repeated Measures

Pattern of Responses	Strategy	Nominal Size of Test
p Distinct	Bonferroni univariate	α/p
	MANOVA	α
	Bonferroni MANOVA c clusters	α/c
p Repeated	Bonferroni univariate	α/p
	MULTIREP	α
	Growth curve	α
	UNIREP	α
	Mixed	α
	Bonferroni for clusters	α/c
p_1 Distinct, repeated p_2	Bonferroni 4–7 for distinct clusters	α/p_2
	Bonferroni MANOVA time clusters	α/p_1
	DMM specific method	α

A collection of univariate analyses with the total test size controlled with a Bonferroni correction may have particular appeal for a modest number of response noncommensurate variables. Additionally, no interest in profiles of response (weighted linear combinations of responses) goes with the desire to consider each variable separately. The approach allows any pattern of missing data and any variety of design matrices across responses. The allocation of test size should reflect the relative scientific interest and importance of the variables.

A collection of multivariate, MULTIREP, UNIREP, or mixed model analyses with total test size controlled by a Bonferroni correction may have particular appeal in the doubly multivariate setting. Toxicologists often study a suite of responses measured repeatedly over time. A separate multivariate could be conducted at each time. Alternately, and most commonly, a separate repeated-measures analysis could be conducted for each type of response variable.

Alternately some methods have been especially developed for the setting. Timm (2002, Section 6.7) provided a good review.

Seemingly unrelated regression (Srivastava and Giles, 1987) and growth curve models (Kshirsagar and Smith, 1995) provide generalizations of the multivariate model. SUR seeks to take advantage of correlations among responses, while allowing for different design matrices. Growth curve models seek to capitalize on modifying a multivariate approach to repeated measures to take advantage of the simplest model that fits the repeated dimension. Timm's book (2002, Chapter 5) contains a detailed introduction to both.

Table 6.4 Linear Model Form Properties for Gaussian Repeated Measures

Approach	Small-N Inference?	Σ Robust?	Additional Plus	Additional Minus
Bonferroni univariate	Good	Yes	Flexible	No Profile, Trend
Bonferroni multivariate	Good	Yes	Flexible	
MANOVA	Good	Yes[1]	Profiles	$\boldsymbol{X}$ Same For All No Missing
MULTIREP	Good	Yes	Trends	$\boldsymbol{X}$ Same For All No Missing
Growth curve	Good	Yes	Trends	$\boldsymbol{X}$ Same For All No Missing
UNIREP	Good[1]	Yes[1]	Trends	$\boldsymbol{X}$ Same No Missing
Mixed	Can be bad	No	Time Vary $\boldsymbol{X}$ OK Missing OK Model Σ	Fragile Limited Inference Limited Diagnostics
EM and adjusted MANOVA, MULTIREP, or UNIREP	Good	Yes	Missing OK In Small N	$\boldsymbol{X}$ Same For All
SUR	Uncertain	Yes	$\boldsymbol{X}$ Varies	No Missing Inference?
DMM	Uncertain	Uncertain	Tailored	No Missing

[1] With appropriate test choice.

Table 6.4 summarizes performance characteristics of the various forms of linear models that might be used. Barton and Cramer (1989) and Catellier and Muller (2000) recommended using the EM algorithm for estimation and adjusted degree of free tests for MANOVA, MULTIREP, and UNIREP tests with missing data. In contrast to a mixed model, the approach always controls test size, even in very small samples. The Appendix (Section A.2) contains a description of free SAS/IML® code which implements the methods and where to retrieve it from the Web.

6.6 CAUTIONS AND RECOMMENDATIONS

The flexibility and generality of the mixed model make it extremely tempting to simply always use it for any linear model analysis. However, just as a skilled carpenter understands and uses many different saws, a skilled data analyst understands and uses many different kinds of linear models. Littel (2003) urged readers to recognize the limitations of the mixed model, especially in terms of accuracy of inference. Some limitations arise from computational difficulties. Numerical problems with currently popular software may badly mislead the user.

The algorithm may fail to converge to a solution, even though a valid and unique answer exists. Muller, Edwards, Simpson, and Taylor (2006) used popular mixed model software to analyze simulated Gaussian data. The observations followed a multivariate linear model with two within-subject factors, each with three levels, giving $p = 9$, and no between-subject factors. In all cases no missing or mistimed data were present and $N \in \{10, 20, 40\}$. Consequently, $\boldsymbol{Y}$ was always $N \times 9$ and $\boldsymbol{X} = \mathbf{1}_N$ (obviously full rank). Any standard multivariate linear model program can compute the unique maximum likelihood and REML estimates for the primary parameters ($\boldsymbol{B}$ and Σ), which are guaranteed to exist, in one step. For mixed model analysis, an unstructured covariance model was always requested, which ensured that the model was valid in the population. Using the default options, the program failed to converge for roughly 2% of the samples. A standard multivariate linear model program was applied to each problematic sample to verify that estimates could be computed. Merely increasing the number of iterations reduced the number of convergence failures, but not completely. Artful tuning of the convergence criteria eliminated some more of the convergence failures, but not all.

Faced with convergence failure, many data analysts would change the request for an unstructured covariance matrix to a request for a compound symmetric matrix. Although doing so might lead to convergence, the strategy will usually inflate test size in small samples. The uncorrected UNIREP test (which assumes compound symmetry) also inflates test size in the same setting but performs better than the mixed model tests in many ways when applicable.

We recommend the following steps to reduce or completely avoid such problems. (1) Use a MULTIREP or UNIREP test and associated model whenever they apply. Current mixed model tests are never better in controlling accuracy of inference. (2) Round the values of time and related predictor values to the smallest number of digits that are scientifically meaningful. In a clinical trial with visits once per month, recording time values in days may greatly destabilize the calculations with no scientific return. Obviously the choice of recording precision must be made jointly with the scientists leading the project. (3) Given the need for a mixed model analysis, begin the process by creating well-conditioned data. Such data have (a) careful scaling, (b) removal of any location differences, (c) full-rank coding schemes for indicator variables, especially effect coding or cell mean style coding, (d) centered or pseudocentered continuous predictors, and (e) design matrices transformed to make them as close to completely orthonormal as needed. A pseudocentered variable has had a scientifically meaningful and convenient value subtracted in order to make the mean approximately zero, such as $D = T - 37$, for T human body temperature in degrees centigrade. Orthogonal or orthonormal polynomial coding for time usually helps greatly when applicable. Both $\boldsymbol{X}_s$ and $\boldsymbol{Z}_s$, as well as $\boldsymbol{y}_s$, should receive the improvements. Chapters 8 and 9 in Muller and Fetterman (2002) contain further discussion in the context of a univariate linear model. (4) Take advantage of the options of the particular program in use to help the program find a solution. An artful choice of starting value estimates is likely to have the most impact. The choice of algorithm and convergence criterion also can greatly affect the ability to find the solution (if one exists). 5) Finally, do not declare the covariance model to be invalid if the limitations of the data disallow estimating it. An alternate analysis method may be the only defensible choice.

Example 6.1 Analyzing data from an observational study with a univariate multiple regression model illustrates the last recommendation. An epidemiologist seeking to build a model of lung function might choose to use smoking status (with three levels, current, previous, never), race (two levels), and gender (two levels) as basic predictors. Obviously interaction variables also have appeal. However, even in a relatively large sample, it might happen that only one white female who never smoked happens to be included. Including the three-way interaction of race by gender by smoking status creates a model with severe collinearity with the intercept and extremely unstable computations. The data do not allow estimating or testing the interaction, a fact revealed by careful attention to regression diagnostics. The data do *not* provide any evidence whatsoever either in favor of or against the existence of such an interaction in the population (no inference is available because no estimate is available). An epidemiologist would report the desire to consider the interaction and the fact that the current study provides no information in either direction. The inability to fit the model in the sample at hand would not be interpreted as evidence against the interaction. Rather it indicates an inconvenient limitation stemming from the finite nature of the observational study design.

Example 6.2 A second example involving a clinical trial comparing three asthma medications also illustrates the last recommendation. Repeated clinic visits and measurements of lung function are scheduled every month for six months. Baseline lung function (measured just before the start of treatment), treatment group, and the interaction of baseline with treatment provide the obvious starting model for between-subject effects. As usual for a repeated-measures study design for living organisms, a covariance pattern more complex than compound symmetry seems necessary. With 100 participants, recording visit time as number of days since baseline will likely create dozens of distinct times of observation. With 1000 participants, recording visit time as number of days since baseline will likely create over 100 distinct time values. In either case, even considering only relatively low order trends across time (linear, quadratic, and cubic), a mixed model analysis with an unstructured covariance pattern may fail to converge. Rounding time to week or half month will help some, as will careful use of orthogonal design coding.

In the example of the epidemiology study, difficulty came from an insufficient $\boldsymbol{X}$ matrix, while in the example of the clinical trial, difficulty came from an insufficient $\boldsymbol{Z}$ matrix. The inability to fit the model for the sample at hand should not be interpreted as evidence against an unstructured covariance pattern or in favor of a simpler model. Evidence for a simpler model might be available. A model assuming autocorrelation or a combination of autocorrelation and compound symmetry may converge *and* provide appropriate and well-behaved regression diagnostics. Without such positive results, some other options still remain viable. One simple approach would be to compute univariate analyses of trend scores and use a Bonferroni correction. Heterogeneity must be treated in a credible way with the approach. Alternately, separate univariate analyses for each time window (clinic visit number), again with a Bonferroni correction, may be preferred. Either approach has many unappealing features. We list them merely to illustrate that the desire to fit a mixed model, or any other model, does not automatically guarantee having enough data to support the model. Defensible inference from a set of data requires a scientifically credible model supported with diagnostic analysis, not merely a model that converges to a numerical solution.

6.7 REVIEW OF LINEAR MODEL NOTATION

Although originating much earlier, the theory and practice of linear models began to flower early in the 1900s. Early work used scalar notation, although proofs often included geometric representations and arguments. Increasing interest in more complex designs, repeated measures, and especially multivariate questions was coupled with a gradually increasing use of matrix notation. The advent and spread of electronic digital computers in the second half of the century accelerated the trend. By the end of the century, nearly all statisticians had access to powerful computers, and nearly all new linear model theory was cast in matrix notation.

Table 6.5 summarizes notation used for univariate, multivariate, and mixed linear models. The scalar equation provides a model statement for a single observation and is rarely used. Current discussion of linear mixed models most often uses the "vector" form for one independent sampling unit (highlighted in gray). In contrast, current discussions of univariate and multivariate models most often use the forms for all observations (highlighted in gray). The disparity between the usual formulations of mixed and other linear models can be a source of confusion. Comparisons between models within a column of the table prove simpler and greatly help understanding. We chose notation for the models in order to facilitate such comparisons.

As discussed in Chapter 4, the growth curve model, as commonly used, can be interpreted as a special case of a restricted multivariate linear model. Although the GCM() would fit naturally into an expanded Table 6.5, the generalized GLM, the GGLM(), would not. The lack of fit arises from the fact that a GGLM() may not have any independent observations.

Table 6.6 summarizes covariance structures for the linear models described in Table 6.5. Comparing the structures for all observations considered together highlights the underlying similarities. In all cases, a block diagonal form occurs, with each diagonal block corresponding to an ISU. Zero values off diagonal reflect the statistical independence. The univariate model has all scalar diagonal blocks, all equal to σ^2 (homogeneity of variance hold). The multivariate model has all diagonal blocks of the same size and value, Σ (homogeneity of covariance holds). The mixed model allows the diagonal blocks to vary in size and value. The covariance elements must be a function of a modest number of parameters (relative to the number of observations) in order to allow computing valid estimates.

Table 6.5 Linear Model Statements
($Np \Leftrightarrow n = \sum_{i=1}^{N} p_i$)

Model	One Observation	Observations for One ISU	All Observations
$\text{GLM}_{N,q}(y_i; \boldsymbol{X}_i\boldsymbol{\beta}, \sigma^2)$	$y_i = \sum_{j=1}^{q} x_{i,j}\beta_j + e_i$	$\underset{1\times 1}{y_i} = \underset{1\times q\times 1}{\boldsymbol{X}_i\boldsymbol{\beta}} + \underset{1\times 1}{e_i}$	$\underset{N\times 1}{\boldsymbol{y}} = \underset{N\times q\times 1}{\boldsymbol{X}\boldsymbol{\beta}} + \underset{N\times 1}{\boldsymbol{e}}$
$\text{GLM}_{N,p,q}(\boldsymbol{Y}_i; \boldsymbol{X}_i\boldsymbol{B}, \boldsymbol{\Sigma})$	$y_{i,k} = \sum_{j=1}^{q} x_{i,j}\beta_{j,k} + e_{i,k}$	$\underset{1\times p}{\boldsymbol{Y}_i} = \underset{1\times q\times p}{\boldsymbol{X}_i\boldsymbol{B}} + \underset{1\times p}{\boldsymbol{E}_i}$	$\underset{N\times p}{\boldsymbol{Y}} = \underset{N\times q\times p}{\boldsymbol{X}\boldsymbol{B}} + \underset{N\times p}{\boldsymbol{E}}$
$\text{LMM}_{N,p_i,q,m}[\boldsymbol{y}_i; \boldsymbol{X}_i\boldsymbol{\beta}, \boldsymbol{\Sigma}_i(\boldsymbol{\tau})]$	$y_{i,k} = \sum_{j=1}^{q} x_{i,j}\beta_{j,k} + \sum_{l=1}^{m} z_{i,l}d_l + e_{i,k}$	$\underset{p_i\times 1}{\boldsymbol{y}_i} = \underset{p_i\times q\times 1}{\boldsymbol{X}_i\boldsymbol{\beta}} + \underset{p_i\times m\times 1}{\boldsymbol{Z}_i\boldsymbol{d}_i} + \underset{p_i\times 1}{\boldsymbol{e}_i}$	$\underset{n\times 1}{\boldsymbol{y}_s} = \underset{n\times q\times 1}{\boldsymbol{X}_s\boldsymbol{\beta}} + \underset{n\times Nm\times 1}{\boldsymbol{Z}_s\boldsymbol{d}_s} + \underset{n\times 1}{\boldsymbol{e}_s}$

Table 6.6 Covariance Structures for Linear Models

Model	One Observation	Observations for One ISU	All Observations
$\text{GLM}_{N,q}(y_i;\boldsymbol{X}_i\boldsymbol{\beta},\sigma^2)$	$\mathcal{V}(y_i) = \sigma^2$	$\mathcal{V}(y_i) = \sigma^2$	$\mathcal{V}(\boldsymbol{y}) = \boldsymbol{I}_N\sigma^2 = \boldsymbol{I}_N \otimes \sigma^2$
$\text{GLM}_{N,p,q}(\boldsymbol{Y}_i;\boldsymbol{X}_i\boldsymbol{B},\boldsymbol{\Sigma})$	$\mathcal{V}(y_{i,k}) = \sigma_{kk}$	$\mathcal{V}(\boldsymbol{Y}_i') = \boldsymbol{\Sigma}$	$\mathcal{V}[\text{vec}(\boldsymbol{Y}')] = \boldsymbol{I}_N \otimes \boldsymbol{\Sigma}$
$\text{LMM}_{N,p_i,q,m}[\boldsymbol{y}_i;\boldsymbol{X}_i\boldsymbol{\beta},\boldsymbol{\Sigma}_i(\boldsymbol{\tau})]$	$\mathcal{V}(y_{i,k}) = \langle\boldsymbol{\Sigma}_i(\boldsymbol{\tau})\rangle_{kk}$	$\mathcal{V}(\boldsymbol{y}_i) = \boldsymbol{\Sigma}_i(\boldsymbol{\tau})$ $= \boldsymbol{Z}_i\boldsymbol{\Sigma}_{di}(\boldsymbol{\tau}_d)\boldsymbol{Z}_i' + \boldsymbol{\Sigma}_{ei}(\boldsymbol{\tau}_e)$	$\mathcal{V}(\boldsymbol{y}_s) = \boldsymbol{\Sigma}_s = \bigoplus_{i=1}^{N}\boldsymbol{\Sigma}_i$

CHAPTER 7

General Theory of Multivariate Distributions

7.1 MOTIVATION

Chapter 7 serves two purposes. First, we present a consistent notation and many basic results for groups of random variables, especially when arranged into vectors. Second, the presentation provides an implicit review of such concepts as moments and generating functions for single random variables. The reader may access the implicit review in a very simple way: Reduce all dimensions to make any matrix or vector a scalar (1×1). Applying the reduction to every result will immediately tell the reader which results are old friends that generalize conveniently and which are new acquaintances requiring extra time to get to know.

Although some results about random vectors apply directly to random matrices, many others do not. If matrix $\boldsymbol{X}$ has random elements, considering $\boldsymbol{y} = \text{vec}(\boldsymbol{X})$ often proves fruitful. For symmetric $\boldsymbol{Z}$, often the form $\boldsymbol{u} = \text{vech}(\boldsymbol{Z})$ is easier to work with. In particular, $\boldsymbol{u} = \text{vech}(\boldsymbol{Z})$ may have a density, while $\text{vec}(\boldsymbol{Z})$ definitely does not. Gupta and Nagar's (2000) book reflects an increasing interest in studying random matrices directly. Johnson and Kotz (1972), Johnson, Kotz, and Balakrishnan (1997), Kotz, Balakrishnan, and Johnson (2000), and Kotz and Nadarajah (2004) contain more traditional treatments of joint distributions.

In the theory of distributions, the adjective "multivariate" describes any collection of more than one random variable, including ones arrayed as vectors or matrices. Clarity requires distinguishing between vector and matrix forms. An example occurs in Chapter 8, which has separate treatments of the vector Gaussian (often called the multivariate Gaussian) and the matrix Gaussian distributions.

As a general principle for the entire book, we provide a carefully stated and technically correct presentation. We rule out most pathological cases by imposing mild regularity assumptions (such as finite variance, $\sigma^2 < \infty$). We do allow for pathologies that occur naturally in practice (such as $\widehat{\sigma}^2 = 0$). Taking advantage of the generality and convenience of characteristic functions involves limited consideration of complex variables. All other contexts involve only real values.

For the sake of brevity, we omit many proofs in the present chapter, especially complicated ones. Most readers will have seen scalar versions of the results in the present chapter. The vector and matrix generalizations are not primarily linear model properties. In all subsequent chapters we (1) provide a proof, (2) provide a reference to a proof, or (3) omit the proofs with the hope that a conscientious student will derive the result.

7.2 NOTATION AND CONCEPTS

The presentation here presumes knowledge of the basic theory of probability and inference. We focus here on multivariate properties of well-defined random variables that have a joint distribution.

Most discussions consider vectors of random variables (random vectors), such as $\boldsymbol{y} = [\, y_1 \; y_2 \; \cdots \; y_n \,]'$, have a joint distribution defined for all $\boldsymbol{y}_* \in S$ (the sample space). Less often, we consider matrices of random variables (random matrices), such as $\boldsymbol{Y} = \{y_{jk}\}$ with a joint distribution defined for all $\boldsymbol{Y}_* \in S$ (the sample space). Implicitly, any result for a random vector always also applies to vec($\boldsymbol{Y}$) or to vech($\boldsymbol{Y}$) if $\boldsymbol{Y} = \boldsymbol{Y}'$. We will review ways of characterizing (uniquely and completely specifying) the distribution of $\boldsymbol{y}$. We will also review some general properties of distributions, such as moments. Although many results apply to any random variable, continuous random variables will be discussed in detail. In contrast, corresponding forms needed for discrete variables will mostly be omitted.

A distribution is an entity which exists apart from any particular characterization. The cumulative distribution function (CDF) is not the distribution. A well-defined distribution always has a CDF and may or may not have a probability density function (PDF). For a given CDF $F_{\boldsymbol{y}}(\boldsymbol{y}_*)$ a corresponding characteristic function (CF) $\phi_{\boldsymbol{y}}(\boldsymbol{t})$ always exists, and is unique (a.e., that is, *almost everywhere*). Conversely, if $\phi_{\boldsymbol{y}}(\boldsymbol{t})$ is the CF of a distribution, a corresponding CDF $F_{\boldsymbol{y}}(\boldsymbol{y}_*)$ always exists and is unique (a.e.). Moreover, mathematical methods allow determining one from the other. Thus a distribution is always completely and uniquely specified (a.e.) by (1) its CDF and (2) its CF.

7.3 FAMILIES OF DISTRIBUTIONS

We often consider a collection of distributions with CDFs which differ, functionally, only in the values of one or more parameters, the elements of $\boldsymbol{\theta} \in \Theta$. In particular, the univariate Gaussian family of distributions is denoted $\{\mathcal{N}(\mu, \sigma^2) : \mu^2 < \infty, 0 \le \sigma^2 < \infty\}$, while the multivariate family is denoted $\{\mathcal{N}_p(\boldsymbol{\mu}, \boldsymbol{\Sigma}) : \boldsymbol{\mu} \in \Re^p, \boldsymbol{\mu}'\boldsymbol{\mu} < \infty, (p \times p)\; \boldsymbol{\Sigma} = \boldsymbol{\Upsilon}\mathrm{Dg}(\boldsymbol{\lambda})\boldsymbol{\Upsilon}', \lambda_j \ge 0\}$. More general families also exist. Members of the *exponential family* have PDFs which are not all of the same functional form but which can all be expressed in the general form:

$$f_{\boldsymbol{y}}(\boldsymbol{y}_*; \boldsymbol{\theta}) = a(\boldsymbol{\theta}) b(\boldsymbol{y}_*) \exp\left[\sum_{j=1}^{p} c_j(\boldsymbol{\theta}) t_j(\boldsymbol{y}_*)\right]. \tag{7.1}$$

Members of the *elliptically symmetric family* have PDFs depending on $\boldsymbol{y}_*$ only through a quadratic form:

$$f_{\boldsymbol{y}}(\boldsymbol{y}_*;\boldsymbol{\theta}) = h\left[(\boldsymbol{y}_* - \boldsymbol{\mu})'\boldsymbol{\Sigma}^{-1}(\boldsymbol{y}_* - \boldsymbol{\mu});\boldsymbol{\theta}\right]. \tag{7.2}$$

Members of the *spherically symmetric family* have PDFs depending on $\boldsymbol{y}$ only through an inner product:

$$f_{\boldsymbol{y}}(\boldsymbol{y}_*;\boldsymbol{\theta}) = h(\boldsymbol{y}_*'\boldsymbol{y}_*;\boldsymbol{\theta}) = h[(\boldsymbol{y}_* - \mathbf{0})'\boldsymbol{I}(\boldsymbol{y}_* - \mathbf{0});\boldsymbol{\theta}]. \tag{7.3}$$

7.4 CUMULATIVE DISTRIBUTION FUNCTION

Definition 7.1 Every distribution is characterized by a *joint cumulative distribution function* (CDF):

$$\begin{aligned} F_{\boldsymbol{y}}(\boldsymbol{y}_*;\boldsymbol{\theta}) &= \Pr\{\boldsymbol{y} \le \boldsymbol{y}_*|\boldsymbol{\theta}\} \\ &= \Pr\left\{\bigcap_{j=1}^{n}(y_j \le y_{j*})|\boldsymbol{\theta}\right\}. \end{aligned} \tag{7.4}$$

The simplified notation $F_{\boldsymbol{y}}(\boldsymbol{y}_*)$ is also used and the CDF is also known as the *distribution function*.

Definition 7.2 A random vector $\boldsymbol{y}$ will be described as *discrete* if and only if it has countable support (smallest S such that $\Pr\{\boldsymbol{y} \in S\} = 1$).

Definition 7.3 $F_{\boldsymbol{y}}(\boldsymbol{y}_*;\boldsymbol{\theta})$ and $\boldsymbol{y}$ are *continuous* if and only if $F_{\boldsymbol{y}}(\boldsymbol{y}_*;\boldsymbol{\theta})$ is a continuous function of $\boldsymbol{y}_*$.

7.5 PROBABILITY DENSITY FUNCTION

Definition 7.4 $F_{\boldsymbol{y}}(\boldsymbol{y}_*;\boldsymbol{\theta})$ and $\boldsymbol{y}$ are *absolutely continuous* (a.c.) if and only if a nonnegative function $f_{\boldsymbol{y}}(\boldsymbol{y}_*;\boldsymbol{\theta})$ exists such that the following n-fold integral over region $\mathrm{S}_{\boldsymbol{y}_0} = \{\boldsymbol{y}_* : \boldsymbol{y}_* \le \boldsymbol{y}_0\}$ exists $\forall \boldsymbol{y}_0 \in \Re^n$:

$$F_{\boldsymbol{y}}(\boldsymbol{y}_0;\boldsymbol{\theta}) \equiv \int_{\mathrm{S}_{\boldsymbol{y}_0}} f_{\boldsymbol{y}}(\boldsymbol{y}_*;\boldsymbol{\theta})\, d\boldsymbol{y}_*. \tag{7.5}$$

The $d\boldsymbol{y}_*$ means $dy_{*1}dy_{*2}\cdots dy_{*n}$ in the n-fold integral.

Definition 7.5 For continuous distributions, the partial derivative

$$f_y(\boldsymbol{y}_*;\boldsymbol{\theta}) = \frac{\partial^{(n)}}{\partial_{y_{*1}}\partial_{y_{*2}}\cdots\partial_{y_{*n}}} F_y(\boldsymbol{y}_*;\boldsymbol{\theta}) \tag{7.6}$$

is the *probability density function* (PDF), or *the density* of the distribution, if it exists $\forall \boldsymbol{y}_* \in S$.

Theorem 7.1 If the PDF $f_y(\boldsymbol{y}_*;\boldsymbol{\theta})$ exists then, with $S_{y_0} = \{\boldsymbol{y}_* : \boldsymbol{y}_* \leq \boldsymbol{y}_0\}$,

$$\begin{aligned} &1.\quad \int_{\Re^n} f_y(\boldsymbol{y}_*;\boldsymbol{\theta})\, d\boldsymbol{y}_* = 1 \\ &2.\quad f_y(\boldsymbol{y}_*;\boldsymbol{\theta}) \geq 0 \ \forall \boldsymbol{y}_* \in \Re^n \\ &3.\quad F_y(\boldsymbol{y}_0;\boldsymbol{\theta}) = \int_{S_{y_0}} f_y(\boldsymbol{y}_*;\boldsymbol{\theta})\, d\boldsymbol{y}_* . \end{aligned} \tag{7.7}$$

Any function satisfying 1 and 2 is a PDF. Condition 3 implies it is a PDF of $F_y(\boldsymbol{y}_*;\boldsymbol{\theta})$.

Theorem 7.2 If $f_y(\boldsymbol{y}_*;\boldsymbol{\theta})$ is a PDF, and function $g(\boldsymbol{y}_*;\boldsymbol{\theta}) = f_y(\boldsymbol{y}_*;\boldsymbol{\theta})$ except at a countable number of points, then $g(\boldsymbol{y}_*;\boldsymbol{\theta})$ is also a PDF of $F_y(\boldsymbol{y}_*;\boldsymbol{\theta})$. Thus we say $f_y(\boldsymbol{y}_*;\boldsymbol{\theta})$ is unique (a.e.).

The nonuniqueness must be considered in defining maximum likelihood estimation.

7.6 FORMULAS FOR PROBABILITIES AND MOMENTS

The formula for a probability or moment changes with the type of random variable. Discrete random variables allow writing the formula in terms of a (possibly infinite) summation, while absolutely continuous random variables use an integral. However, some random variables, such as censored survival time of a person in a clinical trial, are neither discrete nor continuous. Furthermore, some continuous distributions do not have a PDF (although the CDF always exists).

A straightforward generalization of the usual (Riemann) integral solves the problem for all random variables of interest in the present book. The Stieltjes (or Riemann-Stieltjes) integral allows defining probability and moment formulas for nearly any type of random variable (discrete, absolutely continuous, or "neither"). Handling the mathematically exotic random variables not covered by the Stieltjes integral requires measure theory and the methods of Lesbegue integration.

Definition 7.6 Following Weisstein (2003), a Stieltjes integral applies to real functions $f(x)$ and $h(x)$, both bounded on the closed interval $[a, b]$. For a partition $\{a = x_0 < x_1 < \cdots < x_{n-1} < x_n = b\}$ and $x_j < e_j < x_{j+1}$,

$$s = \sum_{j=0}^{n-1} f(e_j)[h(x_{j+1}) - h(x_j)] \tag{7.8}$$

is a Riemann sum. If $s \to s_0$, a fixed number, as $\max(x_{j+1} - x_j) \to 0$, then s_0 is the *Stieltjes integral*, written

$$s_0 = \int_a^b f(x)\, dh(x)\,. \tag{7.9}$$

Considering the complex plane extends the integral to complex variables.

Continuity of $f(x)$ and bounded variation of $h(x)$ over $[a, b]$ ensure the Stieltjes integral exists for $[a, b]$. It fails to exist if $f(x)$ and $h(x)$ are not continuous at a common point. If $h(x)$ has a continuous derivative, the Stieltjes integral reduces to the Riemann integral. Any finite or infinite sum can be expressed as a Stieltjes integral with an appropriate choice of $h(x)$.

7.7 CHARACTERISTIC FUNCTION

Definition 7.7 The *characteristic function* (CF) of the distribution of $\boldsymbol{y}$ is defined for $\mathrm{i} = \sqrt{-1}$ and $\forall \boldsymbol{t} \in \Re^n$ as

$$\begin{aligned}\phi_{\boldsymbol{y}}(\boldsymbol{t}) &= \mathrm{E}[\exp(\mathrm{i}\boldsymbol{t}'\boldsymbol{y})] = \mathrm{E}[\cos(\boldsymbol{t}'\boldsymbol{y})] + \mathrm{i}\mathrm{E}[\sin(\boldsymbol{t}'\boldsymbol{y})] \\ &= \int_{\Re^n} \exp(\mathrm{i}\boldsymbol{t}'\boldsymbol{y}_*)\, dF_{\boldsymbol{y}}(\boldsymbol{y}_*) \\ &= \int_{\Re^n} \exp(\mathrm{i}\boldsymbol{t}'\boldsymbol{y}_*) f_{\boldsymbol{y}}(\boldsymbol{y}_*)\, d\boldsymbol{y}_*\,,\end{aligned} \tag{7.10}$$

with the last line applying only if $f_{\boldsymbol{y}}(\boldsymbol{y}_*)$ exists. The first two always apply if interpreted in terms of complex-variable Stieltjes integration.

In general the integrals in the definition require contour integration methods (taught in a class on complex variables). The last integral is the Fourier transform of $f_{\boldsymbol{y}}(\boldsymbol{y}_*)$. Here $\mathrm{e}^{\mathrm{i}z} \equiv \cos(z) + \mathrm{i} \cdot \sin(z)$ is a complex number. It is represented in two dimensions by the coordinates of a point located on the unit circle, $[\cos(z), \sin(z)]$. Its magnitude is $|\mathrm{e}^{\mathrm{i}z}| = \sqrt{\cos^2(z) + \sin^2(z)} \leq 1$. The mapping of $[\cos(\boldsymbol{t}'\boldsymbol{y}), \sin(\boldsymbol{t}'\boldsymbol{y})]$ has a simple geometric interpretation (Epps, 1993). The mapping wraps the PDF around the unit circle, while $\phi_{\boldsymbol{y}}(\boldsymbol{t}) = \mathrm{E}()$ specifies the location of the center of mass within the circle.

The definition makes it obvious that derivation and manipulation of characteristic functions require complex-variable analysis. Despite that, characteristic functions can provide many powerful and practical results with knowledge of only the most basic results about the complex variables. Most importantly for the developments in later chapters, linear transformations of random variables induce simple changes in characteristic functions of random vectors and matrices. With Gaussian errors, the use of characteristic functions greatly simplifies the derivation of distributions of mean and covariance estimators. The approach has the important advantage of allowing less-than-full-rank coding schemes and population covariance matrices with little extra work.

Theorem 7.3 (a) For any random vector $\boldsymbol{y}$ the characteristic function $\phi_{\boldsymbol{y}}(\boldsymbol{t})$ always exists finitely; specifically, $|\phi_{\boldsymbol{y}}(\boldsymbol{t})| \leq 1\ \forall \boldsymbol{t} \in \Re^n$.
(b) Any multivariate characteristic function $\phi_{\boldsymbol{y}}(\boldsymbol{t})$ is a uniformly continuous function. Equivalently, $\lim_{\boldsymbol{h}\to\boldsymbol{0}} |\phi_{\boldsymbol{y}}(\boldsymbol{t}+\boldsymbol{h}) - \phi_{\boldsymbol{y}}(\boldsymbol{t})| = 0$.
(c) If $\boldsymbol{y}$ $(n \times 1)$ is distributed with CDF $F_{\boldsymbol{y}}(\boldsymbol{y}_*)$, characteristic function $\phi_{\boldsymbol{y}}(\boldsymbol{t})$, and $F_{\boldsymbol{t}'\boldsymbol{y}}(s)$ specifies a distribution symmetric about zero $\forall \boldsymbol{t} \in \Re^n$, then the multivariate characteristic function is real valued and $\phi_{\boldsymbol{y}}(\boldsymbol{t}) \in [-1, 1]$.

Proof of (a). If $\boldsymbol{t} = \boldsymbol{0}$, then $\phi_{\boldsymbol{y}}(\boldsymbol{0}) = \mathrm{E}(\mathrm{e}^{\mathrm{i}\cdot 0}) = \mathrm{E}(1) = 1$. If $\boldsymbol{t} \neq \boldsymbol{0}$, we consider a *fixed* value of $\boldsymbol{t}$. Since $\boldsymbol{t}$ is fixed, $a \in \Re$ exists such that $\phi_{\boldsymbol{y}}(\boldsymbol{t}) = |\phi_{\boldsymbol{y}}(\boldsymbol{t})|\mathrm{e}^{\mathrm{i}a}$ (the "polar form"). Constant a is an unknown function of $\boldsymbol{t}$. For fixed $\{\boldsymbol{t}, a\}$, $\exists \boldsymbol{c}$ such that $a = \boldsymbol{t}'\boldsymbol{c}$. The choice $\boldsymbol{c} = \boldsymbol{t}(a/\boldsymbol{t}'\boldsymbol{t})$ will do. It follows that $\phi_{\boldsymbol{y}}(\boldsymbol{t}) = |\phi_{\boldsymbol{y}}(\boldsymbol{t})|\mathrm{e}^{\mathrm{i}\boldsymbol{t}'\boldsymbol{c}}$ and $|\phi_{\boldsymbol{y}}(\boldsymbol{t})| = \phi_{\boldsymbol{y}}(\boldsymbol{t})\mathrm{e}^{-\mathrm{i}\boldsymbol{t}'\boldsymbol{c}}$. Thus $|\phi_{\boldsymbol{y}}(\boldsymbol{t})| = \mathrm{E}[\exp(\mathrm{i}\boldsymbol{t}'\boldsymbol{y} - \mathrm{i}\boldsymbol{t}'\boldsymbol{c})] = \phi_{(\boldsymbol{y}-\boldsymbol{c})}(\boldsymbol{t})$. Since $|\phi_{\boldsymbol{y}}(\boldsymbol{t})| \in \Re$ by definition, $\phi_{(\boldsymbol{y}-\boldsymbol{c})}(\boldsymbol{t}) = \mathrm{E}[\cos[\boldsymbol{t}'(\boldsymbol{y}-\boldsymbol{c})] + \mathrm{i}\cdot 0$. Necessarily $|\phi_{\boldsymbol{y}}(\boldsymbol{t})| \in [-1, 1]$ since $\cos() \in [-1, 1]$.

Proof of (b) is left as an exercise.

Proof of (c). It is left as an exercise to prove the result holds for $n = 1$. Hence it holds for univariate characteristic functions in the set $\{\phi_{y_j}(s) : s \in \Re, \forall y_j\}$, and for all characteristic functions in the set $\{\phi_{\boldsymbol{t}'\boldsymbol{y}}(s) : \boldsymbol{t} \in \Re^n, s \in \Re\}$. Necessarily $\phi_{\boldsymbol{t}'\boldsymbol{y}}(s) \in [-1, 1]\ \forall \boldsymbol{t} \in \Re^n$, $s \in \Re$, including $s = 1$. Also, $\phi_{\boldsymbol{y}}(\boldsymbol{t}) \in [-1, 1]\ \forall \boldsymbol{t} \in \Re^n$ since $\phi_{\boldsymbol{t}'\boldsymbol{y}}(1) = \phi_{\boldsymbol{y}}(\boldsymbol{t})\ \forall \boldsymbol{t} \in \Re^n$. □

Many properties of characteristic functions of multivariate distributions can be proven via the following lemma and theorems. In the following discussions, sometimes it will be helpful to have $\phi(s; \boldsymbol{t})$ denote the characteristic function of the univariate random variable $\boldsymbol{t}'\boldsymbol{y}$ and write $\phi(s; \boldsymbol{t}) = \phi_{\boldsymbol{t}'\boldsymbol{y}}(s) = \mathrm{E}\{\exp[\mathrm{i}s(\boldsymbol{t}'\boldsymbol{y})]\}$. The notation emphasizes that s is the argument of function ϕ while $\boldsymbol{t}$ is interpreted as a fixed parameter.

Lemma 7.1 If $n \times 1$ $\boldsymbol{y}$ has a multivariate distribution and $\boldsymbol{t}$ is a nonrandom $n \times 1$ vector $\in \Re^n$, then $z = \boldsymbol{t}'\boldsymbol{y} = \sum_{j=1}^{n} t_j y_j$ is a scalar random variable with a univariate distribution.

Theorem 7.4 Knowing $\{\phi_{\boldsymbol{t}'\boldsymbol{y}}(s) : \boldsymbol{t} \in \Re^n, s = 1\}$ determines $\phi_{\boldsymbol{y}}(\boldsymbol{t})$ (Cramér-Wold).

Proof. By definition, the characteristic function of random variable $\boldsymbol{t}'\boldsymbol{y}$ is $\phi_{\boldsymbol{t}'\boldsymbol{y}}(s) = \mathrm{E}\{\exp[\mathrm{i}s(\boldsymbol{t}'\boldsymbol{y})]\}$ $\forall s \in \Re$ and the characteristic function of random vector $\boldsymbol{y}$ is $\phi_{\boldsymbol{y}}(\boldsymbol{t}) = \mathrm{E}[\exp(\mathrm{i}\boldsymbol{t}'\boldsymbol{y})]$ $\forall \boldsymbol{t} \in \Re^n$. By evaluating the univariate characteristic function at $s = 1$ we can determine the value of the multivariate characteristic function $\forall \boldsymbol{t} \in \Re^n$. In fact, $\phi_{\boldsymbol{t}'\boldsymbol{y}}(1) = \phi_{\boldsymbol{y}}(\boldsymbol{t})$ $\forall \boldsymbol{t} \in \Re^n$. □

Theorem 7.5 Knowing $\phi_{\boldsymbol{y}}(\boldsymbol{t})$ determines $\left\{\phi_{\boldsymbol{t}_0'\boldsymbol{y}}(s) : \boldsymbol{t}_0 \in \Re^n, s \in \Re\right\}$.

Proof. Here $\boldsymbol{t}_0$ is any fixed, arbitrary vector in $\Re^n$. By definition the characteristic function of random variable $\boldsymbol{t}_0'\boldsymbol{y}$ is $\phi_{\boldsymbol{t}_0'\boldsymbol{y}}(s) = \mathrm{E}\{\exp[\mathrm{i}s(\boldsymbol{t}_0'\boldsymbol{y})]\}$ $\forall s \in \Re$ and the characteristic function of random vector $\boldsymbol{y}$ is $\phi_{\boldsymbol{y}}(\boldsymbol{t}) = \mathrm{E}[\exp(\mathrm{i}\boldsymbol{t}'\boldsymbol{y})]$ $\forall \boldsymbol{t} \in \Re^n$. By evaluating the multivariate characteristic function at $\boldsymbol{t} = s\boldsymbol{t}_0$ we can determine the value of the univariate characteristic function $\forall s \in \Re$. Merely evaluate $\phi_{\boldsymbol{y}}(s\boldsymbol{t}_0) = \phi_{\boldsymbol{t}_0'\boldsymbol{y}}(s)$ $\forall \boldsymbol{t}_0 \in \Re^n$. □

Theorem 7.6 If the distribution of $\boldsymbol{y}$ $(n \times 1)$ is absolutely continuous, then the PDF is determined by the characteristic function as

$$f_{\boldsymbol{y}}(\boldsymbol{y}_*) = (2\pi)^{-n} \int_{\Re^n} \exp(-\mathrm{i}\boldsymbol{t}'\boldsymbol{y}_*)\phi_{\boldsymbol{y}}(\boldsymbol{t})\, d\boldsymbol{t}. \tag{7.11}$$

Proof. The proof is left as an exercise.

Theorem 7.7 If x is distributed with CDF $F_x(x_*)$ and characteristic function $\phi_x(t)$ while a and b are two points of continuity of $F_x(x_*)$, then $\Pr\{a \leq x \leq b\} \equiv F_x(b) - F_x(a)$ can be expressed as

$$\Pr\{a \leq x \leq b\} = \lim_{T \to \infty} \frac{1}{2\pi} \int_{-T}^{T} \frac{\exp(-\mathrm{i}ta) - \exp(-\mathrm{i}tb)}{-\mathrm{i}t} \phi_x(t)\, dt. \tag{7.12}$$

The result implies $\phi_x(s)$ completely determines $F_x(x_*)$.

Proof. The proof is left as an exercise.

The form in the last theorem does not directly provide the CDF if $\Pr\{x < 0\} > 0$ and $\Pr\{x > 0\} > 0$. Imhof (1961) recommended the following result for such cases.

Theorem 7.8 If random variable x has CDF $F_x(x_*)$ and characteristic function $\phi_x(t)$, while $\Im(z)$ indicates the imaginary part of z, then for x_0 a point of continuity of $F_x(x_0)$,

$$F_x(x_0) = \lim_{T\to\infty} \frac{1}{\pi} \int_0^T \frac{\Im[\exp(-itx_0)\phi_x(t)]}{t}\, dt\,. \tag{7.13}$$

Proof. Gil-Pelaez (1951) included a proof.

Theorem 7.9 If $\boldsymbol{y}$ $(n \times 1)$ is distributed with CDF $F_{\boldsymbol{y}}(\boldsymbol{y}_*)$, characteristic function $\phi_{\boldsymbol{y}}(\boldsymbol{t})$, and $\int_{\Re^n} |\phi_{\boldsymbol{y}}(\boldsymbol{t})|\, d\boldsymbol{t} < \infty$, then $F_{\boldsymbol{y}}(\boldsymbol{y}_*)$ is absolutely continuous and a PDF $f_{\boldsymbol{y}}(\boldsymbol{y}_*)$ exists.

Proof. The proof is left as an exercise.

Theorem 7.10 If random vector $\boldsymbol{x}$ has characteristic function $\phi_{\boldsymbol{x}}(\boldsymbol{t})$ and $\boldsymbol{y} = \boldsymbol{A}\boldsymbol{x} + \boldsymbol{b}$, for conforming constants $\boldsymbol{A}$ and $\boldsymbol{b}$, then

$$\phi_{\boldsymbol{y}}(\boldsymbol{s}) = \exp(\mathrm{i}\boldsymbol{s}'\boldsymbol{b})\phi_{\boldsymbol{x}}(\boldsymbol{A}'\boldsymbol{s})\,. \tag{7.14}$$

Proof. $\phi_{\boldsymbol{y}}(\boldsymbol{s}) = \mathrm{E}\{\exp[\mathrm{i}\boldsymbol{s}'(\boldsymbol{A}\boldsymbol{x}+\boldsymbol{b})]\} = \exp(\mathrm{i}\boldsymbol{s}'\boldsymbol{b})\mathrm{E}[\exp(\mathrm{i}\boldsymbol{s}'\boldsymbol{A}\boldsymbol{x})] =$ $\mathrm{e}(\mathrm{i}\boldsymbol{s}'\boldsymbol{b})\mathrm{E}\{\exp[\mathrm{i}(\boldsymbol{A}'\boldsymbol{s})'\boldsymbol{x}]\} = \exp(\mathrm{i}\boldsymbol{s}'\boldsymbol{b})\phi_{\boldsymbol{x}}(\boldsymbol{A}'\boldsymbol{s})$. □

The notation $x \perp\!\!\!\perp y$ indicates the random variables x and y are statistically independent.

Theorem 7.11 If $\boldsymbol{x}_1 \perp\!\!\!\perp \boldsymbol{x}_2$ and $\boldsymbol{y} = \boldsymbol{x}_1 + \boldsymbol{x}_2$, then $\phi_{\boldsymbol{y}}(\boldsymbol{t}) = \phi_{\boldsymbol{x}_1}(\boldsymbol{t})\phi_{\boldsymbol{x}_2}(\boldsymbol{t})$.

Proof. $\phi_{\boldsymbol{y}}(\boldsymbol{t}) = \mathrm{E}\left[\mathrm{e}^{\mathrm{i}\boldsymbol{t}'(\boldsymbol{x}_1+\boldsymbol{x}_2)}\right] = \mathrm{E}\left(\mathrm{e}^{\mathrm{i}\boldsymbol{t}'\boldsymbol{x}_1}\mathrm{e}^{\mathrm{i}\boldsymbol{t}'\boldsymbol{x}_2}\right) = \mathrm{E}\left(\mathrm{e}^{\mathrm{i}\boldsymbol{t}'\boldsymbol{x}_1}\right)\mathrm{E}\left(\mathrm{e}^{\mathrm{i}\boldsymbol{t}'\boldsymbol{x}_2}\right)$. □

Theorem 7.12 If $\mathrm{E}(|y|^m) < \infty$, then the derivative of order m of $\phi_y(t)$ exists $\forall t$ and is a uniformly continuous function with $\phi_y^{(m)}(t)\Big|_{t=0} = \mathrm{i}^m\mathrm{E}(y^m)$. A converse is true for even but not odd m.

Proof. Feller (1968) provided a proof.

We have given only a brief introduction to characteristic functions. Kendall and Stuart (1977) provided a detailed presentation, including proofs of many of the results not proven here. Lukacs (1983) reviewed subsequent developments. Epps (1993) gave an excellent tutorial on the interpretation and value of characteristic functions based on geometric intuition (in the complex plane).

7.8 MOMENT GENERATING FUNCTION

Definition 7.8 If it exists for all real $t_j \in \{-t_{*j}, t_{*j}\}$ with $t_{*j} > 0$, the *moment generating function* (MGF) of the distribution of random vector $(n \times 1)$ $\boldsymbol{y}$ is

$$m_{\boldsymbol{y}}(\boldsymbol{t}) = \mathrm{E}[\exp(\boldsymbol{t}'\boldsymbol{y})]\,. \tag{7.15}$$

Unlike the characteristic function, the MGF does *not* exist for every distribution. However, when the MGF does exist, the characteristic function is the MGF with $i\boldsymbol{t}$ replacing $\boldsymbol{t}$, namely $\phi_{\boldsymbol{y}}(\boldsymbol{t}) = m_{\boldsymbol{y}}(i\boldsymbol{t})$. A proof requires consideration of the concept of analytic continuation.

Theorem 7.13 If the random vector $\boldsymbol{y}$ has mean $\boldsymbol{\mu}$, dispersion $\boldsymbol{\Sigma}$, and MGF $m_{\boldsymbol{y}}(\boldsymbol{t}) = \mathrm{E}[\exp(\boldsymbol{t}'\boldsymbol{y})]$, then

$$\begin{aligned} \left.\frac{\partial m_{\boldsymbol{y}}(\boldsymbol{t})}{\partial \boldsymbol{t}}\right|_{\boldsymbol{t}=\boldsymbol{0}} &= \mathrm{E}(\boldsymbol{y}) \\ &= \boldsymbol{\mu} \end{aligned} \tag{7.16}$$

and

$$\begin{aligned} \left.\frac{\partial^2 m_{\boldsymbol{y}}(\boldsymbol{t})}{\partial \boldsymbol{t}\,\partial \boldsymbol{t}'}\right|_{\boldsymbol{t}=\boldsymbol{0}} &= \mathrm{E}(\boldsymbol{y}\boldsymbol{y}') \\ &= \boldsymbol{\Sigma} + \boldsymbol{\mu}\boldsymbol{\mu}'\,. \end{aligned} \tag{7.17}$$

When they exist, $\mathrm{E}(\boldsymbol{y}\boldsymbol{y}') = \mathcal{V}(\boldsymbol{y}) + \mathrm{E}(\boldsymbol{y})\mathrm{E}(\boldsymbol{y}')$ and $\partial e^{\mu(\boldsymbol{t})}/\partial \boldsymbol{t} = e^{\mu(\boldsymbol{t})}\,\partial \mu(\boldsymbol{t})/\partial \boldsymbol{t}$.

Proof. The first derivative is an $n \times 1$ vector,

$$\begin{aligned} \frac{\partial m_{\boldsymbol{y}}(\boldsymbol{t})}{\partial \boldsymbol{t}} &= \frac{\partial}{\partial \boldsymbol{t}}\mathrm{E}[\exp(\boldsymbol{t}'\boldsymbol{y})] \\ &= \mathrm{E}\left[\frac{\partial}{\partial \boldsymbol{t}}\exp(\boldsymbol{t}'\boldsymbol{y})\right] = \mathrm{E}[\exp(\boldsymbol{t}'\boldsymbol{y})\boldsymbol{y}]\,. \end{aligned} \tag{7.18}$$

In turn, evaluating $\partial m_{\boldsymbol{y}}(\boldsymbol{t})/\partial \boldsymbol{t}$ at $\boldsymbol{t} = \boldsymbol{0}$ gives $\mathrm{E}(1 \cdot \boldsymbol{y}) = \mathrm{E}(\boldsymbol{y})$.

The second derivative is an $n \times n$ matrix,

$$\begin{aligned} \frac{\partial^2 m_{\boldsymbol{y}}(\boldsymbol{t})}{\partial \boldsymbol{t}\,\partial \boldsymbol{t}'} = \frac{\partial}{\partial \boldsymbol{t}}\left[\frac{\partial m_{\boldsymbol{y}}(\boldsymbol{t})}{\partial \boldsymbol{t}'}\right]' &= \frac{\partial}{\partial \boldsymbol{t}}\,\mathrm{E}[\exp(\boldsymbol{t}'\boldsymbol{y})\boldsymbol{y}'] \\ &= \mathrm{E}\left[\frac{\partial}{\partial \boldsymbol{t}}\,\exp[(\boldsymbol{t}'\boldsymbol{y})\boldsymbol{y}']\right] \\ &= \mathrm{E}\left[\exp(\boldsymbol{t}'\boldsymbol{y})\frac{\partial}{\partial \boldsymbol{t}}\boldsymbol{y}' + \frac{\partial}{\partial \boldsymbol{t}}\exp(\boldsymbol{t}'\boldsymbol{y})\boldsymbol{y}'\right] \\ &= \mathrm{E}[\exp(\boldsymbol{t}'\boldsymbol{y}) \cdot \boldsymbol{0} + \exp(\boldsymbol{t}'\boldsymbol{y})\boldsymbol{y}\boldsymbol{y}']\,, \end{aligned} \tag{7.19}$$

with the matrix $\mathbf{0}$ being $n \times n$. In turn,

$$\left.\frac{\partial^2 m_{\boldsymbol{y}}(\boldsymbol{t})}{\partial \boldsymbol{t}\, \partial \boldsymbol{t}'}\right|_{\boldsymbol{t}=\mathbf{0}} = \mathrm{E}(\mathbf{0} + 1 \cdot \boldsymbol{y}\boldsymbol{y}') = \mathrm{E}(\boldsymbol{y}\boldsymbol{y}') \equiv \mathcal{V}(\boldsymbol{y}) + \mathrm{E}(\boldsymbol{y})\mathrm{E}(\boldsymbol{y}') . \tag{7.20}$$

□

Theorem 7.14 If $\boldsymbol{x}$ is a random vector with MGF $m_{\boldsymbol{x}}(\boldsymbol{t})$ and $\boldsymbol{y} = \boldsymbol{A}\boldsymbol{x} + \boldsymbol{b}$ for conforming constants $\boldsymbol{A}$ and $\boldsymbol{b}$, then

$$m_{\boldsymbol{y}}(\boldsymbol{s}) = \exp(\boldsymbol{s}'\boldsymbol{b}) m_{\boldsymbol{x}}(\boldsymbol{A}'\boldsymbol{s}) . \tag{7.21}$$

Proof. $m_{\boldsymbol{y}}(\boldsymbol{s}) = \mathrm{E}\{\exp[\boldsymbol{s}'(\boldsymbol{A}\boldsymbol{x}+\boldsymbol{b})]\} = \exp(\boldsymbol{s}'\boldsymbol{b})\mathrm{E}[\exp(\boldsymbol{s}'\boldsymbol{A}\boldsymbol{x})] = \exp(\boldsymbol{s}'\boldsymbol{b})\mathrm{E}\big[\exp(\boldsymbol{A}'\boldsymbol{s})'\boldsymbol{x}\big] = \exp^{\boldsymbol{s}'\boldsymbol{b}} m_{\boldsymbol{x}}(\boldsymbol{A}'\boldsymbol{s})$. □

7.9 CUMULANT GENERATING FUNCTION

Definition 7.9 (a) The notation $m_{\boldsymbol{x}}(\boldsymbol{t})$ and $\phi_{\boldsymbol{x}}(\boldsymbol{t})$ indicates the MGF (if it exists) and characteristic function, respectively, of the random vector $\boldsymbol{x}$. If $m_{\boldsymbol{x}}(\boldsymbol{t})$ exists, then

$$c_{\boldsymbol{x}}(\boldsymbol{t}) = \log[m_{\boldsymbol{x}}(\boldsymbol{t})] \tag{7.22}$$

or

$$c_{\boldsymbol{x}}(\boldsymbol{t}) = \log[\phi_{\boldsymbol{x}}(\boldsymbol{t})] \tag{7.23}$$

may be defined to be the *cumulant generating function* (CGF) $c_x(t)$ of the distribution (depending on the author).

(b) If a scalar random variable x has a power series expansion for $\log[m_x(t)]$, then

$$\log[m_x(t)] = \sum_{m=0}^{\infty} \kappa_m \frac{t^m}{m!} . \tag{7.24}$$

The coefficients $\{\kappa_m\}$ are the *cumulants*.

Except for the first cumulant, which equals the mean, all cumulants of the random variable x are also the cumulants of the random variable $x + a$ for constant a (hence the term *semi-invariants*). After taking derivatives, cumulant m equals the derivative of order m of $\log[m(t)]$ evaluated at $t = 0$,

$$\kappa_m = \left.\frac{\partial^{(m)} \log[m_x(t)]}{\partial t^{(m)}}\right|_{t=0} . \tag{7.25}$$

If a distribution is determined by its moments, then it is also determined by its cumulants. Not all distributions are completely determined by their moments.

Moments (and therefore moment and cumulant generating functions) fully characterize one class of random variables, namely all with finite variation. If $\Pr\{-\infty < a \leq x \leq b < \infty\} = 1$ then, $0 \leq \mathrm{E}(|x|^m) < \infty$. For some distributions, cumulants and the CGF are easier to manipulate than moments and the MGF.

A useful application of cumulants occurs in the derivation of moments of certain distributions. A moment about zero is defined as $\mu'_m = \mathrm{E}(x^m)$, while a central moment is $\mu_m = \mathrm{E}\{[x - \mathrm{E}(x)]^m\}$. Although only positive integer values of m are considered for cumulants, moments may exist (or may not, depending on the random variable) for any real m. When the moments and cumulants of order m exist, κ_m may be written as a polynomial of degree m in $\{\mu'_1, \ldots, \mu'_m\}$ or $\{\mu_1, \ldots, \mu_m\}$; alternately, μ_m or μ'_m may be written as a polynomial of degree m in $\{\kappa_1, \ldots, \kappa_m\}$ (Kendall and Stuart, 1977, vol. 1, Section 3.1.4). Given the first four cumulants, the first four moments can easily be found because

$$\kappa_1 = \mu'_1 = \mathrm{E}\big[(x-0)^1\big] = \mu = \mathrm{E}(x) \tag{7.26}$$

$$\kappa_2 = \mu_2 = \mathrm{E}\big[(x-\mu)^2\big] = \sigma^2 = \mathcal{V}(x) \tag{7.27}$$

$$\kappa_3 = \mu_3 = \mathrm{E}\big[(x-\mu)^3\big] \tag{7.28}$$

$$\kappa_4 = \mu_4 - 3\mu^2 = \mathrm{E}\big[(x-\mu)^4\big] - 3\mu^2 . \tag{7.29}$$

Kendall and Stuart (1977), Johnson, Kotz, and Balakrishnan (1994), and Harvey (1972) presented further details.

Theorem 7.15 If random vector $\boldsymbol{y}_j$ $(n \times 1)$ has CGF $c_{\boldsymbol{y}_j}(\boldsymbol{t}_j)$ and m finite cumulants for (finite) $j \in \{1, 2, \ldots, J\}$, with $\{\boldsymbol{y}_j\}$ mutually independent (Section 7.12 contains a precise definition of "mutual independence for vectors") and $\boldsymbol{s} = \sum_{j=1}^{J} \boldsymbol{y}_j$, then

$$c_{\boldsymbol{s}}(\boldsymbol{t}) = \sum_{j=1}^{J} c_{\boldsymbol{y}_j}(\boldsymbol{t}) . \tag{7.30}$$

If $n = 1$, then

$$\kappa_m(\boldsymbol{s}) = \sum_{j=1}^{J} \kappa_m(y_j) . \tag{7.31}$$

The theorem provides some extremely practical and convenient formulas. When the expressions make sense, the CGF of a sum is the sum of the component CGFs. Also, a cumulant of a sum is the sum of the corresponding cumulants. The forms have value in both analytic and numerical work.

7.10 TRANSFORMING RANDOM VARIABLES

Theorem 7.16 If $\boldsymbol{y}$ $(n \times 1)$ is distributed with PDF $f_{\boldsymbol{y}}(\boldsymbol{y}_*)$ and $\boldsymbol{t}(\boldsymbol{y})$ is a vector-valued function defining $(n \times 1)$ $\boldsymbol{z} = \boldsymbol{t}(\boldsymbol{y})$, with PDF $f_{\boldsymbol{z}}(\boldsymbol{z}_*)$ such that the mapping of points in the support of $\boldsymbol{y}$ onto points in the support of $\boldsymbol{z}$ is one-to-one (a.e.), then $\boldsymbol{z} = \boldsymbol{t}(\boldsymbol{y})$ and $\boldsymbol{y} = \boldsymbol{t}^{-1}(\boldsymbol{z})$ exists. Furthermore, with notation as in Searle (1982),

$$\begin{aligned}\mathbf{J}_{\boldsymbol{y}_*\to\boldsymbol{z}_*} &= \left\{\frac{\partial y_{*j}}{\partial z_{*k}}\right\} = \{\mathrm{j}_{jk}\} \\ &= \frac{\partial \boldsymbol{y}_*}{\partial \boldsymbol{z}_*'} = \left(\frac{\partial \boldsymbol{y}_*'}{\partial \boldsymbol{z}_*}\right)', \end{aligned} \tag{7.32}$$

the $(n \times n)$ *Jacobian* matrix for the transformation from $\boldsymbol{y}_*$ to $\boldsymbol{z}_*$, exists. Indicating the absolute value of the determinant of the Jacobian as

$$J_{\boldsymbol{y}_*\to\boldsymbol{z}_*} = ||\mathbf{J}_{\boldsymbol{y}_*\to\boldsymbol{z}_*}|| = \frac{1}{||\mathbf{J}_{\boldsymbol{z}_*\to\boldsymbol{y}_*}||} \tag{7.33}$$

allows writing

$$f_{\boldsymbol{z}}(\boldsymbol{z}_*) = f_{\boldsymbol{y}}\left[\boldsymbol{t}^{-1}(\boldsymbol{z}_*)\right] J_{\boldsymbol{y}_*\to\boldsymbol{z}_*}\,. \tag{7.34}$$

Many authors (including Schott, 2005) define $\mathbf{J}_{\boldsymbol{z}_*\to\boldsymbol{y}_*}$ as the Jacobian. The reciprocal relationship for respective determinants ensures that either definition leads to the same density for the transformed variables.

Definition 7.10 (a) If $\boldsymbol{z}$ $(n \times 1)$ is a random vector and $\boldsymbol{A}$ $(n \times n)$ is a finite constant matrix, then $\boldsymbol{z} = \boldsymbol{A}\boldsymbol{y}$ is a *linear transformation*.
(b) If constant (and finite) $\boldsymbol{b} \neq \boldsymbol{0}$ is $n \times 1$ then $\boldsymbol{z} = \boldsymbol{y} + \boldsymbol{b}$ is a *translation* (shift in origin). Defining $\boldsymbol{x} = \boldsymbol{A}\boldsymbol{y} + \boldsymbol{b}$ indicates a linear transformation with a translation.
(c) A linear transformation is *full rank* (or *nonsingular*) when $\boldsymbol{A}$ is full rank (nonsingular). In such cases $\boldsymbol{A}^{-1}$ exists uniquely and the transformation is one to one and invertible. The transformation is *less than full rank* (or *singular*) when $\boldsymbol{A}$ is less than full rank (singular). In such cases $\boldsymbol{A}^{-1}$ does not exist and the transformation is neither one to one nor invertible.

The definitions reflect precise, mathematical descriptions of transformations. More loosely, it is often convenient to describe a linear transformation with translation as simply a linear transformation. Describing a $\mathrm{GLM}_{N,q}(y_i; \boldsymbol{X}_i\boldsymbol{\beta}, \sigma^2)$ or $\mathrm{GLM}_{N,p,q}(\boldsymbol{Y}_i; \boldsymbol{X}_i\boldsymbol{\beta}, \boldsymbol{\Sigma})$ as a "linear" model because the unknown parameters enter the model equation linearly is consistent with the looser usage.

Lemma 7.2 If $z_* = Ay_* + b$ and A^{-1} $(n \times n)$ exists, then $y_* = A^{-1}(z_* - b)$,

$$J_{y_* \to z_*} = ||\mathbf{J}_{y_* \to z_*}|| = ||\partial y_* / \partial z_*'|| = ||A^{-1}||, \tag{7.35}$$

and

$$J_{z_* \to y_*} = ||\mathbf{J}_{z_* \to y_*}|| = ||\partial z_* / \partial y_*'|| = ||A||. \tag{7.36}$$

Here $||M||$ indicates the absolute value of the determinant of M.

The following theorem illustrates the use of the transformation theorem for a linear transformation. The result is used frequently.

Theorem 7.17 (a) If y is distributed with PDF $f_y(y_*)$, $n \times n$ constant A is full rank and finite, and $n \times 1$ constant b is finite, then the linear transformation $z = Ay + b \equiv t(y)$ is one to one and $y = A^{-1}(z - b) \equiv t^{-1}(z)$.
(b) The Jacobian of the transformation from y to z is $J_{y_* \to z_*} = ||\mathbf{J}_{y_* \to z_*}||$ with $\mathbf{J}_{y_* \to z_*} = \partial y / \partial z' = A^{-1}$.
(c) Also,

$$\int_{\Re^n} f_y(y_*)\, dy_* = \int_{\Re^n} f_y\left[t^{-1}(z_*)\right] J_{y_* \to z_*}\, dz_*. \tag{7.37}$$

(d) Furthermore $z = Ay + b$ is distributed with PDF

$$f_z(z_*) = f_y\left[A^{-1}(z_* - b)\right]||A^{-1}||. \tag{7.38}$$

Here $||A^{-1}||$ indicates the absolute value of the determinant of A^{-1}.

Proof. Left as an exercise.

Theorem 7.18 (The Cramér-Wold Theorem) The distribution of a random vector y $(p \times 1)$, is completely determined by the one-dimensional distributions of all possible linear combinations of the form $t'y$, in which t is a nonstochastic vector. The result does not assume Gaussian distributions!

Proof. Since a distribution is completely determined by its characteristic function, the Cramér-Wold Theorem can be stated $\{\phi_{t'y}(s) : t \in \Re^p\} \Rightarrow \phi_y(t)$. □

The characteristic function of the univariate random variable $t'y$ is $\phi_{t'y}(s) = \mathrm{E}[\exp(ist'y)]$ $\forall s \in \Re$, and the characteristic function of random vector y is $\phi_y(t) = \mathrm{E}[\exp(it'y)]$ $\forall t \in \Re^p$. If $\phi_y(t)$ were unknown, could we determine it from the known set $\{\phi_{t'y}(s) : t \in \Re^p\}$? Yes, evaluating $\phi_{t'y}(s)$ at $s = 1$ gives $\phi_{t'y}(1) = \mathrm{E}[\exp(it'y)] = \phi_y(t)$.

Lemma 7.3 If P is a permutation matrix, then $P^{-1} = P'$. For a given $y \in \Re^n$, $z = Py$ implies $y = P'z$. Also the Jacobian of the transformation is 1.

Theorem 7.19 If $\boldsymbol{y}$ is distributed with CDF $F_{\boldsymbol{y}}(\boldsymbol{y}_*)$, then permutation $\boldsymbol{z} = \boldsymbol{P}\boldsymbol{y}$ is distributed with CDF $F_{\boldsymbol{z}}(\boldsymbol{z}_*) = F_{\boldsymbol{y}}(\boldsymbol{P}\boldsymbol{z}_*)$.

Proof. Here

$$\begin{aligned} F_{\boldsymbol{y}}(\boldsymbol{P}\boldsymbol{z}_0) = \Pr\{\boldsymbol{y} \le \boldsymbol{P}\boldsymbol{z}_0\} &= \Pr\{\boldsymbol{y}_* \le \boldsymbol{y}_0\} \\ &= \Pr\{y_{*1} \le y_{01}, \ldots, y_{*n} \le y_{0n}\} \\ &= \Pr\{z_{*1} \le z_{01}, \ldots, z_{*n} \le z_{0n}\} \\ &= F_{\boldsymbol{z}}(\boldsymbol{z}_0)\,. \end{aligned} \tag{7.39}$$

□

Definition 7.11 The *Box-Cox power transformation* is given by

$$y^{(\lambda)} = \begin{cases} (y^\lambda - 1)/\lambda & \lambda \neq 0 \\ \log(y) & \lambda = 0\,. \end{cases} \tag{7.40}$$

The "standardized" version of $y^{(\lambda)}$ is

$$z_i^{(\lambda)} = \frac{y_i^{(\lambda)}}{\left(\prod_{i=1}^n y_i\right)^{1/n}}\,. \tag{7.41}$$

Box and Cox (1964, 1984) proposed maximum likelihood estimates of λ and $\boldsymbol{\beta}$ based on the model $y_i^{(\lambda)} \sim \mathcal{N}(\boldsymbol{X}_i\boldsymbol{\beta}, \sigma^2)$. Others have studied the properties of the procedure (Carroll and Rupert, 1981; Hinkley and Runger, 1984).

Definition 7.12 The *Bickel-Doksum transformation* ($\lambda > 0$) is given by

$$y^{(\lambda)} = \frac{\text{sign}(y)|y|^\lambda - 1}{\lambda}\,. \tag{7.42}$$

The Bickel-Doksum variation of the Box-Cox transformation can cope with negative y values.

7.11 MARGINAL DISTRIBUTIONS

Definition 7.13 If an ordered set of n random variables has a joint distribution, then any subset of m of them has a joint distribution which is known by any one of the following names: the *marginal*, the *joint marginal*, the *marginal distribution*, or the *joint marginal distribution*.

Permutation matrix properties allow assuming the m variables of interest are the first m variables in the ordered set, without loss of generality. The marginal distribution of the m variables is uniquely (a.e.) characterized by either its CDF or its characteristic function.

Theorem 7.20 If $\boldsymbol{y}$ $(n \times 1)$ has a distribution with CDF $F_{\boldsymbol{y}}(\boldsymbol{y}_*;\boldsymbol{\theta})$ and is partitioned as

$$\boldsymbol{y} = \begin{bmatrix} \boldsymbol{y}_1 \\ \boldsymbol{y}_2 \end{bmatrix} \begin{matrix} m \times 1 \\ (n-m) \times 1 \end{matrix} \tag{7.43}$$

then the marginal joint distribution of the elements in $\boldsymbol{y}_1$ is the distribution with CDF

$$F\boldsymbol{y}_1(\boldsymbol{y}_{*1};\boldsymbol{\theta}) = \lim_{\boldsymbol{y}_{*2}\to\infty} F_{\boldsymbol{y}}(\boldsymbol{y}_{*1},\boldsymbol{y}_{*2};\boldsymbol{\theta})\,. \tag{7.44}$$

Proof. $F_{\boldsymbol{y}}(\boldsymbol{y}_{*1},\boldsymbol{y}_{*2} = \infty;\boldsymbol{\theta}) = \Pr\{\bigcap_{j=1}^{m}(y_j \le y_{j0}), \bigcap_{j=m+1}^{n}(y_j < \infty)|\boldsymbol{\theta}\} = F_{\boldsymbol{y}_1}(\boldsymbol{y}_{*1};\boldsymbol{\theta})\,.$ □

Definition 7.14 For a continuous marginal distribution with CDF $F_1(\boldsymbol{y}_{*1};\boldsymbol{\theta})$, the partial derivative

$$f_{\boldsymbol{y}_1}(\boldsymbol{y}_{*1};\boldsymbol{\theta}) = \frac{\partial^m}{\partial_{y_{*1}}\partial_{y_{*2}}\cdots\partial_{y_{*m}}} F_{\boldsymbol{y}_1}(\boldsymbol{y}_{*1};\boldsymbol{\theta}) \tag{7.45}$$

is *the* PDF of the marginal distribution if it exists $\forall \boldsymbol{y}_{*1} \in S$.

Theorem 7.21 If $f_{\boldsymbol{y}_1}(\boldsymbol{y}_{*1};\boldsymbol{\theta})$ exists (as just defined), then it satisfies, for $\mathrm{S}_{\boldsymbol{y}_0} = \{\boldsymbol{y}_{*1} : \boldsymbol{y}_{*1} \le \boldsymbol{y}_0\}$,

$$f_{\boldsymbol{y}_1}(\boldsymbol{y}_{*1};\boldsymbol{\theta}) = \int_{\Re^{n-m}} f_{\boldsymbol{y}}(\boldsymbol{y}_*;\boldsymbol{\theta})\,d\boldsymbol{y}_{*2} \tag{7.46}$$

$$\int_{\Re^m} f_{\boldsymbol{y}_1}(\boldsymbol{y}_{*1};\boldsymbol{\theta})d\boldsymbol{y}_{*1} = 1 \tag{7.47}$$

$$f_{\boldsymbol{y}_1}(\boldsymbol{y}_{*1};\boldsymbol{\theta}) \ge 0 \;\forall \boldsymbol{y}_{*1} \in \Re^m \tag{7.48}$$

$$F_{\boldsymbol{y}_1}(\boldsymbol{y}_0) = \int_{\mathrm{S}_{\boldsymbol{y}_0}} f_{\boldsymbol{y}_1}(\boldsymbol{y}_{*1};\boldsymbol{\theta})\,d\boldsymbol{y}_{*1}\,. \tag{7.49}$$

Proof. Left as an exercise.

Theorem 7.22 The marginal characteristic function of $\boldsymbol{y}_1$ may be obtained by evaluating the characteristic function of $\boldsymbol{y}' = [\boldsymbol{y}_1' \;\; \boldsymbol{y}_2']$ at $\boldsymbol{t}_2 = \mathbf{0}$,

$$\phi_{\boldsymbol{y}_1}(\boldsymbol{t}_1) = \phi_{\boldsymbol{y}}(\boldsymbol{t})\Big|_{\boldsymbol{t}_2=\mathbf{0}}\,. \tag{7.50}$$

Proof. $\phi_{\boldsymbol{y}_1}(\boldsymbol{t}_1) = \mathrm{E}[\exp(\mathrm{i}\boldsymbol{t}_1'\boldsymbol{y}_1)] = \mathrm{E}[\exp(\mathrm{i}\boldsymbol{t}_1'\boldsymbol{y}_1 + \mathrm{i}\mathbf{0}'\boldsymbol{y}_2)] \equiv \phi_{\boldsymbol{y}}([\boldsymbol{t}_1' \; \mathbf{0}']')$. □

7.12 INDEPENDENCE OF RANDOM VECTORS

Definition 7.15 Any pair of random vectors $\boldsymbol{y}_1$ $(n_1 \times 1)$ and $\boldsymbol{y}_2$ $(n_2 \times 1)$ are *statistically independent* if and only if $\Pr\{\boldsymbol{y}_1 \in S_1, \boldsymbol{y}_2 \in S_2\} = \Pr\{\boldsymbol{y}_1 \in S_1\}\Pr\{\boldsymbol{y}_2 \in S_2\}$ $\forall$ Borel-measurable sets S_1 and S_2.

A Borel set is "a measurable set that can be obtained from closed sets and open sets on the real line by applying the operations of union and intersection repeatedly to countable collections of sets." (Daintith and Nelson, 1989).

Definition 7.16 The members of a set of n random vectors $(n < \infty)$ $\{\boldsymbol{y}_j\ (n_j \times 1), j \in \{1, 2, \ldots, n\}\}$ are *pairwise independent* if and only if any pair of random vectors in the set are *statistically independent.*

The last two definitions do *not* imply the elements of $\boldsymbol{y}_1$ (or the elements within any other $\boldsymbol{y}_j$) are independent. An example illustrates the point. The $\{\boldsymbol{Y}_i'\}$ for $\text{GLM}_{N,p,q}(\boldsymbol{Y}_i; \boldsymbol{X}_i\boldsymbol{B}, \Sigma)$ are pairwise independent. However, within a particular $\boldsymbol{Y}_i'$ the elements are not independent because they have nondiagonal covariance $\boldsymbol{\Sigma}$.

Definition 7.17 The members of a set of n random vectors $(n < \infty)$ $\{\boldsymbol{y}_j, n_j \times 1, j \in \{1, 2, \ldots, n\}\}$ are *mutually independent* if and only if

$$\Pr\left\{\bigcap_{j=1}^{n}(\boldsymbol{y}_j \in S_j)\right\} = \prod_{j=1}^{n}\Pr\{\boldsymbol{y}_j \in S_j\} \tag{7.51}$$

for all Borel-measurable sets $\{S_j\}$. The vectors may also be described as having as *total independence* or just *independence*.

Theorem 7.23 If $n \times 1$ $\boldsymbol{y}' = [\boldsymbol{y}_1'\ \boldsymbol{y}_2']$, with $\boldsymbol{y}_1$ $m \times 1$ and $\boldsymbol{y}_2$ $(n-m) \times 1$, has CDF $F_{\boldsymbol{y}}([\boldsymbol{y}_{*1}'\ \boldsymbol{y}_{*2}']'; \boldsymbol{\theta})$ and characteristic function $\phi_{\boldsymbol{y}}([\boldsymbol{t}_1'\ \boldsymbol{t}_2']'; \boldsymbol{\theta})$ defined $\forall \boldsymbol{y} \in \Re^n, \boldsymbol{t} \in \Re^n$, then $\boldsymbol{y}_1$ and $\boldsymbol{y}_2$ are *statistically independent* $\Leftrightarrow$

$$F_{\boldsymbol{y}}\left(\begin{bmatrix}\boldsymbol{y}_{*1}\\ \boldsymbol{y}_{*2}\end{bmatrix}; \boldsymbol{\theta}\right) = F_{\boldsymbol{y}}\left(\begin{bmatrix}\boldsymbol{y}_{*1}\\ \infty\end{bmatrix}; \boldsymbol{\theta}\right)F_{\boldsymbol{y}}\left(\begin{bmatrix}\infty\\ \boldsymbol{y}_{*2}\end{bmatrix}; \boldsymbol{\theta}\right) \tag{7.52}$$

$\forall \boldsymbol{y} \in \Re^n$. Independence requires the CDF of the joint distribution of $\boldsymbol{y}$ to equal the product of the CDFs of the marginal distributions of $\boldsymbol{y}_1$ and $\boldsymbol{y}_2$. Furthermore, $\boldsymbol{y}_1$ and $\boldsymbol{y}_2$ are *statistically independent* $\Leftrightarrow$

$$\phi_{\boldsymbol{y}}\left(\begin{bmatrix}\boldsymbol{t}_1\\ \boldsymbol{t}_2\end{bmatrix}; \boldsymbol{\theta}\right) = \phi_{\boldsymbol{y}}\left(\begin{bmatrix}\boldsymbol{t}_1\\ \boldsymbol{0}\end{bmatrix}; \boldsymbol{\theta}\right)\phi_{\boldsymbol{y}}\left(\begin{bmatrix}\boldsymbol{0}\\ \boldsymbol{t}_2\end{bmatrix}; \boldsymbol{\theta}\right) \tag{7.53}$$

$\forall \boldsymbol{t} \in \Re^n$. Independence requires the characteristic function of the joint distribution of $\boldsymbol{y}$ to equal the product of the characteristic functions of the marginal distributions of $\boldsymbol{y}_1$ and $\boldsymbol{y}_2$.

Proof. Left as an exercise (Cramér, 1946, p. 266, may be consulted).

Corollary 7.23.1 Similar statements of *mutual independence* can be formulated for a finite number of n subvectors $\{y_1, y_2, \ldots, y_n\}$.

Proof. Left as an exercise.

Corollary 7.23.2 If the characteristic functions are replaced by MGFs, then the second part does *not* hold true for *some* distributions, even though the theorem (as stated in terms of characteristic functions) holds true for *all* distributions.

Proof. Left as an exercise. (*Hint*: Find a distribution.)

Corollary 7.23.3 Any pair of random vectors y_1 $(n_1 \times 1)$, y_2 $(n_2 \times 1)$ are *independent* if and only if $\forall t_1 \in \Re^{n_1}$ and $\forall t_2 \in \Re^{n_2}$ it follows that $t_1'y_1$ is independent of $t_2'y_2$. Symbolically, $y_1 \perp\!\!\!\perp y_2 \Leftrightarrow t_1'y_1 \perp\!\!\!\perp t_2'y_2$ $\forall t_1, \forall t_2$.

Proof. Left as an exercise.

Corollary 7.23.4 The members of a set of n random vectors $(n < \infty)$ $\{y_j\,(n_j \times 1), j \in \{1, 2, \ldots, n\}\}$ are *pairwise independent* if and only if $t_j'y_j$ is independent of $t_k'y_k$ $\forall t_j \in \Re^{n_1}$, $t_k \in \Re^{n_2}$. Symbolically, $\{y_j, j \in \{1, 2, \ldots, n\}\}$ pairwise $\perp\!\!\!\perp$ $\Leftrightarrow$ $\{t_j'y_j \perp\!\!\!\perp t_k'y_k \;\forall j \neq k, t_j, t_k\}$.

Corollary 7.23.5 The members of a set of n random vectors $(n < \infty)$ $\{y_j, n_j \times 1, j \in \{1, 2, \ldots, n\}\}$ are *mutually independent* or just *independent* if and only if the members of a set of n random variables, $\{t_j'y_j : j \in \{1, 2, \ldots, n\}\}$ are pairwise independent for all choices of $t_j \in \Re^{n_j}$.

Theorem 7.24 If $n \times 1$ $y' = [y_1'\, y_2']$, with y_1 $m \times 1$ and y_2 $(n - m) \times 1$, has PDF $f_y(y_*; \theta)$ and the marginal distribution of y_j has PDF $f_j(y_{*j}; \theta)$, $j \in \{1, 2\}$, then y_1 and y_2 are *independent* if and only if $f_y(y_*; \theta) = f_1(y_{1*1}; \theta) f_2(y_{*2}; \theta)$ $\forall y \,\varepsilon\, \Re^n$.

Proof. Left as an exercise.

7.13 CONDITIONAL DISTRIBUTIONS

Definition 7.18 If $n \times 1$ $y' = [y_1'\, y_2']$, with y_1 $m \times 1$ and y_2 $(n - m) \times 1$, has a joint distribution with CDF $F_y(y_*; \theta)$, the marginal of y_1 has CDF $F_{y_1}(y_{*1}; \theta)$, $S \subset \Re^m$ is a Borel-measurable set, and a function $F_{2|1}(y_{*2}|y_{01}; \theta_c)$ exists such that

$$\Pr\{\boldsymbol{y}_1 \in S, \boldsymbol{y}_2 \leq \boldsymbol{y}_{02}\} = \int_S F_{2|1}(\boldsymbol{y}_{*2}|\boldsymbol{y}_{*1};\boldsymbol{\theta}_c)\, dF_{\boldsymbol{y}_1}(\boldsymbol{y}_{*1};\boldsymbol{\theta}) \tag{7.54}$$

$\forall S \subset \Re^p$ and $\boldsymbol{y}_{02} \in \Re^{n-1}$, then $F_{2|1}(\boldsymbol{y}_{*2}|\boldsymbol{y}_{*1};\boldsymbol{\theta}_c)$ is called the *CDF of the conditional distribution* of the random vector $\boldsymbol{y}_2|\boldsymbol{y}_1 = \boldsymbol{y}_{01}$. Furthermore any random vector with CDF $F_{2|1}(\boldsymbol{y}_{*2}|\ \boldsymbol{y}_{01};\boldsymbol{\theta}_c)$ may be called $\boldsymbol{y}_2|\boldsymbol{y}_1 = \boldsymbol{y}_{01}$.

In practice, $\boldsymbol{y}_2|\boldsymbol{y}_1 = \boldsymbol{y}_{01}$ and $\boldsymbol{y}_2|\boldsymbol{y}_1$ are often used interchangeably, although they do not represent precisely the same random vector.

Theorem 7.25 If the joint distribution of $n \times 1$ $\boldsymbol{y}' = [\boldsymbol{y}_1'\ \boldsymbol{y}_2']$, with $\boldsymbol{y}_1$ $m \times 1$ and $\boldsymbol{y}_2$ $(n-m) \times 1$, is absolutely continuous with PDF $f_{\boldsymbol{y}}(\boldsymbol{y}_*;\boldsymbol{\theta})$ and the marginal distribution of $\boldsymbol{y}_1$ has PDF $f_{\boldsymbol{y}_1}(\boldsymbol{y}_{*1};\boldsymbol{\theta})$ with support $S_1 \subseteq \Re^m$ (i.e., $\Pr\{\boldsymbol{y}_1 \in S_1\} = 1$), then
(a) the CDF of the conditional distribution of $\boldsymbol{y}_2|\boldsymbol{y}_1 = \boldsymbol{y}_{01}$ is, with $S_{\boldsymbol{y}_{02}} = \{\boldsymbol{y}_{*2} : \boldsymbol{y}_{*2} \leq \boldsymbol{y}_{02}\}$,

$$F_{2|1}(\boldsymbol{y}_{02}|\boldsymbol{y}_1;\boldsymbol{\theta}_c) = \int_{S_{\boldsymbol{y}_{02}}} \frac{f_{\boldsymbol{y}}(\boldsymbol{y}_*;\boldsymbol{\theta})}{f_{\boldsymbol{y}_1}(\boldsymbol{y}_{*1};\boldsymbol{\theta})}\, d\boldsymbol{y}_{*2} \tag{7.55}$$

and **(b)** the density of the conditional distribution is, $\forall \boldsymbol{y}_{*2} \in S_1$,

$$f_{2|1}(\boldsymbol{y}_{*2}|\boldsymbol{y}_{*1};\boldsymbol{\theta}_c) = \frac{f_{\boldsymbol{y}}(\boldsymbol{y}_*;\boldsymbol{\theta})}{f_{\boldsymbol{y}_1}(\boldsymbol{y}_{*1};\boldsymbol{\theta})}. \tag{7.56}$$

In many cases conditional distributions do not exist. A conditional probability exists only if the probability in the denominator is greater than zero.

7.14 (JOINT) MOMENTS OF MULTIVARIATE DISTRIBUTIONS

Definition 7.19 (a) The *expected value* of random vector $\boldsymbol{y}$ is a vector-valued function of the joint CDF $F_{\boldsymbol{y}}(\boldsymbol{y}_*;\boldsymbol{\theta})$ defined as

$$\mathrm{E}(\boldsymbol{y}) = \begin{bmatrix} \mathrm{E}(y_1) \\ \vdots \\ \mathrm{E}(y_n) \end{bmatrix} \tag{7.57}$$

with

$$\mathrm{E}(y_j) = \begin{cases} \displaystyle\int_{\Re^n} y_{*j}\, dF_{\boldsymbol{y}}(\boldsymbol{y}_*;\boldsymbol{\theta}) & \text{(in general)} \\ \displaystyle\int_{\Re^n} y_{*j} f_{\boldsymbol{y}}(\boldsymbol{y}_*;\boldsymbol{\theta})\, d\boldsymbol{y}_* & \text{(if the PDF exists)}. \end{cases} \tag{7.58}$$

(b) The value of $\mathrm{E}(\boldsymbol{y})$ is an $n \times 1$ vector $\boldsymbol{\mu}$ also known as the *mean vector*. The vector of means exists if all its elements $\{\mu_j\}$ are finite.

Definition 7.20 If $\boldsymbol{y}$ is a random vector which follows a distribution with CDF $F_{\boldsymbol{y}}(\boldsymbol{y};\boldsymbol{\theta})$ and

$$\boldsymbol{h}(\boldsymbol{y}) = \begin{bmatrix} h_1(\boldsymbol{y}) \\ \vdots \\ h_k(\boldsymbol{y}) \end{bmatrix} \tag{7.59}$$

is a vector of functions of $\boldsymbol{y} \in \Re^n$ which are integrable with respect to $F_{\boldsymbol{y}}(\boldsymbol{y};\boldsymbol{\theta})$, then the *expected value* of $\mathrm{E}[\boldsymbol{h}(\boldsymbol{y})]$ is defined as the following elementwise integral operation:

$$\mathrm{E}[\boldsymbol{h}(\boldsymbol{y})] = \begin{bmatrix} \mathrm{E}[h_1(\boldsymbol{y})] \\ \vdots \\ \mathrm{E}[h_k(\boldsymbol{y})] \end{bmatrix} \tag{7.60}$$

with

$$\mathrm{E}[h_j(\boldsymbol{y})] = \begin{cases} \displaystyle\int_{\Re^n} h_j(\boldsymbol{y}_*)\, dF_{\boldsymbol{y}}(\boldsymbol{y}_*;\boldsymbol{\theta}) & \text{(in general)} \\ \displaystyle\int_{\Re^n} h_j(\boldsymbol{y}_*) f_{\boldsymbol{y}}(\boldsymbol{y}_*;\boldsymbol{\theta})\, d\boldsymbol{y}_* & \text{(if the PDF exists)}. \end{cases} \tag{7.61}$$

The moments exist if the integrals are finite.

Definition 7.21 (a) If $\boldsymbol{y}$ follows a distribution with CDF $F_{\boldsymbol{y}}(\boldsymbol{y}_*;\boldsymbol{\theta})$, $\boldsymbol{c} \in \Re^k$, $m > 0$, and

$$\boldsymbol{h}(\boldsymbol{y}) = \begin{bmatrix} h_1(\boldsymbol{y}) \\ \vdots \\ h_k(\boldsymbol{y}) \end{bmatrix} = \begin{bmatrix} (y_1 - c_1)^m \\ \vdots \\ (y_k - c_k)^m \end{bmatrix}, \tag{7.62}$$

then $\mathrm{E}[\boldsymbol{h}(\boldsymbol{y})] = [\,\mathrm{E}[h_1(\boldsymbol{y})] \cdots \mathrm{E}[h_k(\boldsymbol{y})]\,]'$ with

$$\mathrm{E}[h_j(\boldsymbol{y})] = \begin{cases} \displaystyle\int_{\Re^n} (y_{*j} - c_j)^m\, dF_{\boldsymbol{y}}(\boldsymbol{y}_*;\boldsymbol{\theta}) & \text{(in general)} \\ \displaystyle\int_{\Re^n} (y_{*j} - c_j)^m f_{\boldsymbol{y}}(\boldsymbol{y}_*;\boldsymbol{\theta})\, d\boldsymbol{y}_* & \text{(if the PDF exists)}. \end{cases} \tag{7.63}$$

(b) If $\boldsymbol{c} = \boldsymbol{0}$, then $\mathrm{E}[\boldsymbol{h}(\boldsymbol{y})] \equiv [\,\mu_{m',1} \cdots \mu_{m',k}\,]' \equiv \boldsymbol{\mu}_{m'}$ is the set of *moments about* $\boldsymbol{0}$ of order m,
(c) If $\boldsymbol{c} = \boldsymbol{\mu}_{1'}$, then $\mathrm{E}[\boldsymbol{h}(\boldsymbol{y})] \equiv [\,\mu_{m,1} \cdots \mu_{m,k}\,]' \equiv \boldsymbol{\mu}_m$ is the set of *central moments* of order m.

Usually we work with random *vectors* because distributions, covariance matrices, and many related properties of scalar random variables remain well-defined for random vectors. In addition, we sometimes consider random matrices, such as an $n \times p$ data matrix $\boldsymbol{Y}$. Similarly, an estimated covariance (dispersion) matrix is a symmetric random matrix.

Definition 7.22 If $\boldsymbol{y} = \{y_{jk}\}$ is an $n \times p$ array of random variables having a joint distribution and $\mathrm{E}(y_{jk})$ exists (finitely) for all (j, k), the $n \times p$ matrix of expected values is $\mathrm{E}(\boldsymbol{Y}) = \{\mathrm{E}(y_{jk})\}$, with expectation an elementwise operation.

Theorem 7.26 If $\boldsymbol{Y}$ and $\boldsymbol{X}$ are random $n \times p$ arrays and $\boldsymbol{A}$, $\boldsymbol{B}$, and $\boldsymbol{C}$ are constant matrices which conform for the matrix operations implied, then

$$\mathrm{E}(\boldsymbol{X}+\boldsymbol{Y}) = \mathrm{E}(\boldsymbol{X}) + \mathrm{E}(\boldsymbol{Y}) \tag{7.64}$$
$$\mathrm{E}(\boldsymbol{X}+\boldsymbol{C}) = \mathrm{E}(\boldsymbol{X}) + \boldsymbol{C} \tag{7.65}$$
$$\mathrm{E}(\boldsymbol{C}) = \boldsymbol{C} \tag{7.66}$$
$$\mathrm{E}(\boldsymbol{AX}) = \boldsymbol{A}\mathrm{E}(\boldsymbol{X}) \tag{7.67}$$
$$\mathrm{E}(\boldsymbol{XB}) = [\mathrm{E}(\boldsymbol{X})]\boldsymbol{B}\,. \tag{7.68}$$

Definition 7.23 If it exists, the *covariance* or *dispersion* matrix $\mathcal{V}(\boldsymbol{y})$ is

$$\mathcal{V}(\boldsymbol{y}) = \mathrm{E}[(\boldsymbol{y}-\boldsymbol{\mu})(\boldsymbol{y}-\boldsymbol{\mu})'], \tag{7.69}$$

with $\boldsymbol{\mu} = \mathrm{E}(\boldsymbol{y})$ and the (j, k) element a *variance* $\mathcal{V}(y_j)$ if $j = k$ and a *covariance* $\mathcal{V}(y_j, y_k)$ otherwise. In particular,

$$\mathrm{E}[(y_j-\mu_j)(y_k-\mu_k)] = \begin{cases} \displaystyle\int (y_{*j}-\mu_j)(y_{*k}-\mu_k)\, dF_{\boldsymbol{y}}(\boldsymbol{y}_*;\boldsymbol{\theta}) & \text{(in general)} \\ \displaystyle\int (y_{*j}-\mu_j)(y_{*k}-\mu_k) f_{\boldsymbol{y}}(\boldsymbol{y}_*;\boldsymbol{\theta})\, d\boldsymbol{y}_* & \text{(if a.c.)}. \end{cases} \tag{7.70}$$

The array of second-order moments exists if and only if all its elements $\{\sigma_{jk}\}$ have finite values. Although using σ_j^2 in place of σ_{jj} seems tempting, consistent use of σ_{jj} helps prevent confusion.

Theorem 7.27 If random vector $\boldsymbol{y} = [\,y_1 \cdots y_n\,]'$ has a continuous joint distribution, $\mathrm{E}(\boldsymbol{y}) = \boldsymbol{\mu}$, $\mathcal{V}(\boldsymbol{y}) = \boldsymbol{\Sigma}$, and $\sigma_{jj} = \mathcal{V}(y_j) < \infty$ $\forall j \in \{1, 2, \ldots, n\}$, then, for conforming constants $\boldsymbol{A}$, $\boldsymbol{a}$, and $\boldsymbol{t}$, the following all hold.

(a) $(\sigma_{jk})^2 = [\mathcal{V}(y_j, y_k)]^2 \le \mathcal{V}(y_j)\mathcal{V}(y_k) < \infty$
(b) $\mathcal{V}(\boldsymbol{y}+\boldsymbol{a}) = \mathcal{V}(\boldsymbol{y}) = \boldsymbol{\Sigma}\ \forall \boldsymbol{a} \in \Re^n$
(c) $\mathcal{V}(\boldsymbol{Ay}) = \boldsymbol{A}\mathcal{V}(\boldsymbol{y})\boldsymbol{A}' = \boldsymbol{A\Sigma A}'\ \forall \boldsymbol{A} \in \Re^{m\times n}$
(d) $\mathcal{V}(\boldsymbol{t}'\boldsymbol{y}) = \boldsymbol{t}'\mathcal{V}(\boldsymbol{y})\boldsymbol{t} = \boldsymbol{t}'\boldsymbol{\Sigma t} \ge 0\ \forall \boldsymbol{t} \in \Re^n$
(e) $\boldsymbol{\Sigma}$ is either positive definite or positive semidefinite
(f) $\boldsymbol{\Sigma} = \mathrm{E}[(\boldsymbol{y}-\boldsymbol{d})(\boldsymbol{y}-\boldsymbol{d})'] - (\boldsymbol{\mu}-\boldsymbol{d})(\boldsymbol{\mu}-\boldsymbol{d})'\ \forall \boldsymbol{d} \in \Re^n$
(g) $\boldsymbol{\Sigma} = \mathrm{E}(\boldsymbol{yy}') - \boldsymbol{\mu\mu}'$ for $\boldsymbol{d} = \boldsymbol{0}$

Proof. Left as an exercise. *Hints*: The Cauchy-Schwartz inequality, $[\int_a^b f(x)g(x)dx]^2 \leq \int_a^b [f(x)]^2 dx \int_a^b [g(x)]^2 dx$, $\Rightarrow$ **(a)**. The integral definition helps with **(b)**. Also **(c)** $\Rightarrow$ **(d)** $\Leftrightarrow$ **(e)**.

Theorem 7.28 If $\boldsymbol{y}' = [\,\boldsymbol{y}_1'\ \boldsymbol{y}_2'\,]$, with $\boldsymbol{y}_j$ of dimension p_i, and all elements have finite second moments, then

$$\mathcal{V}(\boldsymbol{y}) = \boldsymbol{\Sigma} = \begin{bmatrix} \boldsymbol{\Sigma}_{11} & \boldsymbol{\Sigma}_{12} \\ \boldsymbol{\Sigma}_{21} & \boldsymbol{\Sigma}_{22} \end{bmatrix}, \tag{7.71}$$

with $\boldsymbol{\Sigma}_{jk}$ $(p_j \times p_k)$ described as a covariance matrix. Also $\boldsymbol{\Sigma}_{jj} = \mathcal{V}(\boldsymbol{y}_j)$. Here $\boldsymbol{\Sigma}_{12} = \boldsymbol{\Sigma}_{21}' = \mathrm{E}[(\boldsymbol{y}_1 - \boldsymbol{\mu}_1)(\boldsymbol{y}_2 - \boldsymbol{\mu}_2)']$, often written $\mathcal{V}(\boldsymbol{y}_1, \boldsymbol{y}_2) = [\mathcal{V}(\boldsymbol{y}_2, \boldsymbol{y}_1)]'$. For (conformable) constants $\boldsymbol{A}$ and $\boldsymbol{B}$, $\mathcal{V}()$ is a bilinear operator,

$$\mathcal{V}(\boldsymbol{A}\boldsymbol{y}_1, \boldsymbol{B}\boldsymbol{y}_2) = \boldsymbol{A}\mathcal{V}(\boldsymbol{y}_1, \boldsymbol{y}_2)\boldsymbol{B}', \tag{7.72}$$

and $\mathcal{V}()$ is invariant to adding constants,

$$\mathcal{V}(\boldsymbol{a} + \boldsymbol{A}\boldsymbol{y}_1, \boldsymbol{b} + \boldsymbol{B}\boldsymbol{y}_2) = \boldsymbol{A}\mathcal{V}(\boldsymbol{y}_1, \boldsymbol{y}_2)\boldsymbol{B}'\,. \tag{7.73}$$

Proof. Left as an exercise.

Theorem 7.29 For a random vector $\boldsymbol{z}$ following *any* distribution with finite second moments and $\boldsymbol{T}$ a fixed and conforming matrix, $\mathrm{E}(\boldsymbol{T}\boldsymbol{z}) = \boldsymbol{T}\boldsymbol{\mu}_z$ and $\mathcal{V}(\boldsymbol{T}\boldsymbol{z}) = \boldsymbol{T}\boldsymbol{\Sigma}_z\boldsymbol{T}'$.

Definition 7.24 If $n \times 1$ random vector $\boldsymbol{y}$ has $\mathcal{V}(\boldsymbol{y}) = \boldsymbol{\Sigma} = \{\sigma_{jk}\}$ with $0 < \sigma_{jj} < \infty$, the $n \times n$ *correlation matrix* is

$$\boldsymbol{P} = [\mathrm{Dg}(\{\sigma_{11}, \ldots, \sigma_{nn}\})]^{-1/2}\boldsymbol{\Sigma}[\mathrm{Dg}(\{\sigma_{11}, \ldots, \sigma_{nn}\})]^{-1/2}, \tag{7.74}$$

with (j,k) element $\rho_{jk} = \sigma_{jk}(\sigma_{jj}\sigma_{kk})^{-1/2}$. If $j = k$, then $\rho_{jj} = 1$, and otherwise ρ_{jk} is the *correlation* between y_j and y_k.

Theorem 7.30 If $n \times 1$ random vector $\boldsymbol{y}$ has $\mathcal{V}(\boldsymbol{y}) = \boldsymbol{\Sigma}$, with $0 < \sigma_{jj} < \infty$, then

(a) $\rho_{jk} \in [-1, 1]\ \forall j, k$
(b) $\boldsymbol{P}$ $(n \times n)$ is symmetric and nonnegative definite.
(c) For any given $\boldsymbol{\Sigma}$ the unique value of $\boldsymbol{P}$ is

$$\boldsymbol{P} = [\mathrm{Dg}(\boldsymbol{\Sigma})]^{-1/2}\boldsymbol{\Sigma}[\mathrm{Dg}(\boldsymbol{\Sigma})]^{-1/2}\,. \tag{7.75}$$

(d) For any given $\boldsymbol{P}$ the value of $\boldsymbol{\Sigma}$ depends on knowing $\mathrm{Dg}(\boldsymbol{\Sigma})$ because

$$\boldsymbol{\Sigma} = [\mathrm{Dg}(\boldsymbol{\Sigma})]^{1/2}\boldsymbol{P}[\mathrm{Dg}(\boldsymbol{\Sigma})]^{1/2}\,. \tag{7.76}$$

Proof. **(a)** The Cauchy-Schwartz inequality implies $\sigma_{jk}^2 \leq \sigma_{jj}\sigma_{kk}$, which implies $\rho_{jk} \in [-1, 1]$. **(b)** $\boldsymbol{\Sigma}$ is positive semidefinite and congruent to $\boldsymbol{P}$. Hence

$\boldsymbol{P}$ is also positive semidefinite (by Sylvester's law of inertia). **(c)** is true by definition. **(d)** Any valid choice of standard deviations (strictly positive and finite) yields a valid value for $\boldsymbol{\Sigma}$.

Definition 7.25 For

$$\mathcal{V}\begin{bmatrix} \boldsymbol{y}_1 \\ \boldsymbol{y}_2 \end{bmatrix} = \begin{bmatrix} \boldsymbol{\Sigma}_{11} & \boldsymbol{\Sigma}_{12} \\ \boldsymbol{\Sigma}_{21} & \boldsymbol{\Sigma}_{22} \end{bmatrix}, \tag{7.77}$$

(a) the vectors $\boldsymbol{y}_1$ and $\boldsymbol{y}_2$ are *uncorrelated* $\Leftrightarrow$ $\boldsymbol{\Sigma}_{12} = \boldsymbol{0}$ $\Leftrightarrow$ $\boldsymbol{P}_{12} = \boldsymbol{0}$.
(b) The elements of $\boldsymbol{y}_j$ are *uncorrelated* $\Leftrightarrow$ $\boldsymbol{\Sigma}_{jj}$ is diagonal $\Leftrightarrow$ $\boldsymbol{P}_{jj} = \boldsymbol{I}$.

A lack of correlation does not necessarily imply statistical independence.

7.15 CONDITIONAL MOMENTS OF DISTRIBUTIONS

Here we consider properties of $\boldsymbol{y}|\boldsymbol{x}$. In particular, $\mathrm{E}[\mathrm{E}(\boldsymbol{y}|\boldsymbol{x})] = \mathrm{E}(\boldsymbol{y})$, while $\mathrm{E}[\mathcal{V}(\boldsymbol{y}|\boldsymbol{x})] \leq \mathcal{V}(\boldsymbol{y})$. The second result should seem reasonable because $\mathcal{V}(\boldsymbol{y}|\boldsymbol{x} = \boldsymbol{x}_0) \leq \mathcal{V}(\boldsymbol{y})\ \forall \boldsymbol{x}_0$, with equality whenever $\boldsymbol{y} \perp\!\!\!\perp \boldsymbol{x}$. Knowing the specific realization of $\boldsymbol{x}$ can only reduce uncertainty about the value of $\boldsymbol{y}$.

Lemma 7.4 If $\boldsymbol{y}$ and $\boldsymbol{x}$ have a joint distribution, then the CDF and PDF (if it exists) are

$$\begin{aligned} F_{\boldsymbol{y}}(\boldsymbol{y}_*) &= \mathrm{E}\big[F_{\boldsymbol{y}|\boldsymbol{x}}(\boldsymbol{y}_*|\boldsymbol{x}_*)\big] \quad \text{(always)} \\ f_{\boldsymbol{y}}(\boldsymbol{y}_*) &= \mathrm{E}\big[f_{\boldsymbol{y}|\boldsymbol{x}}(\boldsymbol{y}_*|\boldsymbol{x}_*)\big] \quad \text{(if it exists)}. \end{aligned} \tag{7.78}$$

Proof. Left as an exercise.

Theorem 7.31 If $\boldsymbol{y}$ and $\boldsymbol{x}$ have a joint distribution then $\mathrm{E}(\boldsymbol{y}) = \mathrm{E}[\mathrm{E}(\boldsymbol{y}|\boldsymbol{x})]$, which for clarity may be written $\mathrm{E}(\boldsymbol{y}) = \mathrm{E}_{\boldsymbol{x}}[\mathrm{E}_{\boldsymbol{y}}(\boldsymbol{y}|\boldsymbol{x})]$.

Proof. (For absolutely continuous distributions.)

$$\begin{aligned} \mathrm{E}(\boldsymbol{y}) &= \int \boldsymbol{y}_* f_{\boldsymbol{y}}(\boldsymbol{y}_*)\, d\boldsymbol{y}_* \\ &= \int \boldsymbol{y}_* \left[\int f_{\boldsymbol{y},\boldsymbol{x}}(\boldsymbol{y}_*, \boldsymbol{x}_*) d\boldsymbol{x}_*\right] d\boldsymbol{y}_* \\ &= \int\!\!\int \boldsymbol{y}_* f_{\boldsymbol{y},\boldsymbol{x}}(\boldsymbol{y}_*|\boldsymbol{x}_*) f_{\boldsymbol{x}}(\boldsymbol{x}_*)\, d\boldsymbol{x}_*\, d\boldsymbol{y}_* \\ &= \int \left[\int \boldsymbol{y}_* f_{\boldsymbol{y}|\boldsymbol{x}}(\boldsymbol{y}_*|\boldsymbol{x}_*) d\boldsymbol{y}_*\right] f_{\boldsymbol{x}}(\boldsymbol{x}_*)\, d\boldsymbol{x}_* \\ &= \mathrm{E}[\mathrm{E}(\boldsymbol{y}|\boldsymbol{x})] = \mathrm{E}_{\boldsymbol{x}}[\mathrm{E}_{\boldsymbol{y}}(\boldsymbol{y}|\boldsymbol{x})]. \end{aligned} \tag{7.79}$$

□

Theorem 7.32 If $\boldsymbol{y}$ and $\boldsymbol{x}$ have a joint distribution and the moments exist, then

$$\mathcal{V}(\boldsymbol{y}) = \mathrm{E}[\mathcal{V}(\boldsymbol{y}|\boldsymbol{x})] + \mathcal{V}[\mathrm{E}(\boldsymbol{y}|\boldsymbol{x})]\,. \tag{7.80}$$

Proof. If $\boldsymbol{\mu} = \mathrm{E}(\boldsymbol{y})$ and $\boldsymbol{\mu}_c = \mathrm{E}(\boldsymbol{y}|\boldsymbol{x})$ then

$$\begin{aligned}\mathcal{V}(\boldsymbol{y}) &= \mathrm{E}(\boldsymbol{y}\boldsymbol{y}') - \boldsymbol{\mu}\boldsymbol{\mu}' \\ &= [\mathrm{E}(\boldsymbol{y}\boldsymbol{y}') - \mathrm{E}(\boldsymbol{\mu}_c\boldsymbol{\mu}_c')] + [\mathrm{E}(\boldsymbol{\mu}_c\boldsymbol{\mu}_c)' - \boldsymbol{\mu}\boldsymbol{\mu}']\,. \end{aligned}\tag{7.81}$$

Also

$$\begin{aligned}\mathrm{E}(\boldsymbol{y}\boldsymbol{y}') - \mathrm{E}(\boldsymbol{\mu}_c\boldsymbol{\mu}_c)' &= \mathrm{E}\{\mathrm{E}(\boldsymbol{y}\boldsymbol{y}'|\boldsymbol{x}) - \mathrm{E}(\boldsymbol{y}|\boldsymbol{x})[\mathrm{E}(\boldsymbol{y}|\boldsymbol{x})]'\} \\ &= \mathrm{E}[\mathcal{V}(\boldsymbol{y}|\boldsymbol{x})] \end{aligned}\tag{7.82}$$

and

$$\begin{aligned}\mathrm{E}(\boldsymbol{\mu}_c\boldsymbol{\mu}_c)' - \boldsymbol{\mu}\boldsymbol{\mu}' &= \mathrm{E}(\boldsymbol{\mu}_c\boldsymbol{\mu}_c)' - \mathrm{E}(\boldsymbol{\mu}_c)\mathrm{E}(\boldsymbol{\mu}_c') \\ &= \mathcal{V}(\boldsymbol{\mu}_c) \\ &= \mathcal{V}[(\mathrm{E}(\boldsymbol{y}|\boldsymbol{x})]\,. \end{aligned}\tag{7.83}$$

□

7.16 SPECIAL CONSIDERATIONS FOR RANDOM MATRICES

At first glance, generalizing to matrices introduces no great complication. It obviously makes sense to define, for $\boldsymbol{X} = \{x_{jk}\}$ with x_{jk} random,

$$\mathrm{E}(\boldsymbol{X}) = \{\mathrm{E}(x_{jk})\}\,. \tag{7.84}$$

However, even when $\{x_{jk}\}$ have finite second moments, $\mathcal{V}(\boldsymbol{X})$ has no obvious meaning, although $\mathcal{V}[\mathrm{vec}(\boldsymbol{X})]$ and $\{\mathcal{V}(x_{jk})\}$ are well defined (and typically quite different). For probability calculations and other operations directly involving matrix elements, expressions in terms of the vec() and vech() operators seem most useful. In contrast, it is often advantageous to consider moment calculations in terms of the original matrix form. Consequently the characteristic (and moment and cumulant generating) functions have been generalized.

The characteristic function tends to be used more in the study of random vectors than for scalars. In turn, it plays an even bigger role for random matrices. Scalar (continuous) random variables that are commonly studied rarely fail to have a density. In contrast, singular vector and matrix Gaussian random arrays, which do not have a density, arise naturally in the study of data analysis. Even when a density for a vector or matrix exists, it may not have a convenient closed form. Useful expressions for CDFs are even more rare. Furthermore, even when known, densities for random vectors and matrices often prove difficult to manipulate, for both analytic and computational purposes.

At least for the random vectors and matrices of interest in the present book, the characteristic functions are never more complicated than corresponding densities (when they exist), and often much simpler. In turn, much simpler proofs can often be found in terms of characteristic functions. Most importantly, the proofs often apply to a wider range of distributions by tolerating random vectors and matrices which do not have densities (due to singularities in the distribution function).

The first task is to generalize the definition of the characteristic function of a vector to a matrix. The definition follows the one give by Gupta and Nagar (2000, p45–46, with some notation changed).

Definition 7.26 With $\mathrm{i} = \sqrt{-1}$, $\boldsymbol{T}$ an arbitrary real $n \times p$ matrix, and random $\{x_{jk}\}$ having a well-defined joint distribution, the *characteristic function* of $n \times p$ $\boldsymbol{X} = \{x_{jk}\}$ is

$$\phi_{\boldsymbol{X}}(\boldsymbol{T}) = \mathrm{E}\{\exp[\mathrm{tr}(\mathrm{i}\boldsymbol{T}'\boldsymbol{X})]\}\,. \tag{7.85}$$

Definition 7.27 With $\boldsymbol{T}$ an arbitrary real $n \times p$ matrix and random $\{x_{jk}\}$ having a well-defined joint distribution, the *moment generating function* of $n \times p$ $\boldsymbol{X} = \{x_{jk}\}$, when it exists, is

$$m_{\boldsymbol{X}}(\boldsymbol{T}) = \mathrm{E}\{\exp[\mathrm{tr}(\boldsymbol{T}'\boldsymbol{X})]\}\,. \tag{7.86}$$

Definition 7.28 With $\boldsymbol{T}$ an arbitrary real $n \times p$ matrix, and random $\{x_{jk}\}$ having a well-defined joint distribution, the *cumulant generating function* of $n \times p$ $\boldsymbol{X} = \{x_{jk}\}$, when it exists, is

$$c_{\boldsymbol{X}}(\boldsymbol{T}) = \log[m_{\boldsymbol{X}}(\boldsymbol{T})]\,. \tag{7.87}$$

The following lemma generalizes the result about a linear transformation of a random vector. The matrix form allows simple proofs of many results concerning Gaussian variables and corresponding covariance estimators. Most importantly, the lemma allows less than full rank transformations, even when applied to random variables lacking a density. In applications in later chapters, the resulting characteristic function is often seen to correspond to a known distribution.

Lemma 7.5 If $\boldsymbol{Y} = \boldsymbol{A}\boldsymbol{X}\boldsymbol{B} + \boldsymbol{C}$ for random $\boldsymbol{X}$ and finite constants $\{\boldsymbol{A}, \boldsymbol{B}, \boldsymbol{C}\}$ of *any* rank and conforming size $(n_1 \times p_1) = (n_1 \times n)(n \times p)(p \times p_1) + (n_1 \times p_1)$, then

$$\phi_{\boldsymbol{Y}}(\boldsymbol{T}) = \phi_{\boldsymbol{X}}(\boldsymbol{A}'\boldsymbol{T}\boldsymbol{B}') \cdot \exp[\mathrm{i}\,\mathrm{tr}(\boldsymbol{T}'\boldsymbol{C})]\,. \tag{7.88}$$

Proof. The result is proven by writing

$$\begin{aligned} \mathrm{E}(\exp\{\mathrm{tr}[\mathrm{i}\boldsymbol{T}'(\boldsymbol{A}\boldsymbol{X}\boldsymbol{B} + \boldsymbol{C})]\}) &= \mathrm{E}\{\exp[\mathrm{tr}(\mathrm{i}\boldsymbol{T}'\boldsymbol{A}\boldsymbol{X}\boldsymbol{B}) + \mathrm{tr}(\mathrm{i}\boldsymbol{T}'\boldsymbol{C})]\} \\ &= \mathrm{E}\{\exp[\mathrm{tr}(\mathrm{i}\boldsymbol{B}\boldsymbol{T}'\boldsymbol{A}\boldsymbol{X})]\exp[\mathrm{tr}(\mathrm{i}\boldsymbol{T}'\boldsymbol{C})]\} \qquad (7.89) \\ &= \mathrm{E}\big(\exp\{\mathrm{tr}[\mathrm{i}(\boldsymbol{A}'\boldsymbol{T}\boldsymbol{B}')'\boldsymbol{X}]\}\big)\exp[\mathrm{i}\,\mathrm{tr}(\boldsymbol{T}'\boldsymbol{C})]\,. \end{aligned}$$

□

CHAPTER 8

Scalar, Vector, and Matrix Gaussian Distributions

8.1 MOTIVATION

The Gaussian distribution, which was discovered by de Moivre in 1733, has likely been studied more than any other. The distribution often provides a plausible model in a wide range of applications, primarily because it arises so often as the asymptotic distribution specified in central limit theorems. Equally importantly, Gaussian distributions and distributions of functions of Gaussian variables, such as chi squares, typically provide the backbone of many distribution-free results for large samples. The statement holds for many scalar, vector, and matrix forms.

In the context of linear models, assuming Gaussian error terms typically implies estimators of means and treatment effects (expected-value parameters) often follow Gaussian distributions. Gaussian distributions may provide an appropriate model of a first moment (mean, location) in scalar, vector, or matrix form.

We avoid the term “normal” distribution to avoid the implication that the distribution should be expected to apply or is “usual.” The term originated from a philosophical position about universal laws of nature. Referring to a “Gaussian” distribution helps emphasize it is one particular distribution among many and that the specific choice must be justified in any particular application. We mostly ignore the important practical problem of defending the use of the Gaussian assumption for a particular set of data and proceed under the assumption.

Statisticians have devised many ways of characterizing the Gaussian distribution. The approaches vary in the simplicity of assumptions and presentation, generality of parameters allowed, and ease of learning. We maintain complete generality but sacrifice some simplicity of presentation to make learning easier. We begin with the standard scalar Gaussian and define the general scalar Gaussian as a (scalar) linear transformation of the standard Gaussian. Some transformation constants lead to nonsingular distribution functions (they have continuous derivatives which provide a density) and correspond to nondegenerate random variables. Other transformation constants lead to distribution functions which are singular (they do not have continuous derivatives or a density) and correspond to degenerate random variables.

The same sequence of constructions generalizes from scalars to vectors, beginning with a vector of independent standard Gaussian variables. In turn, many of the convenient and special properties of the vector Gaussian are derived. The chapter concludes with consideration of matrix forms of Gaussian variables. Such forms are particularly useful in the study of multivariate models, especially multivariate linear models.

8.2 THE SCALAR GAUSSIAN DISTRIBUTION

We delay defining the scalar Gaussian until we prove the function we claim is its density is in fact a density. Lemma 8.1 provides the standard Gaussian result.

Lemma 8.1

$$\int_{-\infty}^{\infty} \exp(-z_*^2/2)\, dz_* = (2\pi)^{1/2}. \tag{8.1}$$

Proof. The lemma is true if and only if the square of the integral is 2π. Here

$$\begin{aligned}\left[\int_{-\infty}^{\infty} \exp(-z_*^2/2)\, dz_*\right]^2 &= \left[\int_{-\infty}^{\infty} \exp(-x_*^2/2)\, dx_*\right]\left[\int_{-\infty}^{\infty} \exp(-y_*^2/2)\, dy_*\right] \\ &= \int_{-\infty}^{\infty}\int_{-\infty}^{\infty} \exp\left[-(x_*^2+y_*^2)/2\right] dx_*\, dy_*\,. \end{aligned} \tag{8.2}$$

Switching to polar coordinates gives $x_* = r_*\cos(\theta_*)$ and $y_* = r_*\sin(\theta_*)$, with $r_*^2 = x_*^2 + y_*^2$ and $dx_* dy_* = r_* dr_* d\theta_*$ (r_* is the Jacobian of the transformation). In turn,

$$\begin{aligned}\int_{-\infty}^{\infty}\int_{-\infty}^{\infty} \exp\left[-(x_*^2+y_*^2)/2\right] dx_*\, dy_* &= \int_0^{2\pi}\int_0^{\infty} \exp(-r_*^2/2) r_*\, dr_*\, d\theta_* \\ &= \left(\int_0^{2\pi} d\theta_*\right)\int_0^{\infty} \exp(-r_*^2/2) r_*\, dr_* \\ &= 2\pi\left[-\Big|_{r_*=0}^{\infty} \exp(-r_*^2/2)\right] = 2\pi\,. \end{aligned} \tag{8.3}$$

□

Theorem 8.1 The function $f_z(z_*) = (2\pi)^{-1/2}\exp(-z_*^2/2)$ is the density of an a.c. random variable with domain the real line.

Proof. The function is nonnegative and integrates to 1 over the real line (by the lemma). By Theorem 7.1, the function is a density of a random variable. □

Corollary 8.1.1 For $f_z(z_*) = (2\pi)^{-1/2}\exp(-z_*^2/2)$,
(a) the MGF of z is $m_z(t) = \exp(-t^2/2)$ and
(b) the CF of z is $\phi_z(t) = \exp(-t^2/2)$.
(c) $\mathrm{E}(z) = 0$ and $\mathcal{V}(z) = \mathrm{E}(z^2) = 1$.
(d) All odd moments are zero.
(e) If $m \in \{2, 4, \dots\}$, then $\mathrm{E}(z^m) = (m-1)(m-3)\cdots(3)(1)$.
(f) The cumulants $\{\kappa_m\}$ are all zero for $m > 2$.
Proof. (a)

$$\begin{aligned}
\mathrm{E}[\exp(zt)] &= (2\pi)^{-1/2}\int_{-\infty}^{\infty} \exp(z_*t)\exp(-z_*^2/2)\,dz_* \\
&= (2\pi)^{-1/2}\int_{-\infty}^{\infty} \exp[(-z_*^2 + 2z_*t - t^2 + t^2)/2]\,dz_* \\
&= (2\pi)^{-1/2}\int_{-\infty}^{\infty} \exp[-(z_* - t)^2/2 + t^2/2]\,dz_* \\
&= \exp(t^2/2)(2\pi)^{-1/2}\int_{-\infty}^{\infty} \exp[-(z_* - t)^2/2]\,dz_* \\
&= \exp(t^2/2)(2\pi)^{-1/2}\int_{-\infty}^{\infty} \exp[-(u_*)^2/2]\,du_* \\
&= \exp(t^2/2)\cdot 1\,.
\end{aligned} \tag{8.4}$$

(b) The characteristic function of z is

$$\begin{aligned}
\mathrm{E}[\exp(\mathrm{i}zt)] &= (2\pi)^{-1/2}\int_{-\infty}^{\infty} \exp(\mathrm{i}z_*t)\exp(-z_*^2/2)\,dz_* \\
&= (2\pi)^{-1/2}\int_{-\infty}^{\infty} \exp[(-z_*^2 + 2\mathrm{i}z_*t + t^2 - t^2)/2]\,dz_* \\
&= (2\pi)^{-1/2}\int_{-\infty}^{\infty} \exp[-(z_* - \mathrm{i}t)^2/2 - t^2/2]\,dz_* \\
&= \exp(-t^2/2)(2\pi)^{-1/2}\int_{-\infty}^{\infty} \exp[-(z_* - \mathrm{i}t)^2/2]\,dz_* \\
&\qquad \text{(skipping a complex integration)} \\
&= \exp(-t^2/2)\cdot 1\,.
\end{aligned} \tag{8.5}$$

(c) Moments are easily computed by integration or from derivatives of the MGF.
(d) True due to symmetry about the origin.
(e) and **(f)** are left as exercises (consider induction). □

Corollary 8.1.2 If the random variable z has density $f_z(z_*) = (2\pi)^{-1/2}\exp(-z_*^2/2)$ and domain $\Re^1$, then $y = \sigma z + \mu$,
(a) for (real) constants $-\infty < \mu < \infty$ and $0 < \sigma^2 < \infty$, has domain $\Re^1$ and density

$$f_y(y_*; \mu, \sigma^2) = (2\pi\sigma^2)^{-1/2}\exp[-(y_* - \mu)^2/(2\sigma^2)]\,. \tag{8.6}$$

(b) If $\sigma^2 = 0$ and $-\infty < \mu < \infty$, then $y = \sigma z + \mu$ is a discrete and degenerate random variable, with $\Pr\{y = \mu\} = 1$, and y does not have a density.

Proof. If $0<\sigma^2<\infty$, the function must be nonnegative, integrate to 1 over the domain, and integrate to provide the CDF of y for $0<\sigma^2<\infty$. The function $f_y(y_*;\mu,\sigma^2)$ is clearly nonnegative for all y_* (finite μ, and $0<\sigma^2<\infty$). Also,

$$\begin{aligned}\Pr\{y \le y_0\} &= \Pr\{(\sigma z+\mu) \le y_0\} \\ &= \Pr\{z \le (y_0-\mu)/\sigma\} \\ &= \int_{-\infty}^{(y_0-\mu)/\sigma} (2\pi)^{-1/2}\exp(-z_*^2/2)\,dz_*\,. \end{aligned} \tag{8.7}$$

If $y_0 \to \infty$ then the integral is 1 over the real line (by the lemma). The transformation $t = \sigma z_* + \mu$ gives $dt_* = \sigma\,dz_*$ and $z_* = (t_* - \mu)/\sigma$. In turn,

$$\begin{aligned}\Pr\{y \le y_0\} &= \int_{-\infty}^{y_0} (2\pi)^{-1/2}\exp\big[-(t_*-\mu)^2/(2\sigma^2)\big]\sigma^{-1}\,dt_* \\ &= \int_{-\infty}^{y_0} f_y(t_*;\mu,\sigma^2)\,dt_*\,. \end{aligned} \tag{8.8}$$

If $\sigma^2 = 0$ then $y = 0\cdot z+\mu \equiv \mu$ and $\Pr\{y=\mu\} = 1$. If so, y is a discrete random variable and does not have a density. □

Corollary 8.1.3 For (real) constants $-\infty<\mu<\infty$ and $0\le\sigma^2<\infty$, the characteristic function of $y = \sigma z + \mu$ is

$$\begin{aligned}\phi_y(t) &= \mathrm{e}^{\mathrm{i}t\mu}\phi_z(\sigma t) \\ &= \exp(\mathrm{i}t\mu - \sigma^2t^2/2)\,, \end{aligned} \tag{8.9}$$

and the MGF is $m_y(t) = \exp(t\mu + \sigma^2t^2/2)$. Furthermore, $\mathrm{E}(y) = \mu$ and $\mathcal{V}(y) = \sigma^2$, which fully characterize the distribution.

Definition 8.1 (a) A scalar (real) random variable z with density

$$f_z(z_*) = (2\pi)^{-1/2}\exp(-z_*^2/2) \tag{8.10}$$

is said to follow a *standard Gaussian* distribution, written $z \sim \mathcal{N}(0,1)$.
(b) If μ and $0<\sigma^2$ are finite real constants, then the scalar random variable $y = \sigma z+\mu$ has density

$$f_y(y_*;\mu,\sigma^2) = (2\pi\sigma^2)^{-1/2}\exp\big[-(y_*-\mu)^2/(2\sigma^2)\big] \tag{8.11}$$

and is said to follow a *Gaussian distribution*, written $y \sim \mathcal{N}(\mu,\sigma^2)$.
(c) If $\sigma^2 = 0$ (and μ is finite), then y does not have a density, and follows a *singular Gaussian distribution*, written $y \sim \mathcal{SN}(\mu,0)$, with $\Pr\{y=\mu\}=1$.
(d) Writing $y \sim (\mathcal{S})\mathcal{N}(\mu,\sigma^2)$ indicates y may be either singular or nonsingular (with $0\le\sigma^2<\infty$), which is sometimes written $y \sim \mathcal{N}(\mu,\sigma^2)$.

The adjective "singular" reflects a singularity at $\sigma^2 = 0$ for $f_y(y;\mu,\sigma^2)$, considered as a function of σ^2. The corresponding CDF, considered as a function of y_*, has a singularity at μ. Although $y \sim \mathcal{SN}(\mu, 0)$ does not have a density, its CDF and other properties are well defined. The singular Gaussian can be appropriately treated as the limiting case of a (nonsingular) Gaussian for $\sigma^2 \to 0$.

Other singularities arise if $\mu \to \pm\infty$ or $\sigma^2 \to +\infty$ (separately). Although such conditions arise naturally as limiting cases, it is very common, although not universal, to exclude them from the definition of singular Gaussian, as we do here. In contrast, electrical engineers sometimes find it convenient to describe a theoretical source of random noise as having equal power at all frequencies. Such "white" noise corresponds to a Gaussian random variable with infinite σ^2.

No other values for $\{\mu, \sigma^2\}$ lead to a valid distribution (at least without strong side conditions). In particular, if $-\infty < \sigma^2 < 0$ and $-\infty < \mu < \infty$, then the integral of $\exp[-(y_* - \mu)^2/(2\sigma^2)]$ over the real line does not exist. Furthermore, having μ and σ^2 approach either $+\infty$ or $-\infty$ (at the same time) may lead to undefined forms. Sufficiently strong side conditions, such as having the sequences increase in closely linked and particular ways, are needed to create valid distributions. We leave consideration of such exotic cases to others.

Having carefully defined the scalar Gaussian distribution, it would be natural to derive associated properties, such as the characteristic function, MGF, CGF, moments, and cumulants. In particular, $y \sim (\mathcal{S})\mathcal{N}(\mu, \sigma^2)$ has mean $\mathrm{E}(y) = \mu$ and variance $\mathcal{V}(y) = \sigma^2$ ($-\infty < \mu < \infty$ and $0 \leq \sigma^2 < \infty$). Given that the reader is probably already familiar with properties of the scalar Gaussian, we present the scalar forms only as special cases of vector forms in the next section.

8.3 THE VECTOR ("MULTIVARIATE") GAUSSIAN DISTRIBUTION

Our approach to the vector Gaussian proceeds in three stages. First we construct random vectors as increasingly more complex linear transformations of a set of i.i.d. scalar Gaussian variables $\{z_i\} \sim \mathcal{N}(0, 1)$. Second, we describe the associated characteristic and related generating functions. Third, and finally, we demonstrate how any random vector possessing such a characteristic function can be expressed in terms of an underlying set of i.i.d. scalar Gaussian variables. No claim is made that the underlying variables were constructed directly from independent Gaussian variables. However, we are free to operate as though they were. Consequently properties of *any* vector Gaussian can be expressed in terms of properties of a set of fully independent unit Gaussian variables.

Definition 8.2 (a) An $m \times 1$ random vector $\boldsymbol{z}$ ($1 \leq m < \infty$) with i.i.d. element $z_j \sim \mathcal{N}(0, 1)$ follows a *standard vector (multivariate) Gaussian distribution*, written $\boldsymbol{z} \sim \mathcal{N}_m(\boldsymbol{0}, \boldsymbol{I})$, with $\boldsymbol{0}$ $m \times 1$ and $\boldsymbol{I}$ $m \times m$.
(b) An $m \times 1$ random vector $\boldsymbol{z}$ ($1 \leq m < \infty$) with i.i.d. element $z_j \sim \mathcal{N}(\mu_j, 1)$ follows a *standard noncentral vector (multivariate) Gaussian distribution*, written $\boldsymbol{z} \sim \mathcal{N}_m(\boldsymbol{\mu}, \boldsymbol{I})$, with $\boldsymbol{\mu} = \{\mu_j\}$ $m \times 1$ and $\boldsymbol{I}$ $m \times m$. For clarity, if $\boldsymbol{\mu} = \boldsymbol{0}$, then $\boldsymbol{z}$ may be described as standard *central* vector (multivariate) Gaussian.

(c) If $z \sim \mathcal{N}_m(\mathbf{0}, I)$ and finite constant Φ is $n \times m$ with $1 \leq \text{rank}(\Phi) = m \leq n$, the $n \times 1$ vector $y = \Phi z + \mu$ follows a *vector ("multivariate") Gaussian distribution*, written $y \sim \mathcal{N}_n(\mu, \Sigma)$, with $\Sigma = \Phi\Phi'$ $(n \times n)$ of rank m.
(d) If $m = n$, then Σ is nonsingular and y follows a *nonsingular vector (multivariate) Gaussian distribution.*
(e) If $m < n$, then Σ is singular and y follows a *singular vector (multivariate) Gaussian distribution*, written $y \sim \mathcal{SN}_n(\mu, \Sigma)$.
(f) If $m \leq n$, then Σ may or may not be singular, which may be indicated $y \sim (\mathcal{S})\mathcal{N}_n(\mu_y, \Sigma)$ for clarity, simply as $y \sim \mathcal{N}_n(\mu_y, \Sigma)$.
(g) If $z \sim \mathcal{N}_m(\mathbf{0}, I)$ while finite constant matrix Φ is $n \times m$, $n \geq m \geq 1$, and $\text{rank}(\Phi) = 0$, then $\Phi = \mathbf{0}$ and $n \times 1$ vector $y = \Phi z + \mu$, for μ finite and constant, follows a *degenerate* (or completely singular) *vector (multivariate) Gaussian distribution*, written $y \sim \mathcal{SN}_n(\mu, \mathbf{0})$. Also $\Pr\{y = \mu\} = 1$.

Example 8.1 A $\text{GLM}_{N,q}(y_i; X_i\beta, \sigma^2)$ with Gaussian errors and full rank X has

$$\begin{aligned}\hat{\beta} &= \left[(X'X)^{-1}X'\right]y \\ &\sim \mathcal{N}_q\left[\beta, (X'X)^{-1}\sigma^2\right].\end{aligned} \tag{8.12}$$

With or without full-rank X but with the requirements of full rank of $M = C(X'X)^{-}C'$ and $C = C(X'X)^{-}(X'X)$ (which ensures θ is testable),

$$\begin{aligned}\hat{\theta} - \theta_0 &= C\tilde{\beta} - \theta_0 \\ &\sim \mathcal{N}_a\left\{\theta - \theta_0, [C(X'X)^{-}C']^{-1}\sigma^2\right\}.\end{aligned} \tag{8.13}$$

For $\text{rank}(X) = r \leq q$, $H = X(X'X)^{-}X'$, $\text{rank}(H) = r$, and

$$\begin{aligned}\hat{y} &= Hy \\ &\sim \mathcal{SN}_N(XB, H\sigma^2).\end{aligned} \tag{8.14}$$

Also

$$\begin{aligned}\hat{e} &= (I - H)y \\ &\sim \mathcal{SN}_N[\mathbf{0}, (I - H)\sigma^2],\end{aligned} \tag{8.15}$$

with $\text{rank}(I - H) = N - r$. In contrast, $e \sim \mathcal{N}_N(\mathbf{0}, I\sigma^2)$.

Theorem 8.2 (a) A standard central vector Gaussian $z \sim \mathcal{N}_m(\mathbf{0}, I)$ has characteristic function $\phi_z(t) = \exp(-t't/2)$, moment generating function $m_z(t) = \exp(t't/2)$, mean $\text{E}(z) = \mathbf{0}$ $(m \times 1)$ and covariance $\mathcal{V}(z) = I$ $(m \times m)$.
(b) A standard noncentral vector Gaussian $z \sim \mathcal{N}_m(\mu, I)$ has characteristic function $\phi_z(t) = \exp(it'\mu - t't/2)$, moment generating function $m_z(t) = \exp(t'\mu + t't/2)$, mean $\text{E}(z) = \mu$ $(m \times 1)$, and covariance $\mathcal{V}(z) = I$ $(m \times m)$.

(c) The characteristic function of $\boldsymbol{y} \sim \mathcal{N}_n(\boldsymbol{\mu}, \boldsymbol{\Sigma})$ is $\phi_{\boldsymbol{y}}(\boldsymbol{t}) = \exp(\mathrm{i}\boldsymbol{t}'\boldsymbol{\mu} - \boldsymbol{t}'\boldsymbol{\Sigma}\boldsymbol{t}/2)$, and the moment generating function is $m_{\boldsymbol{y}}(\boldsymbol{t}) = \exp(\boldsymbol{t}'\boldsymbol{\mu} + \boldsymbol{t}'\boldsymbol{\Sigma}\boldsymbol{t}/2)$.
(d) A vector Gaussian $\boldsymbol{y} \sim \mathcal{N}_n(\boldsymbol{\mu}, \boldsymbol{\Sigma})$ has a distribution fully characterized by $\{\boldsymbol{\mu}, \boldsymbol{\Sigma}\}$, which are the mean vector and covariance matrix, respectively.

Proof of (a). Due to independence properties,

$$\begin{aligned}\phi_{\boldsymbol{z}}(\boldsymbol{t}) = \prod_{j=1}^{m}\phi_{z_j}(t_j) &= \prod_{j=1}^{m}\exp(-t_j^2/2)\\ &= \exp\left(-\sum_{j=1}^{m}t_j^2/2\right)\\ &= \exp(-\boldsymbol{t}'\boldsymbol{t}/2)\,. \end{aligned} \tag{8.16}$$

A parallel argument gives $m_{\boldsymbol{z}}(\boldsymbol{t})$.

The first moment is $\mathrm{E}(\boldsymbol{z}) = \{\mathrm{E}(z_j)\} = \{0\}$. Independence among elements gives $\mathcal{V}(z_j, z_{j'}) = 0$. Combining zero covariance with $\mathrm{E}(z_j^2) = 1$ gives $\mathcal{V}(\boldsymbol{z}) = \boldsymbol{I}$.

Proof of (b). Left as an exercise.

Proof of (c). Necessarily $\boldsymbol{\Sigma} = \boldsymbol{\Phi}\boldsymbol{\Phi}'$ for $\boldsymbol{\Phi}$ $m \times n$ of rank m, and $\boldsymbol{y} = \boldsymbol{\Phi}\boldsymbol{z} + \boldsymbol{\mu}$ is a linear transformation of $\boldsymbol{z} \sim \mathcal{N}_m(\boldsymbol{0}, \boldsymbol{I})$. Linear transformation properties give

$$\begin{aligned}\phi_{\boldsymbol{y}}(\boldsymbol{s}) &= \exp(\mathrm{i}\boldsymbol{s}'\boldsymbol{\mu})\phi_{\boldsymbol{z}}(\boldsymbol{\Phi}'\boldsymbol{s})\\ &= \exp(\mathrm{i}\boldsymbol{s}'\boldsymbol{\mu})\exp\left[-(\boldsymbol{\Phi}'\boldsymbol{s})'(\boldsymbol{\Phi}'\boldsymbol{s})/2\right]\\ &= \exp(\mathrm{i}\boldsymbol{s}'\boldsymbol{\mu} - \boldsymbol{s}'\boldsymbol{\Phi}\boldsymbol{\Phi}'\boldsymbol{s}/2)\\ &= \exp(\mathrm{i}\boldsymbol{s}'\boldsymbol{\mu} - \boldsymbol{s}'\boldsymbol{\Sigma}\boldsymbol{s}/2)\,. \end{aligned} \tag{8.17}$$

A parallel approach provides the MGF.

Proof of (d). The characterization by $(\boldsymbol{\mu}, \boldsymbol{\Sigma})$ is obvious from inspection of the characteristic function, coupled with the unique determination of a distribution by a characteristic function. The moments may be found by differentiating the MGF, once for the mean and twice for the second moment about zero, and setting $\boldsymbol{t} = \boldsymbol{0}$. Alternately, $\mathrm{E}(\boldsymbol{y}) = \mathrm{E}(\boldsymbol{\Phi}\boldsymbol{z} + \boldsymbol{\mu}) = \boldsymbol{\Phi}\mathrm{E}(\boldsymbol{z}) + \boldsymbol{\mu} = \boldsymbol{\mu}$ and $\mathcal{V}(\boldsymbol{y}) = \mathcal{V}(\boldsymbol{\Phi}\boldsymbol{z} + \boldsymbol{\mu}_y) = \mathcal{V}(\boldsymbol{\Phi}\boldsymbol{z}) = \boldsymbol{\Phi}\mathcal{V}(\boldsymbol{z})\boldsymbol{\Phi}' = \boldsymbol{\Phi}\boldsymbol{\Phi}' = \Sigma$. □

Lemma 8.2 Any full-rank and orthonormal transformation of a standard vector Gaussian is also standard vector Gaussian of possibly smaller dimension. More specifically, if $\boldsymbol{A}$ is constant, $m \times n$, $1 \leq m \leq n$, $\boldsymbol{A}\boldsymbol{A}' = \boldsymbol{I}_m$ (full row rank and rowwise orthonormal), and $\boldsymbol{z}_1 \sim \mathcal{N}_n(\boldsymbol{0}, \boldsymbol{I})$, then $\boldsymbol{z}_2 = \boldsymbol{A}\boldsymbol{z}_1 \sim \mathcal{N}_m(\boldsymbol{0}, \boldsymbol{I})$.

Proof. As a linear transformation of $\boldsymbol{z}_1$, the characteristic function of $m \times 1$ $\boldsymbol{z}_2$, with $m \times 1$ $\boldsymbol{s}$, is

$$\begin{aligned}\phi_{z_2}(s) = \phi_{Az_1}(s) = e^{is'0} \cdot \phi_{z_1}(A's) \\ = 1 \cdot \exp\left[-(A's)'A's/2\right] \\ = \exp(-s's/2)\,. \qquad (8.18)\end{aligned}$$

Recognizing the last expression as the characteristic function of a standard vector Gaussian (of dimension m), coupled with the unique determination of a distribution by a characteristic function, completes the proof. □

Theorem 8.3 Linear transformations of vector Gaussian variables are Gaussian (the reproductive property). If $y_2 = C_1 y_1 + c_0$ for finite constants C_1 ($n_2 \times n_1$), c_0 ($n_2 \times 1$), and $y_1 \sim \mathcal{N}_{n_1}(\mu_1, \Sigma_1)$, then $y_2 \sim \mathcal{N}_{n_2}(C_1\mu_1 + c_0, C_1\Sigma_1 C_1')$. The special case of $n_2 = 1$ implies all linear combinations of vector Gaussian variables are scalar Gaussian. As for all Gaussian variables, the rank of $\Sigma_2 = C_1\Sigma_1 C_1'$ determines whether y_2 is degenerate, singular, or nonsingular.

Proof. As a linear transformation, the characteristic function may written

$$\begin{aligned}\phi_{y_2}(s) &= \exp(is'c_0)\phi_{y_1}(C_1's) \\ &= \exp(is'c_0)\exp\left[i(C_1's)'\mu_1 - (C_1's)'\Sigma_1(C_1's)/2\right] \\ &= \exp(is'c_0 + is'C_1\mu_1 - s'C_1\Sigma_1C_1's/2) \\ &= \exp[is'(c_0 + C_1\mu_1) - s'C_1\Sigma_1C_1's/2]\,. \qquad (8.19)\end{aligned}$$

The function is the characteristic function of $y_2 \sim \mathcal{N}_{n_2}(C_1\mu_1 + c_0, C_1\Sigma_1C_1')$. □

Theorem 8.4 The central standard vector Gaussian $z \sim \mathcal{N}_n(0, I)$ has density $f_z(z_*) = (2\pi)^{-n/2}\exp(-z_*'z_*/2)$.

Proof. Due to independence properties,

$$\begin{aligned}f_z(z_*) &= \prod_{j=1}^{n} f(z_{*j}) = \prod_{j=1}^{n}\left[(2\pi)^{-1/2}\exp(-z_{*j}^2/2)\right] \\ &= (2\pi)^{-n/2}\exp\left(-\sum_{j=1}^{n} z_{*j}^2/2\right) \qquad (8.20) \\ &= (2\pi)^{-n/2}\exp(-z_*'z_*/2)\,.\end{aligned}$$

□

Corollary 8.4 If $z \sim \mathcal{N}_n(0, I)$ and $y = \Phi z + \mu$, with Φ $n \times n$ and rank n, then $\Sigma = \Phi\Phi'$ is full rank (of n, nonsingular). The nonsingular vector Gaussian $y \sim \mathcal{N}_n(\mu, \Sigma)$ is continuous in n dimensions and absolutely continuous, with density

$$f_y(y_*; \mu, \Sigma) = (2\pi)^{-n/2}|\Sigma|^{-1/2}\exp\left[-(y_* - \mu)'\Sigma^{-1}(y_* - \mu)/2\right]. \qquad (8.21)$$

Proof. Full rank of the $n \times n$ matrix $\boldsymbol{\Sigma} = \boldsymbol{\Phi}\boldsymbol{\Phi}'$ follows from full rank of $n \times n$ $\boldsymbol{\Phi}$. The spectral decomposition allows writing $\boldsymbol{\Sigma} = \boldsymbol{\Upsilon}\text{Dg}(\boldsymbol{\lambda})\boldsymbol{\Upsilon}'$, with $\boldsymbol{\Upsilon}'\boldsymbol{\Upsilon} = \boldsymbol{\Upsilon}\boldsymbol{\Upsilon}' = \boldsymbol{I}_n$. Both $\boldsymbol{\Upsilon}$ and $\text{Dg}(\boldsymbol{\lambda})$ are square ($n \times n$) and full rank, which gives $\boldsymbol{\Sigma}^{-1} = \boldsymbol{\Upsilon}\text{Dg}(\boldsymbol{\lambda})^{-1}\boldsymbol{\Upsilon}'$ and $\boldsymbol{\Phi} = \boldsymbol{\Upsilon}\text{Dg}(\boldsymbol{\lambda})^{1/2}$. In turn,

$$\boldsymbol{y} = \boldsymbol{\Phi}\boldsymbol{z} + \boldsymbol{\mu}, \tag{8.22}$$

and $\boldsymbol{y}$ is a full-rank linear transformation of $\boldsymbol{z} \sim \mathcal{N}_n(\boldsymbol{0}, \boldsymbol{I})$. The theorem ensures it is continuous in n dimensions and absolutely continuous, with density $f_{\boldsymbol{z}}(\boldsymbol{z}_*) = (2\pi)^{-n/2}\exp(-\boldsymbol{z}_*'\boldsymbol{z}_*/2)$. As a smooth and one-to-one function of an a.c. random vector, $\boldsymbol{y}$ is also. By Theorem 7.17

$$\begin{aligned} f_{\boldsymbol{y}}(\boldsymbol{y}_*; \boldsymbol{\mu}, \boldsymbol{\Sigma}) &= \left|\left|\boldsymbol{\Phi}^{-1}\right|\right| f_{\boldsymbol{z}}\left[\boldsymbol{\Phi}^{-1}(\boldsymbol{y}_* - \boldsymbol{\mu})\right] \\ &= \left|\left|\text{Dg}(\boldsymbol{\lambda})^{-1/2}\right|\right| \left|\left|\boldsymbol{\Upsilon}'\right|\right| f_{\boldsymbol{z}_R}\left[\boldsymbol{\Phi}^{-1}(\boldsymbol{y}_* - \boldsymbol{\mu})\right] \\ &= |\boldsymbol{\Sigma}|^{-1/2}(2\pi)^{-n/2}\exp\left\{-\left[\boldsymbol{\Phi}^{-1}(\boldsymbol{y}_* - \boldsymbol{\mu})\right]'\left[\boldsymbol{\Phi}^{-1}(\boldsymbol{y}_* - \boldsymbol{\mu})\right]/2\right\} \\ &= |\boldsymbol{\Sigma}|^{-1/2}(2\pi)^{-n/2}\exp\left[-(\boldsymbol{y}_* - \boldsymbol{\mu})'\left[\boldsymbol{\Phi}^{-t}\boldsymbol{\Phi}^{-1}(\boldsymbol{y}_* - \boldsymbol{\mu})\right]/2\right] \\ &= |\boldsymbol{\Sigma}|^{-1/2}(2\pi)^{-n/2}\exp\left[-(\boldsymbol{y}_* - \boldsymbol{\mu})'\boldsymbol{\Sigma}^{-1}(\boldsymbol{y}_* - \boldsymbol{\mu})/2\right]. \end{aligned} \tag{8.23}$$

□

Theorem 8.5 If $\boldsymbol{y} \sim \mathcal{N}_n(\boldsymbol{\mu}, \boldsymbol{\Sigma})$ and $\boldsymbol{\Sigma} = \boldsymbol{\Phi}\boldsymbol{\Phi}'$, with $\boldsymbol{\Phi}$ $n \times n_1$ of rank n_1 with $0 < n_1 < n$, then $\boldsymbol{y} = \boldsymbol{\Phi}\boldsymbol{z} + \boldsymbol{\mu}$ with $\boldsymbol{z} \sim \mathcal{N}_{n_1}(\boldsymbol{0}, \boldsymbol{I})$, $\text{rank}(\boldsymbol{\Sigma}) = n_1$ (nondegenerate but less than full rank, singular). The vector Gaussian $\boldsymbol{y}$ can also be expressed as the sum of linear transformations of a nonsingular vector Gaussian $\boldsymbol{y}_1$ of dimension n_1 (continuous and a.c.) and a discrete and degenerate random vector $\boldsymbol{y}_2$ of dimension $n - n_1$. Furthermore, the distribution function of $\boldsymbol{y}$ can be stated conveniently in terms of the density of the nonsingular vector Gaussian of dimension n_1. Particular forms for the results may be expressed in terms of the spectral decomposition of the $n \times n$ matrix $\boldsymbol{\Sigma}$. With $\boldsymbol{\lambda}_1$ the $n_1 \times 1$ vector of positive eigenvalues, $n_2 = n - n_1$, $\boldsymbol{0}$ of dimension $n_2 \times 1$, columnwise orthonormal $\boldsymbol{\Upsilon}_1$ of dimension $n \times n_1$, columnwise orthonormal $\boldsymbol{\Upsilon}_2$ of $n \times n_2$, and $\boldsymbol{\Upsilon}_1'\boldsymbol{\Upsilon}_2 = \boldsymbol{0}$,

$$\begin{aligned} \boldsymbol{\Sigma} &= \boldsymbol{\Upsilon}\text{Dg}(\boldsymbol{\lambda}_1, \boldsymbol{0})\boldsymbol{\Upsilon}' \\ &= [\,\boldsymbol{\Upsilon}_1 \quad \boldsymbol{\Upsilon}_2\,]\text{Dg}(\boldsymbol{\lambda}_1, \boldsymbol{0})\begin{bmatrix} \boldsymbol{\Upsilon}_1' \\ \boldsymbol{\Upsilon}_2' \end{bmatrix}. \end{aligned} \tag{8.24}$$

If

$$\begin{bmatrix} \boldsymbol{y}_1 \\ \boldsymbol{y}_2 \end{bmatrix} = \begin{bmatrix} \boldsymbol{\Upsilon}_1' \\ \boldsymbol{\Upsilon}_2' \end{bmatrix}\boldsymbol{y} = \begin{bmatrix} \boldsymbol{\Upsilon}_1'\boldsymbol{y} \\ \boldsymbol{\Upsilon}_2'\boldsymbol{y} \end{bmatrix} \tag{8.25}$$

$$\begin{bmatrix} \boldsymbol{\mu}_1 \\ \boldsymbol{\mu}_2 \end{bmatrix} = \begin{bmatrix} \boldsymbol{\Upsilon}_1' \\ \boldsymbol{\Upsilon}_2' \end{bmatrix}\boldsymbol{\mu} = \begin{bmatrix} \boldsymbol{\Upsilon}_1'\boldsymbol{\mu} \\ \boldsymbol{\Upsilon}_2'\boldsymbol{\mu} \end{bmatrix}, \tag{8.26}$$

then $\boldsymbol{y}_1 \sim \mathcal{N}_{n_1}[\boldsymbol{\mu}_1, \text{Dg}(\boldsymbol{\lambda}_1)]$ is a nonsingular vector Gaussian with density $f_{\boldsymbol{y}_1}[\boldsymbol{y}_{*1}; \boldsymbol{\mu}_1, \text{Dg}(\boldsymbol{\lambda}_1)]$ and $\boldsymbol{y}_2 \sim \mathcal{SN}_{n_2}(\boldsymbol{\mu}_2, \boldsymbol{0})$ is discrete and degenerate. Finally, the distribution function of $\boldsymbol{y}$ can be conveniently expressed in terms of $K_g = \prod_{j=1}^{n_1}(2\pi\lambda_j)^{-1/2}$ and the function

$$\begin{aligned}
g(\boldsymbol{y}_*;\boldsymbol{\mu},\boldsymbol{\Sigma}) &= K_g\exp[-(\boldsymbol{y}_*-\boldsymbol{\mu})'\boldsymbol{\Sigma}^+(\boldsymbol{y}_*-\boldsymbol{\mu})/2] \\
&= K_g\exp\left[-(\boldsymbol{y}_*-\boldsymbol{\mu})'[\boldsymbol{\Upsilon}_1\ \boldsymbol{\Upsilon}_2]\mathrm{Dg}(\boldsymbol{\lambda}_1,\mathbf{0})^+\begin{bmatrix}\boldsymbol{\Upsilon}_1'\\ \boldsymbol{\Upsilon}_2'\end{bmatrix}(\boldsymbol{y}_*-\boldsymbol{\mu})/2\right] \\
&= K_g\exp\left[-\begin{bmatrix}(\boldsymbol{y}_{*1}-\boldsymbol{\mu}_1)\\ (\boldsymbol{y}_{*2}-\boldsymbol{\mu}_2)\end{bmatrix}'\begin{bmatrix}\mathrm{Dg}(\boldsymbol{\lambda}_1)^{-1} & \mathbf{0}\\ \mathbf{0} & \mathbf{0}\end{bmatrix}\begin{bmatrix}(\boldsymbol{y}_{*1}-\boldsymbol{\mu}_1)\\ (\boldsymbol{y}_{*2}-\boldsymbol{\mu}_2)\end{bmatrix}/2\right] \\
&= f_{\boldsymbol{y}_1}[\boldsymbol{y}_{*1};\boldsymbol{\mu}_1,\mathrm{Dg}(\boldsymbol{\lambda}_1)]\,.
\end{aligned} \tag{8.27}$$

Function $g(\boldsymbol{y}_*;\boldsymbol{\mu},\boldsymbol{\Sigma})$ serves as a pseudodensity for $\boldsymbol{y}$ in the following sense. With $n_1 \times 1$ $\boldsymbol{y}_{*1} = \boldsymbol{\Upsilon}_1'\boldsymbol{y}_*$,

$$\begin{aligned}
\Pr\{\boldsymbol{y} \le \boldsymbol{y}_0\} &= \Pr\{\boldsymbol{y}_1 \le \boldsymbol{y}_{0,1}\} \\
&= \int_{-\infty}^{y_{0,1,1}}\cdots\int_{-\infty}^{y_{0,1,n_1}} f_{\boldsymbol{y}_1}[\boldsymbol{y}_{*1};\boldsymbol{\mu}_1,\mathrm{Dg}(\boldsymbol{\lambda}_1)]\,dy_{*1,1}\cdots dy_{*1,n_1}\,.
\end{aligned} \tag{8.28}$$

However, the integration is over $\Re^{n_1} \subset \Re^n$ (and therefore does not meet all requirements for the definition of a density). Explicit and valid integration over $\Re^n$ requires a more general type of integration (Riemann-Stieltjes).

Proof. The outer product $\boldsymbol{\Sigma}$ is symmetric and either positive definite or positive semidefinite. Having $\mathrm{rank}(\boldsymbol{\Sigma}) = n_1$ and $0 < n_1 < n$ ensures $n_0 = n - n_1 > 0$. Therefore the dimensions and nature of the spectral decomposition are as claimed.

Expressing $\boldsymbol{y}$ in terms of a sum of nonsingular and degenerate Gaussian variables is achieved simply by considering three particular characteristic functions. Applying the theorem for the characteristic function of a linear transformation to $\boldsymbol{y}_1 = \boldsymbol{\Upsilon}_1'\boldsymbol{y}$ gives $\phi_{\boldsymbol{y}_1}(\boldsymbol{t}_1) = \exp[\mathrm{i}\boldsymbol{t}_1'\boldsymbol{\mu}_1 - \boldsymbol{t}_1'\mathrm{Dg}(\boldsymbol{\lambda}_1)\boldsymbol{t}_1/2]$ as the characteristic function of the $n_1 \times 1$ random vector $\boldsymbol{y}_1 \sim \mathcal{N}_{n_1}[\boldsymbol{\mu}_1,\mathrm{Dg}(\boldsymbol{\lambda}_1)]$. The full rank of n_1 for $\mathrm{Dg}(\boldsymbol{\lambda}_1)$ ensures $\boldsymbol{y}_1$ is continuous and a.c. (by the previous theorem). Similar reasoning with $\boldsymbol{y}_2 = \boldsymbol{\Upsilon}_2'\boldsymbol{y}$ gives $\phi_{\boldsymbol{y}_2}(\boldsymbol{t}_2) = \exp(\mathrm{i}\boldsymbol{t}_2'\boldsymbol{\mu}_2)$ as the characteristic function of the $n_2 \times 1$ degenerate random vector $\boldsymbol{y}_2 \sim \mathcal{SN}_{n_2}(\boldsymbol{\mu}_2,\mathbf{0})$, with $\Pr\{\boldsymbol{y}_2 = \boldsymbol{\mu}_2\} = 1$ (and therefore discrete). Using the spectral decomposition of $\boldsymbol{\Sigma}$ to expand and then simplify the characteristic function of $\boldsymbol{y}$ gives

$$\begin{aligned}
\phi_{\boldsymbol{y}}(\boldsymbol{t}) &= \exp(\mathrm{i}\boldsymbol{t}'\boldsymbol{\mu} - \boldsymbol{t}'\boldsymbol{\Sigma}\boldsymbol{t}/2) \\
&= \exp\left(\mathrm{i}\boldsymbol{t}'[\boldsymbol{\Upsilon}_1\ \boldsymbol{\Upsilon}_2]\begin{bmatrix}\boldsymbol{\Upsilon}_1'\\ \boldsymbol{\Upsilon}_2'\end{bmatrix}\boldsymbol{\mu} - \boldsymbol{t}'[\boldsymbol{\Upsilon}_1\ \boldsymbol{\Upsilon}_2]\mathrm{Dg}(\boldsymbol{\lambda}_1,\mathbf{0})\begin{bmatrix}\boldsymbol{\Upsilon}_1'\\ \boldsymbol{\Upsilon}_2'\end{bmatrix}\boldsymbol{t}/2\right) \\
&= \exp\left(\mathrm{i}[\boldsymbol{t}_1'\ \boldsymbol{t}_2']\begin{bmatrix}\boldsymbol{\mu}_1\\ \boldsymbol{\mu}_2\end{bmatrix} - [\boldsymbol{t}_1'\ \boldsymbol{t}_2']\mathrm{Dg}(\boldsymbol{\lambda}_1,\mathbf{0})\begin{bmatrix}\boldsymbol{t}_1\\ \boldsymbol{t}_2\end{bmatrix}/2\right) \\
&= \exp[\mathrm{i}(\boldsymbol{t}_2'\boldsymbol{\mu}_2 + \boldsymbol{t}_1'\boldsymbol{\mu}_1) - \boldsymbol{t}_1'\mathrm{Dg}(\boldsymbol{\lambda}_1)\boldsymbol{t}_1/2] \\
&= \exp(\mathrm{i}\boldsymbol{t}_2'\boldsymbol{\mu}_2)\exp[\mathrm{i}\boldsymbol{t}_1'\boldsymbol{\mu}_1 - \boldsymbol{t}_1'\mathrm{Dg}(\boldsymbol{\lambda}_1)\boldsymbol{t}_1/2] \\
&= \phi_{\boldsymbol{y}_2}(\boldsymbol{t}_2)\phi_{\boldsymbol{y}_1}(\boldsymbol{t}_1)\,.
\end{aligned} \tag{8.29}$$

The last form implies $\boldsymbol{y}_1$ and $\boldsymbol{y}_2$ are statistically independent. As eigenvectors of a symmetric matrix (chosen to be orthonormal without loss of generality), $\boldsymbol{\Upsilon} = [\,\boldsymbol{\Upsilon}_1 \quad \boldsymbol{\Upsilon}_2\,]$ is square, full rank, $\boldsymbol{\Upsilon}'\boldsymbol{\Upsilon} = \boldsymbol{\Upsilon}\boldsymbol{\Upsilon}'$, and $(\boldsymbol{\Upsilon})^{-1} = \boldsymbol{\Upsilon}'$. Hence

$$
\begin{aligned}
\boldsymbol{y} &= [\,\boldsymbol{\Upsilon}_1\,\boldsymbol{\Upsilon}_2\,]\begin{bmatrix}\boldsymbol{\Upsilon}_1' \\ \boldsymbol{\Upsilon}_2'\end{bmatrix}\boldsymbol{y} \\
&= [\,\boldsymbol{\Upsilon}_1\,\boldsymbol{\Upsilon}_2\,]\begin{bmatrix}\boldsymbol{y}_1 \\ \boldsymbol{y}_2\end{bmatrix} = \boldsymbol{\Upsilon}_1\boldsymbol{y}_1 + \boldsymbol{\Upsilon}_2\boldsymbol{y}_2\,, \qquad (8.30)
\end{aligned}
$$

which demonstrates $\boldsymbol{y}$ is a full-rank (and invertible) linear transformation of $[\,\boldsymbol{y}_1' \quad \boldsymbol{y}_2'\,]$ and also a sum of linear transformations of $\boldsymbol{y}_1$ and $\boldsymbol{y}_2$.

If $g(\boldsymbol{y}_*;\boldsymbol{\mu},\boldsymbol{\Sigma})$ allows computing the CDF of $\boldsymbol{y}$ as claimed, then it must be the density of $\boldsymbol{y}_1$. If $\boldsymbol{A}$ is $m \times n$, $\boldsymbol{P}$ is $h \times m$ with $\boldsymbol{P}'\boldsymbol{P} = \boldsymbol{I}_m$, and $\boldsymbol{Q}$ is $n \times p$ with $\boldsymbol{Q}'\boldsymbol{Q} = \boldsymbol{I}_p$, then $(\boldsymbol{PAQ})^+ = \boldsymbol{Q}^+\boldsymbol{A}^+\boldsymbol{P}^+ = \boldsymbol{Q}'\boldsymbol{A}^+\boldsymbol{P}'$ (Lemma 1.15). Therefore $\boldsymbol{\Sigma}^+ = \boldsymbol{\Upsilon}\mathrm{Dg}(\boldsymbol{\lambda}_1,\mathbf{0})^+\boldsymbol{\Upsilon}'$. With $K = (2\pi)^{-n_1/2}\prod_{j=1}^{n_1}\lambda_j^{-1/2}$, the function is

$$
\begin{aligned}
f[\boldsymbol{y}_{*1};\boldsymbol{\mu}_1,\mathrm{Dg}(\boldsymbol{\lambda}_1)] &= K\exp[-(\boldsymbol{y}_{*1}-\boldsymbol{\mu}_1)'\mathrm{Dg}(\boldsymbol{\lambda}_1)^{-1}(\boldsymbol{y}_{*1}-\boldsymbol{\mu}_1)/2] \\
&= K\exp\left[-\begin{bmatrix}(\boldsymbol{y}_{*1}-\boldsymbol{\mu}_1) \\ (\boldsymbol{y}_{*2}-\boldsymbol{\mu}_2)\end{bmatrix}'\begin{bmatrix}\mathrm{Dg}(\boldsymbol{\lambda}_1)^{-1} & \mathbf{0} \\ \mathbf{0} & \mathbf{0}\end{bmatrix}\begin{bmatrix}(\boldsymbol{y}_{*1}-\boldsymbol{\mu}_1) \\ (\boldsymbol{y}_{*2}-\boldsymbol{\mu}_2)\end{bmatrix}/2\right] \\
&= K\exp\left[-\begin{bmatrix}(\boldsymbol{\Upsilon}_1'\boldsymbol{y}_{*1}-\boldsymbol{\Upsilon}_1'\boldsymbol{\mu}) \\ (\boldsymbol{\Upsilon}_2'\boldsymbol{y}_{*2}-\boldsymbol{\Upsilon}_2'\boldsymbol{\mu})\end{bmatrix}'\begin{bmatrix}\mathrm{Dg}(\boldsymbol{\lambda}_1)^{-1} & \mathbf{0} \\ \mathbf{0} & \mathbf{0}\end{bmatrix}\begin{bmatrix}(\boldsymbol{\Upsilon}_1'\boldsymbol{y}_{*}-\boldsymbol{\Upsilon}_1'\boldsymbol{\mu}) \\ (\boldsymbol{\Upsilon}_2'\boldsymbol{y}_{*}-\boldsymbol{\Upsilon}_2'\boldsymbol{\mu})\end{bmatrix}/2\right] \\
&= K\exp\left[-(\boldsymbol{y}_*-\boldsymbol{\mu})'[\,\boldsymbol{\Upsilon}_1 \quad \boldsymbol{\Upsilon}_2\,]\mathrm{Dg}(\boldsymbol{\lambda}_1,\mathbf{0})^+\begin{bmatrix}\boldsymbol{\Upsilon}_1' \\ \boldsymbol{\Upsilon}_2'\end{bmatrix}(\boldsymbol{y}_*-\boldsymbol{\mu})/2\right] \\
&= K\exp[-(\boldsymbol{y}_*-\boldsymbol{\mu})'\boldsymbol{\Sigma}^+(\boldsymbol{y}_*-\boldsymbol{\mu})/2] \\
&= g(\boldsymbol{y}_*;\boldsymbol{\mu},\boldsymbol{\Sigma})\,. \qquad (8.31)
\end{aligned}
$$

If $g(\boldsymbol{y}_*;\boldsymbol{\mu},\boldsymbol{\Sigma})$ allows computing the CDF of $\boldsymbol{y}$, it must suffice to consider only $\boldsymbol{y}_1$ in computing the CDF of $\boldsymbol{y}$. Here $\boldsymbol{y}$ and $[\,\boldsymbol{y}_1' \quad \boldsymbol{y}_1'\,]$ are one-to-one functions of each other because $\boldsymbol{y} = \boldsymbol{\Upsilon}[\,\boldsymbol{y}_1' \quad \boldsymbol{y}_2'\,]'$, although the transformation is not smooth. Both are of dimension $n = n_1 + n_2 > n_1$, while $\boldsymbol{y}_1$ has dimension n_1 and $\boldsymbol{y2}$ has dimension n_2. The degenerate nature of $\boldsymbol{y}_2$ gives $\Pr\{\boldsymbol{y}_2 = \boldsymbol{\mu}_0\} = 1$, which ensures $[\,\boldsymbol{y}_1' \quad \boldsymbol{y}_2'\,] = [\,\boldsymbol{y}_1' \quad \boldsymbol{\mu}_2'\,]$, with probability 1. Therefore the n_1 dimensions of $\boldsymbol{y}_1$ capture all of the randomness in $\boldsymbol{y}$, while the n_2 dimensions of $\boldsymbol{y}_2$ affect only the location (the mean of $\boldsymbol{y}$). Intuitively, a constant carries no information about a (truly) random vector.

The argument may be formalized precisely by generalizing the concept of integration to cover computing probabilities for any combination of continuous and discrete random vectors. The interested reader may consult Lindgren (1976) for a more detailed treatment of the basic ideas behind Riemann-Stieltjes integration. □

The dimensions of the matrices change between the first two equations in the sequence, corresponding to a change from $\Re^{n_1}$ to $\Re^{n} = \Re^{n_1} + \Re^{n_0}$. Here the plus sign indicates no overlap between the two subspaces. The form $\boldsymbol{y}_2 - \boldsymbol{\mu}_2$ occurs only with the additional n_2-dimensional subspace and is guaranteed to be

$\boldsymbol{\mu}_2 - \boldsymbol{\mu}_2 = \mathbf{0}$. The expression for $g()$ is not unique because the same result may be computed by replacing $\boldsymbol{\Sigma}^+$ with any choice of $\boldsymbol{\Sigma}^-$ (of which infinitely many exist, except when $\boldsymbol{\Sigma}$ is full rank).

It is often convenient to express $\boldsymbol{y} \sim (\mathcal{S})\mathcal{N}_n(\boldsymbol{\mu}_y, \boldsymbol{\Sigma})$, with $\boldsymbol{\Sigma} = \boldsymbol{\Phi}\boldsymbol{\Phi}'$, explicitly in terms of i.i.d. standard Gaussian variables. If $\text{rank}(\boldsymbol{\Sigma}) = n$, then $\boldsymbol{\Phi}$ is $n \times n$ with full rank and $\boldsymbol{z} = \boldsymbol{\Phi}^{-1}(\boldsymbol{y} - \boldsymbol{\mu}_y) \sim \mathcal{N}_n(\mathbf{0}, \boldsymbol{I})$. In turn, $\boldsymbol{y} = \boldsymbol{\Phi}(\boldsymbol{z} + \boldsymbol{\mu}_z)$ with $\boldsymbol{\mu}_z = \boldsymbol{\Phi}^{-1}\boldsymbol{\mu}_y$ and $(\boldsymbol{z} + \boldsymbol{\mu}_z) \sim \mathcal{N}_n(\boldsymbol{\mu}_z, \boldsymbol{I})$. The following lemma generalizes the decomposition to the singular case. The result is very helpful in deriving the distribution of a general quadratic form, as will be seen in the next chapter.

Lemma 8.3 If $\boldsymbol{y} \sim (\mathcal{S})\mathcal{N}_n(\boldsymbol{\mu}_y, \boldsymbol{\Sigma})$ with finite $\boldsymbol{\mu}_y$ and finite $\boldsymbol{\Sigma}$ of rank $0 < n_1 \leq n$, then $n \times n_1$ finite constant $\boldsymbol{\Phi}_1$ exists such that $\boldsymbol{\Sigma} = \boldsymbol{\Phi}_1\boldsymbol{\Phi}_1'$. If $\boldsymbol{\mu}_z = (\boldsymbol{\Phi}_1'\boldsymbol{\Phi}_1)^{-1}\boldsymbol{\Phi}_1'\boldsymbol{\mu}_y = (\boldsymbol{\Phi}_1^+)'\boldsymbol{\mu}_y$ and $\boldsymbol{z} \sim \mathcal{N}_{n_1}(\mathbf{0}, \boldsymbol{I}_{n_1})$, then $\boldsymbol{y} = \boldsymbol{\Phi}_1(\boldsymbol{z} + \boldsymbol{\mu}_z)$.

Proof. Spectral decomposition gives $\boldsymbol{\Sigma} = \boldsymbol{\Upsilon}_1 \text{Dg}(\boldsymbol{\lambda}_1)\boldsymbol{\Upsilon}_1'$ with $\boldsymbol{\Upsilon}_1'\boldsymbol{\Upsilon}_1 = \boldsymbol{I}_{n_1}$ and $\boldsymbol{\Phi}_1 = \boldsymbol{\Upsilon}_1 \text{Dg}(\boldsymbol{\lambda}_1)^{1/2}$. In turn $\boldsymbol{\Phi}_1^+ = \boldsymbol{\Phi}_1'(\boldsymbol{\Phi}_1'\boldsymbol{\Phi}_1)^{-1} = \text{Dg}(\boldsymbol{\lambda}_1)^{-1/2}\boldsymbol{\Upsilon}_1'$. Furthermore $\mathcal{V}(\boldsymbol{y}) = \boldsymbol{\Phi}_1\boldsymbol{I}_{n_1}\boldsymbol{\Phi}_1' = \boldsymbol{\Sigma}$. Requiring $\boldsymbol{\mu}_y = \boldsymbol{\Phi}_1\boldsymbol{\mu}_z$ implies $(\boldsymbol{\Phi}_1'\boldsymbol{\Phi}_1)^{-1}\boldsymbol{\Phi}_1'\boldsymbol{\mu}_y = \boldsymbol{\mu}_z$. As a linear transformation of $\boldsymbol{z}$, the vector $\boldsymbol{y}$ is Gaussian. □

8.4 MARGINAL DISTRIBUTIONS

Theorem 8.6 If $\boldsymbol{y} \sim \mathcal{N}_{n_1+n_2}(\boldsymbol{\mu}, \boldsymbol{\Sigma})$ with

$$\boldsymbol{y} = \begin{bmatrix} \boldsymbol{y}_1 \\ \boldsymbol{y}_2 \end{bmatrix} \tag{8.32}$$

$$\boldsymbol{\mu} = \begin{bmatrix} \boldsymbol{\mu}_1 \\ \boldsymbol{\mu}_2 \end{bmatrix} \tag{8.33}$$

$$\boldsymbol{\Sigma} = \begin{bmatrix} \boldsymbol{\Sigma}_{11} & \boldsymbol{\Sigma}_{12} \\ \boldsymbol{\Sigma}_{21} & \boldsymbol{\Sigma}_{22} \end{bmatrix}, \tag{8.34}$$

$\boldsymbol{y}_j$ and $\boldsymbol{\mu}_j$ are $n_j \times 1$ for $j \in \{1, 2\}$, while $\boldsymbol{\Sigma}_{jj'}$ is $n_j \times n_{j'}$, then the marginal distribution of $\boldsymbol{y}_j$ is $\mathcal{N}_{n_j}(\boldsymbol{\mu}_j, \boldsymbol{\Sigma}_{jj})$. More generally, all marginal distributions of a vector Gaussian are vector Gaussian.

Proof. The characteristic function of the marginal of $\boldsymbol{y}_1$ is

$$\begin{aligned} \phi_{\boldsymbol{y}_1}(\boldsymbol{t}_1) &= \phi_{\boldsymbol{y}}\left(\begin{bmatrix} \boldsymbol{t}_1 \\ \mathbf{0} \end{bmatrix}\right) \\ &= \exp\left\{ \mathrm{i}[\boldsymbol{t}_1' \ \mathbf{0}'] \begin{bmatrix} \boldsymbol{\mu}_1 \\ \boldsymbol{\mu}_2 \end{bmatrix} - [\boldsymbol{t}_1' \ \mathbf{0}'] \begin{bmatrix} \boldsymbol{\Sigma}_{11} & \boldsymbol{\Sigma}_{21} \\ \boldsymbol{\Sigma}_{21} & \boldsymbol{\Sigma}_{22} \end{bmatrix} \begin{bmatrix} \boldsymbol{t}_1 \\ \mathbf{0} \end{bmatrix} / 2 \right\} \\ &= \exp(\mathrm{i}\boldsymbol{t}_1'\boldsymbol{\mu}_1 - \boldsymbol{t}_1'\boldsymbol{\Sigma}_{11}\boldsymbol{t}_1/2), \end{aligned} \tag{8.35}$$

which is the characteristic function of the $\mathcal{N}_{n_1}(\boldsymbol{\mu}_1, \boldsymbol{\Sigma}_{11})$ distribution. More generally, the result does not depend on the order of the variables. It applies to any rearrangement and therefore to all subsets of $\{y_i\}$. □

Finding a marginal distribution of a vector Gaussian is very easy. One simply selects the corresponding elements of $\boldsymbol{\mu}$ and corresponding rows and columns of $\boldsymbol{\Sigma}$ to get the parameters of the marginal Gaussian. The property does not necessarily hold for other distributions.

The converse of the theorem is not true, because joint distributions do exist which have Gaussian marginal distributions but do *not* have a *joint* Gaussian distribution. Having Gaussian marginal distributions does *not* ensure a Gaussian joint distribution. More specifically, having $\boldsymbol{y}_j \sim \mathcal{N}_{n_j}(\boldsymbol{\mu}_j, \boldsymbol{\Sigma}_{jj})$ for $j \in \{1, 2\}$ and $\boldsymbol{\Sigma} = \mathcal{V}(\boldsymbol{y}) = \mathcal{V}\left(\begin{bmatrix} \boldsymbol{y}_1 \\ \boldsymbol{y}_2 \end{bmatrix}\right) = \begin{bmatrix} \boldsymbol{\Sigma}_{11} & \boldsymbol{\Sigma}_{21} \\ \boldsymbol{\Sigma}_{21} & \boldsymbol{\Sigma}_{22} \end{bmatrix}$ does *not* guarantee $\boldsymbol{y} \sim \mathcal{N}_{n_1+n_2}(\boldsymbol{\mu}, \boldsymbol{\Sigma})$. Proving the result may be a useful exercise.

8.5 INDEPENDENCE

Theorem 8.7 Assuming $\boldsymbol{y} \sim \mathcal{N}_N(\boldsymbol{\mu}, \boldsymbol{\Sigma})$ with $\boldsymbol{y}$, $\boldsymbol{\mu}$, and $\boldsymbol{\Sigma}$ partitioned correspondingly allows writing

$$\boldsymbol{y} = \begin{bmatrix} \boldsymbol{y}_1 \\ \boldsymbol{y}_2 \\ \vdots \\ \boldsymbol{y}_J \end{bmatrix} \sim \mathcal{N}_N \left(\begin{bmatrix} \boldsymbol{\mu}_1 \\ \boldsymbol{\mu}_2 \\ \cdots \\ \boldsymbol{\mu}_J \end{bmatrix}, \begin{bmatrix} \boldsymbol{\Sigma}_{11} & \boldsymbol{\Sigma}_{12} & \cdots & \boldsymbol{\Sigma}_{1J} \\ \boldsymbol{\Sigma}_{21} & \boldsymbol{\Sigma}_{22} & \cdots & \boldsymbol{\Sigma}_{2J} \\ \vdots & \vdots & \ddots & \vdots \\ \boldsymbol{\Sigma}_{J1} & \boldsymbol{\Sigma}_{J2} & \cdots & \boldsymbol{\Sigma}_{JJ} \end{bmatrix} \right), \tag{8.36}$$

with $\boldsymbol{y}_j$ of dimension $n_j \times 1$ and $\sum_{j=1}^J n_j = N$.
The J random vectors $\{\boldsymbol{y}_j, j \in \{1, 2, \ldots, k\}\}$ are mutually (totally) independent
(a) if and only if $\boldsymbol{y}_j$ and $\boldsymbol{y}_{j'}$ are (pairwise) independent for all $j \neq j'$ and
(b) if and only if $\boldsymbol{\Sigma}_{jj'} = \mathbf{0}$ for all $j \neq j'$.
(c) The result does *not* hold for all distributions that are not Gaussian.

Proof. The characteristic function for $\boldsymbol{y} \sim \mathcal{N}_N(\boldsymbol{\mu}, \boldsymbol{\Sigma})$ is defined $\forall \boldsymbol{t} \in \Re^N$ as $\phi_{\boldsymbol{y}}(\boldsymbol{t}) = \exp(\mathrm{i}\boldsymbol{t}'\boldsymbol{\mu} - \boldsymbol{t}'\boldsymbol{\Sigma}\boldsymbol{t}/2)$. Partitioning $\boldsymbol{t}$ to match $\boldsymbol{y}$ gives $\boldsymbol{t}' = [\,\boldsymbol{t}_1' \ \boldsymbol{t}_2' \ \cdots \ \boldsymbol{t}_J'\,]$. Therefore

$$\begin{aligned} \phi_{\boldsymbol{y}}(\boldsymbol{t}) &= \exp\left(\mathrm{i}\sum_{j=1}^J \boldsymbol{t}_j'\boldsymbol{\mu}_j - \sum_{j=1}^J \sum_{j'=1}^J \boldsymbol{t}_j'\boldsymbol{\Sigma}_{jj'}\boldsymbol{t}_{j'}/2 \right) \\ &= \exp\left(\mathrm{i}\sum_{j=1}^J \boldsymbol{t}_j'\boldsymbol{\mu}_j - \sum_{j=1}^J \boldsymbol{t}_j'\boldsymbol{\Sigma}_{jj}\boldsymbol{t}_j/2 - \sum_{j=1}^{J-1} \sum_{j'=j+1}^J \boldsymbol{t}_j'\boldsymbol{\Sigma}_{jj'}\boldsymbol{t}_{j'} \right). \end{aligned} \tag{8.37}$$

The third summation is zero $\forall \boldsymbol{t} \in \Re^N \Leftrightarrow \boldsymbol{\Sigma}_{jj'} = \mathbf{0}\ \forall j \neq j'$. Thus $\boldsymbol{\Sigma}_{jj'} = \mathbf{0}$ $\forall j \neq j'$ if and only if

$$
\begin{aligned}
\phi_{\boldsymbol{y}}(\boldsymbol{t}) &= \exp\left(\mathrm{i}\sum_{j=1}^{J} \boldsymbol{t}_j' \boldsymbol{\mu}_j - \sum_{j=1}^{J} \boldsymbol{t}_j' \boldsymbol{\Sigma}_{jj} \boldsymbol{t}_j / 2 \right) \\
&= \prod_{j=1}^{J} \exp\left(\mathrm{i}\boldsymbol{t}_j' \boldsymbol{\mu}_j - \boldsymbol{t}_j' \boldsymbol{\Sigma}_{jj} \boldsymbol{t}_j / 2\right) \\
&= \prod_{j=1}^{J} \phi_j(\boldsymbol{t}_j) \, . \qquad (8.38)
\end{aligned}
$$

Mutual independence follows from the factorization. Gaussian distribution of each marginal follows from the form of the individual CFs. □

Corollary 8.7.1 If $n_j \equiv 1$ and $\boldsymbol{y} \sim \mathcal{N}_N(\boldsymbol{\mu}, \boldsymbol{\Sigma})$ with $\boldsymbol{\Sigma} = \mathrm{Dg}(\{\sigma_{11}, \ldots, \sigma_{NN}\})$, then the N elements of $\boldsymbol{y}$ are all mutually independent random variables, and $y_j \sim \mathcal{N}_1(\mu_j, \sigma_{jj}) \; \forall j$.

Corollary 8.7.2 If $N = 2$, $n_j \geq 1$, and

$$
\boldsymbol{y} = \begin{bmatrix} \boldsymbol{y}_1 \\ \boldsymbol{y}_2 \end{bmatrix} \sim \mathcal{N}_N\left(\begin{bmatrix} \boldsymbol{\mu}_1 \\ \boldsymbol{\mu}_2 \end{bmatrix}, \begin{bmatrix} \boldsymbol{\Sigma}_{11} & \boldsymbol{\Sigma}_{21} \\ \boldsymbol{\Sigma}_{21} & \boldsymbol{\Sigma}_{22} \end{bmatrix} \right), \qquad (8.39)
$$

then $\boldsymbol{y}_1$ and $\boldsymbol{y}_2$ are independent if and only if $\boldsymbol{\Sigma}_{12} \equiv \boldsymbol{\Sigma}_{21}' = \boldsymbol{0}$.

8.6 CONDITIONAL DISTRIBUTIONS

Theorem 8.8 For $\boldsymbol{y}_j$ and $\boldsymbol{\mu}_j$ of dimension $n_j \times 1$, $\boldsymbol{y} = [\boldsymbol{y}_1' \; \boldsymbol{y}_2']'$, $\boldsymbol{\mu} = [\boldsymbol{\mu}_1' \; \boldsymbol{\mu}_2']'$, while $\boldsymbol{\Sigma}_{jj'}$ is of dimension $n_j \times n_{j'}$ with

$$
\boldsymbol{\Sigma} = \begin{bmatrix} \boldsymbol{\Sigma}_{11} & \boldsymbol{\Sigma}_{12} \\ \boldsymbol{\Sigma}_{21} & \boldsymbol{\Sigma}_{22} \end{bmatrix}, \qquad (8.40)
$$

and $N = n_1 + n_2$. Here $\boldsymbol{y} \sim (\mathcal{S})\mathcal{N}_N(\boldsymbol{\mu}, \boldsymbol{\Sigma})$ with $\mathrm{rank}(\boldsymbol{\Sigma}) = n_+ \leq N$.
(a) The conditional distribution of $(\boldsymbol{y}_1 | \boldsymbol{y}_2 = \boldsymbol{y}_{02})$ is $\mathcal{N}_{n_1}(\boldsymbol{\mu}_{1.2}, \boldsymbol{\Sigma}_{1.2})$ with $\boldsymbol{\mu}_{1.2} = \boldsymbol{\mu}_1 + \boldsymbol{\Sigma}_{12}\boldsymbol{\Sigma}_{22}^{-}(\boldsymbol{y}_{02} - \boldsymbol{\mu}_2)$ and $\boldsymbol{\Sigma}_{1.2} = \boldsymbol{\Sigma}_{11} - \boldsymbol{\Sigma}_{12}\boldsymbol{\Sigma}_{22}^{-}\boldsymbol{\Sigma}_{21}$.
(b) Although $\boldsymbol{\Sigma}_{22}^{-}$ is any generalized inverse of $\boldsymbol{\Sigma}_{22}$, both $\boldsymbol{\mu}_{1.2}$ and $\boldsymbol{\Sigma}_{1.2}$ are invariant to the choice of $\boldsymbol{\Sigma}_{22}^{-}$ (and therefore can be taken to be $\boldsymbol{\Sigma}_{22}^{+}$).
(c) $(\boldsymbol{y}_1 | \boldsymbol{y}_2 = \boldsymbol{y}_{02}) \sim (\mathcal{S})\mathcal{N}_{n_1}[\boldsymbol{\mu}_1 + \boldsymbol{\Sigma}_{12}\boldsymbol{\Sigma}_{22}^{+}(\boldsymbol{y}_{02} - \boldsymbol{\mu}_2), \boldsymbol{\Sigma}_{11} - \boldsymbol{\Sigma}_{12}\boldsymbol{\Sigma}_{22}^{+}\boldsymbol{\Sigma}_{21}]$. (8.41)

Proof. The following are true. (1) A matrix $\boldsymbol{B}$ exists satisfying $\boldsymbol{\Sigma}_{12} = \boldsymbol{B}\boldsymbol{\Sigma}_{22}$. (2) $\boldsymbol{B} = \boldsymbol{\Sigma}_{12}\boldsymbol{\Sigma}_{22}^{-}$ is a solution, and is unique if and only if $\boldsymbol{\Sigma}_{22}$ is nonsingular. (3) $\boldsymbol{\Sigma}_{12} - \boldsymbol{\Sigma}_{12}\boldsymbol{\Sigma}_{22}^{-}\boldsymbol{\Sigma}_{22} = \boldsymbol{0}$.

Proofs of 1 and 2 are left as an exercise. They are consequences of a slight generalization of the Cauchy-Schwartz inequality for inner products based on positive semidefinite matrices.

The proof of 3 starts with the definition of the generalized inverse, $\boldsymbol{\Sigma}_{22} = \boldsymbol{\Sigma}_{22}\boldsymbol{\Sigma}_{22}^{-}\boldsymbol{\Sigma}_{22}$. Substituting the expression into the equation $\boldsymbol{\Sigma}_{12} = \boldsymbol{B}\boldsymbol{\Sigma}_{22}$ implies $\boldsymbol{\Sigma}_{12} = \boldsymbol{B}(\boldsymbol{\Sigma}_{22}\boldsymbol{\Sigma}_{22}^{-}\boldsymbol{\Sigma}_{22}) = (\boldsymbol{B}\boldsymbol{\Sigma}_{22})\boldsymbol{\Sigma}_{22}^{-}\boldsymbol{\Sigma}_{22} = \boldsymbol{\Sigma}_{12}\boldsymbol{\Sigma}_{22}^{-}\boldsymbol{\Sigma}_{22}$.

An important result relates stochastic independence and conditional distribution. If $\boldsymbol{y}_1$ and $\boldsymbol{y}_2$ are random vectors with a joint distribution, then the conditional distribution of $\boldsymbol{y}_1|\boldsymbol{y}_2 = \boldsymbol{y}_{02}$, if it exists, is identical to the marginal distribution of $\boldsymbol{y}_1$ if and only if $\boldsymbol{y}_1$ and $\boldsymbol{y}_2$ are independent. The basic approach is to construct a random variable $\boldsymbol{x} = (\boldsymbol{y}_1|\boldsymbol{y}_2 = \boldsymbol{y}_{02})$ using the three results just stated. If $\boldsymbol{x} = \boldsymbol{A}(\boldsymbol{y} - \boldsymbol{\mu}) + \boldsymbol{c}$ in which

$$\boldsymbol{A} = \begin{bmatrix} \boldsymbol{I}_{n_1} & -\boldsymbol{\Sigma}_{12}\boldsymbol{\Sigma}_{22}^{-} \\ \boldsymbol{0} & \boldsymbol{I}_{n_2} \end{bmatrix} \tag{8.42}$$

$$\boldsymbol{c} = \begin{bmatrix} \boldsymbol{\mu}_1 + \boldsymbol{\Sigma}_{12}\boldsymbol{\Sigma}_{22}^{-}(\boldsymbol{y}_{02} - \boldsymbol{\mu}_2) \\ \boldsymbol{\mu}_2 \end{bmatrix}, \tag{8.43}$$

then $\boldsymbol{x} = [\,\boldsymbol{x}_1'\ \boldsymbol{x}_2'\,]' = \boldsymbol{y} -$ adjustment and

$$\begin{bmatrix} \boldsymbol{x}_1 \\ \boldsymbol{x}_2 \end{bmatrix} = \begin{bmatrix} \boldsymbol{y}_1 + \boldsymbol{\Sigma}_{12}\boldsymbol{\Sigma}_{22}^{-}(\boldsymbol{y}_{02} - \boldsymbol{y}_2) \\ \boldsymbol{y}_2 \end{bmatrix}. \tag{8.44}$$

Given $\boldsymbol{y}_2 = \boldsymbol{y}_{02}$, the adjustment is zero and $\boldsymbol{x} = \boldsymbol{y}$. Furthermore, if $\boldsymbol{y}_2 = \boldsymbol{y}_{02}$ then $\boldsymbol{x}_1 = \boldsymbol{y}_1$ and $(\boldsymbol{y}_1|\boldsymbol{y}_2 = \boldsymbol{y}_{02}) = (\boldsymbol{x}_1|\boldsymbol{x}_2 = \boldsymbol{y}_{02})$. Here $N \times N$ $\boldsymbol{A}$ is full rank of N and $\text{rank}(\boldsymbol{A}\boldsymbol{\Sigma}\boldsymbol{A}') = \text{rank}(\boldsymbol{\Sigma}) = n_+$. As a linear transformation of a vector Gaussian, $\boldsymbol{x}$ is also vector Gaussian with

$$\begin{aligned} \boldsymbol{x} &= \boldsymbol{A}(\boldsymbol{y} - \boldsymbol{\mu}) + \boldsymbol{c} \\ &\sim (S)\mathcal{N}_N(\boldsymbol{c}, \boldsymbol{A}\boldsymbol{\Sigma}\boldsymbol{A}')\,. \end{aligned} \tag{8.45}$$

Using 3 we can prove

$$\boldsymbol{A}\boldsymbol{\Sigma}\boldsymbol{A}' = \begin{bmatrix} \boldsymbol{\Sigma}_{11} - \boldsymbol{\Sigma}_{12}\boldsymbol{\Sigma}_{22}^{-}\boldsymbol{\Sigma}_{21} & \boldsymbol{0} \\ \boldsymbol{0} & \boldsymbol{\Sigma}_{22} \end{bmatrix}. \tag{8.46}$$

Therefore $\boldsymbol{x}_1 \perp\!\!\!\perp \boldsymbol{x}_2$, which ensures the distribution of $(\boldsymbol{x}_1|\boldsymbol{x}_2 = \boldsymbol{y}_{02})$ is identical to the distribution of $\boldsymbol{x}_1$. In turn, the distribution of $(\boldsymbol{y}_1|\boldsymbol{y}_2 = \boldsymbol{y}_{02})$ coincides with the distributions of $(\boldsymbol{x}_1|\boldsymbol{x}_2 = \boldsymbol{y}_{02})$ and therefore $\boldsymbol{x}_1$. Finally, $\boldsymbol{c} = [\,\boldsymbol{\mu}_{1.2}'\ \boldsymbol{\mu}_2'\,]'$ and

$$\boldsymbol{A}\boldsymbol{\Sigma}\boldsymbol{A}' = \begin{bmatrix} \boldsymbol{\Sigma}_{1.2} & \boldsymbol{0} \\ \boldsymbol{0} & \boldsymbol{\Sigma}_{22} \end{bmatrix} \tag{8.47}$$

means $\boldsymbol{x}_1 \sim (S)\mathcal{N}_{n_1}(\boldsymbol{\mu}_{1.2}, \boldsymbol{\Sigma}_{1.2})$, which is also the distribution of $(\boldsymbol{y}_1|\boldsymbol{y}_2 = \boldsymbol{y}_{02})$. □

Corollary 8.8.1 If $\boldsymbol{\Sigma}$ is full rank ($n_+ = n_1 + n_2 = N$), then
(a) $\boldsymbol{\Sigma}_{22}$ has full rank of n_2,
(b) $\boldsymbol{\Sigma}_{1.2}$ has full rank of n_1, and
(c) $(\boldsymbol{y}_1|\boldsymbol{y}_2 = \boldsymbol{y}_{02}) \sim \mathcal{N}_{n_1}(\boldsymbol{\mu}_{1.2}, \boldsymbol{\Sigma}_{1.2})$ has a density with $\boldsymbol{\mu}_{1.2} = \boldsymbol{\mu}_1 + \boldsymbol{\Sigma}_{12}\boldsymbol{\Sigma}_{22}^{-1}(\boldsymbol{y}_{02} - \boldsymbol{\mu}_2)$ and $\boldsymbol{\Sigma}_{1.2} = \boldsymbol{\Sigma}_{11} - \boldsymbol{\Sigma}_{12}\boldsymbol{\Sigma}_{22}^{-1}\boldsymbol{\Sigma}_{21}$.
(d) The dispersion matrix $\boldsymbol{\Sigma}_{1.2}$ is not a function of $\boldsymbol{y}_2$, and equals $\boldsymbol{\Sigma}_{11}$ less an

adjustment which vanishes when $\boldsymbol{\Sigma}_{12} = \mathbf{0}$.
(e) In contrast, mean $\boldsymbol{\mu}_{1.2}$ *is* a linear function of $\boldsymbol{y}_{02}$ and equals $\boldsymbol{\mu}_1$ plus an adjustment which vanishes if $\boldsymbol{\Sigma}_{12} = \mathbf{0}$.

Proof. Both $\boldsymbol{y}_1$ and $\boldsymbol{y}_2$ have marginal densities as well as a joint density. The general form for the PDF of a conditional vector gives, with $\boldsymbol{y}_{*,0} = [\,\boldsymbol{y}'_{*1}\ \boldsymbol{y}'_{02}\,]'$,

$$
\begin{aligned}
f_{1|2}(\boldsymbol{y}_{*1}|\boldsymbol{y}_2 = \boldsymbol{y}_{02}) &= \frac{f_{\boldsymbol{y}}(\boldsymbol{y}_{*,0};\boldsymbol{\mu},\boldsymbol{\Sigma})}{f_{\boldsymbol{y}_2}(\boldsymbol{y}_{02};\boldsymbol{\mu}_2,\boldsymbol{\Sigma}_{22})} \\
&= \frac{(2\pi)^{-N/2}|\boldsymbol{\Sigma}|^{-1/2}\exp[-(\boldsymbol{y}_{*,0}-\boldsymbol{\mu})'\boldsymbol{\Sigma}^{-1}(\boldsymbol{y}_{*,0}-\boldsymbol{\mu})/2]}{(2\pi)^{-n_2/2}|\boldsymbol{\Sigma}_{22}|^{-1/2}\exp[-(\boldsymbol{y}_{02}-\boldsymbol{\mu}_2)'\boldsymbol{\Sigma}_{22}^{-1}(\boldsymbol{y}_{02}-\boldsymbol{\mu}_2)/2]} \\
&\ \ \vdots \qquad \text{(details omitted)} \\
&= (2\pi)^{-n_1/2}|\boldsymbol{\Sigma}_{1.2}|^{-1/2}\exp\left[-(\boldsymbol{y}_{*1}-\boldsymbol{\mu}_{1.2})'\boldsymbol{\Sigma}_{1.2}^{-1}(\boldsymbol{y}_{*1}-\boldsymbol{\mu}_{1.2})/2\right].
\end{aligned}
\tag{8.48}
$$

Details are left as an exercise. □

Corollary 8.8.2 The validity of a multivariate general linear model data analysis with (random) Gaussian predictors depends on *conditional* distribution results. For independent sampling unit $i \in \{1, \ldots, N\}$, the results can be formally stated in terms of $\boldsymbol{u}_i = [\,\boldsymbol{u}'_{i1}\ \boldsymbol{u}'_{i2}\,]' \sim \mathcal{N}_{p+q}(\boldsymbol{\mu}, \boldsymbol{\Sigma})$ with full rank $\boldsymbol{\Sigma}$ and $\boldsymbol{u}_i$ independent of $\boldsymbol{u}_{i'}$ for $i \neq i'$. In turn, $\text{row}_i(\boldsymbol{Y}) = (\boldsymbol{u}_{i1}|\boldsymbol{u}_{i2} = \boldsymbol{u}_{i02})'$ implies $\boldsymbol{Y}$ is $N \times p$, $\text{row}_i(\boldsymbol{X}) = [\,1\ \ (\boldsymbol{u}_{i2}-\boldsymbol{\mu}_{i2})'\,]$ implies $\boldsymbol{X}$ is $N \times q$, and $\boldsymbol{B}$ ($q \times p$) is

$$
\boldsymbol{B} = \begin{bmatrix} \boldsymbol{\mu}'_1 \\ (\boldsymbol{\Sigma}_{12}\boldsymbol{\Sigma}_{22}^{-})' \end{bmatrix}. \tag{8.49}
$$

Under the conditions just stated, $\boldsymbol{Y}$, $\boldsymbol{X}$, and $\boldsymbol{B}$ satisfy the assumptions of the $\text{GLM}_{N,p,q}(\boldsymbol{Y}_i; \boldsymbol{X}_i\boldsymbol{B}, \boldsymbol{\Sigma}_{1.2})$ with Gaussian errors and

$$
\begin{aligned}
\boldsymbol{Y} &= \boldsymbol{X}\boldsymbol{B} \qquad\qquad\qquad + \boldsymbol{E} \\
\begin{bmatrix} (\boldsymbol{u}_{11}|\boldsymbol{u}_{12} = \boldsymbol{u}_{102})' \\ (\boldsymbol{u}_{21}|\boldsymbol{u}_{22} = \boldsymbol{u}_{202})' \\ \vdots \\ (\boldsymbol{u}_{N1}|\boldsymbol{u}_{N2} = \boldsymbol{u}_{N02})' \end{bmatrix} &= \begin{bmatrix} 1 & (\boldsymbol{u}_{12} - \boldsymbol{\mu}_{12})' \\ 1 & (\boldsymbol{u}_{22} - \boldsymbol{\mu}_{22})' \\ \vdots & \vdots \\ 1 & (\boldsymbol{u}_{N2} - \boldsymbol{\mu}_{N2})' \end{bmatrix} \boldsymbol{B} + \boldsymbol{E}.
\end{aligned}
\tag{8.50}
$$

Also $\mathcal{E}[\text{row}_i(\boldsymbol{Y})|\boldsymbol{X}] = \text{row}_i(\boldsymbol{X})\boldsymbol{B}$. If $p = 1$, we have the $\text{GLM}_{N,q}(y_i; \boldsymbol{X}_i\boldsymbol{\beta}, \sigma^2_{1.2})$ with Gaussian errors and $\sigma^2_{1.2} = \sigma_{11} - \boldsymbol{\Sigma}_{12}\boldsymbol{\Sigma}_{22}^{-}\boldsymbol{\Sigma}_{21}$.

Proof. Left as an exercise.

Consequently, if we have $p+q$ random variables with a joint Gaussian distribution, we may analyze p of the variables conditional upon the observed values of the other q variables. We merely use standard general linear models with the observed values of the given variables in the design matrix $\boldsymbol{X}$. The general linear model requires $\boldsymbol{X}$ fixed, constant, and known without (appreciable) error. In a conditional analysis, $\boldsymbol{u}_{i2} = \boldsymbol{u}_{i02}$ and the assumption is met.

The process of *conditioning*, as used here, converts a random variable into a constant, namely the observed value. One can therefore treat a data vector either as a random vector or as a fixed constant vector by conditioning on the realized observation. In the conditional case, one makes inferences about the distribution of $\boldsymbol{u}_1|\boldsymbol{u}_2 = \boldsymbol{u}_{02}$. In linear model notation, one models $\boldsymbol{y}$ given $\boldsymbol{X}$ at its observed value. A different $\boldsymbol{X}$ matrix would produce a different conditional distribution of $\boldsymbol{y}$ and possibly different results and conclusions. However, when one analyzes the unconditional joint distribution of $\boldsymbol{u}_1$ and $\boldsymbol{u}_2$, one treats $\boldsymbol{u}_2$ as a random vector. Subsequently inferences about $\boldsymbol{u}_1$ concern its unconditional distribution, which has mean and variance different from the conditional distribution of $\boldsymbol{u}_1|\boldsymbol{u}_2 = \boldsymbol{u}_{02}$.

Example 8.2 Generally speaking, analysis based on the conditional distribution restricts the scope of inferences one can draw from the data. If $\boldsymbol{u}_1$ and $\boldsymbol{u}_2$ contain heights and weights of adult humans, then the analysis of the marginal distribution of $\boldsymbol{u}_1$ would provide information about the weights of adult humans. In contrast, the analysis of the conditional distribution of $\boldsymbol{u}_1|\boldsymbol{u}_2 = \boldsymbol{u}_{02}$ would provide information about the distribution of adult weights for the given set of heights.

The conditional Gaussian distribution provides the theoretical tools to support conditional analyses. If the data are approximately Gaussian in distribution, then the analysis has a sound theoretical basis. Unfortunately, the simplicity does not extend to the theory and computation of power analysis. Sampson (1974) provided an excellent discussion of the distinctions for the $\text{GLM}_{N,q}(y_i; \boldsymbol{X}_i\boldsymbol{\beta}, \sigma^2)$ and $\text{GLM}_{N,p,q}(\boldsymbol{Y}_i; \boldsymbol{X}_i\boldsymbol{\beta}, \boldsymbol{\Sigma})$. Some discussion is also provided in later chapters on power analysis.

8.7 ASYMPTOTIC PROPERTIES

Theorem 8.9 A central limit theorem may be stated for a set of N i.i.d. length p random vectors $\{\boldsymbol{Y}_i'\}$, with $\text{E}(\boldsymbol{Y}_i') = \boldsymbol{\mu}$ and $\mathcal{V}(\boldsymbol{Y}_i') = \boldsymbol{\Sigma}$, of rank p.

(a) With $\xrightarrow{\mathcal{D}}$ indicating convergence in distribution (law), the $p \times 1$ vector

$$\overline{\boldsymbol{y}}_N = \sum_{i=1}^{N} \boldsymbol{Y}_i'/N = [\boldsymbol{Y}_1' \ \boldsymbol{Y}_2' \ \cdots \ \boldsymbol{Y}_N']\boldsymbol{1}_N/N \tag{8.51}$$

has the property $\sqrt{N}(\overline{\boldsymbol{y}}_N - \boldsymbol{\mu}) \xrightarrow{\mathcal{D}} \mathcal{N}_p(\boldsymbol{0}, \boldsymbol{\Sigma})$ as $N \to \infty$.

(b) From the convergence in distribution, we infer an approximation for large N, namely $\sqrt{N}(\overline{\boldsymbol{y}}_N - \boldsymbol{\mu}) \stackrel{.}{\sim} \mathcal{N}_p(\boldsymbol{0}, \boldsymbol{\Sigma})$.

Proof. Left as an exercise.

Theorem 8.10 For $(q \times 1)$ vector $\boldsymbol{z}_{*N} = \left[f_1(\boldsymbol{y}_{*N})' \cdots f_q(\boldsymbol{y}_{*N}')\right]' = \boldsymbol{f}(\boldsymbol{y}_{*N})$ a vector of real-valued functions of $\boldsymbol{y}_{*N}$ differentiable at $\boldsymbol{y}_{*N} = \boldsymbol{\mu}$,

$$D = \left. \frac{\partial f'}{\partial y_{*N}} \right|_{y_{*N}=\mu} \tag{8.52}$$

exists. Equivalently, $D = \{d_{ij}\}$ and $d_{ij} = \partial f_j(y_{*N})/\partial y_{i*N}$, evaluated at $y_{*N} = \mu$. If the $(p \times 1)$ random vector $y_N \xrightarrow{\mathcal{D}} \mathcal{N}_p(\mu, \Sigma/N)$ has $\text{rank}(\Sigma) = p$, then random vector $z_N = f(y_N)$ is such

$$z_N \xrightarrow{\mathcal{D}} \mathcal{N}_q[f(\mu), D'\Sigma D/N]. \tag{8.53}$$

Proof. Mardia, Kent, and Bibby (1979) provided a proof.

8.8 THE MATRIX GAUSSIAN DISTRIBUTION

Definition 8.3 An $n \times p$ random matrix Y will be said to follow a *general matrix Gaussian* distribution if and only if $\text{vec}(Y) \sim (\mathcal{S})\mathcal{N}_{np}(\mu_{\text{v}}, \Sigma_{\text{v}})$. Necessarily $\mu_{\text{v}} = \text{vec}[\text{E}(Y)]$ and $\mathcal{V}[\text{vec}(Y)] = \Sigma_{\text{v}} = \Sigma_{\text{v}}'$ is n.n.d.

Definition 8.4 The $n \times p$ random matrix Y follows a *direct-product matrix Gaussian* distribution, typically abbreviated *matrix Gaussian* and written $Y \sim \mathcal{N}_{n,p}(M, \Xi, \Sigma)$, if and only if

(a) $\text{vec}(Y) \sim (\mathcal{S})\mathcal{N}_{n,p}[\text{vec}(M), \Sigma \otimes \Xi]$; if and only if

(b) $\text{vec}(Y') \sim (\mathcal{S})\mathcal{N}_{n,p}[\text{vec}(M'), \Xi \otimes \Sigma]$; if and only if

(c) $Y = \Psi Z \Phi' + M$ with $\text{vec}(Z) \sim \mathcal{N}_{n_1 p_1}(0, I)$ and
Ψ $(n \times n_1)$ of rank $n_1 \geq 1$, $\Xi = \Psi\Psi'$,
Φ' $(p_1 \times p)$ of rank $p_1 \geq 1$, $\Sigma = \Phi\Phi'$.

Writing $Y \sim \mathcal{SN}_{n,p}(M, \Xi, \Sigma)$ indicates $n_1 = \text{rank}(\Psi) = \text{rank}(\Xi) < n$, or $p_1 = \text{rank}(\Phi) = \text{rank}(\Sigma) < p$, or both.

Writing $Y \sim (\mathcal{S})\mathcal{N}_{n,p}(M, \Xi, \Sigma)$ emphasizes allowing any combination of $n_1 \leq n$ and $p_1 \leq p$.

Certain direct-product properties help in understanding the definition. As for any direct-product matrix, $\text{rank}(\Xi \otimes \Sigma) = \text{rank}(\Xi) \cdot \text{rank}(\Sigma)$. Hence both Ξ and Σ must be nonsingular for $\Xi \otimes \Sigma$ to be nonsingular (and the distribution to have a density). Furthermore the eigenvalues of $\Xi \otimes \Sigma$ are all products of the eigenvalues of Ξ and Σ of the form $\lambda_{\Xi,j}\lambda_{\Sigma,k}$. Theorem 1.5 gives $\text{vec}(\Psi Z \Phi') = (\Phi \otimes \Psi)\text{vec}(Z)$.

The direct-product matrix Gaussian distribution has not been fully identified because, for any finite constant $a > 0$, one may write $Y \sim \mathcal{N}_{n,p}[M, (1/a)\Xi, a\Sigma]$.

The indeterminacy is not important for most applications because Ξ will be a known matrix and often $\boldsymbol{I}$.

The direct-product matrix Gaussian arises naturally in many places in linear models with Gaussian errors, while the general matrix Gaussian apparently never arises naturally. Therefore, despite the ambiguity, we typically use the abbreviation "matrix Gaussian" to indicate a direct-product matrix Gaussian. The description agrees with Arnold's (1981, p. 310) discussion of the direct product matrix Gaussian as a "matrix normal" and with Gupta and Nagar's (2000), although the latter authors use slightly different notation. In a closely related approach, Mardia, Kent, and Bibby's (1979, p. 64) definition of a "normal data matrix" corresponds to the doubly special case of a direct-product matrix Gaussian with independent rows and homogeneity of mean across rows, namely $\boldsymbol{Y} \sim \mathcal{N}_{n,p}(\mathbf{1}_n\boldsymbol{\mu}', \boldsymbol{I}_n, \Sigma)$ for $\boldsymbol{\mu}$ a $p \times 1$ vector.

Example 8.3 A counterexample demonstrates that not all matrices of jointly Gaussian variables are direct-product matrix Gaussian. If $\boldsymbol{Y}$ is 2×2 with $\text{vec}(\boldsymbol{Y}) \sim \mathcal{N}_4[\mathbf{0}, \text{Dg}(\{1,2,3,4\})]$, then $\boldsymbol{Y}$ being a direct-product matrix Gaussian requires $\{a_1, a_2, b_1, b_2\}$ exists such that $\text{Dg}(\{a_1, a_2\}) \otimes \text{Dg}(b_1, b_2) = \text{Dg}(\{1,2,3,4\})$. If so, then $\{a_1b_1 = 1, a_1b_2 = 2, a_2b_1 = 3, a_2b_2 = 4\}$ implies $a_1 = 1/b_1$ and $b_2/b_1 = 2$. However, $(a_2b_2)/(a_2b_1) = b_2/b_1 = 4/3 \neq 2$, and the equations are inconsistent (they do not have any solution).

Example 8.4 The (direct-product) matrix Gaussian permeates the $\text{GLM}_{N,p,q}(\boldsymbol{Y}_i; \boldsymbol{X}_i\boldsymbol{B}, \Sigma)$ with Gaussian errors. The model equation $\boldsymbol{Y} = \boldsymbol{XB} + \boldsymbol{E}$ has $\boldsymbol{E} \sim \mathcal{N}_{N,p}(\mathbf{0}, \boldsymbol{I}_N, \Sigma)$ and $\boldsymbol{Y} \sim \mathcal{N}_{N,p}(\boldsymbol{XB}, \boldsymbol{I}_N, \Sigma)$. Except for the special cases when $\boldsymbol{B} = \mathbf{0}$ or $\boldsymbol{X} = \mathbf{1}_N$, $\boldsymbol{Y}$ does not meet Mardia, Kent, and Bibby's definition of a normal data matrix, although $\boldsymbol{E}$ always does.

The three parameters of a (direct-product) matrix Gaussian have simple interpretations. Obviously $\text{E}(\boldsymbol{Y}) = \boldsymbol{M}$ is the matrix of expected values. The matrix Σ describes the covariance structure of the columns within a row. The observation may be stated precisely as follows. If $\boldsymbol{d}_{in}$ indicates an $n \times 1$ vector with 1 in position i and 0 everywhere else, then $\text{row}_i'(\boldsymbol{Y}) = \boldsymbol{Y}'\boldsymbol{d}_{in} = \text{vec}(\boldsymbol{d}_{in}'\boldsymbol{Y}) = (\boldsymbol{I}_p \otimes \boldsymbol{d}_{in}')\text{vec}(\boldsymbol{Y})$. The reproductive property under a linear transformation of a vector Gaussian allows writing

$$\begin{aligned} \text{row}_i'(\boldsymbol{Y}) &\sim \mathcal{N}_p\left[(\boldsymbol{I}_p \otimes \boldsymbol{d}_{in}')\text{vec}(\boldsymbol{M}), (\boldsymbol{I}_p \otimes \boldsymbol{d}_{in}')(\Sigma \otimes \Xi)(\boldsymbol{I}_p \otimes \boldsymbol{d}_{in}')'\right] \\ &\sim \mathcal{N}_p[\text{row}_i'(\boldsymbol{M}), (\boldsymbol{I}_p\Sigma\boldsymbol{I}_p) \otimes (\boldsymbol{d}_{in}'\Xi\boldsymbol{d}_{in})] \\ &\sim \mathcal{N}_p[\text{row}_i'(\boldsymbol{M}), \Sigma \otimes \xi_{ii}] \\ &\sim \mathcal{N}_p[\text{row}_i'(\boldsymbol{M}), \Sigma\xi_{ii}]\,. \end{aligned} \tag{8.54}$$

Similarly, the matrix describes the covariance structure of the rows within a column. A particular column may be written as $\text{col}_j(\boldsymbol{Y}) = \left(\boldsymbol{I}_n \otimes \boldsymbol{d}_{jp}'\right)\text{vec}(\boldsymbol{Y}')$ and

$$\begin{aligned}\text{col}_j(\boldsymbol{Y}) &\sim \mathcal{N}_n\Big[(\boldsymbol{I}_n \otimes \boldsymbol{d}'_{jp})\text{vec}(\boldsymbol{M}'), (\boldsymbol{I}_n \otimes \boldsymbol{d}'_{jp})(\boldsymbol{\Xi} \otimes \boldsymbol{\Sigma})(\boldsymbol{I}_n \otimes \boldsymbol{d}'_{jp})'\Big] \\ &\sim \mathcal{N}_n\big[\text{col}_j(\boldsymbol{M}), (\boldsymbol{I}_n \boldsymbol{\Xi} \boldsymbol{I}_n) \otimes (\boldsymbol{d}'_{jp}\boldsymbol{\Sigma}\boldsymbol{d}_{jp})\big] \\ &\sim \mathcal{N}_n[\text{col}_j(\boldsymbol{M}), \boldsymbol{\Xi} \otimes \sigma_{jj}] \\ &\sim \mathcal{N}_n[\text{col}_j(\boldsymbol{M}), \boldsymbol{\Xi}\sigma_{jj}]\,. \end{aligned} \tag{8.55}$$

A useful exercise would be to define $\text{col}_j(\boldsymbol{Y}) = \boldsymbol{Y}\boldsymbol{d}_{jp} = (\boldsymbol{d}'_{jp} \otimes \boldsymbol{I}_n)\text{vec}(\boldsymbol{Y})$ and derive the distribution of the row and column in terms of $\text{vec}(\boldsymbol{Y})$.

Theorem 8.11 (a) With $\boldsymbol{T} = [\,\boldsymbol{t}_1 \cdots \boldsymbol{t}_{p_1}\,]$ an arbitrary real $n_1 \times p_1$ matrix, the characteristic function of $n_1 \times p_1$ $\boldsymbol{Z} \sim \mathcal{N}_{n_1,p_1}(\boldsymbol{0}, \boldsymbol{I}_{n_1}, \boldsymbol{I}_{p_1}) = \{z_{jk}\} = [\,\boldsymbol{z}_1 \cdots \boldsymbol{z}_{p_1}\,]$ with i.i.d. $z_{jk} \sim \mathcal{N}(0,1)$ is $\phi(\boldsymbol{T}; \boldsymbol{Z}) = \text{E}\{\exp[\text{tr}(\text{i}\boldsymbol{T}'\boldsymbol{Z})]\} = \exp[-\text{tr}(\boldsymbol{T}'\boldsymbol{T})/2]$.
(b) For conforming constants $\boldsymbol{\Xi} = \boldsymbol{\Psi}\boldsymbol{\Psi}'$, $\boldsymbol{\Sigma} = \boldsymbol{\Phi}\boldsymbol{\Phi}'$ and $\boldsymbol{M}$, the (direct-product) matrix Gaussian $\boldsymbol{Y} = \boldsymbol{\Psi}\boldsymbol{Z}\boldsymbol{\Phi}' + \boldsymbol{M} \sim (S)\mathcal{N}_{n,p}(\boldsymbol{M}, \boldsymbol{\Xi}, \boldsymbol{\Sigma})$ has characteristic function $\phi_{\boldsymbol{Y}}(\boldsymbol{T}) = \exp[\text{i}\,\text{tr}(\boldsymbol{T}'\boldsymbol{M})]\exp[-\text{tr}(\boldsymbol{T}'\boldsymbol{\Xi}\boldsymbol{T}\boldsymbol{\Sigma})/2]$.

Proof. (a) Independence allows the following exchanges of operations: $\text{E}\{\exp[\text{tr}(\text{i}\boldsymbol{T}'\boldsymbol{Z})]\} = \text{E}\{\exp[\sum_{j=1}^{p_1}(\text{i}\boldsymbol{t}'_j\boldsymbol{z}_j)]\} = \text{E}\{\exp[\sum_{j=1}^{p_1}\sum_{k=1}^{n_1}(\text{i}t_{kj}z_{kj})]\} =$ $\text{E}[\prod_{j=1}^{p_1}\prod_{k=1}^{n_1}\exp(\text{i}t_{kj}z_{kj})] = \prod_{j=1}^{p_1}\prod_{k=1}^{n_1}\text{E}[\exp(\text{i}t_{kj}z_{kj})] =$ $\prod_{j=1}^{p_1}\prod_{k=1}^{n_1}\exp(-t_{kj}^2/2) = \exp[\sum_{j=1}^{p_1}\sum_{k=1}^{n_1}(-t_{kj}^2/2)] = \exp[-\text{tr}(\boldsymbol{T}'\boldsymbol{T})/2]$.

(b) Lemma 7.5 gives

$$\begin{aligned}\phi_{\boldsymbol{Y}}(\boldsymbol{T}) &= \exp[\text{i}\,\text{tr}(\boldsymbol{T}'\boldsymbol{M})]\exp\{-\text{tr}[(\boldsymbol{\Psi}'\boldsymbol{T}\boldsymbol{\Phi})'(\boldsymbol{\Psi}'\boldsymbol{T}\boldsymbol{\Phi})/2]\} \\ &= \exp[\text{i}\,\text{tr}(\boldsymbol{T}'\boldsymbol{M})]\exp[-\text{tr}(\boldsymbol{\Phi}'\boldsymbol{T}'\boldsymbol{\Psi}\boldsymbol{\Psi}'\boldsymbol{T}\boldsymbol{\Phi})/2] \\ &= \exp[\text{i}\,\text{tr}(\boldsymbol{T}'\boldsymbol{M})]\exp[-\text{tr}(\boldsymbol{T}'\boldsymbol{\Psi}\boldsymbol{\Psi}'\boldsymbol{T}\boldsymbol{\Phi}\boldsymbol{\Phi}')/2] \\ &= \exp[\text{i}\,\text{tr}(\boldsymbol{T}'\boldsymbol{M})]\exp[-\text{tr}(\boldsymbol{T}'\boldsymbol{\Xi}\boldsymbol{T}\boldsymbol{\Sigma})/2]\,. \end{aligned} \tag{8.56}$$

□

It would be hard to overstate the convenience and power of matrix Gaussian notation and describing properties of linear models. The following theorem provides one major contribution because it allows quickly deriving the distributions of estimates of expected values. Equally importantly, matrix Gaussian properties allow precisely identifying the distributions of quadratic forms and covariance matrix estimates, which lie at the heart of test statistic theory.

Theorem 8.12 If $\boldsymbol{Y} \sim (S)\mathcal{N}_{n,p}(\boldsymbol{M}, \boldsymbol{\Xi}, \boldsymbol{\Sigma})$ while $\boldsymbol{A} \neq \boldsymbol{0}$ $(n_1 \times n)$, $\boldsymbol{B} \neq \boldsymbol{0}$ $(p \times p_1)$, and $\boldsymbol{C}$ $(n_1 \times p_1)$ are finite constant matrices, then

$$\boldsymbol{A}\boldsymbol{Y}\boldsymbol{B} + \boldsymbol{C} \sim (S)\mathcal{N}_{n_1,p_1}(\boldsymbol{A}\boldsymbol{M}\boldsymbol{B} + \boldsymbol{C}, \boldsymbol{A}\boldsymbol{\Xi}\boldsymbol{A}', \boldsymbol{B}'\boldsymbol{\Sigma}\boldsymbol{B})\,. \tag{8.57}$$

Proof. Lemma 7.5 provides the characteristic function of a linear transformation of a random matrix. Examining the result verifies the reproductive property. □

The theorem holds for Ξ and Σ of any rank. Here $\boldsymbol{A}$ transforms the rows and $\boldsymbol{B}$ transforms the columns. Although the CDF is always well defined, the density exists only if $\text{rank}(\boldsymbol{A}\Xi\boldsymbol{A}') = n_1$ and $\text{rank}(\boldsymbol{B}'\Sigma\boldsymbol{B}) = p_1$. However, every singular form can be expressed in terms of an embedded nonsingular one. A particularly convenient form for the singular case arises from the spectral decomposition of the covariance matrices, as in the next lemma.

Many useful results arise as special cases of the theorem. Choosing $p = p_1 = 1$ provides a standard result about the vector Gaussian. Choosing $\boldsymbol{C} = \boldsymbol{0}$ and one or more other matrices as an identity matrix also produces useful special cases. The theorem helps prove the following lemma, which plays a key role in developing properties of multivariate quadratic forms.

Lemma 8.4 If $\boldsymbol{Y} \sim (\mathcal{S})\mathcal{N}_{n,p}(\boldsymbol{M_Y}, \Xi, \Sigma)$ with $\Xi = \boldsymbol{\Psi\Psi}'$ of rank $n_1 \leq n$, $\Sigma = \boldsymbol{\Phi\Phi}'$ of rank $p_1 \leq p$, and $\boldsymbol{M_Z} = (\boldsymbol{\Psi}'\boldsymbol{\Psi})^{-1}\boldsymbol{\Psi}'\boldsymbol{M_Y}\boldsymbol{\Phi}(\boldsymbol{\Phi}'\boldsymbol{\Phi})^{-1} = \boldsymbol{\Psi}^{+}\boldsymbol{M_Y}\boldsymbol{\Phi}^{+t}$, then (without loss of generality) $\boldsymbol{Y} = \boldsymbol{\Psi}(\boldsymbol{Z} + \boldsymbol{M_Z})\boldsymbol{\Phi}'$ for $\boldsymbol{Z} \sim \mathcal{N}_{n_1,p_1}(\boldsymbol{0}, \boldsymbol{I}_{n_1}, \boldsymbol{I}_{p_1})$. Spectral decomposition gives $\boldsymbol{\Phi} = \boldsymbol{\Upsilon}\text{Dg}(\boldsymbol{\lambda})^{1/2}$ and $\boldsymbol{\Phi}^{+t} = \boldsymbol{\Upsilon}\text{Dg}(\boldsymbol{\lambda})^{-1/2}$.

Proof. Requiring $\text{E}(\boldsymbol{Y}) = \boldsymbol{M_Y} = \boldsymbol{\Psi M_Z \Phi}'$ gives $(\boldsymbol{\Psi}'\boldsymbol{\Psi})^{-1}\boldsymbol{\Psi}'\boldsymbol{M_Y}\boldsymbol{\Phi}(\boldsymbol{\Phi}'\boldsymbol{\Phi})^{-1} = \boldsymbol{M_Z}$. Theorem 8.12 ensures $\boldsymbol{Y}$, as a linear transformation of $\boldsymbol{Z}$, must be matrix Gaussian with the required distribution. □

Example 8.5 The $\text{GLM}_{N,p,q}(\boldsymbol{Y}_i; \boldsymbol{X}_i\boldsymbol{B}, \Sigma)$ with Gaussian errors and full-rank $\boldsymbol{X}$ has

$$\widehat{\boldsymbol{B}} = \left[(\boldsymbol{X}'\boldsymbol{X})^{-1}\boldsymbol{X}'\right]\boldsymbol{Y} \sim \mathcal{N}_{q,p}\left[\boldsymbol{B}, (\boldsymbol{X}'\boldsymbol{X})^{-1}, \Sigma\right]. \tag{8.58}$$

If $p = 1$, then $\widehat{\boldsymbol{\beta}} \sim \mathcal{N}_{q,1}[\boldsymbol{\beta}, (\boldsymbol{X}'\boldsymbol{X})^{-1}, \sigma^2]$ if and only if $\text{vec}(\widehat{\boldsymbol{\beta}}) = \widehat{\boldsymbol{\beta}} \sim \mathcal{N}_q[\boldsymbol{\beta}, (\boldsymbol{X}'\boldsymbol{X})^{-1} \otimes \sigma^2]$, with $(\boldsymbol{X}'\boldsymbol{X})^{-1} \otimes \sigma^2 = (\boldsymbol{X}'\boldsymbol{X})^{-1}\sigma^2$. With or without full-rank $\boldsymbol{X}$ but with the requirements of full rank of $\boldsymbol{M} = \boldsymbol{C}(\boldsymbol{X}'\boldsymbol{X})^{-}\boldsymbol{C}'$ and $\boldsymbol{C} = \boldsymbol{C}(\boldsymbol{X}'\boldsymbol{X})^{-}(\boldsymbol{X}'\boldsymbol{X})$ (which ensures $\boldsymbol{\Theta}$ is testable),

$$\widehat{\boldsymbol{\Theta}} - \boldsymbol{\Theta}_0 = \boldsymbol{C}\widetilde{\boldsymbol{B}}\boldsymbol{U} - \boldsymbol{\Theta}_0 \sim \mathcal{N}_{a,b}\left\{\boldsymbol{\Theta} - \boldsymbol{\Theta}_0, [\boldsymbol{C}(\boldsymbol{X}'\boldsymbol{X})^{-}\boldsymbol{C}']^{-1}, \boldsymbol{U}'\Sigma\boldsymbol{U}\right\}. \tag{8.59}$$

If $\boldsymbol{H} = \boldsymbol{X}(\boldsymbol{X}'\boldsymbol{X})^{-}\boldsymbol{X}'$ and $\text{rank}(\boldsymbol{X}) = r \leq q$, then $\text{rank}(\boldsymbol{H}) = r$ and

$$\widehat{\boldsymbol{Y}} = \boldsymbol{HY} \sim \mathcal{SN}_{N,p}(\boldsymbol{XB}, \boldsymbol{H}, \Sigma). \tag{8.60}$$

Also

$$\hat{\boldsymbol{E}} = (\boldsymbol{I} - \boldsymbol{H})\boldsymbol{Y} \sim \mathcal{SN}_{N,p}[\boldsymbol{0}, (\boldsymbol{I} - \boldsymbol{H}), \boldsymbol{\Sigma}], \tag{8.61}$$

with $\text{rank}(\boldsymbol{I} - \boldsymbol{H}) = N - r$. In contrast, $\boldsymbol{E} \sim \mathcal{N}_{N,p}(\boldsymbol{0}, \boldsymbol{I}, \boldsymbol{\Sigma})$.

Theorem 8.13 The random matrix $\boldsymbol{Y}$ is $N \times p$ with $\text{row}_i(\boldsymbol{Y}) = \boldsymbol{Y}_i$. If $(\boldsymbol{Y}_i)' \sim \mathcal{N}_p(\boldsymbol{\mu}, \boldsymbol{\Sigma})$ is independent of $(\boldsymbol{Y}_{i'})'$ for $i \neq i'$ (i.i.d. rows) and $\text{rank}(\boldsymbol{\Sigma}) = p_1 \leq p$, then $\boldsymbol{Y} \sim \mathcal{N}_{N,p}(\boldsymbol{1}_N\boldsymbol{\mu}', \boldsymbol{I}_N, \boldsymbol{\Sigma})$. In turn, for the special case,
(a) $\text{vec}(\boldsymbol{Y}') = \boldsymbol{y} \sim \mathcal{N}_{Np}(\boldsymbol{1}_N \otimes \boldsymbol{\mu}, \boldsymbol{I}_N \otimes \boldsymbol{\Sigma})$ with $\text{rank}(\boldsymbol{I}_N \otimes \boldsymbol{\Sigma}) = Np_1 \leq Np$,
(b) $\text{vec}(\boldsymbol{Y}) = \boldsymbol{y} \sim \mathcal{N}_{N \cdot p}(\boldsymbol{\mu} \otimes \boldsymbol{1}_N, \boldsymbol{\Sigma} \otimes \boldsymbol{I}_N)$, and
(c) if $p_1 = p$, the density exists and may be computed as

$$\begin{aligned} f_{\boldsymbol{y}}(\boldsymbol{y}_*; \boldsymbol{\mu}, \boldsymbol{\Sigma}) &= (2\pi)^{-Np/2}|\boldsymbol{\Sigma}|^{-N/2}\exp\left[-\sum_{i=1}^{N}(\boldsymbol{Y}'_{*i}-\boldsymbol{\mu})'\,\boldsymbol{\Sigma}^{-1}(\boldsymbol{Y}'_{*i}-\boldsymbol{\mu})/2\right] \\ &= |2\pi\boldsymbol{\Sigma}|^{-N/2}\exp\{-\text{tr}[(\boldsymbol{Y}_* - \boldsymbol{1}_N\boldsymbol{\mu}')\boldsymbol{\Sigma}^{-1}(\boldsymbol{Y}_* - \boldsymbol{1}_N\boldsymbol{\mu}')']/2\} \\ &= |2\pi\boldsymbol{\Sigma}|^{-N/2}\exp\{-\text{tr}[\boldsymbol{\Sigma}^{-1}(\boldsymbol{Y}_* - \boldsymbol{1}_N\boldsymbol{\mu}')'(\boldsymbol{Y}_* - \boldsymbol{1}_N\boldsymbol{\mu}')]/2\}. \end{aligned} \tag{8.62}$$

Proof. Left as an exercise.

Theorem 8.14 If $\boldsymbol{Y} \sim \mathcal{N}_{n,p}(\boldsymbol{M}, \boldsymbol{\Xi}, \boldsymbol{\Sigma})$, $\text{rank}(\boldsymbol{\Xi}) = n$, and $\text{rank}(\boldsymbol{\Sigma}) = p$, then

$$\begin{aligned} f_{\boldsymbol{Y}}(\boldsymbol{Y}_*; \boldsymbol{M}, \boldsymbol{\Xi}, \boldsymbol{\Sigma}) &= \frac{\exp\{-[\text{vec}(\boldsymbol{Y}_*-\boldsymbol{M})]'(\boldsymbol{\Xi}\otimes\boldsymbol{\Sigma})^{-1}\text{vec}(\boldsymbol{Y}_*-\boldsymbol{M})/2\}}{(2\pi)^{np/2}|\boldsymbol{\Xi}\otimes\boldsymbol{\Sigma}|^{1/2}} \\ &= \frac{\exp\{-[\text{vec}(\boldsymbol{Y}_*-\boldsymbol{M})]'(\boldsymbol{\Xi}^{-1}\otimes\boldsymbol{\Sigma}^{-1})\text{vec}(\boldsymbol{Y}_*-\boldsymbol{M})/2\}}{(2\pi)^{np/2}|\boldsymbol{\Xi}|^{1/2}|\boldsymbol{\Sigma}|^{1/2}} \\ &= \frac{\exp\left(-\{\text{vec}[(\boldsymbol{Y}_*-\boldsymbol{M})']\}'(\boldsymbol{\Sigma}\otimes\boldsymbol{\Xi})^{-1}\text{vec}[(\boldsymbol{Y}_*-\boldsymbol{M})']/2\right)}{(2\pi)^{np/2}|\boldsymbol{\Sigma}\otimes\boldsymbol{\Xi}|^{1/2}} \\ &= \frac{\exp\left(-\{\text{vec}[(\boldsymbol{Y}_*-\boldsymbol{M})']\}'(\boldsymbol{\Sigma}^{-1}\otimes\boldsymbol{\Xi}^{-1})\text{vec}[(\boldsymbol{Y}_*-\boldsymbol{M})']/2\right)}{(2\pi)^{np/2}|\boldsymbol{\Sigma}|^{1/2}|\boldsymbol{\Xi}|^{1/2}}. \end{aligned} \tag{8.63}$$

Proof. Left as an exercise.

8.9 ASSESSING MULTIVARIATE GAUSSIAN DISTRIBUTION

Multivariate methods commonly assume the errors follow a Gaussian distribution. Whereas such methods have long been available for univariate data (since circa 1900), tests of the strong assumption of multivariate Gaussian distribution were not developed until relatively recently (circa 1970). Departure from the family of multivariate Gaussian distributions can occur in a great variety of ways. In contrast, departures from univariate Gaussian distribution occur in a relatively small number of ways. Furthermore, obvious alternative univariate distributions often deserve consideration, such as the lognormal or Student's t. Unfortunately, any attempt to generalize the approach encounters a variety of kinds of multivariate t distributions (Kotz and Nadarajah, 2004) or a variety of other elliptical distributions. It is not clear which forms might provide reasonable alternatives. The uncertainty makes it difficult to choose sufficiently broad classes of alternatives to the null hypothesis of a multivariate Gaussian distribution.

Departure from Gaussian distribution can occur in many ways. Many multivariate methods depend on the sample covariance matrix being a good representation of associations among the variables. However, if the dependencies among some or all of the variables are not linear in nature, such as $x_2 = x_1^2 + e$, then the covariances (and associated correlations) can be very poor measures of association. Due to the variety of departures, a variety of detection techniques are needed. The techniques should include descriptive methods, graphical methods, and hypothesis tests.

Any overall test we might formulate will have to examine many features and may therefore have low sensitivity for some. Consequently if the departure from Gaussian distribution involves only a small subset of the variables, then the sensitivity of an overall test may be diluted by most of the variables being jointly Gaussian. On the other hand, any test for a specific feature or small set of features can be powerful for the feature but may not detect the departures in other unexamined features.

D'Agostino and Stephens (1986) surveyed goodness-of-fit techniques, in general. Thode (2002) discussed testing for normality in univariate and multivariate settings. Mecklin and Mundfrom (2005) provided a Monte Carlo comparison of the type I and type II error rates of tests of multivariate normality.

Graphical procedures, some as simple as scatter plots, should always be employed to visually examine the data. Two-dimensional and three-dimensional scatter plots can reveal outliers and other extreme values. Outliers can be misleading as to whether or not the data follow a Gaussian distribution. They can both conceal and falsely mimic departures. On the other hand, extreme values may not be outliers at all in the sense of errors in the data. Rather, they may indicate the need for a transformation to a Gaussian distribution. A typical sample of lognormal data will illustrate the point. Such data are usually highly skewed to the right. A few extreme points almost always occur. In general, it is often difficult to discriminate between outliers (i.e., unacceptable errors) and valid extreme values.

The difficulty of discriminating valid from invalid extreme values suggests modeling, transformation, and outlier detection should be undertaken simultaneously and interactively. In many cases "modeling" should be taken to mean *robust estimation of parameters*. Carroll and Rupert (1985) provided an excellent discussion of the principles for univariate analyses.

The $N \times p$ data matrix $\boldsymbol{Y}$ has $\text{row}_i(\boldsymbol{Y}) = \boldsymbol{Y}_i$ independent of all other rows. In two or three dimensions ($p \leq 3$) we can plot the data to find outliers. Once found, we have several options. In higher dimensions, algorithms become necessary for detecting outliers. The central question is "How far is Y_i' ($p \times 1$) from the center of the cloud of data points?" A measure of distance is needed. However, estimation underlying the measure needs to be robust to the presence of several outliers masking each other.

Definition 8.5 *Masking* occurs when one or more outliers remain undetected due to the presence of other outliers.

Definition 8.6 The *Mahalanobis distance* is

$$d_i^2 = [Y_i' - t(Y)]'[C(Y)]^{-1}[Y_i' - t(Y)], \tag{8.64}$$

with $t(Y) = Y'1_N/N = \overline{y}$, the $p \times 1$ vector of arithmetic means, and $C(Y) = \widehat{\Sigma} = [Y'Y/N - \overline{y}\overline{y}']N/(N-1)$, the sample covariance matrix estimate.

The function suffers from the masking effect because a set of multiple outliers do not necessarily have large d_i values. Furthermore, $t()$ and $\mathcal{V}()$ are not robust because a small cluster of outliers will attract $t()$ and inflate $\mathcal{V}()$ in its direction.

Definition 8.7 The *breakdown point* is the number of outliers (given as a percent of number of independent sampling units) tolerated by a procedure. Larger values are better.

The Mahalanobis distance has a very low breakdown point because a few outliers can mask each other.

Definition 8.8 The *robust distance* based on the minimum volume ellipsoid (MVE) estimator, proposed by Rousseeuw and Van Zomeren (1990), is

$$\mathrm{RD}_i = \left\{[Y_i' - t(Y)]'[S(Y)]^{-1}[Y_i' - t(Y)]\right\}^{1/2}. \tag{8.65}$$

Here $t(Y)$ is the MVE center and $S(Y)$ is the corresponding covariance matrix, with both being high-breakdown estimators. Also $t(Y)$ is the center of the MVE covering half of the observations, and $S(Y)$ is determined by the MVE. It is multiplied by a correction factor to obtain consistency for multivariate Gaussian distributions.

8.10 TESTS FOR GAUSSIAN DISTRIBUTION

The choice of specific tests should target (1) departures anticipated as most likely and (2) departures most detrimental to the particular analysis method being used. It is advantageous if the tests suggest either a transformation which makes the data Gaussian or an alternative data analysis method. We need tests for examining subsets because deviations from Gaussian distribution can be limited to a subset of the variables. It is difficult to check all possible subsets. Choosing subsets most likely to exhibit deviations makes it difficult to determine the true significance level of a test. Essentially three kinds of tests exist: (1) tests for marginal Gaussian distribution, (2) univariate tests of joint Gaussian distribution, and (3) multivariate tests of joint Gaussian distribution.

The simplest tests of marginal Gaussian distribution are achieved by individually testing p subhypotheses $H_{0j} : \{\boldsymbol{y}_j \sim \text{Gaussian}\}$ using standard testing procedures. For the overall test of $H_0 : \{\boldsymbol{y}_j \sim \text{Gaussian}\ \forall j\}$ the null hypothesis is rejected if any one of the subhypotheses is rejected. The significance level of the overall test must be controlled using principles of simultaneous testing. The method is effective in detecting the least Gaussian marginal distribution. However, it may not be powerful for detecting a subtle departure common to many or all of the variables.

Tests of marginal Gaussian distribution designed to detect subtle, common departures have been formulated in terms of measures of skewness and kurtosis. Small (1980) formulated a test in terms of skewness and kurtosis vectors, $\boldsymbol{\beta}_1$ $(p \times 1)$ and $\boldsymbol{\beta}_2$ $(p \times 1)$. Individual elements are the skewness and kurtosis parameters of the marginal distributions. Small derived scalar test statistics, Q_1 and Q_2, each distributed as a multiple of a chi square and "nearly independent."

One approach to detecting departures from joint Gaussian distribution is to examine the Mahalanobis distances. Clustering of observation vectors too far from (or too near to) the sample mean vector is evidence of departure from the assumption of a joint Gaussian distribution. Gnanadesikan and Kettenring (1972) discussed plotting the order statistics $\{d_{(i)}\}$ against their expected values under the null hypothesis of Gaussian distribution. The distribution of the order statistics follows a beta distribution. Small (1988) noted that the order statistics can be converted to normal scores, which should follow a Gaussian distribution under the null hypothesis. A test of univariate Gaussian distribution of the normal scores is, however, equally influenced by clusters of points too close or too far from the sample mean vector.

Another approach to accessing the joint Gaussian distribution derives from the "linear functional" characterization of Gaussian distributions. It arises from the observation that every linear combination of the variates must have a univariate Gaussian distribution.

In principle, we might test every possible linear combination for the univariate Gaussian distribution with the goal of finding the linear combination which maximizes the deviation from the Gaussian distribution. Malkovich and Afifi (1973) investigated the approach using each of three criteria: maximum skewness, maximum kurtosis, and minimum Shapiro-Wilks W statistic. If the number of variables, p, is large, then the approach will be computationally intensive. The alternative is to test specific linear combinations. If the specific linear combinations are suggested by the data, then it is prohibitively difficult to determine the significance levels of the tests.

Cox and Small (1978) suggested yet another approach. An arbitrary transformation to a bivariate distribution, with $\boldsymbol{a}_1$ and $\boldsymbol{a}_2$ $p \times 1$, may be written as

$$\boldsymbol{z}_i = \begin{bmatrix} z_{1i} \\ z_{2i} \end{bmatrix} = \begin{bmatrix} \boldsymbol{a}_1' \\ \boldsymbol{a}_2' \end{bmatrix} \boldsymbol{Y}_i' = \boldsymbol{A}\boldsymbol{Y}_i' \,. \tag{8.66}$$

Under the Gaussian assumption, $\boldsymbol{z}_i$ follows a bivariate Gaussian distribution and the conditional mean is linear in the parameters (rather than quadratic, exponential, etc.). The regression function of interest is

$$\mathrm{E}(z_{1i}|z_{2i}) = \beta_0 + \beta_1 z_{2i} + \beta_2 z_{2i}^{\lambda} \,. \tag{8.67}$$

A test for Gaussian distribution could be obtained by testing $H_0 : \beta_2 = 0$ versus $H_A : \beta_2 \neq 0$. If $\lambda = 2$ and $\eta^2(\boldsymbol{a}_1, \boldsymbol{a}_2)$ is the sum of squares accounted for by the quadratic term in the regression model, we can maximize analytically over $\boldsymbol{a}_1$ to yield $\eta^2(\boldsymbol{a}_2)$, then maximize numerically over $\boldsymbol{a}_2$ to yield $\widehat{\eta}^2_{\max}$. Simulations by Cox and Small (1978) allows concluding that, for $N \geq 50$, $p \leq 6$, and H_0 true, the following holds approximately:

$$\log(\widehat{\eta}^2_{\max}) \sim \mathcal{N}_1\left[\log(5p^2/(8N)), \log^2(0.43 + 3.87/p)\right] . \tag{8.68}$$

Mardia (1970) defined scalar parameters for multivariate skewness and kurtosis, $\beta_{1,p}$ and $\beta_{2,p}$, proposed tests based on estimators of the parameters, and provided tables of critical values for $H_0 : \{(\beta_{1,p} = 0) \wedge [\beta_{2,p} = p(p+2)]\}$. The two subhypotheses are also of interest. Mardia (1975) proved "broadly speaking, in the presence of nonnormality, normal theory tests on *means* are influenced by $\beta_{1,p}$ whereas tests about covariances are influenced by $\beta_{2,p}$."

Definition 8.9 For a set of N vectors, each $p \times 1$, $\boldsymbol{Y}_i'$ i.i.d. with mean vector $\boldsymbol{\mu}$ and covariance $\boldsymbol{\Sigma}$ of rank p, the population parameters for *skewness* and *kurtosis* are, respectively,

$$\beta_{1,p} = \mathrm{E}\big[(\boldsymbol{Y}_i' - \boldsymbol{\mu})'\boldsymbol{\Sigma}^{-1}(\boldsymbol{Y}_{i'}' - \boldsymbol{\mu})\big]^3 \tag{8.69}$$

$$\beta_{2,p} = \mathrm{E}\big[(\boldsymbol{Y}_i' - \boldsymbol{\mu})'\boldsymbol{\Sigma}^{-1}(\boldsymbol{Y}_i' - \boldsymbol{\mu})\big]^2 .$$

The definitions are attributed to Mardia (1970).

Skewness $\beta_{1,p}$ is the expected value of the cube of the angle between vectors $\boldsymbol{Y}_i'$ and $\boldsymbol{Y}_{i'}'$ (weighted by both distances) in the Mahalanobis space with metric $\boldsymbol{\Sigma}$. Kurtosis $\beta_{2,p}$ is the expected value of the squared Mahalanobis distance between vectors $\boldsymbol{Y}_i'$ and $\boldsymbol{Y}_{i'}'$ in the Mahalanobis space with metric $\boldsymbol{\Sigma}$.

Theorem 8.15 For N $\boldsymbol{Y}_i' \sim$ i.i.d. $\mathcal{N}_p(\boldsymbol{\mu}, \boldsymbol{\Sigma})$ with $\text{rank}(\boldsymbol{\Sigma}) = p$, the population parameters for skewness and kurtosis are, respectively,

$$\beta_{1,p} = \mathrm{E}\big[(\boldsymbol{Y}_i' - \boldsymbol{\mu})'\boldsymbol{\Sigma}^{-1}(\boldsymbol{Y}_{i'}' - \boldsymbol{\mu})\big]^3 = \mathrm{E}(d_{ii'}^3) = 0 \tag{8.70}$$

$$\beta_{2,p} = \mathrm{E}\big[(\boldsymbol{Y}_i' - \boldsymbol{\mu})'\boldsymbol{\Sigma}^{-1}(\boldsymbol{Y}_{i'}' - \boldsymbol{\mu})\big]^2 = \mathrm{E}(d_{ii'}^2) = p(2+p)\,. \tag{8.71}$$

Proof. It can be proven that the distribution of $d_{ii'}$ is symmetric about zero. Symmetry with $\mathrm{E}(d_{ii'}) = 0$ implies $\mathrm{E}(d_{ii'}^3) = 0$. Also, $\mathrm{E}(d_i^2) = p(p+2)$ since $d_i^2 \sim \chi^2(p)$. □

Corollary 8.15.1 The scalar (univariate) Gaussian is a special case of the vector Gaussian with $p = 1$. Observation i for variable j, namely $y_{ij} \sim \mathcal{N}_1(\mu_j, \sigma_{jj})$, is i.i.d. for $i \in \{1, 2, \ldots, N\}$. The corresponding population parameters for skewness and kurtosis are, respectively,

$$\beta_1 = \mathrm{E}\big[(y_{ij} - \mu_j)\sigma_{jj}^{-1}(y_{i'j} - \mu_j)\big]^3 = \mathrm{E}(d_{ii'}^3) = 0 \tag{8.72}$$

$$\beta_2 = \mathrm{E}\big[(y_{ij} - \mu_j)^2\sigma_{jj}^{-1}\big]^2 = \mathrm{E}(d_i^2) = 3\,. \tag{8.73}$$

Corollary 8.15.2 If $\boldsymbol{Y}_i' \sim$ i.i.d. $\mathcal{N}_p(\boldsymbol{\mu}, \boldsymbol{\Sigma})$ with $\text{rank}(\boldsymbol{\Sigma}) = p$, $i \in \{1, 2, \ldots, N\}$, the vectors of population parameters for skewness and kurtosis are, respectively, $\boldsymbol{\beta}_1 = \mathbf{0}\ (p \times 1)$ and $\boldsymbol{\beta}_2 = 3 \cdot \mathbf{1}\ (p \times 1)$.

The result is relevant to Small's test for marginal Gaussian distribution discussed above. Andrews, Gnanadesikan, and Warner (1971) discussed a test for multivariate Gaussian distribution in the context of transforming the data to marginal Gaussian distribution. They focused on the class of transformations proposed by Box and Cox (1964), namely $y^{(\lambda)} = (y^\lambda - 1)/\lambda$ for $\lambda \neq 0$ and $y^{(\lambda)} = \log(y)$ for $\lambda = 0$. If $\lambda = 1$, no transformation is needed. Marginal transformations to marginal Gaussian distribution could be obtained as

$$\boldsymbol{Y}_i'^{(\boldsymbol{\lambda})} = \begin{bmatrix} y_{i1}^{(\lambda_1)} & y_{i2}^{(\lambda_2)} & \cdots & y_{ip}^{(\lambda_p)} \end{bmatrix}. \tag{8.74}$$

If the goal is to make every marginal distribution Gaussian, then each λ_i would be estimated separately.

Alternatively, a simultaneous estimation of $\boldsymbol{\lambda}$ can yield a transformation to joint Gaussian distribution. If $\boldsymbol{Y}_i'^{(\boldsymbol{\lambda})}$ is vector Gaussian for some value of $\boldsymbol{\lambda}$, then maximum likelihood methods can be used to obtain $\widehat{\boldsymbol{\lambda}}$, and a likelihood ratio test can be made for $H_0 : \boldsymbol{\lambda} = \mathbf{1}$ versus the general alternative. The maximized log likelihood $L_m(\boldsymbol{\lambda})$ yields test statistic

$$2\Big[L_m(\widehat{\boldsymbol{\lambda}}) - L_m(\mathbf{1})\Big] \xrightarrow{\mathcal{D}} \chi^2(p)\,. \tag{8.75}$$

The MLE method assumes that for some value of $\boldsymbol{\lambda}$ the transformed data are jointly Gaussian. No such value may exist because the Box-Cox transformation has limited flexibility. Even if departures occur, no value of $\boldsymbol{\lambda}$ may significantly improve alignment with a Gaussian distribution. Therefore, failure to reject $H_0 : \boldsymbol{\lambda} = \mathbf{1}$ does not guarantee a Gaussian distribution. However, if the null hypothesis is rejected, then the MLE $\widehat{\boldsymbol{\lambda}}$ indicates a useful transformation.

Andrews et al. (1971) also considered a Box-Cox transformation of a particular projection of the data to univariate dimensions. Specifically, they suggested transforming the least Gaussian projection. First, the projection must be identified. Second, λ must be estimated for the Box-Cox transformation of the corresponding univariate distribution.

Cox and Small (1978) suggested testing $H_0 : \beta_2 = 0$ versus $H_A : \beta_2 \neq 0$ for the bivariate model

$$\mathrm{E}(y_{ij}|y_{ij'}) = \beta_0 + \beta_1 y_{ij} + \beta_2 y_{ij}^2 \tag{8.76}$$

for variables j and j', $j \neq j'$. The test statistic $Q_{jj'}^{(p)}$ follows a Student t distribution. By varying the choice of $j \neq j'$ the set of $p(p+1)/2$ such test statistics could be summarized in several ways. Plotting the ordered $Q_{jj'}^{(p)}$ against the expected values of such order statistics provides an example.

EXERCISES

A vector of 1's will be indicated by $\mathbf{1}$. To be explicit, one would specify the dimension, such as by writing $\mathbf{1}_N$. However, the assumption of conformation for multiplication and addition suffices to determine such dimensions. In some cases, the dimensions of such vectors, which may vary even in the same equation, are omitted in this set of exercises.

8.1 Consider $y_{ij} = a_0 + b_0 t_j + c_0 t_j^2 + a_i + b_i t_j + e_{ij}$ for $j \in \{1, \ldots, p\}$ and $i \in \{1, \ldots, N\}$, with $[\,a_i\; b_i\,]' \sim \mathcal{N}_2(\mathbf{0}, \boldsymbol{D})$, $e_{ij} \sim \mathcal{N}(0, \sigma_e^2)$ i.i.d., elements of $\{t_j, a_0, b_0, c_0\}$ are constants, and $t_j = j$. Assume that the vector $[\,a_i\; b_i\,]'$ is independent of all elements of $\{e_{ij}\}$. Also assume independence between i and i' for $i \neq i'$.
8.1.1 Completely specify the distribution of $\boldsymbol{y}_i = [\,y_{i1}\; \cdots\; y_{ip}\,]'$.
8.1.2 Express $\mathcal{V}(y_{ij})$ as a polynomial in t_j.

8.2 Consider $y_{ij} = \mu + a_i + e_{ij}$ for $j \in \{1, 2, \ldots, m\}$ and $i \in \{1, 2, \ldots, N\}$, i.i.d. $a_i \sim \mathcal{N}(0, \sigma_b^2)$ independent of i.i.d. $e_{ij} \sim \mathcal{N}(0, \sigma_w^2)$ (for all i and j), $\sigma^2 = \sigma_b^2 + \sigma_w^2$, and $\rho = \sigma_b^2/(\sigma_b^2 + \sigma_w^2)$. For $n \times m$ $\boldsymbol{Y}$, $\mathrm{row}_i(\boldsymbol{Y}) = \boldsymbol{y}_i'$ and $\boldsymbol{y}_i = [\,y_{i1}\; \cdots\; y_{im}\,]'$ is $m \times 1$. Also $\boldsymbol{\Sigma}_i = \mathcal{V}(\boldsymbol{y}_i)$ is $m \times m$.
8.2.1 Completely specify the distribution of $\boldsymbol{y}_i$.
8.2.2 Give an interpretation for σ^2.
8.2.3 Give an interpretation for ρ and specify the range of possible values of ρ.
8.2.4 Find $\mathrm{E}(\widehat{\mu})$ and $\mathcal{V}(\widehat{\mu})$ for the "sample mean":
$\widehat{\mu} = \mathbf{1}'\boldsymbol{Y}\mathbf{1}/(nm) = (\mathbf{1}_{nm}'\mathbf{1}_{nm})^{-1}\mathbf{1}_{nm}'[\mathrm{vec}(\boldsymbol{Y})] = (\mathbf{1}_n'\mathbf{1}_n)^{-1}\mathbf{1}'\boldsymbol{Y}\mathbf{1}(\mathbf{1}_m'\mathbf{1}_m)^{-1}$.

8.3 Suppose i.i.d. $\boldsymbol{y}_i \sim \mathcal{N}_{m+1}(\boldsymbol{\mu}_i, \Sigma_i)$ with $\boldsymbol{\mu}_i = \mu\mathbf{1}$ and $\Sigma_i = \sigma^2[\rho\mathbf{1}\mathbf{1}' + (1-\rho)\boldsymbol{I}]$ for $i \in \{1, \ldots, N\}$. Partition $\boldsymbol{y}_i$ into two subvectors, $\boldsymbol{y}_{i1}$ $(m \times 1)$ and y_{i2} (1×1): $\boldsymbol{y}_i = \begin{bmatrix} \boldsymbol{y}_{i1} \\ y_{i2} \end{bmatrix}$.

8.3.1 Completely specify the marginal distributions of $\boldsymbol{y}_{i1}$ and of y_{i2}.

8.3.2 Completely specify the conditional distribution of $y_{i2}|\boldsymbol{y}_{i1} = \boldsymbol{u}_{i10}$.

8.3.3 Describe the behavior of $\mu_{2\cdot1} = \mathrm{E}(y_{i2}|\boldsymbol{y}_{i1} = \boldsymbol{u}_{i10})$ as a function of m, ρ, and σ^2.

8.3.4 (*Optional, noncredit*) Describe the behavior of $\sigma^2_{2\cdot1} = \mathcal{V}(y_{i2}|\boldsymbol{y}_{i1} = \boldsymbol{u}_{i10})$ as a function of m, ρ and σ^2.

8.4 Consider a matrix, $\boldsymbol{Y}$, $N \times p$, with each element of $\boldsymbol{Y}$ marginally Gaussian, $y_{ij} \sim \mathcal{N}(\mu_{ij}, \sigma^2)$, and all elements jointly Gaussian. Any pair of distinct observations has constant correlation ρ [any y_{ij} and $y_{i'j'}$ if (a) $i \neq i'$ (b) $j \neq j'$, or (c) $i \neq i'$ and $j \neq j'$].

8.4.1 What must the correlation be for the only remaining case (d) $i = i'$ and $j = j'$?

8.4.2 Clearly specify the distribution of $\mathrm{vec}(\boldsymbol{Y})$.

8.4.3 Explain why or why not $\boldsymbol{Y}$ is a direct-product matrix Gaussian. If it is, specify an appropriate choice of parameters.

8.5 Consider a matrix $\boldsymbol{X}$ $N \times p$, with each element of $\boldsymbol{X}$ marginally Gaussian, $x_{ij} \sim \mathcal{N}(\mu_{ij}, \sigma^2)$, and all elements jointly Gaussian. Also assume that any pair of distinct observations within a column (any x_{ij} and $x_{i'j}$ if $i \neq i'$) has constant correlation ρ and any pair of distinct observations within a row (any x_{ij} and $x_{ij'}$ if $j \neq j'$) are independent.

8.5.1 Clearly specify the distribution of $\mathrm{vec}(\boldsymbol{X})$.

8.5.2 Explain why or why not $\boldsymbol{X}$ is a direct-product matrix Gaussian. If it is, specify an appropriate choice of parameters.

8.6 Computing assignment, assuming access to SAS/IML, S+, MATLAB, or similar matrix language program. Assume all means are zero.

8.6.1 Use your knowledge of linear transformations as applied to Gaussian variables to sketch a simple algorithm to transform two independent Gaussian variables to Gaussian variables with a correlation coefficient of 0.5.

8.6.2 Based on exercise 8.6.1, generate a sample of size $N = 100$ from a bivariate normal distribution with mean zero, unit variances, and correlation $\rho = 0.5$. Provide a scatter plot of the data.

Hints

Hint 1. Review Section 1.1, especially the suggestions for writing matrices.

Hint 2. The following results may help in exercise 8.3. If $m \times m$ matrix $\boldsymbol{R} = \rho\mathbf{1}\mathbf{1}' + (1-\rho)\boldsymbol{I}$, then the following all hold.

1. The eigenvalues of $\boldsymbol{R}$ are $\lambda_1 = \rho m + (1-\rho)$ and $\lambda_2 = (1-\rho) = \lambda_3 = \cdots = \lambda_m$.

2. $(m \times m)$ $\boldsymbol{R}^{-1} = (\lambda_1\lambda_2)^{-1}(-\rho\mathbf{1}\mathbf{1}' + \lambda_1\boldsymbol{I})$.

3. A set of orthonormal eigenvectors, $\boldsymbol{v}_i$ $(m \times 1)$, for $\boldsymbol{R}$ can be obtained easily by finding the orthonormal polynomial coefficients (Appendix, Section A.3).

4. Chapter 1 has a brief discussion of compound symmetry.
5. The spectral decomposition gives $\boldsymbol{R} = \boldsymbol{V}\text{Dg}(\boldsymbol{\lambda})\boldsymbol{V}' = \sum_{j=1}^{m} \lambda_j \boldsymbol{v}_j \boldsymbol{v}_j'$. The last form is the constituent matrix decomposition, with $\boldsymbol{v}_j \boldsymbol{v}_j' = \boldsymbol{G}_j$ a constituent matrix.
6. Other alternative sets of orthonormal eigenvectors can be used, such as components of a Helmert matrix.

CHAPTER 9

Univariate Quadratic Forms

9.1 MOTIVATION

In the univariate linear model $\boldsymbol{y} = \boldsymbol{X}\boldsymbol{\beta} + \boldsymbol{e}$ estimates of $\boldsymbol{\beta}$ are *linear* transformations of the response vector $\boldsymbol{y}$. In contrast, estimates of second-order moments (including σ^2) are *quadratic* transformations of $\boldsymbol{y}$. More specifically, they are statistical *quadratic forms* in $\boldsymbol{y}$, as defined later in this chapter. Statistics for testing hypotheses are generally scalar-valued functions of quadratic forms in $\boldsymbol{y}$. The present chapter includes some of the more important properties of quadratic forms in Gaussian distributed vectors, with particular emphasis on results with applications in linear models. A series of results have individual importance in general linear univariate models. They combine to allow proving the "ANOVA theorem," which provides the theoretical foundation for hypothesis testing in the Analysis-of-Variance and multiple regression.

9.2 CHI-SQUARE DISTRIBUTIONS

We begin by focusing on the family of central chi-square distributions and the superfamily of noncentral chi square. Noncentral distributions play an important role in power computations for tests of hypotheses. The noncentral chi square is the fundamental distribution.

Definition 9.1 If $\boldsymbol{z} \sim \mathcal{N}_\nu(\boldsymbol{0}, \boldsymbol{I}_\nu)$, then

$$x = \boldsymbol{z}'\boldsymbol{z} = \|\boldsymbol{z}\|^2 \tag{9.1}$$

has a *(central) chi-square distribution* with ν degrees of freedom, indicated $x \sim \chi^2(\nu)$, obviously with $0 \leq x < \infty$.

The definition is unusual. Rather than specifying a density function or some other characteristic of the distribution, we specify a random variable as a function of other random variables and define the distribution to be the distribution of the new random variable. The definition contains very little direct information about the characteristics of the distribution. We do not yet know the CDF, characteristic function, whether it has a density function, or any other characteristic. The primary virtue of the type of definition is its simplicity. The definition also specifies the most important manner in which the distribution arises, namely, a chi square equals the sum of squares of i.i.d. $\mathcal{N}(0,1)$ random variables. The characteristics of the central chi square are easily derived from the characteristics of the univariate Gaussian distribution. The following results are presented in many introductory texts.

Theorem 9.1 If $x \sim \chi^2(\nu)$, then its distribution is completely characterized by the following functions. The probability density function is $(\forall x_* > 0)$

$$f_x(x_*;\nu) = \frac{x_*^{(\nu-2)/2} \mathrm{e}^{-x_*/2}}{2^{\nu/2}\Gamma(\nu/2)} . \tag{9.2}$$

The moment generating function is (for $|t| < 1/2$)

$$m_x(t;\nu) = (1-2t)^{-\nu/2} . \tag{9.3}$$

The characteristic function is $(\forall t)$

$$\phi_x(t;\nu) = (1-2\mathrm{i}t)^{-\nu/2} . \tag{9.4}$$

Proof. Johnson, Kotz, and Balakrishnan (1994) provided a detailed account.

Corollary 9.1.1 The distribution, density, and generating functions remain well-defined for any real $0 < \nu < \infty$.

Proof. Left as an exercise.

Corollary 9.1.2 If $x \sim \chi^2(\nu)$, then moments of the distribution are easily computed, for all real $m > -\nu/2$, as

$$\mathrm{E}(x^m) = 2^m \frac{\Gamma(\nu/2+m)}{\Gamma(\nu/2)} . \tag{9.5}$$

In particular, $\mathrm{E}(x) = \nu$, $\mathcal{V}(x) = 2\nu$, and, if $\nu > 2$, $\mathrm{E}(1/x) = 1/(\nu-2)$.

Proof. Left as an exercise.

Definition 9.2 If $\boldsymbol{z} \sim \mathcal{N}_\nu(\boldsymbol{\mu}, \boldsymbol{I})$, then

$$x = \boldsymbol{z}'\boldsymbol{z} = \sum_{j=1}^{\nu} z_j^2 \tag{9.6}$$

has a *noncentral chi-square distribution* with ν *degrees of freedom* and *noncentrality parameter* $\omega = \boldsymbol{\mu}'\boldsymbol{\mu} \geq 0$, written $x \sim \chi^2(\nu, \omega)$. Alternately, if

$z \sim \mathcal{N}_\nu(\mathbf{1}_\nu\sqrt{\omega/\nu}, \boldsymbol{I}_\nu)$, then $z'z \sim \chi^2(\nu, \omega)$. If $\omega = 0$, the quadratic form reduces to a central chi square.

Warning! Some authors (Searle, 1971; Johnson, Kotz, and Balakrishnan, 1995) use $\omega = \boldsymbol{\mu}'\boldsymbol{\mu}/2$ as the noncentrality parameter, while others (Rao, 1973) use $\omega = \boldsymbol{\mu}'\boldsymbol{\mu}$. Although either choice is valid, constants in the density, generating functions, and moments vary with the choice. When reading material on any function of univariate or multivariate noncentral quadratic forms, one must take care to determine which definition of noncentrality parameter is being used. The warning applies to all noncentral versions of chi square, F, and quadratic forms. The same warning also applies to noncentral versions of multivariate generalizations, including the Wishart random matrix and functions of it.

Theorem 9.2 If $x \sim \chi^2(\nu, \omega)$, then its distribution is completely characterized by the probability density function ($\forall x > 0$),

$$f_x(x_*; \nu, \omega) = \sum_{k=0}^{\infty} \frac{\mathrm{e}^{-\omega/2}\,(\omega/2)^k}{k!} f_{\chi^2}(x_*; \nu + 2k, 0)\,, \tag{9.7}$$

with $f_x(x_*; \nu + 2k, 0) = f_x(x_*; \nu + 2k)$, a central density. The moment generating function (for $|t| < 1/2$), is

$$\begin{aligned} m_x(t; \nu, \omega) &= \sum_{k=0}^{\infty} \frac{\mathrm{e}^{-\omega/2}(\omega/2)^k}{k!} m(t; \nu + 2k, 0) \\ &= (1-2t)^{-\nu/2}\exp\Big\{-\frac{\omega}{2}\big[1-(1-2t)^{-1}\big]\Big\} \\ &= (1-2t)^{-\nu/2}\exp\left(\frac{t\omega}{1-2t}\right), \end{aligned} \tag{9.8}$$

the characteristic function is ($\forall t \in \Re$)

$$\begin{aligned} \phi_x(t; \nu, \omega) &= \sum_{k=0}^{\infty} \frac{\mathrm{e}^{-\omega/2}(\omega/2)^k}{k!} \phi(t; \nu + 2k, 0) \\ &= (1-2\mathrm{i}t)^{-\nu/2}\exp\Big\{-\frac{\omega}{2}\big[1-(1-2\mathrm{i}t)^{-1}\big]\Big\} \\ &= (1-2\mathrm{i}t)^{-\nu/2}\exp\left(\frac{\mathrm{i}t\omega}{1-2\mathrm{i}t}\right), \end{aligned} \tag{9.9}$$

the cumulant generating function is (for $|t| < 1/2$),

$$c_x(t; \nu, \omega) = -(\nu/2)\log(1-2t) + t\omega/(1-2t)\,, \tag{9.10}$$

and cumulant m is

$$\kappa_m(x) = 2^{m-1}(m-1)!(\nu + m\omega)\,. \tag{9.11}$$

Proof. If $\boldsymbol{z} \sim \mathcal{N}_\nu(\boldsymbol{\mu}, \boldsymbol{I}_\nu)$ and $x = \boldsymbol{z}'\boldsymbol{z}$, the MGF is

$$\begin{aligned} m_x(t;\nu,\omega) &= \mathrm{E}[\exp(t\boldsymbol{z}'\boldsymbol{z})] \\ &= (2\pi)^{-\nu/2}\int_{\Re^\nu} \exp[t\boldsymbol{z}_*'\boldsymbol{z}_* - (\boldsymbol{z}_* - \boldsymbol{\mu})'\boldsymbol{I}(\boldsymbol{z}_* - \boldsymbol{\mu})/2]\, d\boldsymbol{z}_* \\ &= (2\pi)^{-\nu/2}\int_{\Re^\nu} \exp[t\boldsymbol{z}_*'\boldsymbol{z}_* - \boldsymbol{z}_*'\boldsymbol{z}_*/2 + \boldsymbol{\mu}'\boldsymbol{z}_* - \boldsymbol{\mu}'\boldsymbol{\mu}/2]\, d\boldsymbol{z}_* \\ &= (2\pi)^{-\nu/2}\int_{\Re^\nu} \exp[-(1-2t)\boldsymbol{z}_*'\boldsymbol{z}_*/2 + \boldsymbol{\mu}'\boldsymbol{z}_* - \boldsymbol{\mu}'\boldsymbol{\mu}/2]\, d\boldsymbol{z}_* . \end{aligned} \tag{9.12}$$

If $\boldsymbol{A}^{-1} = (1-2t)\boldsymbol{I}_\nu$ and $\omega = \boldsymbol{\mu}'\boldsymbol{\mu}$, then

$$\begin{aligned} m_x(t;\nu,\omega) &= (2\pi)^{-\nu/2}\mathrm{e}^{-\omega/2}\int_{\Re^\nu} \exp(-\boldsymbol{z}_*'\boldsymbol{A}^{-1}\boldsymbol{z}_*/2 + \boldsymbol{\mu}'\boldsymbol{z}_*)\, d\boldsymbol{z}_* \\ &= \mathrm{e}^{-\omega/2}|\boldsymbol{A}|^{1/2}\int_{\Re^\nu} \exp(\boldsymbol{\mu}'\boldsymbol{z}_*)\left[(2\pi)^{-\nu/2}|\boldsymbol{A}|^{-1/2}\exp(-\boldsymbol{z}_*'\boldsymbol{A}^{-1}\boldsymbol{z}_*/2)\right] d\boldsymbol{z}_* . \end{aligned} \tag{9.13}$$

The last integral is the MGF of a $\mathcal{N}_\nu(\boldsymbol{0}, \boldsymbol{A})$, with density contained in the brackets (and with $\boldsymbol{\mu}$ replacing the usual $\boldsymbol{t}$). Therefore

$$\begin{aligned} m_x(t;\nu,\omega) &= \mathrm{e}^{-\omega/2}|\boldsymbol{A}|^{1/2}\exp(\boldsymbol{\mu}'\boldsymbol{A}\boldsymbol{\mu}/2) \\ &= (1-2t)^{-\nu/2}\exp\{[(1-2t)^{-1}\boldsymbol{\mu}'\boldsymbol{\mu} - \omega]/2\} \\ &= (1-2t)^{-\nu/2}\exp[t\omega/(1-2t)] . \end{aligned} \tag{9.14}$$

The characteristic function of the distribution with the density shown is

$$\begin{aligned} \phi_x(t;\nu,\omega) &= \mathrm{E}(\mathrm{e}^{itx}) = \int_0^\infty \mathrm{e}^{itx_*} f_x(x_*;\nu,\omega)\, dx_* \\ &= \int_0^\infty \mathrm{e}^{itx_*}\sum_{k=0}^\infty \frac{\mathrm{e}^{-\omega/2}(\omega/2)^k}{k!} f_{x_*}(x_*;\nu+2k,0)\, dx_* . \end{aligned} \tag{9.15}$$

The bounded convergence theorem permits an interchange of the infinite summation and the integral:

$$\begin{aligned} \phi_x(t;\nu,\omega) &= \sum_{k=0}^\infty \frac{\mathrm{e}^{-\omega/2}(\omega/2)^k}{k!}\int_0^\infty \mathrm{e}^{itx_*} f_x(x_*;\nu+2k,0)\, dx_* \\ &= \sum_{k=0}^\infty \frac{\mathrm{e}^{-\omega/2}(\omega/2)^k}{k!}\phi_x(t;\nu+2k,0) \\ &= \sum_{k=0}^\infty \frac{\mathrm{e}^{-\omega/2}(\omega/2)^k}{k!}(1-2\mathrm{i}t)^{-(\nu+2k)/2} \\ &= (1-2\mathrm{i}t)^{-\nu/2}\mathrm{e}^{-\omega/2}\sum_{k=0}^\infty \frac{1}{k!}\left(\frac{\omega/2}{1-2\mathrm{i}t}\right)^k . \end{aligned} \tag{9.16}$$

Recalling $\mathrm{e}^x = \sum_{k=0}^\infty x^k/k!$ and simplifying give the final form. □

The derivation of the characteristic function corresponding to the density can be repeated with $\phi_x(t; \nu, \omega)$ replaced by $m_x(t; \nu, \omega)$ and i omitted to prove the stated PDF corresponds to the stated MGF. Since we have previously derived the MGF from basic principles, the verification proves the stated PDF is correct.

The PDFs in the expression for the density are weighted by Poisson probabilities for $k \in \{0, 1, 2, \dots\}$, which means the density describes a mixture. Similarly, the noncentral MGF and characteristic function are weighted averages of central MGFs and characteristic functions.

Corollary 9.2.1 The distribution, density, and generating functions remain well defined for any real $0 < \nu < \infty$.

Although Siegel (1979) discussed a distribution with zero degrees of freedom, we do not consider it here. Johnson, Kotz, and Balakrishnan (1994) provided related discussions.

Corollary 9.2.2 If $\boldsymbol{z} \sim \mathcal{N}_\nu(\mathbf{0}, \boldsymbol{I}_\nu)$, then $\boldsymbol{z}'\boldsymbol{z} \sim \chi^2(\nu, 0)$, and equivalently $\boldsymbol{z}'\boldsymbol{z} \sim \chi^2(\nu)$, i.e., the central chi-square is a special case of the noncentral chi-square with $\omega = 0$.

Corollary 9.2.3 If $\boldsymbol{y} \sim \mathcal{N}_\nu(\boldsymbol{\mu}, \sigma^2\boldsymbol{I}_\nu)$, then $\boldsymbol{y}'\boldsymbol{y}/\sigma^2 \sim \chi^2(\nu, \boldsymbol{\mu}'\boldsymbol{\mu}/\sigma^2)$.

Corollary 9.2.4 For any finite set of independent chi-square random variables $\{x_j\}$, with $x_j \sim \chi^2(\nu_j, \omega_j)$,

$$\sum_{j=1}^{n} x_j \sim \chi^2\left(\sum_{j=1}^{n} \nu_j, \sum_{j=1}^{n} \omega_j\right). \tag{9.17}$$

Corollary 9.2.5 If $x \sim \chi^2(\nu, \omega)$, then $\mathrm{E}(x) = \nu + \omega$ and $\mathcal{V}(x) = 2\nu + 4\omega$.

Proof. Left as exercises.

9.3 GENERAL PROPERTIES OF QUADRATIC FORMS

As briefly discussed in Chapter 1, in the study of matrix algebra a quadratic form is $q = \boldsymbol{y}'\boldsymbol{A}\boldsymbol{y}$ for any conforming $\boldsymbol{A}$ and $\boldsymbol{y}$. Without loss of generality, $\boldsymbol{A}$ may be assumed to be symmetric (Lemma 1.4) because $q = \boldsymbol{y}'\boldsymbol{A}\boldsymbol{y} = \boldsymbol{y}'\boldsymbol{B}\boldsymbol{y}$ with $\boldsymbol{B} = [(\boldsymbol{A} + \boldsymbol{A}')/2]$. The result allows taking advantage of the special properties of symmetric matrices. Most importantly, we are assured the middle matrix has a spectral decomposition, $\boldsymbol{B} = \boldsymbol{V}_B \mathrm{Dg}(\boldsymbol{\lambda}_B)\boldsymbol{V}_B'$, with square, full-rank, and orthonormal $\boldsymbol{V}_B$.

In studying quadratic forms, it is helpful to remember the eigenvalues of $\boldsymbol{B}$ are necessarily real, but they may be positive, negative, or zero. When no negative eigenvalues occur, the middle matrix in a quadratic form can be expressed as $\boldsymbol{A} = \boldsymbol{M}'\boldsymbol{M}$. Many special properties of inner and outer product matrices are mentioned in Chapter 1. Such products are automatically symmetric. They never have any negative eigenvalues, although some may be zero. In all cases $\text{rank}(\boldsymbol{M}'\boldsymbol{M}) = \text{rank}(\boldsymbol{M}\boldsymbol{M}') = \text{rank}(\boldsymbol{M})$.

Definition 9.3 With $n \times 1$ random vector $\boldsymbol{y}$ and $n \times n$ constant middle matrix $\boldsymbol{A}$, the scalar $q = \boldsymbol{y}'\boldsymbol{A}\boldsymbol{y}$ is a random *quadratic form*.

The next result is true for $\boldsymbol{y}$ following any distribution with finite second moments. For comparison recall $\text{E}(\boldsymbol{y}\boldsymbol{y}') = \boldsymbol{\Sigma} + \boldsymbol{\mu}\boldsymbol{\mu}'$.

Theorem 9.3 If $\boldsymbol{y}$ $(n \times 1)$ is *any* random vector with finite mean $\text{E}(\boldsymbol{y}) = \boldsymbol{\mu}$ $(n \times 1)$ and finite dispersion $\mathcal{V}(\boldsymbol{y}) = \boldsymbol{\Sigma}$ $(n \times n)$, then for any symmetric, finite constant matrix $\boldsymbol{A}$ $(n \times n)$

$$\text{E}(\boldsymbol{y}'\boldsymbol{A}\boldsymbol{y}) = \text{tr}(\boldsymbol{A}\boldsymbol{\Sigma}) + \boldsymbol{\mu}'\boldsymbol{A}\boldsymbol{\mu}\,. \tag{9.18}$$

Proof. The cyclical property of the trace for conforming matrices, $\text{tr}(\boldsymbol{ABC}) = \text{tr}(\boldsymbol{CAB}) = \text{tr}(\boldsymbol{BCA})$, applies. Since the quadratic form is a scalar, it is equal to its trace. Therefore $\text{E}(\boldsymbol{y}'\boldsymbol{A}\boldsymbol{y}) = \text{E}[\text{tr}(\boldsymbol{y}'\boldsymbol{A}\boldsymbol{y})] = \text{E}[\text{tr}(\boldsymbol{A}\boldsymbol{y}\boldsymbol{y}')] = \text{tr}[\boldsymbol{A}\text{E}(\boldsymbol{y}\boldsymbol{y}')] = \text{tr}[\boldsymbol{A}(\boldsymbol{\Sigma} + \boldsymbol{\mu}\boldsymbol{\mu}')] = \text{tr}(\boldsymbol{A}\boldsymbol{\Sigma}) + \boldsymbol{\mu}'\boldsymbol{A}\boldsymbol{\mu}$. □

Corollary 9.3 If $\boldsymbol{A} \neq \boldsymbol{A}'$ the result still holds.

Proof. Although Lemma 1.4 applies, $\text{tr}(\boldsymbol{B}\boldsymbol{\Sigma}) = \text{tr}\{[(\boldsymbol{A}+\boldsymbol{A}')/2]\boldsymbol{\Sigma}\} = [\text{tr}(\boldsymbol{A}\boldsymbol{\Sigma}) + \text{tr}(\boldsymbol{A}'\boldsymbol{\Sigma})]/2$ gives one pause. However, $\text{tr}(\boldsymbol{A}\boldsymbol{\Sigma}) = \text{tr}(\boldsymbol{\Sigma}\boldsymbol{A}') = \text{tr}(\boldsymbol{A}'\boldsymbol{\Sigma})$ proves the result by trace invariance to transposition and permutation. □

9.4 PROPERTIES OF QUADRATIC FORMS IN GAUSSIAN VECTORS

Definition 9.4 A (univariate) *quadratic form in Gaussian variables* is a (scalar) random variable $q = \boldsymbol{y}'\boldsymbol{A}\boldsymbol{y}$ for $n \times n$ constant (finite) $\boldsymbol{A}$ with $\text{rank}(\boldsymbol{A}\boldsymbol{\Sigma}) > 0$ and $\boldsymbol{y} \sim (S)\mathcal{N}_n(\boldsymbol{\mu}, \boldsymbol{\Sigma})$. Without loss of generality $\boldsymbol{A} = \boldsymbol{A}'$.

Example 9.1 Although not obvious, q may be negative. More precisely, although $\Pr\{\boldsymbol{y}'\boldsymbol{y} < 0\} = 0$, depending on the choice of $\boldsymbol{A}$, it may be that $\Pr\{\boldsymbol{y}'\boldsymbol{A}\boldsymbol{y} < 0\} > 0$. If, as an example, $n = 1$, $\boldsymbol{A} = [-1]$, and $\boldsymbol{\Sigma} = 1$, then $q = -\boldsymbol{y}'\boldsymbol{y} = -y_1^2$ and $\Pr\{q < 0\} = 1$.

Example 9.2 If $\boldsymbol{A} = \boldsymbol{I}_n$ and $\boldsymbol{y} \sim \mathcal{N}_n(\boldsymbol{0}, \boldsymbol{I}_n\sigma^2)$, then

$$q = \boldsymbol{y}'\boldsymbol{y} = \sigma^2 \sum_{j=1}^{n} y_j^2/\sigma^2 = \sigma^2 \sum_{j=1}^{n} x_j . \tag{9.19}$$

Here $x_j \sim \chi^2(1,0)$ and $x_j \perp\!\!\!\perp x_{j'}$ if $j \neq j'$ due to the independence of the underlying Gaussian variables. In turn $q/\sigma^2 \sim \chi^2(n,0)$.

Example 9.3 If $\boldsymbol{A} = \boldsymbol{I}_n$ and $\boldsymbol{y} \sim \mathcal{N}_n(\boldsymbol{\mu}, \boldsymbol{I}_n\sigma^2)$, then

$$q = \boldsymbol{y}'\boldsymbol{y} = \sigma^2 \sum_{j=1}^{n} y_j^2/\sigma^2 = \sigma^2 \sum_{j=1}^{n} x_j . \tag{9.20}$$

Here $x_j \sim \chi^2\left(1, \mu_j^2/\sigma^2\right)$ and $x_j \perp\!\!\!\perp x_{j'}$ if $j \neq j'$ due to the independence of the underlying Gaussian variables. In turn $q/\sigma^2 \sim \chi^2(n, \boldsymbol{\mu}'\boldsymbol{\mu}/\sigma^2)$.

The last two examples derive expressions for special quadratic forms in terms of simple sets of underlying Gaussian and chi square variables. In the most general case, similar expressions can be found in terms of weighted sums of possibly noncentral chi-square variables. The following three theorems provide explicit decompositions for increasing more general quadratic forms.

Theorem 9.4 Random $\boldsymbol{y} \sim \mathcal{N}_n(\boldsymbol{\mu}, \sigma^2\boldsymbol{I}_n)$ and $n \times n$ constant (finite) $\boldsymbol{A} = \boldsymbol{A}'$ of rank $0 < n_1 \leq n$ define $q = \boldsymbol{y}'\boldsymbol{A}\boldsymbol{y}$. Spectral decomposition gives $\boldsymbol{A} = \boldsymbol{V}_1 \mathrm{Dg}(\boldsymbol{\lambda}_1)\boldsymbol{V}_1'$, with orthonormal-by-column $\boldsymbol{V}_1 = [\,\boldsymbol{v}_{1,1} \cdots \boldsymbol{v}_{1,n_1}\,]$ $(n \times n_1)$ and $\boldsymbol{\lambda}_1$ $(n_1 \times 1)$.
(a) In any such setting

$$q = \sigma^2 \sum_{k=1}^{n_1} \lambda_{1,k} x_k , \tag{9.21}$$

with $\{x_k\}$ mutually independent, $x_k \sim \chi^2(1, \omega_k)$, and $\omega_k = \left(\boldsymbol{v}_{1,k}'\boldsymbol{\mu}\right)^2/\sigma^2 = \boldsymbol{\mu}'\boldsymbol{v}_{1,k}\boldsymbol{v}_{1,k}'\boldsymbol{\mu}/\sigma^2 = \boldsymbol{v}_{1,k}'\boldsymbol{\mu}\boldsymbol{\mu}'\boldsymbol{v}_{1,k}/\sigma^2$.
(b) With the additional requirement of $\boldsymbol{A} = \boldsymbol{A}^2$ (idempotent with rank n_1), if

$$s = \sum_{k=1}^{n_1} x_k \tag{9.22}$$

$$\omega_+ = \sum_{k=1}^{n_1} \omega_k = \boldsymbol{\mu}'\boldsymbol{V}_1\boldsymbol{V}_1'\boldsymbol{\mu}/\sigma^2 \tag{9.23}$$

then $q = \sigma^2 s$ and $s \sim \chi^2(n_1, \omega_+)$.

Proof. The results are special cases of the next theorem. However, the simpler proof for the special case more clearly illustrates the principles involved. For the special case (a), without loss of generality $\boldsymbol{y} = \sigma(\boldsymbol{z} + \boldsymbol{\mu}/\sigma)$ for $\boldsymbol{z} \sim \mathcal{N}_n(\boldsymbol{0}, \boldsymbol{I}_n)$.

As an orthonormal linear transformation of a unit vector Gaussian, by Lemma 8.2, $\boldsymbol{V}_1'\boldsymbol{z} \sim \mathcal{N}_{n_1}(\boldsymbol{0}, \boldsymbol{I}_{n_1})$. In turn, if $\boldsymbol{\mu}_1 = \boldsymbol{V}_1'\boldsymbol{\mu}/\sigma$, then $\boldsymbol{x} = (\boldsymbol{z}_1 + \boldsymbol{\mu}_1) \sim \mathcal{N}_{n_1}(\boldsymbol{\mu}_1, \boldsymbol{I}_{n_1})$. Furthermore

$$\begin{aligned} q = \boldsymbol{y}'\boldsymbol{A}\boldsymbol{y} &= \sigma(\boldsymbol{z} + \boldsymbol{\mu}/\sigma)'\boldsymbol{V}_1\mathrm{Dg}(\boldsymbol{\lambda}_1)\boldsymbol{V}_1'(\boldsymbol{z} + \boldsymbol{\mu}/\sigma)\sigma \\ &= (\boldsymbol{V}_1'\boldsymbol{z} + \boldsymbol{V}_1'\boldsymbol{\mu}/\sigma)'\mathrm{Dg}(\sigma^2\boldsymbol{\lambda}_1)(\boldsymbol{V}_1'\boldsymbol{z} + \boldsymbol{V}_1'\boldsymbol{\mu}/\sigma) \\ &= (\boldsymbol{z}_1 + \boldsymbol{\mu}_1)'\mathrm{Dg}(\sigma^2\boldsymbol{\lambda}_1)(\boldsymbol{z}_1 + \boldsymbol{\mu}_1) \\ &= \boldsymbol{x}'\mathrm{Dg}(\sigma^2\boldsymbol{\lambda}_1)\boldsymbol{x} \\ &= \sum_{k=1}^{n_1} \sigma^2 \lambda_{1,k} x_k^2 . \end{aligned} \tag{9.24}$$

Here $\{x_k^2\}$ are independent and $x_k^2 \sim \chi^2(1, \omega_k)$.
For the special case (b), $\boldsymbol{A} = \boldsymbol{A}^2$ implies $\lambda_{1,k} \equiv 1$. □

The following theorem contains the preceding two as special cases. It also covers the two examples discussed at the beginning of the present section.

Theorem 9.5 If $\boldsymbol{y} \sim \mathcal{N}_n(\boldsymbol{\mu}_y, \boldsymbol{\Sigma})$ with $\mathrm{rank}(\boldsymbol{\Sigma}) = n$ and $q = \boldsymbol{y}'\boldsymbol{A}\boldsymbol{y}$ for $n \times n$ constant (finite) $\boldsymbol{A} = \boldsymbol{A}'$ of rank $0 < n_1 \leq n$, $\boldsymbol{\Sigma} = \boldsymbol{\Phi}\boldsymbol{\Phi}'$ with $\boldsymbol{\Phi}$ $n \times n$ of rank n, then $\boldsymbol{B} = \boldsymbol{\Phi}'\boldsymbol{A}\boldsymbol{\Phi}$ is $n \times n$, symmetric, and rank n_1. Spectral decomposition gives $\boldsymbol{B} = \boldsymbol{V}_1\mathrm{Dg}(\boldsymbol{\lambda}_1)\boldsymbol{V}_1'$, with columnwise orthonormal $\boldsymbol{V}_1 = [\,\boldsymbol{v}_{1,1} \cdots \boldsymbol{v}_{1,n_1}\,]$ of dimension $n \times n_1$ and $\boldsymbol{\lambda}_1$ $n_1 \times 1$. Furthermore

$$q = \sum_{k=1}^{n_1} \lambda_{1,k} x_k , \tag{9.25}$$

with $\{x_k\}$ mutually independent, $x_k \sim \chi^2(1, \omega_k)$, and $\omega_k = \left(\boldsymbol{v}_{1,k}'\boldsymbol{\Phi}^{-1}\boldsymbol{\mu}_y\right)^2 = \boldsymbol{\mu}_y'\boldsymbol{\Phi}^{-t}\boldsymbol{v}_{1,k}\boldsymbol{v}_{1,k}'\boldsymbol{\Phi}^{-1}\boldsymbol{\mu}_y = \boldsymbol{v}_{1,k}'\boldsymbol{\Phi}^{-1}\boldsymbol{\mu}_y\boldsymbol{\mu}_y'\boldsymbol{\Phi}^{-t}\boldsymbol{v}_{1,k}$.

Proof. The result is a special case of the next theorem. However, the simpler proof for the special case more clearly illustrates the principles involved. For the special case, if $\boldsymbol{\mu}_z = \boldsymbol{\Phi}^{-1}\boldsymbol{\mu}_y$, without loss of generality, $\boldsymbol{y} = \boldsymbol{\Phi}(\boldsymbol{z} + \boldsymbol{\mu}_z)$ with $\boldsymbol{z} \sim \mathcal{N}_n(\boldsymbol{0}, \boldsymbol{I}_n)$. Hence

$$\begin{aligned} q = \boldsymbol{y}'\boldsymbol{A}\boldsymbol{y} &= [\boldsymbol{\Phi}(\boldsymbol{z} + \boldsymbol{\mu}_z)]'\boldsymbol{A}[\boldsymbol{\Phi}(\boldsymbol{z} + \boldsymbol{\mu}_z)] \\ &= (\boldsymbol{z} + \boldsymbol{\mu}_z)\boldsymbol{\Phi}'\boldsymbol{A}\boldsymbol{\Phi}(\boldsymbol{z} + \boldsymbol{\mu}_z) \\ &= (\boldsymbol{z} + \boldsymbol{\mu}_z)\boldsymbol{V}_1\mathrm{Dg}(\boldsymbol{\lambda}_1)\boldsymbol{V}_1'(\boldsymbol{z} + \boldsymbol{\mu}_z) \\ &= (\boldsymbol{V}_1'\boldsymbol{z} + \boldsymbol{V}_1'\boldsymbol{\mu}_z)'\mathrm{Dg}(\boldsymbol{\lambda}_1)(\boldsymbol{V}_1'\boldsymbol{z} + \boldsymbol{V}_1'\boldsymbol{\mu}_z) \\ &= \boldsymbol{x}'\mathrm{Dg}(\boldsymbol{\lambda}_1)\boldsymbol{x} \\ &= \sum_{k=1}^{n_1} \lambda_{1,k} x_k^2 . \end{aligned} \tag{9.26}$$

As an orthonormal linear transformation of a unit vector Gaussian, by Lemma 8.2, $\boldsymbol{V}_1'\boldsymbol{z} \sim \mathcal{N}_{n_1}(\boldsymbol{0}, \boldsymbol{I}_{n_1})$. In turn, $\boldsymbol{x} = (\boldsymbol{V}_1'\boldsymbol{z} + \boldsymbol{V}_1'\boldsymbol{\mu}_z) \sim \mathcal{N}_{n_1}(\boldsymbol{V}_1'\boldsymbol{\mu}_z, \boldsymbol{I}_{n_1})$ is a set of independent Gaussian variables with $\mathrm{E}(x_k) = \boldsymbol{v}'_{1k}\boldsymbol{\mu}_z$. In turn $\{x_k^2\}$ are independent and $x_k^2 \sim \chi^2(1, \omega_k)$, with $\omega_k = (\boldsymbol{v}'_{1k}\boldsymbol{\mu}_z)^2 = (\boldsymbol{v}'_{1k}\boldsymbol{\Phi}^{-1}\boldsymbol{\mu}_y)^2$. □

Theorem 9.6 If $\boldsymbol{y} \sim (\mathcal{S})\mathcal{N}_n(\boldsymbol{\mu}, \boldsymbol{\Sigma})$ with $\mathrm{rank}(\boldsymbol{\Sigma}) = n_1$ for $1 \le n_1 \le n$ and $q = \boldsymbol{y}'\boldsymbol{A}\boldsymbol{y}$ for $n \times n$ constant (finite) $\boldsymbol{A}$, then the following hold.

(a) Without loss of generality, $\boldsymbol{A} = \boldsymbol{A}'$ and $\boldsymbol{\Sigma} = \boldsymbol{\Phi}\boldsymbol{\Phi}'$, with $\boldsymbol{\Phi}$ $n \times n_1$ of rank n_1, while $\boldsymbol{\Phi}^+ = (\boldsymbol{\Phi}'\boldsymbol{\Phi})^{-1}\boldsymbol{\Phi}'$.

(b) The matrix $\boldsymbol{B} = \boldsymbol{\Phi}'\boldsymbol{A}\boldsymbol{\Phi}$ is $n_1 \times n_1$ and symmetric with $\mathrm{rank}(\boldsymbol{B}) = n_2 \le n_1 \le n$. If $n_2 > 0$, then $\boldsymbol{B} = \boldsymbol{V}_2\mathrm{Dg}(\boldsymbol{\lambda}_2)\boldsymbol{V}_2'$, with columnwise orthonormal $\boldsymbol{V}_2 = [\,\boldsymbol{v}_{2,1} \cdots \boldsymbol{v}_{2,n_2}\,]$ of dimension $n_1 \times n_2$ and $\boldsymbol{\lambda}_2$ $n_2 \times 1$.

(c) Furthermore

$$q = \sum_{k=1}^{n_2} \lambda_{2,k} x_k \,, \tag{9.27}$$

with $\{x_k\}$ mutually independent, $x_k \sim \chi^2(1, \omega_k)$, and $\omega_k = \left(\boldsymbol{v}'_{2,k}\boldsymbol{\Phi}^+\boldsymbol{\mu}_y\right)^2 = \boldsymbol{\mu}'_y(\boldsymbol{\Phi}^+)'\boldsymbol{v}_{2,k}\boldsymbol{v}'_{2,k}\boldsymbol{\Phi}^+\boldsymbol{\mu}_y$.

Proof. With $\boldsymbol{z} \sim \mathcal{N}_{n_1}(\boldsymbol{0}, \boldsymbol{I}_{n_1})$, Lemma 8.3 ensures $n_1 \times 1$ constant $\boldsymbol{\mu}_z = \boldsymbol{\Phi}^+\boldsymbol{\mu}_y$ exists such that $\boldsymbol{y} = \boldsymbol{\Phi}(\boldsymbol{z} + \boldsymbol{\mu}_z)$. In turn

$$\begin{aligned} q = \boldsymbol{y}'\boldsymbol{A}\boldsymbol{y} &= (\boldsymbol{z} + \boldsymbol{\mu}_z)'\boldsymbol{\Phi}'\boldsymbol{A}\boldsymbol{\Phi}(\boldsymbol{z} + \boldsymbol{\mu}_z) \\ &= (\boldsymbol{z} + \boldsymbol{\mu}_z)'\boldsymbol{B}(\boldsymbol{z} + \boldsymbol{\mu}_z)\,. \end{aligned} \tag{9.28}$$

Symmetry of $\boldsymbol{A}$ ensures the symmetry of $\boldsymbol{B} = \boldsymbol{\Phi}'\boldsymbol{A}\boldsymbol{\Phi}$. Hence $\boldsymbol{B}$ has $\mathrm{rank}(\boldsymbol{B}) = n_2$ nonzero eigenvalues $\{\lambda_{2,k}\}$ (all real and any mixture of positive and negative values), with $0 < n_2 \le n_1$. Also $\boldsymbol{B}$ has $n_1 - n_2$ zero eigenvalues. The spectral decomposition allows writing

$$\boldsymbol{B} = \boldsymbol{V}_2\mathrm{Dg}(\boldsymbol{\lambda}_2)\boldsymbol{V}_2'\,. \tag{9.29}$$

Here $\mathrm{Dg}(\boldsymbol{\lambda}_2)$ is $n_2 \times n_2$, while $\boldsymbol{V}_2$ is the $n_1 \times n_2$ columnwise orthonormal matrix of corresponding eigenvectors. In turn

$$\begin{aligned} q &= (\boldsymbol{z} + \boldsymbol{\mu}_z)'\boldsymbol{V}_2\mathrm{Dg}(\boldsymbol{\lambda}_2)\boldsymbol{V}_2'(\boldsymbol{z} + \boldsymbol{\mu}_z) \\ &= [\boldsymbol{V}_2'(\boldsymbol{z} + \boldsymbol{\mu}_z)]'\mathrm{Dg}(\boldsymbol{\lambda}_2)[\boldsymbol{V}_2'(\boldsymbol{z} + \boldsymbol{\mu}_z)]\,. \end{aligned} \tag{9.30}$$

The vector

$$\boldsymbol{u} = \boldsymbol{V}_2'(\boldsymbol{z} + \boldsymbol{\mu}_z) = \boldsymbol{V}_2'\boldsymbol{z} + \boldsymbol{V}_2'\boldsymbol{\mu}_z \tag{9.31}$$

is $n_2 \times 1$. Lemma 8.2, with $\boldsymbol{z} \sim \mathcal{N}_{n_1}(\boldsymbol{0}, \boldsymbol{I}_{n_1})$, gives $\boldsymbol{V}_2'\boldsymbol{z} \sim \mathcal{N}_{n_2}(\boldsymbol{0}, \boldsymbol{I}_{n_2})$. In turn

$$\boldsymbol{u} = \boldsymbol{V}_2'\boldsymbol{z} + \boldsymbol{V}_2'\boldsymbol{\mu}_z \sim \mathcal{N}_{n_2}(\boldsymbol{V}_2'\boldsymbol{\mu}_z, \boldsymbol{I}_{n_2})\,. \tag{9.32}$$

Therefore $\{u_k\}$ are mutually independent with $u_k \sim \mathcal{N}_1(\mu_{2,k}, 1)$ and

$$\mu_{2,k} = \boldsymbol{v}'_{2,k}\boldsymbol{\mu}_z = \boldsymbol{v}'_{2,k}\boldsymbol{\Phi}^+\boldsymbol{\mu}_y\,. \tag{9.33}$$

Furthermore

$$q = \boldsymbol{u}'\mathrm{Dg}(\boldsymbol{\lambda}_2)\boldsymbol{u} = \sum_{k=1}^{n_2} \lambda_{2,k} u_k^2, \tag{9.34}$$

with $\{u_k^2\}$ being mutually independent and $u_k^2 \sim \chi^2(1, \mu_{2,k}^2)$. □

Corollary 9.6.1 The characteristic function, moment generating function, cumulant generating function, and cumulant m of q are

$$\begin{aligned}\phi_q(t; \boldsymbol{\lambda}_2, \boldsymbol{\omega}) &= \prod_{k=1}^{n_2}\left[(1 - 2\mathrm{i}\lambda_{2,k}t)^{-1/2}\exp\left(\frac{\mathrm{i}t\lambda_{2,k}\omega_k}{1 - 2\mathrm{i}\lambda_{2,k}t}\right)\right] \\ &= \left[\prod_{k=1}^{n_2}(1 - 2\mathrm{i}\lambda_{2,k}t)^{-1/2}\right]\exp\left(\sum_{k=1}^{n_2}\frac{\mathrm{i}t\lambda_{2,k}\omega_k}{1 - 2\mathrm{i}\lambda_{2,k}t}\right)\end{aligned} \tag{9.35}$$

$$\begin{aligned}m_q(t; \boldsymbol{\lambda}_2, \boldsymbol{\omega}) &= \prod_{k=1}^{n_2}\left[(1 - 2\lambda_{2,k}t)^{-1/2}\exp\left(\frac{t\lambda_{2,k}\omega_k}{1 - 2\lambda_{2,k}t}\right)\right] \\ &= \left[\prod_{k=1}^{n_2}(1 - 2\lambda_{2,k}t)^{-1/2}\right]\exp\left(\sum_{k=1}^{n_2}\frac{t\lambda_{2,k}\omega_k}{1 - 2\lambda_{2,k}t}\right)\end{aligned} \tag{9.36}$$

$$\begin{aligned}c_q(t; \boldsymbol{\lambda}_2, \boldsymbol{\omega}) &= \sum_{k=1}^{n_2}\left[-\frac{1}{2}\log(1 - 2\lambda_{2,k}t) + \left(\frac{t\lambda_{2,k}\omega_k}{1 - 2\lambda_{2,k}t}\right)\right] \\ &= -\frac{1}{2}\sum_{k=1}^{n_2}\log(1 - 2\lambda_{2,k}t) + \sum_{k=1}^{n_2}\frac{t\lambda_{2,k}\omega_k}{1 - 2\lambda_{2,k}t}\end{aligned} \tag{9.37}$$

$$\begin{aligned}\kappa_m(q; \boldsymbol{\lambda}_2, \boldsymbol{\omega}) &= \sum_{k=1}^{n_2}\left[\lambda_{2,k}^m 2^{m-1}(m-1)!(1 + m\omega_k)\right] \\ &= 2^{m-1}(m-1)!\sum_{k=1}^{n_2}\left[\lambda_{2,k}^m(1 + m\omega_k)\right].\end{aligned} \tag{9.38}$$

Proof. The characteristic function of a sum of independent random variables is the product of the individual (marginal) characteristic functions. As a linear transformation of a noncentral chi square, the characteristic function of each x_k can be found by applying the form $\phi_{ay+b}(t) = \exp(\mathrm{i}\,tb)\phi_x(at)$ to the characteristic function of a $\chi^2(1, \omega_k)$. A parallel approach applies to the moment generating function. The cumulant generating function is available by taking a logarithm. The cumulant reflects the simple impact of a linear transformation on cumulants and the fact that the cumulant of a sum of independent random variables is the sum of the individual (marginal) cumulants. □

Corollary 9.6.2 If $\lambda_{2,k} \equiv \lambda_1$, then $q/\lambda_1 \sim \chi^2(n_2, \sum_{k=1}^{n_2}\omega_k)$.

Proof. The characteristic function of q reduces to

$$\begin{aligned}\phi_q(t; \mathbf{1}_{n_2}\lambda_1, \boldsymbol{\omega}) &= \left[\prod_{k=1}^{n_2}(1 - 2\mathrm{i}\lambda_1 t)^{-1/2}\right]\exp\left(\sum_{k=1}^{n_2}\frac{\mathrm{i}t\lambda_1\omega_k}{1 - 2\mathrm{i}\lambda_1 t}\right)\\ &= (1 - 2\mathrm{i}\lambda_1 t)^{-n_2/2}\exp\left(\frac{\mathrm{i}t\lambda_1}{1 - 2\mathrm{i}\lambda_1 t}\sum_{k=1}^{n_2}\omega_k\right). \qquad (9.39)\end{aligned}$$

Observing the effect of transforming q to q/λ_1 completes the proof. □

The theorem characterizes a quadratic form in Gaussian random variables regardless of whether or not the quadratic form has a chi-square distribution. As will be seen in studying estimates of variability in linear models, perhaps the most important univariate quadratic forms are scaled chi-square random variables. However, many important quadratic forms are not scaled chi squares. They occur in a variety of settings, including variance component models and the distributions of test statistics.

Corollary 9.6.3 The mean is

$$\begin{aligned}\mathrm{E}(q) &= \sum_{k=1}^{n_2}\lambda_{2,k} + \sum_{k=1}^{n_2}\lambda_{2,k}\omega_k\\ &= \mathrm{tr}(\boldsymbol{A\Sigma}) + \sum_{k=1}^{n_2}\lambda_{2,k}\boldsymbol{\mu}_y'(\boldsymbol{\Phi}^+)'\boldsymbol{v}_{2,k}\boldsymbol{v}_{2,k}'\boldsymbol{\Phi}^+\boldsymbol{\mu}_y\\ &= \mathrm{tr}(\boldsymbol{A\Sigma}) + \boldsymbol{\mu}_y'(\boldsymbol{\Phi\Phi}^+)'\boldsymbol{A}(\boldsymbol{\Phi\Phi}^+)\boldsymbol{\mu}_y. \qquad (9.40)\end{aligned}$$

If $\boldsymbol{\Sigma}$ has full rank of n, then the last form reduces to $\mathrm{E}(q) = \mathrm{tr}(\boldsymbol{A\Sigma}) + \boldsymbol{\mu}_y'\boldsymbol{A\mu}_y$ and coincides with the result in Theorem 9.3 (which holds whether or not $\boldsymbol{y}$ is Gaussian and whether or not $\boldsymbol{\Sigma}$ is full rank). Also, with $\boldsymbol{B} = \boldsymbol{\Phi}'\boldsymbol{A\Phi}$, $\boldsymbol{\Phi} = \boldsymbol{V}_2\mathrm{Dg}(\boldsymbol{\lambda}_2)^{1/2}$ and $\boldsymbol{\Phi}^+ = \mathrm{Dg}(\boldsymbol{\lambda}_2)^{-1/2}\boldsymbol{V}_2'$, and $\boldsymbol{\Phi\Phi}^+ = \boldsymbol{V}_2\boldsymbol{V}_2'$. Furthermore

$$\begin{aligned}\mathcal{V}(q) &= 2\sum_{k=1}^{n_2}\lambda_{2,k}^2 + 4\sum_{k=1}^{n_2}\lambda_{2,k}^2\omega_k\\ &= 2\mathrm{tr}\big[(\boldsymbol{A\Sigma})^2\big] + 4\sum_{k=1}^{n_2}\lambda_{2,k}^2\boldsymbol{\mu}_y'(\boldsymbol{\Phi}^+)'\boldsymbol{v}_{2,k}\boldsymbol{v}_{2,k}'\boldsymbol{\Phi}^+\boldsymbol{\mu}_y\\ &= 2\mathrm{tr}\big[(\boldsymbol{A\Sigma})^2\big] + 4\boldsymbol{\mu}_y'(\boldsymbol{\Phi}^+)'\sum_{k=1}^{n_2}\lambda_{2,k}^2\boldsymbol{v}_{2,k}\boldsymbol{v}_{2,k}'\boldsymbol{\Phi}^+\boldsymbol{\mu}_y\\ &= 2\mathrm{tr}\big[(\boldsymbol{A\Sigma})^2\big] + 4\boldsymbol{\mu}_y'(\boldsymbol{\Phi}^+)'\boldsymbol{B}^2\boldsymbol{\Phi}^+\boldsymbol{\mu}_y\\ &= 2\mathrm{tr}\big[(\boldsymbol{A\Sigma})^2\big] + 4\boldsymbol{\mu}_y'(\boldsymbol{\Phi\Phi}^+)'\boldsymbol{A\Sigma A}(\boldsymbol{\Phi\Phi}^+)\boldsymbol{\mu}_y. \qquad (9.41)\end{aligned}$$

If $\boldsymbol{\Sigma}$ has full rank of n, then $\mathcal{V}(q) = 2\mathrm{tr}[(\boldsymbol{A\Sigma})^2] + 4\boldsymbol{\mu}_y'\boldsymbol{A\mu}_y$, a very useful result.

Corollary 9.6.4 If Σ has full rank of n and $\boldsymbol{B} = \boldsymbol{\Phi}'\boldsymbol{A}\boldsymbol{\Phi}$ is idempotent ($\lambda_{2,k} = 1$, $\forall k$), then $n_2 = \text{rank}(\boldsymbol{B}) = \text{rank}(\boldsymbol{A})$ and $q = \boldsymbol{y}'\boldsymbol{A}\boldsymbol{y}$ is distributed as the sum of n_2 independent random variables distributed $\chi^2(1, \omega_k)$. Also $q \sim \chi^2\left(n_2, \boldsymbol{\mu}_y'\boldsymbol{A}\boldsymbol{\mu}_y\right)$.

Proof. Partially left as an exercise. Lemma 1.30 is useful ($\boldsymbol{\Phi}'\boldsymbol{A}\boldsymbol{\Phi}$ is idempotent $\Leftrightarrow$ $\Sigma\boldsymbol{A}$ is idempotent). The constituent matrix decomposition of a symmetric and idempotent matrix allows writing

$$\begin{aligned}\sum_{k=1}^{n_2}\omega_k &= \sum_{k=1}^{n_2}\left[\boldsymbol{\mu}_y'(\boldsymbol{\Phi}^+)'\boldsymbol{v}_{2,k}\boldsymbol{v}_{2,k}'\boldsymbol{\Phi}^+\boldsymbol{\mu}_y\right] \\ &= \boldsymbol{\mu}_y'\boldsymbol{\Phi}^{-t}\left(\sum_{k=1}^{n_2}\boldsymbol{v}_{2,k}\boldsymbol{v}_{2,k}'\right)\boldsymbol{\Phi}^{-1}\boldsymbol{\mu}_y \\ &= \boldsymbol{\mu}_y'\boldsymbol{\Phi}^{-t}\left(\boldsymbol{V}\begin{bmatrix}\boldsymbol{I}_{n_2} & \boldsymbol{0} \\ \boldsymbol{0} & \boldsymbol{0}_{n-n_2}\end{bmatrix}\boldsymbol{V}'\right)\boldsymbol{\Phi}^{-1}\boldsymbol{\mu}_y \\ &= \boldsymbol{\mu}_y'\boldsymbol{\Phi}^{-t}(\boldsymbol{\Phi}'\boldsymbol{A}\boldsymbol{\Phi})\boldsymbol{\Phi}^{-1}\boldsymbol{\mu}_y \\ &= \boldsymbol{\mu}_y'\boldsymbol{A}\boldsymbol{\mu}_y\,.\end{aligned} \tag{9.42}$$

□

Theorem 9.7 If $\boldsymbol{y} \sim \mathcal{N}_n(\boldsymbol{\mu}_y, \Sigma)$, $\text{rank}(\Sigma) = n$, and $\boldsymbol{A}_1$ and $\boldsymbol{A}_2$ are conforming constants, then

$$\mathcal{V}(\boldsymbol{y}'\boldsymbol{A}_1\boldsymbol{y}, \boldsymbol{y}'\boldsymbol{A}_2\boldsymbol{y}) = 2\text{tr}(\boldsymbol{A}_1\Sigma\boldsymbol{A}_2\Sigma) + 4\boldsymbol{\mu}_y'\boldsymbol{A}_1\Sigma\boldsymbol{A}_2\boldsymbol{\mu}_y\,. \tag{9.43}$$

Proof. Left as an exercise.

Theorem 9.8 If $q = \boldsymbol{y}'\boldsymbol{A}\boldsymbol{y}$ with $\boldsymbol{y} \sim \mathcal{N}_n(\boldsymbol{\mu}_y, \Sigma)$, $\text{rank}(\Sigma) = n$, $\boldsymbol{A} = \boldsymbol{A}'$ constant, and $\boldsymbol{B}$ ($m \times n$) constant (of any rank), then

$$\mathcal{V}(\boldsymbol{y}, \boldsymbol{y}'\boldsymbol{A}\boldsymbol{y}) = 2\Sigma\boldsymbol{A}\boldsymbol{\mu}_y \tag{9.44}$$

$$\mathcal{V}(\boldsymbol{B}\boldsymbol{y}, \boldsymbol{y}'\boldsymbol{A}\boldsymbol{y}) = 2\boldsymbol{B}\Sigma\boldsymbol{A}\boldsymbol{\mu}_y\,. \tag{9.45}$$

Proof. (Searle, 1971, p56) From the definition, the $n \times 1$ vector can be written

$$\begin{aligned}\mathcal{V}(\boldsymbol{B}\boldsymbol{y}, \boldsymbol{y}'\boldsymbol{A}\boldsymbol{y}) &= \mathrm{E}\left\{(\boldsymbol{B}\boldsymbol{y} - \boldsymbol{B}\boldsymbol{\mu}_y)\left[\boldsymbol{y}'\boldsymbol{A}\boldsymbol{y} - \boldsymbol{\mu}_y'\boldsymbol{A}\boldsymbol{\mu}_y - \text{tr}(\boldsymbol{A}\Sigma)\right]\right\} \\ &= \mathrm{E}\{\boldsymbol{B}(\boldsymbol{y}-\boldsymbol{\mu}_y)[(\boldsymbol{y}-\boldsymbol{\mu}_y)'\boldsymbol{A}(\boldsymbol{y}-\boldsymbol{\mu}_y)+2(\boldsymbol{y}-\boldsymbol{\mu}_y)'\boldsymbol{A}\boldsymbol{\mu}_y-\text{tr}(\boldsymbol{A}\Sigma)]\} \\ &= \boldsymbol{B}\mathrm{E}[(\boldsymbol{y}-\boldsymbol{\mu}_y)(\boldsymbol{y}-\boldsymbol{\mu}_y)'\boldsymbol{A}(\boldsymbol{y}-\boldsymbol{\mu}_y)]+2\boldsymbol{B}\mathrm{E}[(\boldsymbol{y}-\boldsymbol{\mu}_y)(\boldsymbol{y}-\boldsymbol{\mu}_y)']\boldsymbol{A}\boldsymbol{\mu}_y-\boldsymbol{0} \\ &= \boldsymbol{0} + (2\boldsymbol{B}\Sigma\boldsymbol{A}\boldsymbol{\mu}_y) - \boldsymbol{0}\,,\end{aligned} \tag{9.46}$$

because the first and third moments of $\boldsymbol{y} - \boldsymbol{\mu}_y$ are zero. □

In practice, the most frequent use of the following theorem is to infer from the fact $\boldsymbol{A} = \boldsymbol{A}^2$ that q must be chi square regardless of the true value of $\boldsymbol{\mu}_y$.

Theorem 9.9 If $\boldsymbol{y} \sim \mathcal{N}_n(\boldsymbol{\mu}_y, \boldsymbol{I}_n)$, $\boldsymbol{A} = \boldsymbol{A}'$ $(n \times n)$, $\text{rank}(\boldsymbol{A}) = n_1$, and $\omega = \boldsymbol{\mu}_y'\boldsymbol{A}\boldsymbol{\mu}_y$, then $q = \boldsymbol{y}'\boldsymbol{A}\boldsymbol{y} \sim \chi^2(n_1, \omega)\ \forall \boldsymbol{\mu}_y \Leftrightarrow \boldsymbol{A} = \boldsymbol{A}^2$.

Proof. ($\Leftarrow$) The spectral decomposition of $\boldsymbol{A}$ is $\boldsymbol{A} = \boldsymbol{V}\text{Dg}(\boldsymbol{\lambda})\boldsymbol{V}'$ with $\boldsymbol{V}'\boldsymbol{V} = \boldsymbol{V}\boldsymbol{V}' = \boldsymbol{I}_n$. Given $\boldsymbol{A}$ is idempotent we have

$$\boldsymbol{A} = [\,\boldsymbol{V}_1\ \boldsymbol{V}_0\,]\begin{bmatrix}\boldsymbol{I}_\nu & \boldsymbol{0}\\ \boldsymbol{0} & \boldsymbol{0}\end{bmatrix}\begin{bmatrix}\boldsymbol{V}_1'\\ \boldsymbol{V}_0'\end{bmatrix} = \boldsymbol{V}_1\boldsymbol{V}_1'\,. \tag{9.47}$$

Now $q = \boldsymbol{y}'\boldsymbol{A}\boldsymbol{y} = \boldsymbol{y}'\boldsymbol{V}\text{Dg}(\boldsymbol{\lambda})\boldsymbol{V}'\boldsymbol{y}$. If $\boldsymbol{z} = \boldsymbol{V}'\boldsymbol{y}$ then $\boldsymbol{y} = \boldsymbol{V}\boldsymbol{z}$,

$$\boldsymbol{z} = \begin{bmatrix}\boldsymbol{z}_1\\ \boldsymbol{z}_0\end{bmatrix} = \boldsymbol{V}'\boldsymbol{y} = \begin{bmatrix}\boldsymbol{V}_1'\boldsymbol{y}\\ \boldsymbol{V}_0'\boldsymbol{y}\end{bmatrix} \sim \mathcal{N}_n(\boldsymbol{V}'\boldsymbol{\mu}_y, \boldsymbol{I}_n)\,, \tag{9.48}$$

and $q = \boldsymbol{y}'\boldsymbol{A}\boldsymbol{y} = \boldsymbol{z}'\text{Dg}(\boldsymbol{\lambda})\boldsymbol{z} = \boldsymbol{z}_1'\boldsymbol{z}_1$. Having $\boldsymbol{z}_1 \sim \mathcal{N}_{n_1}(\boldsymbol{V}_1'\boldsymbol{\mu}_y, \boldsymbol{I}_\nu)$ implies $q = \boldsymbol{z}_1'\boldsymbol{z}_1 \sim \chi^2(n_1, \omega)$ with $\omega = (\boldsymbol{V}_1'\boldsymbol{\mu}_y)'(\boldsymbol{V}_1'\boldsymbol{\mu}_y) = \boldsymbol{\mu}_y'\boldsymbol{V}_1\boldsymbol{V}_1'\boldsymbol{\mu}_y = \boldsymbol{\mu}_y'\boldsymbol{A}\boldsymbol{\mu}_y$. □

Proof. ($\Rightarrow$) Given $q = \boldsymbol{y}'\boldsymbol{A}\boldsymbol{y} \sim \chi^2(n_1, \omega)$ for every choice of $\boldsymbol{\mu}_y$, in which $\omega = \boldsymbol{\mu}_y'\boldsymbol{A}\boldsymbol{\mu}_y$, and $n_1 = \text{rank}(\boldsymbol{A})$, we must prove $\boldsymbol{A}$ is idempotent. The MGF of the distribution of $\boldsymbol{y}'\boldsymbol{A}\boldsymbol{y}$ is

$$m_{\boldsymbol{y}'\boldsymbol{A}\boldsymbol{y}}(t) = |\boldsymbol{I} - 2t\boldsymbol{A}|^{-1/2}\exp\left\{-\boldsymbol{\mu}_y'\left[\boldsymbol{I} - (\boldsymbol{I} - 2t\boldsymbol{A})^{-1}\right]^{-1}\boldsymbol{\mu}_y/2\right\}, \tag{9.49}$$

while the MGF of $\chi^2(n_1, \omega)$ is

$$m_{\chi^2(n_1,\omega)}(t) = (1 - 2t)^{-\nu/2}\exp\{-(\omega/2)\left[1 - (1 - 2t)^{-1}\right]\}\,. \tag{9.50}$$

By assumption the two MGFs must be equal $\forall \boldsymbol{\mu}_y \in \Re^n$, including $\boldsymbol{\mu}_y = \boldsymbol{0}$. If $\boldsymbol{\mu}_y = \boldsymbol{0}$, then $\omega = 0$, which implies $(1 - 2t)^{-n_1/2} = |\boldsymbol{I} - 2t\boldsymbol{A}|^{-1/2}\ \forall t \in \Re$. For any (square) matrix $|\boldsymbol{I} - u\boldsymbol{A}| = \prod_{k=1}^n (1 - u\lambda_k)$, in which the eigenvalues of $\boldsymbol{A}$ are $\lambda_1 \geq \lambda_2 \geq \cdots \geq \lambda_n$. If $u = 2t$, then raising both sides to the -2 power gives $(1 - u)^{n_1} = \prod_{k=1}^n (1 - u\lambda_k)\ \forall u \in \Re$. The two polynomials are identical and so must be of the same degree. The polynomial on the left is of degree $n_1 < n$, which implies the polynomial on the right is also of degree n_1. In turn, the last $n - n_1$ terms are equal to 1 because $(1 - u\lambda_k) = 1$ for $k \in \{n_1{+}1, n_1{+}2, \ldots, n\}\ \forall u \in \Re$ implies $\lambda_k = 0$ for $k \in \{n_1{+}1, n_1{+}2, \ldots, n\}$. We now have $(1 - u)^{n_1} = \prod_{j=1}^n (1 - u\lambda_k)\ \forall u \in \Re$. Since the polynomials are identical, they must have identical roots. Therefore $\lambda_k = 1$ for $k \in \{1, 2, \ldots, n_1\}$. The matrix $\boldsymbol{A} = \boldsymbol{A}'$ has n_1 eigenvalues of 1 and $n - n_1$ eigenvalues of 0. Finally, $\boldsymbol{A}$ is idempotent of rank n_1. □

Corollary 9.9 If $\boldsymbol{y} \sim \mathcal{N}_n(\boldsymbol{\mu}_y, \boldsymbol{\Sigma})$, $\text{rank}(\boldsymbol{\Sigma}) = n$, $\boldsymbol{A} = \boldsymbol{A}'$ $(n \times n)$, $\text{rank}(\boldsymbol{A}) = n_1$, and $\omega = \boldsymbol{\mu}_y'\boldsymbol{A}\boldsymbol{\mu}_y$, then $\boldsymbol{y}'\boldsymbol{A}\boldsymbol{y} \sim \chi^2(n_1, \omega)\ \forall \boldsymbol{\mu}_y \Leftrightarrow (\boldsymbol{A}\boldsymbol{\Sigma}) = (\boldsymbol{A}\boldsymbol{\Sigma})^2$.

Proof. Factoring the covariance matrix allows transforming the random variables to independence. Spectral decomposition gives $\boldsymbol{\Sigma} = \boldsymbol{\Upsilon}\text{Dg}(\boldsymbol{\lambda})\boldsymbol{\Upsilon}' = \boldsymbol{\Phi}\boldsymbol{\Phi}'$

with $\boldsymbol{\Phi} = \boldsymbol{\Upsilon}\mathrm{Dg}(\boldsymbol{\lambda})^{1/2}$. Alternately use the Cholesky decomposition of $\boldsymbol{\Sigma}$, with $\boldsymbol{\Phi}$ a lower triangular matrix. In general, $\boldsymbol{\Phi}^{-1}$ exists if and only if $\boldsymbol{\Sigma}^{-1}$ exists. For $\mathrm{rank}(\boldsymbol{\Sigma}) = n$ we have $\boldsymbol{\Sigma}^{-1} = \boldsymbol{\Phi}^{-t}\boldsymbol{\Phi}^{-1}$ and $\boldsymbol{\Phi}^{-1}\boldsymbol{\Sigma}\boldsymbol{\Phi}^{-t} = \boldsymbol{I}_n$. Also $\boldsymbol{z} = \boldsymbol{\Phi}^{-1}\boldsymbol{y} \sim \mathcal{N}_n(\boldsymbol{\Phi}^{-1}\boldsymbol{\mu}_y, \boldsymbol{I}_n)$. If $\boldsymbol{\mu}_z = \boldsymbol{\Phi}^{-1}\boldsymbol{\mu}_y$ then $\omega = \boldsymbol{\mu}_z'(\boldsymbol{\Phi}'\boldsymbol{A}\boldsymbol{\Phi})\boldsymbol{\mu}_z = \boldsymbol{\mu}_y'\boldsymbol{A}\boldsymbol{\mu}_y$. Since $\boldsymbol{y} = \boldsymbol{\Phi}\boldsymbol{z}$, we have $q = \boldsymbol{y}'\boldsymbol{A}\boldsymbol{y} = \boldsymbol{z}'\boldsymbol{\Phi}'\boldsymbol{A}\boldsymbol{\Phi}\boldsymbol{z}$. In terms of $\boldsymbol{z}$, $q \sim \chi^2(n_1, \omega)$ if and only if $\boldsymbol{\Phi}'\boldsymbol{A}\boldsymbol{\Phi} = (\boldsymbol{\Phi}'\boldsymbol{A}\boldsymbol{\Phi})^2$. By Lemma 1.30 we conclude $q \sim \chi^2(n_1, \omega)$ if and only if $\boldsymbol{A}\boldsymbol{\Sigma}$ is idempotent. □

The immediately preceding results assumed a nonsingular Gaussian distribution, while Theorem 9.6 and the following theorem provide results for singular and nonsingular Gaussian distributions.

Theorem 9.10 For $\boldsymbol{y} \sim \mathcal{N}_n(\boldsymbol{\mu}, \boldsymbol{\Sigma})$ and $\mathrm{rank}(\boldsymbol{\Sigma}) = n_1 \leq n$, $q = (\boldsymbol{y}'\boldsymbol{A}\boldsymbol{y} + \boldsymbol{b}'\boldsymbol{y} + c)$. If $\boldsymbol{A} = \boldsymbol{A}'$ $(n \times n)$, $\boldsymbol{b}$ $(n \times 1)$, c (1×1), $\nu = \mathrm{tr}(\boldsymbol{A}\boldsymbol{\Sigma})$, and $\omega = (\boldsymbol{A}\boldsymbol{\mu} + \boldsymbol{b}/2)'\boldsymbol{\Sigma}(\boldsymbol{A}\boldsymbol{\mu} + \boldsymbol{b}/2)$ are all fixed constants, then $q = (\boldsymbol{y}'\boldsymbol{A}\boldsymbol{y} + \boldsymbol{b}'\boldsymbol{y} + c) \sim \chi^2(\nu, \omega)$ if and only if all three of the following conditions hold regardless of the value of $\boldsymbol{\mu}$ and $\boldsymbol{\Sigma}$:

1. $\boldsymbol{\Sigma}\boldsymbol{A}\boldsymbol{\Sigma}\boldsymbol{A}\boldsymbol{\Sigma} = \boldsymbol{\Sigma}\boldsymbol{A}\boldsymbol{\Sigma}$
2. $(\boldsymbol{A}\boldsymbol{\mu} + \boldsymbol{b}/2)'\boldsymbol{\Sigma} = (\boldsymbol{A}\boldsymbol{\mu} + \boldsymbol{b}/2)'\boldsymbol{\Sigma}\boldsymbol{A}\boldsymbol{\Sigma}$
3. $\boldsymbol{\mu}'\boldsymbol{A}\boldsymbol{\mu} + \boldsymbol{b}'\boldsymbol{\mu} + c = (\boldsymbol{A}\boldsymbol{\mu} + \boldsymbol{b}/2)'\boldsymbol{\Sigma}(\boldsymbol{A}\boldsymbol{\mu} + \boldsymbol{b}/2)$.

Searle (1971, Section 2.7) presented the theorem and its proof. One can obtain various corollaries by setting some of $\boldsymbol{\mu}$, $\boldsymbol{b}$, and c to zero.

The theorem is easily misinterpreted. If q has a noncentral chi-square distribution, the theorem does *not* imply $\boldsymbol{A}\boldsymbol{\Sigma}$ is idempotent, only that all three of 1–3 are true. Searle (1971, p. 69) commented on problems which may arise from misinterpretation.

9.5 INDEPENDENCE AMONG LINEAR AND QUADRATIC FORMS

The proofs for Theorems 9.11 and 9.14 give independence properties based on factoring $\boldsymbol{A} = \boldsymbol{A}'$ as $\boldsymbol{A} = \boldsymbol{F}\boldsymbol{F}'$. If $\boldsymbol{A}$ has any negative eigenvalues, then complex variables occur in $\boldsymbol{F}$. The complex variables could be avoided with a slight complication of the proofs. With $\boldsymbol{A} = \boldsymbol{V}\mathrm{Dg}(\boldsymbol{\lambda})\boldsymbol{V}'$, if $s(\lambda_j) = 1$ for $\lambda_j \geq 0$ and -1 otherwise, writing $\mathrm{Dg}(\boldsymbol{\lambda}) = \mathrm{Dg}(\{|\lambda_j|\})\mathrm{Dg}(\{s(\lambda_j)\})$ allows treating algebraic sign separately from eigenstructure.

Theorem 9.11 If $\boldsymbol{y} \sim \mathcal{N}_n(\boldsymbol{\mu}, \boldsymbol{\Sigma})$ with $\mathrm{rank}(\boldsymbol{\Sigma}) = n$, $n \times n$ constant $\boldsymbol{A} = \boldsymbol{A}'$ has rank $\nu \leq n$, and $m \times n$ $\boldsymbol{B}$ is constant (and any rank), then $\boldsymbol{y}'\boldsymbol{A}\boldsymbol{y} \perp\!\!\!\perp \boldsymbol{B}\boldsymbol{y}$ $\forall \boldsymbol{\mu} \Leftrightarrow \boldsymbol{B}\boldsymbol{\Sigma}\boldsymbol{A} = \boldsymbol{0}$.

Proof. ($\Leftarrow$) Symmetric $\boldsymbol{A}$ can be written $\boldsymbol{A}=\boldsymbol{F}\boldsymbol{F}'$ with $\boldsymbol{F}$ of dimension $n \times \nu$ and rank ν. One choice is $\boldsymbol{F}=\boldsymbol{V}_1 \text{Dg}(\boldsymbol{\lambda}_1)^{1/2}$, with $\boldsymbol{V}_1$ $n \times \nu$, and the columns the ν eigenvectors corresponding to the nonzero eigenvalues. Full column rank of $\boldsymbol{F}$ ensures $(\boldsymbol{F}'\boldsymbol{F})^{-1}$ exists. Also

$$\begin{aligned} \mathbf{0}=\boldsymbol{B}\boldsymbol{\Sigma}\boldsymbol{A} &= \boldsymbol{B}\boldsymbol{\Sigma}\boldsymbol{F}\boldsymbol{F}' \\ &= \boldsymbol{B}\boldsymbol{\Sigma}\boldsymbol{F}\boldsymbol{F}'[\boldsymbol{F}(\boldsymbol{F}'\boldsymbol{F})^{-1}] \\ &= \boldsymbol{B}\boldsymbol{\Sigma}\boldsymbol{F} \\ &= \mathcal{V}(\boldsymbol{B}\boldsymbol{y}, \boldsymbol{F}'\boldsymbol{y})\,. \end{aligned} \tag{9.51}$$

The last line is true $\Leftrightarrow$ $\boldsymbol{B}\boldsymbol{y} \perp\!\!\!\perp \boldsymbol{F}\boldsymbol{y}'$ (due to the Gaussian distribution assumption), which implies $\boldsymbol{B}\boldsymbol{y} \perp\!\!\!\perp (\boldsymbol{F}'\boldsymbol{y})'(\boldsymbol{F}'\boldsymbol{y}) \equiv \boldsymbol{y}'\boldsymbol{A}\boldsymbol{y}$.

Proof. ($\Rightarrow$) By Theorem 9.8 we know $\mathcal{V}(\boldsymbol{B}\boldsymbol{y}, \boldsymbol{y}'\boldsymbol{A}\boldsymbol{y}) = 2\boldsymbol{B}\boldsymbol{\Sigma}\boldsymbol{A}\boldsymbol{\mu}$. By independence, $\mathbf{0}=2\boldsymbol{B}\boldsymbol{\Sigma}\boldsymbol{A}\boldsymbol{\mu}\ \forall \boldsymbol{\mu}$, which implies $\boldsymbol{B}\boldsymbol{\Sigma}\boldsymbol{A}=\mathbf{0}$. □

The first part of the proof reveals the underlying source of the independence, when it exists. Linear form $\boldsymbol{B}\boldsymbol{y}$ is independent of quadratic form $\boldsymbol{y}'\boldsymbol{A}\boldsymbol{y}$, because the quadratic form can be written as $\boldsymbol{y}'\boldsymbol{A}\boldsymbol{y}=\boldsymbol{y}'\boldsymbol{F}\boldsymbol{F}'\boldsymbol{y}=\boldsymbol{z}'\boldsymbol{z}$ with $\boldsymbol{z}=\boldsymbol{F}'\boldsymbol{y}$ independent of $\boldsymbol{B}\boldsymbol{y}$.

The theorem is both significant and has a familiar result as a special case. A standard result from univariate theory is that $y_i \sim \mathcal{N}(\mu, \sigma^2)$ i.i.d. $\Rightarrow$ $\widehat{\mu}=\overline{y}$ is independent of $\widehat{\sigma}^2$. The following corollary states the familiar result formally.

Corollary 9.11 The sample mean is independent of the sample variance for i.i.d. Gaussian data. Given data $\boldsymbol{y} \sim \mathcal{N}_N(\mu\mathbf{1}, \sigma^2\boldsymbol{I}_N)$, in the notation of the theorem, the usual estimators are $\overline{y}=\boldsymbol{b}'\boldsymbol{y}$ and $\widehat{\sigma}^2=\boldsymbol{y}'\boldsymbol{A}\boldsymbol{y}$ with $1 \times N$ $\boldsymbol{b}'=(\mathbf{1}'\mathbf{1})^{-1}\mathbf{1}'$ and $N \times N$ $\boldsymbol{A}=\left[\boldsymbol{I}-\mathbf{1}(\mathbf{1}'\mathbf{1})^{-1}\mathbf{1}'\right]/(N-1)$. Therefore $\overline{y}$ and $\widehat{\sigma}^2$ are independent.

Proof. It is easy to prove $\boldsymbol{b}'(\sigma^2\boldsymbol{I}_N)\boldsymbol{A}=\mathbf{0}$ because $\mathbf{1}'\boldsymbol{A}=\mathbf{0}$, which allows applying the theorem. □

The following theorem is the foundation for extending the preceding result to the singular Gaussian distribution.

Theorem 9.12 (Good, 1963) For $\boldsymbol{y} \sim \mathcal{N}_n(\mathbf{0}, \boldsymbol{\Sigma})$ with $\text{rank}(\boldsymbol{\Sigma})=n_1 \leq n$, constants $\boldsymbol{A}=\boldsymbol{A}'$ $(n \times n)$, $\boldsymbol{a} \in \Re^n$, and $\boldsymbol{b} \in \Re^n$,
(a) $\boldsymbol{y}'\boldsymbol{A}\boldsymbol{y}$ and $\boldsymbol{b}'\boldsymbol{y}$ are independent $\Leftrightarrow$ $\boldsymbol{\Sigma}\boldsymbol{A}\boldsymbol{\Sigma}\boldsymbol{b}=\mathbf{0}$ and
(b) $\boldsymbol{a}'\boldsymbol{y}$ and $\boldsymbol{b}'\boldsymbol{y}$ are independent $\Leftrightarrow$ $\boldsymbol{a}'\boldsymbol{\Sigma}\boldsymbol{b}=0$.

Proof. Good (1963, Theorem 1C; corrigenda in 1966) provided a proof.

Theorem 9.13 If $\boldsymbol{y} \sim \mathcal{N}_n(\boldsymbol{\mu}, \boldsymbol{\Sigma})$, $\text{rank}(\boldsymbol{\Sigma})=n_1 \leq n$, $\boldsymbol{A}=\boldsymbol{A}'$ $(n \times n)$ and $\boldsymbol{B}$ $(m \times n)$ are constants, then $\boldsymbol{y}'\boldsymbol{A}\boldsymbol{y}$ and $\boldsymbol{B}\boldsymbol{y}$ are independent $\Leftrightarrow$ $\boldsymbol{B}\boldsymbol{\Sigma}\boldsymbol{A}\boldsymbol{\Sigma}=\mathbf{0}$ and $\boldsymbol{B}\boldsymbol{\Sigma}\boldsymbol{A}\boldsymbol{\mu}=\mathbf{0}$.

Proof. (Searle, 1971, p. 70) Necessarily $\boldsymbol{\Sigma} = \boldsymbol{\Phi}\boldsymbol{\Phi}'$, with $\boldsymbol{\Phi}$ $n \times n_1$ and rank n_1. Also $\boldsymbol{y} = \boldsymbol{\mu} + \boldsymbol{\Phi}\boldsymbol{z}$, with $\boldsymbol{z} \sim \mathcal{N}_{n_1}(\boldsymbol{0}, \boldsymbol{I})$. If $\boldsymbol{b}' = \text{row}_i(\boldsymbol{B})$, then $\boldsymbol{y}'\boldsymbol{A}\boldsymbol{y} = \boldsymbol{z}'\boldsymbol{\Phi}'\boldsymbol{A}\boldsymbol{\Phi}\boldsymbol{z} + 2\boldsymbol{\mu}'\boldsymbol{A}\boldsymbol{\Phi}\boldsymbol{z} + \boldsymbol{\mu}'\boldsymbol{A}\boldsymbol{\mu}$ and $\boldsymbol{b}'\boldsymbol{y} = \boldsymbol{b}'\boldsymbol{\Phi}\boldsymbol{z} + \boldsymbol{b}'\boldsymbol{\mu}$. By Good's theorem, $\boldsymbol{z}'\boldsymbol{\Phi}'\boldsymbol{A}\boldsymbol{\Phi}\boldsymbol{z}$ is independent of $\boldsymbol{b}'\boldsymbol{\Phi}\boldsymbol{z} \Leftrightarrow \boldsymbol{I}(\boldsymbol{\Phi}'\boldsymbol{A}\boldsymbol{\Phi})\boldsymbol{I}(\boldsymbol{\Phi}'\boldsymbol{b}) = \boldsymbol{0} \Leftrightarrow \boldsymbol{B}\boldsymbol{\Sigma}\boldsymbol{A}\boldsymbol{\Sigma} = 0$. Also $\boldsymbol{\mu}'\boldsymbol{A}\boldsymbol{\Phi}\boldsymbol{z}$ is independent of $\boldsymbol{b}'\boldsymbol{\Phi}\boldsymbol{z} \Leftrightarrow \boldsymbol{\mu}'\boldsymbol{A}\boldsymbol{\Phi}\boldsymbol{I}\boldsymbol{\Phi}'\boldsymbol{b} = \boldsymbol{0} \Leftrightarrow \boldsymbol{B}\boldsymbol{\Sigma}\boldsymbol{A}\boldsymbol{\mu} = 0$. Combining results, $\boldsymbol{y}'\boldsymbol{A}\boldsymbol{y}$ is independent of $\boldsymbol{B}'\boldsymbol{y} \Leftrightarrow \boldsymbol{B}\boldsymbol{\Sigma}\boldsymbol{A}\boldsymbol{\Sigma} = 0$ and $\boldsymbol{B}\boldsymbol{\Sigma}\boldsymbol{A}\boldsymbol{\mu} = 0$. □

Corollary 9.13 Given the conditions of the theorem, if $\boldsymbol{B}\boldsymbol{\Sigma}\boldsymbol{A} = 0$, then $\boldsymbol{y}'\boldsymbol{A}\boldsymbol{y}$ and $\boldsymbol{B}\boldsymbol{y}$ are independent.

Proof. Immediate from the theorem.

The useful corollary states a sufficient (but not necessary) condition for independence.

Theorem 9.14 If $\boldsymbol{y} \sim \mathcal{N}_n(\boldsymbol{\mu}, \boldsymbol{\Sigma})$ with $\text{rank}(\boldsymbol{\Sigma}) = n$, $\boldsymbol{A} = \boldsymbol{A}'$ $(n \times n)$ and $\boldsymbol{B} = \boldsymbol{B}'$ $(n \times n)$ are constants, then $\boldsymbol{y}'\boldsymbol{A}\boldsymbol{y} \perp\!\!\!\perp \boldsymbol{y}'\boldsymbol{B}\boldsymbol{y}\, \forall \boldsymbol{\mu} \Leftrightarrow \boldsymbol{A}\boldsymbol{\Sigma}\boldsymbol{B} = 0$.

Here $\boldsymbol{0} = \boldsymbol{A}\boldsymbol{\Sigma}\boldsymbol{B} \Leftrightarrow \boldsymbol{0} = (\boldsymbol{A}\boldsymbol{\Sigma}\boldsymbol{B})' = \boldsymbol{B}'\boldsymbol{\Sigma}'\boldsymbol{A}' = \boldsymbol{B}\boldsymbol{\Sigma}\boldsymbol{A}$. It is not necessary for either quadratic form to have a (marginal) chi-square distribution. The theorem is about independence only.

Proof. ($\Leftarrow$) The approach centers on defining a transformation and proving the independence of the two underlying linear forms. If $\boldsymbol{A} = \boldsymbol{F}_A\boldsymbol{F}_A'$ and $\boldsymbol{B} = \boldsymbol{F}_B\boldsymbol{F}_B'$, with $\boldsymbol{F}_A$ and $\boldsymbol{F}_B$ full (column) rank factors, then $(\boldsymbol{F}_A{}'\boldsymbol{F}_A)^{-1}$ and $(\boldsymbol{F}_B{}'\boldsymbol{F}_B)^{-1}$ exist. Furthermore

$$\begin{aligned}
\boldsymbol{0} = \boldsymbol{A}\boldsymbol{\Sigma}\boldsymbol{B} &= \boldsymbol{F}_A\boldsymbol{F}_A{}'\boldsymbol{\Sigma}\boldsymbol{F}_B\boldsymbol{F}_B{}' \\
&= (\boldsymbol{F}_A{}'\boldsymbol{F}_A)^{-1}\boldsymbol{F}_A{}'\boldsymbol{F}_A\boldsymbol{F}_A{}'\boldsymbol{\Sigma}\boldsymbol{F}_B\boldsymbol{F}_B{}'\boldsymbol{F}_B(\boldsymbol{F}_B{}'\boldsymbol{F}_B)^{-1} \\
&= \boldsymbol{F}_A{}'\boldsymbol{\Sigma}\boldsymbol{F}_B \\
&= \mathcal{V}(\boldsymbol{F}_A{}'\boldsymbol{y}, \boldsymbol{F}_B{}'\boldsymbol{y}) = \mathcal{V}(\boldsymbol{x}, \boldsymbol{z}), \qquad (9.52)
\end{aligned}$$

with $\boldsymbol{x} = \boldsymbol{F}_A{}'\boldsymbol{y}$, $\boldsymbol{z} = \boldsymbol{F}_B{}'\boldsymbol{y}$. Under the assumption of Gaussian distribution, $\boldsymbol{x}$ and $\boldsymbol{z}$ are independent. Therefore $\boldsymbol{x}'\boldsymbol{x} = \boldsymbol{y}'\boldsymbol{A}\boldsymbol{y}$ is independent of $\boldsymbol{z}'\boldsymbol{z} = \boldsymbol{y}'\boldsymbol{B}\boldsymbol{y}$.

Proof. ($\Rightarrow$) Assuming independence of $\boldsymbol{y}'\boldsymbol{A}\boldsymbol{y}$ and $\boldsymbol{y}'\boldsymbol{B}\boldsymbol{y}$ allows writing

$$\begin{aligned}
\mathcal{V}(\boldsymbol{y}'\boldsymbol{A}\boldsymbol{y} + \boldsymbol{y}'\boldsymbol{B}\boldsymbol{y}) &= \mathcal{V}(\boldsymbol{y}'\boldsymbol{A}\boldsymbol{y}) + \mathcal{V}(\boldsymbol{y}'\boldsymbol{B}\boldsymbol{y}) \\
&= \mathcal{V}[\boldsymbol{y}'(\boldsymbol{A} + \boldsymbol{B})\boldsymbol{y}]. \qquad (9.53)
\end{aligned}$$

The first part follows from the fact that $\mathcal{V}(\boldsymbol{y}'\boldsymbol{A}\boldsymbol{y}) + \mathcal{V}(\boldsymbol{y}'\boldsymbol{B}\boldsymbol{y}) - \mathcal{V}[\boldsymbol{y}'(\boldsymbol{A} + \boldsymbol{B})\boldsymbol{y}] = 0$. From a previous corollary,

$$\mathcal{V}(\boldsymbol{y}'\boldsymbol{A}\boldsymbol{y}) = 2\text{tr}\left[(\boldsymbol{A}\boldsymbol{\Sigma})^2\right] + 4\boldsymbol{\mu}'\boldsymbol{A}\boldsymbol{\Sigma}\boldsymbol{A}\boldsymbol{\mu} \tag{9.54}$$
$$\mathcal{V}(\boldsymbol{y}'\boldsymbol{B}\boldsymbol{y}) = 2\text{tr}\left[(\boldsymbol{B}\boldsymbol{\Sigma})^2\right] + 4\boldsymbol{\mu}'\boldsymbol{B}\boldsymbol{\Sigma}\boldsymbol{B}\boldsymbol{\mu} \tag{9.55}$$

and

$$\begin{aligned}\mathcal{V}[\boldsymbol{y}'(\boldsymbol{A}+\boldsymbol{B})\boldsymbol{y}] &= 2\text{tr}\{[(\boldsymbol{A}+\boldsymbol{B})\boldsymbol{\Sigma}]^2\} + 4\boldsymbol{\mu}'(\boldsymbol{A}+\boldsymbol{B})\boldsymbol{\Sigma}(\boldsymbol{A}+\boldsymbol{B})\boldsymbol{\mu}\\ &= 2\text{tr}(\boldsymbol{A}\boldsymbol{\Sigma}\boldsymbol{A}\boldsymbol{\Sigma}) + 2\text{tr}(\boldsymbol{A}\boldsymbol{\Sigma}\boldsymbol{B}\boldsymbol{\Sigma}) + 2\text{tr}(\boldsymbol{B}\boldsymbol{\Sigma}\boldsymbol{A}\boldsymbol{\Sigma}) +\\ &\quad 2\text{tr}(\boldsymbol{B}\boldsymbol{\Sigma}\boldsymbol{B}\boldsymbol{\Sigma}) + 4\boldsymbol{\mu}'\boldsymbol{A}\boldsymbol{\Sigma}\boldsymbol{A}\boldsymbol{\mu} + 8\boldsymbol{\mu}'\boldsymbol{A}\boldsymbol{\Sigma}\boldsymbol{B}\boldsymbol{\mu} + 4\boldsymbol{\mu}'\boldsymbol{B}\boldsymbol{\Sigma}\boldsymbol{B}\boldsymbol{\mu}.\end{aligned} \tag{9.56}$$

The difference should be zero $\forall \boldsymbol{\mu}$ and $\boldsymbol{\Sigma}$. Therefore

$$\begin{aligned}0 &= 2\text{tr}(\boldsymbol{A}\boldsymbol{\Sigma}\boldsymbol{B}\boldsymbol{\Sigma}) + 2\text{tr}(\boldsymbol{B}\boldsymbol{\Sigma}\boldsymbol{A}\boldsymbol{\Sigma}) + 8\boldsymbol{\mu}'\boldsymbol{A}\boldsymbol{\Sigma}\boldsymbol{B}\boldsymbol{\mu}\\ &= 4\text{tr}(\boldsymbol{\Sigma}\boldsymbol{A}\boldsymbol{\Sigma}\boldsymbol{B}) + 8\boldsymbol{\mu}'\boldsymbol{A}\boldsymbol{\Sigma}\boldsymbol{B}\boldsymbol{\mu}.\end{aligned} \tag{9.57}$$

Letting $\boldsymbol{\mu} = \boldsymbol{0}$ implies $\text{tr}(\boldsymbol{\Sigma}\boldsymbol{A}\boldsymbol{\Sigma}\boldsymbol{B}) = 0$ which, with the equation above, implies $\boldsymbol{\mu}'\boldsymbol{A}\boldsymbol{\Sigma}\boldsymbol{B}\boldsymbol{\mu} = 0\ \forall \boldsymbol{\mu}$, which implies $\boldsymbol{A}\boldsymbol{\Sigma}\boldsymbol{B} = \boldsymbol{0}$. □

Theorem 9.15 If $\boldsymbol{y} \sim \mathcal{N}_n(\boldsymbol{\mu}, \boldsymbol{\Sigma})$ with $\text{rank}(\boldsymbol{\Sigma}) = n_1 \leq n$, while $n \times n$ matrices $\boldsymbol{A}$ and $\boldsymbol{B}$ are constant (of any rank), then $\boldsymbol{y}'\boldsymbol{A}\boldsymbol{y}$ and $\boldsymbol{y}'\boldsymbol{B}\boldsymbol{y}$ are independent if and only if all three of the following hold:
(1) $\boldsymbol{\mu}'\boldsymbol{A}\boldsymbol{\Sigma}\boldsymbol{B}\boldsymbol{\mu} = 0$, (2) $\boldsymbol{\Sigma}\boldsymbol{A}\boldsymbol{\Sigma}\boldsymbol{B}\boldsymbol{\mu} = \boldsymbol{\Sigma}\boldsymbol{B}\boldsymbol{\Sigma}\boldsymbol{A}\boldsymbol{\mu}$, and (3) $\boldsymbol{\Sigma}\boldsymbol{A}\boldsymbol{\Sigma}\boldsymbol{B}\boldsymbol{\Sigma} = \boldsymbol{0}$.
Without loss of generality, $\boldsymbol{A}$ and $\boldsymbol{B}$ may be assumed symmetric.

Proof. Left as an exercise. The generalization for singular Gaussian vectors is from Searle (1971). The proof relies on Good's theorem.

Corollary 9.15.1 If $\boldsymbol{A}\boldsymbol{\Sigma}\boldsymbol{B} = \boldsymbol{0}$, then $\boldsymbol{y}'\boldsymbol{A}\boldsymbol{y}$ and $\boldsymbol{y}'\boldsymbol{B}\boldsymbol{y}$ are independent.

The useful corollary states a sufficient (but not necessary) condition for independence.

Corollary 9.15.2 If both $\boldsymbol{A}$ and $\boldsymbol{B}$ are positive semidefinite or positive definite, then $\boldsymbol{y}'\boldsymbol{A}\boldsymbol{y}$ and $\boldsymbol{y}'\boldsymbol{B}\boldsymbol{y}$ are independent if and only if $\boldsymbol{A}\boldsymbol{\Sigma}\boldsymbol{B} = \boldsymbol{0}$.

Proof. Shanbhag (1966; Searle, 1971, p. 71).

As one can see, relaxing the assumption of full-rank $\boldsymbol{\Sigma}$ to allow $\boldsymbol{\Sigma}$ to be positive semidefinite complicates the necessary conditions for independence. The basic tool for manipulating the singular Gaussian distribution is a transformation to a full-rank distribution, as follows. For $\boldsymbol{y} \sim \mathcal{N}_n(\boldsymbol{\mu}, \boldsymbol{\Sigma})$, $\text{rank}(\boldsymbol{\Sigma}) = n_1 \leq n$, matrix $\boldsymbol{\Phi}$ exists with full column rank such that $\boldsymbol{\Sigma} = \boldsymbol{\Phi}\boldsymbol{\Phi}'$. In turn, $\boldsymbol{y} = \boldsymbol{\mu} + \boldsymbol{\Phi}\boldsymbol{z}$ with $\boldsymbol{z} \sim \mathcal{N}_{n_1}(\boldsymbol{0}, \boldsymbol{I}_{n_1})$, and $\boldsymbol{z} = (\boldsymbol{\Phi}'\boldsymbol{\Phi})^{-1}\boldsymbol{\Phi}'(\boldsymbol{y} - \boldsymbol{\mu})$. The technique is central to the proof of the last theorem.

9.6 THE ANOVA THEOREM

ANOVA for general linear univariate models partitions the total sum of squares into component sums of squares as

$$SST = \boldsymbol{y}'\boldsymbol{y} = \boldsymbol{y}'\boldsymbol{A}_1\boldsymbol{y} + \boldsymbol{y}'\boldsymbol{A}_2\boldsymbol{y} + \cdots + \boldsymbol{y}'\boldsymbol{A}_n\boldsymbol{y}\,. \tag{9.58}$$

Typically, it is crucial to be able to assume the n sums of squares are totally independent and have marginal chi-square distributions. The ANOVA theorem provides necessary and sufficient conditions for such quadratic forms to have independent chi-square distributions. The theorem is founded on theory for idempotent matrices and has a geometric interpretation. If $\boldsymbol{A}_i = \boldsymbol{A}_i^2$ ($N \times N$), then $\boldsymbol{z} = \boldsymbol{A}_i\boldsymbol{y}$ ($N \times 1$) is a projection onto a subspace, and $\boldsymbol{z}'\boldsymbol{z} = \boldsymbol{y}'\boldsymbol{A}_i\boldsymbol{y}$ is the squared length of the vector. The theorem is about squared lengths of projections of $\boldsymbol{y}$ ($N \times 1$) onto mutually orthogonal subspaces of the sample space.

The proof of the statistical theorem arises directly from a matrix theorem. In turn, Loynes' lemma simplifies proving the matrix theorem. We first prove Loynes' lemma, then derive the matrix decomposition used in the ANOVA theorem, and finally prove the ANOVA theorem itself. The presentation follows Searle (1971, p. 60–64). Early proofs were quite long. Banerjee (1964) produced a shorter proof, which was shortened and improved by Loynes (1966). Therefore Loynes' lemma is a key to a concise proof of the ANOVA theorem.

Lemma 9.1 (Loynes' lemma) If $\boldsymbol{B} = \boldsymbol{B}' = \boldsymbol{B}^2$, $\boldsymbol{Q} = \boldsymbol{Q}'$ is positive definite or positive semidefinite, and $\boldsymbol{I} - \boldsymbol{B} - \boldsymbol{Q}$ is positive definite or positive semidefinite, then $\boldsymbol{BQ} = \boldsymbol{0}$. Furthermore $\boldsymbol{BQ} = \boldsymbol{QB}$.

Proof. We prove $\boldsymbol{QBx} = \boldsymbol{0}$ for all $\boldsymbol{x} \in \Re^N$. It then follows $\boldsymbol{QB} = \boldsymbol{0}$. If $\boldsymbol{y} = \boldsymbol{Bx}$ for arbitrary $\boldsymbol{x} \in \Re^N$ then $\boldsymbol{y}'\boldsymbol{By} = \boldsymbol{y}'\boldsymbol{B}(\boldsymbol{Bx}) = \boldsymbol{y}'\boldsymbol{B}^2\boldsymbol{x} = \boldsymbol{y}'\boldsymbol{B}^1\boldsymbol{x} = \boldsymbol{y}'(\boldsymbol{Bx}) = \boldsymbol{y}'\boldsymbol{y}$. Thus $\boldsymbol{y}'(\boldsymbol{I} - \boldsymbol{B})\boldsymbol{y} = (\boldsymbol{y}'\boldsymbol{Iy} - \boldsymbol{y}'\boldsymbol{By}) = 0$ and $\boldsymbol{y}'(\boldsymbol{I}-\boldsymbol{B}-\boldsymbol{Q})\boldsymbol{y} = -\boldsymbol{y}'\boldsymbol{Qy} \geq 0$. The last inequality follows from the assumption $\boldsymbol{I} - \boldsymbol{B} - \boldsymbol{Q}$ is positive definite or positive semidefinite. The assumption $\boldsymbol{Q}$ is positive definite or positive semidefinite implies by definition that $\boldsymbol{y}'\boldsymbol{Qy}$ is also n.n.d. for all $\boldsymbol{y}$. Hence $\boldsymbol{y}'\boldsymbol{Qy} = 0$. Since $\boldsymbol{Q} = \boldsymbol{Q}'$, $\boldsymbol{F}$ exists such that $\boldsymbol{Q} = \boldsymbol{F}'\boldsymbol{F}$. Therefore $\boldsymbol{y}'\boldsymbol{F}'\boldsymbol{Fy} = 0$ implies $\boldsymbol{Fy} = \boldsymbol{0} = \boldsymbol{F}'\boldsymbol{Fy} = \boldsymbol{Qy} = \boldsymbol{Q}(\boldsymbol{Bx})$ for arbitrary $\boldsymbol{x}$. □

Only $\boldsymbol{B}$ is assumed to be idempotent. In many applications $\boldsymbol{Q}$ is also idempotent. However, the lemma only requires the weaker condition of positive definite or positive semidefinite $\boldsymbol{Q}$. Of course, $\boldsymbol{Q} = \boldsymbol{Q}^2 \Rightarrow \boldsymbol{Q}$ positive definite or positive semidefinite. Similarly, $\boldsymbol{I} - \boldsymbol{B} - \boldsymbol{Q}$ is required to be only positive definite or positive semidefinite rather than idempotent. However, if either $\boldsymbol{Q}$ or $(\boldsymbol{I} - \boldsymbol{B} - \boldsymbol{Q})$ is idempotent, then all of $\boldsymbol{B}$, $\boldsymbol{Q}$, $\boldsymbol{B} + \boldsymbol{Q}$, and $\boldsymbol{I} - \boldsymbol{B} - \boldsymbol{Q}$ are idempotent. Idempotency is easy to prove using the result $\boldsymbol{BQ} = \boldsymbol{0}$.

The matrices $\boldsymbol{B}$ and $\boldsymbol{Q}$ have the same eigenvectors. If $\boldsymbol{A}$ and $\boldsymbol{B}$ are symmetric matrices then $\boldsymbol{AB} = \boldsymbol{BA}$ if and only if $\boldsymbol{A}$ and $\boldsymbol{B}$ have the same eigenvectors

(Theorem 4.17, Schott, 2005). If so, $\boldsymbol{A} = \boldsymbol{V}_A\text{Dg}(\boldsymbol{\lambda}_A)\boldsymbol{V}_A'$ and $\boldsymbol{B} = \boldsymbol{V}_A\text{Dg}(\boldsymbol{\lambda}_B)\boldsymbol{V}_A'$. So $\boldsymbol{AB} = \boldsymbol{BA} = \boldsymbol{0}$ implies $\text{Dg}(\boldsymbol{\lambda}_A)\text{Dg}(\boldsymbol{\lambda}_B) = \boldsymbol{0}$.

Theorem 9.16 (Matrix decomposition used in the ANOVA theorem) If $\boldsymbol{A} = \boldsymbol{A}'$ $(N \times N)$ of rank r is partitioned as $\boldsymbol{A} = \sum_{i=1}^{k}\boldsymbol{A}_i$ with each $\boldsymbol{A}_i = \boldsymbol{A}_i'$ of rank r_i, then the following conditions may be defined.
1. $\boldsymbol{A}_i = \boldsymbol{A}_i^2$ for $i \in \{1, 2, \ldots, k\}$
2. $\boldsymbol{A}_i\boldsymbol{A}_{i'} = \boldsymbol{0} = \boldsymbol{A}_{i'}\boldsymbol{A}_i$ for $i \neq i'$
3. $\boldsymbol{A} = \boldsymbol{A}^2$
4. $r = \sum_{i=1}^{k} r_i$.
With the definitions, it follow that
I. Any two of 1, 2, 3 imply all of 1, 2, 3, and 4.
II. Together, 3 and 4 imply all of 1, 2, 3, and 4.

Proof. Proving I and II require only five steps:
1 and 3 $\Rightarrow$ 2; 2 and 3 $\Rightarrow$ 1; 1 and 2 $\Rightarrow$ 3; 1 and 3 $\Rightarrow$ 4; 3 and 4 $\Rightarrow$ 1.

Proof that 1 and 3 $\Rightarrow$ 2. By 3, $\boldsymbol{A} = \boldsymbol{A}^2 \Rightarrow (\boldsymbol{I} - \boldsymbol{A}) = (\boldsymbol{I} - \boldsymbol{A})^2 \Rightarrow (\boldsymbol{I} - \boldsymbol{A})$ is positive semidefinite (p.s.d.). By 1, $\boldsymbol{A}_k = \boldsymbol{A}_k^2 \Rightarrow \boldsymbol{A}_k$ is p.s.d. $\Rightarrow \sum_{i \neq k \neq i'}\boldsymbol{A}_k$ is p.s.d. $\Rightarrow (\boldsymbol{A} - \boldsymbol{A}_i - \boldsymbol{A}_{i'})$ is p.s.d. Therefore $(\boldsymbol{I} - \boldsymbol{A}) + (\boldsymbol{A} - \boldsymbol{A}_i - \boldsymbol{A}_{i'}) = (\boldsymbol{I} - \boldsymbol{A}_i - \boldsymbol{A}_{i'})$ is p.s.d. By Loynes' lemma, $\boldsymbol{A}_i\boldsymbol{A}_{i'} = \boldsymbol{0}$, which implies 2.

Proof that 2 and 3 $\Rightarrow$ 1. Eigenvector $\boldsymbol{v}$ and corresponding eigenvalue λ of $\boldsymbol{A}_i$ are defined by $\boldsymbol{A}_i\boldsymbol{v} = \boldsymbol{v}\lambda$ or $\lambda^{-1}\boldsymbol{A}_i\boldsymbol{v} = \boldsymbol{v}$ if $\lambda \neq 0$. By 2, $\boldsymbol{A}_k\boldsymbol{v} = \boldsymbol{A}_k\boldsymbol{A}_i\boldsymbol{v}/\lambda = \boldsymbol{0}$ for $k \neq i$ and $\lambda \neq 0$. For any nonzero eigenvalue of $\boldsymbol{A}_i$, $\lambda \neq 0$, and corresponding eigenvector $\boldsymbol{v}$, we have $\boldsymbol{Av} = (\sum_k\boldsymbol{A}_k)\boldsymbol{v} = \boldsymbol{0} + \boldsymbol{A}_i\boldsymbol{v} = \lambda\boldsymbol{v}$. Therefore λ is an eigenvalue of $\boldsymbol{A}$. Since 3 $\Rightarrow$ $\lambda = 1$ or 0, every nonzero eigenvalue of $\boldsymbol{A}_i$ equals 1. Thus 1 holds.

Proof that 1 and 2 $\Rightarrow$ 3. Using 1 and 2 in obvious ways we have 3, $\boldsymbol{A}^2 = \boldsymbol{A}$. Specifically $\boldsymbol{A}^2 = (\sum_k\boldsymbol{A}_k)^2 = \sum_i\sum_{i'}\boldsymbol{A}_i\boldsymbol{A}_{i'} = \sum_k\boldsymbol{A}_k^2 = \sum_k\boldsymbol{A}_k = \boldsymbol{A}$.

Proof that 1 and 3 $\Rightarrow$ 4. Using 1 and 3 in obvious ways we have 4, $r = \sum_k r_k$, as follows: $r = \text{tr}(\boldsymbol{A}) = \text{tr}(\sum_k\boldsymbol{A}_k) = \sum_k\text{tr}(\boldsymbol{A}_k) = \sum_k r_k$.

Proof that 3 and 4 $\Rightarrow$ 1. By 3, $\boldsymbol{A} = \boldsymbol{A}^2 \Rightarrow -(\boldsymbol{A} - \boldsymbol{I}) = (\boldsymbol{A} - \boldsymbol{I})^2 \Rightarrow$ $\text{rank}(\boldsymbol{A} - \boldsymbol{I}) = N - r$. In turn, $(\boldsymbol{A} - \boldsymbol{I})$ has $N - r$ linearly independent columns and so $(\boldsymbol{A} - \boldsymbol{I})\boldsymbol{x} = \boldsymbol{0}$ $(N \times 1)$ is a set of N equations containing $N - r$ linearly independent (LIN) equations. Similarly, $-\boldsymbol{A}_i\boldsymbol{x} = \boldsymbol{0}$ is a set of N equations containing r_i LIN equations. The concatenation of n such sets of equations yields Nn equations,

$$\begin{bmatrix} \boldsymbol{A}-\boldsymbol{I} \\ -\boldsymbol{A}_2 \\ \vdots \\ -\boldsymbol{A}_n \end{bmatrix} \boldsymbol{x} = \boldsymbol{0}. \tag{9.59}$$

The equations contain at most $(N-r)+r_2+r_3+\cdots+r_n$ LIN equations. By 4, a total of $N-r_1$ LIN equations exist. They can be reduced to N equations by adding terms on the left side to yield $(\boldsymbol{A}_1-\boldsymbol{I})\boldsymbol{x}=\boldsymbol{0} \Leftrightarrow \boldsymbol{A}_1\boldsymbol{x}=\boldsymbol{x} \Leftrightarrow \boldsymbol{A}_1\boldsymbol{x}=1\boldsymbol{x}$. At most $N-r_1$ LIN equations exist.

Since the N equations $\boldsymbol{A}_1\boldsymbol{x}=1\boldsymbol{x}$ contain at most $N-r_1$ LIN equations, at least $N-(N-r_1)=r_1$ LIN solutions $\boldsymbol{x}$ exist for the equations. Hence there are at least r_1 eigenvectors for $\boldsymbol{A}_1$ which correspond to eigenvalues equal to 1. Since $\text{rank}(\boldsymbol{A}_1)=r_1$, $\boldsymbol{A}_1$ only has r_1 nonzero eigenvalues. Therefore $\lambda=1$ is an eigenvalue of multiplicity r_1 and $\lambda=0$ is an eigenvalue of multiplicity $N-r_1$. Consequently $\boldsymbol{A}_1$ must be idempotent. The same proof can be applied to any other $\boldsymbol{A}_i$. Thus, by extension, all $\boldsymbol{A}_i$ must be idempotent, which gives 1. □

Theorem 9.17 (The ANOVA theorem) If $\boldsymbol{y}\sim\mathcal{N}_N(\boldsymbol{\mu},\boldsymbol{\Sigma})$ with $\text{rank}(\boldsymbol{\Sigma})=N$, $\boldsymbol{A}=\boldsymbol{A}'$ $(N\times N)$ of rank r with $\boldsymbol{A}=\sum_{i=1}^{k}\boldsymbol{A}_i$, $\boldsymbol{A}_i=\boldsymbol{A}_i'$ with rank r_i, and $q=\boldsymbol{y}'\boldsymbol{A}\boldsymbol{y}$, the following conditions may be defined.
1. $\boldsymbol{A}_i\boldsymbol{\Sigma}=(\boldsymbol{A}_i\boldsymbol{\Sigma})^2$ for $i\in\{1,2,\ldots,k\}$, which is equivalent to $\boldsymbol{A}_i\boldsymbol{\Sigma}\boldsymbol{A}_i=\boldsymbol{A}_i$,
2. $\boldsymbol{A}_i\boldsymbol{\Sigma}\boldsymbol{A}_{i'}=\boldsymbol{0}$ for all $i\neq i'$,
3. $\boldsymbol{A}\boldsymbol{\Sigma}=(\boldsymbol{A}\boldsymbol{\Sigma})^2$, and
4. $r=\sum_{i=1}^{k}r_i$.
For the conditions defined,
(a) $\boldsymbol{y}'\boldsymbol{A}_i\boldsymbol{y}\sim\chi^2(r_i,\boldsymbol{\mu}'\boldsymbol{A}_i\boldsymbol{\mu})$,
(b) $\boldsymbol{y}'\boldsymbol{A}_i\boldsymbol{y}$ is independent of $\boldsymbol{y}'\boldsymbol{A}_{i'}\boldsymbol{y}$ $\forall i\neq i'$, and
(c) $\boldsymbol{y}'\boldsymbol{A}\boldsymbol{y}\sim\chi^2(r,\boldsymbol{\mu}'\boldsymbol{A}\boldsymbol{\mu})$
are all simultaneously true if and only if
I. any two of 1, 2, and 3 are true, or
II. 3 and 4 are both true.

Corollary 9.17.1 (Cochran, 1934) If $\boldsymbol{y}\sim\mathcal{N}_N(\boldsymbol{0},\boldsymbol{I}_N)$ with $\boldsymbol{A}=\boldsymbol{I}_N$ [and $\text{rank}(\boldsymbol{A})=N$] is partitioned as $\boldsymbol{I}=\sum_{i=1}^{k}\boldsymbol{A}_i$ with $\boldsymbol{A}_i=\boldsymbol{A}_i'$ of rank r_i, and $q=\boldsymbol{y}'\boldsymbol{A}\boldsymbol{y}$, then the variates $q_i=\boldsymbol{y}'\boldsymbol{A}_i\boldsymbol{y}$, $i\in\{1,2,\ldots,k\}$, are mutually independent and distributed as $\chi^2(r_i,0) \Leftrightarrow N=\sum_{i=1}^{k}r_i$. Obviously $q\sim\chi^2(N,0)$.

Proof. Left as an exercise.

Example 9.4 A $\text{GLM}_{N,q}(y_i; \boldsymbol{X}_i\boldsymbol{\beta}, \sigma^2)$ with Gaussian distribution has

$$\{\boldsymbol{W}_1, \boldsymbol{W}_2, \boldsymbol{W}_3, \boldsymbol{W}_4\} = \left\{ \begin{bmatrix} 1 \\ 1 \\ 1 \\ 1 \\ 1 \\ 1 \end{bmatrix}, \begin{bmatrix} -1 \\ -1 \\ -1 \\ 1 \\ 1 \\ 1 \end{bmatrix}, \begin{bmatrix} -1 & 1 \\ 0 & -2 \\ 1 & 1 \\ -1 & 1 \\ 0 & -2 \\ 1 & 1 \end{bmatrix}, \begin{bmatrix} 1 & -1 \\ 0 & 2 \\ -1 & -1 \\ -1 & 1 \\ 0 & -2 \\ 1 & 1 \end{bmatrix} \right\}. \tag{9.60}$$

If $\boldsymbol{X} = [\ \boldsymbol{W}_1\ \ \boldsymbol{W}_2\ \ \boldsymbol{W}_3\ ]$ and $\boldsymbol{\beta}' = [\boldsymbol{\beta}_1'\ \boldsymbol{\beta}_2'\ \boldsymbol{\beta}_3']$, then $(\boldsymbol{X}'\boldsymbol{X}) = \text{Dg}(\{6, 6, 4, 12\})$ and $(\boldsymbol{X}'\boldsymbol{X})^{-1} = \text{Dg}(\{1/6, 1/6, 1/4, 1/12\})$. Finally, if

$$\boldsymbol{A}_k = \sigma^{-2}\boldsymbol{W}_k(\boldsymbol{W}_k'\boldsymbol{W}_k)^{-1}\boldsymbol{W}_k', \tag{9.61}$$

then

$$\boldsymbol{A} = \sigma^{-2}\boldsymbol{I} = \sum\nolimits_{k=1}^{4} \boldsymbol{A}_k. \tag{9.62}$$

The $\{\boldsymbol{A}_k\}$ provide projections of $\boldsymbol{y}$ ($N \times 1$) onto mutually orthogonal subspaces of the sample space. The subspace spanned by $(\boldsymbol{A}_1 + \boldsymbol{A}_2 + \boldsymbol{A}_3)$ (with basis $\boldsymbol{X}$) is the *estimation space*. The subspace spanned by $\boldsymbol{A}_4$ (with basis $\boldsymbol{W}_4$) is the *error space*. The corresponding source table is given below. The presence of σ^{-2} in the definition of $\boldsymbol{A}_k$ scales the underlying Gaussian variables to have unit variance, which leads the associated sums of squares to be chi square, as required in the preceding corollary.

ANOVA Table for Example

Source	df	Sum of Squares
Mean	1	$\boldsymbol{y}'\boldsymbol{A}_1\boldsymbol{y}$
A	1	$\boldsymbol{y}'\boldsymbol{A}_2\boldsymbol{y}$
B	2	$\boldsymbol{y}'\boldsymbol{A}_3\boldsymbol{y}$
Residual	2	$\boldsymbol{y}'\boldsymbol{A}_4\boldsymbol{y}$
Total	6	$\boldsymbol{y}'\boldsymbol{A}\boldsymbol{y} = \boldsymbol{y}'\boldsymbol{y}\sigma^{-2}$

Corollary 9.17.2 In a $\text{GLM}_{N,q}\text{FR}(y_i; \boldsymbol{X}_i\boldsymbol{\beta}, \sigma^2)$ with Gaussian errors, the fixed and known constants $\boldsymbol{C}$ and $\boldsymbol{\theta}_0$ define the a priori secondary parameter ($a \times 1$) $\boldsymbol{\theta} = \boldsymbol{C}\boldsymbol{\beta} - \boldsymbol{\theta}_0$, with $\text{rank}(\boldsymbol{C}) = a \leq q$. In turn,

$$F(\boldsymbol{y}) = \frac{(\widehat{\boldsymbol{\theta}} - \boldsymbol{\theta}_0)'[\boldsymbol{C}(\boldsymbol{X}'\boldsymbol{X})^{-1}\boldsymbol{C}']^{-1}(\widehat{\boldsymbol{\theta}} - \boldsymbol{\theta}_0)/a}{\widehat{\sigma}^{\,2}} = \frac{SSH/a}{SSE/(N-q)} \tag{9.63}$$

is a ratio of quadratic forms. Here SSH = sum of squares for the hypothesis is computed from $\widehat{\boldsymbol{\beta}} = (\boldsymbol{X}'\boldsymbol{X})^{-1}\boldsymbol{X}'\boldsymbol{y}$ and $\widehat{\boldsymbol{\theta}} = \boldsymbol{C}\widehat{\boldsymbol{\beta}}$, while

$$\begin{aligned} SSE &= \text{sum of squares for error} \\ &= (\boldsymbol{y} - \boldsymbol{X}\widehat{\boldsymbol{\beta}})'(\boldsymbol{y} - \boldsymbol{X}\widehat{\boldsymbol{\beta}}) \\ &= \boldsymbol{y}'\left[\boldsymbol{I} - \boldsymbol{X}(\boldsymbol{X}'\boldsymbol{X})^{-1}\boldsymbol{X}'\right]\boldsymbol{y}. \end{aligned} \tag{9.64}$$

Hence $F(\boldsymbol{y}) \sim F(a, N-q, \omega)$ for $\omega = (\boldsymbol{\theta} - \boldsymbol{\theta}_0)'[\boldsymbol{C}(\boldsymbol{X}'\boldsymbol{X})^{-1}\boldsymbol{C}']^{-1}(\boldsymbol{\theta} - \boldsymbol{\theta}_0)/\sigma^2$.

Proof. Since $\widehat{\boldsymbol{\beta}}$ and $(\boldsymbol{y} - \boldsymbol{X}\widehat{\boldsymbol{\beta}})$ are statistically independent, *SSE* and *SSH* are statistically independent. Furthermore we can prove $SSE/\sigma^2 \sim \chi^2(N-q)$ and $SSH/\sigma^2 \sim \chi^2(a, \omega)$. Thus $F(\boldsymbol{y})$ is distributed as a ratio of independent chi-square statistics divided by their degrees of freedom. □

We shall see in Chapter 15 that in testing $H_0 = \text{B}(\boldsymbol{\theta} = \boldsymbol{\theta}_0)$ versus $H_A = \text{B}(\boldsymbol{\theta} \neq \boldsymbol{\theta}_0)$ both the likelihood ratio test (LRT) procedure and the union intersection test (UIT) procedure can use $F(\boldsymbol{y})$ as the test statistic.

9.7 RATIOS INVOLVING QUADRATIC FORMS

The F and t distributions play central roles in describing distributions of test statistics for linear models. Both are defined in terms of other random variables. In particular, if $z \sim \mathcal{N}(0, 1)$ and $x \sim \chi^2(\nu)$ are independent, then

$$t = \frac{z + \mu}{\sqrt{x/\nu}}. \tag{9.65}$$

The situation will be indicated by writing $t \sim t(\nu, \mu)$ for the noncentral case or $t \sim t(\nu)$ for the central case. The corresponding CDF is indicated $F_t(t_*; \nu, \mu)$, and the corresponding quantile is $F_t^{-1}(p; \nu, \mu)$. A two-tailed test of size α use $t_{\text{crit}} = F_t^{-1}(1 - \alpha/2; \nu)$, while a one tailed test uses either $F_t^{-1}(1 - \alpha; \nu)$ or $F_t^{-1}(\alpha; \nu)$, depending on the direction (sign of t) required.

If $x_1 \sim \chi^2(\nu_1, \omega_1)$ is independent of $x_2 \sim \chi^2(\nu_2, \omega_2)$, then

$$f = \frac{x_1/\nu_1}{x_2/\nu_2} \tag{9.66}$$

is described as following a doubly noncentral F distribution, $f \sim F(\nu_1, \nu_2, \omega_1, \omega_2)$. The (singly) noncentral F has $\omega_2 = 0$, written $f \sim F(\nu_1, \nu_2, \omega_1)$, and the central has $\omega_1 = \omega_2 = 0$, written $f \sim F(\nu_1, \nu_2)$. Noncentral F has corresponding CDF indicated $F_F(f; \nu_1, \nu_2, \omega_1)$ and quantile $f_{\text{crit}} = F_F^{-1}(p; \nu_1, \nu_2, \omega)$. Equivalently $t^2 \sim F(1, \nu_2, \omega_1)$. A size α test uses $f_{\text{crit}} = F_F^{-1}(1 - \alpha; \nu_1, \nu_2)$, which corresponds to a two-tailed t test if $\nu_1 = 1$. A one-tailed t test may be performed with $F_F^{-1}(1 - 2\alpha; 1, \nu_2)$ while requiring the underlying t to have the correct sign. The notation may be summarized by writing $F_F(f_0; \nu_1, \nu_2, \omega) = \Pr\{f \leq f_0\}$ and, for $\omega = 0$, $\Pr\{f > f_{\text{crit}}\} = \alpha$.

More general ratios of quadratic forms occur naturally in linear models. Such a variable can be written

$$\frac{q_1}{q_2} = \frac{\sum_{j=1}^{J_1} c_j x_j}{\sum_{j=J_1+1}^{J_1+J_2} c_j x_j} \tag{9.67}$$

for constants $\{c_j\}$ and random $x_j \sim \chi^2(\nu_j, \omega_j)$ all fully independent. Typically $c_j > 0$, which implies $\Pr\{q_1/q_2 > 0\} = 1$. For $r_0 > 0$ a simple transformation gives

$$\begin{aligned}\Pr\{q_1/q_2 \le r_0\} &= \Pr\Big\{\sum_{j=1}^{J_1} c_j x_j - r_0 \sum_{j=J_1+1}^{J_1+J_2} c_j x_j \le 0\Big\}\\ &= \Pr\Big\{\sum_{j=1}^{J_1+J_2} d_j x_j \le 0\Big\}.\end{aligned} \tag{9.68}$$

Although $s = \sum_{j=1}^{J_1+J_2} d_j x_j$ has a simple and known characteristic function, computing $\Pr\{s \le 0\} = \Pr\{q_1/q_2 \le r_0\}$ proves difficult. With $\{d_j\}$ known, Davies' (1980) algorithm allows computing $\Pr\{s \le 0\}$ and more general results with specifiable precision. The method uses numerical inversion of the characteristic function (a numerical integration). Interest in special cases had led to the development of many alternative algorithms, typically based on series expansions. Johnson and Kotz (1970, Chapter 29, p. 169–173) summarized many issues of theory and computational practice in the long history of the problem. Johnson, Kotz, and Balakrishnan (1994) provided some additional information.

In many settings, an approximation provides sufficient accuracy and can be much faster to compute. For $c_j > 0$ and $\omega_j \equiv 0$, a Satterthwaite (1946) approximation matches the mean and variance of q_k to q_{*k}, with $q_{*k}/\lambda_{*k} \sim \chi^2(\nu_{*k}, 0)$. In turn, $\Pr\{q_1/q_2 \le r_0\} \approx F_F[r_0(\lambda_{*2}\nu_{*2})/(\lambda_{*k}\nu_{*1}); \nu_{*1}, \nu_{*2}]$. Kim, Gribbin, Muller and Taylor (2005) generalized the approximation to allow $\omega_j > 0$ for $j \le J_1$ by using $q_{*1}/\lambda_{*1} \sim \chi^2(\nu_{*1}, \omega_{*1})$.

EXERCISES

9.1 Suppose $\boldsymbol{y} \sim \mathcal{N}_n(\mathbf{1}_n\mu, \boldsymbol{\Sigma})$ and $\boldsymbol{\Sigma} = \sigma^2 \begin{bmatrix} 1 & \rho & . & \rho \\ \rho & 1 & . & \rho \\ . & . & . & . \\ \rho & \rho & . & 1 \end{bmatrix}$. Thus $\mathrm{E}(y_i) = \mu$ for all i, $\mathcal{V}(y_i) = \sigma^2$ for all i, and $\mathcal{V}(y_i, y_j) = \sigma^2\rho$ for all $i \ne j$; that is, the y's are equicorrelated. Equivalently, $\boldsymbol{\Sigma} = \sigma^2[(1-\rho)\boldsymbol{I} + \rho\mathbf{1}\mathbf{1}']$.

9.1.1 Show that $\sum_{i=1}^{n}(y_i - \overline{y})^2/[\sigma^2(1-\rho)]$ is $\chi^2(n-1)$.

9.1.2 Given that $\boldsymbol{V}_T$ is an $n \times (n-1)$ matrix which is columnwise orthonormal such that $\boldsymbol{V}_T'\mathbf{1}_n = \mathbf{0}$, find the distribution of $\boldsymbol{y}_T = \boldsymbol{V}_T'\boldsymbol{y}$.

9.1.3 Explicitly specify the distribution of $Q = \boldsymbol{y}_T'\boldsymbol{y}_T$.

9.2 Suppose $\boldsymbol{y} \sim \mathcal{N}_3(\boldsymbol{\mu}, \boldsymbol{\Sigma})$, with $\boldsymbol{\mu} = \begin{bmatrix} 2 \\ -1 \\ 3 \end{bmatrix}$, $\boldsymbol{\Sigma} = \begin{bmatrix} 4 & 1 & 0 \\ 1 & 2 & 1 \\ 0 & 1 & 3 \end{bmatrix}$, and

$$\boldsymbol{A} = \begin{bmatrix} 1 & -3 & -8 \\ -3 & 2 & -6 \\ -8 & -6 & 3 \end{bmatrix}.$$

9.2.1 Find $\mathrm{E}(\boldsymbol{y}'\boldsymbol{A}\boldsymbol{y})$.
9.2.2 Find $\mathcal{V}(\boldsymbol{y}'\boldsymbol{A}\boldsymbol{y})$/
9.2.3 Fully specify the exact distribution of $\boldsymbol{y}'\boldsymbol{A}\boldsymbol{y}$ (including all parameters and dimensions). Explain why the distribution you claim provides the correct answer. You may use IML or any other matrix language for numerical calculations. If you do, please show both the code and final results to help the grader.
9.2.4 If $\boldsymbol{\Sigma} = \sigma^2\boldsymbol{I}$, does $\boldsymbol{y}'\boldsymbol{A}\boldsymbol{y}/\sigma^2$ have a χ^2 distribution?

9.3 Assume $x_1/\sigma^2 \sim \chi^2(\nu_1, \omega)$, independent of x_2, and $x_2/\sigma^2 \sim \chi^2(\nu_2)$. Define $r = (x_1/\nu_1)/(x_2/\nu_2)$, $c = \nu_2/(\nu_1 r_0)$ for $r_0 > 0$, and $s = cx_1 - x_2$.
9.3.1 Prove that $\Pr\{r \leq r_0\} = \Pr\{s \leq 0\}$.
9.3.2 What is $\mathrm{E}(s)$?
9.3.3 What is $\mathcal{V}(s)$?
9.3.4 What is the moment generating function of s?

CHAPTER 10

Multivariate Quadratic Forms

10.1 THE WISHART DISTRIBUTION

With $\boldsymbol{A}$ constant and $\boldsymbol{y}$ Gaussian, $q = \boldsymbol{y}'\boldsymbol{A}\boldsymbol{y}$ is a univariate quadratic form and equals a weighted sum of chi squares. With equal weights, q is a scaled chi square. The necessary and sufficient conditions are detailed in the previous chapter. Replacing vector Gaussian $\boldsymbol{y}$ by matrix Gaussian leads to the following definition.

Definition 10.1 The $N \times N$ constant $\boldsymbol{A} = \boldsymbol{A}'$ and $\boldsymbol{Y} \sim (\mathcal{S})\mathcal{N}_{N,p}(\boldsymbol{M}, \boldsymbol{\Xi}, \boldsymbol{\Sigma})$ create the $p \times p$ matrix $\boldsymbol{Q} = \boldsymbol{Y}'\boldsymbol{A}\boldsymbol{Y}$, a *multivariate quadratic form.*

In the multivariate case we must always assume $\boldsymbol{A} = \boldsymbol{A}'$. Idempotent $\boldsymbol{A}$ leads to a special distribution (chi square) in the univariate case. Similarly, idempotent $\boldsymbol{A}$ and $\boldsymbol{\Xi}$ leads to $\boldsymbol{Q}$ having a Wishart distribution, which is one multivariate generalization (of many) of the chi square. Wishart (1928), Johnson and Kotz (1972), Arnold (1981), Muirhead (1984), Gupta and Nagar (2000), and Anderson (2004) included related treatments. The nomenclature and approach to many proofs used here closely follow the presentation in Muller and Chi (2006).

Definition 10.2 (a) If $\boldsymbol{Y} \sim \mathcal{N}_{\nu,p}(\boldsymbol{0}, \boldsymbol{I}_\nu, \boldsymbol{\Sigma})$, then $\boldsymbol{Y}'\boldsymbol{Y} \sim \mathcal{W}_p(\nu, \boldsymbol{\Sigma})$ indicates $\boldsymbol{Y}'\boldsymbol{Y}$ follows a *central (integer) Wishart* distribution with (integer) $\nu > 0$ degrees of freedom.
(b) If $\boldsymbol{Y} \sim \mathcal{N}_{\nu,p}(\boldsymbol{M}, \boldsymbol{I}_\nu, \boldsymbol{\Sigma})$, then $\boldsymbol{Y}'\boldsymbol{Y} \sim \mathcal{W}_p(\nu, \boldsymbol{\Sigma}, \boldsymbol{M}'\boldsymbol{M})$ indicates $\boldsymbol{Y}'\boldsymbol{Y}$ follows a *noncentral (integer) Wishart* distribution with (integer) $\nu > 0$ degrees of freedom, *shift* $\boldsymbol{\Delta} = \boldsymbol{M}'\boldsymbol{M}$, and *noncentrality* $\boldsymbol{\Omega} = \boldsymbol{M}'\boldsymbol{M}\boldsymbol{\Sigma}^+$.
(c) Singular $\boldsymbol{\Sigma}$ may be emphasized by writing $\mathcal{SW}_p(\nu, \boldsymbol{\Sigma})$ or $\mathcal{SW}_p(\nu, \boldsymbol{\Sigma}, \boldsymbol{\Delta})$.
(d) Writing $(\mathcal{S})\mathcal{W}_p(\nu, \boldsymbol{\Sigma})$ or $(\mathcal{S})\mathcal{W}_p(\nu, \boldsymbol{\Sigma}, \boldsymbol{\Delta})$ indicates possibly singular $\boldsymbol{\Sigma}$.

In parallel to a chi square, a Wishart is defined as a quadratic form of independent standard Gaussian random vectors. The Wishart definition also includes scale. The shift and noncentrality parameters reflect the nature of the underlying Gaussian variables. If (and only if) $\boldsymbol{M} = \boldsymbol{0}$, then $\boldsymbol{\Delta} = \boldsymbol{M}'\boldsymbol{M} = \boldsymbol{0}$ and $\boldsymbol{\Omega} = \boldsymbol{0}$, which reduces a noncentral Wishart to a central. Eigenvalues of the noncentrality matrix $\mathbf{\Omega}$ are invariant to any full-rank transformation of the columns of $\boldsymbol{Y}$. Therefore they are scale free, in the sense of being invariant to multiplying each variable by a possibly distinct nonzero constant.

Historically, most discussions of the Wishart have assumed full-rank Σ, which implies $\mathbf{\Omega} = \boldsymbol{M}'\boldsymbol{M}\Sigma^{-1}$ is unique, and led to the notation $\mathcal{W}_p(\nu, \Sigma, \mathbf{\Omega})$. For singular Σ, some authors define noncentrality as $\boldsymbol{M}'\boldsymbol{M}\Sigma^{-}$ or $\boldsymbol{M}'\boldsymbol{M}\Sigma^{+}$. In many applications, only functions of $\boldsymbol{M}'\boldsymbol{M}\Sigma^{-}$ invariant to the choice of Σ^{-} occur, which allows using $\boldsymbol{M}'\boldsymbol{M}\Sigma^{+}$ without loss of generality.

Defining the Wishart in terms of $\mathbf{\Delta} = \boldsymbol{M}'\boldsymbol{M}$ not only avoids the ambiguity in singular cases but also is consistent with chi-square notation. If $z \sim \mathcal{N}(\mu_z, 1)$, then $y = z\sigma \sim \mathcal{N}(\mu_y, \sigma^2)$ with $\mu_y = \sigma\mu_z$. In turn, $y^2/\sigma^2 = z^2 \sim \chi^2(1, \mu_z^2)$ and $z^2 \sim \mathcal{W}_1(1, \mu_z^2)$, with noncentrality $\mu_z^2 = \mu_y^2/\sigma^2 = \omega$, while $y^2 \sim \mathcal{W}_1(1, \mu_y^2)$. If $\boldsymbol{Z} \sim \mathcal{N}_{\nu,p}(\boldsymbol{M}_Z, \boldsymbol{I}_\nu, \boldsymbol{I}_p)$ and $\Sigma = \mathbf{\Phi}\mathbf{\Phi}'$, then $\boldsymbol{Y} = \boldsymbol{Z}\mathbf{\Phi}' \sim \mathcal{N}_{\nu,p}(\boldsymbol{M}_Y, \boldsymbol{I}_\nu, \Sigma)$ with $\boldsymbol{M}_Y = \boldsymbol{M}_Z\mathbf{\Phi}'$. In turn, $\mathbf{\Phi}^{-1}\boldsymbol{Y}'\boldsymbol{Y}\mathbf{\Phi}^{-t} = \boldsymbol{Z}'\boldsymbol{Z} \sim \mathcal{W}_p(\nu, \boldsymbol{I}_p, \boldsymbol{M}_Z'\boldsymbol{M}_Z)$ with noncentrality $\boldsymbol{M}_Z'\boldsymbol{M}_Z = \mathbf{\Phi}^{-1}\boldsymbol{M}_Y'\boldsymbol{M}_Y\mathbf{\Phi}^{-t} = \mathbf{\Omega}$, while $\boldsymbol{Y}'\boldsymbol{Y} \sim \mathcal{W}_p(\nu, \Sigma, \boldsymbol{M}_Y'\boldsymbol{M}_Y)$.

Table 10.1 Impact of Rank Conditions on Eigenvalue Estimation: Central Wishart, $\boldsymbol{S} = \nu\widehat{\Sigma} \sim \mathcal{W}_p(\nu, \Sigma)$

	Σ Singular **rank(Σ)$=p_1<p$**	**Σ Nonsingular** **rank(Σ)$=p_1=p$**	$\forall\lambda_j \neq 0$ *Estimable?*
rank($\widehat{\Sigma}$)$=\nu<p_1$	$0<\nu<p_1$ $<p$	$0<\nu<p_1=p$	*No*
rank($\widehat{\Sigma}$)$=p_1\leq\nu$	0 $<p_1$ $<p\leq\nu$	0 $<p_1=p\leq\nu$	*Yes*
	0 $<p_1\leq\nu<p$		*Yes*

Arnold (1981, p. 317) noted that various authors have used the term "singular" or "pseudo" if Σ is singular or if $\nu < p$. Such approaches fail to describe all possible combinations in Table 10.1. We suggest the following terms. The distinction between *population singular* and *population nonsingular* specifies rank(Σ) $= p$ or rank(Σ) $= p_1 < p$. In turn, $\widehat{\Sigma}$ may be singular due to $\nu < p$ or rank(Σ) $= p_1 < p$. The contrast between *sample-rank sufficient* (to estimate all nonzero population eigenvalues) and *sample-rank insufficient* fully captures the necessary distinction. Estimated eigenvalues are roots of the scalar polynomial $|\widehat{\Sigma} - \widehat{\lambda}\boldsymbol{I}_p| = 0$. Hence rank($\widehat{\Sigma}$) determines the number of eigenvalues that can be estimated. As long as rank($\widehat{\Sigma}$) = rank(Σ) all population eigenvalues can be estimated, which reflects whether $\nu \geq p_1$ or $\nu < p_1$. Although four of five cases have singular $\widehat{\Sigma}$ in Table 10.1, only two of five are sample-rank insufficient.

Since $\boldsymbol{S} = \boldsymbol{S}'$, only $p(p+1)/2$ distinct elements exist, and the p^2 elements are not functionally independent. Saying "$\boldsymbol{S}$ follows a Wishart distribution" always refers to the distribution of $\boldsymbol{z} = \text{vech}(\boldsymbol{S})$. Only Wishart matrices that are both population and sample nonsingular have a density. In parallel to the singular Gaussian, a population-nonsingular Wishart can be extracted from a population-singular one to represent all of the information available. The statement holds whether or not the Wishart is sample-rank sufficient or insufficient. In contrast, sample-rank insufficiency can *not* be converted to sample-rank-sufficiency, except for special cases of simple covariance structure.

Example 10.1 A professor interested in predicting Graduate Record Examination (GRE) scores from undergraduate grade point might examine the residual covariance matrix for the scores on Verbal, Quantitative and Analytic sections of the test. The test construction process makes it reasonable to assume the data are multivariate Gaussian. With data from $N = 20$ students, the covariance estimate $\widehat{\boldsymbol{\Sigma}}_1$ would be such that $\boldsymbol{S}_1 = 19\widehat{\boldsymbol{\Sigma}}_1 \sim \mathcal{W}_3(20-1, \boldsymbol{\Sigma}_1)$, which is population nonsingular and sample-rank sufficient.

If the professor includes Total = Verbal + Quantitative + Analytic, then $\boldsymbol{S}_2 = 19\widehat{\boldsymbol{\Sigma}}_2 \sim \mathcal{SW}_4(20-1, \boldsymbol{\Sigma}_2)$, which is population-singular and sample-rank-sufficient. Infinitely many 4×3 transformation matrices can transform $\boldsymbol{S}_2$ into a population-nonsingular Wishart. Choices include

$$\boldsymbol{T}_1 = \begin{bmatrix} 1 & 0 & 0 \\ 0 & 1 & 0 \\ 0 & 0 & 1 \\ 0 & 0 & 0 \end{bmatrix} = \begin{bmatrix} \boldsymbol{I}_3 \\ \boldsymbol{0} \end{bmatrix} \tag{10.69}$$

$$\boldsymbol{T}_2 = \begin{bmatrix} 1 & 0 & 0 \\ 0 & 1 & 0 \\ 0 & 1 & 0 \\ 0 & 0 & 1 \end{bmatrix}. \tag{10.70}$$

As will be proven later in the chapter, $\boldsymbol{T}_j'\boldsymbol{S}_2\boldsymbol{T}_j \sim \mathcal{W}_3(20-1, \boldsymbol{T}_j'\boldsymbol{\Sigma}_2\boldsymbol{T}_j)$, which corresponds to studying $\boldsymbol{Y}\boldsymbol{T}_j$ with dimensions $(20 \times 4)(4 \times 3)$.

Example 10.2 Medical imaging and genetic scientists often encounter data with more variables than participants. Pizer et al. (2003) compared human and computer segmentations from CT images of $N = 12$ kidneys. They modeled the surfaces at 88 points, giving $p = 88 \cdot 3 = 264$ (x, y, z) variables, describing location in three dimensions. Gaussian data lead to the assumption of an observed covariance matrix $\widehat{\boldsymbol{\Sigma}}$ such that $\boldsymbol{S} = 11\widehat{\boldsymbol{\Sigma}} \sim \mathcal{W}_{264}(12-1, \boldsymbol{\Sigma})$. Here $\boldsymbol{S}$ is population nonsingular and sample-rank insufficient.

10.2 THE CHARACTERISTIC FUNCTION OF THE WISHART

The following lemma summarizes properties of any covariance matrix. The notation helps describe the characteristic function of the Wishart.

Lemma 10.1 A $p \times p$ covariance matrix is symmetric with no negative eigenvalues, $\{\lambda_j\}$, which allows writing $\boldsymbol{\Sigma} = \boldsymbol{\Upsilon}\mathrm{Dg}(\boldsymbol{\lambda})\boldsymbol{\Upsilon}' = \boldsymbol{\Phi}\boldsymbol{\Phi}'$, with $\boldsymbol{\Upsilon}$, $\mathrm{Dg}(\boldsymbol{\lambda})$, and $\boldsymbol{\Phi} = \boldsymbol{\Upsilon}\mathrm{Dg}(\boldsymbol{\lambda})^{1/2}$ all $p \times p$. Considering p_1 columns in the first of two partitions, with orthonormal $\boldsymbol{\Upsilon}$, gives $\boldsymbol{\Upsilon} = [\,\boldsymbol{\Upsilon}_1\ \boldsymbol{\Upsilon}_0\,]$, $\boldsymbol{\lambda} = [\,\boldsymbol{\lambda}_1'\ \boldsymbol{\lambda}_0'\,]'$, $\boldsymbol{\Phi} = [\,\boldsymbol{\Phi}_1\ \boldsymbol{\Phi}_0\,]$, $\boldsymbol{\Phi}_1 = \boldsymbol{\Upsilon}_1 \mathrm{Dg}(\boldsymbol{\lambda}_1)^{1/2}$, $\boldsymbol{\Phi}_0 = \boldsymbol{\Upsilon}_0 \mathrm{Dg}(\boldsymbol{\lambda}_0)^{1/2}$, and

$$\begin{aligned}\boldsymbol{\Sigma} &= [\,\boldsymbol{\Upsilon}_1\ \boldsymbol{\Upsilon}_0\,]\begin{bmatrix}\mathrm{Dg}(\boldsymbol{\lambda}_1) & \mathbf{0}\\ \mathbf{0} & \mathrm{Dg}(\boldsymbol{\lambda}_0)\end{bmatrix}\begin{bmatrix}\boldsymbol{\Upsilon}_1'\\ \boldsymbol{\Upsilon}_0'\end{bmatrix}\\ &= [\,\boldsymbol{\Phi}_1\ \boldsymbol{\Phi}_0\,]\begin{bmatrix}\boldsymbol{\Phi}_1'\\ \boldsymbol{\Phi}_0'\end{bmatrix}\\ &= \boldsymbol{\Phi}_1\boldsymbol{\Phi}_1' + \boldsymbol{\Phi}_0\boldsymbol{\Phi}_0'\,. \end{aligned} \tag{10.1}$$

If $\mathrm{rank}(\boldsymbol{\Sigma}) = p_1 \leq p$, then $\boldsymbol{\Sigma} = \boldsymbol{\Upsilon}_1 \mathrm{Dg}(\boldsymbol{\lambda}_1)\boldsymbol{\Upsilon}_1' = \boldsymbol{\Phi}_1\boldsymbol{\Phi}_1'$ and $\boldsymbol{\Phi}_1^+ = (\boldsymbol{\Phi}_1'\boldsymbol{\Phi}_1)^{-1}\boldsymbol{\Phi}_1' = \mathrm{Dg}(\boldsymbol{\lambda}_1)^{-1/2}\boldsymbol{\Upsilon}_1'$.

Proof. Eigenanalysis and partitioned matrix properties give the results.

Theorem 10.1 (a) For $\nu > 0$, $\boldsymbol{\Sigma} = \boldsymbol{\Phi}\boldsymbol{\Phi}' = \boldsymbol{\Phi}_1\boldsymbol{\Phi}_1'$, $p \times p$ $\boldsymbol{\Phi}$, $p \times p_1$ $\boldsymbol{\Phi}_1$, $\mathrm{rank}(\boldsymbol{\Phi}) = \mathrm{rank}(\boldsymbol{\Phi}_1) = \mathrm{rank}(\boldsymbol{\Sigma}) = p_1 \leq p$, $\mathrm{i} = (-1)^{1/2}$, $p \times p$ real $\boldsymbol{U} = \boldsymbol{U}'$, $\langle \boldsymbol{T}\rangle_{jj} = u_{jj}$, and $\langle \boldsymbol{T}\rangle_{jk} = u_{jk}/2$, the characteristic function of $\boldsymbol{S} \sim (\mathcal{S})\mathcal{W}_p(\nu, \boldsymbol{\Sigma})$ is

$$\phi_{\boldsymbol{S}}(\boldsymbol{T}) = |\boldsymbol{I}_p - 2\mathrm{i}\boldsymbol{T}\boldsymbol{\Sigma}|^{-\nu/2} \tag{10.2}$$

$$= |\boldsymbol{I}_p - 2\mathrm{i}\boldsymbol{T}\boldsymbol{\Phi}\boldsymbol{\Phi}'|^{-\nu/2} \tag{10.3}$$

$$= |\boldsymbol{I}_p - 2\mathrm{i}\boldsymbol{T}\boldsymbol{\Phi}_1\boldsymbol{\Phi}_1'|^{-\nu/2} \tag{10.4}$$

$$= |\boldsymbol{I}_p - 2\mathrm{i}\boldsymbol{\Phi}'\boldsymbol{T}\boldsymbol{\Phi}|^{-\nu/2} \tag{10.5}$$

$$= |\boldsymbol{I}_{p_1} - 2\mathrm{i}\boldsymbol{\Phi}_1'\boldsymbol{T}\boldsymbol{\Phi}_1|^{-\nu/2}\,. \tag{10.6}$$

(b) The function $\phi_{\boldsymbol{S}}(\boldsymbol{T})$ is a valid characteristic function for all real $\nu > 0$.

Proof. We use the approach of Muller and Chi (2006), who generalized earlier work to cover all population-singular and sample-rank-insufficient cases.

Proof of (a). For $\mathrm{rank}(\boldsymbol{\Sigma}) = p_1 = p$, Muirhead (1984) proved (10.2) for positive integer ν. The proof of Theorem 10.2 includes (10.2) as a special case ($\boldsymbol{\Omega} = \mathbf{0}$) for $p_1 \leq p$. Lemma 10.1 gives (10.3). Equation (10.5) follows from $|\boldsymbol{I}_p - 2\mathrm{i}\boldsymbol{T}\boldsymbol{\Phi}\boldsymbol{\Phi}'|^{-\nu/2} = (|\boldsymbol{\Phi}^{-t} - 2\mathrm{i}\boldsymbol{T}\boldsymbol{\Phi}\boldsymbol{\Phi}'||\boldsymbol{\Phi}'|)^{-\nu/2} = (|\boldsymbol{\Phi}'||\boldsymbol{\Phi}^{-t} - 2\mathrm{i}\boldsymbol{T}\boldsymbol{\Phi}\boldsymbol{\Phi}'|)^{-\nu/2}$. If $\mathrm{rank}(\boldsymbol{\Sigma}) = p = p_1$, then $\boldsymbol{\Phi}' = \boldsymbol{\Phi}_1$ gives (10.4) from (10.3) and (10.6) from (10.4).

For $p_1 < p$ and $\boldsymbol{\lambda}_1 > 0$, there exists $p_1 \times p_1$ $\boldsymbol{S}_1 \sim \mathcal{W}_{p_1}[\nu, \mathrm{Dg}(\boldsymbol{\lambda}_1)]$ with $\phi_{\boldsymbol{S}_1}(\boldsymbol{T}_1) = |\boldsymbol{I}_{p_1} - 2\mathrm{i}\boldsymbol{T}_1\mathrm{Dg}(\boldsymbol{\lambda}_1)|^{-\nu/2}$ and $\boldsymbol{S} = \boldsymbol{\Upsilon}_1\boldsymbol{S}_1\boldsymbol{\Upsilon}_1' \sim \mathcal{SW}_p[\nu, \boldsymbol{\Upsilon}_1\mathrm{Dg}(\boldsymbol{\lambda}_1)\boldsymbol{\Upsilon}_1']$. With $p \times p$ real $\boldsymbol{T} = \boldsymbol{T}'$ as defined in the theorem, Lemma 7.5 gives

$$\begin{aligned}\phi_{\boldsymbol{S}}(\boldsymbol{T}) &= |\boldsymbol{I}_{p_1} - 2\mathrm{i}\boldsymbol{\Upsilon}_1'\boldsymbol{T}(\boldsymbol{\Upsilon}_1')'\mathrm{Dg}(\boldsymbol{\lambda}_1)|^{-\nu/2}\\ &= |\mathrm{Dg}^{-1/2}(\boldsymbol{\lambda}_1) - 2\mathrm{i}\boldsymbol{\Upsilon}_1'\boldsymbol{T}\boldsymbol{\Upsilon}_1\mathrm{Dg}^{1/2}(\boldsymbol{\lambda}_1)|^{-\nu/2}|\mathrm{Dg}^{1/2}(\boldsymbol{\lambda}_1)|^{-\nu/2}\\ &= |\boldsymbol{I}_{p_1} - 2\mathrm{i}\mathrm{Dg}^{1/2}(\boldsymbol{\lambda}_1)\boldsymbol{\Upsilon}_1'\boldsymbol{T}\boldsymbol{\Upsilon}_1\mathrm{Dg}^{1/2}(\boldsymbol{\lambda}_1)|^{-\nu/2}\,, \end{aligned} \tag{10.7}$$

which gives equation (10.6) for $\mathrm{rank}(\boldsymbol{\Sigma}) = p_1 < p$. Equation (10.5) follows from

$$\begin{aligned}
\phi_{S}(T) &= (|I_{p_1} - 2\mathrm{i}\Phi_1' T\Phi_1||I_{p-p_1}|)^{-\nu/2} \\
&= \left|\begin{bmatrix} I_{p_1} - 2\mathrm{i}\Phi_1' T\Phi_1 & 0 \\ 0 & I_{p-p_1}\end{bmatrix}\right|^{-\nu/2} \\
&= \left|I_p - 2\mathrm{i}[\,\Phi_1\ 0\,]' T[\,\Phi_1\ \ 0\,]\right|^{-\nu/2} \\
&= \left|I_p - 2\mathrm{i}[\,\Phi_1\ \Phi_0\,]' T[\,\Phi_1\ \Phi_0\,]\right|^{-\nu/2}. \qquad (10.8)
\end{aligned}$$

With $\phi_S(T) = |I_{p_1} - 2\mathrm{i}\Upsilon_1' T\Upsilon_1 \mathrm{Dg}(\lambda_1)|^{-\nu/2}$, Theorem 7.2 in Schott (2005) gives

$$\begin{aligned}
\phi_S(T) &= \left|\begin{bmatrix} I_{p_1} - 2\mathrm{i}\Upsilon_1' T\Upsilon_1 \mathrm{Dg}(\lambda_1) & 0 \\ -2\mathrm{i}\Upsilon_0' T\Upsilon_1 \mathrm{Dg}(\lambda_1) & I_{p-p_1}\end{bmatrix}\right|^{-\nu/2} \\
&= \left|I_p - 2\mathrm{i}\begin{bmatrix} \Upsilon_1' T\Upsilon_1 & \Upsilon_1' T\Upsilon_0 \\ \Upsilon_0' T\Upsilon_1 & \Upsilon_0' T\Upsilon_0\end{bmatrix}\begin{bmatrix} \mathrm{Dg}(\lambda_1) & 0 \\ 0 & 0\end{bmatrix}\right|^{-\nu/2} \\
&= |I_p - 2\mathrm{i}\Upsilon' T\Upsilon \mathrm{Dg}(\lambda_1, 0)|^{-\nu/2} \\
&= |\Upsilon\Upsilon' - 2\mathrm{i}T\Upsilon \mathrm{Dg}(\lambda_1, 0)\Upsilon'|^{-\nu/2} \\
&= |I_p - 2\mathrm{i}T\Upsilon_1 \mathrm{Dg}(\lambda_1)\Upsilon_1'|^{-\nu/2}. \qquad (10.9)
\end{aligned}$$

Equation (10.4) and then (10.3) and (10.2) are seen to hold for $\mathrm{rank}(\Sigma) = p_1 < p$.

Proof of (b). Conditions guaranteeing a valid characteristic function (Kendall and Stuart, 1977, p. 105) may be verified directly for noninteger ν. □

Theorem 10.2 (a) For $\nu > 0$, $\mathrm{rank}(\Sigma) = \mathrm{rank}(\Phi_1) = p_1 \le p$, $\Sigma = \Phi_1\Phi_1'$, $p\times p_1$ Φ_1, $\mathrm{i} = (-1)^{1/2}$, $p\times p$ real $U = U'$, $\langle T\rangle_{jj} = u_{jj}$, and $\langle T\rangle_{jk} = u_{jk}/2$, the characteristic function of $S = Y'Y \sim (\mathcal{S})\mathcal{W}_p(\nu, \Sigma, \Delta)$ for $\Delta = M'M$ and $Y \sim \mathcal{N}_{\nu,p}(M, I_\nu, \Sigma)$ is

$$\begin{aligned}
\phi_{S_Y}(T) &= |I_p - 2\mathrm{i}T\Sigma|^{-\nu/2}\exp\{\mathrm{i}\,\mathrm{tr}[T\Sigma(I_p - 2\mathrm{i}T\Sigma)^{-1}\Delta]\} \\
&= |I_p - 2\mathrm{i}T\Sigma|^{-\nu/2}\exp\{\mathrm{i}\,\mathrm{tr}[T\Sigma(I_p - 2\mathrm{i}T\Sigma)^{+}\Delta]\} \\
&= |I_{p_1} - 2\mathrm{i}\Phi_1' T\Phi_1|^{-\nu/2}\exp\left\{\mathrm{tr}\left[\mathrm{i}T\Phi_1(I_{p_1} - 2\mathrm{i}\Phi_1' T\Phi_1)^{-1}\Phi_1^{+}\Delta\right]\right\}. \qquad (10.10)
\end{aligned}$$

(b) The function $\phi_S(T)$ is a valid characteristic function for all real $\nu > 0$.

Proof of (a). First the result for $\Sigma = I_p$ is proven. If $Z \sim \mathcal{N}_{\nu,p}(0, I_\nu, I_p)$, then $S_Z = (Z + M_Z)'(Z + M_Z) \sim \mathcal{W}_p(\nu, I_\nu, M_Z' M_Z)$. Symmetry of S_Z restricts attention to $T = V\mathrm{Dg}(t)V'$ with $V'V = VV' = I_p$. In turn

$$\begin{aligned}
\phi_{S_Z}(T) &= \mathrm{E}\{\exp[\mathrm{tr}(\mathrm{i}T'S_Z)]\} \\
&= \mathrm{E}\big(\exp\{\mathrm{tr}[V\mathrm{Dg}(t)V'(Z+M_Z)'(Z+M_Z)]\}\big) \\
&= \mathrm{E}\big(\exp\{\mathrm{i}\,\mathrm{tr}[\mathrm{Dg}(t)(ZV+M_ZV)'(ZV+M_ZV)]\}\big) \\
&= \mathrm{E}\big(\exp\{\mathrm{i}\,\mathrm{tr}[\mathrm{Dg}(t)(U+Q)'(U+Q)]\}\big) \\
&= \mathrm{E}\left\{\exp\left[\mathrm{i}\sum_{j=1}^{p} t_j(u_j+q_j)'(u_j+q_j)\right]\right\} \\
&= \mathrm{E}\left\{\exp\left[\mathrm{i}\sum_{j=1}^{p} t_j\sum_{k=1}^{\nu}(u_{kj}+q_{kj})^2\right]\right\} = \mathrm{E}\left\{\prod_{j=1}^{p}\prod_{k=1}^{\nu}\exp\left[\mathrm{i}\,t_j(u_{kj}+q_{kj})^2\right]\right\}. \qquad (10.11)
\end{aligned}$$

Here $U = ZV \sim \mathcal{N}_{\nu,p}(\mathbf{0}, I_\nu, I_p)$ and $Q = M_Z V$ is constant. Hence $\{u_{kj} + q_{kj}\}$ and $\{(u_{kj} + q_{kj})^2\}$ are independent. In turn $u_{kj} + q_{kj} \sim \mathcal{N}(q_{kj}, 1)$ implies $(u_{kj} + q_{kj})^2 \sim \chi^2(1, q_{kj}^2)$. Independence and chi-square properties give

$$\begin{aligned}
\phi_{S_Z}(T) &= \prod_{j=1}^{p}\prod_{k=1}^{\nu}\mathrm{E}\{\exp[\mathrm{i}t_j(u_{kj}+q_{kj})^2]\} \\
&= \prod_{j=1}^{p}\prod_{k=1}^{\nu}\left\{(1-2\mathrm{i}t_j)^{-1/2}\exp\left[\mathrm{i}t_j q_{kj}^2(1-2\mathrm{i}t_j)^{-1}\right]\right\} \\
&= \left[\prod_{j=1}^{p}(1-2\mathrm{i}t_j)^{-\nu/2}\right]\exp\left[\mathrm{i}\sum_{j=1}^{p}t_j(1-2\mathrm{i}t_j)^{-1}\sum_{k=1}^{\nu}q_{kj}^2\right] \\
&= |I-2\mathrm{i}T|^{-\nu/2}\exp\left[\mathrm{i}\sum_{j=1}^{p}t_j(1-2\mathrm{i}t_j)^{-1}q_j'q_j\right] \\
&= |I-2\mathrm{i}T|^{-\nu/2}\exp\left(\mathrm{i}\,\mathrm{tr}\{\mathrm{Dg}(t)[\mathrm{Dg}(1-2\mathrm{i}t_j)]^{-1}V'M_Z'M_ZV\}\right) \\
&= |I-2\mathrm{i}T|^{-\nu/2}\exp\left(\mathrm{i}\,\mathrm{tr}\{V\mathrm{Dg}(t)V'V[\mathrm{Dg}(1-2\mathrm{i}t_j)]^{-1}V'M_Z'M_Z\}\right) \\
&= |I-2\mathrm{i}T|^{-\nu/2}\exp\{\mathrm{tr}[\mathrm{i}T(I-2\mathrm{i}T)^{-1}M_Z'M_Z]\}. \qquad (10.12)
\end{aligned}$$

The last form is the CF of a noncentral Wishart with covariance I_p.

For clarity in generalizing to $\Sigma \neq I_p$ with $\mathrm{rank}(\Sigma) = p_1 \leq p$, in the remainder of the proof S_Y replaces S and $\Delta = M_Y'M_Y$. Also $Z \sim \mathcal{N}_{\nu,p_1}(\mathbf{0}, I_\nu, I_{p_1})$ has $p_1 \leq p$ columns in the remainder of the proof. For $Y \sim (\mathcal{S})\mathcal{N}_{\nu,p}(M_Y, I_\nu, \Sigma)$, $p \times p_1$ Φ_1, $\Sigma = \Phi_1\Phi_1'$, $\nu \times p_1$ M_Z, and $Z \sim \mathcal{N}_{\nu,p_1}(\mathbf{0}, I_\nu, I_{p_1})$, Lemma 8.4 gives $Y = (Z + M_Z)\Phi_1'$, with $M_Y = M_Z\Phi_1'$ and $M_Z = M_Y\Phi_1^{+t}$. In turn

$$S_Y = Y'Y = \Phi_1(Z + M_Z)'(Z + M_Z)\Phi_1' = \Phi_1 S_Z \Phi_1'. \qquad (10.13)$$

If $\Phi_1 = \Upsilon_1\mathrm{Dg}(\lambda_1)^{1/2}$, then $\Phi_1^{+t} = \Upsilon_1\mathrm{Dg}(\lambda_1)^{-1/2}$. Lemma 7.5 gives

$$\begin{aligned}
\phi_{S_Y}(T) &= |I-2\mathrm{i}\Phi_1'T\Phi_1|^{-\nu/2}\exp\left\{\mathrm{tr}\left[\mathrm{i}\Phi_1'T\Phi_1(I_{p_1}-2\mathrm{i}\Phi_1'T\Phi_1)^{-1}M_Z'M_Z\right]\right\} \\
&= |I_p-2\mathrm{i}T\Sigma|^{-\nu/2}\exp\left\{\mathrm{tr}\left[\mathrm{i}T\Phi_1(I_{p_1}-2\mathrm{i}\Phi_1'T\Phi_1)^{-1}M_Z'M_Y\right]\right\}. \qquad (10.14)
\end{aligned}$$

The following equalities arise from repeated use of (1) $\boldsymbol{\Phi}_1 = \boldsymbol{\Upsilon}_1 \mathrm{Dg}(\boldsymbol{\lambda}_1)^{1/2}$ in Lemma 10.1, (2) $(\boldsymbol{AB})^{-1} = \boldsymbol{B}^{-1}\boldsymbol{A}^{-1}$ for square and full-rank matrices, and (3) similar generalized inverse results in parts (a) and (b) of Lemma 1.15.

$$\begin{aligned}
&\boldsymbol{T\Phi}_1\left[\mathrm{Dg}(\boldsymbol{\lambda}_1)^{-1/2} - 2\mathrm{i}\boldsymbol{\Upsilon}_1'\boldsymbol{T\Phi}_1\right]^{-1}\mathrm{Dg}(\boldsymbol{\lambda}_1)^{-1/2}\boldsymbol{M}_Z'\boldsymbol{M}_Y = \\
&\boldsymbol{T\Phi}_1\left[\boldsymbol{\Upsilon}_1'\boldsymbol{\Upsilon}_1\mathrm{Dg}(\boldsymbol{\lambda}_1)^{-1/2} - 2\mathrm{i}\boldsymbol{\Upsilon}_1'\boldsymbol{T\Phi}_1\right]^{-1}\mathrm{Dg}(\boldsymbol{\lambda}_1)^{-1/2}\boldsymbol{M}_Z'\boldsymbol{M}_Y = \\
&\boldsymbol{T\Phi}_1\left[\boldsymbol{\Upsilon}_1\mathrm{Dg}(\boldsymbol{\lambda}_1)^{-1/2} - 2\mathrm{i}\boldsymbol{T\Phi}_1\right]^{+}\boldsymbol{\Upsilon}_1'\mathrm{Dg}(\boldsymbol{\lambda}_1)^{-1/2}\boldsymbol{M}_Z'\boldsymbol{M}_Y = \\
&\boldsymbol{T\Phi}_1\left\{[\boldsymbol{\Upsilon}_1 - 2\mathrm{i}\boldsymbol{T\Upsilon}_1\mathrm{Dg}(\boldsymbol{\lambda}_1)]\mathrm{Dg}(\boldsymbol{\lambda}_1)^{-1/2}\right\}^{+}\boldsymbol{M}_Y'\boldsymbol{M}_Y = \\
&\boldsymbol{T\Upsilon}_1\mathrm{Dg}(\boldsymbol{\lambda}_1)\{[\boldsymbol{\Upsilon}_1 - 2\mathrm{i}\boldsymbol{T\Upsilon}_1\mathrm{Dg}(\boldsymbol{\lambda}_1)]\}^{+}\boldsymbol{\Delta} = \\
&\boldsymbol{T\Upsilon}_1\mathrm{Dg}(\boldsymbol{\lambda}_1)[\boldsymbol{\Upsilon}_1 - 2\mathrm{i}\boldsymbol{T\Upsilon}_1\mathrm{Dg}(\boldsymbol{\lambda}_1)\boldsymbol{\Upsilon}_1'\boldsymbol{\Upsilon}_1]^{+}\boldsymbol{\Delta} = \\
&\boldsymbol{T\Upsilon}_1\mathrm{Dg}(\boldsymbol{\lambda}_1)\boldsymbol{\Upsilon}_1'[\boldsymbol{I}_p - 2\mathrm{i}\boldsymbol{T\Upsilon}_1\mathrm{Dg}(\boldsymbol{\lambda}_1)\boldsymbol{\Upsilon}_1']^{+}\boldsymbol{\Delta} = \boldsymbol{T\Sigma}(\boldsymbol{I}_p - 2\mathrm{i}\boldsymbol{T\Sigma})^{+}\boldsymbol{\Delta}. \qquad (10.15)
\end{aligned}$$

For $p \times p$ $\boldsymbol{\Upsilon}$, $(\boldsymbol{I}_p - 2\mathrm{i}\boldsymbol{T\Sigma})$ is similar to $\boldsymbol{A} = [\boldsymbol{I}_p - 2\mathrm{i}\boldsymbol{\Upsilon}'\boldsymbol{T\Upsilon}\mathrm{Dg}(\boldsymbol{\lambda}_1, \boldsymbol{0})]$. For $\boldsymbol{D} = \mathrm{Dg}(\boldsymbol{\lambda}_1, \boldsymbol{1})^{1/2}$ and $\boldsymbol{T}$ partitioned like $\boldsymbol{\Upsilon}$, $(\boldsymbol{I}_p - 2\mathrm{i}\boldsymbol{T\Sigma})$ and $\boldsymbol{A}$ are similar to

$$\boldsymbol{DAD}^{-1} = \begin{bmatrix} \boldsymbol{I}_{p_1} - 2\mathrm{i}\boldsymbol{\Phi}_1'\boldsymbol{T}_{1,1}\boldsymbol{\Phi}_1 & \boldsymbol{0} \\ \boldsymbol{T}_{2,1}\boldsymbol{\Phi}_1 & \boldsymbol{I}_{p-p_1} \end{bmatrix}. \qquad (10.16)$$

The triangular form ensures $\mathrm{rank}(\boldsymbol{DAD}^{-1}) = \mathrm{rank}(\boldsymbol{I}_{p_1} - 2\mathrm{i}\boldsymbol{\Phi}_1'\boldsymbol{T}_{1,1}\boldsymbol{\Phi}_1) + (p - p_1)$. In turn, $\boldsymbol{\Phi}_1'\boldsymbol{T}_{1,1}\boldsymbol{\Phi}_1 = \boldsymbol{V}\mathrm{Dg}(\boldsymbol{a})\boldsymbol{V}'$ and $(\boldsymbol{I}_{p_1} - 2\mathrm{i}\boldsymbol{\Phi}_1'\boldsymbol{T}_{1,1}\boldsymbol{\Phi}_1) = \boldsymbol{V}[\boldsymbol{I}_{p_1} - 2\mathrm{i}\mathrm{Dg}(\boldsymbol{a})]\boldsymbol{V}'$. Consequently $\mathrm{rank}(\boldsymbol{I}_p - 2\mathrm{i}\boldsymbol{T\Sigma}) < p$ only if $(1 - 2\mathrm{i}a_k) = 0$ for one or more a_k, which never happens because a_k is always real. Furthermore $(\boldsymbol{I}_p - 2\mathrm{i}\boldsymbol{T\Sigma})^{+} = (\boldsymbol{I}_p - 2\mathrm{i}\boldsymbol{T\Sigma})^{-1}$. A parallel analysis for the moment generating function gives $(1 - 2a_k) > 0$ as sufficient to ensure existence and full-rank $(\boldsymbol{I}_p - 2\mathrm{i}\boldsymbol{T\Sigma})$.

Proof of (b). Conditions guaranteeing a function is a valid characteristic function (Kendall and Stuart, 1977, p. 105) may be verified for noninteger ν. □

10.3 PROPERTIES OF THE WISHART

Theorem 10.3 For $\nu \times p$ $\boldsymbol{Y} = [\boldsymbol{y}_1\, \boldsymbol{y}_2 \cdots \boldsymbol{y}_p] \sim \mathcal{N}_{\nu,p}(\boldsymbol{0}, \boldsymbol{I}_\nu, \boldsymbol{\Sigma})$, $\mathrm{rank}(\boldsymbol{\Sigma}) = p$, $\nu \geq p$, the $p \times p$ matrix $\boldsymbol{S} = \boldsymbol{Y}'\boldsymbol{Y}$ has a central nonsingular Wishart distribution with parameters ν and $\boldsymbol{\Sigma}$. The joint PDF of the $p(p+1)/2$ distinct elements of $\boldsymbol{S}$ is (Gupta and Nagar, 2000)

$$f_{\boldsymbol{S}}(\boldsymbol{S}_*) = \frac{|\boldsymbol{S}_*|^{(\nu-p-1)/2}\exp[-\mathrm{tr}(\boldsymbol{\Sigma}^{-1}\boldsymbol{S}_*)/2]}{2^{\nu p/2}\pi^{p(p-1)/4}|\boldsymbol{\Sigma}|^{\nu/2}\prod_{j=1}^{p}\Gamma[(\nu+1-j)/2]}. \qquad (10.17)$$

The j, j' element of $\boldsymbol{S}$ is a bilinear form for $j \neq j'$ and a univariate quadratic form if $j = j'$. If $\nu < p$ the density does not exist.

With $\nu = N - \text{rank}(\boldsymbol{X})$, the central Wishart describes the distribution of $\nu\widehat{\boldsymbol{\Sigma}}$ in the multivariate general linear model. The noncentral form applies to the hypothesis sum-of-squares matrix under the alternative.

Theorem 10.4 If $\boldsymbol{S} \sim (\mathcal{S})\mathcal{W}_p(\nu, \boldsymbol{\Sigma}, \boldsymbol{\Delta})$ and $\boldsymbol{T}$ is *any* $p \times p_1$ constant, then

$$\boldsymbol{T}'\boldsymbol{S}\boldsymbol{T} \sim (\mathcal{S})\mathcal{W}_{p_1}(\nu, \boldsymbol{T}'\boldsymbol{\Sigma}\boldsymbol{T}, \boldsymbol{T}'\boldsymbol{\Delta}\boldsymbol{T}). \tag{10.18}$$

Proof. Left as an exercise. *Hints*: characteristic function, linear transformation.

Many authors consider only full-rank $\boldsymbol{\Sigma}$ and $\boldsymbol{T}$ (and $p_1 \leq p$). The result holds for any sort of Wishart and any conforming constant $\boldsymbol{T}$. The transformation is equivalent to having transformed the underlying Gaussian (which can be the basis of a proof). If $\boldsymbol{S} = \boldsymbol{Y}'\boldsymbol{Y}$ with $\boldsymbol{Y} \sim (\mathcal{S})\mathcal{N}_{\nu,p}(\boldsymbol{M}, \boldsymbol{I}_\nu, \boldsymbol{\Sigma})$, then

$$\boldsymbol{T}'\boldsymbol{S}\boldsymbol{T} = \boldsymbol{T}'\boldsymbol{Y}'\boldsymbol{Y}\boldsymbol{T} = \boldsymbol{Y}_1'\boldsymbol{Y}_1 \tag{10.19}$$

and

$$\boldsymbol{Y}_1 = \boldsymbol{Y}\boldsymbol{T} \sim \mathcal{N}_{\nu,p_1}(\boldsymbol{M}\boldsymbol{T}, \boldsymbol{I}_\nu, \boldsymbol{T}'\boldsymbol{\Sigma}\boldsymbol{T}). \tag{10.20}$$

Theorem 10.5 If $\boldsymbol{S} \sim \mathcal{W}_p(\nu, \boldsymbol{\Sigma}, \boldsymbol{\Delta})$ and $\boldsymbol{t}$ is a vector of constants, then $\boldsymbol{t}'\boldsymbol{S}\boldsymbol{t}/\boldsymbol{t}'\boldsymbol{\Sigma}\boldsymbol{t} \sim \chi^2(\nu, \omega)$, with $\omega = \boldsymbol{t}'\boldsymbol{\Delta}\boldsymbol{t}/\boldsymbol{t}'\boldsymbol{\Sigma}\boldsymbol{t}$.

Proof. With $\boldsymbol{S} = \sum_{i=1}^{\nu} \boldsymbol{Y}_i'\boldsymbol{Y}_i$ for $[\text{row}_i(\boldsymbol{Y})]' = \boldsymbol{Y}_i' \sim \mathcal{N}_\nu(\boldsymbol{\mu}_i, \boldsymbol{\Sigma})$ it follows that

$$\boldsymbol{t}'\boldsymbol{S}\boldsymbol{t} = \boldsymbol{t}'\left(\sum_{i=1}^{\nu} \boldsymbol{Y}_i'\boldsymbol{Y}_i\right)\boldsymbol{t} = \sum_{i=1}^{\nu} \boldsymbol{t}'\boldsymbol{Y}_i'\boldsymbol{Y}_i\boldsymbol{t} = \sum_{i=1}^{\nu} (\boldsymbol{t}'\boldsymbol{Y}_i')^2. \tag{10.21}$$

Also $\boldsymbol{t}'\boldsymbol{Y}_i \sim \mathcal{N}_1(\boldsymbol{t}'\boldsymbol{\mu}_i, \boldsymbol{t}'\boldsymbol{\Sigma}\boldsymbol{t})$ allows concluding

$$\boldsymbol{t}'\boldsymbol{Y}_i/(\boldsymbol{t}'\boldsymbol{\Sigma}\boldsymbol{t})^{1/2} \sim \mathcal{N}_1\left[\boldsymbol{t}'\boldsymbol{\mu}_i/(\boldsymbol{t}'\boldsymbol{\Sigma}\boldsymbol{t})^{1/2}, 1\right]. \tag{10.22}$$

The result follows because a squared unit Gaussian is a noncentral chi square. □

The result can be proven more generally as a special case of the previous theorem. The converse is *not* true (Mitra, 1970). Theorem 10.9 provides a form of a converse, based on a far stronger condition.

Theorem 10.6 Principal diagonal blocks of a Wishart are Wishart and independent if and only if the corresponding interblock covariance is zero. If

$$\boldsymbol{S} = \begin{bmatrix} \boldsymbol{S}_{11} & \boldsymbol{S}_{12} \\ \boldsymbol{S}_{21} & \boldsymbol{S}_{22} \end{bmatrix} \sim (\mathcal{S})\mathcal{W}_p(\nu, \boldsymbol{\Sigma}, \boldsymbol{\Delta}), \tag{10.23}$$

with $\boldsymbol{\Sigma}$ and $\boldsymbol{\Delta}$ partitioned to match, $\boldsymbol{S}_{11}$ $(q \times q)$, then

$$S_{11} \sim (\mathcal{S})\mathcal{W}_q(\nu, \Sigma_{11}, \Delta_{11}) \tag{10.24}$$
$$S_{22} \sim (\mathcal{S})\mathcal{W}_{p-q}(\nu, \Sigma_{22}, \Delta_{22}) \tag{10.25}$$

$$S_{11} \perp\!\!\!\perp S_{22} \Leftrightarrow \Sigma_{12} = \mathbf{0}. \tag{10.26}$$

Proof. Highly recommended as an exercise.

Theorem 10.7 If $S_1 \sim (\mathcal{S})\mathcal{W}_p(\nu_1, \Sigma, \Delta_1) \perp\!\!\!\perp S_2 \sim (\mathcal{S})\mathcal{W}_p(\nu_2, \Sigma, \Delta_2)$ then $S_1 + S_2 \sim (\mathcal{S})\mathcal{W}_p(\nu_1 + \nu_2, \Sigma, \Delta_1 + \Delta_2)$.

Proof. Highly recommended as an exercise.

Theorem 10.8 If $Y \sim (\mathcal{S})\mathcal{N}_{N,p}(M, I_N, \Sigma)$, $\text{rank}(\Sigma) = p_1 \le p$ with $N \ge p_1$, $N \times N$ A and B are constants, the following hold.
(a) AY and BY are independent if and only if $AB' = \mathbf{0}$.
(b) If $A = A'$ is positive definite or positive semidefinite and $AB' = \mathbf{0}$, then BY and $Y'AY$ are independent.
(c) If $A = A'$ and $B = B'$ are positive definite or positive semidefinite and $BA = \mathbf{0}$, then $Y'AY$ and $Y'BY$ are independent.
(d) If $A = A' = A^2$, then $\nu = \text{rank}(A) = \text{tr}(A)$ and

$$S = Y'AY \sim (\mathcal{S})\mathcal{W}_p(\nu, \Sigma, M'AM). \tag{10.27}$$

Proof. The basic approach is as follows. Here $Y \sim (\mathcal{S})\mathcal{N}_{N,p}(M, I_N, \Sigma)$ gives

$$\begin{bmatrix} Y \\ Y \end{bmatrix} \sim (\mathcal{S})\mathcal{N}_{2N,p}\left(\begin{bmatrix} M \\ M \end{bmatrix}, [(\mathbf{1}_2\mathbf{1}_2') \otimes I_N], \Sigma\right) \tag{10.28}$$

$$\begin{bmatrix} AY \\ BY \end{bmatrix} \sim (\mathcal{S})\mathcal{N}_{(\nu_a+\nu_b),p}\left(\begin{bmatrix} M \\ M \end{bmatrix}, \begin{bmatrix} A & \mathbf{0} \\ \mathbf{0} & B \end{bmatrix}[(\mathbf{1}_2\mathbf{1}_2') \otimes I_N]\begin{bmatrix} A' & \mathbf{0} \\ \mathbf{0} & B' \end{bmatrix}, \Sigma\right) \tag{10.29}$$

$$\begin{bmatrix} A & \mathbf{0} \\ \mathbf{0} & B \end{bmatrix}\begin{bmatrix} I_N & I_N \\ I_N & I_N \end{bmatrix}\begin{bmatrix} A' & \mathbf{0} \\ \mathbf{0} & B' \end{bmatrix} = \begin{bmatrix} AA' & AB' \\ BA' & BB' \end{bmatrix}. \tag{10.30}$$

Part (a) follows immediately. If $A = A'$ is nonnegative definite, then $A = FF'$ and $Y'AY = Y'FF'Y$. Parts (b) and (c) follow from independence of the underlying Gaussian matrices, as seen from part (a). In part (d), $A = A' = A^2$ implies $A = VV'$ with $V'V = I_\nu$, $Y_A = V'Y \sim (\mathcal{S})\mathcal{N}_{N,\nu}(V'M, I_\nu, \Sigma)$, and $S = Y_A'Y_A \sim (\mathcal{S})\mathcal{W}_p(\nu, \Sigma, M'VV'M)$. □

What does the theorem tell us about the independence of the usual sample covariance and sample mean vector?

Corollary 10.8.1 If $p = 1$ and $\sigma^2 > 0$, then $\boldsymbol{y} \sim \mathcal{N}_{n,1}(\boldsymbol{\mu}, \boldsymbol{I}_n, \sigma^2) \Leftrightarrow \boldsymbol{y} \sim \mathcal{N}_n(\boldsymbol{\mu}, \boldsymbol{I}_n\sigma^2)$ and

$$s = \boldsymbol{y}'\boldsymbol{A}\boldsymbol{y} \sim \mathcal{W}_1(\nu, \sigma^2, \boldsymbol{\mu}'\boldsymbol{A}\boldsymbol{\mu}). \tag{10.31}$$

Equivalently

$$s/\sigma^2 \sim \chi^2(\nu, \boldsymbol{\mu}'\boldsymbol{A}\boldsymbol{\mu}/\sigma^2). \tag{10.32}$$

Proof. Left as an exercise.

Corollary 10.8.2 The matrix $\boldsymbol{Y}'\boldsymbol{A}\boldsymbol{Y}$ is Wishart if and only if $\boldsymbol{A} = \boldsymbol{A}' = \boldsymbol{A}^2$ (idempotent).

Proof. Only the statement "$\boldsymbol{Y}'\boldsymbol{A}\boldsymbol{Y}$ is Wishart implies $\boldsymbol{A}$ is idempotent" must be proven. Spectral decomposition gives $\boldsymbol{A} = \boldsymbol{V}_1\mathrm{Dg}(\boldsymbol{\lambda}_1)\boldsymbol{V}_1'$ with $\boldsymbol{V}_1$ $N \times \nu$ and $\boldsymbol{V}_1'\boldsymbol{V}_1 = \boldsymbol{I}_\nu$. In turn $\boldsymbol{Y}_1 = \boldsymbol{V}_1'\boldsymbol{Y} \sim \mathcal{N}_{\nu,p}(\boldsymbol{V}_1'\boldsymbol{M}, \boldsymbol{I}_\nu, \boldsymbol{\Sigma})$ and

$$\boldsymbol{Y}'\boldsymbol{A}\boldsymbol{Y} = \boldsymbol{Y}'\boldsymbol{V}_1\mathrm{Dg}(\boldsymbol{\lambda}_1)\boldsymbol{V}_1'\boldsymbol{Y} = \boldsymbol{Y}_1'\mathrm{Dg}(\boldsymbol{\lambda}_1)\boldsymbol{Y}_1. \tag{10.33}$$

The last equation can only be guaranteed to exist if $\boldsymbol{A} = \boldsymbol{A}'$. By the definition of a Wishart, $\boldsymbol{Y}_1'\boldsymbol{Y}_1 \sim \mathcal{W}_p(\nu, \boldsymbol{\Sigma}, \boldsymbol{M}'\boldsymbol{V}_1\boldsymbol{V}_1'\boldsymbol{M})$. Referring again to the definition of a Wishart, the $\nu \times \nu$ matrix $\mathrm{Dg}(\boldsymbol{\lambda}_1)$ must equal $\boldsymbol{I}_\nu$ for $\boldsymbol{Y}_1'\mathrm{Dg}(\boldsymbol{\lambda}_1)\boldsymbol{Y}_1 = \boldsymbol{Y}'\boldsymbol{A}\boldsymbol{Y}$ to be a Wishart. Requiring $\boldsymbol{A} = \boldsymbol{V}_1\boldsymbol{I}_\nu\boldsymbol{V}_1'$ is equivalent to requiring $\boldsymbol{A} = \boldsymbol{A}' = \boldsymbol{A}^2$. □

If all conditions for part (a) of the last theorem are met except $\boldsymbol{A} \neq \boldsymbol{A}^2$, then $\boldsymbol{Y}'\boldsymbol{A}\boldsymbol{Y}$ is a general multivariate quadratic form. As a generalization of results for univariate quadratic forms, a constituent matrix decomposition of $\boldsymbol{A}$ gives

$$\boldsymbol{Y}'\boldsymbol{A}\boldsymbol{Y} = \sum_{k=1}^{\nu} \lambda_{1k}\boldsymbol{Y}'\boldsymbol{v}_{1k}\boldsymbol{v}_{1k}'\boldsymbol{Y} \tag{10.34}$$

with i.i.d. $\boldsymbol{Y}'\boldsymbol{v}_{1k}\boldsymbol{v}_{1k}'\boldsymbol{Y} \sim (\mathcal{S})\mathcal{W}_p(1, \boldsymbol{\Sigma}, \boldsymbol{M}\boldsymbol{v}_{1k}\boldsymbol{v}_{1k}'\boldsymbol{M})$. We leave consideration of such forms for another venue.

Theorem 10.9 For $N \times p$ $\boldsymbol{Y} = [\boldsymbol{y}_1\ \boldsymbol{y}_2 \cdots \boldsymbol{y}_p] \sim \mathcal{N}_{N,p}(\boldsymbol{M}, \boldsymbol{I}_N, \boldsymbol{\Sigma})$, $\mathrm{rank}(\boldsymbol{\Sigma}) = p$, $N \geq p$, and $\boldsymbol{Y}_i = \mathrm{row}_i(\boldsymbol{Y})$. The constant $N \times N$ matrix $\boldsymbol{A} = \boldsymbol{A}'$ has $\nu = \mathrm{rank}(\boldsymbol{A})$ and $\boldsymbol{\Delta} = \boldsymbol{M}'\boldsymbol{A}\boldsymbol{M}$. For the conditions given,

$$\boldsymbol{S} = \boldsymbol{Y}'\boldsymbol{A}\boldsymbol{Y} \sim \mathcal{W}_p(\nu, \boldsymbol{\Sigma}, \boldsymbol{\Delta}) \tag{10.35}$$

if and only if for all $\boldsymbol{t} \in \Re^p$

$$\frac{\boldsymbol{t}'\boldsymbol{Y}'\boldsymbol{A}\boldsymbol{Y}\boldsymbol{t}}{\boldsymbol{t}'\boldsymbol{\Sigma}\boldsymbol{t}} \sim \chi^2(\nu, \boldsymbol{t}'\boldsymbol{\Delta}\boldsymbol{t}/\boldsymbol{t}'\boldsymbol{\Sigma}\boldsymbol{t}). \tag{10.36}$$

Furthermore $\nu = \mathrm{tr}(\boldsymbol{A})$.

Proof. The fact $t'Y'AYt/t'\Sigma t$ is chi square may be proven by combining two previous theorems. The proof for the other direction begins with defining $N \times 1$ $x = Yt$ for a fixed nonzero t and $Yt \sim \mathcal{N}_{N,1}(Mt, I_N, t'\Sigma t)$. Equivalently $x \sim \mathcal{N}_N(Mt, I_N t'\Sigma t)$. By assumption $t'Y'AYt/(t'\Sigma t) = x'Ax/\mathcal{V}(x) \sim \chi^2(\nu, t'\Delta t/t'\Sigma t)$. By Theorem 9.6, for $x'Ax/\mathcal{V}(x) \sim \chi^2(\nu, t'\Delta t/t'\Sigma t)$ to hold, the eigenvalues of A must all be 1. Therefore A is idempotent, which suffices, with Theorem 10.8, to ensure $S \sim \mathcal{W}_p(\nu, \Sigma, \Delta)$. □

Lemma 10.2 (a) If $S \sim (\mathcal{S})\mathcal{W}_p(\nu, \Sigma, \Delta)$, then, without loss of generality, it may be assumed $S = Y_1'Y_1$ with $Y_1 \sim (\mathcal{S})\mathcal{N}_{\nu,p}(M, I_\nu, \Sigma)$.
(b) If $A = A' = A^2$, then $A = V_1V_1'$ with V_1 $N \times \nu$ and $V_1'V_1 = I_\nu$.
If $Y \sim (\mathcal{S})\mathcal{N}_{N,p}(M, I_N, \Sigma)$, then $Y_1 = V_1'Y \sim (\mathcal{S})\mathcal{N}_{\nu,p}(M, I_\nu, \Sigma)$ and $Y'AY = Y_1'Y_1 \sim (\mathcal{S})\mathcal{W}_p(\nu, \Sigma, M'V_1V_1'M)$.

Proof. (a) By definition of a Wishart. **(b)** Shown by construction. □

Theorem 10.10 For $N \times p$ $Y = [\, y_1\, y_2 \cdots y_p\,] \sim \mathcal{N}_{N,p}(M, I_N, \Sigma)$, $N \geq p$, $\text{rank}(\Sigma) = p$, and $Y_i = \text{row}_i(Y)$. The constant $N \times N$ matrix $A = A'$ has $\nu = \text{rank}(A)$ and $\Delta = M'AM$. Given the definitions,

$$\text{E}(Y'AY) = \text{tr}(A)\Sigma + \Delta\,. \tag{10.37}$$

The result remains true with the weaker assumption of i.i.d. rows with finite second moments, even without Gaussian variables.

The ANOVA theorem generalizes directly from univariate to multivariate quadratic forms. The result is the MANOVA theorem.

Theorem 10.11 (MANOVA theorem) The $N \times p$ matrix $Y = [\, y_1\, y_2 \cdots y_p\,]$, with $Y_i = \text{row}_i(Y)$ and $N \geq p$, is matrix Gaussian, $Y \sim \mathcal{N}_{N,p}(M, I_N, \Sigma)$, with $\text{rank}(\Sigma) = p$. Also, with $k \in \{1, 2, \ldots, d\}$ and $N \times N$ matrix $A_k = A_k'$ of rank $\nu_k > 0$, the matrix $A_0 = \sum_{k=1}^{d} A_k$, with rank of ν_0, is symmetric and $N \times N$. The assumptions allow defining five conditions.
1. A_k is idempotent for $k \in \{1, 2, \ldots, d\}$;
2. $A_kA_{k'} = 0\ \forall k \neq k'$ with $k \in \{1, 2, \ldots, d\}$ and $k' \in \{1, 2, \ldots, d\}$;
3. A_0 is idempotent;
4. $\nu_0 = \sum_{k=1}^{d} \nu_k$, which is equivalent to $\text{rank}(\sum_{k=1}^{d} A_k) = \sum_{k=1}^{d} \text{rank}(A_k)$; and
5. A_k is idempotent for $k \in \{1, \ldots, d-1\}$ *and* A_d is nonnegative definite.
Given the assumptions,
(a) $Y'A_kY \sim \mathcal{W}_p(\nu_k, \Sigma, M'A_kM)$ for $k \in \{0, 1, 2, \ldots, d\}$ *and*
(b) $\{Y'A_kY\}$ are mutually independent for $k \in \{1, 2, \ldots, d\}$ and $k \neq 0$
if and only if
I. any two of conditions 1, 2 and 3 are true, or
II. conditions 3 and 4 are true, or

III. conditions 3 and 5 are true.
Furthermore, if *any* of I, II, or III, holds, then *all* hold.

Proof. Mostly left as an exercise. Theorem 10.9 combines with the ANOVA theorem and Theorem 5 in Searle (1971, Section 2.5).

Theorem 10.12 The conditional sum-of-squares matrix from a Wishart is Wishart. If

$$\boldsymbol{S} = \begin{bmatrix} \boldsymbol{S}_{11} & \boldsymbol{S}_{12} \\ \boldsymbol{S}_{21} & \boldsymbol{S}_{22} \end{bmatrix} \sim \mathcal{W}_p(\nu, \boldsymbol{\Sigma}), \tag{10.38}$$

with $\boldsymbol{\Sigma}$ partitioned to match, full-rank $\boldsymbol{S}_{22}$, and $\boldsymbol{S}_{1.2} = \boldsymbol{S}_{11} - \boldsymbol{S}_{12}\boldsymbol{S}_{22}^{-1}\boldsymbol{S}_{21}$, then

$$\boldsymbol{S}_{1.2} \sim \mathcal{W}_q(\nu - p + q, \boldsymbol{\Sigma}_{1.2}). \tag{10.39}$$

Proof. Highly recommended as an exercise.

Corollary 10.12 For any combination of singular $\boldsymbol{\Sigma}_{22}$ and $\nu \leq p$, it follows that $\boldsymbol{S}_{1.2} = \boldsymbol{S}_{11} - \boldsymbol{S}_{12}\boldsymbol{S}_{22}^{-}\boldsymbol{S}_{21} = \boldsymbol{S}_{11} - \boldsymbol{S}_{12}\boldsymbol{S}_{22}^{+}\boldsymbol{S}_{21}$ and $\boldsymbol{S}_{1.2} \sim (\mathcal{S})\mathcal{W}_q(\nu - p + q, \boldsymbol{\Sigma}_{1.2})$.

Theorem 10.13 (Wijsman, 1959; Odell and Feiveson, 1966) If $\boldsymbol{S} \sim \mathcal{W}_p(\nu, \boldsymbol{I}, \boldsymbol{0})$ and $\nu \geq p$, then a $p \times p$ lower triangular matrix $\boldsymbol{T}$ exists such that $\boldsymbol{S} = \boldsymbol{T}\boldsymbol{T}'$ and
1. $\boldsymbol{t}_{jj} \sim \chi(\nu - j + 1, 0)$ for the diagonal elements,
2. $\boldsymbol{t}_{jj'} \sim \mathcal{N}_1(0, 1)$ for $j > j'$, the lower off-diagonal elements, and
3. all elements of $\boldsymbol{T}$ are statistically independent.

The result, the *Bartlett decomposition*, also holds for any (finite) real $\nu > p$.

Corollary 10.13 **(a)** If $\boldsymbol{\Sigma} = \boldsymbol{\Phi}\boldsymbol{\Phi}'$ is positive definite, $\boldsymbol{T}_1$ has the properties described in the theorem, $\boldsymbol{S}_1 = \boldsymbol{T}_1\boldsymbol{T}_1' \sim \mathcal{W}_p(\nu, \boldsymbol{I})$, $\nu \geq p$, and $\boldsymbol{T}_2 = \boldsymbol{\Phi}\boldsymbol{T}_1$, then $\boldsymbol{S}_2 = \boldsymbol{T}_2\boldsymbol{T}_2' \sim \mathcal{W}_p(\nu, \boldsymbol{\Sigma})$.
(b) If $\nu \geq p+1$ and $\boldsymbol{y} \sim \mathcal{N}_p(\boldsymbol{\mu}, \boldsymbol{\Sigma})$, then a noncentral Wishart, $\boldsymbol{S}_+ \sim \mathcal{W}_p(\nu, \boldsymbol{\Sigma}, \boldsymbol{\mu}\boldsymbol{\mu}')$, may be generated as $\boldsymbol{S}_+ = \boldsymbol{S}_1 + \boldsymbol{S}_2$, with $\boldsymbol{S}_1 \sim \mathcal{W}_p(\nu - 1, \boldsymbol{\Sigma}, \boldsymbol{0})$ and $\boldsymbol{S}_2 = \boldsymbol{y}\boldsymbol{y}' \sim \mathcal{W}_p(1, \boldsymbol{\Sigma}, \boldsymbol{\mu}\boldsymbol{\mu}')$. Furthermore, as long as $\nu > p+1$, the approach also allows directly generating noncentral pseudo-random Wishart matrices with fractional degrees of freedom.

The corollary leads to a simulation algorithm for generating pseudo-random Wishart matrices that can be dramatically faster than computing the inner product of a matrix of pseudo-random Gaussian variables. As long as $\nu > p$, the approach also allows generating pseudo-random Wishart matrices with fractional ν.

The steps are as follows.
1. Compute $\boldsymbol{\Phi}$ via a Cholesky algorithm or a spectral decomposition. If $\boldsymbol{\Sigma} = \boldsymbol{\Upsilon}\mathrm{Dg}(\boldsymbol{\lambda})\boldsymbol{\Upsilon}'$ then $\boldsymbol{\Phi} = \boldsymbol{\Upsilon}\mathrm{Dg}(\boldsymbol{\lambda})^{1/2}$.
2. Generate the required $p(p+1)/2$ nonzero elements for matrix $\boldsymbol{T}_1$ using pseudo-

random number generators.
3. Compute $\boldsymbol{T}_2 = \boldsymbol{\Phi T}_1$, then $\boldsymbol{T}_2\boldsymbol{T}_2'$, a pseudo-random realization.
Repeat steps 2 and 3 to create the desired number of replications. For fixed p, as ν becomes large, the approach is faster than generating a $\nu \times p$ matrix of psuedo-random Gaussian variables and taking the inner product.

10.4 THE INVERSE WISHART

The inverse of a Wishart matrix occurs in many forms of multivariate test statistics. As a special case, the F statistic for testing a general linear hypothesis in the univariate GLM may be written, with $r = \text{rank}(\boldsymbol{X})$,

$$F(\boldsymbol{y}) = \frac{SSH/a}{SSE/(N-r)} = SSH(SSE)^{-1}[(N-r)/a]\,. \tag{10.40}$$

Here $SSE = \sigma^2 x_e \sim \mathcal{W}_1(N-r, \sigma^2, 0)$ and $x_e \sim \chi^2(N-r, 0)$.

Definition 10.3 For $\nu \geq p = \text{rank}(\boldsymbol{\Sigma})$, the $p \times p$ matrix $\boldsymbol{T}$ has the inverse (central) Wishart distribution, denoted $\boldsymbol{T} \sim \mathcal{W}_p^{-1}(\nu, \boldsymbol{\Sigma}^{-1})$, if and only if $\boldsymbol{T}^{-1} = \boldsymbol{S} \sim \mathcal{W}_p(\nu, \boldsymbol{\Sigma})$.

Certain features of the definition should be noted. Most importantly, the matrices $\boldsymbol{T}$ and $\boldsymbol{\Sigma}$ are implicitly assumed to be symmetric, full rank, and positive definite. Furthermore, the definition leaves open the possibility of extending the concept to noncentral Wishart matrices. Finally, the random matrix is defined in terms of another random matrix, with no reference to the distribution function or other properties. Given the conditions of the definition, $\boldsymbol{T}$ has the density

$$f(\boldsymbol{T}; \nu, \boldsymbol{\Sigma}^{-1}) = \frac{|\boldsymbol{T}|^{-(\nu+p+1)/2}\exp[-\text{tr}(\boldsymbol{\Sigma}^{-1}\boldsymbol{T}^{-1})/2]}{2^{\nu p/2}\pi^{p(p-1)/4}|\boldsymbol{\Sigma}|^{\nu/2}\prod_{j=1}^{p}\Gamma[(\nu+1-j)/2]}\,. \tag{10.41}$$

Mardia, Kent, and Bibby (1979, p. 85) gave a derivation. It is simple to prove $\text{E}(\boldsymbol{T}) = [\boldsymbol{\Sigma}(\nu - p - 1)]^{-1}$. Sampson (1974) stated the following two lemmas.

Lemma 10.3 If $\boldsymbol{T} \sim \mathcal{W}_p^{-1}(\nu, \boldsymbol{\Sigma}^{-1})$ and constant $\boldsymbol{A}$ is full rank and $p \times p$, then $\boldsymbol{A}'\boldsymbol{T}\boldsymbol{A} \sim \mathcal{W}_p^{-1}(\nu, \boldsymbol{A}'\boldsymbol{\Sigma}^{-1}\boldsymbol{A})$.

Proof. Left as an exercise.

The result can be generalized, at least to $p \times p_1$ $\boldsymbol{A}$ of rank p_1 with $p_1 < p$.

Lemma 10.4 If $\boldsymbol{T}_{jk}$ and $\boldsymbol{\Sigma}_{jk}$ are $p_j \times p_k$ with $p_j + p_k = p$ and also

$$\boldsymbol{T} = \begin{bmatrix} \boldsymbol{T}_{11} & \boldsymbol{T}_{12} \\ \boldsymbol{T}_{21} & \boldsymbol{T}_{22} \end{bmatrix} \sim \mathcal{W}_p^{-1}\left(\nu, \begin{bmatrix} \boldsymbol{\Psi}_{11} & \boldsymbol{\Psi}_{12} \\ \boldsymbol{\Psi}_{21} & \boldsymbol{\Psi}_{22} \end{bmatrix}\right), \tag{10.42}$$

then $\boldsymbol{T}_{11} \sim \mathcal{W}_{p_1}^{-1}(\nu - p_2, \boldsymbol{\Psi}_{11})$.

Proof. Left as an exercise. *Hint*: Use theorems about the inverse of a partitioned matrix and $\boldsymbol{S}_{1.2} = \boldsymbol{S}_{11} - \boldsymbol{S}_{12}\boldsymbol{S}_{22}^{-1}\boldsymbol{S}_{21}$.

10.5 RELATED DISTRIBUTIONS

The Wishart has useful relationships to other distributions. We have earlier noted $\boldsymbol{S} \sim \mathcal{W}_p(\nu, \boldsymbol{\Sigma}, \boldsymbol{\Delta}) \Leftrightarrow q_t = \boldsymbol{t}'\boldsymbol{S}\boldsymbol{t}/(\boldsymbol{t}'\boldsymbol{\Sigma}\boldsymbol{t}) \sim \chi^2[\nu, \boldsymbol{t}'\boldsymbol{\Delta}\boldsymbol{t}/(\boldsymbol{t}'\boldsymbol{\Sigma}\boldsymbol{t})] \quad \forall \boldsymbol{t} \in \Re^p$. The trace and determinant of a Wishart have simple distributions.

Theorem 10.14 Idempotent matrix $\boldsymbol{A}$ of rank $\nu > 0$ is constant and $N \times N$. Also $p \times p$ $\boldsymbol{\Sigma}$ is symmetric, positive semidefinite, rank $p_1 \leq p$, with spectral decomposition $\boldsymbol{\Sigma} = \boldsymbol{\Upsilon}\mathrm{Dg}(\boldsymbol{\lambda})\boldsymbol{\Upsilon}'$, and $\boldsymbol{\Upsilon}'\boldsymbol{\Upsilon} = \boldsymbol{I}_p$. If $\boldsymbol{\lambda}_1$ is the $p_1 \times 1$ vector of strictly positive eigenvalues, then, without loss of generality,

$$\boldsymbol{\Sigma} = [\boldsymbol{\Upsilon}_1 \; \boldsymbol{\Upsilon}_0]\mathrm{Dg}(\boldsymbol{\lambda}_1, \boldsymbol{0})\begin{bmatrix} \boldsymbol{\Upsilon}_1' \\ \boldsymbol{\Upsilon}_0' \end{bmatrix}. \tag{10.43}$$

We indicate a $p \times 1$ vector with a 1 in row k and 0 elsewhere as $\boldsymbol{d}_k$. If $\boldsymbol{Y} = (\mathcal{S})\mathcal{N}_{N,p}(\boldsymbol{M}, \boldsymbol{I}_N, \boldsymbol{\Sigma})$, $\boldsymbol{m}_k = \boldsymbol{M}\boldsymbol{\Upsilon}\boldsymbol{d}_k$ and $x_k \sim \chi^2(\nu, \boldsymbol{m}_k'\boldsymbol{A}\boldsymbol{m}_k/\lambda_k)$ are independent, then

$$\begin{aligned} \mathrm{tr}(\boldsymbol{Y}'\boldsymbol{A}\boldsymbol{Y}) &= \sum_{k=1}^{p_1}\lambda_k x_k + \sum_{k=p_1+1}^{p} \boldsymbol{m}_k'\boldsymbol{A}\boldsymbol{m}_k \\ &= \sum_{k=1}^{p_1}\lambda_k x_k + \mathrm{tr}(\boldsymbol{\Upsilon}_0'\boldsymbol{M}'\boldsymbol{A}\boldsymbol{M}\boldsymbol{\Upsilon}_0). \end{aligned} \tag{10.44}$$

If $p_1 = p$, then $\mathrm{tr}(\boldsymbol{\Upsilon}_0'\boldsymbol{M}'\boldsymbol{M}\boldsymbol{\Upsilon}_0) = 0$. If $\boldsymbol{M} = \boldsymbol{0}$, then $\mathrm{tr}(\boldsymbol{\Upsilon}_0'\boldsymbol{M}'\boldsymbol{M}\boldsymbol{\Upsilon}_0) = 0$ and $x_k \sim \chi^2(\nu, 0)$.

Proof. Glueck and Muller (1998) proved the result.

Corollary 10.14.1 The reproductive property of the matrix Gaussian allow concluding

$$\begin{aligned} [\boldsymbol{Y}_1 \; \boldsymbol{Y}_0] &= \boldsymbol{Y}[\boldsymbol{\Upsilon}_1 \; \boldsymbol{\Upsilon}_0] \\ &= \boldsymbol{Y}_1\boldsymbol{\Upsilon}_1' + \boldsymbol{Y}_0\boldsymbol{\Upsilon}_0' \sim (\mathcal{S})\mathcal{N}_{N,p}[\boldsymbol{M}\boldsymbol{\Upsilon}, \boldsymbol{I}_N, \mathrm{Dg}(\boldsymbol{\lambda}_1, \boldsymbol{0})]. \end{aligned} \tag{10.45}$$

Therefore $\boldsymbol{Y}_1$ and $\boldsymbol{Y}_0$ are statistically independent with

$$Y_1 \sim \ \mathcal{N}_{N,p_1}[M\Upsilon_1, I_N, \text{Dg}(\lambda_1)] \tag{10.46}$$
$$Y_0 \sim \mathcal{SN}_{N,p-p_1}[M\Upsilon_0, I_N, \text{Dg}(\mathbf{0})] \,. \tag{10.47}$$

In turn

$$Y_1'AY_1 \sim \mathcal{W}_{p_1}[\nu, \text{Dg}(\lambda_1), \Upsilon_1'M'AM\Upsilon_1] \tag{10.48}$$
$$Y_0'AY_0 \sim \mathcal{W}_{p-p_1}[\nu, \text{Dg}(\mathbf{0}), \Upsilon_0'M'AM\Upsilon_0]\,, \tag{10.49}$$

with $Y_1'AY_1$ independent of (degenerate and discrete) $Y_0'AY_0$. In fact, $\Pr\{Y_0'AY_0 = \Upsilon_0'M'AM\Upsilon_0\} = 1$. Furthermore

$$\text{tr}(Y'AY) = \text{tr}(Y_1'AY_1) + \text{tr}(Y_0'AY_0)\,. \tag{10.50}$$

Corollary 10.14.2 The distribution of the trace and other properties of $S \sim \mathcal{W}_p(\nu, \Sigma, \Delta)$ can be described by choosing $A = I_N$. In particular, if $Y = (\mathcal{S})\mathcal{N}_{N,p}(M, I_N, \Sigma)$, with $\Sigma = \Upsilon \text{Dg}(\lambda)\Upsilon'$ and $\Upsilon = [\,\Upsilon_1\ \Upsilon_0\,]$, then

$$\text{tr}(Y'Y) = \sum_{k=1}^{p_1} \lambda_k x_k + \text{tr}(\Upsilon_0'M'M\Upsilon_0)\,, \tag{10.51}$$

with independent $x_k \sim \chi^2(\nu, m_k'm_k/\lambda_k)$. If $p_1 = p$ then $\text{tr}(\Upsilon_0'M'M\Upsilon_0) = 0$. If $M = \mathbf{0}$, then $\text{tr}(\Upsilon_0'M'M\Upsilon_0) = 0$ and $x_k \sim \chi^2(\nu, 0)$. If $p_1 = p$ or $M = \mathbf{0}$, then $\text{tr}(\Upsilon_0'M'M\Upsilon_0) = 0$ and $x_k \sim \chi^2(\nu, 0)$.

Theorem 10.15 If $S \sim \mathcal{W}_p(\nu, \Sigma, \mathbf{0})$ and $\nu \geq p$, then $|S||\Sigma|^{-1} \sim \prod_{j=1}^p x_j$ with $x_j \sim \chi^2(\nu + 1 - j) \perp\!\!\!\perp x_{j'} \sim \chi^2(\nu + 1 - j')\, \forall j \neq j'$.

Proof. Here $|S||\Sigma|^{-1} = |S||\Phi\Phi'|^{-1} = |S|(|\Phi||\Phi'|)^{-1} = |S||\Phi|^{-1}|\Phi'|^{-1} = |S||\Phi^{-1}||\Phi^{-t}| = |\Phi^{-t}||S||\Phi^{-1}| = |\Phi^{-1}S\Phi^{-t}|$. By a previous theorem, $|\Phi^{-1}S\Phi^{-t}| \sim \mathcal{W}_p(\nu, I, \mathbf{0})$. The Bartlett decomposition theorem and corollary give $|\Phi^{-1}S\Phi^{-t}| = |TT'| = |T||T'| = |T|^2$. Since T is triangular, its determinant is the product of its diagonals,

$$|T|^2 = \left(\prod_{j=1}^p t_{jj}\right)^2 = \prod_{j=1}^p t_{jj}^2\,. \tag{10.52}$$

The proof is completed by describing the joint distribution of $\{t_{jj}^2\}$ in terms of the distribution of $\{t_{jj}\}$ given in the triangular decomposition. □

EXERCISES

10.1 Prove that if $S = \begin{bmatrix} S_{11} & S_{12} \\ S_{21} & S_{22} \end{bmatrix} \sim \mathcal{W}_p(\nu, \Sigma, \Delta)$ with $S_{11}(q \times q)$ and Σ and Δ are partitioned to match, then $S_{11} \sim \mathcal{W}_q(\nu, \Sigma_{11}, \Delta_{11})$.

10.2 Prove directly, without using the theorem for which the following is a (very) special case:

If $\boldsymbol{Y} \sim \mathcal{N}_{n,p}(\mathbf{0}, \boldsymbol{I}_n\gamma, \boldsymbol{I}_p)$, then $\text{tr}(\boldsymbol{Y}'\boldsymbol{Y})/\gamma \sim \chi^2(np)$.

10.3 Consider the following SAS/IML program.

```
START GAUSS1(N,MUMAT,FSIGMAT,SEED);
*Function returns matrix Gaussian N(MUMAT,I(N),SIGMA);
*SIGMA=FSIGMAT`*FSIGMAT;
*SEED is the random # generator initial value;
Y=MUMAT + NORMAL(J(N,NCOL(FSIGMAT),SEED))*FSIGMAT;
RETURN(Y);
FINISH GAUSS1;
```

The J(r,c,x) function in IML creates a matrix with r rows, c columns, and all elements equal to x. Also, A`*B indicates matrix A transposed and multiplied by B.

10.3.1 If SIGMA is $p \times p$, what must the dimension of MUMAT be?

10.3.2 What will be the dimension of the matrix Y?

10.3.3 Fully specify the distribution of data created by NORMAL(J(N,NCOL(FSIGMAT),SEED)). The Normal function returns pseudo-random $\mathcal{N}(0, 1)$ values. In SAS/IML, as with many pseudo-random number generators, the seed value is essentially ignored except for the very first call. Furthermore successive invocations create independent values.

10.3.4 Indicate which theorem is being used to compute Y and use the theorem to fully specify the distribution of Y. The program may be used to help answer the following exercise.

10.4 There are two common ways to generate a Wishart matrix in simulations. The simplest way is to generate Gaussian data and compute the Wishart as a function of the Gaussian data. The second approach is often much faster and uses the Bartlett decomposition. Here the simpler method is acceptable.

10.4.1 Generate 100 pseudo-random samples following the $\mathcal{W}_3(r, \boldsymbol{\Sigma}, \mathbf{0})$ distribution with $r = 10$, and $\boldsymbol{\Sigma} = \sigma^2[(1-\rho)\boldsymbol{I}_3 + \rho\mathbf{1}_3\mathbf{1}_3']$, with $\mathbf{1}_m$ an $m \times m$ matrix filled with 1's.

Assume $\sigma^2 = 2.0$ and $\rho = 0.50$. The result should be a set of random matrices that follow the specified Wishart distribution: $\{\boldsymbol{S}_i : \boldsymbol{S}_i \sim \mathcal{W}_3(r, \boldsymbol{\Sigma}, \mathbf{0})\}$ for $i \in \{1, \ldots, 100\}$.

10.4.2 Print $\boldsymbol{\Sigma}$, $\boldsymbol{S}_1$, $\boldsymbol{S}_2$, $\boldsymbol{S}_3$, and a copy of your program.

10.4.3 Compute the largest eigenvalue λ_{1i} of each of the $N = 100$ random matrices.

10.4.4 Report the mean and variance of the set of largest eigenvalues.

10.4.5 Display a frequency histogram for the distribution of the set of largest eigenvalues.

CHAPTER 11

Estimation for Univariate and Weighted Linear Models

11.1 MOTIVATION

Linear model estimation theory ranks as among the most beautiful in statistics. The derivations illustrate classical techniques for obtaining good estimators. The resulting explicit linear functions of the data also have nice geometric interpretations, which helps understand the estimators. We focus on the univariate linear model in the present chapter and leave multivariate generalizations to the next chapter. Separate treatments allows seeing the proof techniques in two slightly different settings. Although the approach creates some redundancy, taking more but shorter steps gives a faster path to understanding the multivariate GLM.

The present chapter centers on deriving estimators for $\boldsymbol{\beta}$ and σ^2 with a variety of optimal properties. Most of the properties do not require any particular distribution for the responses and are exact, even in small samples. In contrast, testing hypotheses uses test statistics. Finding exact test statistic distributions for small samples usually requires explicitly specifying the data distribution.

11.2 STATEMENT OF THE PROBLEM

Definition 11.1 (a) The vector $\boldsymbol{\beta}$ $(q \times 1)$ and scalar σ^2 are the *primary parameters* of a $\text{GLM}_{N,q}(y_i; \boldsymbol{X}_i\boldsymbol{\beta}, \sigma^2)$, with or without Gaussian errors.
(b) Any (finite) known constant $\boldsymbol{C}$ $(a \times q)$ and (finite) known constant $\boldsymbol{\theta}_0$ $(a \times 1)$ define *secondary parameter* $\boldsymbol{\theta} = \boldsymbol{C}\boldsymbol{\beta} + \boldsymbol{\theta}_0$ $(a \times 1)$.

Definition 11.2 (a) *Estimators* of primary parameters with good properties, especially unbiasedness, are indicated by $\widehat{\boldsymbol{\beta}}$ and $\widehat{\sigma}^2$, while ones with distinct and possibly fewer desirable properties are indicated by $\widetilde{\boldsymbol{\beta}}$ and $\widetilde{\sigma}^2$.
(b) Estimators of secondary parameters take the form $\widehat{\boldsymbol{\theta}} = \boldsymbol{C}\widehat{\boldsymbol{\beta}} + \boldsymbol{\theta}_0$, or $\widetilde{\boldsymbol{\theta}} = \boldsymbol{C}\widetilde{\boldsymbol{\beta}} + \boldsymbol{\theta}_0$, or $\widehat{\boldsymbol{\theta}} = \boldsymbol{C}\widetilde{\boldsymbol{\beta}} + \boldsymbol{\theta}_0$. The third form is used only when $\widehat{\boldsymbol{\theta}}$ possesses a desirable property not shared by $\widetilde{\boldsymbol{\beta}}$, such as unbiasedness.
(c) Covariance matrices such as $\mathcal{V}(\widehat{\boldsymbol{\theta}})$ are also secondary parameters.

Overall, four classes of $\text{GLM}_{N,q}(y_i; \boldsymbol{X}_i\boldsymbol{\beta}, \sigma^2)$ are defined by allowing either FR or LTFR designs, combined with either Gaussian or unspecified distributions for responses. For $\text{GLM}_{N,q}(y_i; \boldsymbol{X}_i\boldsymbol{\beta}|\boldsymbol{R\beta} = \boldsymbol{a}, \sigma^2)$, with or without Gaussian errors, estimation theory for primary parameter $\boldsymbol{\beta}$ and secondary parameter $\boldsymbol{\theta}$ is closely tied to $r = \text{rank}([\,\boldsymbol{X}'\ \boldsymbol{R}'\,]')$.

The distinction between FR and LTFR describes properties of the columns of $[\,\boldsymbol{X}'\ \boldsymbol{R}'\,]'$, which correspond directly to properties of the rows of $\boldsymbol{\beta}$. In the absence of restrictions on $\boldsymbol{\beta}$ the classification depends solely on $r = \text{rank}(\boldsymbol{X})$. Full rank models have an unbiased estimator for $\boldsymbol{\beta}$ and for all $\boldsymbol{\theta}$, while LTFR models never have an unbiased estimator for $\boldsymbol{\beta}$ and have an unbiased estimator only for some $\boldsymbol{\theta}$. Whether or not the model is FR, unbiased estimators of σ^2 and related properties are always available [as long as $N > r = \text{rank}([\,\boldsymbol{X}'\ \boldsymbol{R}'\,]')$].

11.3 (UNRESTRICTED) LINEARLY EQUIVALENT LINEAR MODELS

The basic theory of transformations between linearly equivalent univariate linear models is contained in the present section and in Section 11.15. The results allow defining two important types of linear equivalence, namely (1) equivalence between a LTFR and FR model and (2) equivalence between an explicitly and an implicitly restricted model. The value of the concept lies in being able to work with a model that is simpler (often due to involving fewer parameters and variables) while being assured of not losing access to any information from the original model.

Definition 11.3 $\text{GLM}_{N,q_1}(y_i; \boldsymbol{X}_{i,1}\boldsymbol{\beta}_1, \sigma^2)$ and $\text{GLM}_{N,q_2}(y_i; \boldsymbol{X}_{i,2}\boldsymbol{\beta}_2, \sigma^2)$ are *linearly equivalent* whenever (1) for any $\boldsymbol{\beta}_1$ there exists $\boldsymbol{\beta}_2$ such that $\boldsymbol{X}_1\boldsymbol{\beta}_1 = \boldsymbol{X}_2\boldsymbol{\beta}_2$ *and* (2) for any $\boldsymbol{\beta}_2$ there exists $\boldsymbol{\beta}_1$ such that $\boldsymbol{X}_1\boldsymbol{\beta}_1 = \boldsymbol{X}_2\boldsymbol{\beta}_2$.

Linear equivalence describes the expected values, the means, of $\{y_i\}$, because $\text{E}(\boldsymbol{y}) = \boldsymbol{X}_1\boldsymbol{\beta}_1$ and $\text{E}(\boldsymbol{y}) = \boldsymbol{X}_2\boldsymbol{\beta}_2$. The definition implicitly requires $\boldsymbol{X}_1$ and $\boldsymbol{X}_2$ have the same number of rows and $\text{rank}(\boldsymbol{X}_1) = \text{rank}(\boldsymbol{X}_2)$. For a given $\boldsymbol{\beta}_1$ the required $\boldsymbol{\beta}_2$ need not be unique, while, given $\boldsymbol{\beta}_2$, the required $\boldsymbol{\beta}_1$ need not be unique.

Lemma 11.1 (a) Models $\text{GLM}_{N,q_1}(y_i; \boldsymbol{X}_{i,1}\boldsymbol{\beta}_1, \sigma^2)$ and $\text{GLM}_{N,q_2}(y_i; \boldsymbol{X}_{i,2}\boldsymbol{\beta}_2, \sigma^2)$ are linearly equivalent if and only if $\boldsymbol{X}_1$ and $\boldsymbol{X}_2$ (1) have N rows *and* (2) their columns span the same subspace of $\Re^N$.
(b) Alternately, $\text{GLM}_{N,q_1}(y_i; \boldsymbol{X}_{i,1}\boldsymbol{\beta}_1, \sigma^2)$ and $\text{GLM}_{N,q_2}(y_i; \boldsymbol{X}_{i,2}\boldsymbol{\beta}_2, \sigma^2)$ are linearly equivalent if and only if (1) $\boldsymbol{X}_2 = \boldsymbol{X}_1\boldsymbol{T}_1$ *and* (2) $\boldsymbol{X}_1 = \boldsymbol{X}_2\boldsymbol{T}_2$.

Proof. Left as an exercise.

A clearer understanding of equivalence may be achieved by considering the SVDs of the two design matrices. Necessarily the number of rows and ranks of $\boldsymbol{X}_1$ and $\boldsymbol{X}_2$ must be the same (otherwise one would contain information not present in the other). With $\text{Dg}(\boldsymbol{s}_{1,j})$ containing only the strictly positive singular values of dimension $\text{rank}(\boldsymbol{X}_j)$, it helps to write the SVDs (using Lemma 1.29) as

$$\begin{aligned} \boldsymbol{X}_j &= \boldsymbol{L}_j \begin{bmatrix} \text{Dg}(\boldsymbol{s}_j) \\ \boldsymbol{0} \end{bmatrix} \boldsymbol{R}_j' \\ &= \boldsymbol{L}_{1,j}\text{Dg}(\boldsymbol{s}_{1,j})\boldsymbol{R}_{1,j}' \end{aligned} \tag{11.1}$$

and define

$$\begin{aligned} \boldsymbol{\beta}_{1,j} &= \boldsymbol{L}_{1,j}'\boldsymbol{X}_j\boldsymbol{\beta}_j \\ &= \text{Dg}(\boldsymbol{s}_{1,j})\boldsymbol{R}_{1,j}'\boldsymbol{\beta}_j\,. \end{aligned} \tag{11.2}$$

Here the 1 subscript indicates the component of the original matrix corresponding to positive singular values and the full-rank basis. Although $\boldsymbol{\beta}_1$ and $\boldsymbol{\beta}_2$ may have different dimensions, $\boldsymbol{\beta}_{1,1}$ and $\boldsymbol{\beta}_{1,2}$ must have the same dimension, because otherwise $\boldsymbol{X}_1$ and $\boldsymbol{X}_2$ would have different ranks. Requiring that for any $\boldsymbol{\beta}_1$ there exists $\boldsymbol{\beta}_2$ such that $\boldsymbol{X}_1\boldsymbol{\beta}_1 = \boldsymbol{X}_2\boldsymbol{\beta}_2$ implies

$$\boldsymbol{X}_1\boldsymbol{\beta}_1 = \boldsymbol{X}_2\boldsymbol{\beta}_2 \tag{11.3}$$

$$\boldsymbol{L}_{1,1}\text{Dg}(\boldsymbol{s}_{1,1})\boldsymbol{R}_{1,1}'\boldsymbol{\beta}_1 = \boldsymbol{L}_{1,2}\text{Dg}(\boldsymbol{s}_{1,2})\boldsymbol{R}_{1,2}'\boldsymbol{\beta}_2 \tag{11.4}$$

$$\boldsymbol{L}_{1,1}\boldsymbol{\beta}_{1,1} = \boldsymbol{L}_{1,2}\boldsymbol{\beta}_{1,2}\,. \tag{11.5}$$

If $\boldsymbol{X}_1$ and $\boldsymbol{X}_2$ span the same column space, then $\boldsymbol{L}_{1,1} = \boldsymbol{L}_{1,2}\boldsymbol{T}$ with $\boldsymbol{T}$ square and FR. Therefore

$$\boldsymbol{L}_{1,2}\boldsymbol{T}\boldsymbol{\beta}_{1,1} = \boldsymbol{L}_{1,2}\boldsymbol{\beta}_{1,2} \tag{11.6}$$

$$\boldsymbol{L}_{1,2}'\boldsymbol{L}_{1,2}\boldsymbol{T}\boldsymbol{\beta}_{1,1} = \boldsymbol{L}_{1,2}'\boldsymbol{L}_{1,2}\boldsymbol{\beta}_{1,2} \tag{11.7}$$

$$\boldsymbol{T}\boldsymbol{\beta}_{1,1} = \boldsymbol{\beta}_{1,2}\,. \tag{11.8}$$

In turn, any parameters corresponding to a basis of one design matrix are full rank transformations of corresponding parameters in a linearly equivalent model (even though $\boldsymbol{\beta}_1$ and $\boldsymbol{\beta}_2$ need not be).

The definition does not directly address the possibility of equivalence between an unrestricted model (1) and a restricted model (2) or equivalence between two restricted models. The first case is covered by results in Section 11.13 which allows finding an unrestricted model (3) which is linearly equivalent to model 2. The question is then reduced to comparing models 1 and 3. Similarly, two restricted models may be compared by first transforming each separately to an unrestricted model and then comparing the results.

The guaranteed existence of equivalencies between FR and LTFR models and between restricted and unrestricted models allows deriving many of the desired results in the unrestricted and FR case. Naturally care must be taken in differentiating between properties that do generalize (from FR to LTFR and from unrestricted to restricted) from properties that do not.

Traditionally, the most important transformations between *linearly equivalent* linear models were from LTFR models to FR models. The transformations allow computing in the FR setting rather than in the LTFR setting. The following theorem formally establishes that one is never required to work with a LTFR model. One can always transform to a linearly equivalent FR model and do all estimation (theory and data analysis) in the FR setting.

Theorem 11.1 Every LTFR model has a linearly equivalent FR model. In particular, for $\text{GLM}_{N,q}\text{LTFR}(y_i; \boldsymbol{X}_i\boldsymbol{\beta}, \sigma^2)$ and $r = \text{rank}(\boldsymbol{X})$ there exists $\text{GLM}_{N,r}\text{FR}(y_i; \boldsymbol{X}_{i1}\boldsymbol{\beta}_1, \sigma^2)$ which is linearly equivalent. The FR model provides a *reparameterization* of the LTFR.

Proof. With the subscript 1 indicating the components corresponding to positive singular values, the SVD results summarized in Lemma 1.29 give

$$\begin{aligned} \boldsymbol{X} &= [\boldsymbol{L}_1 \;\; \boldsymbol{L}_0] \begin{bmatrix} \text{Dg}(\boldsymbol{s}_1) & \boldsymbol{0} \\ \boldsymbol{0} & \boldsymbol{0} \end{bmatrix} \begin{bmatrix} \boldsymbol{R}_1' \\ \boldsymbol{R}_0' \end{bmatrix} \\ &= \boldsymbol{L}_1 \text{Dg}(\boldsymbol{s}_1) \boldsymbol{R}_1' \,, \end{aligned} \tag{11.9}$$

which allows defining

$$\boldsymbol{X}_1 = \boldsymbol{L}_1 \text{Dg}(\boldsymbol{s}_1) \tag{11.10}$$

$$\boldsymbol{\beta}_1 = \boldsymbol{R}_1' \boldsymbol{\beta} \,. \tag{11.11}$$

Furtheremore

$$\begin{aligned} \boldsymbol{y} &= \boldsymbol{X}\boldsymbol{\beta} \;\; + \boldsymbol{e} \\ &= \boldsymbol{X}_1\boldsymbol{\beta}_1 + \boldsymbol{e} \,, \end{aligned} \tag{11.12}$$

with $\boldsymbol{X}_1$ having full rank. Although especially elegant and well behaved, the model is merely one choice among infinitely many linearly equivalent models. □

11.4 ESTIMABILITY AND CRITERIA FOR CHECKING IT

Definition 11.4 In a $\text{GLM}_{N,q}(y_i; \boldsymbol{X}_i\boldsymbol{\beta}, \sigma^2)$ or $\text{GGLM}_{N,q}(\boldsymbol{y}; \boldsymbol{X}\boldsymbol{\beta}, \boldsymbol{\Upsilon})$, $\boldsymbol{\beta}$ is the primary expected-value parameter while $\boldsymbol{\theta} = \boldsymbol{C}\boldsymbol{\beta}$ is a secondary (expected-value) parameter, as is $\boldsymbol{\theta} + \boldsymbol{\theta}_0$.
(a) Primary parameter $\boldsymbol{\beta}$ is *estimable* if and only if a $q \times N$ constant matrix $\boldsymbol{A}_1$ exists such that $\text{E}(\boldsymbol{A}_1\boldsymbol{y}) = \boldsymbol{\beta}$.
(b) Secondary parameter $\boldsymbol{\theta} = \boldsymbol{C}\boldsymbol{\beta}$ is *estimable* if and only if a constant matrix $\boldsymbol{A}_2$ $(a \times N)$ exists such that $\text{E}(\boldsymbol{A}_2\boldsymbol{y}) = \boldsymbol{\theta}$.
(c) For known, fixed $\boldsymbol{\theta}_0$, $\boldsymbol{\theta} + \boldsymbol{\theta}_0$ is *estimable* if and only if $\boldsymbol{\theta}$ is *estimable.*

Part **(a)** is essentially redundant since it is a special case of part **(b)**, with $\boldsymbol{\beta}$ a secondary parameter and $\boldsymbol{C} = \boldsymbol{I}_q$.

Although estimability and parameter definition stem from different issues, Helms (1988a) proved a secondary parameter is estimable if and only if it is well defined. In the present chapter, understanding *estimability*, apart from the issue of being well defined, suffices for the study of estimation. The next theorem leads to the conclusion that nonestimability is a problem *only in* LTFR models.

Theorem 11.2 Primary parameter $\boldsymbol{\beta}$ and secondary parameter $\boldsymbol{\theta} = \boldsymbol{C\beta}$, defined by known constant $\boldsymbol{C}$ may or may not be estimable in $\text{GLM}_{N,q}(y_i; \boldsymbol{X}_i\boldsymbol{\beta}, \sigma^2)$.
(a) If $\text{rank}(\boldsymbol{X}) = q$, then $\boldsymbol{\beta}$ and $\boldsymbol{\theta}$ are always estimable in $\text{GLM}_{N,q}\text{FR}(y_i; \boldsymbol{X}_i\boldsymbol{\beta}, \sigma^2)$.
(b) If $\text{rank}(\boldsymbol{X}) = r < q$, then $\boldsymbol{\beta}$ is never estimable in $\text{GLM}_{N,q}\text{LTFR}(y_i; \boldsymbol{X}_i\boldsymbol{\beta}, \sigma^2)$, while $\boldsymbol{\theta}$ may or may not be estimable.

Proof of (a). The FR assumption $q = \text{rank}(\boldsymbol{X}) = \text{rank}(\boldsymbol{X}'\boldsymbol{X})$ implies $(\boldsymbol{X}'\boldsymbol{X})^{-1}$ exists. If $\boldsymbol{A}_1 = (\boldsymbol{X}'\boldsymbol{X})^{-1}\boldsymbol{X}'$ $(q \times N)$ and $\boldsymbol{A}_2 = \boldsymbol{C}\boldsymbol{A}_1$ $(a \times N)$, then $\widehat{\boldsymbol{\beta}} = \boldsymbol{A}_1\boldsymbol{y}$ and $\widehat{\boldsymbol{\theta}} = \boldsymbol{A}_2\boldsymbol{y}$ are linear estimators, which gives $\text{E}(\widehat{\boldsymbol{\beta}}) = \boldsymbol{A}_1\text{E}(\boldsymbol{y}) = (\boldsymbol{X}'\boldsymbol{X})^{-1}\boldsymbol{X}'\boldsymbol{X}\boldsymbol{\beta} = \boldsymbol{\beta}$ and $\text{E}(\widehat{\boldsymbol{\theta}}) = \boldsymbol{C}\boldsymbol{A}_1\text{E}(\boldsymbol{y}) = \boldsymbol{C\beta} = \boldsymbol{\theta}$. We have described linear unbiased estimators. Therefore both $\boldsymbol{\beta}$ and $\boldsymbol{\theta}$ are *estimable.*

Proof of (b). By contradiction. If $\boldsymbol{A}$ exists defining estimator $\widehat{\boldsymbol{\beta}} = \boldsymbol{Ay}$ such that $\text{E}(\widehat{\boldsymbol{\beta}}) = \text{E}(\boldsymbol{Ay}) = \boldsymbol{AX\beta} = \boldsymbol{\beta}$ $\forall \boldsymbol{\beta}$, then $\boldsymbol{AX} = \boldsymbol{I}_q$, which implies $\text{rank}(\boldsymbol{AX}) = q$. However, $\text{rank}(\boldsymbol{AX}) \leq \min\{\text{rank}(\boldsymbol{A}), \text{rank}(\boldsymbol{X})\} \leq r < q$. The assumption $\boldsymbol{A}$ exists has led to a contradiction. Therefore no such $\boldsymbol{A}$ exists. □

Theorem 11.3 If constant $\boldsymbol{C}$ defines $\boldsymbol{\theta} = \boldsymbol{C\beta}$ for $\text{GLM}_{N,q}\text{LTFR}(y_i; \boldsymbol{X}_i\boldsymbol{\beta}, \sigma^2)$, then $\boldsymbol{\theta}$ is estimable, i.e., $\boldsymbol{T}$ exists such that $\text{E}(\boldsymbol{Ty}) = \boldsymbol{\theta}$,
$\Leftrightarrow$ $\boldsymbol{A}$ exists such that $\boldsymbol{C} = \boldsymbol{AX}$,
$\Leftrightarrow$ rows of $\boldsymbol{C}$ are linear combinations of rows of $\boldsymbol{X}$, and
$\Leftrightarrow$ rows of $\boldsymbol{C}$ are in the space spanned by the rows of $\boldsymbol{X}$.

Proof. $(\Rightarrow)$ If $\boldsymbol{A}$ exists such that $\boldsymbol{C} = \boldsymbol{AX}$ then the estimator defined by $\widehat{\boldsymbol{\theta}} = \boldsymbol{Ay}$ is unbiased, $\text{E}(\widehat{\boldsymbol{\theta}}) = \text{E}(\boldsymbol{Ay}) = \boldsymbol{AX\beta} = \boldsymbol{C\beta} = \boldsymbol{\theta}$.
$(\Leftarrow)$ If $\boldsymbol{\theta}$ is estimable, then, by definition, matrix $\boldsymbol{T}$ exists such that $\text{E}(\boldsymbol{TY}) = \boldsymbol{TX\beta} = \boldsymbol{\theta} = \boldsymbol{C\beta}$ for any $\boldsymbol{\beta}$. Thus $\boldsymbol{TX\beta} = \boldsymbol{C\beta}$ $\forall \boldsymbol{\beta}$ implies $\boldsymbol{TX} = \boldsymbol{C}$ and $\boldsymbol{A} = \boldsymbol{T}$ exists, as required, although is not necessarily unique. □

For checking estimability, the preceding theorem has the disadvantage of involving $\boldsymbol{X}$, which has N rows. In practice, slow and inaccurate computations can make determining estimability difficult. The following theorems use $a \times q$ matrices, which are usually much smaller, with $q \ll N$. Computationally, if $(\boldsymbol{X}'\boldsymbol{X})^-$ is available (as it typically will be), then the next theorem provides the easiest check for estimability. Computer arithmetic reduces the task to checking $\max\{\text{abs}[\boldsymbol{C} - \boldsymbol{C}(\boldsymbol{X}'\boldsymbol{X})^-(\boldsymbol{X}'\boldsymbol{X})]\} \leq \{\max[\text{abs}(\boldsymbol{C})]\}\epsilon$ for ϵ a value judged to be numeric zero, such as 10^{-12} in many contemporary software environments. The abs() operator gives the matrix of absolute values of the original elements. The next theorem also simplifies derivations of moments and distributions of estimators in the univariate and multivariate GLM. The subsequent two theorems can serve the same purposes, although perhaps less conveniently.

Theorem 11.4 If $\text{GLM}_{N,q}\text{LTFR}(y_i; \boldsymbol{X}_i\boldsymbol{\beta}, \sigma^2)$ has secondary parameter $\boldsymbol{\theta} = \boldsymbol{C\beta}$, with $(\boldsymbol{X}'\boldsymbol{X})^-$ any particular generalized inverse, then $\boldsymbol{\theta}$ is estimable $\Leftrightarrow$ $\boldsymbol{C} = \boldsymbol{C}(\boldsymbol{X}'\boldsymbol{X})^-(\boldsymbol{X}'\boldsymbol{X})$.

Proof. $(\Rightarrow)$ If $\boldsymbol{\theta}$ is estimable, then constant $\boldsymbol{A}$ $(a \times N)$ exists such that $\boldsymbol{C} = \boldsymbol{AX}$. In turn

$$\begin{aligned} \boldsymbol{C}\,[(\boldsymbol{X}'\boldsymbol{X})^-\boldsymbol{X}'\boldsymbol{X}] &= \boldsymbol{AX}[(\boldsymbol{X}'\boldsymbol{X})^-\boldsymbol{X}'\boldsymbol{X}] \\ &= \boldsymbol{A}[\boldsymbol{X}(\boldsymbol{X}'\boldsymbol{X})^-\boldsymbol{X}'\boldsymbol{X}] \\ &= \boldsymbol{AX} = \boldsymbol{C}\,. \end{aligned} \tag{11.13}$$

Searle (1971, p. 20) proved $\boldsymbol{X} = \boldsymbol{X}(\boldsymbol{X}'\boldsymbol{X})^-\boldsymbol{X}'\boldsymbol{X}$, which is in Theorem 1.15.

$(\Leftarrow)$ Given $\boldsymbol{C} = \boldsymbol{C}(\boldsymbol{X}'\boldsymbol{X})^-(\boldsymbol{X}'\boldsymbol{X})$, if $\boldsymbol{A} = \boldsymbol{C}(\boldsymbol{X}'\boldsymbol{X})^-\boldsymbol{X}'$, then $\boldsymbol{C} = \boldsymbol{AX}$ and $\boldsymbol{\theta}$ is estimable by Theorem 11.3. □

Corollary 11.4 Secondary parameter $\boldsymbol{\theta}$ is estimable $\Leftrightarrow$ $\boldsymbol{C} = \boldsymbol{C}(\boldsymbol{X}'\boldsymbol{X})^+(\boldsymbol{X}'\boldsymbol{X})$.

Proof. $(\Rightarrow)$ Estimable implies $\boldsymbol{C} = \boldsymbol{C}(\boldsymbol{X}'\boldsymbol{X})^-(\boldsymbol{X}'\boldsymbol{X})$, which clearly implies $\boldsymbol{C} = \boldsymbol{C}(\boldsymbol{X}'\boldsymbol{X})^+(\boldsymbol{X}'\boldsymbol{X})$.

$(\Leftarrow)$ If $\boldsymbol{C} = \boldsymbol{C}(\boldsymbol{X}'\boldsymbol{X})^+(\boldsymbol{X}'\boldsymbol{X})$, then $\boldsymbol{C} = \boldsymbol{C}(\boldsymbol{X}'\boldsymbol{X})^-(\boldsymbol{X}'\boldsymbol{X})$. The result follows by the theorem. □

Theorem 11.5 A $\text{GLM}_{N,q}\text{LTFR}(y_i; \boldsymbol{X}_i\boldsymbol{\beta}, \sigma^2)$ may have $N \times (r+s)$ $\boldsymbol{X}$ partitioned as $\boldsymbol{X} = [\,\boldsymbol{X}_1\, \boldsymbol{X}_2\,]$ with $N \times r$ $\boldsymbol{X}_1$ and $N \times s$ $\boldsymbol{X}_2$, while $s = q - r$. If $(\boldsymbol{X}_1'\boldsymbol{X}_1)$ is nonsingular and $\boldsymbol{C} = [\,\boldsymbol{C}_1\, \boldsymbol{C}_2\,]$ with $\boldsymbol{C}_1$ $a \times r$ and $\boldsymbol{C}_2$ $a \times s$, then $\boldsymbol{\theta} = \boldsymbol{C\beta}$ is estimable if and only if $\boldsymbol{C}_2 = \boldsymbol{C}_1(\boldsymbol{X}_1'\boldsymbol{X}_1)^{-1}\boldsymbol{X}_1'\boldsymbol{X}_2$.

Proof. (Roy, 1957). Left as an exercise.

Since $\text{rank}(\boldsymbol{X}) = r$, $\boldsymbol{X}$ has r linearly independent columns. It may be necessary to permute the columns of $\boldsymbol{X}$ to make the columns the first r columns. The rows of $\boldsymbol{\beta}$ and then $\boldsymbol{C}$ must be permuted to match.

Theorem 11.6 If $\text{GLM}_{N,q}\text{LTFR}(y_i; \boldsymbol{X}_i\boldsymbol{\beta}, \sigma^2)$ has $\boldsymbol{\theta} = \boldsymbol{C\beta}$, then $\boldsymbol{\theta}$ is estimable $\Leftrightarrow$ $\exists \boldsymbol{D}$ $(a \times q)$ such that $\boldsymbol{C} = \boldsymbol{D}(\boldsymbol{X}'\boldsymbol{X})$. If $\boldsymbol{D}$ exists it is unique.

Proof. $(\Rightarrow)$ If $\boldsymbol{D}_1 = \boldsymbol{C}(\boldsymbol{X}'\boldsymbol{X})^-$ $(a \times q)$, with $(\boldsymbol{X}'\boldsymbol{X})^-$ any particular generalized inverse, and $\boldsymbol{\theta}$ is estimable, then $\boldsymbol{C} = \boldsymbol{C}(\boldsymbol{X}'\boldsymbol{X})^-(\boldsymbol{X}'\boldsymbol{X})$, which allows writing $\boldsymbol{C} = \boldsymbol{D}_1(\boldsymbol{X}'\boldsymbol{X})$ and ensures $\boldsymbol{D}$ exists.

$(\Leftarrow)$ If $\boldsymbol{D}_2$ exists such that $\boldsymbol{C} = \boldsymbol{D}_2(\boldsymbol{X}'\boldsymbol{X})$, then $\boldsymbol{A} = \boldsymbol{D}_2\boldsymbol{X}'$ exists such that $\boldsymbol{C} = \boldsymbol{AX}$. Therefore $\boldsymbol{\theta}$ is estimable.

Proving uniqueness is left as an exercise. □

Example 11.1 Two different model formulations for the same simple situation illustrate some of the complications arising in considering estimability for LTFR models. For $N = 3$, two experimental units receive treatment 1, and one experimental unit receives treatment 2. Model 1, a cell-mean model, with $q = 2$, has scalar equation $y_{ij} = \mu_i + e_{ij}$ and corresponds to a $\text{GLM}_{N,2}\text{FR}(y_i; \boldsymbol{X}_{i,1}\boldsymbol{\beta}_1, \sigma^2)$ with Gaussian errors. Model 2, a classical ANOVA model, with $q = 3$, has scalar equation $y_{ij} = \mu + \alpha_i + e_{ij}$ and corresponds to a $\text{GLM}_{N,3}\text{LTFR}(y_i; \boldsymbol{X}_{i,2}\boldsymbol{\beta}, \sigma^2)$ with Gaussian errors. In the setting described,

$$\begin{aligned} y_{ig} &= \text{observation } i \text{ subject to treatment } g \\ \mu_g &= \text{E}(y_{ig}) \\ \mu &= \text{an ``overall mean''} \\ \alpha_g &= \text{``the effect of treatment } g\text{''} \\ e_{ig} &\sim \mathcal{N}_1(0, \sigma^2) \text{ i.i.d.} \end{aligned}$$

Also

$$\begin{array}{ccccc} \boldsymbol{y} & = & \boldsymbol{X}_1\boldsymbol{\beta}_1 & + & \boldsymbol{e} \\ \begin{bmatrix} y_{11} \\ y_{12} \\ y_{21} \end{bmatrix} & = & \begin{bmatrix} 1 & 0 \\ 1 & 0 \\ 0 & 1 \end{bmatrix} \begin{bmatrix} \mu_1 \\ \mu_2 \end{bmatrix} & + & \begin{bmatrix} e_{11} \\ e_{12} \\ e_{21} \end{bmatrix} \end{array} \tag{11.14}$$

and

$$\begin{array}{ccccc} \boldsymbol{y} & = & \boldsymbol{X}_2\boldsymbol{\beta}_2 & + & \boldsymbol{e} \\ \begin{bmatrix} y_{11} \\ y_{12} \\ y_{21} \end{bmatrix} & = & \begin{bmatrix} 1 & 1 & 0 \\ 1 & 1 & 0 \\ 1 & 0 & 1 \end{bmatrix} \begin{bmatrix} \mu \\ \alpha_1 \\ \alpha_2 \end{bmatrix} & + & \begin{bmatrix} e_{11} \\ e_{12} \\ e_{21} \end{bmatrix} \end{array} . \tag{11.15}$$

It is easily verified that

$$\widehat{\boldsymbol{\beta}}_1 = \begin{bmatrix} \overline{y}_{1.} \\ y_{21} \end{bmatrix} = \begin{bmatrix} \widehat{\mu}_1 \\ \widehat{\mu}_2 \end{bmatrix} \tag{11.16}$$

is unbiased, $\text{E}(\widehat{\boldsymbol{\beta}}_1) = \boldsymbol{\beta}_1$. Furthermore any linear combination of μ_1 and μ_2, say $\boldsymbol{C}\boldsymbol{\beta}_1 = c_1\mu_1 + c_2\mu_2$, is estimable since $\boldsymbol{C}\widehat{\boldsymbol{\beta}}_1$ is a linear unbiased estimator of $\boldsymbol{C}\boldsymbol{\beta}_1$.

In contrast, $\boldsymbol{\beta}_2$ is not estimable (no linear unbiased estimator of $\boldsymbol{\beta}_2$ exits), and element j of $\boldsymbol{\beta}_2$ is not estimable. In particular, no individual linear estimators with expectation equal to μ, or to α_1, or to α_2 exist. They are not individually estimable, even as secondary parameters ($\theta = \boldsymbol{c}_j\boldsymbol{\beta}_2 = \beta_{j,2}$, with $\boldsymbol{c}_j$ $1 \times q$ with all zero elements except for a 1 at position j).

For model 2 the normal equations are

$$(\boldsymbol{X}_2'\boldsymbol{X}_2)\widehat{\boldsymbol{\beta}}_2 = \boldsymbol{X}_2'\boldsymbol{y}$$

$$= \begin{bmatrix} 3 & 2 & 1 \\ 2 & 2 & 0 \\ 1 & 0 & 1 \end{bmatrix} \begin{bmatrix} \widehat{\mu} \\ \widehat{\alpha}_1 \\ \widehat{\alpha}_2 \end{bmatrix}$$

$$= \begin{bmatrix} 1 & 1 & 1 \\ 1 & 1 & 0 \\ 0 & 0 & 1 \end{bmatrix} \boldsymbol{y}. \tag{11.17}$$

Subtracting multiples of the third row from the first two rows gives the equivalent equations

$$\begin{bmatrix} 0 & 2 & -2 \\ 0 & 2 & -2 \\ 1 & 0 & 1 \end{bmatrix} \begin{bmatrix} \widehat{\mu} \\ \widehat{\alpha}_1 \\ \widehat{\alpha}_2 \end{bmatrix} = \begin{bmatrix} 1 & 1 & -2 \\ 1 & 1 & -2 \\ 0 & 0 & 1 \end{bmatrix} \boldsymbol{y}. \tag{11.18}$$

Since two of the equations are identical, we have a system of two equations in three unknowns. Infinitely many solutions exist for such a system. If $\widehat{\boldsymbol{\beta}}_2$ is a solution, then so is $\widetilde{\boldsymbol{\beta}}_2 = \widehat{\boldsymbol{\beta}}_2 + \boldsymbol{z}$ for any $\boldsymbol{z} = [-z_0 \; z_0 \; z_0]'$, because

$$\boldsymbol{X}_2\boldsymbol{z} = \begin{bmatrix} 1 & 1 & 0 \\ 1 & 1 & 0 \\ 1 & 0 & 1 \end{bmatrix} \begin{bmatrix} -z_0 \\ z_0 \\ z_0 \end{bmatrix} = \boldsymbol{0}. \tag{11.19}$$

Particular solutions are obtained by placing an additional linear restriction on $\boldsymbol{\beta}_2$, specifically $\boldsymbol{r}'\boldsymbol{\beta}_2 = a$, to supply a third equation. Requiring $\mu = 0$ corresponds to $\boldsymbol{r}' = [1\,0\,0]$ and $a = 0$. The choice implies three equations in three unknowns,

$$\begin{bmatrix} 1 & 0 & 0 \\ 0 & 2 & -2 \\ 1 & 0 & 1 \end{bmatrix} \begin{bmatrix} \widehat{\mu} \\ \widehat{\alpha}_1 \\ \widehat{\alpha}_2 \end{bmatrix} = \begin{bmatrix} 0 & 0 & 0 \\ 1 & 1 & -2 \\ 0 & 0 & 1 \end{bmatrix} \boldsymbol{y}. \tag{11.20}$$

Further manipulation yields the solution

$$\widehat{\boldsymbol{\beta}}_2 = \begin{bmatrix} \widehat{\mu} \\ \widehat{\alpha}_1 \\ \widehat{\alpha}_2 \end{bmatrix} = \begin{bmatrix} 0 \\ (y_{11} + y_{12})/2 \\ y_{21} \end{bmatrix}. \tag{11.21}$$

It is interesting to look at the expected values of the estimators. They are

$$\mathrm{E}\left(\widehat{\boldsymbol{\beta}}_2\right) = \mathrm{E}\left(\begin{bmatrix} \widehat{\mu} \\ \widehat{\alpha}_1 \\ \widehat{\alpha}_2 \end{bmatrix}\right) = \begin{bmatrix} 0 \\ \mu + \alpha_1 \\ \mu + \alpha_2 \end{bmatrix} \neq \begin{bmatrix} \mu \\ \alpha_1 \\ \alpha_2 \end{bmatrix} = \boldsymbol{\beta}_2. \tag{11.22}$$

We can also examine the expected value of all other solutions since they must be of the form $\widetilde{\boldsymbol{\beta}}_2 = \widehat{\boldsymbol{\beta}}_2 + \boldsymbol{z}$. No known quantity z_0 exists such that

$$\mathrm{E}(\widehat{\boldsymbol{\beta}}_2 + \boldsymbol{z}) = \begin{bmatrix} 0 \\ \mu + \alpha_1 \\ \mu + \alpha_2 \end{bmatrix} + \mathrm{E}\left(\begin{bmatrix} -z_0 \\ z_0 \\ z_0 \end{bmatrix}\right) = \begin{bmatrix} \mu \\ \alpha_1 \\ \alpha_2 \end{bmatrix} = \boldsymbol{\beta}_2 . \tag{11.23}$$

Regardless of model specifications, $\mu_{ij} = \mathrm{E}(y_{ij})$ is always estimable, which is a trivial result since y_{ij} is a linear unbiased estimator of $\mu_{ij} = \mathrm{E}(y_{ij})$. In the example, the fundamental parameters are $\mathrm{E}(y_{11}) = \mu + \alpha_1 = \mathrm{E}(y_{12})$ and $\mathrm{E}(y_{21}) = \mu + \alpha_2$. Since the parameters are estimable, linear combinations of them are also estimable, including $\alpha_1 - \alpha_2 = (\mu + \alpha_1) - (\mu + \alpha_2)$. However, no linear combination of the fundamental estimable parameter has expectation equal to μ, or to α_1, or to α_2. They are not estimable, even as secondary parameters. The proofs of the assertions are recommended as exercises.

In spite of the estimability problems, LTFR models are popular and can be useful in some circumstances. If one restricts interest to estimable secondary parameters, such parameters may be estimated equally well via either FR or LTFR models (aside from certain computational problems for LTFR models). The FR models have an advantage when defining a secondary parameter because one need not be concerned with whether the parameter is estimable.

Since it is embarrassing to present an estimate of a nonestimable parameter, when using LTFR models one must check each secondary parameter for estimability, which can be a nuisance. Verifying the condition $\boldsymbol{C} = \boldsymbol{C}(\boldsymbol{X}'\boldsymbol{X})^{-}(\boldsymbol{X}'\boldsymbol{X})$ seems the simplest.

The following theorem formally justifies taking advantage of the convenience of working with a linearly equivalent model. Analyzing a full-rank model which is linearly equivalent to a less-than-full-rank model avoids the need to check for estimability and loses no information available in the original model.

Theorem 11.7 Any primary or secondary expected-value parameter estimable in $\mathrm{GLM}_{N,q}(y_i; \boldsymbol{X}_i\boldsymbol{\beta}, \sigma^2)$ is also estimable in a linearly equivalent model.

Proof. Considering estimable $\boldsymbol{\theta} = \boldsymbol{C}\boldsymbol{\beta}$ first allows treating $\boldsymbol{\beta}$ as a special case. Direct comparison of the definitions of estimable and linearly equivalent provides the basis of the result. Details are left as an exercise.

11.5 CODING SCHEMES AND THE ESSENCE MATRIX

Even simple design matrices can be coded in a variety of ways. The choice affects the definitions and estimability of expected-value parameters $\boldsymbol{\beta}$ and $\boldsymbol{\theta} = \boldsymbol{C}\boldsymbol{\beta}$. Readers with limited knowledge of coding schemes will find it profitable to consult Chapters 12–16 in Muller and Fetterman (2002). They provided extensive guidance concerning practical applications in ANOVA and regression. They also explicitly describe equivalencies among models using such coding schemes in one- and two-way designs.

In the univariate linear model $\text{GLM}_{N,q}(y_i; \boldsymbol{X}_i\boldsymbol{\beta}, \sigma^2)$, with corresponding model equation for all observations $\boldsymbol{y} = \boldsymbol{X}\boldsymbol{\beta} + \boldsymbol{e}$, the design matrix $\boldsymbol{X}$ represents between-subject information (distinctions among independent sampling units, ISUs). Most often, data analysts use one of six coding schemes to code distinctions among G groups of observations: reference cell (regression), cell mean, effect, classical ANOVA, natural polynomial, and orthogonal polynomial.

Example 11.2 A $G = 3$ group ANOVA design, with N_g independent observations in group g, may be written with reference cell coding as

$$\begin{aligned}\boldsymbol{y} &= \boldsymbol{X}_1\boldsymbol{\beta} + \boldsymbol{e} \\ &= \begin{bmatrix} \mathbf{1} & \mathbf{0} & \mathbf{0} \\ \mathbf{1} & \mathbf{1} & \mathbf{0} \\ \mathbf{1} & \mathbf{0} & \mathbf{1} \end{bmatrix} \boldsymbol{\beta} + \boldsymbol{e} \\ &= \begin{bmatrix} \mathbf{1}_{N_1} \otimes [1 \;\; 0 \;\; 0] \\ \mathbf{1}_{N_2} \otimes [1 \;\; 1 \;\; 0] \\ \mathbf{1}_{N_3} \otimes [1 \;\; 0 \;\; 1] \end{bmatrix} \boldsymbol{\beta} + \boldsymbol{e}. \end{aligned} \tag{11.24}$$

If $N_1 = N_2 = N_3$ (a balanced design), then $N_g = N/G$ and

$$\boldsymbol{X}_1 = \mathbf{1}_{N_g} \otimes \begin{bmatrix} 1 & 0 & 0 \\ 1 & 1 & 0 \\ 1 & 0 & 1 \end{bmatrix} = \mathbf{1}_{N_g} \otimes \boldsymbol{X}_{E,1}. \tag{11.25}$$

Cell mean coding gives

$$\begin{aligned}\boldsymbol{y} &= \boldsymbol{X}_2\boldsymbol{\beta} + \boldsymbol{e} \\ &= \begin{bmatrix} \mathbf{1} & \mathbf{0} & \mathbf{0} \\ \mathbf{0} & \mathbf{1} & \mathbf{0} \\ \mathbf{0} & \mathbf{0} & \mathbf{1} \end{bmatrix} \boldsymbol{\beta} + \boldsymbol{e} \\ &= \begin{bmatrix} \mathbf{1}_{N_1} \otimes [1 \;\; 0 \;\; 0] \\ \mathbf{1}_{N_2} \otimes [0 \;\; 1 \;\; 0] \\ \mathbf{1}_{N_3} \otimes [0 \;\; 0 \;\; 1] \end{bmatrix} \boldsymbol{\beta} + \boldsymbol{e}. \end{aligned} \tag{11.26}$$

If the design is balanced, then

$$\boldsymbol{X}_2 = \mathbf{1}_{N_g} \otimes \begin{bmatrix} 1 & 0 & 0 \\ 0 & 1 & 0 \\ 0 & 0 & 1 \end{bmatrix} = \mathbf{1}_{N_g} \otimes \boldsymbol{X}_{E,2}. \tag{11.27}$$

Obviously $\boldsymbol{X}_{E,2} = \boldsymbol{I}_3$.

Definition 11.5 The *essence matrix* (Helms, 1988a, b) simplifies any discussion of coding schemes. With N observations and q predictors, an $N \times q$ design matrix, $\boldsymbol{X}$, has $G \times q$ essence matrix $\text{Es}(\boldsymbol{X})$, which contains one and only one copy of each unique row of $\boldsymbol{X}$.

With G rows in $\text{Es}(\boldsymbol{X})$ each row $\boldsymbol{X}_g = \text{row}_g[\text{Es}(\boldsymbol{X})]$ identifies one of G distinct groups of sampling units. Observations which share the same $\text{row}_i(\boldsymbol{X})$ and hence the same $\boldsymbol{X}_g$ are described as being in the same cell, which corresponds to the same treatment combination for an experiment. With N_g observations in group g and an unbalanced design,

$$\boldsymbol{X} = \begin{bmatrix} \boldsymbol{X}_1 \otimes \mathbf{1}_{N_1} \\ \boldsymbol{X}_2 \otimes \mathbf{1}_{N_2} \\ \vdots \\ \boldsymbol{X}_G \otimes \mathbf{1}_{N_G} \end{bmatrix} . \tag{11.28}$$

A balanced design has an equal number of observations in each cell so $N_g \equiv N/G$ and $\boldsymbol{X} = \text{Es}(\boldsymbol{X}) \otimes \mathbf{1}_{N/G}$.

For a one-way ANOVA design, cell-mean coding has $\text{Es}(\boldsymbol{X}) = \boldsymbol{I}_G$. Reference-cell, effect, and polynomial coding schemes also have $G \times G$ and full-rank $\text{Es}(\boldsymbol{X})$ for $N \times G$ and full rank $\boldsymbol{X}$. Classical ANOVA coding has less-than-full-rank $\text{Es}(\boldsymbol{X}) = [\,\mathbf{1}_G\ \boldsymbol{I}_G\,]$, while deleting any of the last G columns creates a reference-cell coding.

Questions of parameter definition and estimability can be answered in terms of the essence matrix. The easily proven fact that $\text{rank}(\boldsymbol{X}) = [\text{Es}(\boldsymbol{X})]$ gives some hint of value in considering the essence matrix. Furthermore, linear equivalence of two models can be assessed conveniently in terms of essence matrices.

11.6 UNRESTRICTED MAXIMUM LIKELIHOOD ESTIMATION OF $\boldsymbol{\beta}$

The following two lemmas support the derivation of the likelihood estimators.

Lemma 11.2 If $f(\boldsymbol{x})$ is a positive valued function of $\boldsymbol{x}$ on $\Re^p$, then $\boldsymbol{x}_0$ is the location of a local maximum of $f(\boldsymbol{x})$ if and only if $\boldsymbol{x}_0$ is the location of a local maximum of $\log[f(\boldsymbol{x})]$.

Proof. ($\Rightarrow$) Proving $f(\boldsymbol{x}_0)$ is a local maximum is equivalent to proving $\exists\delta > 0$ such that $\| \boldsymbol{x} - \boldsymbol{x}_0 \| < \delta \Rightarrow f(\boldsymbol{x}) \leq f(\boldsymbol{x}_0)$. The monotonicity of the log() function allows writing $f(\boldsymbol{x}) \leq f(\boldsymbol{x}_0) \Rightarrow \log[f(\boldsymbol{x})] \leq \log[f(\boldsymbol{x}_0)]$. Therefore we know $\| \boldsymbol{x} - \boldsymbol{x}_0 \| < \delta \Rightarrow \log[f(\boldsymbol{x})] \leq \log[f(\boldsymbol{x}_0)]$.

($\Leftarrow$) Proving $\log[f(\boldsymbol{x}_0)]$ is a local maximum is equivalent to proving $\exists\delta > 0$ such that $\| \boldsymbol{x} - \boldsymbol{x}_0 \| < \delta \Rightarrow \log[f(\boldsymbol{x})] \leq \log[f(\boldsymbol{x}_0)] \Rightarrow f(\boldsymbol{x}) \leq f(\boldsymbol{x}_0)$ due to the monotonicity of exp(). □

Lemma 11.3 For $\text{GLM}_{N,q}(y_i; \boldsymbol{X}_i\boldsymbol{\beta}, \sigma^2)$, the system of equations $(\boldsymbol{X}'\boldsymbol{X})\boldsymbol{\beta} = \boldsymbol{X}'\boldsymbol{y}$ is consistent (Definition 1.34) when solving for unknown $\boldsymbol{\beta}$ in terms of known $\boldsymbol{X}$ and $\boldsymbol{y}$.

Proof. $\boldsymbol{a}'(\boldsymbol{X}'\boldsymbol{X}) = \mathbf{0} \Rightarrow \boldsymbol{a}'(\boldsymbol{X}'\boldsymbol{X})\boldsymbol{a} = 0 \Rightarrow \boldsymbol{a}'\boldsymbol{X}' = \mathbf{0} \Rightarrow \boldsymbol{a}'\boldsymbol{X}'\boldsymbol{y} = 0.$ □

Any model with less-than-full-rank design matrix disallows finding a unique estimate for $\boldsymbol{\beta}$. Infinitely many different estimates provide a valid solution, which could each be termed a supremum likelihood estimate. However, even when a unique estimate for $\boldsymbol{\beta}$ cannot be found, a unique estimate of σ^2 will still exist. We follow convention in describing the estimates for $\{\boldsymbol{\beta}, \sigma^2\}$ as the maximum likelihood estimates, even with a less-than-full-rank design.

Theorem 11.8 For $\text{GLM}_{N,q}(y_i; \boldsymbol{X}_i\boldsymbol{\beta}, \sigma^2)$ with Gaussian errors and $r = \text{rank}(\boldsymbol{X}) \leq q$, the joint supremum (for $\boldsymbol{\beta}$ if $r < q$) or maximum (if $r = q$ for $\boldsymbol{\beta}$; always for σ^2) likelihood estimators of $\boldsymbol{\beta}$ and σ^2 are

$$\widetilde{\boldsymbol{\beta}} = (\boldsymbol{X}'\boldsymbol{X})^{-}\boldsymbol{X}'\boldsymbol{y} \tag{11.29}$$

$$\widetilde{\sigma}^2 = (\boldsymbol{y} - \boldsymbol{X}\widetilde{\boldsymbol{\beta}})'(\boldsymbol{y} - \boldsymbol{X}\widetilde{\boldsymbol{\beta}})/N = \boldsymbol{y}'[\boldsymbol{I}_N - \boldsymbol{X}(\boldsymbol{X}'\boldsymbol{X})^{-}\boldsymbol{X}']\boldsymbol{y}/N. \tag{11.30}$$

Here $\widetilde{\boldsymbol{\beta}}$ is any solution of $(\boldsymbol{X}'\boldsymbol{X})\widetilde{\boldsymbol{\beta}} = \boldsymbol{X}'\boldsymbol{y}$, with infinitely many for $r < q$, and one unique solution if $r = q$ and $(\boldsymbol{X}'\boldsymbol{X})^{-} = (\boldsymbol{X}'\boldsymbol{X})^{-1}$. The value of $\widetilde{\sigma}^2$ is invariant to the choice of $(\boldsymbol{X}'\boldsymbol{X})^{-}$. It is customary to use $\widehat{\sigma}^2 = \widetilde{\sigma}^2 N/(N-r)$, which is unbiased. If $\boldsymbol{X}$ $(N \times q)$ is full rank, then $(\boldsymbol{X}'\boldsymbol{X})^{-} = (\boldsymbol{X}'\boldsymbol{X})^{-1}$, and $\widetilde{\boldsymbol{\beta}} = \widehat{\boldsymbol{\beta}} = (\boldsymbol{X}'\boldsymbol{X})^{-1}\boldsymbol{X}'\boldsymbol{y}$ is unique and unbiased.

Proof. The density function of $\boldsymbol{y}$ $(N \times 1)$ with $\boldsymbol{\mu} = \boldsymbol{X}\boldsymbol{\beta}$, $\boldsymbol{\Sigma} = \sigma^2\boldsymbol{I}$ is

$$f_y(\boldsymbol{y}_*) = (2\pi)^{-N/2}|\boldsymbol{\Sigma}|^{-1/2}\exp\left[-(\boldsymbol{y}_* - \boldsymbol{\mu})'\boldsymbol{\Sigma}^{-1}(\boldsymbol{y}_* - \boldsymbol{\mu})/2\right]. \tag{11.31}$$

We actually maximize the log likelihood (using Lemma 11.2), which is

$$\begin{aligned} \log L(\boldsymbol{y}_*; \boldsymbol{\beta}, \sigma^2) &= \log[f_y(\boldsymbol{y}_*)] \\ &= -\frac{N}{2}\log(2\pi) - \frac{N}{2}\log(\sigma^2) - \frac{1}{2}(\boldsymbol{y}_* - \boldsymbol{X}\boldsymbol{\beta})'(\boldsymbol{y}_* - \boldsymbol{X}\boldsymbol{\beta})/\sigma^2 \\ &= c - \frac{N}{2}\log(\sigma^2) - (\boldsymbol{y}_*'\boldsymbol{y}_* - 2\boldsymbol{y}_*'\boldsymbol{X}\boldsymbol{\beta} + \boldsymbol{\beta}'\boldsymbol{X}'\boldsymbol{X}\boldsymbol{\beta})/(2\sigma^2). \end{aligned} \tag{11.32}$$

Step 1. Critical points are found by finding zeros of partial derivatives of $\log L$ with respect to $\boldsymbol{\tau} = [\boldsymbol{\beta}' \;\; \sigma^2]'$. Here $\partial \log L/\partial\boldsymbol{\beta} = (2\sigma^2)^{-1}[2\boldsymbol{X}'\boldsymbol{y} - 2(\boldsymbol{X}'\boldsymbol{X})\boldsymbol{\beta}]$ is $q \times 1$. Evaluating $\boldsymbol{0} = \partial\log L/\partial\boldsymbol{\beta}$ at $\boldsymbol{\beta} = \widetilde{\boldsymbol{\beta}}$ implies $(\boldsymbol{X}'\boldsymbol{X})\widetilde{\boldsymbol{\beta}} = \boldsymbol{X}'\boldsymbol{y}$. Equivalently, $\widetilde{\boldsymbol{\beta}} = (\boldsymbol{X}'\boldsymbol{X})^{-}\boldsymbol{X}'\boldsymbol{y}_*$ since the equations are consistent by Lemma 1.17. Also,

$$\frac{\partial \log L}{\partial \sigma^2} = -\frac{N}{2\sigma^2} + \frac{1}{2(\sigma^2)^2}[(\boldsymbol{y}_* - \boldsymbol{X}\boldsymbol{\beta})'(\boldsymbol{y}_* - \boldsymbol{X}\boldsymbol{\beta})]. \tag{11.33}$$

Evaluating $0 = \partial\log L/\partial\sigma^2$ at $\sigma^2 = \widetilde{\sigma}^2$ implies $\widetilde{\sigma}^2 = (\boldsymbol{y} - \boldsymbol{X}\widetilde{\boldsymbol{\beta}})'(\boldsymbol{y} - \boldsymbol{X}\widetilde{\boldsymbol{\beta}})/N$ or $\widetilde{\sigma}^2 = \boldsymbol{y}'(\boldsymbol{I} - \boldsymbol{X}(\boldsymbol{X}'\boldsymbol{X})^{-}\boldsymbol{X}')\boldsymbol{y}/N$. The estimator is invariant to the choice of generalized inverse by Theorem 1.15.

Step 2. Is the critical point a maximum, minimum, or saddle point? Deciding requires proving the Hessian matrix

$$\frac{\partial^2 \log L}{\partial \tau\, \partial \tau'} = \begin{bmatrix} \dfrac{\partial^2 \log L}{\partial \sigma^2\, \partial \sigma^2} & \dfrac{\partial^2 \log L}{\partial \sigma^2\, \partial \boldsymbol{\beta}'} \\ \dfrac{\partial^2 \log L}{\partial \boldsymbol{\beta}\, \partial \sigma^2} & \dfrac{\partial^2 \log L}{\partial \boldsymbol{\beta}\, \partial \boldsymbol{\beta}'} \end{bmatrix} \tag{11.34}$$

is negative definite if $(\boldsymbol{X}'\boldsymbol{X})^{-1}$exists and negative semidefinite otherwise. Further details are left as an exercise. □

Lemma 11.4 The solution to the equations $(\boldsymbol{X}'\boldsymbol{X})\boldsymbol{\beta} = \boldsymbol{X}'\boldsymbol{y}$ gives predicted values orthogonal to the residuals, which leads to referring to $(\boldsymbol{X}'\boldsymbol{X})\boldsymbol{\beta} = \boldsymbol{X}'\boldsymbol{y}$ as the "normal" equations.

Proof.

$$\begin{aligned} \widehat{\boldsymbol{y}}'\widehat{\boldsymbol{e}} &= [\boldsymbol{X}(\boldsymbol{X}'\boldsymbol{X})^{-}\boldsymbol{X}'\boldsymbol{y}]'[\boldsymbol{y} - \boldsymbol{X}(\boldsymbol{X}'\boldsymbol{X})^{-}\boldsymbol{X}'\boldsymbol{y}] \\ &= \boldsymbol{y}'\boldsymbol{X}(\boldsymbol{X}'\boldsymbol{X})^{-}\boldsymbol{X}'[\boldsymbol{I} - \boldsymbol{X}(\boldsymbol{X}'\boldsymbol{X})^{-}\boldsymbol{X}']\boldsymbol{y} \\ &= \boldsymbol{y}'\boldsymbol{H}(\boldsymbol{I} - \boldsymbol{H})\boldsymbol{y} \\ &= \boldsymbol{y}'(\boldsymbol{H} - \boldsymbol{H}^2)\boldsymbol{y} \\ &= 0\,. \end{aligned} \tag{11.35}$$

□

The adjective "normal" in the lemma refers to a perpendicular property and applies whether or not the data are Gaussian, as does the following lemma.

Lemma 11.5 (a) Predicted values and residuals from *any* $\mathrm{GLM}_{N,q}(y_i; \boldsymbol{X}_i\boldsymbol{\beta}, \sigma^2)$ have zero covariance and zero correlation.
(b) With Gaussian errors the predicted values are statistically independent of the residuals.

Proof.

$$\begin{aligned} \mathcal{V}(\widehat{\boldsymbol{y}}, \widehat{\boldsymbol{e}}) &= \mathcal{V}[\boldsymbol{H}\boldsymbol{y}, (\boldsymbol{I} - \boldsymbol{H})\boldsymbol{y}] \\ &= \mathrm{E}\{(\boldsymbol{H}\boldsymbol{y})[(\boldsymbol{I} - \boldsymbol{H})\boldsymbol{y}]'\} - \mathrm{E}(\boldsymbol{H}\boldsymbol{y})\mathrm{E}\{[(\boldsymbol{I} - \boldsymbol{H})\boldsymbol{y}]'\} \\ &= \boldsymbol{H}[\mathrm{E}(\boldsymbol{y}\boldsymbol{y}')](\boldsymbol{I} - \boldsymbol{H}) - \boldsymbol{H}\mathrm{E}(\boldsymbol{y})\mathrm{E}(\boldsymbol{y}')(\boldsymbol{I} - \boldsymbol{H}) \\ &= \boldsymbol{H}[\mathrm{E}(\boldsymbol{y}\boldsymbol{y}') - \mathrm{E}(\boldsymbol{y})\mathrm{E}(\boldsymbol{y}')](\boldsymbol{I} - \boldsymbol{H}) \\ &= \boldsymbol{H}\mathcal{V}(\boldsymbol{y})(\boldsymbol{I} - \boldsymbol{H}) \\ &= \boldsymbol{H}\sigma^2\boldsymbol{I}_N(\boldsymbol{I} - \boldsymbol{H}) = \boldsymbol{0}_{N\times N}\,. \end{aligned} \tag{11.36}$$

Jointly expressing predicted values and residuals as a linear transformation of a vector Gaussian implies they are jointly vector Gaussian:

$$\begin{bmatrix} \widehat{\boldsymbol{y}} \\ \widehat{\boldsymbol{e}} \end{bmatrix} = \begin{bmatrix} \boldsymbol{H} \\ \boldsymbol{I}-\boldsymbol{H} \end{bmatrix} \boldsymbol{y} \sim \mathcal{SN}_{2N}\left\{ \begin{bmatrix} \boldsymbol{X}\boldsymbol{\beta} \\ 0 \end{bmatrix}, \sigma^2 \begin{bmatrix} \boldsymbol{H} & 0 \\ 0 & \boldsymbol{I}-\boldsymbol{H} \end{bmatrix} \right\}. \tag{11.37}$$

Zero covariance among Gaussian vectors implies statistical independence. □

Lemma 11.6 (a) The $\boldsymbol{\beta}$ estimator and the residuals from *any* $\text{GLM}_{N,q}(y_i; \boldsymbol{X}_i\boldsymbol{\beta}, \sigma^2)$ have zero covariance and zero correlation.
(b) With Gaussian errors the $\boldsymbol{\beta}$ estimator and the residuals are statistically independent of each other.

Proof. For $\boldsymbol{H} = \boldsymbol{X}(\boldsymbol{X}'\boldsymbol{X})^-\boldsymbol{X}'$, Theorem 1.15 gives $[(\boldsymbol{X}'\boldsymbol{X})^-\boldsymbol{X}']\boldsymbol{H} = (\boldsymbol{X}'\boldsymbol{X})^-\boldsymbol{X}'$, which implies $[(\boldsymbol{X}'\boldsymbol{X})^-\boldsymbol{X}'](\boldsymbol{I}_N - \boldsymbol{H}) = \boldsymbol{0}$. Proceeding as in the proof of the last lemma gives

$$\begin{bmatrix} \widetilde{\boldsymbol{\beta}} \\ \widehat{\boldsymbol{e}} \end{bmatrix} = \begin{bmatrix} (\boldsymbol{X}'\boldsymbol{X})^-\boldsymbol{X}' \\ (\boldsymbol{I}_N - \boldsymbol{H}) \end{bmatrix} \boldsymbol{y} \tag{11.38}$$

$$\sim \mathcal{SN}_{q+N}\left\{ \begin{bmatrix} (\boldsymbol{X}'\boldsymbol{X})^-\boldsymbol{X}'\boldsymbol{X}\boldsymbol{\beta} \\ \boldsymbol{0} \end{bmatrix}, \sigma^2 \begin{bmatrix} (\boldsymbol{X}'\boldsymbol{X})^- & \boldsymbol{0} \\ \boldsymbol{0} & (\boldsymbol{I}_N - \boldsymbol{H}) \end{bmatrix} \right\}. \qquad \square$$

11.7 UNRESTRICTED BLUE ESTIMATION OF $\boldsymbol{\beta}$

Definition 11.6 If at least one function of a set of data provides a linear unbiased estimator of a parameter, one may be best in the sense of having minimum variance (among unbiased estimators). Such a *best linear unbiased estimator* (BLUE) is also known as a *(linear) uniformly minimum variance unbiased estimator* (UMVUE).

Theorem 11.9 For a $\text{GLM}_{N,q}\text{FR}(y_i; \boldsymbol{X}_i\boldsymbol{\beta}, \sigma^2)$, the class of linear unbiased estimators is $\mathbb{C}_{\text{LUE}} = \{\widetilde{\boldsymbol{\beta}} : \widetilde{\boldsymbol{\beta}} = \boldsymbol{A}\boldsymbol{y}, \boldsymbol{A}\ (q \times N), \text{E}(\widetilde{\boldsymbol{\beta}}) = \boldsymbol{\beta}\}$ for $\boldsymbol{A}$ constant. The BLUE can be identified in two ways. (1) The BLUE of $\boldsymbol{\beta}$ is $\widehat{\boldsymbol{\beta}} = \boldsymbol{A}_0\boldsymbol{y} \Leftrightarrow \mathcal{V}(\widetilde{\boldsymbol{\beta}}) - \mathcal{V}(\widehat{\boldsymbol{\beta}}) \equiv \mathcal{V}(\boldsymbol{A}\boldsymbol{y}) - \mathcal{V}(\boldsymbol{A}_0\boldsymbol{y})$ is *positive semidefinite* for all $\widetilde{\boldsymbol{\beta}} \in \mathbb{C}_{\text{LUE}}$. (2) For $\widetilde{\theta}_C = \boldsymbol{C}\widetilde{\boldsymbol{\beta}}$ and $\widehat{\theta}_C = \boldsymbol{C}\widehat{\boldsymbol{\beta}}$ scalar $\mathcal{V}(\widetilde{\theta}_C) - \mathcal{V}(\widehat{\theta}_C) \equiv \mathcal{V}(\boldsymbol{C}\widetilde{\boldsymbol{\beta}}) - \mathcal{V}(\boldsymbol{C}\widehat{\boldsymbol{\beta}})$ is nonnegative for all $\boldsymbol{C}' \in \Re^q$ and $\widetilde{\boldsymbol{\beta}} \in \mathbb{C}_{\text{LUE}}$. Condition 1 holds $\Leftrightarrow$ condition 2 holds.

Proof. For any $\widetilde{\boldsymbol{\beta}}, \widehat{\boldsymbol{\beta}} \in \mathbb{C}_{\text{LUE}}$, the matrix $\mathcal{V}(\widetilde{\boldsymbol{\beta}}) - \mathcal{V}(\widehat{\boldsymbol{\beta}})$ is *positive semidefinite* if and only if the scalar $\mathcal{V}(\boldsymbol{C}\widetilde{\boldsymbol{\beta}}) - \mathcal{V}(\boldsymbol{C}\widehat{\boldsymbol{\beta}})$ is nonnegative $\forall \boldsymbol{C}' \in \Re^q$ because

$$\mathcal{V}(\boldsymbol{C}\widetilde{\boldsymbol{\beta}}) - \mathcal{V}(\boldsymbol{C}\widehat{\boldsymbol{\beta}}) = \boldsymbol{C}[\mathcal{V}(\widetilde{\boldsymbol{\beta}}) - \mathcal{V}(\widehat{\boldsymbol{\beta}})]\boldsymbol{C}'\,. \tag{11.39}$$

Matrix theory guarantees the matrix (the left side of the equation) is positive definite or positive semidefinite if and only if the quadratic form (the right side of the equation) is nonnegative $\forall \boldsymbol{C}$. $\square$

Theorem 11.10 For a $\text{GLM}_{N,q}\text{FR}(y_i; \boldsymbol{X}_i\boldsymbol{\beta}, \sigma^2)$ the unique BLUE of $\boldsymbol{\beta}$ is $\widehat{\boldsymbol{\beta}} = (\boldsymbol{X}'\boldsymbol{X})^{-1}\boldsymbol{X}'\boldsymbol{y}$. The result is known as the *Gauss-Markov theorem.*

Proof. If $\boldsymbol{A}_0 = (\boldsymbol{X}'\boldsymbol{X})^{-1}\boldsymbol{X}'$ $(q \times N)$, then $\widehat{\boldsymbol{\beta}} = \boldsymbol{A}_0\boldsymbol{y}$ is in the class of linear unbiased estimators, $\widehat{\boldsymbol{\beta}} \in \mathbb{C}_{\text{LUE}} = \{\widetilde{\boldsymbol{\beta}} : \widetilde{\boldsymbol{\beta}} = \boldsymbol{A}\boldsymbol{y} \text{ for some } \boldsymbol{A} \text{ and } \mathrm{E}(\widetilde{\boldsymbol{\beta}}) = \boldsymbol{\beta}\}$. An arbitrary LUE $\widetilde{\boldsymbol{\beta}} = \boldsymbol{A}\boldsymbol{y}$ has $\mathrm{E}(\widetilde{\boldsymbol{\beta}}) = \mathrm{E}(\boldsymbol{A}\boldsymbol{y}) = \boldsymbol{A}\boldsymbol{X}\boldsymbol{\beta} = \boldsymbol{\beta}$ for arbitrary $\boldsymbol{\beta}$. Here $\boldsymbol{A}\boldsymbol{X}\boldsymbol{\beta} = \boldsymbol{\beta}$ for all $\boldsymbol{\beta}$ if and only if $\boldsymbol{A}\boldsymbol{X} = \boldsymbol{I}$. The special property $\boldsymbol{A}$ $(q \times N)$ must have in relation to $\boldsymbol{X}$ $(N \times q)$ is used to establish the fact that the matrix

$$\mathcal{V}(\widetilde{\boldsymbol{\beta}}) - \mathcal{V}(\widehat{\boldsymbol{\beta}}) = \mathcal{V}(\widetilde{\boldsymbol{\beta}} - \widehat{\boldsymbol{\beta}}) \tag{11.40}$$

is a covariance matrix and therefore must be positive definite or positive semidefinite. It then follows $\widehat{\boldsymbol{\beta}}$ is best in $\mathbb{C}_{\text{LUE}}$. The details are as follows. Beginning with

$$\begin{aligned}\mathcal{V}(\widetilde{\boldsymbol{\beta}}) &= \mathcal{V}[(\widetilde{\boldsymbol{\beta}} - \widehat{\boldsymbol{\beta}}) + \widehat{\boldsymbol{\beta}}] \\ &= \mathcal{V}(\widetilde{\boldsymbol{\beta}} - \widehat{\boldsymbol{\beta}}) + \mathcal{V}(\widehat{\boldsymbol{\beta}}) + 2\mathcal{V}\left[\widehat{\boldsymbol{\beta}}, (\widetilde{\boldsymbol{\beta}} - \widehat{\boldsymbol{\beta}})\right]\end{aligned} \tag{11.41}$$

leads to examining the $(q \times q)$ matrix

$$\begin{aligned}\mathcal{V}[\widehat{\boldsymbol{\beta}}, (\widetilde{\boldsymbol{\beta}} - \widehat{\boldsymbol{\beta}})] &= \mathcal{V}[\boldsymbol{A}_0\boldsymbol{y}, (\boldsymbol{A} - \boldsymbol{A}_0)\boldsymbol{y}] \\ &= \boldsymbol{A}_0(\sigma^2\boldsymbol{I})(\boldsymbol{A} - \boldsymbol{A}_0)' \\ &= \sigma^2(\boldsymbol{X}'\boldsymbol{X})^{-1}\boldsymbol{X}'\left[\boldsymbol{A}' - \boldsymbol{X}(\boldsymbol{X}'\boldsymbol{X})^{-1}\right] \\ &= \sigma^2(\boldsymbol{X}'\boldsymbol{X})^{-1}\left[\boldsymbol{X}'\boldsymbol{A}' - \boldsymbol{X}'\boldsymbol{X}(\boldsymbol{X}'\boldsymbol{X})^{-1}\right] \\ &= \sigma^2(\boldsymbol{X}'\boldsymbol{X})^{-1}(\boldsymbol{I} - \boldsymbol{I}) = \boldsymbol{0}.\end{aligned} \tag{11.42}$$

In turn, $\mathcal{V}(\widetilde{\boldsymbol{\beta}}) - \mathcal{V}\left(\widehat{\boldsymbol{\beta}}\right) = \mathcal{V}(\widetilde{\boldsymbol{\beta}} - \widehat{\boldsymbol{\beta}}) + 2 \cdot \boldsymbol{0}$. Since the difference matrix is a covariance matrix, the difference matrix must be positive definite or positive semidefinite. Thus $\widehat{\boldsymbol{\beta}}$ is BLUE.

The vector $\widehat{\boldsymbol{\beta}}$ is also the unique BLUE, which can be proven by contradiction. Suppose $\widetilde{\boldsymbol{\beta}} = \boldsymbol{A}\boldsymbol{y}$ and $\widehat{\boldsymbol{\beta}} = \boldsymbol{A}_0\boldsymbol{y}$ are both BLUE with $\boldsymbol{A}_0 = (\boldsymbol{X}'\boldsymbol{X})^{-1}\boldsymbol{X}' \neq \boldsymbol{A}$. If both are BLUE, $[\mathcal{V}(\widetilde{\boldsymbol{\beta}}) - \mathcal{V}(\widehat{\boldsymbol{\beta}})]$ and $[\mathcal{V}(\widehat{\boldsymbol{\beta}}) - \mathcal{V}(\widetilde{\boldsymbol{\beta}})]$ are positive definite or positive semidefinite, which gives $[\mathcal{V}(\widetilde{\boldsymbol{\beta}}) - \mathcal{V}(\widehat{\boldsymbol{\beta}})] = \boldsymbol{0}$. Here $[\mathcal{V}(\widetilde{\boldsymbol{\beta}}) - \mathcal{V}(\widehat{\boldsymbol{\beta}})] = [\mathcal{V}(\widehat{\boldsymbol{\beta}}) - \mathcal{V}(\widetilde{\boldsymbol{\beta}})]$ implies $\mathcal{V}(\widetilde{\boldsymbol{\beta}} - \widehat{\boldsymbol{\beta}}) = \boldsymbol{0}$. In turn $\mathcal{V}[(\boldsymbol{A} - \boldsymbol{A}_0)\boldsymbol{y}] \equiv (\boldsymbol{A} - \boldsymbol{A}_0)\sigma^2\boldsymbol{I}(\boldsymbol{A} - \boldsymbol{A}_0)' = \boldsymbol{0}$ for all σ^2, which implies $(\boldsymbol{A} - \boldsymbol{A}_0) = \boldsymbol{0}$, which is a contradiction and implies the assumption $(\boldsymbol{A} \neq \boldsymbol{A}_0)$ is false. □

11.8 UNRESTRICTED LEAST SQUARES ESTIMATION OF $\boldsymbol{\beta}$

Here we consider building a model for a vector of observations, $\boldsymbol{y}$ $(N \times 1)$. Also $\widehat{\boldsymbol{y}} = \boldsymbol{t}(\boldsymbol{\beta})$ models $\boldsymbol{y}$, with $\boldsymbol{t}()$ a (vector-valued) transformation from $\boldsymbol{\beta} \in \Re^q$ to $\widehat{\boldsymbol{y}} \in \Re^N$. Here $\mathbf{B}$ indicates the parameter space of $\boldsymbol{\beta}$, with $\boldsymbol{\beta} \in \mathbf{B} \subset \Re^q$.

The choice of estimator may be restricted to a certain class of estimators $\mathbb{C}$ which satisfy certain restrictions or conditions, such as $\mathbb{C}_{\text{LUE}}$. Goodness of fit for such a model is assessed by examining the vector of estimated errors, the residuals $(\boldsymbol{y} - \widehat{\boldsymbol{y}}) = \boldsymbol{r}(\widetilde{\boldsymbol{\beta}}) = \widehat{\boldsymbol{e}}$ for $\widetilde{\boldsymbol{\beta}}$, an estimator of $\boldsymbol{\beta}$. The value of the residuals depend on the choice of $\widetilde{\boldsymbol{\beta}}$. The study of decision theory has led to many criteria for "small error," corresponding to many vector norms.

Definition 11.7 (a) The *squared error loss function* is $SSE(\widetilde{\boldsymbol{\beta}}) = \widetilde{\boldsymbol{e}}'\widetilde{\boldsymbol{e}} = \sum_{i=1}^{N} \widetilde{e}_i^2 = ||\widetilde{\boldsymbol{e}}||^2$.
(b) If $\widetilde{\boldsymbol{y}} = \boldsymbol{t}(\boldsymbol{\beta})$ is a model for $\boldsymbol{y}$, and $\widehat{\boldsymbol{\beta}} \in \mathbb{C}$ for $\mathbb{C}$ a class of estimators of $\boldsymbol{\beta} \in \mathbf{B}$, then $\widehat{\boldsymbol{\beta}}$ is a *least squares estimator* of $\boldsymbol{\beta}$ if and only if $SSE(\widehat{\boldsymbol{\beta}}) \leq SSE(\widetilde{\boldsymbol{\beta}})$ for all $\widetilde{\boldsymbol{\beta}} \in \mathbb{C}$.

Such an estimator may or may not be unique, biased, or linear.

The Gauss-Markov theorem for the full-rank GLM provides a unique linear least-squares estimator $\widehat{\boldsymbol{\beta}}$ a representation which is linear in parameters, $\boldsymbol{t}(\boldsymbol{\beta}) = \boldsymbol{X}\boldsymbol{\beta}$. For each observation, the parameter estimates imply a predicted value, as in $\widehat{y}_i = \widehat{\beta}_1 x_{i1} + \widehat{\beta}_2 x_{i2} + \widehat{\beta}_3 x_{i3}$, and error estimates, the residuals, $\widehat{e}_i = y_i - \widehat{y}_i$. In contrast, we may seek a least squares estimator in the context of an inherently nonlinear model such as $\widehat{y}_i = \widehat{\beta}_1 x_{i1}^{\widehat{\beta}_2} + \widehat{\beta}_3 x_{i2}$.

A less-than-full-rank GLM corresponds to an inconsistent system of equations without a unique solution. Each nonunique set of parameter estimates does imply predicted values and error estimates. The following theorem, and other results later in the chapter, characterize important properties of such models.

Definition 11.8 A $\text{GLM}_{N,q}\text{LTFR}(y_i; \boldsymbol{X}_i\boldsymbol{\beta}, \sigma^2)$ with $r = \text{rank}(\boldsymbol{X}) \leq q$, $\boldsymbol{\beta} \in \Re^q$, $\widetilde{\boldsymbol{\beta}} \in \mathbb{C}_{\text{LE}} \equiv \{\widetilde{\boldsymbol{\beta}} : \widetilde{\boldsymbol{\beta}} = \boldsymbol{A}\boldsymbol{y} \text{ for some } \boldsymbol{A}\ (q \times N)\}$, has $\widetilde{\boldsymbol{e}} = \boldsymbol{y} - \boldsymbol{X}\widetilde{\boldsymbol{\beta}}$ and $SSE(\widetilde{\boldsymbol{\beta}}) = \widetilde{\boldsymbol{e}}'\widetilde{\boldsymbol{e}} = (\boldsymbol{y} - \boldsymbol{X}\widetilde{\boldsymbol{\beta}})'(\boldsymbol{y} - \boldsymbol{X}\widetilde{\boldsymbol{\beta}})$. The vector $\widehat{\boldsymbol{\beta}} \in \mathbb{C}_{\text{LE}}$ is a (linear) least squares estimator of $\boldsymbol{\beta}$ if and only if $SSE(\widehat{\boldsymbol{\beta}}) \leq SSE(\widetilde{\boldsymbol{\beta}})$ $\forall \widetilde{\boldsymbol{\beta}} \in \mathbb{C}_{\text{LE}}$.

Theorem 11.11 The following hold for $\text{GLM}_{N,q}(y_i; \boldsymbol{X}_i\boldsymbol{\beta}, \sigma^2)$.
(a) Any solution $\widetilde{\boldsymbol{\beta}}$ of the normal equations $(\boldsymbol{X}'\boldsymbol{X})\widetilde{\boldsymbol{\beta}} = \boldsymbol{X}'\boldsymbol{y}$ is a least squares estimator for $\boldsymbol{\beta}$. The solutions are of the form $\widetilde{\boldsymbol{\beta}} = (\boldsymbol{X}'\boldsymbol{X})^{-}\boldsymbol{X}'\boldsymbol{y}$ with $SSE(\widetilde{\boldsymbol{\beta}}) = \boldsymbol{y}'[\boldsymbol{I} - \boldsymbol{X}(\boldsymbol{X}'\boldsymbol{X})^{-}\boldsymbol{X}']\boldsymbol{y}$, which is invariant to the choice of generalized inverse.
(b) If $\boldsymbol{X}$ $(N \times q)$ has rank q, then $(\boldsymbol{X}'\boldsymbol{X})^{-} = (\boldsymbol{X}'\boldsymbol{X})^{-1}$, and $\widetilde{\boldsymbol{\beta}} = \widehat{\boldsymbol{\beta}} = (\boldsymbol{X}'\boldsymbol{X})^{-1}\boldsymbol{X}'\boldsymbol{y}$ is unique and unbiased. Uniqueness of $\widehat{\boldsymbol{\beta}}$ ensures $SSE(\widehat{\boldsymbol{\beta}}) < SSE(\widetilde{\boldsymbol{\beta}})$ for all $\widetilde{\boldsymbol{\beta}}$ not identical to $\widehat{\boldsymbol{\beta}}$.

Proof. We want to minimize $SSE(\boldsymbol{b}) = (\boldsymbol{y} - \boldsymbol{X}\boldsymbol{b})'(\boldsymbol{y} - \boldsymbol{X}\boldsymbol{b}) = \boldsymbol{y}'\boldsymbol{y} - 2\boldsymbol{b}'\boldsymbol{X}'\boldsymbol{y} + \boldsymbol{b}'\boldsymbol{X}'\boldsymbol{X}\boldsymbol{b}$. Critical points are found by finding zeros of partial derivatives of *SSE* with respect to $\boldsymbol{b}$. The $q \times 1$ vector of derivatives is

$$\partial SSE/\partial \boldsymbol{b} = -2\boldsymbol{X}'\boldsymbol{y} + 2(\boldsymbol{X}'\boldsymbol{X})\boldsymbol{b}\,. \tag{11.43}$$

Evaluating $\partial SSE/\partial \boldsymbol{b} = \boldsymbol{0}$ at $\boldsymbol{b} = \widetilde{\boldsymbol{\beta}}$ implies $(\boldsymbol{X}'\boldsymbol{X})\widetilde{\boldsymbol{\beta}} = \boldsymbol{X}'\boldsymbol{y}$, which implies

$\widetilde{\beta} = (X'X)^{-}X'y$ because the equations are consistent. Is the critical point a minimum, maximum, or saddle point? The matrix of second derivatives,

$$\frac{\partial^{(2)} SSE}{\partial b\, \partial b'} = 2X'X, \tag{11.44}$$

is at least positive semidefinite. Having $X'X$ positive definite $\Rightarrow$ a minimum point; $X'X$ negative definite $\Rightarrow$ a maximum point; $X'X$ $\pm$ semidefinite $\Rightarrow$ a minimum, maximum, or saddle point; and $X'X$ indefinite $\Rightarrow$ a saddle point.

If X has full rank, then $\widetilde{\beta} = \widehat{\beta} = (X'X)^{-1}X'y$ corresponds to a minimum and is unique and unbiased.

For the LTFR case, it must be determined whether or not $\widetilde{\beta} = (X'X)^{-}X'y$ corresponds to a minimum. The result can be determined by evaluating *SSE* for other points in the neighborhood of $\widetilde{\beta}$, say $\widetilde{\beta} + h$, for h a vector of any length in any direction. Here $SSE(\widetilde{\beta} + h) = ||y - X(\widetilde{\beta} + h)||^2$ can be written as

$$(\widetilde{\beta} + h) = ||y - X\widetilde{\beta}||^2 + ||Xh||^2 + 2h'\left[(X'X)\widetilde{\beta} - X'y\right], \tag{11.45}$$

a Taylor series expansion. Thus $SSE(\widetilde{\beta} + h) = SSE(\widetilde{\beta}) + h'(X'X)h + 0$. Now, $X'X$ is positive definite $\Leftrightarrow$ $h'(X'X)h > 0\ \forall h$. Also, $X'X$ is positive semidefinite $\Leftrightarrow$ $h'(X'X)h \geq 0\ \forall h$ and $h'(X'X)h = 0$ for some h. Therefore, in the LTFR case *SSE* has a level valley which is minimal.

Invariance of $SSE(\widetilde{\beta})$ follows from uniqueness of $X(X'X)^{-}X'$ (Theorem 1.15). □

Theorem 11.12 For a $\text{GLM}_{N,q}\text{FR}(y_i; X_i\beta, \sigma^2)$ with Gaussian errors,

$$\widehat{\beta} = (X'X)^{-1}X'y, \tag{11.46}$$

$$\widehat{\sigma}^2 = \frac{1}{N-q} y'\left[I - X\left(X'X\right)^{-1}X'\right]y \tag{11.47}$$

are **(a)** unbiased, **(b)** consistent, **(c)** efficient, **(d)** complete, **(e)** sufficient, and **(f)** UMVUE. Furthermore **(g)** $\widehat{\beta}$ and $\widehat{\sigma}^2$ are mutually independent, **(h)** $\widehat{\beta} \sim N_q\left[\beta, \sigma^2(X'X)^{-1}\right]$, and **(i)** $\widehat{\sigma}^2(N-q)/\sigma^2 \sim \chi^2(N-q)$.

Proof. Proofs are left to the reader.

11.9 UNRESTRICTED MAXIMUM LIKELIHOOD ESTIMATION OF θ

Least squares estimation does not apply directly to estimation of secondary parameters θ except when MLE results apply to secondary parameter estimation. Even then, some additional details must be considered.

Theorem 11.13 For $\mathrm{GLM}_{N,q}(y_i; \boldsymbol{X}_i\boldsymbol{\beta}, \sigma^2)$ with Gaussian errors and $r = \mathrm{rank}(\boldsymbol{X}) \leq q$, $\boldsymbol{\theta} = \boldsymbol{C\beta}$, constant $\boldsymbol{C}$ $(q \times q)$ of rank q, the joint supremum (for $\boldsymbol{\theta}$ if $\boldsymbol{\theta}$ is not estimable) or maximum (for $\boldsymbol{\theta}$ if $\boldsymbol{\theta}$ is estimable; always for σ^2) likelihood estimators of $\boldsymbol{\theta}$ and σ^2 are

$$\widetilde{\boldsymbol{\theta}} = \boldsymbol{C}\widetilde{\boldsymbol{\beta}} \tag{11.48}$$

$$\widetilde{\sigma}^2 = (\boldsymbol{y} - \boldsymbol{X}\widetilde{\boldsymbol{\beta}})'(\boldsymbol{y} - \boldsymbol{X}\widetilde{\boldsymbol{\beta}})/N = \boldsymbol{y}'[\boldsymbol{I}_N - \boldsymbol{X}(\boldsymbol{X}'\boldsymbol{X})^{-}\boldsymbol{X}']\boldsymbol{y}/N\,. \tag{11.49}$$

Here $\widetilde{\boldsymbol{\beta}}$ is one of infinitely many solutions of $(\boldsymbol{X}'\boldsymbol{X})\widetilde{\boldsymbol{\beta}} = \boldsymbol{X}'\boldsymbol{y}$ for $r < q$ and the unique solution if $r = q$ and $(\boldsymbol{X}'\boldsymbol{X})^{-} = (\boldsymbol{X}'\boldsymbol{X})^{-1}$. Estimable $\boldsymbol{\theta}$ gives $\widetilde{\boldsymbol{\theta}}$ invariant to the choice of $(\boldsymbol{X}'\boldsymbol{X})^{-}$,while $\widetilde{\sigma}^2$ is always invariant to $(\boldsymbol{X}'\boldsymbol{X})^{-}$.

Proof. The invariance of $\widetilde{\boldsymbol{\theta}}$ for estimable $\boldsymbol{\theta}$ derives from two facts. First, estimability $\Rightarrow$ $\boldsymbol{C} = \boldsymbol{AX}$ for some $\boldsymbol{A}$. Second, $\boldsymbol{X}(\boldsymbol{X}'\boldsymbol{X})^{-}\boldsymbol{X}'$ is invariant to choice of generalized inverse (Theorem 1.15). Here $\widetilde{\boldsymbol{\theta}} = \widehat{\boldsymbol{\theta}} = \boldsymbol{C}\widetilde{\boldsymbol{\beta}} = \boldsymbol{AX}(\boldsymbol{X}'\boldsymbol{X})^{-}\boldsymbol{X}'\boldsymbol{y}$. For nonestimable $\boldsymbol{\theta}$, infinitely many MLEs exist.

It remains to be proven that MLE estimators are of the stated form, namely $\widetilde{\boldsymbol{\theta}} = \boldsymbol{C}\widetilde{\boldsymbol{\beta}}$ and $\widetilde{\boldsymbol{\beta}} = (\boldsymbol{X}'\boldsymbol{X})^{-}\boldsymbol{X}'\boldsymbol{y}$. Since matrix $\boldsymbol{C}$ is nonsingular and $q \times q$, $\mathrm{E}(\boldsymbol{y}) \equiv \boldsymbol{X\beta} = (\boldsymbol{XC}^{-1})(\boldsymbol{C\beta}) = \boldsymbol{Z\theta}$ with $(N \times q)$ $\boldsymbol{Z} = \boldsymbol{XC}^{-1}$ and $\boldsymbol{\theta} = \boldsymbol{C\beta}$. Thus a linearly equivalent model with $\boldsymbol{\theta}$ as the primary expected-value parameter is $\mathrm{GLM}_{N,q}(y_i; \boldsymbol{Z}_i\boldsymbol{\theta}, \sigma^2)$ with Gaussian errors and $r = \mathrm{rank}(\boldsymbol{Z})$. The problem is thus reduced to one already solved. □

Corollary 11.13 For a $\mathrm{GLM}_{N,q}(y_i; \boldsymbol{X}_i\boldsymbol{\beta}, \sigma^2)$ with Gaussian errors and $r = \mathrm{rank}(\boldsymbol{X}) \leq q$, secondary parameter $\boldsymbol{\theta}_i = \boldsymbol{C}_i\boldsymbol{\beta} \in \Re^{a_i}$ for $i \in \{1, 2, \ldots, t\}$ may be defined by $\boldsymbol{C}_i$ $(a_i \times q)$ such that $\mathrm{rank}(\boldsymbol{C}_i) = a_i \leq q = \sum_{i=1}^{t} a_i$ and

$$\boldsymbol{\theta} = \begin{bmatrix} \boldsymbol{\theta}_1 \\ \boldsymbol{\theta}_2 \\ \vdots \\ \boldsymbol{\theta}_t \end{bmatrix} = \begin{bmatrix} \boldsymbol{C}_1 \\ \boldsymbol{C}_2 \\ \vdots \\ \boldsymbol{C}_t \end{bmatrix} \boldsymbol{\beta} = \boldsymbol{C\beta}\,. \tag{11.50}$$

(a) A joint MLE of $\boldsymbol{\theta}_i$ $(a_i \times 1)$ is $\widetilde{\boldsymbol{\theta}}_i = \boldsymbol{C}_i\widetilde{\boldsymbol{\beta}}$, in which $\widetilde{\boldsymbol{\beta}} = (\boldsymbol{X}'\boldsymbol{X})^{-}\boldsymbol{X}'\boldsymbol{y}$.
(b) If $\boldsymbol{\theta}_i$ is estimable, then $\widetilde{\boldsymbol{\theta}}_i$ is invariant to the choice of $(\boldsymbol{X}'\boldsymbol{X})^{-}$.
(c) $\widetilde{\boldsymbol{\theta}}_i$ is invariant to the choice of $\boldsymbol{C}_{i'}$ $\forall i' \neq i$.

Proof. The results follow from the theorem since $\boldsymbol{C}$ $(q \times q)$ is nonsingular. The rows of the $\boldsymbol{C}_i$ are linearly independent, as are, collectively, the rows of $\boldsymbol{C}$. □

Although the $\boldsymbol{C}_i$ of interest may not span the estimation space, one can always find additional rows to create nonsingular $\boldsymbol{C}$ $(q \times q)$. It is then convenient to find the joint MLE of all the secondary parameters, $\{\boldsymbol{\theta}_1, \boldsymbol{\theta}_2, \ldots, \boldsymbol{\theta}_t\}$, and define the joint MLE of $\{\boldsymbol{\theta}_1, \boldsymbol{\theta}_2, \ldots, \boldsymbol{\theta}_k\}$, $k < t$, to be the same values as the MLE of all of $\boldsymbol{\theta}$ $(q \times 1)$. The result is invariant to the choice of the additional rows making $\boldsymbol{C}$ invertible.

If one attempts maximum likelihood estimation of $\boldsymbol{\theta}_i$, $i \in \{1, 2, \ldots, t-1\}$ (or fewer), then the problem is essentially indeterminate. The elements of $\boldsymbol{\theta}_t$ make an appearance as nuisance parameters, which makes the maximization depend on the unknown value of $\boldsymbol{\theta}_t$.

11.10 UNRESTRICTED BLUE ESTIMATION OF θ

Definition 11.9 (a) For $\text{GLM}_{N,q}(y_i; \boldsymbol{X}_i\boldsymbol{\beta}, \sigma^2)$ with $r = \text{rank}(\boldsymbol{X}) \leq q$, any $a \times 1$ estimable secondary parameter $\boldsymbol{\theta} = \boldsymbol{C\beta}$ has an associated class $\mathbb{C}_{\text{LUE}} = \{\widetilde{\boldsymbol{\theta}} : \widetilde{\boldsymbol{\theta}} = \boldsymbol{Ay} \text{ for some } \boldsymbol{A}\ (a \times N) \text{ and } \text{E}(\widetilde{\boldsymbol{\theta}}) = \boldsymbol{\theta}\}$ of linear unbiased estimators.
(b) One element of $\mathbb{C}_{\text{LUE}}$ may have minimum variance and therefore gives a BLUE.
(c) Such an estimator is also known as the (linear) *uniformly minimum variance unbiased estimator* (UMVUE).

Theorem 11.14 (a) The BLUE of $\boldsymbol{\theta}$ is $\widehat{\boldsymbol{\theta}} = \boldsymbol{A}_0\boldsymbol{y}$ such that $[\mathcal{V}((\widetilde{\boldsymbol{\theta}})) - \mathcal{V}(\widehat{\boldsymbol{\theta}})] \equiv [\mathcal{V}(\boldsymbol{Ay}) - \mathcal{V}(\boldsymbol{A}_0\boldsymbol{y})]$ is nonnegative definite $\forall \widetilde{\boldsymbol{\theta}} \in \mathbb{C}_{\text{LUE}}$.
(b) Equivalently, (scalar) $[\mathcal{V}(\boldsymbol{t}'\widetilde{\boldsymbol{\theta}}) - \mathcal{V}(\boldsymbol{t}'\widehat{\boldsymbol{\theta}})] \geq 0\ \forall \boldsymbol{t} \in \Re^a$ and $\widetilde{\boldsymbol{\theta}} \in \mathbb{C}_{\text{LUE}}$.

Proof. For any $\widetilde{\boldsymbol{\theta}} \in \mathbb{C}_{\text{LUE}}$ and any $\widehat{\boldsymbol{\theta}} \in \mathbb{C}_{\text{LUE}}$, the matrix $[\mathcal{V}(\widetilde{\boldsymbol{\theta}}) - \mathcal{V}(\widehat{\boldsymbol{\theta}})]$ is nonnegative definite $\Leftrightarrow$ the scalar $[\mathcal{V}(\boldsymbol{t}'\widetilde{\boldsymbol{\theta}}) - \mathcal{V}(\boldsymbol{t}'\widehat{\boldsymbol{\theta}})]$ is nonnegative $\forall \boldsymbol{t} \in \Re^a$ because $[\mathcal{V}(\boldsymbol{t}'\widetilde{\boldsymbol{\theta}}) - \mathcal{V}(\boldsymbol{t}'\widehat{\boldsymbol{\theta}})] = \boldsymbol{t}'[\mathcal{V}(\widetilde{\boldsymbol{\theta}}) - \mathcal{V}(\widehat{\boldsymbol{\theta}})]\boldsymbol{t}$. □

Example 11.3 With $2N$ i.i.d. observations $y_i \sim \mathcal{N}(\mu, \sigma^2)$, the estimators

$$\widetilde{\mu}_1 = \sum_{i=1}^{N} y_i / N \tag{11.51}$$

$$\widetilde{\mu}_2 = \sum_{i=N+1}^{2N} y_i / N \tag{11.52}$$

$$\widehat{\mu} = \sum_{i=1}^{2N} y_i / (2N) = (\widetilde{\mu}_1 + \widetilde{\mu}_2)/2 \tag{11.53}$$

are all unbiased. However, $\mathcal{V}(\widetilde{\mu}_1) = \mathcal{V}(\widetilde{\mu}_2) = 2\mathcal{V}(\widehat{\mu})$.

Corollary 11.14 For a $\text{GLM}_{N,q}(y_i; \boldsymbol{X}_i\boldsymbol{\beta}, \sigma^2)$ with $r = \text{rank}(\boldsymbol{X}) \leq q$, the BLUE of estimable $\boldsymbol{\theta} = \boldsymbol{C\beta}$ $(a \times 1)$ is $\widehat{\boldsymbol{\theta}}$ $\Leftrightarrow$ the BLUE of $\boldsymbol{t}'\boldsymbol{\theta}$ is $\boldsymbol{t}'\widehat{\boldsymbol{\theta}}$ $\forall \boldsymbol{t} \in \Re^a$.

Proof. The BLUE properties linearity, $\widehat{\boldsymbol{\theta}} = \boldsymbol{Ty} \Leftrightarrow \boldsymbol{t}'\widehat{\boldsymbol{\theta}} = \boldsymbol{t}'\boldsymbol{Ty}\ \forall \boldsymbol{t} \in \Re^a$, and unbiasedness give $\text{E}(\widehat{\boldsymbol{\theta}}) = \boldsymbol{\theta} \Leftrightarrow \boldsymbol{t}'\text{E}(\widehat{\boldsymbol{\theta}}) = \boldsymbol{t}'\boldsymbol{\theta}\ \forall \boldsymbol{t} \in \Re^a$. To prove the variance is a minimum, we proceed as follows. The matrix $\mathcal{V}(\widehat{\boldsymbol{\theta}})$ is minimal (as in BLUE version a) $\Leftrightarrow$ $\forall \boldsymbol{t} \in \Re^a$ scalar $\mathcal{V}(\boldsymbol{t}'\widehat{\boldsymbol{\theta}})$ is minimal. □

Theorem 11.15 For a $\text{GLM}_{N,q}(y_i; \boldsymbol{X}_i\boldsymbol{\beta}, \sigma^2)$ with $r = \text{rank}(\boldsymbol{X}) \leq q$, the unique BLUE of an estimable $\boldsymbol{\theta} = \boldsymbol{C\beta}$ $(a \times 1)$ is $\widehat{\boldsymbol{\theta}} = \boldsymbol{C}\widetilde{\boldsymbol{\beta}}$, with $\widetilde{\boldsymbol{\beta}} = (\boldsymbol{X}'\boldsymbol{X})^{-}\boldsymbol{X}'\boldsymbol{y}$ the least squares estimator of $\boldsymbol{\beta}$.

The assumptions guarantee $\widehat{\boldsymbol{\theta}}$ is unbiased, is invariant to the choice of generalized inverse defining $\widetilde{\boldsymbol{\beta}}$, has minimum variance among all LUEs, and is unique in that it is the only estimator satisfying all of the requirements.

Proof. It is sufficient to prove $\forall \boldsymbol{t} \in \Re^q$ the BLUE of $\boldsymbol{t}'\boldsymbol{\theta}$ is $\boldsymbol{t}'\widehat{\boldsymbol{\theta}}$. If $\boldsymbol{c}' = \boldsymbol{t}'\boldsymbol{C}$, then by Lemma 11.7 the BLUE of $\theta_t = \boldsymbol{c}'\boldsymbol{\beta} = \boldsymbol{t}'\boldsymbol{C\beta} = \boldsymbol{t}'\boldsymbol{\theta}$ is $\widehat{\theta}_t = (\boldsymbol{t}'\boldsymbol{C})(\boldsymbol{X}'\boldsymbol{X})^{-}\boldsymbol{X}'\boldsymbol{y}$. □

Corollary 11.15 For the special case of a $\text{GLM}_{N,q}\text{FR}(y_i; \boldsymbol{X}_i\boldsymbol{\beta}, \sigma^2)$ with estimable $\boldsymbol{\theta} = \boldsymbol{C\beta}$ the unique BLUE is $\widehat{\boldsymbol{\theta}} = \boldsymbol{C}\widehat{\boldsymbol{\beta}}$ in which $\widehat{\boldsymbol{\beta}} = (\boldsymbol{X}'\boldsymbol{X})^{-1}\boldsymbol{X}'\boldsymbol{y}$.

11.11 RELATED DISTRIBUTIONS

It is definitely worth repeating that nearly all results about estimation presented earlier in the present chapter, except for likelihood properties, are distribution free. As long as the GLM assumptions hold, the random variables can have any distribution with finite second moments. The special case of Gaussian errors leads to simple forms of distributions for many estimators.

Theorem 11.16 (a) For $\text{GLM}_{N,q}\text{LTFR}(y_i; \boldsymbol{X}_i\boldsymbol{\beta}, \sigma^2)$ with Gaussian errors and two-condition inverse $(\boldsymbol{X}'\boldsymbol{X})^{-}$,

$$\widetilde{\boldsymbol{\beta}} \sim \mathcal{SN}_q\left[(\boldsymbol{X}'\boldsymbol{X})^{-}(\boldsymbol{X}'\boldsymbol{X})\boldsymbol{\beta}, \sigma^2(\boldsymbol{X}'\boldsymbol{X})^{-}\right]. \tag{11.54}$$

(b) A $\text{GLM}_{N,q}\text{FR}(y_i; \boldsymbol{X}_i\boldsymbol{\beta}, \sigma^2)$ has

$$\widehat{\boldsymbol{\beta}} \sim \mathcal{N}_q\left[\boldsymbol{\beta}, \sigma^2(\boldsymbol{X}'\boldsymbol{X})^{-1}\right]. \tag{11.55}$$

(c) For $\text{rank}(\boldsymbol{X}) \leq q$ and any one-condition inverse, estimable $\boldsymbol{\theta} = \boldsymbol{C\beta}$ ensures

$$\widehat{\boldsymbol{\theta}} - \boldsymbol{\theta}_0 \sim (\mathcal{S})\mathcal{N}_a\left[\boldsymbol{\theta} - \boldsymbol{\theta}_0, \sigma^2\boldsymbol{C}(\boldsymbol{X}'\boldsymbol{X})^{-}\boldsymbol{C}'\right]. \tag{11.56}$$

(d) For any one-condition inverse, $\boldsymbol{H} = \boldsymbol{X}(\boldsymbol{X}'\boldsymbol{X})^{-}\boldsymbol{X}' = \boldsymbol{X}(\boldsymbol{X}'\boldsymbol{X})^{+}\boldsymbol{X}'$ and

$$\begin{aligned}\widehat{\boldsymbol{y}} &= \boldsymbol{X}\widetilde{\boldsymbol{\beta}} = \boldsymbol{Hy} \\ &\sim \mathcal{SN}_N(\boldsymbol{X\beta}, \sigma^2\boldsymbol{H}).\end{aligned} \tag{11.57}$$

(e) Furthermore

$$\begin{aligned}\widehat{\boldsymbol{e}} &= \boldsymbol{y} - \widehat{\boldsymbol{y}} = (\boldsymbol{I}_N - \boldsymbol{H})\boldsymbol{y} \\ &\sim \mathcal{SN}_N[\boldsymbol{X\beta}, \sigma^2(\boldsymbol{I}_N - \boldsymbol{H})].\end{aligned} \tag{11.58}$$

Proof. Left as an exercise. *Hints*: linear transformations of Gaussian variables, Theorem 1.15, generalized inverse conditions, and estimability condition.

Theorem 11.17 In a $\text{GLM}_{N,q}(y_i; \boldsymbol{X}_i\boldsymbol{\beta}, \sigma^2)$ with Gaussian errors, $\widetilde{\boldsymbol{\beta}}$ and $\widehat{\boldsymbol{e}}$ are jointly Gaussian with zero covariance and correlation and are statistically independent.

Proof. The joint distribution of the estimators is established by expressing the two vectors as a single linear transformation of a vector Gaussian:

$$\begin{bmatrix} \widetilde{\boldsymbol{\beta}} \\ \widehat{\boldsymbol{e}} \end{bmatrix} = \begin{bmatrix} (\boldsymbol{X}'\boldsymbol{X})^{-}\boldsymbol{X}' \\ (\boldsymbol{I} - \boldsymbol{H}) \end{bmatrix} \boldsymbol{y} = \boldsymbol{T}\boldsymbol{y}. \tag{11.59}$$

In turn

$$\begin{aligned} \mathcal{V}(\widetilde{\boldsymbol{\beta}}, \widehat{\boldsymbol{e}}) &= \mathrm{E}(\widetilde{\boldsymbol{\beta}}\widehat{\boldsymbol{e}}') - \mathrm{E}(\widetilde{\boldsymbol{\beta}})\mathrm{E}(\widehat{\boldsymbol{e}}') \\ &= \mathrm{E}\{(\boldsymbol{X}'\boldsymbol{X})^{-}\boldsymbol{X}'\boldsymbol{y}[(\boldsymbol{I} - \boldsymbol{H})\boldsymbol{y}]'\} - \mathrm{E}(\widetilde{\boldsymbol{\beta}})\mathbf{0} \\ &= (\boldsymbol{X}'\boldsymbol{X})^{-}\boldsymbol{X}'\mathrm{E}(\boldsymbol{y}\boldsymbol{y}')(\boldsymbol{I} - \boldsymbol{H}) \\ &= (\boldsymbol{X}'\boldsymbol{X})^{-}\boldsymbol{X}'(\sigma^2\boldsymbol{I}_N)(\boldsymbol{I} - \boldsymbol{H}) \\ &= \sigma^2(\boldsymbol{X}'\boldsymbol{X})^{-}[\boldsymbol{X}' - \boldsymbol{X}'\boldsymbol{X}(\boldsymbol{X}'\boldsymbol{X})^{-}\boldsymbol{X}'] = \sigma^2(\boldsymbol{X}'\boldsymbol{X})^{-}[\mathbf{0}]. \end{aligned} \tag{11.60}$$

The last step is true by Theorem 1.15 and gives $\mathcal{V}(\widetilde{\boldsymbol{\beta}}, \widehat{\boldsymbol{e}}) = \mathbf{0}$. □

11.12 FORMULATIONS OF EXPLICIT RESTRICTIONS OF $\boldsymbol{\beta}$ AND $\boldsymbol{\theta}$

The notation $\text{GLM}_{N,q}(y_i; \boldsymbol{X}_i\boldsymbol{\beta}, \sigma^2)$ describes a situation in which no restrictions have been placed on relationships among the elements of $\boldsymbol{\beta}$. In some cases, data analysts wish to ensure some explicit linear relationships hold among the elements of $\boldsymbol{\beta}$. The classical approach to one-way ANOVA coding for G groups uses the model $y_{ij} = \mu + \alpha_j + e_{ij}$ subject to the restriction $\sum_{j=1}^{G}\alpha_j = 0$. The use of the restriction overcomes the inherently LTFR nature of the classical coding. It leads to a method for identifying a set of G estimable parameters (from among the set of $G+1$ nonestimable original parameters).

The notation may be extended to incorporate explicit linear restrictions on the elements of $\boldsymbol{\beta}$ by writing $\text{GLM}_{N,q}(y_i; \boldsymbol{X}_i\boldsymbol{\beta}, |\boldsymbol{R}\boldsymbol{\beta} = \boldsymbol{a}, \sigma^2)$ for $\boldsymbol{R}$ $(k \times q)$ and $\boldsymbol{a}$ $(k \times 1)$ conforming and known constants with $k \leq q$. Most often, $\boldsymbol{a} = \mathbf{0}$. Without restrictions, $\boldsymbol{\beta} \in \Re^q$, whereas, with the explicit restrictions, $\boldsymbol{R}\boldsymbol{\beta} = \boldsymbol{a}$, $\boldsymbol{\beta}$ is required to lie in a subspace. Such linear restrictions on (the parameters space for) $\boldsymbol{\beta}$ can be imposed on both FR and LTFR models. They are particularly important for LTFR models because such additional restrictions can lead to a model with *estimable* primary expected-value parameters. They will also be seen to be important in the formulation of the general linear hypothesis test.

Example 11.4 The classical ANOVA coding for a $G = 3$ group one-way ANOVA provides a simple example. In matrix notation, the model is

$$
\begin{aligned}
\boldsymbol{y} = \quad & \boldsymbol{X\beta} \quad + \boldsymbol{e} \\
= & \begin{bmatrix} 1 & 1 & 0 & 0 \\ 1 & 0 & 1 & 0 \\ 1 & 0 & 0 & 1 \end{bmatrix} \begin{bmatrix} \mu \\ \alpha_1 \\ \alpha_2 \\ \alpha_3 \end{bmatrix} + \boldsymbol{e}.
\end{aligned} \tag{11.61}
$$

The "sum to zero" zero constraints may be stated as

$$
\begin{aligned}
\boldsymbol{R\beta} &= \boldsymbol{a} \\
\begin{bmatrix} 0 & 1 & 1 & 1 \end{bmatrix} \begin{bmatrix} \mu \\ \alpha_1 \\ \alpha_2 \\ \alpha_3 \end{bmatrix} &= [0].
\end{aligned} \tag{11.62}
$$

Since the usual estimation procedures involve optimization, estimation subject to linear restrictions can be performed via constrained optimization, such as through the use of Lagrange multipliers. However, the following two theorems allow restating the restrictions and in turn restating the restricted model as an unrestricted model.

Theorem 11.18 A restricted $\text{GLM}_{N,q}(y_i; \boldsymbol{X}_i\boldsymbol{\beta}|\boldsymbol{R\beta} = \boldsymbol{a}, \sigma^2)$ has $\boldsymbol{R}$ $(k \times q)$ and $\boldsymbol{a}$ $(k \times 1)$ known constants, $k \leq q$, and consistent equations $\boldsymbol{R\beta} = \boldsymbol{a}$ (with at least one solution).

(a) If $\text{rank}(\boldsymbol{R}) \leq k$ and $\boldsymbol{R}^-$ $(q \times k)$ is any particular generalized inverse (i.e., $\boldsymbol{RR}^-\boldsymbol{R} = \boldsymbol{R}$), then any value $\boldsymbol{\beta}_R$ which satisfies the restrictions may be written, for some $\boldsymbol{\gamma} \in \Re^q$, as

$$
\boldsymbol{\beta}_R = \boldsymbol{R}^-\boldsymbol{a} + (\boldsymbol{R}^-\boldsymbol{R} - \boldsymbol{I}_q)\boldsymbol{\gamma}. \tag{11.63}
$$

(b) If $\text{rank}(\boldsymbol{R}) = k$, then the singular value decomposition allows writing

$$
\begin{aligned}
\boldsymbol{R} &= \boldsymbol{P}[\boldsymbol{\Lambda}\,\boldsymbol{0}]\boldsymbol{Q}' \\
&= \boldsymbol{P}[\boldsymbol{\Lambda}\,\boldsymbol{0}][\boldsymbol{Q}_1\,\boldsymbol{Q}_2]' \\
&= \boldsymbol{P\Lambda Q}_1',
\end{aligned} \tag{11.64}
$$

the (unique) Moore-Penrose (four-condition) generalized inverse is

$$
\boldsymbol{R}^+ = \boldsymbol{Q}_1\boldsymbol{\Lambda}^{-1}\boldsymbol{P}', \tag{11.65}
$$

and $\boldsymbol{\beta}$ satisfies the restrictions if and only if, for some $\boldsymbol{\tau} \in \Re^{q-k}$,

$$
\boldsymbol{\beta} = \boldsymbol{R}^+\boldsymbol{a} + \boldsymbol{Q}_2\boldsymbol{\tau}. \tag{11.66}
$$

Here $\boldsymbol{P}$ is $k \times k$, $\boldsymbol{Q}$ is $q \times q$, $\boldsymbol{Q}_1$ is $q \times k$, and $\boldsymbol{Q}_2$ with $q \times (q-k)$, with all being columnwise orthonormal.

(c) If $\text{rank}(\boldsymbol{R}) = k$, then, with $s = q - k$ and columns of $\boldsymbol{R}$ and corresponding rows of $\boldsymbol{\beta}$ permuted prior to partitioning if necessary,

$$R\beta = [\,R_1\;R_2\,]\begin{bmatrix}\beta_1\\ \beta_2\end{bmatrix}, \tag{11.67}$$

with R_1 $k \times k$ and full rank, which ensures R_1^{-1} exists. Also β_1 is $k \times 1$, R_2 is $k \times s$, and β_2 is $s \times 1$. Only β_2 is free to vary and β satisfies $R\beta = a \Leftrightarrow$

$$\beta = \begin{bmatrix}\beta_1\\ \beta_2\end{bmatrix} = \begin{bmatrix}R_1^{-1}\\ 0_{s\times k}\end{bmatrix} a + \begin{bmatrix}-R_1^{-1}R_2\\ I_s\end{bmatrix}\beta_2 . \tag{11.68}$$

Proof of (a). ($\Leftarrow$) Any such β_R is a solution.

($\Rightarrow$) If β is *any* solution, then choosing $\gamma = (R^- R - I_q)\beta$ ensures β satisfies the relationship required of β_R. In turn,

$$\begin{aligned}
R^- a + (R^- R - I_q)\gamma &= R^- a + (R^- R - I_q)(R^- R - I_q)\beta \\
&= R^- a + (R^- R R^- R - R^- R - R^- R + I_q)\beta \\
&= R^- a + (I_q - R^- R)\beta \\
&= R^-(R\beta) + (I_q - R^- R)\beta \\
&= \beta + 0 .
\end{aligned} \tag{11.69}$$

Proof of (b). If $R^- = R^+$ and $\tau = Q_2'\gamma$, applying part (a) gives

$$\begin{aligned}
R^- a + (R^- R - I_q)\gamma &= R^+ a + (Q_1 Q_1' - I_q)\gamma \\
&= R^+ a + (Q_2 Q_2')\gamma \\
&= R^+ a + Q_2(\tau) .
\end{aligned} \tag{11.70}$$

Proof of (c). $R_1\beta_1 + R_2\beta_2 = a$ implies $\beta_1 = R_1^{-1}a - R_1^{-1}R_2\beta_2$. □

Corollary 11.18 The matrix $[\,X'\;R'\,]$ has full rank $\Leftrightarrow$ $Z = (X_2 - X_1 R_1^{-1} R_2)$ has full rank.

Proof. Left as an exercise.

11.13 RESTRICTED ESTIMATION VIA EQUIVALENT MODELS

Definition 11.10 (a) An *explicitly restricted linear model* is indicated $\mathrm{GLM}_{N,q}(y_i; X_i\beta | R\beta = a, \sigma^2)$, with $R\beta = a$ indicating restrictions on β. (b) A linearly equivalent unrestricted form is an *implicitly restricted linear model.*

The following theorem provides explicit methods for finding an unrestricted model for any given restricted model. Hence the theorem guarantees a linearly equivalent unrestricted form always exists for any restricted model.

Theorem 11.19 A restricted $\text{GLM}_{N,q}(y_i; \boldsymbol{X}_i\boldsymbol{\beta}|\boldsymbol{R}\boldsymbol{\beta} = \boldsymbol{a}, \sigma^2)$ has $\boldsymbol{R}$ $(k \times q)$ and $\boldsymbol{a}$ $(k \times 1)$ known constants, with $k \leq q$, $r = \text{rank}(\boldsymbol{X}) \leq q$ and $s = q - k$. Consistent equations $\boldsymbol{R}\boldsymbol{\beta} = \boldsymbol{a}$ give three distinct ways to create and analyze a linearly equivalent unrestricted model, depending on $\text{rank}(\boldsymbol{R})$.

(a) Having $\text{rank}(\boldsymbol{R}) \leq k$ allows the following method.
1. Compute $\boldsymbol{R}^-$ $(q \times k)$ such that $\boldsymbol{R}\boldsymbol{R}^-\boldsymbol{R} = \boldsymbol{R}$.
2. Compute transformed responses $\boldsymbol{u} = \boldsymbol{y} - \boldsymbol{X}\boldsymbol{R}^-\boldsymbol{a}$ $(N \times 1)$.
3. Compute transformed predictors $\boldsymbol{Z} = \boldsymbol{X}(\boldsymbol{R}^-\boldsymbol{R} - \boldsymbol{I}_q)$ $(N \times q)$.
4. Analyze unrestricted model (I) $\text{GLM}_{N,q}(u_i; \boldsymbol{Z}_i\boldsymbol{\gamma}, \sigma^2)$ with $\text{rank}(\boldsymbol{Z}) \leq s$.

(b) Having $\text{rank}(\boldsymbol{R}) = k$ allows the following method.
1. Compute the SVD $\boldsymbol{R} = \boldsymbol{P}[\boldsymbol{\Lambda}\,\boldsymbol{0}][\boldsymbol{Q}_1\,\boldsymbol{Q}_2]'$ with $\boldsymbol{P}'\boldsymbol{P} = \boldsymbol{P}\boldsymbol{P}' = \boldsymbol{I}_k$, $\boldsymbol{Q}_1$ $(q \times k)$, $\boldsymbol{Q}_1'\boldsymbol{Q}_1 = \boldsymbol{I}_k$, and $\boldsymbol{Q}_2$ $(q \times s)$, $\boldsymbol{Q}_2'\boldsymbol{Q}_2 = \boldsymbol{I}_s$.
2. Compute $\boldsymbol{R}^+ = \boldsymbol{Q}_1\boldsymbol{\Lambda}^{-1}\boldsymbol{P}'$.
3. Compute transformed responses $\boldsymbol{u} = \boldsymbol{y} - \boldsymbol{X}\boldsymbol{R}^+\boldsymbol{a}$ $(N \times 1)$.
4. Compute transformed (and reduced) predictors $\boldsymbol{Z} = \boldsymbol{X}\boldsymbol{Q}_2$ $(N \times s)$.
5. Analyze unrestricted model (II) $\text{GLM}_{N,s}(u_i; \boldsymbol{Z}_i\boldsymbol{\tau}, \sigma^2)$ with $\text{rank}(\boldsymbol{Z}) \leq s$.

(c) Having $\text{rank}(\boldsymbol{R}) = k$ also allows the following method.
1. Permute columns of $\boldsymbol{X}$ and corresponding rows of $\boldsymbol{\beta}$ to have $\boldsymbol{R} = [\boldsymbol{R}_1\,\boldsymbol{R}_2]$, with $\boldsymbol{R}_1$ $(k \times k)$ full rank and $\boldsymbol{X} = [\boldsymbol{X}_1\,\boldsymbol{X}_2]$ partitioned similarly.
2. Compute transformed responses $\boldsymbol{u} = \boldsymbol{y} - \boldsymbol{X}_1\boldsymbol{R}_1^{-1}\boldsymbol{a}$ $(N \times 1)$.
3. Compute transformed (and reduced) predictors $(N \times s)$

$$\boldsymbol{Z} = \boldsymbol{X}\begin{bmatrix} -\boldsymbol{R}_1^{-1}\boldsymbol{R}_2 \\ \boldsymbol{I}_s \end{bmatrix}. \tag{11.71}$$

4. Analyze unrestricted model (III) $\text{GLM}_{N,s}(u_i; \boldsymbol{Z}_i\boldsymbol{\beta}_2, \sigma^2)$ with $\text{rank}(\boldsymbol{Z}) \leq s$.

Here $(\boldsymbol{u} - \boldsymbol{Z}\boldsymbol{\gamma}) = (\boldsymbol{u} - \boldsymbol{Z}\boldsymbol{\tau}) = (\boldsymbol{u} - \boldsymbol{Z}\boldsymbol{\beta}_2) = (\boldsymbol{y} - \boldsymbol{X}\boldsymbol{\beta})$.

Proof. If $\boldsymbol{y} = \boldsymbol{X}\boldsymbol{\beta} + \boldsymbol{\epsilon}$ given $\boldsymbol{R}\boldsymbol{\beta} = \boldsymbol{a}$, then Theorem 11.18 (a) implies

$$\boldsymbol{y} = \boldsymbol{X}\boldsymbol{R}^-\boldsymbol{a} + \boldsymbol{X}(\boldsymbol{R}^-\boldsymbol{R} - \boldsymbol{I}_q)\boldsymbol{\gamma} + \boldsymbol{\epsilon}. \tag{11.72}$$

Hence the restricted model and model I are linearly equivalent. Theorem 11.18 (b) implies

$$\boldsymbol{y} = \boldsymbol{X}\boldsymbol{R}^+\boldsymbol{a} + \boldsymbol{X}\boldsymbol{Q}_2\boldsymbol{\tau} + \boldsymbol{\epsilon}. \tag{11.73}$$

Hence the restricted model and model II are linearly equivalent. Theorem 11.18 (c) implies

$$\boldsymbol{y} = \boldsymbol{X}\boldsymbol{R}_1^{-1}\boldsymbol{a} + \boldsymbol{X}\begin{bmatrix} -\boldsymbol{R}_1^{-1}\boldsymbol{R}_2 \\ \boldsymbol{I}_s \end{bmatrix}\boldsymbol{\beta}_2 + \boldsymbol{\epsilon}. \tag{11.74}$$

Hence the restricted model and model III are linearly equivalent. □

Theorem 11.20 With the notation and assumptions of Theorem 11.18 (b), $\boldsymbol{\theta} = \boldsymbol{C\beta}$ $(a \times 1)$ is estimable. Here $\boldsymbol{C} = \boldsymbol{I}$ is allowed $\Leftrightarrow$ rank$(\boldsymbol{X}) = q$. The BLUE of $\boldsymbol{\theta}$ is $\widehat{\boldsymbol{\theta}}_R = \boldsymbol{C}(\boldsymbol{R}^+\boldsymbol{a} + \boldsymbol{Q}_2\widetilde{\boldsymbol{\tau}})$, in which $\widetilde{\boldsymbol{\tau}}$ is the BLUE of $\boldsymbol{\tau}$ $(s \times 1)$ in the linearly equivalent unrestricted model II, namely $\text{GLM}_{N,s}(u_i; \boldsymbol{Z}_i\boldsymbol{\tau}, \sigma^2)$ with rank$(\boldsymbol{Z}) \le s$, $\boldsymbol{u} = \boldsymbol{y} - \boldsymbol{XR}^+\boldsymbol{a}$, and $\boldsymbol{Z} = \boldsymbol{XQ}_2$ $(N \times s)$.

Proof. Subject to the restrictions $\boldsymbol{R\beta} = \boldsymbol{a}$, $\boldsymbol{\theta} = \boldsymbol{C\beta}$ is still estimable because $\boldsymbol{\theta} = \boldsymbol{CR}^+\boldsymbol{a} + \boldsymbol{CQ}_2\boldsymbol{\tau}$ implies $\boldsymbol{\theta} = \text{constant} + (\boldsymbol{CQ}_2)\boldsymbol{\tau}$. Also, $\boldsymbol{CQ}_2\boldsymbol{\tau}$ is estimable in unrestricted model II because $\boldsymbol{C\beta}$ is estimable in the restricted model. Therefore, by Theorem 11.15, the BLUE of $\boldsymbol{CQ}_2\boldsymbol{\tau}$ is $\boldsymbol{CQ}_2\widetilde{\boldsymbol{\tau}}$, and the desired result follows. □

Corollary 11.20 (a) If rank$(\boldsymbol{X}) = q$, then the BLUE of $\boldsymbol{\beta}$ is

$$\begin{aligned} \widehat{\boldsymbol{\beta}}_R &= (\boldsymbol{R}^+\boldsymbol{a} + \boldsymbol{Q}_2\widetilde{\boldsymbol{\tau}}) \\ &= \widehat{\boldsymbol{\beta}} - (\boldsymbol{X}'\boldsymbol{X})^{-1}\boldsymbol{R}'\left[\boldsymbol{R}(\boldsymbol{X}'\boldsymbol{X})^{-1}\boldsymbol{R}'\right]^{-1}(\boldsymbol{R}\widehat{\boldsymbol{\beta}} - \boldsymbol{a}), \end{aligned} \tag{11.75}$$

in which $\widehat{\boldsymbol{\beta}} = (\boldsymbol{X}'\boldsymbol{X})^{-1}\boldsymbol{X}'\boldsymbol{y}$.
(b) An unbiased estimator of σ^2 is

$$\begin{aligned} \widehat{\sigma}^2 &= (\boldsymbol{y} - \boldsymbol{X}\widehat{\boldsymbol{\beta}}_R)'(\boldsymbol{y} - \boldsymbol{X}\widehat{\boldsymbol{\beta}}_R)/[(N - q + k)] \\ &= (SSE + SSH)/[(N - q + k)], \end{aligned} \tag{11.76}$$

with

$$SSE = \boldsymbol{y}'\left[\boldsymbol{I} - \boldsymbol{X}(\boldsymbol{X}'\boldsymbol{X})^{-1}\boldsymbol{X}'\right]\boldsymbol{y} = (\boldsymbol{y} - \boldsymbol{X}\widehat{\boldsymbol{\beta}})'(\boldsymbol{y} - \boldsymbol{X}\widehat{\boldsymbol{\beta}}) \tag{11.77}$$

and

$$SSH = (\boldsymbol{R}\widehat{\boldsymbol{\beta}} - \boldsymbol{a})'\left[\boldsymbol{R}(\boldsymbol{X}'\boldsymbol{X})^{-1}\boldsymbol{R}'\right]^{-1}(\boldsymbol{R}\widehat{\boldsymbol{\beta}} - \boldsymbol{a}). \tag{11.78}$$

Proof. The first form for $\widehat{\boldsymbol{\beta}}_R$ may be found by considering $\boldsymbol{\theta} = \boldsymbol{\beta}$ and applying the original theorem. In the special case of rank$(\boldsymbol{X}) = q$ the parameter $\boldsymbol{\theta} = \boldsymbol{C\beta}$ is estimable for all $\boldsymbol{C}$ including $\boldsymbol{C} = \boldsymbol{I}$.

The second form for $\widehat{\boldsymbol{\beta}}_R$ may be found by using Theorem 11.18 (a) to help define a linearly equivalent unrestricted model, $\text{GLM}_{N,q}(u_i; \boldsymbol{Z}_i\boldsymbol{\gamma}, \sigma^2)$ with $\boldsymbol{Z}$ $N \times q$ with rank$(\boldsymbol{Z}) \le q$, $\boldsymbol{R}$ $q \times k$, $\boldsymbol{a}$ $k \times 1$, $\boldsymbol{u} = \boldsymbol{y} - \boldsymbol{XR}^-\boldsymbol{a}$, $\boldsymbol{Z} = \boldsymbol{X}(\boldsymbol{R}^-\boldsymbol{R} - \boldsymbol{I}_q)$, and $\boldsymbol{\beta}_R = \boldsymbol{R}^-\boldsymbol{a} + (\boldsymbol{R}^-\boldsymbol{R} - \boldsymbol{I}_q)\boldsymbol{\gamma}$. Here $\boldsymbol{\beta}_R$ is an estimable parameter, $\boldsymbol{\beta}_R = \boldsymbol{C\gamma} + \boldsymbol{R}^-\boldsymbol{a}$, since $\boldsymbol{C} = (\boldsymbol{R}^-\boldsymbol{R} - \boldsymbol{I}_q) = (\boldsymbol{X}'\boldsymbol{X})^{-1}\boldsymbol{X}'\boldsymbol{Z}$ is a linear combination of the rows of $\boldsymbol{Z}$, and $\boldsymbol{R}^-\boldsymbol{a}$ is a constant. By Theorem 11.15 the unique BLUE of $\boldsymbol{\beta}_R$ is $\widehat{\boldsymbol{\beta}}_R = \boldsymbol{C}\widetilde{\boldsymbol{\gamma}} + \boldsymbol{R}^-\boldsymbol{a}$ in which $\widetilde{\boldsymbol{\gamma}} = (\boldsymbol{Z}'\boldsymbol{Z})^-\boldsymbol{Z}'\boldsymbol{U}$. The estimator can be written

$$\begin{aligned}\hat{\beta}_R &= (R^-R - I_q)(Z'Z)^-Z'(Y - XR^-a) + R^-a \\ &= [(R^-R-I_q)(Z'Z)^-Z']y+\{[I_k-(R^-R-I_q)(Z'Z)^-Z'X]R^-\}a \\ &= T_{1y}y + T_{1a}a\,. \end{aligned} \tag{11.79}$$

The estimator can also be written

$$\begin{aligned}\hat{\beta}_R &= \hat{\beta} - (X'X)^{-1}R'\left[R(X'X)^{-1}R'\right]^{-1}(R\hat{\beta} - a) \\ &= (\{I_q - (X'X)^{-1}R'[R(X'X)^{-1}R']^{-1}R\}(X'X)^{-1}X')y + \\ &\qquad\qquad \{(X'X)^{-1}R'[R(X'X)^{-1}R']^{-1}\}a \\ &= T_{2y}y + T_{2a}a\,. \end{aligned} \tag{11.80}$$

The equality of the expressions can be proved by first equating the coefficients of y and second equating the coefficients of a.

The fact that R^- can be chosen to ensure $(Z'Z)^-Z' = (R^-R - I_q)(X'X)^{-1}X'$ helps accomplish the two equating tasks. The forms

$$\begin{aligned} Z^- &= (Z'Z)^-Z' && \text{(i)} \\ Z^- &= (R^-R - I_q)(X'X)^{-1}X' && \text{(ii)} \end{aligned}$$

are both valid representations for a generalized inverse of Z. The representation (i) is contained in Theorem 1.15. The validity of (ii) follows directly from verifying $ZZ^-Z = Z$ for $Z = X(R^-R - I_q)$. All possible generalized inverses of Z are represented by each of the two expressions. Therefore, for any particular generalized inverse of $Z'Z$ a particular generalized inverse of R exists which makes (i) and (ii) identical.

Applying the equality, we have

$$\begin{aligned} T_{1y} &= (R^-R - I_q)(R^-R - I_q)(X'X)^{-1}X' \\ &= (I_q - R^-R)(X'X)^{-1}X'\,, \end{aligned} \tag{11.81}$$

compared with $T_{2y} = \{I_q - (X'X)^{-1}R'\left[R(X'X)^{-1}R'\right]^{-1}R\}(X'X)^{-1}X'$. Hence it will suffice to prove $R^-R = (X'X)^{-1}R'[R(X'X)^{-1}R']^{-1}R$. Here

$$\begin{aligned} R(X'X)^{-1}R' &= R[(X'X)^{-1}R'] && \Downarrow \\ R^-[R(X'X)^{-1}R'] &= (X'X)^{-1}R' && \Downarrow \\ R^- &= (X'X)^{-1}R'[R(X'X)^{-1}R']^{-1} && \Downarrow \\ R^-R &= (X'X)^{-1}R'[R(X'X)^{-1}R']^{-1}R && \Downarrow \\ T_{1y} &= T_{2y}\,. \end{aligned} \tag{11.82}$$

Given $T_{1y} = T_{2y}$, it follows almost immediately that $T_{1a} = T_{2a}$ since

$$\begin{aligned} T_{1a} &= (I - T_{1y}X)R^- \\ T_{2a}R &= I - T_{2y}X\,. \end{aligned} \tag{11.83}$$

Multiplying the first equation by a and the second equation by R^-a gives

$$\begin{aligned} T_{1a}a &= (I - T_{1y}X)R^{-}a \\ T_{2a}RR^{-}a &= (I - T_{2y}X)R^{-}a\,. \end{aligned} \tag{11.84}$$

Since $RR^{-}a = a$ and $T_{1y} = T_{2y}$, it follows that $T_{1a}a = T_{2a}a$ for all a. Thus $T_{1y} = T_{2y}$ and $T_{1a} = T_{2a}$.

It is not necessary to assume Gaussian errors to formulate a reasonable estimator of σ^2. By Theorem 11.18, a linearly equivalent unrestricted model is $\text{GLM}_{N,q}(u_i; Z_i\gamma, \sigma^2)$ with $\text{rank}(Z) \leq q$, $u = y - XR^{-}a$, $Z = X(R^{-}R - I_q)$, and $\beta_R = R^{-}a + (R^{-}R - I_q)\gamma$. For any real symmetric A $\text{E}(u'Au) = \text{tr}(A\sigma^2 I_N) + (Z\tau)'A(Z\tau)$. If $A = [I - Z(Z'Z)^{-}Z] \equiv (I - P_z)$, then $\text{E}(u'Au) = \sigma^2\text{rank}(A) + 0 = \sigma^2[N-(q-k)]$. Hence an unbiased estimator for σ^2 is $\widehat{\sigma}^2 = u'(I - P_z)u/(N-q+k) = (u - Z\widetilde{\tau})'(u - Z\widetilde{\tau})/(N - q + r)$. As noted in Theorem 11.18, $(u - Z\widetilde{\tau}) = (y - X\widehat{\beta}_R)$ and hence the desired result follows.

The second form for $\widehat{\sigma}^2$ may be found in terms of the following expressions:

$$(\widehat{\beta} - \widehat{\beta}_R) = (X'X)^{-1}R'[R(X'X)^{-1}R']^{-1}(R\widehat{\beta} - a) \tag{11.85}$$

$$SSE = (y - X\widehat{\beta})'(y - X\widehat{\beta}) \tag{11.86}$$

$$SSH = (R\widehat{\beta} - a)'\left[R(X'X)^{-1}R'\right]^{-1}(R\widehat{\beta} - a)\,. \tag{11.87}$$

Simple manipulations give

$$\begin{aligned} \widehat{\sigma}^2(N - q + k) &= (y - X\widehat{\beta}_R)'(y - X\widehat{\beta}_R) \\ &= \left[y - X\widehat{\beta} + X(\widehat{\beta} - \widehat{\beta}_R)\right]'\left[y - X\widehat{\beta} + X(\widehat{\beta} - \widehat{\beta}_R)\right] \\ &= (y - X\widehat{\beta})'(y - X\widehat{\beta}) + (\widehat{\beta} - \widehat{\beta}_R)'(X'X)(\widehat{\beta} - \widehat{\beta}_R) \\ &= SSE + SSH\,. \end{aligned} \tag{11.88}$$

□

11.14 FITTING PIECEWISE POLYNOMIAL MODELS VIA SPLINES

Definition 11.11 A collection of m piecewise linear functions connected at m points, known as *knots*, together define a *spline* function.

In some cases, the study design and scientific setting dictate that no single model applies to the entire range of the predictor. The complication could arise in modeling the yield of a chemical manufacturing process as a function of total volume of ingredients. The batch size z_i (a model predictor) can vary only within a limited range for each of $m = 3$ container sizes. Using regression splines allows specifying a distinct polynomial model for each container size, as does the obvious alternative of fitting three different models. However, not only do splines allow creating a smooth model for the entire range of interest, they also provide much greater precision by using a pooled estimator of variance. Smith (1979) provided a useful introduction to regression splines.

A statistical model for $\boldsymbol{y}$ $(N \times 1)$ specifying the mean as a piecewise polynomial function with $m = 3$ parts provides an example:

$$\mathrm{E}(y_i|z_i) = f(z_i;\boldsymbol{\beta}) = \begin{cases} a_1 + b_1 z_i & z_i \in (k_0, k_1] \\ a_2 + b_2 z_i + c_2 z_i^2 & z_i \in (k_1, k_2] \\ a_3 + b_3 z_i & z_i \in (k_2, k_3]\,. \end{cases} \tag{11.89}$$

The points of connection $\boldsymbol{k}$ $(m \times 1)$ are the knots. The m values are assumed to be fixed and known. Usually the values are chosen by the analyst and are evenly spaced with $m \leq N^{1/5}$. With the knots fixed and known, the only remaining parameters of $f()$ are the regression coefficients $\boldsymbol{\beta} = [\,a_1\ b_1\ a_2\ b_2\ c_2\ a_3\ b_3\,]'$. The fact that $\mathrm{E}(y_i|z_i) = f(z_i;\boldsymbol{\beta}) = \boldsymbol{X}_i\boldsymbol{\beta}$ verifies the mean is linear in $\boldsymbol{\beta}$. If $\boldsymbol{X}_{1i} = [\,1\ z_i\,]$, $\delta_{1i} = \mathbb{B}\{z_i \in (k_0, k_1]\}$, $\boldsymbol{X}_{2i} = [1\ z_i\ z_i^2]$, $\delta_{2i} = \mathbb{B}\{z_i \in (k_1, k_2]\}$, $\boldsymbol{X}_{3i} = [\,1\ z_i\,]$, $\delta_{3i} = \mathbb{B}\{z_i \in (k_2, k_3]\}$, $\boldsymbol{\beta}_1 = [\,a_1\ b_1\,]'$, $\boldsymbol{\beta}_2 = [\,a_2\ b_2\ c_2\,]'$, and $\boldsymbol{\beta}_3 = [\,a_3\ b_3\,]'$, then equivalent expressions for $f(z_i;\boldsymbol{\beta})$ are

$$\begin{aligned} f(z_i;\boldsymbol{\beta}) &= \begin{cases} \boldsymbol{X}_{1i}\boldsymbol{\beta}_1 & z_i \in (k_0, k_1] \\ \boldsymbol{X}_{2i}\boldsymbol{\beta}_2 & z_i \in (k_1, k_2] \\ \boldsymbol{X}_{3i}\boldsymbol{\beta}_3 & z_i \in (k_2, k_3] \end{cases} \\ &= \sum_{j=1}^{3} \delta_{ji}\boldsymbol{x}_{ji}\boldsymbol{\beta}_j \\ &= [\,\delta_{1i}\boldsymbol{X}_{1i}\quad \delta_{2i}\boldsymbol{X}_{2i}\quad \delta_{3i}\boldsymbol{X}_{3i}\,] \begin{bmatrix} \boldsymbol{\beta}_1 \\ \boldsymbol{\beta}_2 \\ \boldsymbol{\beta}_3 \end{bmatrix} \\ &= \boldsymbol{X}_i\boldsymbol{\beta}\,. \end{aligned} \tag{11.90}$$

It follows $\mathrm{E}(\boldsymbol{y}|\boldsymbol{z}) = f(\boldsymbol{z};\boldsymbol{\beta}) = \boldsymbol{X}\boldsymbol{\beta}$. (Can you say how $\boldsymbol{X}$ is defined?)

Theorem 11.21 Regression splines allow defining a valid univariate GLM. We assume $\mathrm{E}(y_i|z_i)$ is a piecewise linear function of $\boldsymbol{\beta}$, namely

$$\mathrm{E}(y_i|z_i) = \begin{cases} \boldsymbol{X}_{i,1}\boldsymbol{\beta}_1 & z_i \in (k_0, k_1] \\ \boldsymbol{X}_{i,2}\boldsymbol{\beta}_2 & z_i \in (k_1, k_2] \\ \vdots & \\ \boldsymbol{X}_{i,m}\boldsymbol{\beta}_m & z_i \in (k_{m-1}, k_m] \end{cases} \tag{11.91}$$

with known points of connection (knots) $\boldsymbol{k}$ $(m \times 1)$. If

$$d_{i,j} = \begin{cases} 1 & z_i \in (k_{j-1}, k_j] \\ 0 & \text{otherwise}\,, \end{cases} \tag{11.92}$$

$$\boldsymbol{X}_i = [\,d_{i1}\boldsymbol{X}_{i,1}\ \cdots\ d_{i,m}\boldsymbol{X}_{i,m}\,] \tag{11.93}$$

and

$$\boldsymbol{\beta} = \begin{bmatrix} \boldsymbol{\beta}_1 \\ \vdots \\ \boldsymbol{\beta}_m \end{bmatrix}, \tag{11.94}$$

then $\boldsymbol{y}$ satisfies $\text{GLM}_{N,q}(y_i; \boldsymbol{X}_i\boldsymbol{\beta}, \sigma^2)$, with q the number of columns in $\{\boldsymbol{X}_j\}$.

Proof. Left as an exercise.

Generally, the polynomial pieces $\{f_j(z_i; \boldsymbol{\beta}), j \in \{1, \ldots, m\}\}$ do not connect and will not form a smooth continuous curve unless some constraints are placed on them. To make $f(z_i; \boldsymbol{\beta}) = \sum_{j=1}^{m} \delta_{ji} f_j(z_i; \boldsymbol{\beta})$ continuous at knot k_j we must require $f_j(z; \boldsymbol{\beta})\big|_{z=k_j} = f_{j+1}(z; \boldsymbol{\beta})\big|_{z=k_j}$. Even with the requirement, the pieces may not connect to form a smooth curve. A degree of smoothness is obtained by requiring the derivatives from the left and right to be equal,

$$\left.\frac{\partial f_j(z; \boldsymbol{\beta})}{\partial z}\right|_{z=k_j} = \left.\frac{\partial f_{j+1}(z; \boldsymbol{\beta})}{\partial z}\right|_{z=k_j}. \tag{11.95}$$

Higher degrees of smoothness are obtained by specifying requirements for higher order derivatives such as

$$\left.\frac{\partial^2 f_j(z; \boldsymbol{\beta})}{\partial z^2}\right|_{z=k_j} = \left.\frac{\partial^2 f_{j+1}(z; \boldsymbol{\beta})}{\partial z^2}\right|_{z=k_j}. \tag{11.96}$$

Each constraint at each knot constitutes a single linear constraint on $\boldsymbol{\beta}$. Collectively the constraints can be expressed in the usual form, $\boldsymbol{R\beta} = \boldsymbol{a}$. The resulting constrained model can be transformed to a linearly equivalent unconstrained model.

Corollary 11.21 If smoothness constraints are placed on $f(z_i; \boldsymbol{\beta})$, then $\boldsymbol{y}$ satisfies $\text{GLM}_{N,q}(y_i; \boldsymbol{X}_i\boldsymbol{\beta} | \boldsymbol{R\beta} = \boldsymbol{a}, \sigma^2)$.

Proof. Left as an exercise.

Some scientific settings require estimating the join points rather than having them fixed. Gallant and Fuller (1973) provided a useful discussion of splines.

11.15 ESTIMATION FOR THE GGLM: WEIGHTED LEAST SQUARES

All results presented so far in the chapter apply not only to the GLM but also to the GGLM. The generality arises due to most of the results depending only on properties of the design matrix as they relate to expected values.

Definition 11.12 Exactly as for two GLMs, models $\mathrm{GGLM}_{N,q_1}(\boldsymbol{y}; \boldsymbol{X}_1\boldsymbol{\beta}_1, \boldsymbol{\Upsilon})$ and $\mathrm{GGLM}_{N,q_2}(\boldsymbol{y}; \boldsymbol{X}_2\boldsymbol{\beta}_2, \boldsymbol{\Upsilon})$ are *linearly equivalent* whenever (1) for any $\boldsymbol{\beta}_1$ there exists $\boldsymbol{\beta}_2$ such that $\boldsymbol{X}_1\boldsymbol{\beta}_1 = \boldsymbol{X}_2\boldsymbol{\beta}_2$ *and* (2) for any $\boldsymbol{\beta}_2$ there exists $\boldsymbol{\beta}_1$ such that $\boldsymbol{X}_1\boldsymbol{\beta}_1 = \boldsymbol{X}_2\boldsymbol{\beta}_2$.

It is useful to recognize that a GGLM may be linearly equivalent to a univariate GLM. Theorems about weighted least squares later in the chapter use such an equivalence. Similar results lie at the heart of some derivations of the "univariate" approach to repeated measures. The approach depends on the assumption of compound symmetric covariance among a set of p observations on each of N independent sampling units.

Theorem 11.22 For any $\mathrm{GGLM}_{N,q}\mathrm{LTFR}(\boldsymbol{y}; \boldsymbol{X\beta}, \boldsymbol{\Upsilon})$ with $r = \mathrm{rank}(\boldsymbol{X})$, linearly equivalent $\mathrm{GGLM}_{N,r}\mathrm{FR}(\boldsymbol{y}; \boldsymbol{X}_1\boldsymbol{\beta}_1, \boldsymbol{\Upsilon})$ always exists.

Proof. Left as an exercise.

Although results about restricted models are stated in terms of the univariate GLM, they also apply to the $\mathrm{GGLM}_{N,q}(\boldsymbol{y}; \boldsymbol{X\beta}|\boldsymbol{R\beta} = \boldsymbol{a}, \boldsymbol{\Upsilon})$. The generalization is valid because the constraints apply only to the expected- value portion of the model.

Definition 11.13 Estimators satisfying the least squares criterion for the GGLM are described as *exact weighted least squares* estimators.

The BLUE and likelihood results involve second-moment properties. Hence the forms of the result differ between the GLM and the GGLM and require separate proofs.

Lemma 11.7 For a $\mathrm{GGLM}_{N,q}(\boldsymbol{y}; \boldsymbol{X\beta}, \sigma^2\boldsymbol{D})$ with $r = \mathrm{rank}(\boldsymbol{X}) \leq q$, $\boldsymbol{D} = \boldsymbol{D}'$ known, positive definite and $N \times N$, and estimable scalar $\theta = \boldsymbol{c}'\boldsymbol{\beta}$, the unique BLUE of θ is $\widehat{\theta} = \boldsymbol{c}'\widetilde{\boldsymbol{\beta}}$ in which $\widetilde{\boldsymbol{\beta}} = (\boldsymbol{X}'\boldsymbol{D}^{-1}\boldsymbol{X})^{-}\boldsymbol{X}'\boldsymbol{D}^{-1}\boldsymbol{y}$.

Proof. The approach is to construct and discover $\widehat{\theta}$ which must be a linear estimator, namely $\widehat{\theta} = \boldsymbol{a}'\boldsymbol{y}$ for some $\boldsymbol{a} \in \Re^N$. The estimator $\widehat{\theta}$ must be unbiased, $\mathrm{E}(\widehat{\theta}) = \theta$. If $\mathrm{E}(\boldsymbol{a}'\boldsymbol{y}) = \boldsymbol{c}'\boldsymbol{\beta}\ \forall\boldsymbol{\beta}$, then $\boldsymbol{a}'\boldsymbol{X\beta} = \boldsymbol{c}'\boldsymbol{\beta}\ \forall\boldsymbol{\beta}$, which implies $\boldsymbol{a}'\boldsymbol{X} = \boldsymbol{c}'$ and gives q linear restrictions on $\boldsymbol{a}$. That θ is estimable implies $\boldsymbol{c}' = \boldsymbol{AX}$ for some $\boldsymbol{A}$ $(1 \times N)$. Hence the equations $\boldsymbol{a}'\boldsymbol{X} = \boldsymbol{c}'$ are consistent and there exists a value of $\boldsymbol{a}'$ which makes $\widehat{\theta}$ unbiased. Also, $\widehat{\theta}$ must have minimum variance, so $\mathcal{V}(\widehat{\theta}) \equiv \sigma^2\boldsymbol{a}'\boldsymbol{D}\boldsymbol{a}$ must be minimized subject to consistent restrictions $\boldsymbol{a}'\boldsymbol{X} = \boldsymbol{c}'$. The solution is found with the Lagrange function $s^2 = \sigma^2\boldsymbol{a}'\boldsymbol{D}\boldsymbol{a} - 2(\boldsymbol{a}'\boldsymbol{X} - \boldsymbol{c}')\boldsymbol{\lambda}$ for $\boldsymbol{\lambda} \in \Re^q$. The stationary point (value of $\boldsymbol{a}$) is specified by the requirements $\partial s^2/\partial\boldsymbol{\lambda} = \mathbf{0}$ and $\partial s^2/\partial\boldsymbol{a} = \mathbf{0}$, which imply

$$a'X = c' \tag{11.97}$$
$$X\lambda = \sigma^2 Da . \tag{11.98}$$

In turn, substituting $a = \sigma^{-2}D^{-1}X\lambda$ into the first system gives

$$\lambda'(X'D^{-1}X)\sigma^{-2} = c' , \tag{11.99}$$

which is a consistent system. Therefore solutions are of the form

$$\lambda' = \sigma^{-2}c'(X'D^{-1}X)^{-} . \tag{11.100}$$

Substituting the λ solution back into the original form and solving for a give

$$a' = c'(X'D^{-1}X)^{-}X'D^{-1} . \tag{11.101}$$

Hence λ and a satisfy the original conditions, and $(\sigma^{-2})^{-1}$ cancels σ^{-2}.

It must now be determined that the stationary point corresponds to a minimum rather than a saddle point, which may be done by comparing $\widehat{\theta} = a'y$ with possible competitors. If $\widetilde{\theta} = \alpha'y$ is also a LUE, then α satisfies the restrictions $\alpha'X = c'$. The variance of $\widetilde{\theta}$ can be expressed as

$$\begin{aligned} \mathcal{V}(\widetilde{\theta}) &= \mathcal{V}\left[(\widetilde{\theta} - \widehat{\theta}) + \widehat{\theta}\right] \\ &= \mathcal{V}(\widetilde{\theta} - \widehat{\theta}) + \mathcal{V}(\widehat{\theta}) + 2\mathcal{V}\left[\widehat{\theta}, (\widetilde{\theta} - \widehat{\theta})\right], \end{aligned} \tag{11.102}$$

in which

$$\begin{aligned} \mathcal{V}\left[\widehat{\theta}, (\widetilde{\theta} - \widehat{\theta})\right] &= \mathcal{V}[a'y, (\alpha' - a')y] \\ &= \sigma^2 a'D(\alpha - a) \\ &= \left[c'(X'D^{-1}X)^{-}X'D^{-1}\right]D(\alpha - a)\sigma^2 \\ &= c'(X'D^{-1}X)^{-}(X'\alpha - X'a)\sigma^2 \\ &= c'(X'D^{-1}X)^{-}(c - c)\sigma^2 = 0 . \end{aligned} \tag{11.103}$$

Hence $\widehat{\theta}$ has minimum variance because $\mathcal{V}(\widetilde{\theta}\,) - \mathcal{V}(\widehat{\theta}) = \mathcal{V}(\widetilde{\theta} - \widehat{\theta}) \geq 0$. □

Theorem 11.23 Model 1, $\text{GGLM}_{N,q}(y; X\beta, \sigma^2 D)$ with Gaussian errors, has $r = \text{rank}(X) \leq q$, $D = D'$ known and positive definite, and $D^{-1} = LL'$ for nonsingular L $(N \times N)$. Model 2, $\text{GLM}_{N,q}[\text{row}_i(L'y); \text{row}_i(L'X)\beta, \sigma^2]$, has $\text{rank}(L'X) = r$, is linearly equivalent, and satisfies the least squares and Gaussian assumptions of the GLM.

Proof. $y = X\beta + \epsilon \sim \mathcal{N}_N(X\beta, \sigma^2 D)$ implies $L'y = L'X\beta + L'\epsilon \sim \mathcal{N}_N(L'X\beta, \sigma^2 I)$. □

Theorem 11.24 $\text{GGLM}_{N,q}(y; X\beta, \sigma^2 D)$ with Gaussian errors, $r = \text{rank}(X) \leq q$, D known, positive definite, and symmetric, has estimable secondary parameter

$\boldsymbol{\theta} = \boldsymbol{C\beta}$. Joint MLEs of $\boldsymbol{\tau}' = [\boldsymbol{\beta}' \; \sigma^2]$ and $\boldsymbol{\theta}$ are

$$\widetilde{\boldsymbol{\beta}} = (\boldsymbol{X}'\boldsymbol{D}^{-1}\boldsymbol{X})^{-}\boldsymbol{X}'\boldsymbol{D}^{-1}\boldsymbol{y} \tag{11.104}$$

$$\widetilde{\sigma}^2 = (\boldsymbol{y} - \boldsymbol{X}\widetilde{\boldsymbol{\beta}})'\boldsymbol{D}^{-1}(\boldsymbol{y} - \boldsymbol{X}\widetilde{\boldsymbol{\beta}})/N \tag{11.105}$$

and

$$\widehat{\boldsymbol{\theta}} = \boldsymbol{C}\widetilde{\boldsymbol{\beta}}, \tag{11.106}$$

with $\widehat{\boldsymbol{\theta}}$ and $\widetilde{\sigma}^2$ invariant to $\widetilde{\boldsymbol{\beta}}$. Unbiased $\widehat{\sigma}^2 = \widetilde{\sigma}^2 N/(N-r)$ is usually preferred.

Proof. Left as an exercise.

Theorem 11.25 For $\text{GGLM}_{N,q}(\boldsymbol{y}; \boldsymbol{X\beta}, \sigma^2\boldsymbol{D})$, if $\boldsymbol{D} = \boldsymbol{D}'$ is known and positive definite while $\boldsymbol{\theta} = \boldsymbol{C\beta}$ is estimable, the following hold.
(a) If $\text{rank}(\boldsymbol{X}) = q$, then the unique BLUE of $\boldsymbol{\beta}$ is $\widehat{\boldsymbol{\beta}} = (\boldsymbol{X}'\boldsymbol{D}^{-1}\boldsymbol{X})^{-1}\boldsymbol{X}'\boldsymbol{D}^{-1}\boldsymbol{y}$.
(b) If $\text{rank}(\boldsymbol{X}) < q$, then the class of LUEs of $\boldsymbol{\beta}$ is empty and $\boldsymbol{\beta}$ is not estimable.
(c) The unique BLUE of $\boldsymbol{\theta}$ is $\widehat{\boldsymbol{\theta}} = \boldsymbol{C}\widetilde{\boldsymbol{\beta}}$ in which $\widetilde{\boldsymbol{\beta}} = (\boldsymbol{X}'\boldsymbol{D}^{-1}\boldsymbol{X})^{-}\boldsymbol{X}'\boldsymbol{D}^{-1}\boldsymbol{y}$.

Proof. Left as an exercise.

Theorem 11.26 Weighted least squares (WLS) estimators, also known as generalized least squares (GLS) estimators, may be found for $\text{GGLM}_{N,q}(\boldsymbol{y}; \boldsymbol{X\beta}, \sigma^2\boldsymbol{D})$ with $r = \text{rank}(\boldsymbol{X}) \leq q$ and known positive-definite, matrix $\boldsymbol{D}$ such that $\boldsymbol{D}^{-1} = \boldsymbol{LL}'$.
(a) Any solution $\widetilde{\boldsymbol{\beta}}$ of the WLS equations $(\boldsymbol{X}'\boldsymbol{D}^{-1}\boldsymbol{X})\widetilde{\boldsymbol{\beta}} = \boldsymbol{X}'\boldsymbol{D}^{-1}\boldsymbol{y}$ is a WLS estimator for $\boldsymbol{\beta}$. The solutions are of the form

$$\widetilde{\boldsymbol{\beta}} = (\boldsymbol{X}'\boldsymbol{D}^{-1}\boldsymbol{X})^{-}\boldsymbol{X}'\boldsymbol{D}^{-1}\boldsymbol{y} \tag{11.107}$$

with

$$\begin{aligned} SSE(\widetilde{\boldsymbol{\beta}}) &= \boldsymbol{y}'[\boldsymbol{I} - \boldsymbol{L}'\boldsymbol{X}(\boldsymbol{X}'\boldsymbol{D}^{-1}\boldsymbol{X})^{-}\boldsymbol{X}'\boldsymbol{L}]\boldsymbol{y} \\ &= \boldsymbol{y}'\boldsymbol{L}[\boldsymbol{I} - \boldsymbol{L}'\boldsymbol{X}(\boldsymbol{X}'\boldsymbol{D}^{-1}\boldsymbol{X})^{-}\boldsymbol{X}'\boldsymbol{L}]\boldsymbol{L}'\boldsymbol{y}, \end{aligned} \tag{11.108}$$

which is invariant to the choice of generalized inverse.
(b) If $\text{rank}(\boldsymbol{X}) = q$, then $(\boldsymbol{X}'\boldsymbol{D}^{-1}\boldsymbol{X})^{-} = (\boldsymbol{X}'\boldsymbol{D}^{-1}\boldsymbol{X})^{-1}$ and $\widetilde{\boldsymbol{\beta}} = \widehat{\boldsymbol{\beta}} = (\boldsymbol{X}'\boldsymbol{D}^{-1}\boldsymbol{X})^{-1}\boldsymbol{X}'\boldsymbol{D}^{-1}\boldsymbol{y}$ is unique and unbiased.

Proof. Left as an exercise.

Theorem 11.27 Ordinary least squares (OLS) and WLS estimators may coincide. If

$$\widehat{\boldsymbol{\beta}}_D = (\boldsymbol{X}'\boldsymbol{D}^{-1}\boldsymbol{X})^{-1}\boldsymbol{X}'\boldsymbol{D}^{-1}\boldsymbol{y}, \tag{11.109}$$

$$\widehat{\boldsymbol{\beta}} = (\boldsymbol{X}'\boldsymbol{X})^{-1}\boldsymbol{X}'\boldsymbol{y}, \tag{11.110}$$

$$\widehat{\sigma}^2_D = (\boldsymbol{y} - \boldsymbol{X}\widehat{\boldsymbol{\beta}}_D)'\boldsymbol{D}^{-1}(\boldsymbol{y} - \boldsymbol{X}\widehat{\boldsymbol{\beta}}_D)/N\,, \tag{11.111}$$

and

$$\widehat{\sigma}^2 = (\boldsymbol{y} - \boldsymbol{X}\widehat{\boldsymbol{\beta}})'\boldsymbol{D}^{-1}(\boldsymbol{y} - \boldsymbol{X}\widehat{\boldsymbol{\beta}})/N\,, \tag{11.112}$$

then $\widehat{\boldsymbol{\beta}}_D = \widehat{\boldsymbol{\beta}} \Leftrightarrow \exists \boldsymbol{B}^{-1}$ such that $\boldsymbol{D}\boldsymbol{X}\boldsymbol{B}^{-1} = \boldsymbol{X}$. If so, then $\widehat{\sigma}^2 = \widehat{\sigma}^2_D$.

Proof. $(\Rightarrow)$ The fact that $\widehat{\boldsymbol{\beta}}_D = \widehat{\boldsymbol{\beta}}$ $(\forall \boldsymbol{y})$ implies

$$\begin{aligned}(\boldsymbol{X}'\boldsymbol{D}^{-1}\boldsymbol{X})^{-1}\boldsymbol{X}'\boldsymbol{D}^{-1} &= (\boldsymbol{X}'\boldsymbol{X})^{-1}\boldsymbol{X}' \\ \boldsymbol{X}' &= \left[(\boldsymbol{X}'\boldsymbol{D}^{-1}\boldsymbol{X})(\boldsymbol{X}'\boldsymbol{X})^{-1}\right]\boldsymbol{X}'\boldsymbol{D} \\ \boldsymbol{B}^{-1} &= \left[(\boldsymbol{X}'\boldsymbol{D}^{-1}\boldsymbol{X})(\boldsymbol{X}'\boldsymbol{X})^{-1}\right]'.\end{aligned} \tag{11.113}$$

$(\Leftarrow)$ If such $\boldsymbol{B}$ exists, then together $\boldsymbol{D}^{-1}\boldsymbol{X} = \boldsymbol{X}\boldsymbol{B}^{-1}$ and $(\boldsymbol{X}'\boldsymbol{D}^{-1}\boldsymbol{X})^{-1} = \boldsymbol{B}(\boldsymbol{X}'\boldsymbol{X})^{-1}$ give $\boldsymbol{D}^{-1}\boldsymbol{X}(\boldsymbol{X}'\boldsymbol{D}^{-1}\boldsymbol{X})^{-1} = \boldsymbol{X}\boldsymbol{B}^{-1}\boldsymbol{B}(\boldsymbol{X}'\boldsymbol{X})^{-1}$. In turn, $(\boldsymbol{X}'\boldsymbol{D}^{-1}\boldsymbol{X})^{-1}\boldsymbol{X}'\boldsymbol{D}^{-1} = (\boldsymbol{X}'\boldsymbol{X})^{-1}\boldsymbol{X}'$, or $\widehat{\boldsymbol{\beta}}_D = \widehat{\boldsymbol{\beta}}\ \forall \boldsymbol{y}$. □

The special case of compound symmetric covariance illustrates the simplification that can occur with known covariance in a GGLM. The importance of the next lemma lies in the "univariate" approach to repeated measures and is discussed in detail in the context of hypothesis testing.

Lemma 11.8 A $\mathrm{GGLM}_{N,q}(\boldsymbol{y}; \boldsymbol{X}\boldsymbol{\beta}, \boldsymbol{D})$ may have $\boldsymbol{D}$ which is compound symmetric (Lemma 1.33 summarizes properties),

$$\begin{aligned}\boldsymbol{D} &= \sigma^2[\rho\boldsymbol{1}_N\boldsymbol{1}_N' + (1-\rho)\boldsymbol{I}_N] \\ &= \sigma^2\begin{bmatrix}1 & \rho & \cdots & \rho \\ \rho & 1 & \cdots & \rho \\ \vdots & \vdots & \ddots & \vdots \\ \rho & \rho & \cdots & 1\end{bmatrix} \\ &= \boldsymbol{V}\mathrm{Dg}(\boldsymbol{\lambda})\boldsymbol{V}'\,,\end{aligned} \tag{11.114}$$

with unknown $\sigma^2 > 0$ and $-1/(p-1) < \rho < 1$. Such models allow an exact transformation to a model with uncorrelated but heteroscedastic observations, $\mathrm{GGLM}_{N,q}[\boldsymbol{V}'\boldsymbol{y}; \boldsymbol{V}'\boldsymbol{X}\boldsymbol{\beta}, \mathrm{Dg}(\boldsymbol{\lambda})]$.

Proof. The restrictions on the unknown parameters are necessary and sufficient for $\boldsymbol{D}$ to be positive definite. The eigenvalues are $\lambda_1 = \sigma^2(1 - \rho + p\rho)$ and, $\forall j \neq 1,\ \lambda_j = \lambda_2 = \sigma^2(1-\rho)$. By Lemma 1.33, the eigenvectors $\boldsymbol{V} = [\,\boldsymbol{v}_0\ \boldsymbol{v}_1 \cdots \boldsymbol{v}_{N-1}\,]$ may be taken to be $\boldsymbol{v}_0 = p^{-1/2}\boldsymbol{1}_N$ and $[\,\boldsymbol{v}_2\ \cdots\ \boldsymbol{v}_{N-1}\,] = \boldsymbol{V}_T$, the set of $N-1$ normalized trends for N measurements. Hence $\boldsymbol{V}$ may be specified without knowing σ^2 or ρ. The result gives an exact transformation of $\boldsymbol{y} = \boldsymbol{X}\boldsymbol{\beta} + \boldsymbol{e}$ to a model with uncorrelated but heteroscedastic observations, $\boldsymbol{V}'\boldsymbol{y} = \boldsymbol{V}'\boldsymbol{X}\boldsymbol{\beta} + \boldsymbol{V}'\boldsymbol{e}$. The new model has $\boldsymbol{V}'\boldsymbol{e} \sim \mathcal{N}_N[\boldsymbol{0}, \mathrm{Dg}(\boldsymbol{\lambda})]$, corresponding to $\mathrm{GGLM}_{N,q}[\boldsymbol{V}'\boldsymbol{y}; \boldsymbol{V}'\boldsymbol{X}\boldsymbol{\beta}, \mathrm{Dg}(\boldsymbol{\lambda})]$. □

EXERCISES

Prove Theorem 11.9. For a $\mathrm{GLM}_{N,q}\mathrm{FR}(y_i; \boldsymbol{X}_i\boldsymbol{\beta}, \sigma^2)$ with Gaussian errors, estimators

$$\widehat{\boldsymbol{\beta}} = (\boldsymbol{X}'\boldsymbol{X})^{-1}\boldsymbol{X}'\boldsymbol{y} \tag{11.49}$$

$$\widehat{\sigma}^2 = \frac{1}{N-q}\boldsymbol{y}'\left[\boldsymbol{I} - \boldsymbol{X}\,(\boldsymbol{X}'\boldsymbol{X})^{-1}\boldsymbol{X}'\right]\boldsymbol{y} \tag{11.49}$$

are
11.1 consistent,
11.2 efficient,
11.3 unbiased,
11.4 complete,
11.5 sufficient, and
11.6 UMVUE.
11.7 Furthermore $\widehat{\boldsymbol{\beta}}$ and $\widehat{\sigma}^2$ are mutually independent,
11.8 $\widehat{\boldsymbol{\beta}} \sim \mathcal{N}_q[\boldsymbol{\beta}, \sigma^2(\boldsymbol{X}'\boldsymbol{X})^{-1}]$ and
11.9 $\widehat{\sigma}^2(N-q)/\sigma^2 \sim \chi^2(N-q)$.

Hints. You may prove the results in any order you wish.

Among asymptotically unbiased estimators, one with minimum variance in large samples is called an efficient estimator or simply efficient.

You may cite results in Chapters 1–11 for the exercises. When you cite a result from Chapters 1–11, clearly indicate the number of the theorem, corollary, or lemma that you are using.

For your own enlightenment (but not for the exercises), you may wish to consider proving the results without using any results from Chapter 11.

CHAPTER 12

Estimation for Multivariate Linear Models

12.1 ALTERNATE FORMULATIONS OF THE MODEL

The univariate general linear model concerns an $N \times 1$ vector of responses $\boldsymbol{y}$ with all observations independent. The multivariate general linear model allows generalizing the response vector to an $N \times p$ matrix $\boldsymbol{Y}$ with some observations independent and some dependent (correlated). Only particular patterns of correlation meet the assumptions. *For data with correlated observations, distinguishing between the observational unit and the independent sampling unit often provides the key first step in choosing a valid model.*

Depending on the task at hand, it may be more convenient to express $N \times p\, \boldsymbol{Y}$ in terms of its N rows $\{\boldsymbol{Y}_i\}$ or its p columns $\{\boldsymbol{y}_j\}$. Each row, sometimes indicated $\boldsymbol{Y}_i = \text{row}_i(\boldsymbol{Y})$, is $1 \times p$, while each column, say $\boldsymbol{y}_j = \text{col}_j(\boldsymbol{Y})$, is $N \times 1$. Each row corresponds to a particular independent sampling unit, with elements within a row being observations for the sampling unit. In a study with (unrelated) human participants, each row of $\boldsymbol{Y}$ contains the data for one person, while different columns contain different response variables, which might be repeated measures of a single variable. It helps to remember that

$$\boldsymbol{Y} = \begin{bmatrix} \boldsymbol{y}_1 \, \boldsymbol{y}_2 \cdots \boldsymbol{y}_p \end{bmatrix} = \begin{bmatrix} \boldsymbol{Y}_1 \\ \boldsymbol{Y}_2 \\ \vdots \\ \boldsymbol{Y}_N \end{bmatrix}. \tag{12.1}$$

The first form decomposes $\boldsymbol{Y}$ into p variables and the second into N independent sampling units.

The present chapter centers on deriving estimators for $\boldsymbol{B}$ and Σ which satisfy a variety of optimal properties. The great majority of such properties do not depend on a particular choice of distribution function for the responses and are exact, even in small samples. In contrast, the desire to test hypotheses leads to describing distributions of test statistics. Finding exact test distributions for small samples usually requires explicit and particular specification of the distribution of the data. The most common choice involves assuming each row of errors independently follows a multivariate Gaussian distribution.

Describing the model in vector form helps explain the many relationships among the elements. The vector form helps in understanding the connections and differences among the multivariate and univariate GLM, mixed models, and more general forms. It also provides a convenient basis for many proofs of multivariate GLM properties. In parallel to expressions for $\boldsymbol{Y}$ in terms of rows or columns,

$$\boldsymbol{X} = [\,\boldsymbol{x}_1\ \boldsymbol{x}_2\ \cdots\ \boldsymbol{x}_q\,] = \begin{bmatrix} \boldsymbol{X}_1 \\ \boldsymbol{X}_2 \\ \vdots \\ \boldsymbol{X}_N \end{bmatrix}, \tag{12.2}$$

$$\boldsymbol{B} = [\,\boldsymbol{\beta}_1\ \boldsymbol{\beta}_2\ \cdots\ \boldsymbol{\beta}_p\,] = \begin{bmatrix} \boldsymbol{B}_1 \\ \boldsymbol{B}_2 \\ \vdots \\ \boldsymbol{B}_q \end{bmatrix}, \tag{12.3}$$

and

$$\boldsymbol{E} = [\,\boldsymbol{e}_1\ \boldsymbol{e}_2\ \cdots\ \boldsymbol{e}_p\,] = \begin{bmatrix} \boldsymbol{E}_1 \\ \boldsymbol{E}_2 \\ \vdots \\ \boldsymbol{E}_N \end{bmatrix}. \tag{12.4}$$

As discussed in Chapter 3, a multivariate linear model is often written as $\boldsymbol{Y} = \boldsymbol{X}\boldsymbol{B} + \boldsymbol{E}$. The notation just described allows stacking the data *by variable* (column of $\boldsymbol{Y}$, $\boldsymbol{B}$, and $\boldsymbol{E}$) to give a single column of responses and errors:

$$\begin{aligned}
\operatorname{vec}(\boldsymbol{Y}) &= \operatorname{vec}(\boldsymbol{X}\boldsymbol{B}) + \operatorname{vec}(\boldsymbol{E}) \\
\begin{bmatrix} \boldsymbol{y}_1 \\ \boldsymbol{y}_2 \\ \vdots \\ \boldsymbol{y}_p \end{bmatrix} &= \begin{bmatrix} \boldsymbol{X}\boldsymbol{\beta}_1 \\ \boldsymbol{X}\boldsymbol{\beta}_2 \\ \vdots \\ \boldsymbol{X}\boldsymbol{\beta}_p \end{bmatrix} + \begin{bmatrix} \boldsymbol{e}_1 \\ \boldsymbol{e}_2 \\ \vdots \\ \boldsymbol{e}_p \end{bmatrix} \\
\begin{bmatrix} \boldsymbol{y}_1 \\ \boldsymbol{y}_2 \\ \vdots \\ \boldsymbol{y}_p \end{bmatrix} &= (\boldsymbol{I}_p \otimes \boldsymbol{X})\operatorname{vec}(\boldsymbol{B}) + \begin{bmatrix} \boldsymbol{e}_1 \\ \boldsymbol{e}_2 \\ \vdots \\ \boldsymbol{e}_p \end{bmatrix} \\
\begin{bmatrix} \boldsymbol{y}_1 \\ \boldsymbol{y}_2 \\ \vdots \\ \boldsymbol{y}_p \end{bmatrix} &= (\boldsymbol{I}_p \otimes \boldsymbol{X}) \begin{bmatrix} \boldsymbol{\beta}_1 \\ \boldsymbol{\beta}_2 \\ \vdots \\ \boldsymbol{\beta}_p \end{bmatrix} + \begin{bmatrix} \boldsymbol{e}_1 \\ \boldsymbol{e}_2 \\ \vdots \\ \boldsymbol{e}_p \end{bmatrix} \\
\begin{bmatrix} \boldsymbol{y}_1 \\ \boldsymbol{y}_2 \\ \vdots \\ \boldsymbol{y}_p \end{bmatrix} &= \begin{bmatrix} \boldsymbol{X} & \boldsymbol{0} & \cdots & \boldsymbol{0} \\ \boldsymbol{0} & \boldsymbol{X} & \cdots & \boldsymbol{0} \\ \vdots & \vdots & \ddots & \vdots \\ \boldsymbol{0} & \boldsymbol{0} & \cdots & \boldsymbol{X} \end{bmatrix} \begin{bmatrix} \boldsymbol{\beta}_1 \\ \boldsymbol{\beta}_2 \\ \vdots \\ \boldsymbol{\beta}_p \end{bmatrix} + \begin{bmatrix} \boldsymbol{e}_1 \\ \boldsymbol{e}_2 \\ \vdots \\ \boldsymbol{e}_p \end{bmatrix}.
\end{aligned} \tag{12.5}$$

Stacking the data by variable allows concluding that the submodel for response variable j is a (valid) GLM corresponding to $\boldsymbol{y}_j = \boldsymbol{X}\boldsymbol{\beta}_j + \boldsymbol{e}_j$, with a common design matrix and distinct parameters $\text{GLM}_{N,q}(y_{ij}; \boldsymbol{X}_i\boldsymbol{\beta}_j, \sigma_j^2)$. Elements of $\boldsymbol{X}$ and $\boldsymbol{\beta}_j$, combine to describe differences *between* sampling units in response variable j.

Alternatively, stacking the data by independent sampling unit (row of $\boldsymbol{Y}$, $\boldsymbol{X}$, and $\boldsymbol{E}$) focuses on differences *within* sampling unit, between correlated response variables, as captured by $\boldsymbol{B}_k$. The approach gives

$$\begin{aligned}
\text{vec}(\boldsymbol{Y}') &= \text{vec}\left[(\boldsymbol{X}\boldsymbol{B})'\right] + \text{vec}(\boldsymbol{E}') \\
\begin{bmatrix} \boldsymbol{Y}_1' \\ \boldsymbol{Y}_2' \\ \vdots \\ \boldsymbol{Y}_N' \end{bmatrix} &= \begin{bmatrix} (\boldsymbol{X}_1\boldsymbol{B})' \\ (\boldsymbol{X}_2\boldsymbol{B})' \\ \vdots \\ (\boldsymbol{X}_N\boldsymbol{B})' \end{bmatrix} + \begin{bmatrix} \boldsymbol{E}_1' \\ \boldsymbol{E}_2' \\ \vdots \\ \boldsymbol{E}_N' \end{bmatrix} \\
\begin{bmatrix} \boldsymbol{Y}_1' \\ \boldsymbol{Y}_2' \\ \vdots \\ \boldsymbol{Y}_N' \end{bmatrix} &= (\boldsymbol{X} \otimes \boldsymbol{I}_p)\text{vec}(\boldsymbol{B}') + \begin{bmatrix} \boldsymbol{E}_1' \\ \boldsymbol{E}_2' \\ \vdots \\ \boldsymbol{E}_N' \end{bmatrix} \\
\begin{bmatrix} \boldsymbol{Y}_1' \\ \boldsymbol{Y}_2' \\ \vdots \\ \boldsymbol{Y}_N' \end{bmatrix} &= (\boldsymbol{X} \otimes \boldsymbol{I}_p)\begin{bmatrix} \boldsymbol{B}_1' \\ \boldsymbol{B}_2' \\ \vdots \\ \boldsymbol{B}_q' \end{bmatrix} + \begin{bmatrix} \boldsymbol{E}_1' \\ \boldsymbol{E}_2' \\ \vdots \\ \boldsymbol{E}_N' \end{bmatrix} \\
\begin{bmatrix} \boldsymbol{Y}_1' \\ \boldsymbol{Y}_2' \\ \vdots \\ \boldsymbol{Y}_N' \end{bmatrix} &= \begin{bmatrix} x_{11}\boldsymbol{I}_p & x_{12}\boldsymbol{I}_p & \cdots & x_{1q}\boldsymbol{I}_p \\ x_{21}\boldsymbol{I}_p & x_{22}\boldsymbol{I}_p & \cdots & x_{2q}\boldsymbol{I}_p \\ \vdots & \vdots & \ddots & \vdots \\ x_{N1}\boldsymbol{I}_p & x_{N2}\boldsymbol{I}_p & \cdots & x_{Nq}\boldsymbol{I}_p \end{bmatrix} \begin{bmatrix} \boldsymbol{B}_1' \\ \boldsymbol{B}_2' \\ \vdots \\ \boldsymbol{B}_q' \end{bmatrix} + \begin{bmatrix} \boldsymbol{E}_1' \\ \boldsymbol{E}_2' \\ \vdots \\ \boldsymbol{E}_N' \end{bmatrix}.
\end{aligned} \tag{12.6}$$

The corresponding equation for sampling unit i, namely

$$\boldsymbol{Y}_i' = (\boldsymbol{X}_i \otimes \boldsymbol{I}_p)\text{vec}(\boldsymbol{B}') + \boldsymbol{E}_i', \tag{12.7}$$

does not describe a valid univariate GLM because $\mathcal{V}(\boldsymbol{Y}_i') = \mathcal{V}(\boldsymbol{E}_i') = \boldsymbol{\Sigma}$. However, it does implicitly define $\text{GGLM}_{p,q}[\boldsymbol{Y}_i'; (\boldsymbol{X}_i \otimes \boldsymbol{I}_p)\text{vec}(\boldsymbol{B}'), \boldsymbol{\Sigma}]$ for independent sampling unit i and $\text{GGLM}_{Np,q}[\text{vec}(\boldsymbol{Y}'); (\boldsymbol{X} \otimes \boldsymbol{I}_p)\text{vec}(\boldsymbol{B}'), \boldsymbol{I}_p \otimes \boldsymbol{\Sigma}]$ for all of the data. Later discussions of mixed models will demonstrate that the equation also defines a corresponding particular mixed model.

As a special case of a mixed model, the defining characteristics of a multivariate GLM are "Kronecker design" and "Kronecker covariance." The first requires a common design matrix for all response variables (columns of $\boldsymbol{Y}$, which may be repeated measures), and the second requires a common covariance matrix for all independent sampling units (rows of $\boldsymbol{Y}$, which may be persons). Many current approaches to mixed models were developed largely to allow relaxing the restrictive assumptions of homogenous design and homogenous covariance inherent in the multivariate GLM.

Definition 12.1 (a) The matrices $\boldsymbol{B}$ ($q \times p$) and $\boldsymbol{\Sigma}$ ($p \times p$) are the *primary parameters* of a multivariate $\text{GLM}_{N,p,q}(\boldsymbol{Y}_i; \boldsymbol{X}_i\boldsymbol{B}, \boldsymbol{\Sigma})$ with or without Gaussian errors.
(b) Any (finite) known constant $\boldsymbol{C}$ ($a \times q$), (finite) known constant $\boldsymbol{U}$ ($p \times b$), and (finite) known constant $\boldsymbol{\Theta}_0$ ($a \times b$) define *secondary parameter* $\boldsymbol{\Theta} = \boldsymbol{CBU} + \boldsymbol{\Theta}_0$ ($a \times b$).

Definition 12.2 (a) *Estimators* of primary parameters with good properties, especially unbiasedness, are indicated by $\widehat{\boldsymbol{B}}$ and $\widehat{\boldsymbol{\Sigma}}$, while ones with distinct and possibly fewer desirable properties are indicated by $\widetilde{\boldsymbol{B}}$ and $\widetilde{\boldsymbol{\Sigma}}$.
(b) Estimators of secondary parameters take the form $\widehat{\boldsymbol{\Theta}} = \boldsymbol{C}\widehat{\boldsymbol{B}}\boldsymbol{U} + \boldsymbol{\Theta}_0$, or $\widetilde{\boldsymbol{\Theta}} = \boldsymbol{C}\widetilde{\boldsymbol{B}}\boldsymbol{U} + \boldsymbol{\Theta}_0$, or $\widehat{\boldsymbol{\Theta}} = \boldsymbol{C}\widetilde{\boldsymbol{B}}\boldsymbol{U} + \boldsymbol{\Theta}_0$. The third form is used only when $\widehat{\boldsymbol{\Theta}}$ possesses a desirable property not shared by $\widetilde{\boldsymbol{B}}$, such as unbiasedness.
(c) Covariance matrices such as $\mathcal{V}[\text{vec}(\widehat{\boldsymbol{\Theta}})]$ are also secondary parameters.

Definition 12.3 (a) With $\boldsymbol{T}_1$ ($m_1 \times p$) and $\boldsymbol{T}_2$ ($m_2 \times p$) known constant matrices, $\text{GLM}_{N,p,q_1}(\boldsymbol{Y}_i; \boldsymbol{X}_{i1}\boldsymbol{B}_1\boldsymbol{T}_1, \boldsymbol{\Sigma})$ and $\text{GLM}_{N,p,q_2}(\boldsymbol{Y}_i; \boldsymbol{X}_{i2}\boldsymbol{B}_2\boldsymbol{T}_2, \boldsymbol{\Sigma})$ are *linearly equivalent between subjects* whenever the *columns* of $\boldsymbol{X}_1$ and $\boldsymbol{X}_2$ span the same subspace of $\Re^q$.
(b) The two models are *linearly equivalent within subject* whenever the *rows* of $\boldsymbol{T}_1$ and $\boldsymbol{T}_2$ span the same subspace of $\Re^p$.
(c) If both conditions **(a)** and **(b)** hold, then the two models are simply *linearly equivalent.*

Linear equivalence describes the expected values, the means, of $\{y_{ij}\}$, because $\text{E}(\boldsymbol{Y}) = \boldsymbol{X}_1\boldsymbol{B}_1\boldsymbol{T}_1$ and $\text{E}(\boldsymbol{Y}) = \boldsymbol{X}_2\boldsymbol{B}_2\boldsymbol{T}_2$. While $\boldsymbol{X}_1$ and $\boldsymbol{X}_2$ provide between-subject information, $\boldsymbol{T}_1$ and $\boldsymbol{T}_2$ provide within-subject information. Most often, one or both of $\boldsymbol{T}_1$ and $\boldsymbol{T}_2$ are simply $\boldsymbol{I}_p$ (invisible). With linearly equivalent between-subject models, (1) for any $\boldsymbol{B}_1$ there exists $\boldsymbol{B}_2$ such that $\boldsymbol{X}_1\boldsymbol{B}_1 = \boldsymbol{X}_2\boldsymbol{B}_2$ and (2) for any $\boldsymbol{B}_2$ there exists $\boldsymbol{B}_1$ such that $\boldsymbol{X}_1\boldsymbol{B}_1 = \boldsymbol{X}_2\boldsymbol{B}_2$. With linearly equivalent within-subject models, (1) for any $\boldsymbol{B}_1$ there exists $\boldsymbol{B}_2$ such that $\boldsymbol{B}_1\boldsymbol{T}_1 = \boldsymbol{B}_2\boldsymbol{T}_2$ and (2) for any $\boldsymbol{B}_2$ there exists $\boldsymbol{B}_1$ such that $\boldsymbol{B}_1\boldsymbol{T}_1 = \boldsymbol{B}_2\boldsymbol{T}_2$.

Obviously, if $\boldsymbol{T}_1 = \boldsymbol{T}_2$, then no attention must be paid to the question of linear equivalence within subjects. Often we have $\boldsymbol{T}_1 = \boldsymbol{T}_2 = \boldsymbol{I}_p$. In such cases, verifying linear equivalence of two multivariate models ($p > 1$) reduces to considering only between-subject properties. The multivariate setting requires the univariate requirement to apply simultaneously to p parameter vector pairs. Here

$$\boldsymbol{Y} = [\boldsymbol{y}_1\, \boldsymbol{y}_2 \cdots \boldsymbol{y}_p] \tag{12.8}$$

$$\boldsymbol{B}_1 = [\boldsymbol{\beta}_{1,1}\, \boldsymbol{\beta}_{2,1} \cdots \boldsymbol{\beta}_{p,1}] \tag{12.9}$$

$$\boldsymbol{B}_2 = [\boldsymbol{\beta}_{1,2}\, \boldsymbol{\beta}_{2,2} \cdots \boldsymbol{\beta}_{p,2}]\,. \tag{12.10}$$

Linear equivalence of $\boldsymbol{Y} = \boldsymbol{X}_1\boldsymbol{B}_1 + \boldsymbol{E}_1$ and $\boldsymbol{Y} = \boldsymbol{X}_2\boldsymbol{B}_2 + \boldsymbol{E}_2$ corresponds to linear equivalence of all p univariate model pairs; $\boldsymbol{y}_j = \boldsymbol{X}_1\boldsymbol{\beta}_{j,1} + \boldsymbol{e}_{j,1}$ must be linearly equivalent to $\boldsymbol{y}_j = \boldsymbol{X}_2\boldsymbol{\beta}_{j,2} + \boldsymbol{e}_{j,2}$.

If $\boldsymbol{X}_1 = \boldsymbol{X}_2$, then no question arises about linear equivalence between subjects. Similarly, if $\boldsymbol{T}_1 = \boldsymbol{T}_2$, then no question arises about linear equivalence within subjects. If both conditions hold, the term "linear equivalence" can be discussed without ambiguity.

Example 12.1 A growth curve model provides one application of $\boldsymbol{T}_1 \neq \boldsymbol{I}_p$, with the columns of $\boldsymbol{Y}$ containing repeated measures and $\boldsymbol{T}$ defining the (within-subject) predictor values of interest as a function of time. With $p = 3$ times and no between-subject factor, a simple example is

$$\begin{aligned}\boldsymbol{Y} &= \boldsymbol{XBT} + \boldsymbol{E} \\ &= \boldsymbol{1}_N[\,\beta_0\ \beta_1\ \beta_2\,]\begin{bmatrix}1 & 1 & 1\\ t_1 & t_2 & t_3\\ t_1^2 & t_2^2 & t_3^2\end{bmatrix} + \boldsymbol{E}\,. \end{aligned} \tag{12.11}$$

The corresponding scalar form is $y_{ij} = \beta_0 + \beta_1 t_1 + \beta_2 t_1^2 + e_{ij}$. The model may be converted to one without $\boldsymbol{T}$:

$$\begin{aligned}\boldsymbol{YT}^{-1} &= \boldsymbol{XBT}^{-1} + \boldsymbol{ET}^{-1} \\ \boldsymbol{Y}_T &= \boldsymbol{XB}_T \quad + \boldsymbol{E}_T\,. \end{aligned} \tag{12.12}$$

If $\boldsymbol{T}$ has $m \leq p$ rows, eliminating $\boldsymbol{T}$ requires multiplying by $\boldsymbol{T}'(\boldsymbol{TT}')^{-1} = \boldsymbol{T}^{+}$.

Overall, four classes of multivariate GLMs are defined by allowing either FR or LTFR designs, combined with either Gaussian or unspecified distributions for responses. For $\text{GLM}_{N,p,q}(\boldsymbol{Y}_i; \boldsymbol{X}_i\boldsymbol{B}|\boldsymbol{R}_x\boldsymbol{B}\boldsymbol{R}_y = \boldsymbol{A}, \Sigma)$, with or without Gaussian errors, estimation theory for primary parameter $\boldsymbol{B}$ and secondary parameter $\boldsymbol{\Theta}$ is closely tied to $r = \text{rank}([\,\boldsymbol{X}'\ \boldsymbol{R}_x'\,]')$.

The distinction between FR and LTFR describes properties of the columns of $[\,\boldsymbol{X}'\ \boldsymbol{R}_x'\,]'$, which correspond directly to properties of the rows of $\boldsymbol{B}$. The form of the model statement implies the properties apply only to every element of each row and hence to each column in $\boldsymbol{B}$ considered separately. Just as for the univariate models considered in the preceding chapter, every multivariate GLM constrained by $\boldsymbol{R}_x \neq \boldsymbol{I}_q$ has a linearly equivalent unconstrained model.

The presence of $\boldsymbol{R}_y \neq \boldsymbol{I}_p$ describes constraints among the columns of $\boldsymbol{Y}$, which correspond directly to properties of the columns of $\boldsymbol{B}$. Such constraints may either introduce or eliminate singularity of the error covariance matrix. Not surprisingly, every multivariate GLM constrained by $\boldsymbol{R}_y \neq \boldsymbol{I}_p$ has a linearly equivalent unconstrained model.

In the absence of $\boldsymbol{R}_x$, which gives row restrictions on $\boldsymbol{B}$, the distinction between FR and LTFR depends solely on $r = \text{rank}(\boldsymbol{X})$. FR models have an unbiased estimator for $\boldsymbol{B}$ and for all $\boldsymbol{\Theta}$, while LTFR models never have an unbiased estimator for $\boldsymbol{B}$ and have an unbiased estimator for only some $\boldsymbol{\Theta}$.

Whether or not the model is FR, unbiased estimators of Σ and related properties are always available [as long as $N > r = \text{rank}([\boldsymbol{X}' \ \boldsymbol{R}_x']')$]. The presence or absence of $\boldsymbol{R}_y$ has no effect on bias of estimators. However, it may introduce singularity and hence cause distributions to degenerate. In turn, hypothesis tests may be undefined and require modification. Just as an equivalent FR model can be found for any LTFR model, an equivalent "nonsingular" model can be found for any "singular" model.

Theorem 12.1 Every multivariate GLM with LTFR design matrix has a linearly equivalent model with a full-rank design matrix. In particular, for $\text{GLM}_{N,p,q}\text{LTFR}(\boldsymbol{Y}_i; \boldsymbol{X}_i\boldsymbol{B}, \Sigma)$ with $\text{rank}(\boldsymbol{X}) = r < q$, there exists $\text{GLM}_{N,p,r}\text{FR}(\boldsymbol{Y}_i; \boldsymbol{X}_{i,1}\boldsymbol{B}_1, \Sigma)$ which is linearly equivalent with $\text{rank}(\boldsymbol{X}_1) = r$.

Proof. With the subscript 1 indicating the components corresponding to positive singular values, the SVD allows writing

$$\begin{aligned} \boldsymbol{X} &= [\, \boldsymbol{L}_1 \ \boldsymbol{L}_0 \,] \begin{bmatrix} \text{Dg}(\boldsymbol{s}_1) & 0 \\ 0 & 0 \end{bmatrix} \begin{bmatrix} \boldsymbol{R}_1' \\ \boldsymbol{R}_0' \end{bmatrix} \\ &= \boldsymbol{L}_1 \text{Dg}(\boldsymbol{s}_1) \boldsymbol{R}_1' \end{aligned} \tag{12.13}$$

and also allows defining

$$\boldsymbol{X}_1 = \boldsymbol{L}_1 \text{Dg}(\boldsymbol{s}_1) \tag{12.14}$$

$$\boldsymbol{B}_1 = \boldsymbol{R}_1' \boldsymbol{B} \,. \tag{12.15}$$

In turn

$$\begin{aligned} \boldsymbol{Y} &= \boldsymbol{X}\boldsymbol{B} \ + \boldsymbol{E} \\ &= \boldsymbol{X}_1 \boldsymbol{B}_1 + \boldsymbol{E} \,, \end{aligned} \tag{12.16}$$

with full rank $\boldsymbol{X}_1$. Although well behaved and elegant, the full-rank model is merely one particular choice among infinitely many equivalent models. □

12.2 ESTIMABILITY IN THE MULTIVARIATE GLM

For secondary parameter $\Theta = \boldsymbol{C}\boldsymbol{B}\boldsymbol{U}$, the within-subject contrast matrix $\boldsymbol{U}$ helps define Θ. It also plays a key role in determining (within-subject) linear equivalence as well as whether or not a valid hypothesis test exists. However, it has no effect on estimability.

Definition 12.4 A $\text{GLM}_{N,p,q}(\boldsymbol{Y}_i; \boldsymbol{X}_{i1}\boldsymbol{B}, \Sigma)$ has $\Theta = \boldsymbol{C}\boldsymbol{B}\boldsymbol{U}$.
(a) Primary parameter $\boldsymbol{B}$ is *estimable* if and only if a $q \times N$ constant matrix $\boldsymbol{A}_1$ exists such that $\text{E}(\boldsymbol{A}_1\boldsymbol{Y}) = \boldsymbol{B}$.
(b) Secondary parameter Θ is *estimable* if and only if constant matrix $\boldsymbol{A}_2$ $(a \times N)$ exists such that $\text{E}(\boldsymbol{A}_2\boldsymbol{Y}\boldsymbol{U}) = \Theta$.
(c) For known, fixed Θ_0, $\Theta + \Theta_0$ is *estimable* if and only if Θ is *estimable*.

Part (a) of the definition is essentially redundant since it is a special case of part (b). Choosing $\boldsymbol{C} = \boldsymbol{I}_q$ defines $\boldsymbol{B}$ as a secondary parameter.

Theorem 12.2 In $\text{GLM}_{N,p,q}\text{FR}(\boldsymbol{Y}_i; \boldsymbol{X}_i\boldsymbol{B}, \Sigma)$ with $\boldsymbol{\Theta} = \boldsymbol{CBU}$ defined by constants $\boldsymbol{C}$ and $\boldsymbol{U}$, both $\boldsymbol{B}$ and $\boldsymbol{\Theta}$ are estimable.

Proof. The proof is essentially the same as for univariate models. Only the properties of the design matrix $\boldsymbol{X}$ are involved in the issue of estimability. □

Theorem 12.3 For $\text{GLM}_{N,p,q}\text{LTFR}(\boldsymbol{Y}_i; \boldsymbol{X}_i\boldsymbol{B}, \Sigma)$ with $\boldsymbol{\Theta} = \boldsymbol{CBU}$, the following hold.

(a) With $\boldsymbol{A}_1$ $q \times N$ and constant, *no* linear transformation of the data, say $\widehat{\boldsymbol{B}} = \boldsymbol{A}_1\boldsymbol{Y}$, exists such that $\text{E}(\widehat{\boldsymbol{B}}) = \boldsymbol{B}$.

(b) For $a \times b$ secondary parameter $\boldsymbol{\Theta}$, with constants $\boldsymbol{A}_2$ $a \times N$ and $\boldsymbol{U}$ $p \times b$, a linear transformation of the data, say $\widehat{\boldsymbol{\Theta}} = \boldsymbol{A}_2\boldsymbol{YU}$, *may* or *may not* exist such that $\text{E}(\widehat{\boldsymbol{\Theta}}) = \boldsymbol{\Theta}$.

(c) If $\boldsymbol{X}_i = \text{row}_i(\boldsymbol{X})$, $\boldsymbol{E}_i = \text{row}_i(\boldsymbol{I}_N)$, and $\theta_{ij} = \text{E}(\boldsymbol{E}_i\boldsymbol{Y}\boldsymbol{E}_j') = \boldsymbol{E}_i\boldsymbol{XB}\boldsymbol{E}_j' = \boldsymbol{X}_i\boldsymbol{B}\boldsymbol{E}_j$, then θ_{ij} is always *estimable*.

(d) The within-subject contrast matrix $\boldsymbol{U}$ plays no role in determining estimability of $\boldsymbol{\Theta}$, only $\boldsymbol{C}$ does. Secondary parameters $\boldsymbol{\Theta} = \boldsymbol{CBU}$ and $\boldsymbol{\Theta} - \boldsymbol{\Theta}_0$ are estimable if and only if $\boldsymbol{CB}$ is estimable.

Proof. Essentially the same as for univariate models, by considering individual columns of $\boldsymbol{U}$. Only the properties of $\boldsymbol{X}$ are involved in the issue of estimability.

Theorem 12.4 Any primary or secondary expected-value parameter estimable in $\text{GLM}_{N,p,q}(\boldsymbol{Y}_i; \boldsymbol{X}_i\boldsymbol{B}, \Sigma)$ is also estimable in a linearly equivalent model.

Proof. Considering estimable $\boldsymbol{\Theta} = \boldsymbol{CBU}$ allows treating $\boldsymbol{B}$ as a special case. Extending the known result for $p = 1$ (the univariate result) is easy due to the unimportance of $\boldsymbol{U}$ in determining estimability. Details are left as an exercise.

Estimators for the univariate GLM arise naturally as special cases of the multivariate forms. However, multivariate models require expanding the concepts and results surrounding estimability and restricted models. Furthermore, differences in development are sufficient to warrant separate treatment in some cases, especially for maximum likelihood estimation. We have omitted separate development of least squares estimators for the multivariate case. The task is straightforward and can easily be done by the interested reader. Although distribution free, the key mathematical forms coincide with or closely resemble the forms needed for deriving maximum likelihood estimators.

12.3 UNRESTRICTED LIKELIHOOD ESTIMATION

Derivation of the maximum likelihood estimators of $\boldsymbol{B}$ and Σ under $\text{GLM}_{N,p,q}(\boldsymbol{Y}_i; \boldsymbol{X}_i\boldsymbol{B}, \Sigma)$ assumptions with Gaussian variables is lengthy but straightforward. In formulating the likelihood function it helps to keep in mind the close relationship between the univariate and multivariate GLM.

Lemma 12.1 Model 1, $\text{GLM}_{N,p,q}(\boldsymbol{Y}_i; \boldsymbol{X}_i\boldsymbol{B}, \Sigma)$ with constant $\boldsymbol{C}$ has estimable $a \times p$ secondary parameter $\Theta = \boldsymbol{C}\boldsymbol{B}$. The assumptions and definitions of the model can also be expressed as specified in models 2 and 3 below.
Model 2, $\text{GGLM}_{Np,q}[\text{vec}(\boldsymbol{Y}); (\boldsymbol{I}_p \otimes \boldsymbol{X})\text{vec}(\boldsymbol{B}), \Sigma \otimes \boldsymbol{I}_N]$ and estimable $ap \times 1$ secondary parameter

$$\begin{aligned} \boldsymbol{\tau} &= \text{vec}(\Theta) \\ &= \text{vec}(\boldsymbol{C}\boldsymbol{B}) \\ &= (\boldsymbol{I}_p \otimes \boldsymbol{C})\text{vec}(\boldsymbol{B}) . \end{aligned} \tag{12.17}$$

Model 3, $\text{GGLM}_{Np,q}[\text{vec}(\boldsymbol{Y}'); (\boldsymbol{X} \otimes \boldsymbol{I}_p)\text{vec}(\boldsymbol{B}'), (\boldsymbol{I}_N \otimes \Sigma)]$ and estimable $ap \times 1$ secondary parameter

$$\begin{aligned} \boldsymbol{\tau}_T &= \text{vec}(\Theta') \\ &= \text{vec}(\boldsymbol{B}'\boldsymbol{C}') \\ &= (\boldsymbol{C} \otimes \boldsymbol{I}_p)\text{vec}(\boldsymbol{B}') , \end{aligned} \tag{12.18}$$

with $\boldsymbol{\tau}_T$ and $\boldsymbol{\tau}$ differing only by being a permutation of each other.

Proof. Results follow directly from the definitions of the multivariate GLM, the GGLM, the vec() operator, and the Kronecker product $\boldsymbol{A} \otimes \boldsymbol{B}$. In particular, Theorem 1.5 gives $\text{vec}(\boldsymbol{ABC}) = (\boldsymbol{C}' \otimes \boldsymbol{A})\text{vec}(\boldsymbol{B})$. □

Lemma 12.2 For a $\text{GLM}_{N,p,q}(\boldsymbol{Y}_i; \boldsymbol{X}_i\boldsymbol{B}, \Sigma)$ with Gaussian errors and $\text{rank}(\boldsymbol{X}) \leq q$, the log likelihood may be written

$$\log L(\boldsymbol{B}, \Sigma; \boldsymbol{Y}_*) = -N\log|2\pi\Sigma|/2 - \text{tr}\left[\Sigma^{-1}(\boldsymbol{Y}_* - \boldsymbol{X}\boldsymbol{B})'(\boldsymbol{Y}_* - \boldsymbol{X}\boldsymbol{B})\right]/2 . \tag{12.19}$$

Proof. With $\boldsymbol{Y}_i$ $1 \times p$ and $\boldsymbol{X}_i$ $1 \times q$, the joint density function for $\text{vec}(\boldsymbol{Y})$ is

$$\begin{aligned} L = f_{\boldsymbol{Y}}(\boldsymbol{Y}_*) &= \prod_{i=1}^{N} |2\pi\Sigma|^{-1/2}\exp\left[-(\boldsymbol{Y}_{i*} - \boldsymbol{X}_i\boldsymbol{B})\Sigma^{-1}(\boldsymbol{Y}_{i*} - \boldsymbol{X}_i\boldsymbol{B})'/2\right] \\ &= |2\pi\Sigma|^{-N/2}\exp\left\{-\text{tr}\left[(\boldsymbol{Y}_* - \boldsymbol{X}\boldsymbol{B})\Sigma^{-1}(\boldsymbol{Y}_* - \boldsymbol{X}\boldsymbol{B})'\right]/2\right\} . \end{aligned} \tag{12.20}$$

In turn $\log L = -N\log|2\pi\Sigma|/2 - \text{tr}[(\boldsymbol{Y}_* - \boldsymbol{X}\boldsymbol{B})\Sigma^{-1}(\boldsymbol{Y}_* - \boldsymbol{X}\boldsymbol{B})']/2$. The final form follows from the cyclical property of the trace function. □

Lemma 12.3 For a $\text{GLM}_{N,p,q}(\boldsymbol{Y}_i; \boldsymbol{X}_i\boldsymbol{B}, \Sigma)$ with Gaussian errors and $\text{rank}(\boldsymbol{X}) \leq q$, if $\boldsymbol{y}_s = \text{vec}(\boldsymbol{Y})$, $\boldsymbol{X}_s = \boldsymbol{I}_p \otimes \boldsymbol{X}$, $\boldsymbol{\beta}_s = \text{vec}(\boldsymbol{B})$, $\Sigma_s = (\Sigma \otimes \boldsymbol{I}_N)$,

$\boldsymbol{e}_s = \text{vec}(\boldsymbol{E})$, and $\boldsymbol{E} = (\boldsymbol{Y} - \boldsymbol{XB})$, then the log likelihood is

$$\begin{aligned}\log L(\boldsymbol{\beta}_s, \boldsymbol{\Sigma}_s; \boldsymbol{y}_{s*}) &= -\log|2\pi\boldsymbol{\Sigma}_s|/2 - (\boldsymbol{y}_{s*} - \boldsymbol{X}_s\boldsymbol{\beta}_s)'\boldsymbol{\Sigma}_s^{-1}(\boldsymbol{y}_{s*} - \boldsymbol{X}_s\boldsymbol{\beta}_s)/2 \quad (12.21)\\ &= -\log|2\pi\boldsymbol{\Sigma}_s|/2 - \text{tr}(\boldsymbol{\Sigma}_s^{-1}\boldsymbol{e}_{s*}\boldsymbol{e}_{s*}')/2\,.\end{aligned}$$

Proof. The joint density function of $\boldsymbol{y}_s$ can be written as

$$f_{\boldsymbol{y}_s}(\boldsymbol{y}_{s*}) = |2\pi\boldsymbol{\Sigma}_s|^{-1/2}\exp\left[-(\boldsymbol{y}_{s*} - \boldsymbol{X}_s\boldsymbol{\beta}_s)'\boldsymbol{\Sigma}_s^{-1}(\boldsymbol{y}_{s*} - \boldsymbol{X}_s\boldsymbol{\beta}_s)/2\right]. \quad (12.22)$$

Further details are left to the reader. □

The following lemmas are in numerous matrix theory books and will be used in the proof of the next theorem. Proofs are left as exercises.

Lemma 12.4 (a) If $\boldsymbol{A}$ is symmetric and nonsingular, then $\partial\log|\boldsymbol{A}|/\partial\boldsymbol{A} = 2\boldsymbol{A}^{-1} - \text{Dg}(\{\langle\boldsymbol{A}^{-1}\rangle_{jj}\})$.
(b) If $\boldsymbol{X}$ is symmetric and $\boldsymbol{a}$ is fixed, then $\partial(\boldsymbol{a}'\boldsymbol{X}\boldsymbol{a})/\partial\boldsymbol{X} = 2\boldsymbol{a}\boldsymbol{a}' - \text{Dg}(\boldsymbol{a}\boldsymbol{a}')$.

Lemma 12.5 (a) If $\boldsymbol{A}$, $\boldsymbol{B}$, $\boldsymbol{C}$, and $\boldsymbol{D}$ conform to the operations, then $(\boldsymbol{A}\otimes\boldsymbol{B})(\boldsymbol{C}\otimes\boldsymbol{D}) = (\boldsymbol{AC}\otimes\boldsymbol{BD})$.
(b) If $\boldsymbol{X}$ and $\boldsymbol{B}$ conform to the operations, then $\text{vec}(\boldsymbol{XB}) = (\boldsymbol{I}\otimes\boldsymbol{X})\text{vec}(\boldsymbol{B})$ and $\text{vec}(\boldsymbol{B}'\boldsymbol{X}') = (\boldsymbol{X}\otimes\boldsymbol{I})\text{vec}(\boldsymbol{B}')$.
(c) If $\boldsymbol{E}$ and $\boldsymbol{S}$ conform to the operations, then $[\text{vec}(\boldsymbol{E})]'(\boldsymbol{S}\otimes\boldsymbol{I})[\text{vec}(\boldsymbol{E})] = [\text{vec}(\boldsymbol{E}')]'(\boldsymbol{I}\otimes\boldsymbol{S}')\text{vec}(\boldsymbol{E}')$.
(d) $(\boldsymbol{A}\otimes\boldsymbol{B})^- = \boldsymbol{A}^-\otimes\boldsymbol{B}^-$ (which is just one of infinitely many).

Theorem 12.5 For $\text{GLM}_{N,p,q}(\boldsymbol{Y}_i; \boldsymbol{X}_i\boldsymbol{B}, \boldsymbol{\Sigma})$ with Gaussian errors, $r = \text{rank}(\boldsymbol{X}) \le q$, and $a \times b$ estimable secondary parameter $\boldsymbol{\Theta} = \boldsymbol{CBU}$, the joint supremum (for $\boldsymbol{B}$ if $r < q$) or maximum (if $r = q$ for $\boldsymbol{B}$; always for $\boldsymbol{\Sigma}$ and estimable $\boldsymbol{\Theta}$) likelihood estimators of $\boldsymbol{B}$, $\boldsymbol{\Sigma}$, and $\boldsymbol{\Theta}$ are

$$\widetilde{\boldsymbol{B}} = (\boldsymbol{X}'\boldsymbol{X})^-\boldsymbol{X}'\boldsymbol{Y} \quad (12.23)$$
$$\widetilde{\boldsymbol{\Sigma}} = (\boldsymbol{Y} - \boldsymbol{X}\widetilde{\boldsymbol{B}})'(\boldsymbol{Y} - \boldsymbol{X}\widetilde{\boldsymbol{B}})/N = \boldsymbol{Y}'[\boldsymbol{I}_N - \boldsymbol{X}(\boldsymbol{X}'\boldsymbol{X})^-\boldsymbol{X}']\boldsymbol{Y}/N \quad (12.24)$$
$$\widehat{\boldsymbol{\Theta}} = \boldsymbol{C}\widetilde{\boldsymbol{B}}\boldsymbol{U}\,, \quad (12.25)$$

with $\widehat{\boldsymbol{\Theta}}$ invariant to $\widetilde{\boldsymbol{B}}$ for estimable $\boldsymbol{\Theta}$. Here $\widetilde{\boldsymbol{B}}$ is any solution of $(\boldsymbol{X}'\boldsymbol{X})\widetilde{\boldsymbol{B}} = \boldsymbol{X}'\boldsymbol{Y}$, with infinitely many for $r < q$, and one unique solution if $r = q$ and $(\boldsymbol{X}'\boldsymbol{X})^- = (\boldsymbol{X}'\boldsymbol{X})^{-1}$. The value of $\widetilde{\boldsymbol{\Sigma}}$ is always invariant to $(\boldsymbol{X}'\boldsymbol{X})^-$. It is customary to use $\widehat{\boldsymbol{\Sigma}} = \widetilde{\boldsymbol{\Sigma}}N/(N-r)$, which is unbiased.

Proof. To obtain the solution equations, we differentiate the log likelihood with respect to the elements of $\boldsymbol{B}$ and $\boldsymbol{\Sigma}$, set the derivatives to zero, and solve for $\boldsymbol{B}$ and $\boldsymbol{\Sigma}$. Next $\boldsymbol{y}_s = \text{vec}(\boldsymbol{Y})$ is $Np\times 1$, $\boldsymbol{X}_s = \boldsymbol{I}_p\otimes\boldsymbol{X}$ $(Np\times pq)$, $\boldsymbol{\beta}_s = \text{vec}(\boldsymbol{B})$, $\boldsymbol{\Sigma}_s = \boldsymbol{\Sigma}\otimes\boldsymbol{I}_N$, $\boldsymbol{e}_s = \text{vec}(\boldsymbol{E}) = (\boldsymbol{y}_s - \boldsymbol{X}_s\boldsymbol{\beta}_s)$, and $\boldsymbol{E} = \boldsymbol{Y} - \boldsymbol{XB}$. By Lemma 12.3

$$\log L(\boldsymbol{B},\boldsymbol{\Sigma};\boldsymbol{Y}_*) = -\log|2\pi\boldsymbol{\Sigma}_s|/2 - (\boldsymbol{y}_{s*}-\boldsymbol{X}_s\boldsymbol{\beta}_s)\boldsymbol{\Sigma}_s^{-1}(\boldsymbol{y}_{s*}-\boldsymbol{X}_s\boldsymbol{\beta}_s)'/2\,. \quad (12.26)$$

If $t_0 = -(Np/2)\log(2\pi)$, $t_1(\boldsymbol{\Sigma}) = -(1/2)\log|\boldsymbol{\Sigma}_s|$, and $t_2(\boldsymbol{\beta},\boldsymbol{\Sigma}) = (\boldsymbol{y}_{s*}-\boldsymbol{X}_s\boldsymbol{\beta}_s)'\boldsymbol{\Sigma}_s^{-1}(\boldsymbol{y}_{s*}-\boldsymbol{X}_s\boldsymbol{\beta}_s)$, then $\log L(\boldsymbol{B},\boldsymbol{\Sigma};\boldsymbol{Y}_*) = t_0 + t_1(\boldsymbol{\Sigma}) - t_2(\boldsymbol{B},\boldsymbol{\Sigma})/2$. Also,

$$\begin{aligned} t_1(\boldsymbol{\Sigma}) &= -(1/2)\log|\boldsymbol{\Sigma}\otimes\boldsymbol{I}_N| \\ &= (1/2)\log|\boldsymbol{\Sigma}\otimes\boldsymbol{I}_N|^{-1} \\ &= (1/2)\log\left(|\boldsymbol{\Sigma}^{-1}|^N|\boldsymbol{I}_N|^p\right) \\ &= (N/2)\log|\boldsymbol{\Sigma}^{-1}|\,. \end{aligned} \quad (12.27)$$

Using direct-product properties in Lemma 12.5,

$$\begin{aligned} t_2(\boldsymbol{\beta}_s,\boldsymbol{\Sigma}) &= \boldsymbol{y}_{s*}'(\boldsymbol{\Sigma}\otimes\boldsymbol{I}_N)^{-1}\boldsymbol{y}_{s*} - 2\boldsymbol{\beta}_s'(\boldsymbol{I}_p\otimes\boldsymbol{X}')(\boldsymbol{\Sigma}\otimes\boldsymbol{I}_N)^{-1}\boldsymbol{y}_{s*} + \\ &\qquad \boldsymbol{\beta}_s'(\boldsymbol{I}_p\otimes\boldsymbol{X}')(\boldsymbol{\Sigma}\otimes\boldsymbol{I}_N)^{-1}(\boldsymbol{I}_p\otimes\boldsymbol{X})\boldsymbol{\beta}_s \\ &= \boldsymbol{y}_{s*}'(\boldsymbol{\Sigma}^{-1}\otimes\boldsymbol{I}_N)\boldsymbol{y}_{s*} - 2\boldsymbol{\beta}_s'(\boldsymbol{\Sigma}^{-1}\otimes\boldsymbol{X}')\boldsymbol{y}_{s*} + \boldsymbol{\beta}_s'(\boldsymbol{\Sigma}^{-1}\otimes\boldsymbol{X}'\boldsymbol{X})\boldsymbol{\beta}_s\,. \end{aligned} \quad (12.28)$$

Differentiating $\log L$ with respect to $\boldsymbol{\beta}_s$ gives

$$\begin{aligned} \partial\log L/\partial\boldsymbol{\beta}_s &= \partial[-t_2(\boldsymbol{\beta}_s,\boldsymbol{\Sigma})/2]/\partial\boldsymbol{\beta}_s \\ &= \boldsymbol{0} + (\boldsymbol{\Sigma}^{-1}\otimes\boldsymbol{X}')\boldsymbol{y}_{s*} - (\boldsymbol{\Sigma}^{-1}\otimes\boldsymbol{X}'\boldsymbol{X})\boldsymbol{\beta}_s\,. \end{aligned} \quad (12.29)$$

Setting the expression to zero (and using Lemma 12.5) gives the MLE for $\boldsymbol{\beta}_s$,

$$\begin{aligned} \widetilde{\boldsymbol{\beta}}_s &= (\boldsymbol{\Sigma}^{-1}\otimes\boldsymbol{X}'\boldsymbol{X})^-(\boldsymbol{\Sigma}^{-1}\otimes\boldsymbol{X}')\boldsymbol{y}_s \\ &= [\boldsymbol{\Sigma}\otimes(\boldsymbol{X}'\boldsymbol{X})^-](\boldsymbol{\Sigma}^{-1}\otimes\boldsymbol{X}')\boldsymbol{y}_s \\ &= [\boldsymbol{I}_p\otimes(\boldsymbol{X}'\boldsymbol{X})^-\boldsymbol{X}']\boldsymbol{y}_s\,. \end{aligned} \quad (12.30)$$

Surprisingly, and very conveniently, $\boldsymbol{\Sigma}$ cancels! Equivalently, $\text{vec}(\widetilde{\boldsymbol{B}}) = [\boldsymbol{I}_p\otimes(\boldsymbol{X}'\boldsymbol{X})^-\boldsymbol{X}']\text{vec}(\boldsymbol{Y})$. By Lemma 12.4, $\widetilde{\boldsymbol{B}} = (\boldsymbol{X}'\boldsymbol{X})^-\boldsymbol{X}'\boldsymbol{Y}$. In summary, $\widetilde{\boldsymbol{B}} = \left[\widetilde{\boldsymbol{\beta}}_1 \cdots \widetilde{\boldsymbol{\beta}}_p\right]$ with $\widetilde{\boldsymbol{\beta}}_j = (\boldsymbol{X}'\boldsymbol{X})^-\boldsymbol{X}'\boldsymbol{y}_j$ and

$$\widetilde{\boldsymbol{\beta}}_s = \text{vec}(\widetilde{\boldsymbol{B}}) = \begin{bmatrix} \widetilde{\boldsymbol{\beta}}_1 \\ \widetilde{\boldsymbol{\beta}}_2 \\ \vdots \\ \widetilde{\boldsymbol{\beta}}_p \end{bmatrix}. \quad (12.31)$$

The cancellation of $\boldsymbol{\Sigma}$ in the derivation of $\boldsymbol{\beta}_s$ is extremely important. The "disappearance" allows estimating $\boldsymbol{\beta}_s$ (and thus $\boldsymbol{B}$) in a noniterative fashion. If $\boldsymbol{\Sigma}$ did not disappear from the likelihood equations, one would be required to know or estimate $\boldsymbol{\Sigma}$ before estimating $\boldsymbol{B}$. The strong structural assumptions allow $\boldsymbol{\Sigma}$ to cancel. Rows of $\boldsymbol{Y}$ are independent, each row of $\boldsymbol{Y}$ has the same covariance matrix, and each column of $\boldsymbol{Y}$ has the same design.

Since $\boldsymbol{\Sigma}$ and $\boldsymbol{\Sigma}^{-1}$ have a one-to-one correspondence, maximization with respect to $\boldsymbol{\Sigma}^{-1}$ is the same as maximization with respect to $\boldsymbol{\Sigma}$. It is more convenient to take derivatives of $\log L$ with respect to $\boldsymbol{\Sigma}^{-1}$:

$$\frac{\partial \log L}{\partial \Sigma^{-1}} = \frac{\partial}{\partial \Sigma^{-1}} t_1(\Sigma) - \frac{1}{2} \frac{\partial}{\partial \Sigma^{-1}} t_2(\boldsymbol{\beta}_s, \Sigma). \tag{12.32}$$

By Lemma 12.4

$$\frac{\partial}{\partial \Sigma^{-1}} t_1(\Sigma) = \frac{N}{2}[2\Sigma - \mathrm{Dg}(\Sigma)]. \tag{12.33}$$

For differentiating with respect to Σ^{-1}, it is more convenient to express t_2 as a sum of quadratic forms in Σ^{-1}, which corresponds to working with the likelihood function of $\mathrm{vec}(\boldsymbol{Y}')$ (i.e., $\boldsymbol{Y}$ stacked by rows) rather than $\mathrm{vec}(\boldsymbol{Y})$. With $\boldsymbol{E}_* = (\boldsymbol{Y}_* - \boldsymbol{X}\boldsymbol{B})$ and $\boldsymbol{E}_{i*} = \mathrm{row}_i(\boldsymbol{E}_*)$, direct-product properties in Lemma 12.5 give

$$\begin{aligned} t_2(\boldsymbol{\beta}, \Sigma) &= (\boldsymbol{y}_{s*} - \boldsymbol{X}_s\boldsymbol{\beta}_s)'\Sigma_s^{-1}(\boldsymbol{y}_{s*} - \boldsymbol{X}_s\boldsymbol{\beta}_s) \\ &= [\mathrm{vec}(\boldsymbol{E}_*)]'(\Sigma^{-1} \otimes \boldsymbol{I}_N)[\mathrm{vec}(\boldsymbol{E}_*)] \\ &= [\mathrm{vec}(\boldsymbol{E}_*')]'(\boldsymbol{I}_N \otimes \Sigma^{-1})[\mathrm{vec}(\boldsymbol{E}_*')] \\ &= \sum_{i=1}^{N} \boldsymbol{E}_{i*}\Sigma^{-1}\boldsymbol{E}_{i*}'. \end{aligned} \tag{12.34}$$

Using Lemma 12.4 we have

$$\begin{aligned} \frac{\partial}{\partial \Sigma^{-1}} t_2(\boldsymbol{\beta}, \Sigma) &= \sum_{i=1}^{N} \frac{\partial}{\partial \Sigma^{-1}} \boldsymbol{E}_{i*}\Sigma^{-1}\boldsymbol{E}_{i*}' \\ &= \sum_{i=1}^{N} [2\boldsymbol{E}_{i*}'\boldsymbol{E}_{i*} - \mathrm{Dg}(\{\langle \boldsymbol{E}_{i*}'\boldsymbol{E}_{i*}\rangle_{jj}\})] \\ &= 2\boldsymbol{E}_*'\boldsymbol{E}_* - \mathrm{Dg}(\{\langle \boldsymbol{E}_*'\boldsymbol{E}_*\rangle_{jj}\}) \\ &= 2(\boldsymbol{Y}_* - \boldsymbol{X}\boldsymbol{\beta})'(\boldsymbol{Y}_* - \boldsymbol{X}\boldsymbol{\beta}) - \mathrm{Dg}(\{\langle (\boldsymbol{Y}_* - \boldsymbol{X}\boldsymbol{B})'(\boldsymbol{Y}_* - \boldsymbol{X}\boldsymbol{B})\rangle_{jj}\}). \end{aligned} \tag{12.35}$$

The MLEs for $\boldsymbol{B}$ and Σ are produced by combining results and setting the derivatives equal to zero. Considering $\partial \log L(\boldsymbol{B}, \Sigma, \boldsymbol{Y}_{s*})/\partial \Sigma^{-1} = \boldsymbol{0}$ implies

$$\left[2\widetilde{\Sigma} - \mathrm{Dg}(\widetilde{\Sigma})\right]N/2 = \left[2\widehat{\boldsymbol{E}}'\widehat{\boldsymbol{E}}/N - \mathrm{Dg}(\{\langle \boldsymbol{E}'\boldsymbol{E}\rangle_{jj}\})/N\right]N/2, \tag{12.36}$$

in which $\widehat{\boldsymbol{E}} = (\boldsymbol{Y} - \boldsymbol{X}\widetilde{\boldsymbol{B}})$. Hence $\widetilde{\Sigma} = \widehat{\boldsymbol{E}}'\widehat{\boldsymbol{E}}/N = (\boldsymbol{Y} - \boldsymbol{X}\widetilde{\boldsymbol{B}})'(\boldsymbol{Y} - \boldsymbol{X}\widetilde{\boldsymbol{B}})/N$.

Differentiating the log likelihood function and setting the results to zeros gave $\widetilde{\boldsymbol{B}}$ and $\widetilde{\Sigma}$ as solutions of the likelihood equations. The uniqueness of the solutions implies that *if* the likelihood function has an extremum (minimum, maximum, or saddle point), then our estimators provide the coordinates of the extremum.

The likelihood function, being a density, is nonnegative for all values of the variable ($\boldsymbol{Y}_*$) and the parameters. For a fixed value of $\boldsymbol{Y}_*$ (say $\boldsymbol{Y}_0$), one can make the likelihood arbitrarily close to zero by choosing $\Sigma = \boldsymbol{I}_p$ and $\boldsymbol{\beta}_s$ such that $[\mathrm{vec}(\boldsymbol{Y}_0) - \boldsymbol{X}_s\boldsymbol{\beta}_s] \to \pm\infty$. Thus, the likelihood function has an infimum at zero but no minimum. The only possibilities are that $\widetilde{\boldsymbol{B}}$ and $\widetilde{\Sigma}$ locate a maximum or a saddle point.

To demonstrate the extremum is a maximum for $\text{rank}(\boldsymbol{X}) = r = q$ and supremum for $r < q$, one must prove

$$\boldsymbol{H} = \begin{bmatrix} \partial^2 \text{log}L/(\partial\boldsymbol{\beta}_s \partial\boldsymbol{\beta}_s') & \partial^2 \text{log}L/(\partial\boldsymbol{\beta}_s \partial\boldsymbol{\Sigma}^{-1}) \\ \partial^2 \text{log}L(\partial\boldsymbol{\Sigma}^{-1} \partial\boldsymbol{\beta}_s') & \partial^2 \text{log}L/(\partial\boldsymbol{\Sigma}^{-1} \partial\boldsymbol{\Sigma}^{-1}) \end{bmatrix} \tag{12.37}$$

is negative definite at $\widehat{\boldsymbol{B}}$ and $\widehat{\boldsymbol{\Sigma}}$. Details will be left to the reader. □

Theorem 12.6 For $\text{GLM}_{N,p,q}(\boldsymbol{Y}_i; \boldsymbol{X}_i\boldsymbol{B}, \boldsymbol{\Sigma})$ with $r = \text{rank}(\boldsymbol{X}) \leq q$ and $\boldsymbol{\beta}_s = \text{vec}(\boldsymbol{B})$ $(pq \times 1)$ the following results hold.
(a) If $r < q$, then no BLUE exists and $\boldsymbol{B}$ is not estimable.
(b) If $r = q$, then $\text{BLUE}(\boldsymbol{B}) = \widehat{\boldsymbol{B}} = (\boldsymbol{X}'\boldsymbol{X})^{-1}\boldsymbol{X}'\boldsymbol{Y}$.
(c) Equivalently $\text{BLUE}(\boldsymbol{\beta}_s) = \widehat{\boldsymbol{\beta}}_s = \text{vec}(\widehat{\boldsymbol{B}}) = \left[\boldsymbol{I} \otimes (\boldsymbol{X}'\boldsymbol{X})^{-1}\boldsymbol{X}'\right]\text{vec}(\boldsymbol{Y})$. (12.38)

Proof. By Lemma 12.1, the assumptions of $\text{GLM}_{N,p,q}\text{FR}(\boldsymbol{Y}_i; \boldsymbol{X}_i\boldsymbol{B}, \boldsymbol{\Sigma})$ may be expressed as $\text{GGLM}_{Np,q}\text{FR}(\boldsymbol{y}_s; \boldsymbol{X}_s\boldsymbol{B}_s, \boldsymbol{\Sigma}_s)$, in which $\boldsymbol{y}_s$ is $Np \times 1$, $\boldsymbol{\Sigma}$ is $p \times p$ and $\boldsymbol{X}$ is $N \times q$, $\boldsymbol{y}_s = \text{vec}(\boldsymbol{Y})$, $\boldsymbol{X}_s = (\boldsymbol{I}_p \otimes \boldsymbol{X})$, $\boldsymbol{\beta}_s = \text{vec}(\boldsymbol{B})$, and $\boldsymbol{\Sigma}_s = (\boldsymbol{\Sigma} \otimes \boldsymbol{I}_N)$. By Theorem 11.25 on weighted estimation, the BLUE of $\boldsymbol{\beta}_s$ is

$$\begin{aligned}
\widehat{\boldsymbol{\beta}}_s &= (\boldsymbol{X}_s'\boldsymbol{\Sigma}_s^{-1}\boldsymbol{X}_s)^{-1}\boldsymbol{X}_s\boldsymbol{\Sigma}_s^{-1}\boldsymbol{y}_s \\
&= \left[(\boldsymbol{I}_p \otimes \boldsymbol{X}')'(\boldsymbol{\Sigma} \otimes \boldsymbol{I}_N)^{-1}(\boldsymbol{I}_p \otimes \boldsymbol{X})\right]^{-1}(\boldsymbol{I}_p \otimes \boldsymbol{X}')(\boldsymbol{\Sigma} \otimes \boldsymbol{I}_N)^{-1}\boldsymbol{y}_s \\
&= \left[(\boldsymbol{\Sigma}^{-1} \otimes \boldsymbol{X}')(\boldsymbol{I}_p \otimes \boldsymbol{X})\right]^{-1}(\boldsymbol{\Sigma}^{-1} \otimes \boldsymbol{X}')\boldsymbol{y}_s \\
&= (\boldsymbol{\Sigma}^{-1} \otimes \boldsymbol{X}'\boldsymbol{X})^{-1}(\boldsymbol{\Sigma}^{-1} \otimes \boldsymbol{X}')\boldsymbol{y}_s \\
&= \left[\boldsymbol{\Sigma} \otimes (\boldsymbol{X}'\boldsymbol{X})^{-1}\right](\boldsymbol{\Sigma}^{-1} \otimes \boldsymbol{X}')\boldsymbol{y}_s \\
&= \left[\boldsymbol{I} \otimes (\boldsymbol{X}'\boldsymbol{X})^{-1}\boldsymbol{X}'\right]\boldsymbol{y}_s \\
&= \begin{bmatrix} (\boldsymbol{X}'\boldsymbol{X})^{-1}\boldsymbol{X}'\boldsymbol{y}_1 \\ \vdots \\ (\boldsymbol{X}'\boldsymbol{X})^{-1}\boldsymbol{X}'\boldsymbol{y}_p \end{bmatrix}.
\end{aligned} \tag{12.39}$$

Hence $\widehat{\boldsymbol{B}}$ must satisfy $\widehat{\boldsymbol{\beta}}_s = \text{vec}(\widehat{\boldsymbol{B}}) = \left[\widehat{\boldsymbol{\beta}}_1' \, \widehat{\boldsymbol{\beta}}_2' \cdots \widehat{\boldsymbol{\beta}}_p'\right]'$. □

Here $\boldsymbol{\Sigma}$ "drops out" in the derivation of $\widehat{\boldsymbol{B}}$. Thus, unlike the situation in weighted least squares, one need not know the value of $\boldsymbol{\Sigma}$ in order to determine the BLUE in the multivariate GLM.

Corollary 12.6 The covariance matrix of $\text{BLUE}(\boldsymbol{B})$ is $\mathcal{V}(\widehat{\boldsymbol{\beta}}_s) = \boldsymbol{\Sigma} \otimes (\boldsymbol{X}'\boldsymbol{X})^{-1}$.

Proof.

$$\begin{aligned}\mathcal{V}(\widehat{\boldsymbol{\beta}}_s) &= [\boldsymbol{I}_p \otimes (\boldsymbol{X}'\boldsymbol{X})^{-1}\boldsymbol{X}']\mathcal{V}(\boldsymbol{y}_s)[\boldsymbol{I}_p \otimes (\boldsymbol{X}'\boldsymbol{X})^{-1}\boldsymbol{X}']' \\ &= [\boldsymbol{I}_p \otimes (\boldsymbol{X}'\boldsymbol{X})^{-1}\boldsymbol{X}'](\boldsymbol{\Sigma} \otimes \boldsymbol{I}_N)[\boldsymbol{I}_p \otimes \boldsymbol{X}(\boldsymbol{X}'\boldsymbol{X})^{-1}] \\ &= [\boldsymbol{\Sigma} \otimes (\boldsymbol{X}'\boldsymbol{X})^{-1}\boldsymbol{X}'][\boldsymbol{I}_p \otimes \boldsymbol{X}(\boldsymbol{X}'\boldsymbol{X})^{-1}] \\ &= \boldsymbol{\Sigma} \otimes (\boldsymbol{X}'\boldsymbol{X})^{-1}\boldsymbol{X}'\boldsymbol{X}(\boldsymbol{X}'\boldsymbol{X})^{-1} \\ &= \boldsymbol{\Sigma} \otimes (\boldsymbol{X}'\boldsymbol{X})^{-1}. \end{aligned} \tag{12.40}$$

□

12.4 ESTIMATION OF SECONDARY PARAMETERS

The commonly described format of secondary parameters in the multivariate GLM is the $a \times b$ form $\boldsymbol{\Theta} = \boldsymbol{CBU}$, with $\boldsymbol{C}$ and $\boldsymbol{U}$ usually full-rank matrices, $a = \text{rank}(\boldsymbol{C}) \leq q$ and $b = \text{rank}(\boldsymbol{U}) \leq p$. Subtraction of a constant matrix $\boldsymbol{\Theta}_0$ has no effect on estimation and precision and we may ignore $\boldsymbol{\Theta}_0$ without loss of generality when discussing nearly all estimation properties.

An alternative form is $\boldsymbol{\tau} = \boldsymbol{L\beta}_s = \boldsymbol{L}\text{vec}(\boldsymbol{B})$, with $\boldsymbol{L}$ constant. For convenience we assume $\ell = \text{rank}(\boldsymbol{L}) \leq pq$. An additive constant may be ignored without loss of generality. Although the $\boldsymbol{\Theta} = \boldsymbol{CBU}$ form is more common, the $\boldsymbol{\tau} = \boldsymbol{L\beta}_s$ form is also valid and more general in that some secondary parameters can be defined in the $\boldsymbol{L\beta}_s$ form but cannot be defined in the $\boldsymbol{CBU}$ form.

One can easily verify the above statements as follows. Each element of $\boldsymbol{\Theta} = \boldsymbol{CBU}$ is a linear combination of the elements of $\boldsymbol{B}$ and, therefore, of the elements of $\boldsymbol{\beta}_s$. Any linear combinations of the elements of $\boldsymbol{B}$ can be written in the form $\boldsymbol{L\beta}_s$. It is easy to find counterexamples to the converse.

Theorem 12.7 Having $\boldsymbol{B} \in \Re^{q \times p}$ allows defining $\boldsymbol{\beta}_s = \text{vec}(\boldsymbol{B})$ and considering fixed constants $\boldsymbol{C}, \boldsymbol{U}$, and $\boldsymbol{L}$.
(a) If $\boldsymbol{\Theta} = \boldsymbol{CBU}$ and $\boldsymbol{\tau} = \boldsymbol{L\beta}_s$, then $\text{vec}(\boldsymbol{\Theta}) = \boldsymbol{\tau}\ \forall \boldsymbol{B} \Leftrightarrow \boldsymbol{L} = (\boldsymbol{U}' \otimes \boldsymbol{C})$.
(b) If $\boldsymbol{\Theta} = \boldsymbol{CBU}$ and $\boldsymbol{\tau} = \text{vec}(\boldsymbol{\Theta})$, then $\boldsymbol{L}$ exists such that $\boldsymbol{\tau} = \boldsymbol{L\beta}_s\ \forall \boldsymbol{B}$.
(c) If $\boldsymbol{\tau} = \boldsymbol{L\beta}_s$, then $\{\boldsymbol{C}, \boldsymbol{U}\}$ may or may not exist such that $\boldsymbol{\tau} = \text{vec}(\boldsymbol{CBU})\ \forall \boldsymbol{B}$.

Proof. (a) follows directly from the properties of vec() and $\boldsymbol{A} \otimes \boldsymbol{B}$.
For **(b)**, choosing $\boldsymbol{L} = (\boldsymbol{U}' \otimes \boldsymbol{C})$ satisfies the claim.

(c) A counterexample is $\boldsymbol{L} = [1\ 0\ 0\ {-1}]$, $\boldsymbol{B} = \begin{bmatrix} \beta_{11} & \beta_{12} \\ \beta_{21} & \beta_{22} \end{bmatrix}$, $\boldsymbol{\beta}_s = [\beta_{11}\ \beta_{21}\ \beta_{12}\ \beta_{22}]'$, and $\tau = \boldsymbol{L\beta}_s = (\beta_{11} - \beta_{22})$. If $1 \times q$ $\boldsymbol{C}$ and $p \times 1$ $\boldsymbol{U}$ exist such that $\tau = \boldsymbol{L\beta}_s = \boldsymbol{CBU}$, then $\boldsymbol{CBU} = \sum_{i=1}^{2}\sum_{i=1}^{2} C_i \beta_{ij} U_j = (\beta_{11} - \beta_{22})$ must hold $\forall \boldsymbol{B}$. The last equation requires the impossible, that (a) some elements of $\{\boldsymbol{C}, \boldsymbol{U}\}$ are zero, and (b) none of the elements of $\{\boldsymbol{C}, \boldsymbol{U}\}$ are zero. The supposition that $\{\boldsymbol{C}, \boldsymbol{U}\}$ exists has led to a contradiction. □

Lemma 12.6 If $\boldsymbol{\theta}$ $(a \times 1)$ is an estimable secondary (or primary) parameter for a $\text{GLM}_{N,p,q}(\boldsymbol{Y}_i; \boldsymbol{X}_i\boldsymbol{B}, \Sigma)$ and $\widehat{\boldsymbol{\theta}}$ is the BLUE of $\boldsymbol{\theta}$, then a tertiary (or secondary) parameter vector is $\boldsymbol{\gamma} = \boldsymbol{T\theta}$, with $g = \text{rank}(\boldsymbol{T}) \leq a$. The BLUE of $\boldsymbol{\gamma}$ is $\widehat{\boldsymbol{\gamma}} = \boldsymbol{T}\widehat{\boldsymbol{\theta}}$.

Proof. If $\widehat{\boldsymbol{\theta}} = \text{BLUE}(\boldsymbol{\theta})$, then (1) $\widehat{\boldsymbol{\theta}}$ is an LUE, which ensures $\exists \boldsymbol{A}$ such that $\widehat{\boldsymbol{\theta}} = \boldsymbol{Ay}$ and $\text{E}(\boldsymbol{Ay}) = \boldsymbol{\theta}$. Hence $\widehat{\boldsymbol{\gamma}}$ is a LUE since $\text{E}(\widehat{\boldsymbol{\gamma}}) = \text{E}[(\boldsymbol{TA})\boldsymbol{y}] = \boldsymbol{T}\text{E}(\boldsymbol{Ay}) = \boldsymbol{T\theta} = \boldsymbol{\gamma}$. Also, $\widehat{\boldsymbol{\theta}} = \text{BLUE}(\boldsymbol{\theta})$ implies (2) $\widehat{\boldsymbol{\theta}}$ is the "best" LUE. For any other LUE, say, $\widehat{\boldsymbol{\theta}}_1 = \boldsymbol{A}_1\boldsymbol{y}$ $(\neq \boldsymbol{A})$, $\mathcal{V}(\boldsymbol{s}'\boldsymbol{Ay}) = \boldsymbol{s}'\mathcal{V}(\boldsymbol{Ay})\boldsymbol{s}$ and $\mathcal{V}(\boldsymbol{s}'\boldsymbol{A}_1\boldsymbol{y}) = \boldsymbol{s}'\mathcal{V}(\boldsymbol{A}_1\boldsymbol{y})\boldsymbol{s}$. By Theorem 11.9, $\mathcal{V}(\boldsymbol{A}_1\boldsymbol{y}) - \mathcal{V}(\boldsymbol{Ay})$ is *positive semidefinite. The definition of "positive semidefinite" immediately implies* $\mathcal{V}(\boldsymbol{s}'\boldsymbol{A}_1\boldsymbol{y}) \geq \mathcal{V}(\boldsymbol{s}'\boldsymbol{Ay})$ $\forall \boldsymbol{s} \in \Re^a$. For $\boldsymbol{s} \in \{\boldsymbol{s} : \boldsymbol{s}' = \boldsymbol{k}'\boldsymbol{T}$ with $\boldsymbol{k} \in \Re^g\} \subseteq \Re^a$ it follows that $\mathcal{V}(\boldsymbol{s}'\boldsymbol{A}_1\boldsymbol{y}) \geq \mathcal{V}(\boldsymbol{s}'\boldsymbol{Ay})$ for all such $\boldsymbol{s}$ and $\mathcal{V}(\boldsymbol{k}'\boldsymbol{TA}_1\boldsymbol{y}) \geq \mathcal{V}(\boldsymbol{k}'\boldsymbol{TAy})$ $\forall \boldsymbol{k} \in \Re^g$. □

Theorem 12.8 For $\text{GLM}_{N,p,q}(\boldsymbol{Y}_i; \boldsymbol{X}_i\boldsymbol{B}, \Sigma)$, $\boldsymbol{\beta}_s = \text{vec}(\boldsymbol{B})$, $r = \text{rank}(\boldsymbol{X}) \leq q$, and estimable secondary parameter $\boldsymbol{\tau} = \boldsymbol{L\beta}_s$ with $\boldsymbol{\tau}$ $\ell \times 1$, $\ell = \text{rank}(\boldsymbol{L}) \leq pq$, if $\widehat{\boldsymbol{\beta}}_s$ is the BLUE of $\boldsymbol{\beta}_s$, then $r = q$ and the BLUE of $\boldsymbol{\tau}$ is $\widehat{\boldsymbol{\tau}} = \boldsymbol{L}\widehat{\boldsymbol{\beta}}_s$.

Proof. The theorem follows from the form of the likelihood and from a univariate estimation theorem. The details are left to the reader.

Corollary 12.8 For estimable secondary parameter matrix $\boldsymbol{\Theta} = \boldsymbol{CBU}$ with $a = \text{rank}(\boldsymbol{C}) \leq q$ and $b = \text{rank}(\boldsymbol{U}) \leq p$, the BLUE is $\widehat{\boldsymbol{\Theta}} = \boldsymbol{C}\widehat{\boldsymbol{B}}\boldsymbol{U}$.

Proof. The result follows immediately from Theorem 11.14 and the preceding theorems in the present chapter. The details are left to the reader as an exercise.

Theorem 12.9 (a) For $\text{GLM}_{N,p,q}\text{LTFR}(\boldsymbol{Y}_i; \boldsymbol{X}_i\boldsymbol{B}, \Sigma)$ with Gaussian errors and two-condition inverse $(\boldsymbol{X}'\boldsymbol{X})^{-}$,

$$\widehat{\boldsymbol{B}} \sim \mathcal{SN}_{p,q}[(\boldsymbol{X}'\boldsymbol{X})^{-}(\boldsymbol{X}'\boldsymbol{X})\boldsymbol{B}, (\boldsymbol{X}'\boldsymbol{X})^{-}, \Sigma]. \tag{12.41}$$

(b) A $\text{GLM}_{N,p,q}\text{FR}(\boldsymbol{Y}_i; \boldsymbol{X}_i\boldsymbol{B}, \Sigma)$ has

$$\widehat{\boldsymbol{B}} \sim \mathcal{N}_{p,q}\left[\boldsymbol{B}, (\boldsymbol{X}'\boldsymbol{X})^{-1}, \Sigma\right]. \tag{12.42}$$

(c) For $\text{rank}(\boldsymbol{X}) \leq q$ and any one-condition inverse, estimable $\boldsymbol{\Theta} = \boldsymbol{CBU}$ gives

$$\widehat{\boldsymbol{\Theta}} - \boldsymbol{\Theta}_0 \sim (\mathcal{S})\mathcal{N}_a[\boldsymbol{\Theta} - \boldsymbol{\Theta}_0, \boldsymbol{C}(\boldsymbol{X}'\boldsymbol{X})^{-}\boldsymbol{C}', \boldsymbol{U}'\Sigma\boldsymbol{U}]. \tag{12.43}$$

Proof. Left as an exercise.

12.5 ESTIMATION WITH MULTIVARIATE RESTRICTIONS

Writing $\text{GLM}_{N,p,q}(\boldsymbol{Y}_i; \boldsymbol{X}_i\boldsymbol{B}|\boldsymbol{R}_x\boldsymbol{B}\boldsymbol{R}_y = \boldsymbol{A}, \Sigma)$ indicates an *explicitly* restricted multivariate model. The $(q \times p)$ matrix $\boldsymbol{B}$ has between-ISU (independent sampling unit, "subject") restrictions defined by known and constant matrices $\boldsymbol{R}_x$ of dimension $a \times q$ with $a \leq q$, $\text{rank}(\boldsymbol{R}_x) = a$. The $(q \times p)$ matrix $\boldsymbol{B}$ has within-subject restrictions defined by $\boldsymbol{R}_y$ of dimension $p \times b$ with $p \geq b$ and $\text{rank}(\boldsymbol{R}_y) = b$. The $a \times b$ constant $\boldsymbol{A}$ plays a role in both sorts of restrictions. The matrices $\boldsymbol{R}_x$, $\boldsymbol{R}_y$, and $\boldsymbol{A}$ must all be chosen without knowledge of the data, prior to data collection. As in univariate models, matrix $\boldsymbol{R}_x$ implicitly specifies restrictions about $\boldsymbol{X}$, the between-subjects design matrix. The matrix $\boldsymbol{R}_y$ implicitly specifies restrictions about $\boldsymbol{Y}$ and can be thought of as indirectly defining a within-subject design matrix. Naturally all results about restrictions in univariate models occur in the special case of the multivariate model with $p = 1$.

Example 12.2 A set of *ipsative* measures add to a known constant. Such variables occur naturally in the study of allocation of behavior and many other areas. The pandemic of obesity in the United States has led scientists to assess how much time Americans allocate to sleeping, eating, watching television, walking, etc. Each person's allocations always add to 24 hours. The implicit constraint creates a singular covariance matrix (Σ), and difficulties in defining scientifically appealing parameter estimators which account for the constraint.

Example 12.3 Pan (2003) evaluated models of land use allocation among rural farmers in the Amazon Basin of South America. With farm as the independent sampling unit, the multivariate response of interest was % land in crops, % land in pasture, % land fallow. The formulation corresponds to the restrictions $\boldsymbol{B}\boldsymbol{R}_y = \boldsymbol{A}$, with $\boldsymbol{R}_y = [\,1\;1\;1\,]$ and $\boldsymbol{A} = [\,100\,]$.

Example 12.4 The same constraint arises in the study of diets as predictors of cancer. Nutritional epidemiologists seek to build models of % calories from fat, % calories from protein, % calories from carbohydrates, and % calories from alcohol.

Having restrictions only between subjects requires $\boldsymbol{R}_y = \boldsymbol{I}_p$ and simplifies the model statement of an explicitly restricted linear model to the form $\text{GLM}_{N,p,q}(\boldsymbol{Y}_i; \boldsymbol{X}_i\boldsymbol{B}|\boldsymbol{R}_x\boldsymbol{B} = \boldsymbol{A}, \Sigma)$. Any such model can be transformed to a linearly equivalent *unrestricted* linear model $\text{GLM}_{N,p,q}(\boldsymbol{Y}_{iu}; \boldsymbol{X}_{iu}\boldsymbol{B}_u, \Sigma_u)$ with all dimensions the same. Adding within-subject restrictions corresponds to choosing $\boldsymbol{R}_y \neq \boldsymbol{I}_p$ and writing $\text{GLM}_{N,p,q}(\boldsymbol{Y}_i; \boldsymbol{X}_i\boldsymbol{B}|\boldsymbol{R}_x\boldsymbol{B}\boldsymbol{R}_y = \boldsymbol{A}, \Sigma)$.

The same statement does not hold, except in special cases, for a model having within-subject restrictions, $\text{GLM}_{N,p,q}(\boldsymbol{Y}_i; \boldsymbol{X}_i\boldsymbol{B}|\boldsymbol{R}_x\boldsymbol{B}\boldsymbol{R}_y = \boldsymbol{A}, \Sigma)$. Without loss of generality, for the purposes of the present argument, we may assume $\boldsymbol{R}_x = \boldsymbol{I}_q$ and discuss $\text{GLM}_{N,p,q}(\boldsymbol{Y}_i; \boldsymbol{X}_i\boldsymbol{B}|\boldsymbol{B}\boldsymbol{R}_y = \boldsymbol{A}, \Sigma)$. A linearly equivalent unrestricted model can be guaranteed to exist in such a model only if $b = p = \text{rank}(\boldsymbol{R}_y)$.

However, in practice, the restriction statement often implicitly defines the only parameters of interest. If so, analysis of the (within-subject) restricted model only loses access to scientifically irrelevant parameters. Equally importantly, and not surprisingly, properties of the restricted model suffice to fully specify the distributions of estimators and tests available in the restricted model.

The next theorem has many uses. The results provide ways of defining linearly equivalent multivariate GLMs and GGLMs in the presence of restrictions. Looking ahead, testing a general linear hypothesis may be cast as specifying a set of restrictions. The various forms help simplify derivations of estimators, tests, and the associated distributions.

Example 12.5 A paired data t test can be cast as a test in a multivariate linear model with $p = 2$ or as a univariate model with $p = 1$ and the responses being difference scores. In both cases, $\boldsymbol{X} = \boldsymbol{1}_N$. The following theorem allows expressing the $p = 2$ model as a restriction of the $p = 1$ model. The restriction matrix is $\boldsymbol{U} = [\,1\ -1\,]'$.

Theorem 12.10 **Model 1** is $\text{GLM}_{N,p,q}(\boldsymbol{Y}_i; \boldsymbol{X}_i\boldsymbol{B}|\boldsymbol{R}_x\boldsymbol{B}\boldsymbol{R}_y = \boldsymbol{A}, \boldsymbol{\Sigma})$, with $\boldsymbol{R}_x$ $(a \times q)$ rank $a \leq q$ and $\boldsymbol{R}_y$ $(p \times b)$ rank $b \leq p$. If $\boldsymbol{R}_{y\perp}$ is any $p \times (p - b)$ matrix such that $\boldsymbol{T} = [\,\boldsymbol{R}_y\ \boldsymbol{R}_{y\perp}\,]$ is $p \times p$ and full rank, it defines

$$\boldsymbol{\Gamma} = \boldsymbol{B}\boldsymbol{T} = [\,\boldsymbol{B}\boldsymbol{R}_y\ \boldsymbol{B}\boldsymbol{R}_{y\perp}\,] = [\ \underset{q\times b}{\boldsymbol{\Gamma}_1}\quad \underset{q\times(p-b)}{\boldsymbol{\Gamma}_2}\ ]\,. \tag{12.44}$$

Here $\boldsymbol{B} = [\,\boldsymbol{\Gamma}_1\,\boldsymbol{\Gamma}_2\,]\boldsymbol{T}^{-1}$. If $\boldsymbol{R}_x\boldsymbol{\Gamma}_1 = \boldsymbol{A}$ is consistent, then $\boldsymbol{\Gamma}_1$ satisfies the restrictions if and only if, for some $q \times b$ $\boldsymbol{\Delta}$ (with sign adjustable),

$$\boldsymbol{\Gamma}_1 = \boldsymbol{R}_x^-\boldsymbol{A} \pm (\boldsymbol{R}_x^-\boldsymbol{R}_x - \boldsymbol{I}_q)\boldsymbol{\Delta}\,. \tag{12.45}$$

In turn, $\boldsymbol{R}_x\boldsymbol{\Gamma}_1 = \boldsymbol{A}$ is equivalent to the original restrictions, and models 2, 3, 4, and 5, defined below, satisfy the restrictions. Models 1, 2, 4, and 5 are always linearly equivalent to each other, and linearly equivalent to model 3 if $\boldsymbol{R}_y$ is square and full rank.
Model 2 is $\text{GLM}_{N,p,q}(\boldsymbol{Y}_i\boldsymbol{T}, \boldsymbol{X}_i[\,\boldsymbol{\Gamma}_1\,\boldsymbol{\Gamma}_2\,]|\boldsymbol{R}_x\boldsymbol{\Gamma}_1 = \boldsymbol{A}, \boldsymbol{T}'\boldsymbol{\Sigma}\boldsymbol{T})$.
Model 3 is $\text{GLM}_{N,b,q}(\boldsymbol{Y}_{i3}, \boldsymbol{X}_{i3}\boldsymbol{\Delta}, \boldsymbol{R}_y'\boldsymbol{\Sigma}\boldsymbol{R}_y)$, with $\boldsymbol{Y}_3 = \boldsymbol{Y}\boldsymbol{R}_y - \boldsymbol{X}\boldsymbol{R}_x^-\boldsymbol{A}$ and $\boldsymbol{X}_3 = \boldsymbol{X}(\boldsymbol{R}_x^-\boldsymbol{R}_x - \boldsymbol{I}_q)$ and ignores $\boldsymbol{\Gamma}_2$. If $b = p = \text{rank}(\boldsymbol{R}_y)$, it is linearly equivalent.
Model 4 is $\text{GGLM}_{Np,q}(\boldsymbol{y}_s; \boldsymbol{X}_s\boldsymbol{\beta}_s|\boldsymbol{R}_s\boldsymbol{\beta}_s = \boldsymbol{a}_s, \boldsymbol{\Sigma} \otimes \boldsymbol{I}_N)$, with $\boldsymbol{y}_s = \text{vec}(\boldsymbol{Y})$, $\boldsymbol{X}_s = (\boldsymbol{I}_p \otimes \boldsymbol{X})$, $\boldsymbol{\beta}_s = \text{vec}(\boldsymbol{B})$, $\boldsymbol{R}_s$ $(ab \times q)$, $\boldsymbol{a}_s$ $(ab \times 1)$, $\boldsymbol{R}_s = (\boldsymbol{T}' \otimes \boldsymbol{R})$, and $\boldsymbol{a}_s = \text{vec}(\boldsymbol{A})$.
Model 5 is $\text{GGLM}_{Np,q}(\boldsymbol{u}; \boldsymbol{Z}\boldsymbol{\gamma}_s, \boldsymbol{\Sigma} \otimes \boldsymbol{I}_N)$, with $\boldsymbol{u} = \boldsymbol{y}_s - \boldsymbol{X}_s\boldsymbol{R}_s^-\boldsymbol{a}_s$, $\boldsymbol{Z} = \boldsymbol{X}_s(\boldsymbol{R}_s^-\boldsymbol{R}_s - \boldsymbol{I}_{qp})$, $\boldsymbol{\gamma}_s$ of dimension $qp \times 1$, and $\boldsymbol{Z}$ of dimension $Np \times qp$.

Proof. Expressions for $\boldsymbol{\Gamma}_1$ in terms of $\boldsymbol{\Delta}$ may be derived in terms of an arbitrary column j of $\boldsymbol{\Gamma}_1$, say $\boldsymbol{\gamma}_{1j}$ $(q \times 1)$. Here $\boldsymbol{R}_x\boldsymbol{\gamma}_{1j} = \boldsymbol{a}_i$ $(a \times 1)$ $\Leftrightarrow$ $\boldsymbol{\gamma}_{1j} = \boldsymbol{R}_x^-\boldsymbol{A}_j + (\boldsymbol{R}_x^-\boldsymbol{R}_x - \boldsymbol{I}_q)\boldsymbol{\delta}_j$ for some $\boldsymbol{\delta}_j$ $(q \times 1)$. Results on univariate models ensure the desired properties hold. The remainder of the proof is also based on univariate results and is left to the reader as an exercise. □

12.6 UNRESTRICTED ESTIMATION WITH COMPOUND SYMMETRY: THE "UNIVARIATE" APPROACH TO REPEATED MEASURES

As detailed in Lemma 1.33, a $p \times p$ compound symmetric covariance matrix may be written

$$\begin{aligned}\boldsymbol{\Sigma} &= \sigma^2\left[\mathbf{1}_p\mathbf{1}_p'\rho + \boldsymbol{I}_p(1-\rho)\right] \\ &= \boldsymbol{V}\mathrm{Dg}(\boldsymbol{\lambda})\boldsymbol{V}' \\ &= \begin{bmatrix}\boldsymbol{v}_{\mathrm{B}} & \boldsymbol{V}_{\mathrm{W}}\end{bmatrix}\begin{bmatrix}\lambda_1 & \mathbf{0} \\ \mathbf{0} & \lambda_2\boldsymbol{I}_{p-1}\end{bmatrix}\begin{bmatrix}\boldsymbol{v}_{\mathrm{B}}' \\ \boldsymbol{V}_{\mathrm{W}}'\end{bmatrix}.\end{aligned} \tag{12.46}$$

The $p \times p$ matrix $\mathrm{Dg}(\boldsymbol{\lambda}) = \mathrm{Dg}(\lambda_1, \lambda_2\mathbf{1}_{p-1})$ has $\lambda_1 = \sigma^2[1+(p-1)\rho]$ and $\lambda_2 = \sigma^2(1-\rho)$, with known eigenvectors.

Theorem 12.11 Model 1, $\mathrm{GLM}_{N,p,q}(\boldsymbol{Y}_i; \boldsymbol{X}_i\boldsymbol{B}, \boldsymbol{\Sigma})$, has positive definite compound symmetric covariance $\boldsymbol{\Sigma} = \boldsymbol{V}\mathrm{Dg}(\boldsymbol{\lambda})\boldsymbol{V}'$. The eigenvectors provide a *known, constant and exact* transformation to a model with uncorrelated observations, model 2, $\mathrm{GLM}_{N,p,q}[\boldsymbol{Y}_i\boldsymbol{V}; \boldsymbol{X}_i\boldsymbol{B}\boldsymbol{V}, \mathrm{Dg}(\boldsymbol{\lambda})]$. With Gaussian data the columns of $\boldsymbol{YV}$, as well as the rows, are independent. Column 1 has variance $\lambda_1 = \sigma^2[1+(p-1)\rho]$ and the other $p-1$ columns have variance $\lambda_2 = \sigma^2(1-\rho)$.

Proof. Lemma 1.33 provides most results needed. Model transformation gives

$$\begin{aligned}\boldsymbol{Y} &= \boldsymbol{XB} + \boldsymbol{E} \\ \boldsymbol{YV} &= \boldsymbol{XBV} + \boldsymbol{EV},\end{aligned} \tag{12.47}$$

$\mathrm{E}[\mathrm{row}_i(\boldsymbol{EV})] = \{\mathrm{E}[\mathrm{row}_i(\boldsymbol{E})]\}\boldsymbol{V} = \mathbf{0}$, and $\mathcal{V}\{[\mathrm{row}_i(\boldsymbol{EV})]'\} = \boldsymbol{V}'\boldsymbol{\Sigma}\boldsymbol{V} = \mathrm{Dg}(\boldsymbol{\lambda})$. □

Corollary 12.11 If the original data are Gaussian, then all observations split into one set of N observations which exactly follow model 2B (between), a *univariate* $\mathrm{GLM}_{N,q}()$, and a second set of $N(p-1)$ observations which exactly follow model 2W (within), a *univariate* $\mathrm{GLM}_{N(p-1),q}()$.

Proof. If $\boldsymbol{y}_{\mathrm{B}} = \boldsymbol{Y}\boldsymbol{v}_{\mathrm{B}}$ $(N \times 1)$ and $\boldsymbol{Y}_{\mathrm{W}} = \boldsymbol{Y}\boldsymbol{V}_{\mathrm{W}}$ $[N \times (p-1)]$, then

$$\begin{aligned}\boldsymbol{Y}\begin{bmatrix}\boldsymbol{v}_{\mathrm{B}} & \boldsymbol{V}_{\mathrm{W}}\end{bmatrix} &= \boldsymbol{XB}\begin{bmatrix}\boldsymbol{v}_{\mathrm{B}} & \boldsymbol{V}_{\mathrm{W}}\end{bmatrix} + \boldsymbol{E}\begin{bmatrix}\boldsymbol{v}_{\mathrm{B}} & \boldsymbol{V}_{\mathrm{W}}\end{bmatrix} \\ \begin{bmatrix}\boldsymbol{y}_{\mathrm{B}} & \boldsymbol{Y}_{\mathrm{W}}\end{bmatrix} &= \boldsymbol{X}\begin{bmatrix}\boldsymbol{B}\boldsymbol{v}_{\mathrm{B}} & \boldsymbol{B}\boldsymbol{V}_{\mathrm{W}}\end{bmatrix} + \begin{bmatrix}\boldsymbol{e}_{\mathrm{B}} & \boldsymbol{E}_{\mathrm{W}}\end{bmatrix}.\end{aligned} \tag{12.48}$$

If $\boldsymbol{\beta}_{\mathrm{B}} = \boldsymbol{B}\boldsymbol{v}_{\mathrm{B}}$ $(q \times 1)$, extracting the model (2B) for the first column gives

$$\boldsymbol{y}_{\mathrm{B}} = \boldsymbol{X}\boldsymbol{\beta}_{\mathrm{B}} + \boldsymbol{e}_{\mathrm{B}}\,. \tag{12.49}$$

Furthermore

$$\boldsymbol{e}_{\mathrm{B}} \sim \mathcal{N}_N(\mathbf{0}, \lambda_1\boldsymbol{I}_N) \tag{12.50}$$

$$\boldsymbol{y}_{\mathrm{B}} \sim \mathcal{N}_N(\boldsymbol{X}\boldsymbol{\beta}_{\mathrm{B}}, \lambda_1\boldsymbol{I}_N)\,. \tag{12.51}$$

Hence model 2B meets the assumptions of $\mathrm{GLM}_{N,q}(y_{\mathrm{B}i}; \boldsymbol{X}_i\boldsymbol{\beta}_{\mathrm{B}}, \lambda_1)$ with Gaussian errors. With similar notation, $\boldsymbol{B}_{\mathrm{W}} = \boldsymbol{B}\boldsymbol{V}_{\mathrm{W}}$ is $q \times (p-1)$ and

$$\boldsymbol{Y}_{\mathrm{W}} = \boldsymbol{X}\boldsymbol{B}_{\mathrm{W}} + \boldsymbol{E}_{\mathrm{W}}\,. \tag{12.52}$$

Here

$$\begin{aligned}\boldsymbol{E}_{\mathrm{W}} &\sim \mathcal{N}_{N,(p-1)}(\boldsymbol{0}, \boldsymbol{I}_N, \lambda_2 \boldsymbol{I}_{p-1}) \Leftrightarrow \\ \boldsymbol{e}_{\mathrm{W}} &= \mathrm{vec}(\boldsymbol{E}_{\mathrm{W}}) \sim \mathcal{N}_{N(p-1)}\big(\boldsymbol{0}, \lambda_2 \boldsymbol{I}_{N(p-1)}\big)\,,\end{aligned} \tag{12.53}$$

$$\begin{aligned}\boldsymbol{Y}_{\mathrm{W}} &\sim \mathcal{N}_{N,(p-1)}(\boldsymbol{X}\boldsymbol{B}_{\mathrm{W}}, \boldsymbol{I}_N, \lambda_2 \boldsymbol{I}_{p-1}) \Leftrightarrow \\ \boldsymbol{y}_{\mathrm{W}} &= \mathrm{vec}(\boldsymbol{Y}_{\mathrm{W}}) \sim \mathcal{N}_{N(p-1)}\big[\mathrm{vec}(\boldsymbol{X}\boldsymbol{B}_{\mathrm{W}}), \lambda_2 \boldsymbol{I}_{N(p-1)}\big]\,.\end{aligned} \tag{12.54}$$

Observing

$$\begin{aligned}\mathrm{vec}(\boldsymbol{X}\boldsymbol{B}_{\mathrm{W}}) &= (\boldsymbol{I}_{p-1} \otimes \boldsymbol{X})\mathrm{vec}(\boldsymbol{B}_{\mathrm{W}}) \\ &= \boldsymbol{X}_{\mathrm{W}}\boldsymbol{\beta}_{\mathrm{W}}\end{aligned} \tag{12.55}$$

allows writing

$$\begin{aligned}\mathrm{vec}(\boldsymbol{Y}_{\mathrm{W}}) &= (\boldsymbol{I}_{p-1} \otimes \boldsymbol{X})\mathrm{vec}(\boldsymbol{B}_{\mathrm{W}}) + \mathrm{vec}(\boldsymbol{E}_{\mathrm{W}}) \Leftrightarrow \\ \boldsymbol{y}_{\mathrm{W}} &= \qquad \boldsymbol{X}_{\mathrm{W}}\boldsymbol{\beta}_{\mathrm{W}} \qquad + \quad \boldsymbol{e}_{\mathrm{W}}\,.\end{aligned} \tag{12.56}$$

Finally, the last equation satisfies univariate $\mathrm{GLM}_{N(p-1),q}[y_{\mathrm{W},i}; \mathrm{row}_i(\boldsymbol{X}_{\mathrm{W}})\boldsymbol{\beta}_{\mathrm{W}}, \lambda_2]$ with Gaussian errors, model 2W. □

Model 2B describes differences between subjects (ISUs), while model 2W describes differences within subjects. The formulation provides a basis of the theory underlying the "univariate" approach to repeated measures and the "uncorrected" UNIREP test. Muller and Barton (1989) included many details.

EXERCISES

Here $\boldsymbol{e}_1$ is a vector with a 1 as the first element and zeros as all other elements: $\boldsymbol{e}_1 = [\,1\;0\cdots0\,]'$. In different exercises $\boldsymbol{e}_1$ can have different dimensions.

True/False Exercises. Several propositions are stated below. For each choose one of the following: T if the proposition is true and F if the proposition is false. Be careful! Some propositions may be tricky. Most are either correct theorems or a slight modification of correct theorems (in which case the proposition is false).

For each proposition you mark as "false," give either (1) a brief counterexample or (2) a brief remark on why the proposition fails to be true. This should make clear which aspects of the proposition you reject. In addition to intentional errors in the propositions, there is always the possibility of unintentional typographical errors. Therefore, your brief remarks are important. If you mark a proposition "true" when it is in fact false, it will be inferred that your proof would contain one or more errors. Therefore, do not attach proofs or comments for propositions marked true.

The following assumption and definitions apply to exercises 12.1–12.6. Assume $\mathrm{GLM}_{N,p,q}(\boldsymbol{Y}_i; \boldsymbol{X}_i\boldsymbol{B}, \boldsymbol{\Sigma})$ with Gaussian errors and rank($\boldsymbol{X}$) $= r \le q$. Also $\widehat{\boldsymbol{B}} = (\boldsymbol{X}'\boldsymbol{X})^{-}\boldsymbol{X}'\boldsymbol{Y} = [\widehat{\boldsymbol{\beta}}_1\ \widehat{\boldsymbol{\beta}}_2 \cdots \widehat{\boldsymbol{\beta}}_p]$, $\widehat{\boldsymbol{\Sigma}} = \boldsymbol{Y}'[\boldsymbol{I} - \boldsymbol{X}(\boldsymbol{X}'\boldsymbol{X})^{-}\boldsymbol{X}']\boldsymbol{Y}/(N-r)$, $\widehat{\boldsymbol{\beta}}_1 = (\boldsymbol{X}'\boldsymbol{X})^{-}\boldsymbol{X}'\boldsymbol{Y}\boldsymbol{e}_1$, and $\widehat{\sigma}_1^2 = \boldsymbol{e}_1'\boldsymbol{Y}'[\boldsymbol{I} - \boldsymbol{X}(\boldsymbol{X}'\boldsymbol{X})^{-}\boldsymbol{X}']\boldsymbol{Y}\boldsymbol{e}_1/(N-r)$.

T F 12.1 *Proposition*: $\widehat{\boldsymbol{\beta}}_1 \sim \mathcal{N}_q[\boldsymbol{B}\boldsymbol{e}_1, \boldsymbol{e}_1'\boldsymbol{\Sigma}\boldsymbol{e}_1(\boldsymbol{X}'\boldsymbol{X})^{-}]$ of rank $r \le q$.

T F 12.2 *Proposition*: $\widehat{\sigma}_1^2/(\boldsymbol{e}_1'\boldsymbol{\Sigma}\boldsymbol{e}_1) \sim \chi^2(N-r, 0)$.

T F 12.3 *Proposition*: $\widehat{\boldsymbol{\beta}}_1$ and $\widehat{\sigma}_1^2$ are uncorrelated.

T F 12.4 Additionally assume $\boldsymbol{\theta} = \boldsymbol{C}\boldsymbol{B}\boldsymbol{e}_1$ and $\widehat{\boldsymbol{\theta}} = \boldsymbol{C}\widehat{\boldsymbol{B}}\boldsymbol{e}_1$ are $a \times 1$, $a \le q$. *Proposition*: $\widehat{\boldsymbol{\theta}} \sim \mathcal{N}_a[\boldsymbol{C}\boldsymbol{B}\boldsymbol{e}_1, \boldsymbol{e}_1'\boldsymbol{\Sigma}\boldsymbol{e}_1\boldsymbol{C}(\boldsymbol{X}'\boldsymbol{X})^{-}\boldsymbol{C}']$ and rank$[\mathcal{V}(\widehat{\boldsymbol{\theta}})] = a$.

The following assumptions and notation apply to exercises 12.5–12.8. Assume $\mathrm{GLM}_{N,p,q}\mathrm{FR}(\boldsymbol{Y}_i; \boldsymbol{X}_i\boldsymbol{B}, \boldsymbol{\Sigma})$ with Gaussian errors, $N \gg q$, and rank($\boldsymbol{\Sigma}$) $= p$. Also $\widehat{\boldsymbol{B}} = (\boldsymbol{X}'\boldsymbol{X})^{-1}\boldsymbol{X}'\boldsymbol{Y}$, $\boldsymbol{S} = \boldsymbol{Y}'[\boldsymbol{I} - \boldsymbol{X}(\boldsymbol{X}'\boldsymbol{X})^{-}\boldsymbol{X}']\boldsymbol{Y}/(N-q)$, $\boldsymbol{y}_s = \mathrm{vec}(\boldsymbol{Y})$, $\boldsymbol{\beta}_s = \mathrm{vec}(\boldsymbol{B})$, and $\widehat{\boldsymbol{\beta}}_s = \mathrm{vec}(\widehat{\boldsymbol{B}})$.

T F 12.5 *Proposition*: $\widehat{\boldsymbol{\beta}}_s \sim \mathcal{N}_{pq}[\boldsymbol{\beta}_s, \boldsymbol{\Sigma} \otimes (\boldsymbol{X}'\boldsymbol{X})^{-1}]$ and rank$[\mathcal{V}(\widehat{\boldsymbol{\beta}}_s)] = pq$.

T F 12.6 *Proposition*: $\boldsymbol{S} \sim \mathcal{W}_p(N-q, \boldsymbol{\Sigma}, \boldsymbol{0})$.

T F 12.7 *Proposition*: $\widehat{\boldsymbol{B}}$ and $\boldsymbol{S}$ are independent.

T F 12.8 *Proposition*: $\mathrm{E}(\boldsymbol{S}) = \boldsymbol{\Sigma}$ and $\mathcal{V}(s_{ij}) = (N-q)\sigma_{ij}^2 + (N-q)^2\sigma_{ii}\sigma_{jj}$.

12.9 Three models may be written for the same data as follows: For model 1

$$\begin{aligned}\boldsymbol{Y} &= (\boldsymbol{1}_n \otimes \boldsymbol{I}_3)\boldsymbol{B}_1 + \boldsymbol{E}\\ \boldsymbol{Y} &= \boldsymbol{X}_1\boldsymbol{B}_1 + \boldsymbol{E}.\end{aligned}$$

For model 2

$$\begin{aligned}\boldsymbol{Y} &= \left(\boldsymbol{1}_n \otimes \begin{bmatrix}1&0&0\\1&1&0\\1&0&1\end{bmatrix}\right)\boldsymbol{B}_2 + \boldsymbol{E}\\ \boldsymbol{Y} &= \boldsymbol{X}_2\boldsymbol{B}_2 + \boldsymbol{E}.\end{aligned}$$

For model 3

$$\begin{aligned}\boldsymbol{Y} &= \left(\boldsymbol{1}_n \otimes \begin{bmatrix}1&1&0&0&-1&-1\\1&0&1&0&1&0\\1&0&0&1&0&1\end{bmatrix}\right)\boldsymbol{B}_3 + \boldsymbol{E}\\ \boldsymbol{Y} &= \boldsymbol{X}_3\boldsymbol{B}_3 + \boldsymbol{E}.\end{aligned}$$

Each meets the assumptions of a $\mathrm{GLM}_{N,p,q_k}(\boldsymbol{Y}_i; \boldsymbol{X}_{i,k}\boldsymbol{B}_k, \boldsymbol{\Sigma})$.

12.9.1 Prove that model 1 and model 2 are linearly equivalent.

12.9.2 Prove that model 1 and model 3 are linearly equivalent.

12.9.3 matrices $\boldsymbol{R}_x = [\boldsymbol{0}\ \boldsymbol{I}_3\ \boldsymbol{0}]$ and $\boldsymbol{A} = \boldsymbol{0}$ specify model 4 $\mathrm{GLM}_{N,p,q_3}(\boldsymbol{Y}_i; \boldsymbol{X}_{i,3}\boldsymbol{B}_3 | \boldsymbol{R}_x\boldsymbol{B}_3 = \boldsymbol{A}, \boldsymbol{\Sigma})$ as a restriction of model 3. Specify the dimensions of $\boldsymbol{A}$ and submatrices in $\boldsymbol{R}_x$.

12.9.4 Find a full-rank and unrestricted model 5, equivalent to model 4, with the additional restriction that all elements of $\boldsymbol{B}_5$ are also elements of $\boldsymbol{B}_3$. This may be done by inspection (no proof needed).

CHAPTER 13

Estimation for Generalizations of Multivariate Models

13.1 MOTIVATION

As detailed in the preceding two chapters, the "least squares" assumptions of the univariate and multivariate GLM allow finding unbiased estimators of the primary parameters and broad classes of secondary parameters. The additional assumption of Gaussian errors leads to the estimators also satisfying likelihood principles. In all cases the estimates can be computed in closed form, with a noniterative algorithm for solving a system of simultaneous linear equations.

Do more general linear models allow such nice results? For a range of generalizations of the multivariate linear model, the answer is essentially "Yes." Although usually much less convenient to compute, estimates satisfying generalized least squares or likelihood criteria can be found. Furthermore, the expected value parameter estimators are usually unbiased or nearly so. Covariance parameter estimators can range from unbiased to substantially biased.

As discussed in Chapter 4, models for growth curves, seemingly unrelated regressions, multiple design matrix settings, and doubly multivariate settings can be cast as generalizations of the GLM. The references given there contain details about particular techniques.

13.2 CRITERIA AND ALGORITHMS

Given Gaussian errors, ordinary least squares (OLS) estimators coincide with the likelihood estimators in the univariate and multivariate GLM. The same holds true for *exact* weighted least squares (WLS), as seen in results in the previous two chapters for the GGLM.

> **Definition 13.1 (a)** Using estimated weights (covariance matrix) in weighted least squares (WLS) formulas defines *approximate* weighted least squares (AWLS) estimators.
> **(b)** Using WLS formulas to iteratively estimate expected values and covariance defines *iterative approximate least squares* (ITAWLS).

AWLS requires an initial computation of an OLS estimate for the expected-value parameters, followed by computation of covariance parameter estimates in terms of the residuals. Except in very special cases (e.g., Gaussian errors in special patterns), the two-step method does not produce maximum likelihood estimates.

Although much more efficient algorithms can often be found, iterating the two-step AWLS method (ITAWLS) yields maximum likelihood estimates for a variety of models with Gaussian errors. The covariance estimate from the second step allows computing an updated estimate of the expected-value parameters. In turn, a new covariance estimate may be computed, etc.

13.3 WEIGHTED ESTIMATION OF B AND Σ

Weighted estimation in a multivariate model implies considering $\boldsymbol{Y} \sim \mathcal{N}_{N,p}(\boldsymbol{XB}, \boldsymbol{D}, \boldsymbol{\Sigma})$, with *known* $\boldsymbol{X}$, full-rank $\boldsymbol{D} = \boldsymbol{F}_D\boldsymbol{F}_D'$, and unknown $\boldsymbol{B}$ and $\boldsymbol{\Sigma}$. Given the assumptions, $\text{vec}(\boldsymbol{Y}') \sim \mathcal{N}_{Np}[(\boldsymbol{X} \otimes \boldsymbol{I}_p)\text{vec}(\boldsymbol{B}'), \boldsymbol{D} \otimes \boldsymbol{\Sigma}]$. Multiplication by $\boldsymbol{F}_D^{-1}$ gives $\boldsymbol{Y}_D = \boldsymbol{F}_D^{-1}\boldsymbol{Y} \sim \mathcal{N}_{N,p}(\boldsymbol{F}_D^{-1}\boldsymbol{XB}, \boldsymbol{I}, \boldsymbol{\Sigma})$. If $\boldsymbol{X}_D = \boldsymbol{F}_D^{-1}\boldsymbol{X}$, then $\boldsymbol{Y}_D$ corresponds to $\text{GLM}_{N,p,q}[\text{row}_i(\boldsymbol{Y}_D), \text{row}_i(\boldsymbol{X}_D)\boldsymbol{B}, \boldsymbol{\Sigma}] \Leftrightarrow \text{GLM}_{N,p,q}(\boldsymbol{F}_D^{-1}\boldsymbol{Y}_i, \boldsymbol{F}_D^{-1}\boldsymbol{X}_i\boldsymbol{B}, \boldsymbol{\Sigma})$.

Theorem 13.1 Matrices $\boldsymbol{Y}$ $(N \times p)$, $\boldsymbol{X}$ $(N \times q)$, $\text{rank}(\boldsymbol{X}) \leq q$, positive definite $\boldsymbol{D} = \boldsymbol{D}'$ $(N \times N)$, and positive definite or positive semidefinite $\boldsymbol{\Sigma} = \boldsymbol{\Sigma}'$ $(p \times p)$ define $\text{GGLM}_{Np,q}[\text{vec}(\boldsymbol{Y}'); (\boldsymbol{X} \otimes \boldsymbol{I}_p)\text{vec}(\boldsymbol{B}'), \boldsymbol{D} \otimes \boldsymbol{\Sigma}]$, with Gaussian errors. The same assumptions give $\text{GGLM}_{Np,q}[\text{vec}(\boldsymbol{Y}'\boldsymbol{L}); (\boldsymbol{L}'\boldsymbol{X} \otimes \boldsymbol{I}_p)\text{vec}(\boldsymbol{B}'), \boldsymbol{I}_N \otimes \boldsymbol{\Sigma}]$. The latter model has a corresponding multivariate GLM satisfying the least squares assumptions with Gaussian errors and $\text{rank}(\boldsymbol{L}'\boldsymbol{X}) = \text{rank}(\boldsymbol{X})$.

Proof. With $\boldsymbol{Y}_i = \text{row}_i(\boldsymbol{Y})$ and $\boldsymbol{Y}'$ $p \times N$, $\text{vec}(\boldsymbol{Y}')$ is the "vertical concatenation of rows." Hence

$$\begin{aligned} \mathcal{V}[\text{vec}(\boldsymbol{Y}')] &= \boldsymbol{D} \otimes \boldsymbol{\Sigma} \\ &= \begin{bmatrix} d_{11}\boldsymbol{\Sigma} & d_{12}\boldsymbol{\Sigma} & \cdots & d_{1N}\boldsymbol{\Sigma} \\ d_{21}\boldsymbol{\Sigma} & d_{22}\boldsymbol{\Sigma} & \cdots & d_{2N}\boldsymbol{\Sigma} \\ \vdots & \vdots & \ddots & \vdots \\ d_{N1}\boldsymbol{\Sigma} & d_{N2}\boldsymbol{\Sigma} & \cdots & d_{NN}\boldsymbol{\Sigma} \end{bmatrix}, \end{aligned} \tag{13.1}$$

and $\text{vec}(\boldsymbol{L}'\boldsymbol{Y}) = (\boldsymbol{I}_p \otimes \boldsymbol{L}')\text{vec}(\boldsymbol{Y})$ has covariance $(\boldsymbol{I}_p \otimes \boldsymbol{L}')(\boldsymbol{\Sigma} \otimes \boldsymbol{D})(\boldsymbol{I}_p \otimes \boldsymbol{L}) = \boldsymbol{\Sigma} \otimes \boldsymbol{I}_N$, while $\text{vec}(\boldsymbol{Y}'\boldsymbol{L}) = (\boldsymbol{L}' \otimes \boldsymbol{I}_p)\text{vec}(\boldsymbol{Y}')$ has covariance $(\boldsymbol{L}' \otimes \boldsymbol{I}_p)(\boldsymbol{D} \otimes \boldsymbol{\Sigma})(\boldsymbol{L} \otimes \boldsymbol{I}_p) = \boldsymbol{I}_N \otimes \boldsymbol{\Sigma}$ and mean $(\boldsymbol{L}' \otimes \boldsymbol{I}_p)(\boldsymbol{X} \otimes \boldsymbol{I}_p)\text{vec}(\boldsymbol{B}') = (\boldsymbol{L}'\boldsymbol{X} \otimes \boldsymbol{I}_p)\text{vec}(\boldsymbol{B}')$. The proof is completed by appealing to properties of a direct-product matrix Gaussian. □

Theorem 13.2 For $\text{GGLM}_{Np,q}[\text{vec}(\boldsymbol{Y}'); (\boldsymbol{X} \otimes \boldsymbol{I}_p)\text{vec}(\boldsymbol{B}'), \boldsymbol{D} \otimes \Sigma]$ with Gaussian errors, $r = \text{rank}(\boldsymbol{X}) \leq q$, $\boldsymbol{\Theta} = \boldsymbol{CB}$ is estimable, unknown $\Sigma = \Sigma'$ $(p \times p)$ is positive definite, while $\boldsymbol{D} = \boldsymbol{D}'$ $(N \times N)$ is *known* and positive definite. The joint MLEs of $\boldsymbol{B}, \Sigma$, and $\boldsymbol{\Theta}$ are

$$\widetilde{\boldsymbol{B}} = (\boldsymbol{X}'\boldsymbol{D}^{-1}\boldsymbol{X})^{-}\boldsymbol{X}'\boldsymbol{D}^{-1}\boldsymbol{Y} \tag{13.2}$$

$$\widetilde{\Sigma} = (\boldsymbol{Y} - \boldsymbol{X}\widetilde{\boldsymbol{B}})'\boldsymbol{D}^{-1}(\boldsymbol{Y} - \boldsymbol{X}\widetilde{\boldsymbol{B}})/N \tag{13.3}$$

$$\widehat{\boldsymbol{\Theta}} = \boldsymbol{C}\widetilde{\boldsymbol{B}}. \tag{13.4}$$

Both $\widetilde{\Sigma}$ and $\widehat{\boldsymbol{\Theta}}$ are invariant to $\widetilde{\boldsymbol{B}}$, which is unique if $r = q$. The unbiased restricted maximum likelihood estimator $\widehat{\Sigma} = \widetilde{\Sigma}N/(N - r)$ is typically used.

Proof. The proof is left as an exercise.

13.4 TRANSFORMATIONS AMONG GROWTH CURVE DESIGNS

Often a $\text{GCM}_{N,p,q,m}(\boldsymbol{Y}_i; \boldsymbol{X}_i\boldsymbol{BT}, \Sigma)$ leads to interest in a reduced model for the within-subject dimension. As in all our discussions of the GCM, for convenience of exposition we assume that columns of $\boldsymbol{Y}_i$ differ (only) in regard to the time a repeated measure was collected. With polynomial coding, the reduced (within-subject) model can typically be created from the full (within-subject) model by deleting higher-order terms corresponding to trailing columns of $\boldsymbol{B}$. If so, the expected value for the full model may be written

$$\begin{aligned} \text{E}(\boldsymbol{Y}|\boldsymbol{X},\boldsymbol{T}) &= \boldsymbol{XBT} \\ &= \boldsymbol{X}[\boldsymbol{B}_1 \;\; \boldsymbol{B}_2]\begin{bmatrix} \boldsymbol{T}_1 \\ \boldsymbol{T}_2 \end{bmatrix} \\ &= \boldsymbol{XB}_1\boldsymbol{T}_1 + \boldsymbol{XB}_2\boldsymbol{T}_2, \end{aligned} \tag{13.5}$$

with corresponding expected value for the reduced model of

$$\begin{aligned} \text{E}(\boldsymbol{Y}|\boldsymbol{X},\boldsymbol{T},\boldsymbol{B}_2 = \boldsymbol{0}) &= \boldsymbol{X}[\boldsymbol{B}_1 \;\; \boldsymbol{0}]\begin{bmatrix} \boldsymbol{T}_1 \\ \boldsymbol{T}_2 \end{bmatrix} \\ &= \boldsymbol{XB}_1\boldsymbol{T}_1. \end{aligned} \tag{13.6}$$

With $m \times p$ $\boldsymbol{T}$ for $1 \leq m \leq p$ and $m_1 \times p$ $\boldsymbol{T}_1$, necessarily $\boldsymbol{T}_2$ must be $(m - m_1) \times p$ for $1 \leq m_1 \leq m \leq p$. The fully saturated (within-subject) model has $m = p$ and $\boldsymbol{T} = \{d_j^{i-1}\}$. Equivalently, for ISU i the expected values are written $\text{E}(\boldsymbol{Y}_i|\boldsymbol{X}_i,\boldsymbol{T}) = \boldsymbol{X}_i\boldsymbol{BT} = \boldsymbol{X}_i\boldsymbol{B}_1\boldsymbol{T}_1 + \boldsymbol{X}_i\boldsymbol{B}_2\boldsymbol{T}_2$ and $\text{E}(\boldsymbol{Y}_i|\boldsymbol{X}_i,\boldsymbol{T},\boldsymbol{B}_2 = \boldsymbol{0}) = \boldsymbol{X}_i\boldsymbol{B}_1\boldsymbol{T}_1$.

A similar construction relating a full and reduced model for between-subject design applies when the columns of $\boldsymbol{X}$ contain polynomials for a continuous predictor. A scientist using natural polynomial coding and a participant's age as the basic predictor can use $\boldsymbol{X}_i = \{x_{ij}\}$ with $x_{ij} = \text{age}_i^{j-1}$, giving $\boldsymbol{X}_i = [1 \;\; \text{age}_i \;\; \text{age}_i^2 \;\; \cdots \;\; \text{age}_i^{q-1}]$. The strong constraints relating the univariate GLM to all variations of the multivariate GLM, including the GCM, guarantee univariate techniques for coding $\boldsymbol{X}$ and reducing models apply directly to $\boldsymbol{X}$ in the multivariate setting. Hence we do not discuss the topic here. Muller and Fetterman (2002, especially Chapters 9, 11, and 16), as well as many other authors of univariate texts, provided detailed discussions of polynomial models.

Natural polynomial models seem easy to handle and interpret. Disadvantages include potential numerical problems and a corresponding lack of statistical independence among the columns of $\widehat{\boldsymbol{B}}_1$ (which contains the estimators assumed to not be zero, as in the last equation). In turn, dependencies among terms can greatly complicate interpretations (a fact often overlooked). Muller and Fetterman (2002, Chapter 9) addressed the difficulty in the context of comparing added-last and added-in-order tests and corresponding correlations. Orthogonal or orthonormal polynomial models avoid the disadvantages. Within-subject design matrix $\boldsymbol{T}$ $(m \times p)$ is assumed to be rank $m \leq p$. It is orthogonal (by rows) if $\boldsymbol{T}\boldsymbol{T}'$ is diagonal and orthonormal (by rows) if $\boldsymbol{T}\boldsymbol{T}' = \boldsymbol{I}_m$.

Definition 13.2 In discussing growth curve models, $\boldsymbol{T}_{\text{NAT}}$ indicates a matrix with $t_{ij} = (d_j)^{i-1}$, a natural polynomial. Similarly, $\boldsymbol{T}_{\text{ORT}}$ indicates $\boldsymbol{T}_{\text{ORT}}\boldsymbol{T}'_{\text{ORT}} = \boldsymbol{D}$, a diagonal matrix, which implies $\boldsymbol{T}_{\text{ORT}}$ is orthogonal by rows. Also, $\boldsymbol{T}_{\text{ORN}}$ indicates a matrix orthonormal by rows, $\boldsymbol{T}_{\text{ORN}}\boldsymbol{T}'_{\text{ORN}} = \boldsymbol{I}$.

Example 13.1 The weight of each of $N = 50$ children is measured in kilograms at 5, 6, and 7 years of age. The resulting $p = 3$ ordered responses for child i is $\boldsymbol{Y}_i = [\,\text{WT}_{i5}\ \text{WT}_{i6}\ \text{WT}_{i7}\,]$ and $\boldsymbol{d}' = [\,5\ 6\ 7\,] = \{d_j\}$. The cohort is assumed to be a single homogeneous group, and consequently the between-individual design matrix has one column, $\boldsymbol{X} = \mathbf{1}_N = [\,1\ \cdots\ 1\,]'$. Without loss of generality, any $\boldsymbol{X}$ matrix could be used, but to keep the example simple, we assume no additional predictor variables are needed. We further suppose the data were collected with the understanding the growth curve for the children would be well approximated by a low-order polynomial. Therefore a polynomial matrix $\boldsymbol{T}$ is desired and may be defined in terms of a natural polynomial:

$$\boldsymbol{B}_{\text{NAT}} = [\,\beta_0\ \beta_1\ \beta_2\,] \tag{13.7}$$

and

$$\boldsymbol{T}_{\text{NAT}} = \begin{bmatrix} 1 & 1 & 1 \\ 5 & 6 & 7 \\ 25 & 36 & 49 \end{bmatrix} = \{d_j^{i-1}\}. \tag{13.8}$$

Element (i, j) of $\boldsymbol{T} = [\,\boldsymbol{t}_1\ \cdots\ \boldsymbol{t}_p\,]$ is a monomial. In turn, the elements of the parameter matrix $\boldsymbol{B}$ are the coefficients of the natural polynomial. The expected weight of child i at time j is

$$\mathrm{E}(y_{ij}|\boldsymbol{X}, \boldsymbol{T}) = 1\boldsymbol{B}_{\text{NAT}}\boldsymbol{t}_j = [\,\beta_0\ \beta_1\ \beta_2\,]\begin{bmatrix} 1 \\ d_j \\ d_j^2 \end{bmatrix}. \tag{13.9}$$

If we assume $\beta_2 = 0$, then $\boldsymbol{B}$ is partitioned such that $\boldsymbol{B}_1 = [\beta_0\ \beta_1]$ and $\boldsymbol{B}_2 = [\beta_2]$. Correspondingly, $\boldsymbol{T}$ is partitioned with $\boldsymbol{T}_1 = \begin{bmatrix} 1 & 1 & 1 \\ 5 & 6 & 7 \end{bmatrix}$ and $\boldsymbol{T}_2 = [25\ 36\ 49]$.

Definition 13.3 One-to-one linear transformations exist among natural, orthogonal, and orthonormal polynomial design matrices. A set of matrices, all $p \times p$, may be defined as follows:

$$\boldsymbol{A}_{\text{ORTNAT}} = \boldsymbol{T}_{\text{ORT}}\boldsymbol{T}_{\text{NAT}}^{-1} \tag{13.10}$$

$$\boldsymbol{A}_{\text{ORNNAT}} = \boldsymbol{T}_{\text{ORN}}\boldsymbol{T}_{\text{NAT}}^{-1} \tag{13.11}$$

$$\boldsymbol{A}_{\text{ORNORT}} = \boldsymbol{T}_{\text{ORN}}\boldsymbol{T}_{\text{ORT}}^{-1}\,. \tag{13.12}$$

Although $(\boldsymbol{T}_{\text{NAT}})^{-1}$ always exists, the ratio of its eigenvalues, λ_1/λ_p, approaches zero as $p \to \infty$. Numerical problems will be encountered when trying to invert large natural polynomial matrices represented in computer arithmetic with typical finite precision. By contrast, inverting the orthonormal polynomial design matrix is perfectly stable and a trivial operation: $(\boldsymbol{T}_{\text{ORN}})^{-1} = (\boldsymbol{T}_{\text{ORN}})'$.

Lemma 13.1 One-to-one linear transformations among natural, orthogonal, and orthonormal polynomial regression coefficients exist. The following one-to-one relationships hold (with all $\boldsymbol{B}$ of dimension $q \times p$ and all $\boldsymbol{A}$ $p \times p$:

$$\boldsymbol{B}_{\text{NAT}} = \boldsymbol{B}_{\text{ORT}}\boldsymbol{A}_{\text{ORTNAT}} \tag{13.13}$$

$$\boldsymbol{B}_{\text{NAT}} = \boldsymbol{B}_{\text{ORN}}\boldsymbol{A}_{\text{ORNNAT}} \tag{13.14}$$

$$\boldsymbol{B}_{\text{ORN}} = \boldsymbol{B}_{\text{ORT}}\boldsymbol{A}_{\text{ORNORT}}\,. \tag{13.15}$$

Proof. Left as an exercise.

Definition 13.4 If the elements of $\boldsymbol{d}$ are equally spaced, then $d_{j+1} - d_j$ is a constant $\forall i$, and an *orthogonal design matrix* $\boldsymbol{T}_{\text{ORT}}$ exists with every element an integer. The elements of $\boldsymbol{T}_{\text{ORT}}$ can be found in tables of orthogonal polynomial coefficients (in the Appendix, Section A.3).

Example 13.2 For $p = 3$

$$\boldsymbol{T}_{\text{ORT}} = \begin{bmatrix} 1 & 1 & 1 \\ -1 & 0 & 1 \\ 1 & -2 & 1 \end{bmatrix} \begin{matrix} \leftarrow \text{constant} \\ \leftarrow \text{linear} \\ \leftarrow \text{quadratic} \end{matrix} \tag{13.16}$$

For $p = 4$

$$\boldsymbol{T}_{\mathrm{ORT}} = \begin{bmatrix} 1 & 1 & 1 & 1 \\ -3 & -1 & 1 & 3 \\ 1 & -1 & -1 & 1 \\ -1 & 3 & -3 & 1 \end{bmatrix} \begin{matrix} \leftarrow \text{constant} \\ \leftarrow \text{linear} \\ \leftarrow \text{quadratic} \\ \leftarrow \text{cubic} \end{matrix} \qquad (13.17)$$

Dividing t_{ij} by $\left(\sum_{j=1}^{p} t^2\right)^{1/2}$ creates the corresponding orthonormal matrix. Premultiplying by $\mathrm{Dg}\left[\left(\sum_{j=1}^{p} t_{ij}^2\right)^{-1/2}\right]$ accomplishes the task.

Example 13.3 For the example,

$$\boldsymbol{T}_{\mathrm{ORN}} = \begin{bmatrix} 0.577350 & 0.577350 & 0.577350 \\ -0.707107 & 0.000000 & 0.707107 \\ 0.408248 & -0.816497 & 0.408248 \end{bmatrix} \begin{matrix} \leftarrow \text{constant} \\ \leftarrow \text{linear} \\ \leftarrow \text{quadratic} \end{matrix} \qquad (13.18)$$

In the example of children's weights ($N = 50$, $m = 2$, $p = 3$, $q = 1$), the children's weights were measured at equally spaced time intervals. Therefore an orthogonal matrix exists which has integer entries:

$$\boldsymbol{T}_{\mathrm{ORT}} = \begin{bmatrix} 1 & 1 & 1 \\ -1 & 0 & 1 \\ 1 & -2 & 1 \end{bmatrix}. \qquad (13.19)$$

In turn, the expected value of the weight of child i at time j is

$$\mathrm{E}(y_{ij}|\boldsymbol{X} = \mathbf{1}, \boldsymbol{T}) = \boldsymbol{B}_{\mathrm{ORT}}\boldsymbol{T}_{j\mathrm{ORT}} = [\,\beta_0 \;\; \beta_1 \;\; \beta_2\,] \begin{bmatrix} 1 \\ -6 + d_j \\ 106 - 36d_j + 3d_j^2 \end{bmatrix}$$

$$= [\,\beta_0 \;\; \beta_1 \;\; \beta_2\,] \begin{bmatrix} 1 & 0 & 0 \\ -6 & 1 & 0 \\ 106 & -36 & 3 \end{bmatrix} \begin{bmatrix} 1 \\ d_j \\ d_j^2 \end{bmatrix}. \qquad (13.20)$$

The $\boldsymbol{B}$ matrices and $\boldsymbol{T}$ matrices of the natural and orthogonal models are linearly related. The linear transformation matrix for the two designs is

$$\boldsymbol{A}_{\mathrm{ORTNAT}} = \begin{bmatrix} 1 & 0 & 0 \\ -6 & 1 & 0 \\ 106 & -36 & 3 \end{bmatrix} = \boldsymbol{T}_{\mathrm{ORT}}\boldsymbol{T}_{\mathrm{NAT}}^{-1}. \qquad (13.21)$$

Thus we have $\boldsymbol{B}_{\mathrm{NAT}} = \boldsymbol{B}_{\mathrm{ORT}}\boldsymbol{A}_{\mathrm{ORTNAT}}$,

$$[\,\beta_0 \;\; \beta_1 \;\; \beta_2\,]_{\mathrm{NAT}} = [\,\beta_0 \;\; \beta_1 \;\; \beta_2\,]_{\mathrm{ORT}} \begin{bmatrix} 1 & 0 & 0 \\ -6 & 1 & 0 \\ 106 & -36 & 3 \end{bmatrix}, \qquad (13.22)$$

and $\boldsymbol{T}_{\mathrm{ORT}} = \boldsymbol{A}_{\mathrm{ORTNAT}}\boldsymbol{T}_{\mathrm{NAT}}$,

$$\begin{bmatrix} 1 & 1 & 1 \\ -1 & 0 & 1 \\ 1 & -2 & 1 \end{bmatrix} = \begin{bmatrix} 1 & 0 & 0 \\ -6 & 1 & 0 \\ 106 & -36 & 3 \end{bmatrix} \begin{bmatrix} 1 & 1 & 1 \\ 5 & 6 & 7 \\ 25 & 36 & 49 \end{bmatrix}. \tag{13.23}$$

We suppose mean weight is approximately linear in time over the ages from 5 to 7 years. If so, it is appropriate to assume $\beta_2 = 0$ in the natural polynomial model, or $\beta_2 = 0$ in the orthogonal polynomial model. The resulting model in terms of orthogonal polynomials is $\mathrm{E}(\boldsymbol{Y}|\boldsymbol{X},\boldsymbol{T}) = [\beta_0 \quad \beta_1]_{\mathrm{ORT}}\, \boldsymbol{T}_{1,\mathrm{ORT}}$, in which

$$\boldsymbol{T}_{1\mathrm{ORT}} = \begin{bmatrix} 1 & 1 & 1 \\ -1 & 0 & 1 \end{bmatrix}. \tag{13.24}$$

The possibility of specifying a GCM using one of several kinds of polynomials raises the question of equivalence among models more generally in terms of any specifications of the within-individual design matrix $\boldsymbol{T}$.

Definition 13.5 Two models specified by their design matrices and parameter spaces, $\mathrm{GCM}(\boldsymbol{Y};\boldsymbol{X}_1\boldsymbol{B}_1\boldsymbol{T}_1,\boldsymbol{\Sigma})$, $\boldsymbol{B}_1 \in \Omega_1$, and $\mathrm{GCM}(\boldsymbol{Y};\boldsymbol{X}_2\boldsymbol{B}_2\boldsymbol{T}_2,\boldsymbol{\Sigma})$, $\boldsymbol{B}_2 \in \Omega_2$, are *linearly equivalent* $\Leftrightarrow$
(a) $\forall \boldsymbol{B}_1 \in \Omega_1$ there exists $\boldsymbol{B}_2 \in \Omega_2$ such that $\boldsymbol{X}_1\boldsymbol{B}_1\boldsymbol{T}_1 = \boldsymbol{X}_2\boldsymbol{B}_2\boldsymbol{T}_2$ and
(b) $\forall \boldsymbol{B}_2 \in \Omega_2$ there exists $\boldsymbol{B}_1 \in \Omega_1$ such that $\boldsymbol{X}_1\boldsymbol{B}_1\boldsymbol{T}_1 = \boldsymbol{X}_2\boldsymbol{B}_2\boldsymbol{T}_2$.

Having fitted a GCM, it is desirable to create graphical representations of the estimated growth curve. As for all polynomial models, computations are best conducted in orthogonal or orthonormal terms, although displayed in terms of natural polynomials.

Definition 13.6 The *estimated polynomial growth curve*, or dose-response curve, is the estimator of mean response as a function of time, or dose, d of the form $\widehat{\boldsymbol{\mu}}(d;\boldsymbol{C}) = \boldsymbol{C}\widehat{\boldsymbol{B}}\boldsymbol{u}_d$ $(1 \times q \times m \times 1)$ with $\boldsymbol{u}_d = [1 \; d \; \cdots \; d^{m-1}]'$ $(m \times 1)$. A set of coordinates $\{[d, \widehat{\boldsymbol{\mu}}(d;\boldsymbol{C})] : d \in [d_1, d_p]\}$ define an estimated growth curve.

Theorem 13.3 For $\mathrm{GCM}_{N,p,q,m}(\boldsymbol{Y}_i;\boldsymbol{X}_i\boldsymbol{B}_{\mathrm{ORT}}\boldsymbol{T}_{\mathrm{ORT}},\boldsymbol{\Sigma})$ and any value of d in the range of the data, a growth curve can be conveniently and plausibly estimated. With $\boldsymbol{C}$ $(1 \times q)$ and $\boldsymbol{u}_d$ $(m \times 1)$ such that $\{[d, \widehat{\mu}(d;\boldsymbol{C})] : d \in [d_1, d_p]\}$ gives

$$\widehat{\mu}(d;\boldsymbol{C}) = \boldsymbol{C}\widehat{\boldsymbol{B}}_{\mathrm{ORT}}\boldsymbol{A}_{\mathrm{ORTNAT}}\boldsymbol{u}_d \tag{13.25}$$

with $m \times 1$

$$\boldsymbol{u}_d = [1 \; d \; \cdots \; d^{m-1}]'. \tag{13.26}$$

The result allows plotting growth curves (or dose-response curves) at arbitrarily many points in terms of the orthogonal polynomial regression parameters.

Proof. The natural polynomial model and the orthogonal polynomial model are linearly equivalent (with all corresponding matrices of the same dimensions). In particular, $\mathrm{GCM}(\boldsymbol{Y};\boldsymbol{X}\boldsymbol{B}_{\mathrm{ORT}}\boldsymbol{T}_{\mathrm{ORT}},\boldsymbol{\Sigma})$ with $\boldsymbol{B}_{2\mathrm{ORT}}=\mathbf{0}$ is equivalent to $\mathrm{GCM}(\boldsymbol{Y};\boldsymbol{X}\boldsymbol{B}_{\mathrm{NAT}}\boldsymbol{T}_{\mathrm{NAT}},\boldsymbol{\Sigma})$ with $\boldsymbol{B}_{2\mathrm{NAT}}=\mathbf{0}$. The linear relationship between natural and orthogonal polynomial coefficients gives $\widehat{\boldsymbol{B}}_{\mathrm{NAT}}=\widehat{\boldsymbol{B}}_{\mathrm{ORT}}\boldsymbol{A}_{\mathrm{ORTNAT}}$. In turn $\widehat{\mu}(d;\boldsymbol{C})=\boldsymbol{C}\widehat{\boldsymbol{B}}_{\mathrm{NAT}}\boldsymbol{u}_d=\boldsymbol{C}\left(\widehat{\boldsymbol{B}}_{\mathrm{ORT}}\boldsymbol{A}_{\mathrm{ORTNAT}}\right)\boldsymbol{u}_d$. □

Extrapolating a growth curve, or any other model, outside the range of data can be quite misleading. Muller and Fetterman (2002, Chapter 9) provided examples. As with most issues of practical data analysis and interpretation, we leave the topic for consideration in other settings.

Example 13.4 The integer-valued orthogonal polynomial model for $d\in\{1,2,3,4\}$ is

$$\begin{aligned}\mathrm{E}(\boldsymbol{Y}|\boldsymbol{X},\boldsymbol{T})&=\boldsymbol{X}\boldsymbol{B}_{\mathrm{ORT}}\boldsymbol{T}_{\mathrm{ORT}}\\&=\boldsymbol{X}[\,\beta_0\;\;\beta_1\;\;\beta_2\;\;\beta_3\,]\begin{bmatrix}1&1&1&1\\-3&-1&1&3\\1&-1&-1&1\\-1&3&-3&1\end{bmatrix}.\end{aligned}\tag{13.27}$$

If we require $\beta_1=0$ and $\beta_3=0$, then $\boldsymbol{B}_{\mathrm{ORT}}=[\,\beta_0\;\;0\;\;\beta_2\;\;0\,]$. The corresponding natural polynomial regression coefficients are $\boldsymbol{B}_{\mathrm{NAT}}=[\,\beta_0\;\;\beta_1\;\;\beta_2\;\;0\,]$. The result is obtained from $\boldsymbol{B}_{\mathrm{NAT}}=\boldsymbol{B}_{\mathrm{ORT}}\boldsymbol{A}_{\mathrm{ORTNAT}}$, with

$$\boldsymbol{A}_{\mathrm{ORTNAT}}=\begin{bmatrix}1&0&0&0\\-5&2&0&0\\5&-5&1&0\\-35&55.66&-25&3.33\end{bmatrix}.\tag{13.28}$$

The example natural polynomial has $m=3$ nonzero coefficients, $\boldsymbol{B}_{1,\mathrm{NAT}}=[\,\beta_0\,\beta_1\,\beta_2\,]$, while the orthogonal polynomial has only $m_1=2$ nonzero coefficients, $\boldsymbol{B}_{1,\mathrm{ORT}}=[\,\beta_0\,\beta_2\,]$. The quadratic term of the integer-valued orthogonal polynomial is $\beta_2(5-d+d^2)$, which includes a linear component. It is important to remember that different kinds of polynomials (of the same order) for a given model may have different numbers of nonzero regression coefficients.

The notation $\boldsymbol{A}[\![\,\cdot\,,\,\cdot\,]\!]$ indicates a submatrix of $\boldsymbol{A}$ created by selecting all elements of rows in a list and all elements of columns in a list. We use $\boldsymbol{J}_1$ $(1\times m)$ to indicate which of p columns are assumed to be *not* zero. With $\boldsymbol{J}_1=[\,1\;3\,]$ and $m_1=\max\{\boldsymbol{J}_1\}=3$, $\boldsymbol{A}_{\mathrm{ORTNAT}}[\![\boldsymbol{J}_1,[\,1\;2\cdots m_1\,]]\!]$ has m rows and m_1 columns and $\widehat{\boldsymbol{B}}_{1\mathrm{NAT}}=\widehat{\boldsymbol{B}}_{1\mathrm{ORT}}\boldsymbol{A}_{\mathrm{ORTNAT}}[\![\boldsymbol{J}_1,[\,1\;2\cdots m_1\,]]\!]$. The $\boldsymbol{J}_1$ matrix selects the rows and columns of $\boldsymbol{A}_{\mathrm{ORTNAT}}$ needed to transform orthogonal polynomial results into results in terms of natural polynomials. Similarly $\boldsymbol{J}_2=[\,2\;4\,]$ $(1\times p-m)$ indicates columns 2 and 4 of the coefficient matrix *are* assumed to be zero.

Theorem 13.4 For $\text{GCM}_{N,p,q,m}(\boldsymbol{Y}_i; \boldsymbol{X}_i\boldsymbol{B}\boldsymbol{T}, \Sigma)$ with $\boldsymbol{B}$ and $\boldsymbol{T}$ in terms of an orthogonal or orthonormal polynomial, growth curves can be plotted in terms of $\boldsymbol{J}_1$ $(1 \times m)$, the list of columns of $\boldsymbol{B}$ assumed *not* to be zero. At time d the estimated value is specified by $\boldsymbol{C}$ $(1 \times q)$, $\widehat{\boldsymbol{B}}_1$ $(q \times m)$, and $\boldsymbol{A}$ $(m \times m_1)$ as

$$\widehat{\mu}(d; \boldsymbol{C}) = \boldsymbol{C}\widehat{\boldsymbol{B}}_1\boldsymbol{A}\boldsymbol{u}_d \tag{13.29}$$

with $\boldsymbol{u}_d = [\,1 \;\; d \;\; d^2 \;\; \cdots \;\; d^{m-1}\,]'$ and

$$\boldsymbol{A} = \boldsymbol{A}_{\text{ORTNAT}}[\![\boldsymbol{J}_1, [\,1 \;\; 2 \;\; \cdots \;\; m_1\,]]\!], \tag{13.30}$$

in which $m_1 = \max\{\boldsymbol{J}_1\}$.

Proof. Left as an exercise.

Corollary 13.4.1 A set of $s > p$ points on a growth curve can be computed with a natural polynomial design matrix with more columns than rows, $\boldsymbol{U}$ $(p \times s)$, the first p rows of $\boldsymbol{T}_{\text{NAT}}$ $(s \times s)$. For $\boldsymbol{C}$ $(1 \times q)$, $\boldsymbol{B}$ $(q \times p)$, $\boldsymbol{\mu}'$ $(1 \times s)$, $\boldsymbol{d}$ $(s \times 1)$,

$$\widehat{\boldsymbol{\mu}}'(\boldsymbol{d}; \boldsymbol{C}) = \boldsymbol{C}\widehat{\boldsymbol{B}}_{\text{ORN}}\boldsymbol{A}_{\text{ORNNAT}}\boldsymbol{U}\,. \tag{13.31}$$

The set cannot be computed with an orthogonal design matrix with $s > p$ columns (the dimension of the fitted model). In particular, $\widehat{\boldsymbol{\mu}}(\boldsymbol{d}; \boldsymbol{C}) \neq \boldsymbol{C}\widehat{\boldsymbol{B}}_{\text{ORN}}\boldsymbol{U}_{\boldsymbol{T}}$ if $\boldsymbol{U}_{\boldsymbol{T}}$ $(p \times s)$ is the first p rows of $\boldsymbol{T}_{\text{ORN}}$ $(s \times s)$.

Proof. Left as an exercise.

Corollary 13.4.2 Under the same assumptions, estimated growth curve values can be computed as

$$\widehat{\mu}(d; \boldsymbol{C}) = \boldsymbol{C}\widehat{\boldsymbol{B}}_{\text{ORN}}\,\boldsymbol{A}_{\text{ORNNAT}}\,\boldsymbol{u}_d\,, \tag{13.32}$$

with $\boldsymbol{u}_d = [\,1 \;\; d \;\; d^2 \;\; \cdots \;\; d^{m-1}\,]'$, in which d is any value.

Proof. Left as an exercise.

The example of children's weights ($N = 50$, $q = 1$, $p = 3$, $m = 2$) allows illustrating the process. Having estimated $\boldsymbol{B}_{1\text{ORT}} = [\,\beta_0 \;\; \beta_1\,]$, we can plot the estimated growth curve. Suppose we only want to plot three points on the curve corresponding to $\boldsymbol{t}' = [\,d_1 \;\; d_2 \;\; d_3\,]$. With $\boldsymbol{C} = 1$ and $\boldsymbol{T}_{\text{ORT}}$ (3×3) we have

$$\begin{aligned} \widehat{\mu}(d; \boldsymbol{C}) &= \boldsymbol{C}\widehat{\boldsymbol{B}}_{\text{ORT}}\boldsymbol{T}_{\text{ORT}} \\ &= 1[\,\widehat{\beta}_0 \;\; \widehat{\beta}_1 \;\; 0\,]\boldsymbol{T}_{\text{ORT}}\,. \end{aligned} \tag{13.33}$$

Here $\boldsymbol{C} = 1$ because the participants form a single homogeneous group. The three points correspond to the observed ages (5, 6, and 7 years). Corollary 13.4.1 answers the question "How would additional intermediate points be plotted?"

Example 13.5 We seek to compute five points within the range of the data, corresponding to ages $[5\ 5.5\ 6.0\ 6.5\ 7.0]' = \boldsymbol{t}_1$. With $\boldsymbol{C} = 1$, $\boldsymbol{A}$ (3×3), $\boldsymbol{U}$ (3×5), the points would be computed as

$$\begin{aligned}
\widehat{\mu}(\boldsymbol{t}_1; \boldsymbol{C}) &= \boldsymbol{C}\widehat{\boldsymbol{B}}_{\text{ORT}}\boldsymbol{A}_{\text{ORTNAT}}\boldsymbol{U} \\
&= [\widehat{\beta}_0\ \widehat{\beta}_1\ 0]\begin{bmatrix} 1 & 0 & 0 \\ -6 & 1 & 0 \\ 106 & -36 & 3 \end{bmatrix}\begin{bmatrix} 1 & 1 & 1 & 1 & 1 \\ 5 & 5.50 & 6 & 6.50 & 7 \\ 25 & 30.25 & 36 & 42.25 & 49 \end{bmatrix} \\
&= [\widehat{\beta}_0\ \widehat{\beta}_1]\begin{bmatrix} 1 & 0 \\ -6 & 1 \end{bmatrix}\begin{bmatrix} 1 & 1 & 1 & 1 & 1 \\ 5 & 5.5 & 6 & 6.5 & 7 \end{bmatrix} \\
&= [\widehat{\beta}_0\ \widehat{\beta}_1]\begin{bmatrix} 1 & 1 & 1 & 1 & 1 \\ -1 & -0.5 & 0 & -0.5 & 1 \end{bmatrix}.
\end{aligned} \tag{13.34}$$

Matrix $\boldsymbol{U}$ can be interpreted as being $\boldsymbol{T}_{\text{NAT}}$ (3×3) augmented with two additional columns. It can also be interpreted as being the first three rows of $\boldsymbol{T}_{\text{NAT}}$ $(s \times s)$. Arbitrarily large numbers of points on the curve may be plotted by augmenting the $\boldsymbol{T}_{\text{NAT}}$ matrix with additional columns. It is usually desirable (but not necessary) to choose evenly spaced points.

13.5 WITHIN-INDIVIDUAL DESIGN MATRICES

Two of the most desirable attributes of the $\boldsymbol{T}$ matrix are full rank and orthogonality. The creative analyst can draw from the great variety of orthogonal designs to formulate an appropriate GCM. Any full-rank nonorthogonal design matrix can be reparameterized as an orthogonal design. After applying the new design, hypotheses about the original nonorthogonal design are easily tested.

The process may be described in terms of $\boldsymbol{T}$, a $p \times p$ arbitrary full-rank design matrix. Since $\boldsymbol{TT}'$ is full rank, $p \times p$, and symmetric, a lower triangular matrix $\boldsymbol{L}$ $(p \times p)$ exists such that $\boldsymbol{TT}' = \boldsymbol{LL}'$ (the Cholesky decomposition). Lower triangular $\boldsymbol{L}^{-1}$ maps $\boldsymbol{T}$ onto an orthogonal matrix $\boldsymbol{T}_{\text{new}} = \boldsymbol{L}^{-1}\boldsymbol{T}$. If a model which assumes $\text{E}(\boldsymbol{Y}|\boldsymbol{X}, \boldsymbol{T}_{\text{new}}) = \boldsymbol{XB}_{\text{new}}\boldsymbol{T}_{\text{new}}$ is fitted, the original hypothesis $H_0 : \boldsymbol{CBU} = \boldsymbol{0}$ $(a \times b)$ can be tested in the form $H_0 : \boldsymbol{CB}_{\text{new}}\boldsymbol{U}_{\text{new}} = \boldsymbol{0}$ $(a \times b)$, in which $\boldsymbol{U}_{\text{new}} = \boldsymbol{L}^{-1}\boldsymbol{U}$. If some columns of $\boldsymbol{B}$ $(q \times p)$ are assumed to be zero, then $\boldsymbol{T}$ $(p \times p)$ is replaced by $\boldsymbol{T}_1$ $(m \times p)$ in the above equations.

Polynomials are useful when little is known about the mechanisms underlying the ordered responses. If more is known about the underlying process, then a formulation which uses the additional information likely will be better. As an example, if the underlying process is known to be periodic, then the design matrix $\boldsymbol{T}$ might be formulated in terms of sine and cosine functions of time.

Definition 13.7 It may be convenient to define transformation matrix $\boldsymbol{T}$ in a GCM in terms of a *discrete Fourier transform* (DFT). If p is even, then $p_c = (p-0)/2$ and $p_b = (p-2)/2$. If p is odd, then $p_a = p_b = (p-1)/2$. For time d_j, $c_{ij} = \cos(2\pi i d_j f)$ and $s_{ij} = \sin(2\pi i t_j f)$, with f the *fundamental frequency* of the process and $1/f$ the *fundamental period.* In turn,

$$\boldsymbol{T}_{\text{DFT}} = \begin{bmatrix} \mathbf{1}' \\ \boldsymbol{T}_c \\ \boldsymbol{T}_s \end{bmatrix} \begin{matrix} \leftarrow \text{constants} & (1 \times p) \\ \leftarrow \text{cosines} & (p_c \times p) \\ \leftarrow \text{sines} & (p_b \times p) \end{matrix} \tag{13.35}$$

in which $\boldsymbol{T}_c = \{c_{ij}\}$ with $1 \le i \le p_a$ and $\boldsymbol{T}_s = \{s_{ij}\}$ with $1 \le i \le p_b$. Larger values of i indicate the higher order terms. The terms with $i = 1$ represent the lowest frequency component. The terms with $i = 2$ represent the *first harmonic*, while $i = 3$ terms give the *second harmonic*.

Theorem 13.5 If $\boldsymbol{d}' = [1\ 2\ \cdots\ p] = \{d_j\} = \{j\}$ and $f = 1/p$, then the DFT design matrix $\boldsymbol{T}_{\text{DFT}}$ is a $p \times p$ orthogonal matrix.

Proof. Left as an exercise.

So far we have assumed the measurement made at time j is a function of d_j, which implies the ordered columns of $\boldsymbol{Y}$ $(N \times p)$ are functionally related. The design matrix $\boldsymbol{T}$ can, when necessary, be formulated in terms of a polynomial in more than one variable.

Example 13.6 Temporal changes in level of pain (Y) indexed by time (t) in hours can be observed for several doses (d) of an analgesic drug via a multi-period crossover design; i.e., the individual sequentially experiences the different doses in different periods of the study and longitudinal observations are recorded within each period. In other human studies, simultaneous administration of different doses of a drug can occur, for example, when two different doses of a drug are applied to the left and right eyes respectively. Other examples include allergy skin testing (various patches of skin are exposed to different allergens) and in vitro studies [e.g., a blood sample is drawn, divided into multiple subsamples of equal volume (aliquots), each aliquot is treated with one of the dose levels of interest, and response in each aliquot is then observed longitudinally].

For illustration, suppose two within-subject factors, time (t) and dose (d), are present, and further suppose a response y is observed at $p = 3$ times for each of 3 doses. In terms of a two-variable natural polynomial the mean response for the 3×3 factorial experiment is of the form

$$\begin{aligned} \mu_{ijk} &= \mathrm{E}[y_i(t_j, d_k)|\boldsymbol{X}, \boldsymbol{T}]_j \\ &= \boldsymbol{X}_i \boldsymbol{B} \boldsymbol{t}_j, \end{aligned} \tag{13.36}$$

in which $\boldsymbol{T}$ is the Kronecker product of two single-variable design matrices:

$$\boldsymbol{T}(t,d) = \boldsymbol{T}(t) \otimes \boldsymbol{T}(d)\,. \tag{13.37}$$

If the participants are a homogeneous group, then $\boldsymbol{X} = \mathbf{1}$ $(N \times 1)$. If the two-variable polynomial is of maximal order, then

$$\mu_{ijk} = \beta_{00}+\beta_{10}t+\beta_{20}t^2+\beta_{01}d+\beta_{11}td+\beta_{21}t^2d+\beta_{02}d^2+\beta_{12}td^2+\beta_{22}t^2d^2\,. \tag{13.38}$$

For $\boldsymbol{T}(t,d)$ to be an orthogonal matrix, a Kronecker product of orthogonal $\boldsymbol{T}$ matrices is used: $\boldsymbol{T}(t,d) = \boldsymbol{T}(t) \otimes \boldsymbol{T}(d)$ is (9×9) and

$$\boldsymbol{T}(t) = \boldsymbol{T}(d) = \begin{bmatrix} 1 & 1 & 1 \\ -1 & 0 & 1 \\ 1 & -2 & 1 \end{bmatrix}. \tag{13.39}$$

Similarly the design matrix for a polynomial in three variables is

$$\boldsymbol{T}(t,s,u) = \boldsymbol{T}(t) \otimes \boldsymbol{T}(s) \otimes \boldsymbol{T}(u)\,. \tag{13.40}$$

In general, the $\boldsymbol{T}$ matrix for a polynomial in K variables is the Kronecker product of K matrices. The inverse of $\boldsymbol{T}$ is the Kronecker product of the individual inverses. If $\boldsymbol{T}(t,d) = \boldsymbol{T}(t) \otimes \boldsymbol{T}(d)$, then $\boldsymbol{T}^{-1}(t,d) = \boldsymbol{T}^{-1}(t) \otimes \boldsymbol{T}^{-1}(d)$.

In the GCM, the definition of matrix $\boldsymbol{X}$ $(N \times q)$ does not require any additional considerations beyond the ones made in the multivariate GLM. Therefore the between-individual design matrix $\boldsymbol{X}$ may be thought of as the usual $\boldsymbol{X}$ matrix in a general linear model. Given a matrix of responses $\boldsymbol{Y}$ $(N \times p)$, the definition of matrix $\boldsymbol{T}$ $(p \times p)$ does not place any restrictions on the definition of $\boldsymbol{X}$.

13.6 ESTIMATION METHODS

Under the assumptions of $\text{GCM}_{N,p,q,m}(\boldsymbol{Y}_i; \boldsymbol{X}_i\boldsymbol{B}\boldsymbol{T}, \boldsymbol{\Sigma})$, the correct model for a single individual's responses is

$$\text{E}(\boldsymbol{Y}_i'|\boldsymbol{X}_i,\boldsymbol{T}) = (\boldsymbol{\theta}_i'\boldsymbol{T})'\,, \tag{13.41}$$

with $m \times 1$ $\boldsymbol{\theta}_i = (\boldsymbol{X}_i\boldsymbol{B})'$ and $\mathcal{V}(\boldsymbol{Y}_i'|\boldsymbol{X}_i,\boldsymbol{T}) = \boldsymbol{\Sigma}$ $(p \times p)$. Multiplying both sides of the regression equation yields

$$\text{E}[\boldsymbol{Y}_i\boldsymbol{V}^{-1}\boldsymbol{T}'(\boldsymbol{T}\boldsymbol{V}^{-1}\boldsymbol{T}')^{-1}|\boldsymbol{X}_i,\boldsymbol{T}] = \boldsymbol{\theta}_i' = \boldsymbol{X}_i\boldsymbol{B}\,. \tag{13.42}$$

The vector $\widehat{\boldsymbol{\theta}}_i = (\boldsymbol{T}\boldsymbol{V}^{-1}\boldsymbol{T}')^{-1}\boldsymbol{T}\boldsymbol{V}^{-1}\boldsymbol{Y}_i'$ is an $m \times 1$ weighted-least-squares estimator (unweighted if $\boldsymbol{V} = \boldsymbol{I}_p$) for the individual-specific model $\text{E}(\boldsymbol{Y}_i'|\boldsymbol{X}_i,\boldsymbol{T}) = \boldsymbol{T}'\boldsymbol{\theta}_i$. Given $\boldsymbol{\theta}_i' = \boldsymbol{X}_i\boldsymbol{B}$, with $\boldsymbol{X}_i\boldsymbol{B}$ invariant to the choice of $\boldsymbol{V}$, it follows $\boldsymbol{\theta}_i$ is invariant to the choice of $\boldsymbol{V}$. Also

$$\begin{aligned}\mathrm{E}(\widehat{\boldsymbol{\theta}}_i) &= \mathrm{E}\left[(\boldsymbol{T}\boldsymbol{V}^{-1}\boldsymbol{T}')^{-1}\boldsymbol{T}\boldsymbol{V}^{-1}\boldsymbol{Y}_i'\right] \\ &= (\boldsymbol{T}\boldsymbol{V}^{-1}\boldsymbol{T}')^{-1}\boldsymbol{T}\boldsymbol{V}^{-1}\mathrm{E}(\boldsymbol{Y}_i') \\ &= (\boldsymbol{T}\boldsymbol{V}^{-1}\boldsymbol{T}')^{-1}\boldsymbol{T}\boldsymbol{V}^{-1}(\boldsymbol{\theta}_i'\boldsymbol{T})' \\ &= \boldsymbol{\theta}_i\,. \end{aligned} \tag{13.43}$$

Although $\widehat{\boldsymbol{\theta}}_i'$ is unbiased for all choices of $\boldsymbol{V}$, its precision is optimal if $\boldsymbol{V} = \boldsymbol{\Sigma}$. Unfortunately $\boldsymbol{\Sigma}$ is unknown.

If a separate univariate GLM model is fitted for each individual, it would then be reasonable to fit a multivariate GLM for $\widehat{\boldsymbol{\Theta}} = \left[\widehat{\boldsymbol{\theta}}_1 \cdots \widehat{\boldsymbol{\theta}}_N\right]' = \boldsymbol{Y}\boldsymbol{V}^{-1}\boldsymbol{T}(\boldsymbol{T}\boldsymbol{V}^{-1}\boldsymbol{T}')^{-1}$ $(N \times m)$, with $N \times q \times m$ mean

$$\mathrm{E}(\widehat{\boldsymbol{\theta}}|\boldsymbol{X}) = \boldsymbol{X}\boldsymbol{B} \tag{13.44}$$

and $\mathcal{V}[\mathrm{vec}(\widehat{\boldsymbol{\theta}})|\boldsymbol{X}] = \boldsymbol{\Omega} \otimes \boldsymbol{I}$. Here $\boldsymbol{\Omega} = \boldsymbol{G}'\boldsymbol{\Sigma}\boldsymbol{G}$ with $\boldsymbol{G} = \boldsymbol{V}^{-1}\boldsymbol{T}(\boldsymbol{T}\boldsymbol{V}^{-1}\boldsymbol{T}')^{-1}$ gives estimator

$$\widehat{\boldsymbol{B}} = (\boldsymbol{X}'\boldsymbol{X})^{-1}\boldsymbol{X}'\widehat{\boldsymbol{\Theta}} = (\boldsymbol{X}'\boldsymbol{X})^{-1}\boldsymbol{X}'\boldsymbol{Y}\boldsymbol{V}^{-1}\boldsymbol{T}(\boldsymbol{T}\boldsymbol{V}^{-1}\boldsymbol{T}')^{-1}\,. \tag{13.45}$$

The variance of the estimator,

$$\mathcal{V}[\mathrm{vec}(\widehat{\boldsymbol{B}})|\boldsymbol{X}] = (\boldsymbol{G}'\boldsymbol{\Sigma}\boldsymbol{G}) \otimes (\boldsymbol{X}'\boldsymbol{X})^{-1}\,, \tag{13.46}$$

depends on the choice of $\boldsymbol{V}$. Obvious choices for $\boldsymbol{V}$ include $\widetilde{\boldsymbol{\Sigma}} = \boldsymbol{Y}'[\boldsymbol{I} - \boldsymbol{X}(\boldsymbol{X}'\boldsymbol{X})^{-1}\boldsymbol{X}'])\boldsymbol{Y}/N$, the MLE, and the unbiased estimator $\widehat{\boldsymbol{\Sigma}} = \widetilde{\boldsymbol{\Sigma}}N/[N - \mathrm{rank}(\boldsymbol{X})]$.

The results just stated could be derived in terms of a transformation of the columns of an ordinary multivariate GLM using $\boldsymbol{G} = \boldsymbol{T}^{-1}$ $(p \times p)$, combined with the fact that conditioning on $\boldsymbol{T}$ is the same as conditioning on $\boldsymbol{G}$. In particular,

$$\mathrm{E}(\,\boldsymbol{Y}\boldsymbol{G}|\boldsymbol{X},\boldsymbol{G}) = \boldsymbol{X}\boldsymbol{B} \tag{13.47}$$

and

$$\mathcal{V}(\boldsymbol{Y}\boldsymbol{G}|\boldsymbol{X},\boldsymbol{G}) = (\boldsymbol{G}'\boldsymbol{\Sigma}\boldsymbol{G}) \otimes \boldsymbol{I}_N\,. \tag{13.48}$$

The transformation yields new dependent variables, $\boldsymbol{Y}\boldsymbol{G}$, and provides a GLM setting for estimation and inference. The partitioned model

$$\begin{aligned}\mathrm{E}(\boldsymbol{Y}|\boldsymbol{X},\boldsymbol{T}) &= \boldsymbol{X}\boldsymbol{B}\boldsymbol{T} \\ &= \boldsymbol{X}[\,\boldsymbol{B}_1\ \ \boldsymbol{B}_2\,]\begin{bmatrix}\boldsymbol{T}_1\\ \boldsymbol{T}_2\end{bmatrix}\end{aligned} \tag{13.49}$$

induces the partitioning $\boldsymbol{Y}\boldsymbol{G} = [\,\boldsymbol{Y}\boldsymbol{G}_1\,\boldsymbol{Y}\boldsymbol{G}_2\,] = [\,\boldsymbol{Y}_1\ \ \boldsymbol{Y}_2\,]$. Constrained by $\boldsymbol{B}_2 = \boldsymbol{0}$, here $\mathrm{E}(\boldsymbol{Y}_2|\boldsymbol{X},\boldsymbol{T}) = \boldsymbol{0}$ $[N \times (p-q)]$ while $\mathrm{E}(\boldsymbol{Y}_1|\boldsymbol{X},\boldsymbol{T}) = \boldsymbol{X}\boldsymbol{B}_1$. The columns of $\boldsymbol{G}_1$ are usually the first few columns of $\boldsymbol{G}$. In any case, the rows of $\boldsymbol{T}$ can always be permuted (along with columns of $\boldsymbol{B}$ and columns of $\boldsymbol{G}$) so no generality is lost by assuming $\boldsymbol{G}_1$ contains the first few columns of $\boldsymbol{G}$ $(p \times p)$.

Restricting some columns of $\boldsymbol{B}$ to be zero is achieved by omitting Y_2 $(N \times p - q)$ from the model. The test of $H_0 : \boldsymbol{B}_2 = \boldsymbol{0}$ serves as a goodness-of-fit test.

If all of the $\boldsymbol{Y}_2$ variables are omitted from the GLM in order to restrict $\boldsymbol{B}$ $(q \times p)$, then the following model is fitted:

$$\mathrm{E}(\boldsymbol{Y}_1|\boldsymbol{X}, \boldsymbol{G}_1) = \boldsymbol{X}\boldsymbol{B}_1 \tag{13.50}$$

$$\mathcal{V}(\boldsymbol{Y}_1|\boldsymbol{X}, \boldsymbol{G}_1) = \boldsymbol{G}_1'\Sigma\,\boldsymbol{G}_1 \otimes \boldsymbol{I}\,. \tag{13.51}$$

Rao (1965) and Khatri (1966) described how to reduce a GCM to an ordinary multivariate GLM. They proved some information is discarded if the $\boldsymbol{Y}_2$ variables are ignored. Furthermore the use of some or all of $\boldsymbol{Y}_2$ as independent covariates can improve the power of tests and reduce the widths of confidence intervals. Including none of the $\boldsymbol{Y}_2$ variables as covariates yields the unweighted estimator

$$\widetilde{\boldsymbol{B}}_1 = (\boldsymbol{X}'\boldsymbol{X})^{-1}\boldsymbol{X}'\boldsymbol{Y}\boldsymbol{T}_1'(\boldsymbol{T}_1\boldsymbol{T}_1')^{-1}\,, \tag{13.52}$$

while using all of the $\boldsymbol{Y}_2$ variables as covariates yields the weighted estimator

$$\widehat{\boldsymbol{B}}_1 = (\boldsymbol{X}'\boldsymbol{X})^{-1}\boldsymbol{X}'\boldsymbol{Y}\widehat{\boldsymbol{\Sigma}}^{-1}\boldsymbol{T}_1'(\boldsymbol{T}_1\widehat{\boldsymbol{\Sigma}}^{-1}\boldsymbol{T}_1')^{-1}\,, \tag{13.53}$$

in which $\widehat{\boldsymbol{\Sigma}}$ is the MLE of $\boldsymbol{\Sigma}$. Using all $\boldsymbol{Y}_2$ variables requires fitting the model

$$\mathrm{E}(\boldsymbol{Y}_1|\boldsymbol{X}, \boldsymbol{G}_1, \boldsymbol{Y}_2) = \boldsymbol{X}\boldsymbol{B}_1 + \boldsymbol{Y}_2\boldsymbol{\Gamma} \tag{13.54}$$

$$\mathcal{V}(\boldsymbol{Y}_1|\boldsymbol{X}, \boldsymbol{G}_1, \boldsymbol{Y}_2) = \boldsymbol{G}_1'\boldsymbol{\Sigma}\boldsymbol{G}_1 \otimes \boldsymbol{I}\,, \tag{13.55}$$

which is equivalent to computing $\widehat{\boldsymbol{B}}_1 = (\boldsymbol{X}'\boldsymbol{X})^{-1}\boldsymbol{X}'\boldsymbol{Y}\boldsymbol{V}^{-1}\boldsymbol{T}(\boldsymbol{T}\boldsymbol{V}^{-1}\boldsymbol{T}')^{-1}$ with $\boldsymbol{V} = \widehat{\boldsymbol{\Sigma}}$. Grizzle and Allen (1969) derived the variance of $\widehat{\boldsymbol{B}}_1$:

$$\mathcal{V}\left[\mathrm{vec}(\widehat{\boldsymbol{B}})|\boldsymbol{X}\right] = \boldsymbol{G}'\boldsymbol{\Sigma}\boldsymbol{G} \otimes (\boldsymbol{X}'\boldsymbol{X})^{-1}(N-1)/[N-(p-q)-1]\,. \tag{13.56}$$

Conditioning on some or all of the $\boldsymbol{Y}_2$ variables uses additional degrees of freedom. Since some of the $\boldsymbol{Y}_2$ variables may be redundant of one another in the information they provide, Rao (1965) and Grizzle and Allen (1969) proposed balancing the gain of information against the use of degrees of freedom by including only a few well-chosen $\boldsymbol{Y}_2$ covariates. As criteria for including or excluding variables, Rao proposed examining the resulting widths of confidence intervals for elements of $\widehat{\boldsymbol{B}}$. Grizzle and Allen proposed relying on a measure of generalized variance, the determinant of the covariance matrix of estimator $\widehat{\boldsymbol{B}}$. For confirmatory hypothesis testing, the conservative approach is to include all of the $\boldsymbol{Y}_2$ covariates.

Berger (1986) conducted simulations evaluating various strategies for growth curve analysis. His results indicate that using observed properties of the data to help model the covariance inflates the type I error rate. Hence he recommended avoiding the approach unless an appropriate correction could be determined.

13.7 RELATIONSHIPS TO THE UNIVARIATE AND MIXED MODELS

The model $\boldsymbol{Y}=\boldsymbol{XB}+\boldsymbol{E}$ has the corresponding equation for sampling unit i:

$$\boldsymbol{Y}_i' = (\boldsymbol{X}_i \otimes \boldsymbol{I}_p)\text{vec}(\boldsymbol{B}') + \boldsymbol{E}_i' . \tag{13.57}$$

For $p \times m$ $\boldsymbol{T}$ the model $\boldsymbol{Y} = \boldsymbol{XBT} + \boldsymbol{E}$ has corresponding equation for sampling unit i:

$$\boldsymbol{Y}_i' = (\boldsymbol{X}_i \otimes \boldsymbol{I}_m)\text{vec}(\boldsymbol{T}'\boldsymbol{B}') + \boldsymbol{E}_i' . \tag{13.58}$$

From Theorem 1.5, $\text{vec}(\boldsymbol{ABC}) = (\boldsymbol{C}' \otimes \boldsymbol{A})\text{vec}(\boldsymbol{B})$. With $\boldsymbol{T}'\boldsymbol{B}' = \boldsymbol{T}'\boldsymbol{B}'\boldsymbol{I}_q$,

$$\boldsymbol{Y}_i' = (\boldsymbol{X}_i \otimes \boldsymbol{I}_m)(\boldsymbol{I}_q \otimes \boldsymbol{T}')\text{vec}(\boldsymbol{B}') + \boldsymbol{E}_i' . \tag{13.59}$$

Hence

$$\boldsymbol{Y}_i' = (\boldsymbol{X}_i \otimes \boldsymbol{T}')\text{vec}(\boldsymbol{B}') + \boldsymbol{E}_i' . \tag{13.60}$$

$$(m \times 1) = [(1 \times q) \otimes (m \times p)](pq \times 1) + (m \times 1)$$

$$(m \times 1) = (m \times qp)(pq \times 1) + (m \times 1)$$

Here $\boldsymbol{X}_i$ contains between-subject design information, while $\boldsymbol{T}$ contains within-subject design information. The last equation corresponds directly to a mixed model or a GGLM. In contrast to a GCM, a mixed model allows elements of $\boldsymbol{T}$ and m to vary across subjects.

It can be extremely helpful in understanding an incomplete design to examine the corresponding complete (factorial design). Considering restrictions or variable deletions can then be used to specify the incomplete design in terms of the complete design. The model form just described is intended to help the reader in such an endeavor in the context of mixed models.

EXERCISES

13.1 An investigator gathered n_i observations on each of N sampling units for $i \in \{1, \ldots, N\}$ from two different groups. A total of N_1 sampling units were selected from group 1 and N_2 sampling units were from group 2 ($N_1 + N_2 = N$). The n_i observations from each sampling unit are assumed to be independent of the observations from another sampling unit and follow the linear model $\boldsymbol{y}_i = \boldsymbol{X}_i\boldsymbol{\gamma} + \boldsymbol{e}_{iy}$ with $\boldsymbol{X}_i = [\,\boldsymbol{1}_{n_i} \quad \boldsymbol{0}\,]$ if sampling unit i belongs to group 1 and $\boldsymbol{X}_i = [\,\boldsymbol{0} \quad \boldsymbol{1}_{n_i}\,]$ if sampling unit i belongs to group 2. Also $\boldsymbol{\gamma} = [\,\gamma_1 \quad \gamma_2\,]'$ is fixed and unknown, while $\boldsymbol{e}_{iy} \sim N_{n_i}(\boldsymbol{0}, \sigma^2\boldsymbol{I}_{n_i})$ with σ^2 unknown. The original data, y_{ij}, are not available, but we do know $\Sigma_{j=1}^{n_i} y_{ij}$ for each sampling unit. From the sums, we calculate $m_i = \Sigma_{j=1}^{n_i} y_{ij}/n_i$ and consider the following linear model: $\boldsymbol{m} = \{m_i\} = \boldsymbol{Z\beta} + \boldsymbol{e}$, with $\boldsymbol{\beta} = [\,\beta_1 \quad \beta_2\,]'$ and $\boldsymbol{Z} = \begin{bmatrix} \boldsymbol{1}_{N_1} & \boldsymbol{0} \\ \boldsymbol{0} & \boldsymbol{1}_{N_2} \end{bmatrix}$.

Hint for some parts of the exercise: Summation notation may help simplify.

13.1.1 Completely specify the distribution of $\boldsymbol{e}$.

13.1.2 Find an expression for the OLS estimator $\widehat{\boldsymbol{\beta}}$. Simplify the expression, and very briefly describe the nature of the elements of $\widehat{\boldsymbol{\beta}}$ in terms understandable to a scientific collaborator.
13.1.3 Find $\mathrm{E}(\widehat{\boldsymbol{\beta}})$. Simplify the expression.
13.1.4 Find $\mathcal{V}(\widehat{\boldsymbol{\beta}})$. Simplify the expression.
13.1.5 Enough is known to allow using (exact) weighted least squares. Given what is known, provide an expression for an appropriate WLS estimator $\widetilde{\boldsymbol{\beta}}$ and simplify it.
13.1.6 Specify the model for $\boldsymbol{m}$ associated with 13.1.5 as a GGLM.
13.1.7 Create a linearly equivalent GLM for the GGLM in 13.1.6.
13.1.8 Find $\mathrm{E}(\widetilde{\boldsymbol{\beta}})$. Simplify the expression.
13.1.9 Find $\mathcal{V}(\widetilde{\boldsymbol{\beta}})$. Simplify the expression.
13.1.10 Compare $\mathcal{V}(\widehat{\boldsymbol{\beta}})$ with $\mathcal{V}(\widetilde{\boldsymbol{\beta}})$. In particular, is $\mathcal{V}(\widehat{\boldsymbol{\beta}}) - \mathcal{V}(\widetilde{\boldsymbol{\beta}})$ always positive definite, negative definite, nonnegative definite, or nonpositive definite? Explain why briefly. Based on your results in the exercises, can either or both be ruled out as a candidate for being the BLUE?

CHAPTER 14

Estimation for Linear Mixed Models

14.1 MOTIVATION

Given the background of the previous chapters, the theory of estimation may be stated quite simply for the general linear mixed model. Without the need to assume any particular distribution, the method of approximate weighted least squares provides parameter estimates. With the additional assumption of Gaussian errors, iterating the process yields maximum likelihood estimates.

Although the mixed model theory of estimation takes simple forms, successfully finding numerical solutions satisfying the theory sometimes proves difficult with real data. The difficulty often has the unfortunate effect of encouraging practitioners to choose a simple covariance model without data to support the choice. Instead, the analyst should first round the within-subject predictor values to the most coarse resolution scientifically acceptable (often time has been recorded with much numerical precision) before attempting analysis. Second, the analyst should improve the scaling, centering, and coding of the data, as discussed in Chapters 8, 9, and 12 in Muller and Fetterman (2000). If a model with an appropriate and defensible covariance structure will not converge, a different analysis strategy, other than oversimplifying the covariance structure, should be considered. In particular, compound symmetry does not seem to be a plausible model for many types of repeated measures in time.

Throughout most of our discussion of linear models, we make two strong assumptions: a valid model and a sample size sufficient to compute estimates. In contrast to univariate and multivariate model properties, the two assumptions do not suffice to guarantee optimal estimators for mixed models. We must also require symmetry of distributions to guarantee unbiased estimates of expected-value parameters. More importantly, unbiased estimates of covariance parameters are available only with complete and balanced designs (no missing or mistimed data). In practice, such models very often correspond to multivariate models, which can and should be analyzed with multivariate techniques in order to use the best available methods for estimation and inference. Some limitations of the mixed model gradually diminish as sample size increases.

Given that relatively little is known about the finite-sample properties of mixed model methods, we should consider them to be "large-sample" methods. The question naturally arises: How big is "large?" We speculate that serious concerns about accuracy may be present with fewer than $N = 100$ independent sampling units (ISUs), and no more than a modest number of observations for each (perhaps all $p_i < 10$). It should be emphasized that, except in very special cases, simply having a large number of *observations* ($n = \sum_{i=1}^{N} p_i$) is not reassuring. The ratio of N to n, as well as the absolute size, plays a role in the performance of mixed models.

14.2 STATEMENT OF THE GENERAL LINEAR MIXED MODEL

The definition and notation of the mixed model was introduced in Chapter 5.

14.3 ESTIMATION AND ESTIMABILITY

Except for various special cases, the mixed model does not have closed-form expressions for estimators for any criterion. Gaussian theory estimation procedures for the mixed model with incomplete and unbalanced data include maximum likelihood (ML), restricted maximum likelihood (REML), moment estimators, and general linear model (ANOVA) estimators. Hocking (1985) discussed estimation for each. Harville (1977) gave a comprehensive review of ML and REML procedures, along with computational techniques. Laird and Ware (1982) discussed the Bayesian approach to variance component estimation, its relationship to REML estimation, and the application of the EM algorithm. Fairclough and Helms (1986) and Andrade and Helms (1986) explored ML estimation for the mixed model with linear covariance structure. More recently, Vonesh and Chinchilli (1997), Verbeke and Molenberghs (2000), and Demidenko (2004) provided book-length treatments of mixed models.

Conditional on knowing $\{\Sigma_i\}$ *exactly*, the theory of exact weighted least squares applies, which implies the theory of estimability and linearly equivalent models developed in previous chapters applies. Our treatment of estimability and linear equivalence concerned only the expected-value parameters $\{\beta_j\} = \boldsymbol{\beta}$ (the fixed effect parameters). The possibility of ambiguous parameter sets caused by (1) purposefully or accidentally LTFR between-subject ($\boldsymbol{X}$) design matrices or (2) explicit restrictions on rows of $\boldsymbol{\beta}$ motivated the developments for univariate models. Multivariate and related growth curve models introduce the additional concerns of (3) purposefully or accidentally LTFR within-subject design matrices $(\boldsymbol{T}, \boldsymbol{U})$ or (4) explicit restrictions on columns of the multivariate form of the expected value parameters, $\boldsymbol{B} = [\boldsymbol{\beta}_1 \cdots \boldsymbol{\beta}_p]$. The mixed model form may be created by vertically concatenating all of the columns of $\boldsymbol{B}$ to give $\boldsymbol{\beta}_{\text{mixed}} = \text{vec}(\boldsymbol{B}) = [\boldsymbol{\beta}_1' \cdots \boldsymbol{\beta}_p']'$. All aspects of the same estimability issues arise in mixed models, albeit with between-subject and within-subject design and parameters jumbled together.

The simplicity of the GLM covariance model essentially eliminates any need to study the analog of estimability for covariance parameters. The (univariate)

$\text{GLM}_{N,q}(y_i; \boldsymbol{X}_i\boldsymbol{\beta}, \sigma^2)$ describes a model for N ISUs and 1 observation per ISU. Stacking all of the data together gives an $N \times 1$ vector $\boldsymbol{y}$ with $\boldsymbol{y}' = [\, y_1' \; \cdots \; y_N' \,]$. The associated covariance model matrix for the entire set of data describes a total of $n = N$ observations,

$$\begin{aligned}\mathcal{V}(\boldsymbol{y}) &= \boldsymbol{I}_N\sigma^2 = \boldsymbol{I}_N \otimes \sigma^2 \\ &= \begin{bmatrix} \sigma^2 & & 0 \\ & \ddots & \\ 0 & & \sigma^2 \end{bmatrix} . \end{aligned} \tag{14.1}$$

The (multivariate) $\text{GLM}_{N,p,q}(\boldsymbol{Y}_i; \boldsymbol{X}_i\boldsymbol{B}, \boldsymbol{\Sigma})$ describes a model for N ISUs and p observations per ISU. Stacking all of the data together by participant gives an $(Np) \times 1$ vector, $\text{vec}(\boldsymbol{Y}')$ with $\text{vec}(\boldsymbol{Y}') = [\, \boldsymbol{Y}_1 \cdots \boldsymbol{Y}_N \,]'$. The associated covariance model for all data describes a total of $n = Np$ observations,

$$\begin{aligned}\mathcal{V}[\text{vec}(\boldsymbol{Y}')] &= \boldsymbol{I}_N \otimes \boldsymbol{\Sigma} \\ &= \begin{bmatrix} \boldsymbol{\Sigma} & & \boldsymbol{0} \\ & \ddots & \\ \boldsymbol{0} & & \boldsymbol{\Sigma} \end{bmatrix} . \end{aligned} \tag{14.2}$$

Although rarely considered, the multivariate GLM theory of estimation can tolerate a singular $\boldsymbol{\Sigma}$, if handled carefully. The calculation of primary parameter estimates $(\widehat{\boldsymbol{B}}, \widehat{\boldsymbol{\Sigma}})$ does also. An appropriate choice of $\boldsymbol{U}$ matrix or multivariate restrictions $(\boldsymbol{R}_x\boldsymbol{B}\boldsymbol{R}_y = \boldsymbol{A})$ avoids any difficulties in testing associated with defining a singular error covariance matrix $\boldsymbol{\Sigma}_* = \boldsymbol{U}'\boldsymbol{\Sigma}\boldsymbol{U}$.

In contrast to the univariate and multivariate GLM, the general linear mixed model $\text{LMM}_{N,p_i,q,m}[\boldsymbol{y}_i; \boldsymbol{X}_i\boldsymbol{\beta}, \boldsymbol{Z}_i\boldsymbol{\Sigma}_{di}(\boldsymbol{\tau}_d)\boldsymbol{Z}_i' + \boldsymbol{\Sigma}_{ei}(\boldsymbol{\tau}_e)]$ describes a model for N ISUs and p_i observations per ISU. The presence of subscript i allows the number of observations to vary across ISU. Stacking all data together by participant gives, with $n = \sum_{i=1}^{n} p_i$, an $n \times 1$ vector $\boldsymbol{y}_s$ with $\boldsymbol{y}_s' = [\, \boldsymbol{y}_1' \cdots \boldsymbol{y}_N' \,]$. The associated covariance model for the entire set of data describes a total of n observations,

$$\begin{aligned}\boldsymbol{\Sigma}_s = \mathcal{V}(\boldsymbol{y}_s) &= \bigoplus_{i=1}^{N} (\boldsymbol{Z}_i\boldsymbol{\Sigma}_{di}\boldsymbol{Z}_i' + \boldsymbol{\Sigma}_{ei}) \\ &= \bigoplus_{i=1}^{N} \boldsymbol{\Sigma}_i \\ &= \begin{bmatrix} \boldsymbol{\Sigma}_1 & & \boldsymbol{0} \\ & \ddots & \\ \boldsymbol{0} & & \boldsymbol{\Sigma}_N \end{bmatrix} . \end{aligned} \tag{14.3}$$

Edwards, Stewart, Muller, and Helms (2001) described how linear restrictions can be defined which allow identifying a corresponding unrestricted and linearly equivalent general mixed model. Their results apply to linear constraints providing parallel restrictions on the fixed effect (expected-value) parameters and random effect (covariance) parameters and the class of two-stage mixed models. Hence much work remains to be done to extend the results to the full range of mixed models and constraints.

Heterogeneity among $\{\Sigma_i\}$ leads to heterogeneity among $\{\widehat{\Sigma}_i\}$, which serve as weights for $\widehat{\beta}$ during the *iterative* calculation of estimates. Despite having full-rank $\{\Sigma_i\}$, either $\{\widehat{\Sigma}_i\}$ or expressions based on the estimates may have difficulty maintaining their intended ranks due to finite precision computer arithmetic. Imprecision may arise due to limitations of the sample size, timing of observations, poor scaling, lack of centering, or poor predictor coding. Estimability may fail at any iteration. Much remains to be learned about how to overcome such local difficulties.

14.4 SOME SPECIAL TYPES OF MODELS

Practical use of the linear mixed model requires further assumptions and constraints to reduce the number of covariance parameters to a manageable level. For stationary processes we often assume that the j, k element of the covariance matrix, $\sigma_{jk} = \langle \Sigma_i(\boldsymbol{\tau}) \rangle_{jk}$, is a simple function of the elapsed times between the various pairs of observations. In particular, an autoregressive covariance model of order 1, AR(1), assumes $\langle \Sigma_i(\boldsymbol{\tau}) \rangle_{jk} = \sigma^2 \rho^{|t_j - t_k|}$ for observation j at time t_j and observation k at time t_k. The AR(1) structure specifies a nonlinear model for the variances and covariances. It has two parameters, $\boldsymbol{\tau}' = [\sigma^2 \ \rho]$. In some cases, an inherently linear model for the covariances can be defined.

Definition 14.1 A LMM$_{N,p_i,q,m}[\boldsymbol{y}_i; \boldsymbol{X}_i\boldsymbol{\beta}, \boldsymbol{Z}_i\Sigma_{di}(\boldsymbol{\tau}_d)\boldsymbol{Z}_i' + \Sigma_{ei}(\boldsymbol{\tau}_e)]$ with

$$\begin{aligned} \Sigma_i(\boldsymbol{\tau}) &= \Sigma_{di}(\boldsymbol{\tau}_d)\boldsymbol{Z}_i' + \Sigma_{ei}(\boldsymbol{\tau}_e) \\ &= \sum_{k=1}^{K} \tau_k \boldsymbol{G}_{ik} \end{aligned} \tag{14.4}$$

and all $\{\boldsymbol{G}_{ik}\}$ known constants which may vary with i has *linear covariance structure.*

A wide variety of useful covariance models can be expressed as a linear structure. Andrade and Helms (1986) developed estimators and test statistics for linear hypotheses on $\boldsymbol{\beta}$ and $\boldsymbol{\tau}$ under the assumption of linear covariance structure.

Example 14.1 If $\boldsymbol{Z}_i = \boldsymbol{1}_{p_i}$ while $\Sigma_{di}(\boldsymbol{\tau}_d) = \tau_1$ and $\Sigma_{ei}(\boldsymbol{\tau}_e) = \tau_2\boldsymbol{I}_{p_i}$, then $\Sigma_i(\boldsymbol{\tau}) = \tau_1\boldsymbol{1}_{p_i}\boldsymbol{1}_{p_i}' + \tau_2\boldsymbol{I}_{p_i}$ exhibits linear covariance structure. The covariance model corresponds to assuming compound symmetry for each independent sampling unit. What are ρ and σ^2 as functions of $\{\tau_1, \tau_2\}$?

Definition 14.2 A LMM$_{N,p_i,q,m}[\boldsymbol{y}_i; \boldsymbol{X}_i\boldsymbol{\beta}, \boldsymbol{Z}_i\Sigma_{di}(\boldsymbol{\tau}_d)\boldsymbol{Z}_i' + \Sigma_{ei}(\boldsymbol{\tau}_e)]$ with full-rank $\boldsymbol{X}_i = \boldsymbol{Z}_i = \boldsymbol{X}_1$ is a *balanced random coefficient model.*

Demidenko (2004, p. 62) noted that a balanced random coefficient model corresponds exactly to a particular form of a growth curve model. Hence the machinery for the $\text{GLM}_{N,p}()$ multivariate model can provide noniterative and optimal estimators as well as exact size-α tests in small samples. Except for (even more) special cases, the commonly used mixed model tests will not coincide with the optimal (multivariate) tests. Although we do not pursue the topic here, we note that a wider class of mixed models may be cast and analyzed as multivariate models and an even wider set as multivariate models with missing data.

14.5 ML ESTIMATION

As discussed in Chapter 5, the mixed model for ISU i may be expressed as $\boldsymbol{y}_i = \boldsymbol{X}_i\boldsymbol{\beta} + \boldsymbol{Z}_i\boldsymbol{d}_i + \boldsymbol{e}_i$, with $\mathrm{E}(\boldsymbol{y}_i) = \boldsymbol{X}_i\boldsymbol{\beta}$ and $\mathcal{V}(\boldsymbol{y}_i) = \boldsymbol{\Sigma}_i(\boldsymbol{\tau})$ abbreviated as $\boldsymbol{\Sigma}_i$. In turn,

$$\begin{aligned} \boldsymbol{e}_{+i} &= \boldsymbol{y}_i - \boldsymbol{X}_i\boldsymbol{\beta} \\ &= \boldsymbol{Z}_i\boldsymbol{d}_i + \boldsymbol{e}_i \,. \end{aligned} \tag{14.5}$$

Recalling that $n = \sum_{i=1}^{N} p_i$, the joint log likelihood is

$$\log L(\boldsymbol{\beta},\boldsymbol{\tau}) = -\frac{1}{2}n\log(2\pi) - \frac{1}{2}\sum_{i=1}^{N}\left[\log|\boldsymbol{\Sigma}_i| + (\boldsymbol{y}_i - \boldsymbol{X}_i\boldsymbol{\beta})'\boldsymbol{\Sigma}_i^{-1}(\boldsymbol{y}_i - \boldsymbol{X}_i\boldsymbol{\beta})\right] \tag{14.6}$$

or

$$-2\log L(\boldsymbol{\beta},\boldsymbol{\tau}) = n\log(2\pi) + \sum_{i=1}^{N}\left[\log|\boldsymbol{\Sigma}_i| + \mathrm{tr}(\boldsymbol{e}_{+i}\boldsymbol{e}_{+i}'\boldsymbol{\Sigma}_i^{-1})\right] . \tag{14.7}$$

Here $\partial \mathrm{vec}[\boldsymbol{\Sigma}_i]/\partial\boldsymbol{\beta} = \mathbf{0}$. For any fixed $\boldsymbol{\tau}$ the value of $\boldsymbol{\beta}$ which maximizes the likelihood is the weighted least squares estimate. Using $\widehat{\boldsymbol{\tau}}$ leads to an approximate weighted least squares estimator,

$$\begin{aligned} \boldsymbol{\beta}(\widehat{\boldsymbol{\tau}}) &= \left(\boldsymbol{X}_s'\widehat{\boldsymbol{\Sigma}}_s^{-1}\boldsymbol{X}_s\right)^{-1}\left(\boldsymbol{X}_s'\widehat{\boldsymbol{\Sigma}}_s^{-1}\boldsymbol{y}_s\right) \\ &= \left(\sum_{i=1}^{N}\boldsymbol{X}_i'\widehat{\boldsymbol{\Sigma}}_i^{-1}\boldsymbol{X}_i\right)^{-1}\left(\sum_{i=1}^{N}\boldsymbol{X}_i'\widehat{\boldsymbol{\Sigma}}_i^{-1}\boldsymbol{y}_i\right) . \end{aligned} \tag{14.8}$$

Hence the task reduces to maximizing the profile likelihood, with $\widehat{\boldsymbol{e}}_{+i} = \boldsymbol{y}_i - \boldsymbol{X}_i\boldsymbol{\beta}(\widehat{\boldsymbol{\tau}})$,

$$\log L_{\mathrm{ML}}(\widehat{\boldsymbol{\tau}}) = -\frac{1}{2}n\log(2\pi) - \frac{1}{2}\sum_{i=1}^{N}\left[\log|\boldsymbol{\Sigma}_i| + \mathrm{tr}(\boldsymbol{\Sigma}_i^{-1}\widehat{\boldsymbol{e}}_{+i}\widehat{\boldsymbol{e}}_{+i}')\right] . \tag{14.9}$$

Finding the MLE of $\boldsymbol{\tau}$ requires solving $\partial \mathrm{L}(\widehat{\boldsymbol{\tau}})/\partial\widehat{\boldsymbol{\tau}} = \mathbf{0}$. The most useful representations of the system of estimating equations depend on features of the model. If $\boldsymbol{\Sigma}_i$ has linear structure $[\mathcal{V}(\boldsymbol{y}_i) = \boldsymbol{\Sigma}_i(\boldsymbol{\tau}) = \sum_g \tau_g \boldsymbol{G}_{ig}$ with $\{\boldsymbol{G}_{ig}\}$ known$]$, then the MLE of $\boldsymbol{\tau}$ is

$$\widehat{\boldsymbol{\tau}} = \left[\left\langle \sum_{i=1}^{N} \text{tr}(\widehat{\boldsymbol{\Sigma}}_i^{-1} \boldsymbol{G}_{ig} \widehat{\boldsymbol{\Sigma}}_i^{-1} \boldsymbol{G}_{ih}) \right\rangle_{gh}\right]^{-1} \left[\left\langle \sum_{i=1}^{N} \text{tr}(\widehat{\boldsymbol{e}}_{+i} \widehat{\boldsymbol{e}}_{+i}' \widehat{\boldsymbol{\Sigma}}_i^{-1} \boldsymbol{G}_{ig} \widehat{\boldsymbol{\Sigma}}_i^{-1}) \right\rangle_{g}\right]. \quad (14.10)$$

Here $[\langle\,\rangle_g]$ is a vector and the expression in $\langle\,\rangle$ is the value of its element g. Matrix $[\langle\,\rangle_{gh}]$ is similarly defined and has the indicated value for the element in row g and column h. Advantageous simplifications of the estimating equations can depend on whether the data are balanced, whether the repeated measurements are consistently timed, or whether compound symmetry is assumed.

Except for special cases, the above equations for $\widehat{\boldsymbol{\beta}}$ and $\widehat{\boldsymbol{\tau}}$ must be iterated to achieve a maximum. Computation requires solution of simultaneous nonlinear equations via algorithms such as the Newton-Raphson, the Method of Scoring, or the EM algorithm. Jennrich and Schluchter (1985) and Fairclough and Helms (1986) compared performances of the algorithms. Lindstrom and Bates (1988) provided arguments favoring the use of the Newton-Raphson approach. The MIXED procedure in the SAS® System software uses a ridge-stabilized Newton-Raphson algorithm and offers, optionally, Method of Scoring steps for the initial iterations. Wolfinger, Tobias, and Sall (1994) reported many further details.

All three algorithms are iterative. Given results from iteration $t \in \{1, 2, \dots\}$, all three algorithms compute a new estimate of the form

$$\widehat{\boldsymbol{\theta}}_{t+1} = \widehat{\boldsymbol{\theta}}_t + \lambda_t \boldsymbol{H}_t^{-1} \partial \log L / \partial \boldsymbol{\theta} \Big|_{\boldsymbol{\theta} = \widehat{\boldsymbol{\theta}}_t}, \quad (14.11)$$

with $\widehat{\boldsymbol{\theta}}_{t+1}$ being the next value of the parameter estimates, $\widehat{\boldsymbol{\theta}} = [\widehat{\boldsymbol{\beta}}' \; \widehat{\boldsymbol{\tau}}']'$. The length of the next step taken towards the MLE is indicated by scalar $\lambda_t \in [0, 1]$. Vector $\boldsymbol{H}_t^{-1} \partial \log L / \partial \boldsymbol{\theta}$ specifies the direction of step t. In the Newton-Raphson algorithm $\boldsymbol{H}_t$ is the negative of the observed information matrix,

$$\boldsymbol{\mathcal{I}}(\widehat{\boldsymbol{\theta}}) = -\frac{\partial^2}{\partial \boldsymbol{\theta} \partial \boldsymbol{\theta}'} \log L(\boldsymbol{\theta}) \Big|_{\boldsymbol{\theta} = \widehat{\boldsymbol{\theta}}}. \quad (14.12)$$

In the Fisher Scoring algorithm $\boldsymbol{H}_t$ is the negative of the expected information matrix. At convergence, the two algorithms yield a computed value $\boldsymbol{H}_t^{-1}$ which is an estimate of the asymptotic covariance matrix of $\widehat{\boldsymbol{\theta}}_t$. Maximum likelihood theory gives $\boldsymbol{H}_t^{1/2}(\widehat{\boldsymbol{\theta}}_t - \boldsymbol{\theta}) \overset{\text{asy}}{\sim} \mathcal{N}(\boldsymbol{0}, \boldsymbol{I})$. Hence the matrix is useful computing the estimated asymptotic standard errors of $\widehat{\boldsymbol{\theta}}_t$, with sufficiently large N.

The EM algorithm has advantages of simplicity, positive definite $\boldsymbol{H}_t$ $\forall t$, and assured increments in the likelihood at each step. However, its iterations provide only the MLEs of $\widehat{\boldsymbol{\beta}}$ and $\boldsymbol{\tau}$ and do not provide any standard errors. Thus, additional computations are needed to estimate standard errors of $\widehat{\boldsymbol{\beta}}$ and $\widehat{\boldsymbol{\tau}}$. Of several proposed methods, the supplemental EM algorithm (SEM) of Meng and Rubin (1991) seems best. The SEM derives necessary information about the asymptotic standard errors via evaluation of the rate of convergence of the EM algorithm.

The key parts of conventional maximum likelihood estimation are 1) specification of the likelihood (which arises from assumptions about the model), 2) numerical computation of the MLEs of $\widehat{\boldsymbol{\beta}}$ and $\widehat{\boldsymbol{\tau}}$, 3) computation of estimates of asymptotically correct approximations of the standard errors of $\widehat{\boldsymbol{\beta}}$ and $\widehat{\boldsymbol{\tau}}$, and 4) subsequent computation of confidence intervals and test statistics. Under certain regularity conditions the ML estimators have the desirable properties of being consistent, asymptotically Gaussian, and efficient (Harville, 1977; Magnus, 1978). As noted earlier, Kackar and Harville (1984) proved $\widehat{\boldsymbol{\beta}}$ is unbiased, while variance estimators are optimistically biased (too small).

14.6 REML ESTIMATION

Restricted maximum likelihood (REML) estimators of variance are less biased because they take into account the loss of degrees of freedom due to the estimation of $\boldsymbol{\beta}$. For purposes of estimation (as distinct from hypothesis testing), the method seems preferable to ML methods.

In the simplest $\text{GLM}_{N,q}(y_i; \boldsymbol{X}_i\boldsymbol{\beta}, \sigma^2)$ with N i.i.d. errors, the estimators of $\boldsymbol{\beta}$ and σ^2 have very desirable properties. With Gaussian data, Lemma 11.6 guarantees the independence of $\widetilde{\boldsymbol{\beta}}$ and the residuals $\widehat{\boldsymbol{e}}$ in the $\text{GLM}_{N,q}(y_i; \boldsymbol{X}_i\boldsymbol{\beta}, \sigma^2)$. It follows immediately that the ML estimators $\widetilde{\boldsymbol{\beta}}$ and $\widetilde{\sigma}^2 = \widehat{\boldsymbol{e}}'\widehat{\boldsymbol{e}}/N$ are statistically independent. However, the ML estimator of variance is optimistically small. Hence it is customary to replace the MLE $\widetilde{\sigma}^2$ with $\widehat{\sigma}^2 = \widetilde{\sigma}^2 N/(N-r)$. Here N is the number of ISUs and $r = \text{rank}(\boldsymbol{X})$. The advantage of $\widehat{\sigma}^2$ over $\widetilde{\sigma}^2$ lies in the fact that $\text{E}(\widehat{\sigma}^2) = \sigma^2$, while $\text{E}(\widetilde{\sigma}^2) = \sigma^2(N-r)/N \leq \sigma^2$.

In seeking estimators with similar properties for mixed models, Patterson and Thompson (1971) recommended transforming the data to functionally separate computation of covariance and expected-value parameter estimators. With

$$\boldsymbol{y}_s' = [\,\boldsymbol{y}_1' \cdots \boldsymbol{y}_N'\,] \tag{14.13}$$

$$\boldsymbol{X}_s' = [\,\boldsymbol{X}_1' \cdots \boldsymbol{X}_N'\,], \tag{14.14}$$

they considered the transformation

$$\widehat{\boldsymbol{e}}_s = [\boldsymbol{I}_n - \boldsymbol{X}_s(\boldsymbol{X}_s'\boldsymbol{X}_s)^- \boldsymbol{X}_s']\boldsymbol{y}_s\,. \tag{14.15}$$

As always, $n = \sum_{i=1}^{N} p_i$ indicates the total number of observations (not the number of ISUs). The appeal of the transformation arises from properties of the $\text{GLM}_{N,q}(y_i; \boldsymbol{X}_i\boldsymbol{\beta}, \sigma^2)$ with i.i.d. Gaussian errors, as summarized in Lemmas 11.5 (independence of $\widehat{\boldsymbol{y}}$ and $\widehat{\boldsymbol{e}}$) and 11.6 (independence of $\widetilde{\boldsymbol{\beta}}$ and $\widehat{\boldsymbol{e}}$). Harville (1974) discussed the concept in the context of estimating variance components. Applying the transformation to the mixed model gives the reduced-profile log likelihood:

$$-2\log L_{\text{REML}}(\widehat{\boldsymbol{\tau}}) = (n-q)\log(2\pi) + \sum_{i=1}^{N}(\log|\boldsymbol{\Sigma}_i| + \widehat{\boldsymbol{e}}_{+i}'\boldsymbol{\Sigma}_i^{-1}\widehat{\boldsymbol{e}}_{+i}) + \log\left|\sum_{i=1}^{N}\boldsymbol{X}_i'\boldsymbol{\Sigma}_i^{-1}\boldsymbol{X}_i\right|. \tag{14.16}$$

The REML estimates are found by maximizing the last equation. The REML and ML forms differ only by

$$-2[\log L_{\text{REML}}(\widehat{\boldsymbol{\tau}}) - \log L_{\text{ML}}(\widehat{\boldsymbol{\tau}})] = -q\log(2\pi) + \log\left|\sum_{i=1}^{N}\boldsymbol{X}_i'\boldsymbol{\Sigma}_i^{-1}\boldsymbol{X}_i\right|. \tag{14.17}$$

Lindstrom and Bates (1988) provided helpful discussion about algorithms, as did Demidenko (2004).

14.7 SMALL-SAMPLE PROPERTIES OF ESTIMATORS

Kackar and Harville (1984) proved $\widehat{\boldsymbol{\beta}}$ from iterated approximate least squares (sometimes called estimated generalized least squares, among other names) is unbiased. However, covariance parameter estimators, at least in small samples, typically are biased. Littel (2003) provided an excellent overview.

Theorem 14.1 (a) In the $\text{LMM}_{N,p_i,q,m}[\boldsymbol{y}_i; \boldsymbol{X}_i\boldsymbol{\beta}, \boldsymbol{Z}_i\boldsymbol{\Sigma}_{di}(\boldsymbol{\tau}_d)\boldsymbol{Z}_i' + \boldsymbol{\Sigma}_{ei}(\boldsymbol{\tau}_e)]$ with Gaussian errors and full-rank $\boldsymbol{X}_s = [\boldsymbol{X}_1' \; \boldsymbol{X}_2' \cdots \boldsymbol{X}_N']'$, the ML estimator of $\boldsymbol{\beta}$, namely $\widehat{\boldsymbol{\beta}}$, is unbiased.
(b) If $\boldsymbol{d}_i$ and $\boldsymbol{e}_i$ have symmetric but not necessarily Gaussian distributions and all other assumptions are met, the result still holds.
(c) If $\boldsymbol{d}_i$ and $\boldsymbol{e}_i$ have symmetric but not necessarily Gaussian distributions, and all other assumptions are met, the REML, method-of-moments (MM) and MINQUE estimators of $\boldsymbol{\beta}$ are also unbiased.
(d) In general, ML, REML, estimators of $\boldsymbol{\tau}$, and functions thereof, including $\boldsymbol{\Sigma}_{di}(\boldsymbol{\tau}_d)$ and $\boldsymbol{\Sigma}_{ei}(\boldsymbol{\tau}_e)$ are biased.

Proof. A surprisingly simple proof arises from the combination of the symmetry of distribution assumption and a variance estimator expressible as an even function. Demidenko (2004, Section 3.6) provided detailed proofs.

Theorem 14.2 (a) In the *general* $\text{LMM}_{N,p_i,q,m}[\boldsymbol{y}_i; \boldsymbol{X}_i\boldsymbol{\beta}, \boldsymbol{Z}_i\boldsymbol{\Sigma}_{di}(\boldsymbol{\tau}_d)\boldsymbol{Z}_i' + \boldsymbol{\Sigma}_{ei}(\boldsymbol{\tau}_e)]$, with or without Gaussian errors, the ML, REML, MM, and MINQUE variance estimators of $\boldsymbol{\tau}$ and functions thereof, including $\boldsymbol{\Sigma}_{di}(\boldsymbol{\tau}_d)$ and $\boldsymbol{\Sigma}_{ei}(\boldsymbol{\tau}_e)$, are biased.
(b) In the special case of a balanced random-coefficient model, the REML, MM, and MINQUE variance estimators (1) coincide, (2) are unbiased, and (3) differ from corresponding ML estimators only by a scaling constant, such as $N/(N-1)$.

Proof. Dimedenko's (2004, p. 140) proof of his Theorem 14 and surrounding discussion contain the desired results.

14.8 LARGE-SAMPLE PROPERTIES OF VARIANCE ESTIMATORS

Given the preceding theorem, it is not surprising that only asymptotic results are available for properties of variance parameter estimators $\widehat{\boldsymbol{\tau}}$ and $\boldsymbol{\Sigma}_i(\widehat{\boldsymbol{\tau}})$. Most such results provide only very low order approximations. As often happens with variance estimation, any bias is nearly always optimistic (estimates too small). In turn, estimated confidence intervals and tests are also optimistic. Much useful work remains to be done.

14.9 CONDITIONAL ESTIMATION OF d_i AND BLUP PREDICTION

In addition to estimation of $\widehat{\boldsymbol{\beta}}$ and $\widehat{\boldsymbol{\tau}}$, prediction of $\{\boldsymbol{d}_i\}$ (collectively $\boldsymbol{d}_s$) is often of interest in applications of the general linear mixed model. Henderson (1963) popularized the use of the "best linear unbiased predictor" (BLUP) of $\boldsymbol{d}_i$, namely $\boldsymbol{d}_i = \boldsymbol{\Sigma}_{di}\boldsymbol{Z}_i'\boldsymbol{\Sigma}_i^{-1}(\boldsymbol{y}_i - \boldsymbol{X}_i\boldsymbol{\beta})$, with $\boldsymbol{\Sigma}_{di}$ and $\boldsymbol{\Sigma}_i$ as in Definition 5.1. For Gaussian data the BLUP is easily proven to be the expected value of $\boldsymbol{d}_i$ *conditional on* the observed value of $\boldsymbol{y}_i$. Substituting the MLEs for the unknown parameters yields the empirical BLUP (eBLUP):

$$\widehat{\boldsymbol{d}}_i = \widehat{\boldsymbol{\Sigma}}_{di}\boldsymbol{Z}_i'\widehat{\boldsymbol{\Sigma}}_i^{-1}(\boldsymbol{y}_i - \boldsymbol{X}_i\widehat{\boldsymbol{\beta}}). \tag{14.18}$$

Collectively, with $\boldsymbol{\Sigma}_s$, $\boldsymbol{\Sigma}_{ds}$, and $\boldsymbol{\Sigma}_{es}$ as defined in equation 5.9,

$$\widehat{\boldsymbol{d}}_s = \widehat{\boldsymbol{\Sigma}}_{ds}\boldsymbol{Z}_s'\widehat{\boldsymbol{\Sigma}}_s^{-1}(\boldsymbol{y}_s - \boldsymbol{X}_s\widehat{\boldsymbol{\beta}}). \tag{14.19}$$

Depending on $\widehat{\boldsymbol{\Sigma}}_i^{-1}$ and the dimension and magnitude of $(\boldsymbol{y}_i - \boldsymbol{X}_i\widehat{\boldsymbol{\beta}})$, the value of $\widehat{\boldsymbol{d}}_i$ is subject to shrinkage toward zero. If p_i is small, then conditioning $\boldsymbol{d}_i$ on $\boldsymbol{y}_i$ is not highly informative and the shrinkage toward zero will be substantial. In turn, the eBLUP of the ISU central tendency, namely,

$$\widehat{\boldsymbol{y}}_i = \boldsymbol{X}_i\widehat{\boldsymbol{\beta}} + \boldsymbol{Z}_i\widehat{\boldsymbol{d}}_i \tag{14.20}$$

shrinks toward the estimated population central tendency, $\boldsymbol{X}_i\widehat{\boldsymbol{\beta}}$.

Computing the expression for $\widehat{\boldsymbol{d}}_i$ above would require inversion of a $p_i \times p_i$ matrix $\widehat{\boldsymbol{\Sigma}}_i^{-1}$, which can be problematic for large p_i. Henderson (1963) offered equivalent mixed model equations of the form

$$\begin{bmatrix} \boldsymbol{X}_s'\widehat{\boldsymbol{\Sigma}}_{es}^{-1}\boldsymbol{X}_s & \boldsymbol{X}_s'\widehat{\boldsymbol{\Sigma}}_{es}^{-1}\boldsymbol{Z}_s \\ \boldsymbol{Z}_s'\widehat{\boldsymbol{\Sigma}}_{es}^{-1}\boldsymbol{X}_s & \widehat{\boldsymbol{\Sigma}}_{ds}^{-1} + \boldsymbol{Z}_s'\widehat{\boldsymbol{\Sigma}}_{es}^{-1}\boldsymbol{Z}_s \end{bmatrix} \begin{bmatrix} \widehat{\boldsymbol{\beta}} \\ \widehat{\boldsymbol{d}} \end{bmatrix} = \begin{bmatrix} \boldsymbol{X}'\widehat{\boldsymbol{\Sigma}}_{es}^{-1}\boldsymbol{y} \\ \boldsymbol{Z}'\widehat{\boldsymbol{\Sigma}}_{es}^{-1}\boldsymbol{y} \end{bmatrix}, \tag{14.21}$$

which involve the more manageable inversion of $\widehat{\boldsymbol{\Sigma}}_{es}$ and $\widehat{\boldsymbol{\Sigma}}_{ds}$ (as defined in equation 5.9). The greatest advantage accrues with diagonal $\widehat{\Sigma}_{es}$ and block diagonal $\widehat{\Sigma}_{ds}$.

Harville (1990) proved that

$$\mathcal{V}\left(\begin{bmatrix} \widehat{\boldsymbol{\beta}} \\ \widehat{\boldsymbol{d}}_s - \boldsymbol{d}_s \end{bmatrix}\right) = \begin{bmatrix} \boldsymbol{X}_s'\boldsymbol{\Sigma}_{es}^{-1}\boldsymbol{X}_s & \boldsymbol{X}_s'\boldsymbol{\Sigma}_{es}^{-1}\boldsymbol{Z}_s \\ \boldsymbol{Z}_s'\boldsymbol{\Sigma}_{es}^{-1}\boldsymbol{X}_s & \boldsymbol{\Sigma}_{ds}^{-1} + \boldsymbol{Z}_s'\boldsymbol{\Sigma}_{es}^{-1}\boldsymbol{Z}_s \end{bmatrix}^{-}, \tag{14.22}$$

in which both $\widehat{\boldsymbol{d}}_s$ and $\boldsymbol{d}_s$ are random vectors. The MLEs must be substituted for the unknown parameters $\boldsymbol{\Sigma}_{es}$ and $\boldsymbol{\Sigma}_{ds}$ to use the expression computing estimated standard errors, confidence intervals, and prediction intervals.

EXERCISES

14.1 The $\text{LMM}_{N,p_i,q,m}[\boldsymbol{y}_i; \boldsymbol{X}_i\boldsymbol{\beta}, \boldsymbol{Z}_i\boldsymbol{\Sigma}_{di}(\boldsymbol{\tau}_d)\boldsymbol{Z}_i' + \boldsymbol{\Sigma}_{ei}(\boldsymbol{\tau}_e)]$ has $\boldsymbol{Z}_i = \boldsymbol{0}$, $p_i = p = 3$, $\boldsymbol{\Sigma}_{ei}(\boldsymbol{\tau}_e) = \begin{bmatrix} \sigma_1^2 & 0 & 0 \\ 0 & \sigma_2^2 & 0 \\ 0 & 0 & \sigma_3^2 \end{bmatrix}$, and $\boldsymbol{\tau} = [\sigma_1^2 \;\; \sigma_2^2 \;\; \sigma_3^2]'$.

14.1.1 Show that $\boldsymbol{\Sigma}_i(\boldsymbol{\tau})$ has a linear structure.
14.1.2 Assuming $\widehat{\boldsymbol{\tau}}$ is given, provide an appropriate expression for $\widehat{\boldsymbol{\Sigma}}_i$.
14.1.3 Use equation (14.8) to find a slightly simplified expression for $\boldsymbol{\beta}(\boldsymbol{\tau})$.
14.1.4 Use equation (14.10) to provide an expression for $\widehat{\boldsymbol{\tau}} = [\widehat{\sigma}_1^2 \;\; \widehat{\sigma}_2^2 \;\; \widehat{\sigma}_3^2]'$. Simplify the expression.

14.2 Give an example of a covariance matrix that does *not* have a linear structure.

CHAPTER 15

Tests for Univariate Linear Models

15.1 MOTIVATION

In the present chapter, we always assume Gaussian errors in considering a univariate linear model, a $\text{GLM}_{N,q}(y_i; \boldsymbol{X}_i\boldsymbol{\beta}, \sigma^2)$ with $\text{rank}(\boldsymbol{X}) = r \leq q$, or a GGLM. Throughout the chapter, $\boldsymbol{\theta} = \boldsymbol{C}\boldsymbol{\beta}$ is $a \times 1$ and $\text{rank}(\boldsymbol{C}) = a \leq q$. With $\boldsymbol{\theta}_0$ an $a \times 1$ vector of known constants, the associated general linear hypothesis (GLH) may be stated

$$H_0 : \boldsymbol{\theta} = \boldsymbol{\theta}_0 \,, \tag{15.1}$$

with corresponding alternative $H_A : \boldsymbol{\theta} \neq \boldsymbol{\theta}_0$. In terms of Boolean variables, with $\mathbb{B}() \in \{0, 1\}$, the hypothesis may be written $H_0 = \mathbb{B}(\boldsymbol{\theta} = \boldsymbol{\theta}_0)$ *versus* $H_A = \mathbb{B}(\boldsymbol{\theta} \neq \boldsymbol{\theta}_0)$. The first issue addressed is testability. We will restrict attention to well-defined hypotheses and then examine the properties of the resulting tests. We will demonstrate that reasonable hypothesis testing procedures do not exist for nontestable hypotheses.

Another important issue is whether $\boldsymbol{\theta}$ is truly a fixed constant. If the model and its parameters are defined prior to collection of the data, then $\boldsymbol{\theta}$ is an unknown, fixed constant. However, in the course of an analysis one frequently discovers interesting aspects of the data which were not considered prior to data collection and which are directly suggested by the results at hand. In such a case the dimensions and the definition of $\boldsymbol{\theta}$ (implied by the choice of $\boldsymbol{C}$) or the model itself may have been influenced by the observed value of $\boldsymbol{y}$. Hypotheses suggested by the data ($\boldsymbol{y}$) raise issues of multiplicity which may or may not be intractable.

Definition 15.1 (a) Parameters defined without regard to $\boldsymbol{y}$ are *a priori* parameters.
(b) Other parameters, including ones which are suggested by the data, are *post hoc* parameters.

As indicated by the status of the parameter, the corresponding hypothesis is either an a priori hypothesis or a post hoc hypothesis. Different statistical tests are required for the two different kinds of hypotheses. When appropriate methods are used, tests of a priori hypotheses are more powerful than tests of post hoc hypotheses.

15.2 TESTABILITY OF UNIVARIATE HYPOTHESES

In the theoretical formulation of any hypothesis test procedure, the hypothesis is always manipulated in terms of its linearly independent (LIN) components. Hence one requires the rows of $\boldsymbol{C}$ to be LIN, with rank$(\boldsymbol{C}) = a \leq q$. In the following developments, we first address the issue of testability for full-rank $\boldsymbol{C}$ matrices. Later we consider less-than-full-rank $\boldsymbol{C}$. With full-rank $\boldsymbol{C}$, the formal definition of a testable hypothesis implies questions of testability that arise only in LTFR models.

Definition 15.2 Under the assumptions of $\text{GLM}_{N,q}(y_i; \boldsymbol{X}_i\boldsymbol{\beta}, \sigma^2)$ with rank$(\boldsymbol{X}) = r \leq q$ and $\boldsymbol{\theta} = \boldsymbol{C}\boldsymbol{\beta}$ $a \times 1$, the hypothesis $H_0 = \mathbb{B}(\boldsymbol{\theta} = \boldsymbol{\theta}_0)$ *versus* $H_A = \mathbb{B}(\boldsymbol{\theta} \neq \boldsymbol{\theta}_0)$ is *testable* if and only if $\boldsymbol{\theta}$ is estimable *and* rank$(\boldsymbol{C}) = a \leq q$.

In FR models, $\boldsymbol{\beta}$ and $\boldsymbol{\theta}$ are always estimable and corresponding tests with full-rank $\boldsymbol{C}$ are always testable. In LTFR models $H_0 = \mathbb{B}(\boldsymbol{\beta} = \boldsymbol{0})$ *versus* $H_A = \mathbb{B}(\boldsymbol{\beta} \neq \boldsymbol{0})$ is never testable, while $\boldsymbol{\theta}$ may or may not be estimable and hence may or may not be testable.

The great majority of all results presented about properties of hypothesis tests depend on the assumption of Gaussian errors. In contrast, in estimation theory, many first- and second-moment properties are distribution free in the sense that the particular likelihood need not be specified (finite second moments and appropriate independence and homogeneity suffice). The distribution-free feature carries over to testability, even though the forms of the distributions of test statistics studied here depend strongly on the Gaussian assumption.

The following argument helps explain the relationship between estimable parameters and testable hypotheses. We will find that test statistics are functions of $\widehat{\boldsymbol{\theta}} = \boldsymbol{C}\widetilde{\boldsymbol{\beta}}$, with $\widetilde{\boldsymbol{\beta}} = \widehat{\boldsymbol{\beta}}$ in the FR case. The quantity $\widehat{\boldsymbol{\theta}}$ needs to be invariant to the choice of $\widetilde{\boldsymbol{\beta}}$ in order for the test to be invariant. We do not want the test result to depend on our choice of generalized inverse! We have previously proven $\widehat{\boldsymbol{\theta}}$ is invariant if and only if $\boldsymbol{\theta}$ is estimable. A series of theorems were presented in Chapter 10 which give estimability criteria. In particular, $\boldsymbol{C}(\boldsymbol{X}'\boldsymbol{X})^{-}(\boldsymbol{X}'\boldsymbol{X}) = \boldsymbol{C}$ if and only if $\boldsymbol{C}\boldsymbol{\beta}$ is estimable.

Lemma 15.1 For $a \times q$ $\boldsymbol{C}$, the matrix $\boldsymbol{M} = \boldsymbol{C}(\boldsymbol{X}'\boldsymbol{X})^{-}\boldsymbol{C}'$ is $a \times a$.

(a) If $\boldsymbol{X}$ is full rank, then rank$(\boldsymbol{M}) = a$ provides a necessary and sufficient condition for testability.

(b) If $\boldsymbol{X}$ is less than full rank, then the condition rank$(\boldsymbol{M}) = a$ provides a necessary but not sufficient condition for testability.

(c) If $\boldsymbol{X}$ is less than full rank, then $\text{rank}(\boldsymbol{M}) = a$ combined with the requirement $\boldsymbol{C} = \boldsymbol{C}(\boldsymbol{X}'\boldsymbol{X})^{-}(\boldsymbol{X}'\boldsymbol{X})$, or any other condition which guarantees estimability, provide a necessary and sufficient set of conditions for testability.

Proof. Left as an exercise.

Lemma 15.2 Estimable $\boldsymbol{\theta} = \boldsymbol{C\beta}$ $(a \times 1)$ with $\text{rank}(\boldsymbol{C}) = a$ implies $\boldsymbol{M}$ is symmetric, and unique and $\text{rank}(\boldsymbol{M}) = a$. Also $\boldsymbol{C} = \boldsymbol{AX}$ with $\text{rank}(\boldsymbol{A}) = a$.

Proof. Estimability allows writing $\boldsymbol{C} = \boldsymbol{AX}$ (Theorem 11.3). In turn,

$$\begin{aligned} \boldsymbol{M} &= \boldsymbol{C}(\boldsymbol{X}'\boldsymbol{X})^{-}\boldsymbol{C}' \\ &= \boldsymbol{AX}(\boldsymbol{X}'\boldsymbol{X})^{-}\boldsymbol{X}'\boldsymbol{A}' \\ &= \boldsymbol{AX}(\boldsymbol{X}'\boldsymbol{X})^{+}\boldsymbol{X}'\boldsymbol{A}' \\ &= \boldsymbol{AHA}'\,. \end{aligned} \tag{15.2}$$

The matrix $\boldsymbol{H} = \boldsymbol{X}(\boldsymbol{X}'\boldsymbol{X})^{+}\boldsymbol{X}'$ is symmetric, idempotent, unique and of rank $r = \text{rank}(\boldsymbol{X})$ (Theorem 1.15). Hence $\boldsymbol{M}$ is symmetric whether or not $(\boldsymbol{X}'\boldsymbol{X})^{-}$ is symmetric, and $\boldsymbol{M}$ is unique.

Lemma 1.29 gives $\boldsymbol{X} = \boldsymbol{L}_1\text{Dg}(\boldsymbol{s}_1)\boldsymbol{R}_1'$, with $\boldsymbol{R}_1'\boldsymbol{R}_1 = \boldsymbol{I}_r$, $\text{Dg}(\boldsymbol{s}_1)$ $r \times r$ of rank r, $\boldsymbol{L}_1$ $N \times r$ of rank r, $\boldsymbol{L}_1'\boldsymbol{L}_1 = \boldsymbol{I}_r$, and $\boldsymbol{H} = \boldsymbol{L}_1\boldsymbol{L}_1'$. Hence $\boldsymbol{M} = (\boldsymbol{AL}_1)(\boldsymbol{AL}_1)'$. Having $\boldsymbol{C} = \boldsymbol{AX}$ implies $\boldsymbol{CC}' = \boldsymbol{AXX}'\boldsymbol{A}' = \boldsymbol{AL}_1\text{Dg}(\boldsymbol{s}_1)^2\boldsymbol{L}_1'\boldsymbol{A}' = \boldsymbol{FF}'$ for $\boldsymbol{F} = [\boldsymbol{AL}_1\text{Dg}(\boldsymbol{s}_1)]$ and $a = \text{rank}(\boldsymbol{C}) = \text{rank}(\boldsymbol{CC}') = \text{rank}(\boldsymbol{F}) = \text{rank}(\boldsymbol{AL}_1) = \text{rank}(\boldsymbol{M})$. With $\boldsymbol{C} = \boldsymbol{AX}$ of rank $a \leq r$, Lemma 1.6 gives $\text{rank}(\boldsymbol{AX}) \leq \min[\text{rank}(\boldsymbol{A}), \text{rank}(\boldsymbol{X})]$, which implies $a \leq \min[\text{rank}(\boldsymbol{A}), r]$ and $a \leq \text{rank}(\boldsymbol{A})$. Also $\boldsymbol{A}$ is $a \times N$, with $a \leq N$, which implies $\text{rank}(\boldsymbol{A}) \leq a$. Hence $\text{rank}(\boldsymbol{A}) = a$. □

Example 15.1 The theory of linearly equivalent models provides many insights into the structure underlying a testable hypothesis. For any testable $a \times 1$ $\boldsymbol{\theta} = \boldsymbol{C\beta}$, the last lemma and the notation developed in its proof allow defining full-rank $N \times r$ matrix $\boldsymbol{X}_1 = \boldsymbol{XR}_1 = \boldsymbol{L}_1\text{Dg}(\boldsymbol{s}_1)$ and $r \times 1$ matrix $\boldsymbol{\beta}_1 = \boldsymbol{R}_1'\boldsymbol{\beta}$. The corresponding model equation may be written

$$\begin{aligned} \boldsymbol{y} &= \boldsymbol{X}\boldsymbol{\beta} + \boldsymbol{e} \\ &= \boldsymbol{L}_1\text{Dg}(\boldsymbol{s}_1)\boldsymbol{R}_1'\boldsymbol{\beta} + \boldsymbol{e} \\ &= \boldsymbol{X}_1\boldsymbol{\beta}_1 + \boldsymbol{e}\,. \end{aligned} \tag{15.3}$$

Choosing the $a \times r$ matrix $\boldsymbol{C}_1 = \boldsymbol{CR}_1$ implies $\boldsymbol{C}_1\boldsymbol{R}_1' = \boldsymbol{C}$ and $\boldsymbol{\theta}_1 = \boldsymbol{C}_1\boldsymbol{\beta}_1 = \boldsymbol{C}_1\boldsymbol{R}_1'\boldsymbol{\beta} = \boldsymbol{C\beta}$. Orthonormal $\boldsymbol{P}$ and $\boldsymbol{Q}$ ensure $(\boldsymbol{PAQ}')^{+} = \boldsymbol{QA}^{+}\boldsymbol{P}'$. Also

$$\begin{aligned} \boldsymbol{M}_1 = \boldsymbol{C}_1(\boldsymbol{X}_1'\boldsymbol{X}_1)^{-1}\boldsymbol{C}_1' &= \boldsymbol{CR}_1(\boldsymbol{X}_1'\boldsymbol{X}_1)^{-1}\boldsymbol{R}_1'\boldsymbol{C}' \\ &= \boldsymbol{C}(\boldsymbol{R}_1\boldsymbol{X}_1'\boldsymbol{X}_1\boldsymbol{R}_1')^{+}\boldsymbol{C}' \\ &= \boldsymbol{M}\,. \end{aligned} \tag{15.4}$$

Hence with a less-than-full-rank model, a testable hypothesis corresponds to finding the linearly equivalent full-rank model with predictors $\boldsymbol{X}_1 = \boldsymbol{XR}_1$, the principal component scores. In turn, C_1 expresses the original contrasts in terms of the component score parameters, namely β_1.

Theorem 15.1 For $\text{GLM}_{N,q}(y_i; X_i\beta, \sigma^2)$ with $\text{rank}(X) = r \leq q$, $\theta = C\beta$ $a \times 1$, and $\text{rank}(C) = a \leq q$, if θ is *not* estimable then $H_0 = \mathbb{B}(\theta = \theta_0)$ *versus* $H_A = \mathbb{B}(\theta \neq \theta_0)$ is not testable. Consequently no reasonable test exists for the hypothesis; i.e., any test procedure will have size $\alpha = 0$ and power $1 - \beta = 0$. Furthermore the test will not be invariant to changes in irrelevant quantities.

Proof. (Searle, 1971, p. 193–194) Intuitively, tests involve a comparison of the model with and without the linear restrictions $C\beta = \theta_0$. In the case of nontestable hypotheses the model seems to fit equally well with or without the restrictions. If (a) θ is not estimable, (b) $\widetilde{\beta}$ satisfies the restricted normal equations $X'X\widetilde{\beta} + C'\lambda = X'y$, and (c) $C\widetilde{\beta} = \theta_0$ (in which λ is the vector of "Lagrange multipliers"), then $\widetilde{\beta}$ also satisfies the unrestricted normal equations $X'X\widetilde{\beta} = X'y$. For a nontestable hypothesis, *SSE* is the same with or without restrictions! In terms of goodness of fit, the unrestricted model and the restricted model are indistinguishable; we have no reason to prefer one to the other. Thus, any test has size $\alpha \equiv \Pr\{\text{rejecting } H_0 | H_0 = \text{TRUE}\} = 0$ and power $1 - \beta \equiv \Pr\{\text{reject } H_0 | H_A = \text{TRUE}\} = 0$. □

Corollary 15.1 If $\theta_s = [\theta_1' \;\; \theta_2']'$ and θ_1 is estimable but θ_2 is not, then tests for θ_s are indistinguishable from tests for θ_1.

Proof. For $\theta_0 = [\theta_{0,1}' \;\; \theta_{0,2}']'$ the model subject to the restrictions $C_1\beta = \theta_{0,1}$ is linearly equivalent to some unrestricted model. It then follows from the proof for Theorem 11.18 (with restrictions not necessarily of full rank) that, in terms of goodness of fit via *SSE*, the model constrained by both $\theta_1 \equiv C_1\beta - \theta_{0,1} = \mathbf{0}$ and $\theta_2 \equiv C_2\beta - \theta_{0,2} = \mathbf{0}$ cannot be distinguished from the model constrained only by $\theta_1 \equiv C_1\beta - \theta_{0,1} = \mathbf{0}$. Therefore either of the two constrained models may be compared with the original, unrestricted model; the results in terms of difference in *SSE* would be the same. □

Test statistics are functions of $\widehat{\delta} = (C\widetilde{\beta} - \theta_0)'[C(X'X)^- C']^{-1}(C\widetilde{\beta} - \theta_0)$ (with $\widetilde{\beta} = \widehat{\beta}$ and $X'X$ nonsingular in the FR case) and the test statistics depend on C and θ_0 only through $\widehat{\delta}$. The ability to compute $\widehat{\delta}$ depends on the existence of $M^{-1} = [C(X'X)^- C']^{-1}$ and can be computed if and only if $M = [C(X'X)^- C'] \neq \mathbf{0}$ has full rank. Having testable θ ensures full-rank M.

In many cases it is possible to compute $\widehat{\delta}$ (and therefore the test statistic) even though θ is *not* estimable and the hypothesis is *not* testable. The result is formalized in the following theorem. In turn, the next theorem answers the question "What hypothesis is the test statistic addressing?" in such cases.

Theorem 15.2 If $\text{GLM}_{N,q}(y_i; \boldsymbol{X}_i\boldsymbol{\beta}, \sigma^2)$ has $\text{rank}(\boldsymbol{X}) = r \leq q$, $\boldsymbol{\theta} = \boldsymbol{C}\boldsymbol{\beta}$ $(a \times 1)$, *and* $\text{rank}(\boldsymbol{C}) = a \leq q$, then
(a) $\boldsymbol{\theta}$ is estimable implies $\boldsymbol{M} = \boldsymbol{C}(\boldsymbol{X}'\boldsymbol{X})^{-}\boldsymbol{C}'$ has full rank, although
(b) $\boldsymbol{M} = \boldsymbol{C}(\boldsymbol{X}'\boldsymbol{X})^{-}\boldsymbol{C}'$ having full rank does *not* imply $\boldsymbol{\theta}$ is estimable.

Proof of (a). (Compare with Searle, 1971, p. 189–190) It is assumed $a \times q$ $\boldsymbol{C}$ has (full) $\text{rank}(\boldsymbol{C}) = a \leq q$. Estimable $\boldsymbol{C}\boldsymbol{\beta}$ insures there exists unique matrix $\boldsymbol{A}$ such that $\boldsymbol{C} = \boldsymbol{A}(\boldsymbol{X}'\boldsymbol{X})$ which implies $\boldsymbol{C}(\boldsymbol{X}'\boldsymbol{X})^{-}\boldsymbol{C}' = \boldsymbol{A}(\boldsymbol{X}'\boldsymbol{X})\boldsymbol{A}'$. Therefore $\text{rank}[\boldsymbol{C}(\boldsymbol{X}'\boldsymbol{X})^{-}\boldsymbol{C}'] = \text{rank}[(\boldsymbol{A}\boldsymbol{X}')(\boldsymbol{X}\boldsymbol{A}')] \equiv \text{rank}(\boldsymbol{A}\boldsymbol{X}') \leq a$.

It suffices to prove $\text{rank}(\boldsymbol{A}\boldsymbol{X}') = a$. Matrix theory gives $a \equiv \text{rank}(\boldsymbol{C}) \equiv \text{rank}[\boldsymbol{A}(\boldsymbol{X}'\boldsymbol{X})] \leq \min\{\text{rank}(\boldsymbol{A}\boldsymbol{X}'), \text{rank}(\boldsymbol{X})\}$. Therefore $\text{rank}(\boldsymbol{A}\boldsymbol{X}') \geq a$ and $\text{rank}(\boldsymbol{X}) \equiv r \geq a$. Hence $\text{rank}(\boldsymbol{A}\boldsymbol{X}') = a$ since $\text{rank}(\boldsymbol{A}\boldsymbol{X}')$ cannot exceed a. We can similarly prove $\text{rank}(\boldsymbol{A}) = a$ since $a \equiv \text{rank}(\boldsymbol{C}) \equiv \text{rank}[\boldsymbol{A}(\boldsymbol{X}'\boldsymbol{X})] \leq \min\{\text{rank}(\boldsymbol{A}), \text{rank}(\boldsymbol{X}'\boldsymbol{X})\}$.

Proof of (b). By counterexample, with $p = 1, q = N = 3$:

$$\boldsymbol{X}\boldsymbol{\beta} = \begin{bmatrix} 1 & 1 & 0 \\ 1 & 1 & 0 \\ 1 & 0 & 1 \end{bmatrix} \begin{bmatrix} \mu \\ \alpha_1 \\ \alpha_2 \end{bmatrix} \tag{15.5}$$

$$(\boldsymbol{X}'\boldsymbol{X})^{-} = \left(\begin{bmatrix} 3 & 2 & 1 \\ 2 & 2 & 0 \\ 1 & 0 & 1 \end{bmatrix} \right)^{-} = \frac{1}{2} \begin{bmatrix} 2 & -2 & 0 \\ -2 & 3 & 0 \\ 0 & 0 & 0 \end{bmatrix} \tag{15.6}$$

$$\boldsymbol{\theta} = \boldsymbol{C}\boldsymbol{\beta} = \begin{bmatrix} 1 & 0 & 0 \\ 0 & 1 & 0 \end{bmatrix} \begin{bmatrix} \mu \\ \alpha_1 \\ \alpha_2 \end{bmatrix} = \begin{bmatrix} \mu \\ \alpha_1 \end{bmatrix} \tag{15.7}$$

$$\boldsymbol{M}^{-1} = \begin{bmatrix} 3 & 2 \\ 2 & 2 \end{bmatrix} \tag{15.8}$$

$$\boldsymbol{C}(\boldsymbol{X}'\boldsymbol{X})^{-}(\boldsymbol{X}'\boldsymbol{X}) = \begin{bmatrix} 1 & 0 & 1 \\ 0 & 1 & -1 \end{bmatrix} \neq \boldsymbol{C}. \tag{15.9}$$

Hence the result is true by counterexample. □

Theorem 15.3 For $\text{GLM}_{N,q}(y_i; \boldsymbol{X}_i\boldsymbol{\beta}, \sigma^2)$ with $\text{rank}(\boldsymbol{X}) = r \leq q$, $\boldsymbol{\theta} = \boldsymbol{C}\boldsymbol{\beta}$ $a \times 1$, and $\text{rank}(\boldsymbol{C}) = a \leq q$, even though $\boldsymbol{M}^{-1} = [\boldsymbol{C}(\boldsymbol{X}'\boldsymbol{X})^{-}\boldsymbol{C}']^{-1}$ exists, $\boldsymbol{\theta}$ may not be estimable. If so, test statistics based on $\widehat{\delta} = (\boldsymbol{C}\widetilde{\boldsymbol{\beta}} - \boldsymbol{\theta}_0)'\boldsymbol{M}^{-1}(\boldsymbol{C}\widetilde{\boldsymbol{\beta}} - \boldsymbol{\theta}_0)$ can be computed and the testable hypothesis actually tested is $H_0 = \mathbb{B}(\boldsymbol{\theta}_G = \boldsymbol{0})$ *versus* $H_A = \mathbb{B}(\boldsymbol{\theta}_G \neq \boldsymbol{0})$ with $\boldsymbol{\theta}_G = \boldsymbol{C}(\boldsymbol{X}'\boldsymbol{X})^{-}(\boldsymbol{X}'\boldsymbol{X})\boldsymbol{\beta} - \boldsymbol{\theta}_0$.

Proof. Searle (1971, p. 195) provided a proof.

Here $\boldsymbol{\theta}_G$ is not necessarily invariant to the choice of $(\boldsymbol{X}'\boldsymbol{X})^-$. For each choice of generalized inverse a potentially different testable hypothesis is tested! The special case of $r = q$ has all $\boldsymbol{\theta}$ estimable. In the special case, full rank of $\boldsymbol{M} = [\boldsymbol{C}(\boldsymbol{X}'\boldsymbol{X})^-\boldsymbol{C}']$ guarantees a testable hypothesis, which corresponds to a unique parameter and a well-defined test.

Example 15.2 The following example illustrates a poorly defined hypothesis. If $p = 1, q = N = 3, \boldsymbol{\theta}_0 = \boldsymbol{0}$,

$$\boldsymbol{X} = \begin{bmatrix} 1 & 1 & 0 \\ 1 & 1 & 0 \\ 1 & 0 & 1 \end{bmatrix} \tag{15.10}$$

$$(\boldsymbol{X}'\boldsymbol{X})^- = \left(\begin{bmatrix} 3 & 2 & 1 \\ 2 & 2 & 0 \\ 1 & 0 & 1 \end{bmatrix}\right)^- = \frac{1}{2}\begin{bmatrix} 2 & -2 & 0 \\ -2 & 3 & 0 \\ 0 & 0 & 0 \end{bmatrix} \tag{15.11}$$

$$\boldsymbol{C} = \begin{bmatrix} 1 & 0 & 0 \\ 0 & 1 & 0 \end{bmatrix} \tag{15.12}$$

$$\boldsymbol{C}(\boldsymbol{X}'\boldsymbol{X})^-\boldsymbol{X}'\boldsymbol{X} = \boldsymbol{C}\begin{bmatrix} 1 & 0 & 1 \\ 0 & 1 & -1 \\ 0 & 0 & 0 \end{bmatrix}, \tag{15.13}$$

$\boldsymbol{\beta} = [\mu \quad \alpha_1 \quad \alpha_2]'$, $\boldsymbol{\theta} = [\mu \quad \alpha_1]'$, and $\boldsymbol{\theta}_G = [\mu + \alpha_2 \quad \alpha_1 - \alpha_2]'$. More generally, all possible generalized inverses are given by

$$(\boldsymbol{X}'\boldsymbol{X})^- = \frac{1}{2}\begin{bmatrix} 2 & -2 & 0 \\ -2 & 3 & 0 \\ 0 & 0 & 0 \end{bmatrix} + \begin{bmatrix} 0 & 0 & -1 \\ 0 & 0 & 1 \\ 0 & 0 & 1 \end{bmatrix}\boldsymbol{T} + \boldsymbol{S}\begin{bmatrix} 0 & 0 & 0 \\ 0 & 0 & 0 \\ -1 & 1 & 1 \end{bmatrix}, \tag{15.14}$$

in which $\boldsymbol{T}$ and $\boldsymbol{S}$ are arbitrary. Choosing

$$(\boldsymbol{X}'\boldsymbol{X})^- = \frac{1}{2}\begin{bmatrix} 2 & -2 & -2 \\ -2 & 3 & 2 \\ 0 & 0 & 2 \end{bmatrix} \tag{15.15}$$

would give $\boldsymbol{\theta}_G = [0 \quad \mu + \alpha_1]'$.

We now consider more general forms of $\boldsymbol{C}$. Previously we have assumed $a \times q$ $\boldsymbol{C}$ with $1 \leq \text{rank}(\boldsymbol{C}) = a \leq q$. Can the requirements be relaxed? We are occasionally interested in less-than-full-rank $\boldsymbol{C}$. In any case we will need $\boldsymbol{C} = \boldsymbol{A}\boldsymbol{X}$ (for some $\boldsymbol{A}$) to have any hope of formulating a testable hypothesis. Otherwise $\boldsymbol{\theta}$ will not be estimable. Estimability of $\boldsymbol{\theta}$ does not require $\boldsymbol{C}$ to have full rank. In particular, if $\boldsymbol{C}\boldsymbol{\beta} = \boldsymbol{\theta}$ is estimable, then so is

$$\begin{bmatrix} C \\ C \end{bmatrix} \beta = C_2\beta = \begin{bmatrix} \theta \\ \theta \end{bmatrix} = \theta_2 \,. \tag{15.16}$$

Here $1 \leq \text{rank}(C_2) < 2a$. Also, $H_0 = \mathbb{B}(\theta_2 = \mathbf{0}) = \mathbb{B}[(\theta = \mathbf{0}) \cap (\theta = \mathbf{0})]$ is overstated and is identical to $H_{0,1} = \mathbb{B}(\theta = \mathbf{0})$. We should test subhypothesis $H_{0,1}$ rather than H_0 itself.

Theorem 15.4 A test with a LTFR C matrix is indistinguishable from a test with C replaced by a full (row) rank matrix with rows that span all rows of C. More specifically, model $\text{GLM}_{N,q}(y_i; X_i\beta, \sigma^2)$ has $\text{rank}(X) = r \leq q$, estimable $(a \times 1)$

$$\theta = C\beta = \begin{bmatrix} C_1 \\ C_2 \end{bmatrix} \beta = \begin{bmatrix} \theta_1 \\ \theta_2 \end{bmatrix}, \tag{15.17}$$

and $\theta_0 = [\,\theta_{0,1}' \;\; \theta_{0,2}'\,]'$. Also C_j, θ_j, and $\theta_{0,j}$ have a_j rows for $j \in \{1, 2\}$, $\text{rank}(C) = \text{rank}(C_1) = a_1 < \min\{a, q\}$.
(a) Tests about θ are indistinguishable from tests about θ_1; i.e., $\theta = \theta_0$ iff $\theta_1 = \theta_{0,1}$, and $H_0 = \mathbb{B}(\theta = \theta_0) = \mathbb{B}(\theta_1 = \theta_{0,1})$.
(b) Tests about $\theta_1 = C_1\beta$ are testable; i.e., C_1 has full row rank and θ_1 is estimable.

Proof. The rows of C_2 are linear combinations of the rows of C_1; i.e., $C_2 = AC_1$ for some fixed matrix A. Hence $\Theta_2 = AC_1\beta = A\theta_1$. Also, $\theta = \theta_0$ iff $\theta_1 = \theta_{0,1}$ and

$$C\beta = \begin{bmatrix} I \\ A \end{bmatrix} C_1\beta = \theta = \begin{bmatrix} \theta_1 \\ \theta_2 \end{bmatrix} = \begin{bmatrix} I \\ A \end{bmatrix} \theta_1 \,. \tag{15.18}$$

Estimable θ implies θ_1 is estimable; combining estimability with C_1 having full row rank implies $\widehat{\delta}_1$ exists. While $\widehat{\delta} = (\widehat{\theta} - \theta_0)'[C(X'X)^{-}C']^{-1}(\widehat{\theta} - \theta_0)$ does *not* exist and cannot be computed, $\widehat{\delta}_1 = (\widehat{\theta}_1 - \theta_{0,1})'[C_1(X'X)^{-}C_1']^{-1}(\widehat{\theta}_1 - \theta_{0,1})$ does exist and can be computed. □

The following theorem justifies using a simpler model which is linearly equivalent. The approach allows avoiding some of the pitfalls with LTFR models.

Theorem 15.5 Any primary or secondary expected-value parameter testable in $\text{GLM}_{N,q}(y_i; X_i\beta, \sigma^2)$ is also testable in a linearly equivalent model.

Proof. Follows almost directly from the parallel result about estimability.

15.3 TESTS OF A PRIORI HYPOTHESES

Theorem 15.6 The likelihood ratio test in the univariate full rank GLM may be expressed exactly in terms of an F. For a $\text{GLM}_{N,q}\text{FR}(y_i; X_i\beta, \sigma^2)$ with

Gaussian errors, interest centers on a priori and estimable $\boldsymbol{\theta} = \boldsymbol{C\beta}$ ($a \times 1$) with $\text{rank}(\boldsymbol{C}) = a \leq q$, $\boldsymbol{M} = \boldsymbol{C}(\boldsymbol{X}'\boldsymbol{X})^{-1}\boldsymbol{C}'$, and $\boldsymbol{C}$ and $\boldsymbol{\theta}_0$ known constants.
(a) For testing $H_0 = \mathbb{B}(\boldsymbol{\theta} = \boldsymbol{\theta}_0)$ *versus* $H_A = \mathbb{B}(\boldsymbol{\theta} \neq \boldsymbol{\theta}_0)$, with $\Pr\{F(\nu_1, \nu_2, \omega) > f_{\text{crit}}\} = \alpha$, the *likelihood ratio test* (LRT) of size α is $\phi(\boldsymbol{y}) = \mathbb{B}[F(\boldsymbol{y}) > f_{\text{crit}}]$, in which

$$\widehat{\boldsymbol{\beta}} = (\boldsymbol{X}'\boldsymbol{X})^{-1}\boldsymbol{X}'\boldsymbol{y} \tag{15.19}$$
$$\widehat{\boldsymbol{\theta}} = \boldsymbol{C}\widehat{\boldsymbol{\beta}} \tag{15.20}$$
$$\widehat{\sigma}^2 = \boldsymbol{y}'[\boldsymbol{I} - \boldsymbol{X}(\boldsymbol{X}'\boldsymbol{X})^{-1}\boldsymbol{X}']\boldsymbol{y}/(N-q) \tag{15.21}$$
$$F(\boldsymbol{y}) = \left[(\widehat{\boldsymbol{\theta}} - \boldsymbol{\theta}_0)'\boldsymbol{M}^{-1}(\widehat{\boldsymbol{\theta}} - \boldsymbol{\theta}_0)/a\right]/\widehat{\sigma}^2 . \tag{15.22}$$

(b) Under H_0 $F(\boldsymbol{y}) \sim F(a, N-q, 0)$.
(c) In general, $F(\boldsymbol{y}) \sim F(a, N-q, \omega)$, with

$$\omega = (\boldsymbol{\theta} - \boldsymbol{\theta}_0)'\boldsymbol{M}^{-1}(\boldsymbol{\theta} - \boldsymbol{\theta}_0)/\sigma^2 . \tag{15.23}$$

(d) Testing $H_0 = \mathbb{B}(\boldsymbol{\theta} = \boldsymbol{\theta}_0)$ versus $H_A = \mathbb{B}(\boldsymbol{\theta} \neq \boldsymbol{\theta}_0)$ is equivalent to testing $H_0 = \mathbb{B}(\omega = 0)$ versus $H_A = \mathbb{B}(\omega \neq 0)$.
(e) Test statistic $F(\boldsymbol{y})$ can be written as a ratio of chi-square statistics (scaled sums of squares) divided by their degrees of freedom,

$$F(\boldsymbol{y}) = \frac{SSH/a}{SSE/(N-q)} , \tag{15.24}$$

with SSH= sum of squares for the hypothesis and SSE= sum of squares for error.

Proof. Results in Chapter 11 on univariate linear model estimation provide the basis for the proof, which is cast in terms of the more general concept of supremum, rather than maximum. Doing so simplifies generalizing the proof to LTFR models because FR models lead to a unique estimator of $\boldsymbol{\beta}$, while LTFR models do not.

Part 1. The model has primary parameters $\boldsymbol{\tau} = [\boldsymbol{\beta}' \ \sigma^2]'$. Unrestricted estimation finds values in the set $\boldsymbol{\tau}_A = \{\boldsymbol{\tau} : \boldsymbol{\beta} \in \Re^q, \sigma^2 \geq 0\}$. The null hypothesis restricts attention to the set $\boldsymbol{\tau}_0 = \{\boldsymbol{\tau} : \boldsymbol{\theta} = \boldsymbol{\theta}_0, \boldsymbol{\beta} \in \Re^q, \sigma^2 \geq 0\} < \boldsymbol{\tau}_A$.

Unrestricted maximum likelihood estimators are derived as follows. The maximum likelihood estimate is any particular value of $\boldsymbol{\tau}$, say $\widehat{\boldsymbol{\tau}}$, for which the likelihood

$$L(\boldsymbol{\tau}; \boldsymbol{y}_*) = (2\pi\sigma^2)^{-N/2}\exp\left[-\sigma^{-2}(\boldsymbol{y}_* - \boldsymbol{X\beta})'(\boldsymbol{y}_* - \boldsymbol{X\beta})/2\right] , \tag{15.25}$$

or equivalently

$$\log L(\boldsymbol{\tau}; \boldsymbol{y}_*) = -N\log(2\pi) - N\log(\sigma^2) - \sigma^{-2}(\boldsymbol{y}_* - \boldsymbol{X\beta})'(\boldsymbol{y}_* - \boldsymbol{X\beta})/2 , \tag{15.26}$$

achieves its supremum. The LRT statistic is

$$\gamma(\boldsymbol{y}) = \frac{\sup\limits_{\boldsymbol{\tau}\in\mathcal{T}_0} L(\boldsymbol{\beta}, \sigma^2; \boldsymbol{y}_*)}{\sup\limits_{\boldsymbol{\tau}\in\mathcal{T}_A} L(\boldsymbol{\beta}, \sigma^2; \boldsymbol{y}_*)} = \frac{L(\widehat{\boldsymbol{\tau}}_0)}{L(\widehat{\boldsymbol{\tau}}_A)}. \tag{15.27}$$

Here $L(\widehat{\boldsymbol{\tau}}_0)$ is the restricted supremum of L over all $\boldsymbol{\tau}$ which satisfy H_0, while $L(\widehat{\boldsymbol{\tau}}_A)$ is the unrestricted supremum of L over all $\boldsymbol{\tau}$ which satisfy the model.

For the ful-rank case, the unrestricted supremum is obtained at $\widehat{\boldsymbol{\tau}}_A$ specified by

$$\widehat{\boldsymbol{\beta}} = (\boldsymbol{X}'\boldsymbol{X})^{-1}\boldsymbol{X}'\boldsymbol{y}_* \tag{15.28}$$

$$\widetilde{\sigma}^2 = \left(\boldsymbol{y}_* - \boldsymbol{X}\widehat{\boldsymbol{\beta}}\right)'\left(\boldsymbol{y}_* - \boldsymbol{X}\widehat{\boldsymbol{\beta}}\right)/N. \tag{15.29}$$

Hence

$$\begin{aligned} L(\widehat{\boldsymbol{\tau}}_A) &= \left(2\pi\widetilde{\sigma}^2\right)^{-N/2}\exp\left[-(\boldsymbol{y}_* - \boldsymbol{X}\widehat{\boldsymbol{\beta}})'(\boldsymbol{y}_* - \boldsymbol{X}\widehat{\boldsymbol{\beta}})/(2\widetilde{\sigma}^2)\right] \\ &= \left(2\pi\widetilde{\sigma}^2\right)^{-N/2}\exp\left(-\frac{N}{2}\frac{\widetilde{\sigma}^2}{\widetilde{\sigma}^2}\right) \\ &= \left(2\pi\widetilde{\sigma}^2 \mathrm{e}\right)^{-N/2}. \end{aligned} \tag{15.30}$$

Part 2. The technique of Lagrangian multipliers allows finding the supremum of $\log L(\boldsymbol{\tau}; \boldsymbol{y}_*)$ subject to the restrictions of $H_0 : \boldsymbol{\theta} = \boldsymbol{0}$. In the following, $\boldsymbol{\lambda}$ is an $a \times 1$ vector of LaGrangian multipliers. The restricted optimization with respect to $\boldsymbol{\tau}$ is achieved by undertaking unrestricted optimization with respect to $[\boldsymbol{\tau}'\ \boldsymbol{\lambda}']'$ for the objective function

$$h(\boldsymbol{\tau}, \boldsymbol{\lambda}; \boldsymbol{y}_*) = \log L(\boldsymbol{\tau}; \boldsymbol{y}_*) - (\boldsymbol{C}\boldsymbol{\beta} - \boldsymbol{\theta}_0)'\boldsymbol{\lambda}. \tag{15.31}$$

Here $h() = \log L()$ for all values of $\boldsymbol{\beta}$ satisfying the restriction. For taking derivatives with respect to $\boldsymbol{\beta}$ it is useful to write

$$\partial h/\partial \boldsymbol{\lambda} = \boldsymbol{C}\boldsymbol{\beta} - \boldsymbol{\theta}_0 \tag{15.32}$$

$$\partial h/\partial \boldsymbol{\beta} = \frac{2}{2\sigma^2}\boldsymbol{X}'\boldsymbol{y}_* - \frac{2}{2\sigma^2}(\boldsymbol{X}'\boldsymbol{X})\boldsymbol{\beta} - \boldsymbol{C}'\boldsymbol{\lambda} \tag{15.33}$$

$$\partial h/\partial \sigma^2 = \frac{-N}{2\sigma^2} + \frac{1}{2\sigma^4}(\boldsymbol{y}_* - \boldsymbol{X}\boldsymbol{\beta})'(\boldsymbol{y}_* - \boldsymbol{X}\boldsymbol{\beta}). \tag{15.34}$$

Setting each derivative to zero and simplifying gives

$$\boldsymbol{C}\overline{\boldsymbol{\beta}} = \boldsymbol{\theta}_0 \tag{15.35}$$

$$\boldsymbol{X}'\boldsymbol{X}\overline{\boldsymbol{\beta}} + \boldsymbol{C}'\boldsymbol{\lambda}\sigma^2 = \boldsymbol{X}'\boldsymbol{y}_* \tag{15.36}$$

$$\overline{\sigma}^2 = (\boldsymbol{y}_* - \boldsymbol{X}\overline{\boldsymbol{\beta}})'(\boldsymbol{y}_* - \boldsymbol{X}\overline{\boldsymbol{\beta}})/N. \tag{15.37}$$

The second equation implies

$$\overline{\boldsymbol{\beta}} = \widehat{\boldsymbol{\beta}} - (\boldsymbol{X}'\boldsymbol{X})^{-1}\boldsymbol{C}'\boldsymbol{\lambda}\sigma^2. \tag{15.38}$$

Combining the last result with the first equation gives

$$\theta_0 = C\overline{\beta} = C\widehat{\beta} - C(X'X)^{-1}C'\lambda\sigma^2 . \tag{15.39}$$

Hence

$$\lambda = [C(X'X)^{-1}C']^{-1}(C\widehat{\beta} - \theta_0)/\sigma^2 . \tag{15.40}$$

The last two equations together give

$$\overline{\beta} = \widehat{\beta} - (X'X)^{-1}C'[C(X'X)^{-1}C']^{-1}(C\widehat{\beta} - \theta_0) . \tag{15.41}$$

Hence $\overline{\beta}$ satisfies the restriction $C\overline{\beta} = \theta_0$ and is in fact the restricted MLE of β.

Part 3. Demonstrating $\overline{\beta}$ is the supremum of the function, as desired, may be approached in various ways. A direct approach would be to begin by applying Theorem 9.15 in Schott (2005), which allows verifying a local maximum was achieved. Complete verification also requires excluding boundary values of the parameter space, the set over which maximization was performed, as possible solutions. Alternately, monotonicity properties of the likelihood could be used. Finally, results for the restricted model could be cast in terms of a linearly equivalent unrestricted model.

Part 4. The last form allows writing the restricted maximum of the likelihood as

$$\begin{aligned} L(\widehat{\tau}_0) &= \left(2\pi\overline{\sigma}^2\right)^{-N/2}\exp\left[-(y_* - X\overline{\beta})'(y_* - X\overline{\beta})/(2\overline{\sigma}^2)\right] \\ &= \left(2\pi\overline{\sigma}^2\right)^{-N/2}\exp\left(\frac{-N}{2}\frac{\overline{\sigma}^2}{\overline{\sigma}^2}\right) \\ &= \left(2\pi\overline{\sigma}^2 e\right)^{-N/2} . \end{aligned} \tag{15.42}$$

In turn, the LRT statistic is

$$\begin{aligned} \gamma(y) = \frac{L(\widehat{\tau}_0)}{L(\widehat{\tau}_A)} &= \frac{[(y_* - X\overline{\beta})'(y_* - X\overline{\beta})]^{-N/2}}{[(y_* - X\widehat{\beta})'(y_* - X\widehat{\beta})]^{-N/2}} \\ &= \left(\overline{\sigma}^2/\widehat{\sigma}^2\right)^{-N/2} . \end{aligned} \tag{15.43}$$

Part 5. The fact that $0 \leq L(\widehat{\tau}_0) \leq L(\widehat{\tau}_A)$ allows concluding $\gamma(y) \in [0,1]$. If $\|C\beta - \theta_0\| = (C\beta - \theta_0)'(C\beta - \theta_0) \approx 0$, then $L(\widehat{\tau}_0) \approx L(\widehat{\tau}_A)$ and $\gamma \approx 1$, while if $\|C\beta - \theta_0\| \gg 0$, then $L(\widehat{\tau}_0) \ll L(\widehat{\tau}_A)$ and $\gamma(y) \ll 1$. Hence reject H_0 for small values of $\gamma(y)$ with the decision function

$$\phi_{\text{LRT}}(y) = \mathbb{B}[\gamma(y) < \gamma_\alpha] . \tag{15.44}$$

We indicate the density of $\gamma(y)$ by $g(\gamma;\tau)$ [which exists due to $\gamma(y)$ being a well-behaved and smooth function of a.c. random variables]. For tests of size α, the appropriate critical value is γ_α as specified by

$$\int_0^{\gamma_\alpha} g(\gamma;\tau|\theta = \theta_0)\, d\gamma = \alpha . \tag{15.45}$$

The critical region is $[0, \gamma_\alpha]$.

Part 6. The complexity of the density, $g(\gamma)$, led statisticians to work with a different statistic. Using the fact

$$\overline{\boldsymbol{\beta}} = \widehat{\boldsymbol{\beta}} - (\boldsymbol{X}'\boldsymbol{X})^{-1}\boldsymbol{C}'[\boldsymbol{C}(\boldsymbol{X}'\boldsymbol{X})^{-1}\boldsymbol{C}']^{-1}(\boldsymbol{C}\widehat{\boldsymbol{\beta}} - \boldsymbol{\theta}_0) \tag{15.46}$$

allows writing

$$\boldsymbol{y}_* - \boldsymbol{X}\overline{\boldsymbol{\beta}} = (\boldsymbol{y}_* - \boldsymbol{X}\widehat{\boldsymbol{\beta}}) + \boldsymbol{X}(\boldsymbol{X}'\boldsymbol{X})^{-1}\boldsymbol{C}'[\boldsymbol{C}(\boldsymbol{X}'\boldsymbol{X})^{-1}\boldsymbol{C}']^{-1}(\widehat{\boldsymbol{\theta}} - \boldsymbol{\theta}_0)\,. \tag{15.47}$$

Hence

$$\begin{aligned}(\boldsymbol{y}_* - \boldsymbol{X}\overline{\boldsymbol{\beta}})'(\boldsymbol{y}_* - \boldsymbol{X}\overline{\boldsymbol{\beta}}) &= (\boldsymbol{y}_* - \boldsymbol{X}\widehat{\boldsymbol{\beta}})'(\boldsymbol{y}_* - \boldsymbol{X}\widehat{\boldsymbol{\beta}}) + \\ &\quad (\widehat{\boldsymbol{\theta}} - \boldsymbol{\theta}_0)'[\boldsymbol{C}(\boldsymbol{X}'\boldsymbol{X})^{-1}\boldsymbol{C}']^{-1}(\widehat{\boldsymbol{\theta}} - \boldsymbol{\theta}_0) \\ &= SSE + SSH,\end{aligned} \tag{15.48}$$

in which $\widehat{\boldsymbol{\theta}} = \boldsymbol{C}\widehat{\boldsymbol{\beta}}$,

$$\begin{aligned}SSE &= (\boldsymbol{y}_* - \boldsymbol{X}\widehat{\boldsymbol{\beta}})'(\boldsymbol{y}_* - \boldsymbol{X}\widehat{\boldsymbol{\beta}}) \\ &= \boldsymbol{y}_*'[\boldsymbol{I} - \boldsymbol{X}(\boldsymbol{X}'\boldsymbol{X})^{-1}\boldsymbol{X}']\boldsymbol{y}_*\,,\end{aligned} \tag{15.49}$$

and

$$SSH = (\widehat{\boldsymbol{\theta}} - \boldsymbol{\theta}_0)'[\boldsymbol{C}(\boldsymbol{X}'\boldsymbol{X})^{-1}\boldsymbol{C}']^{-1}(\widehat{\boldsymbol{\theta}} - \boldsymbol{\theta}_0)\,. \tag{15.50}$$

Thus

$$\begin{aligned}\gamma(\boldsymbol{y}) &= \left[\frac{(\boldsymbol{y}_* - \boldsymbol{X}\overline{\boldsymbol{\beta}})'(\boldsymbol{y}_* - \boldsymbol{X}\overline{\boldsymbol{\beta}})}{(\boldsymbol{y}_* - \boldsymbol{X}\widehat{\boldsymbol{\beta}})'(\boldsymbol{y}_* - \boldsymbol{X}\widehat{\boldsymbol{\beta}})}\right]^{-N/2} \\ &= \left(\frac{SSE + SSH}{SSE}\right)^{-N/2} \\ &= \left(1 + \frac{SSH}{SSE}\right)^{-N/2}.\end{aligned} \tag{15.51}$$

Part 7. Statistical independence of $\widehat{\boldsymbol{\beta}}$ and $(\boldsymbol{y} - \boldsymbol{X}\widehat{\boldsymbol{\beta}})$ guarantees independence of quadratic forms SSE and SSH. Furthermore $SSE/\sigma^2 \sim \chi^2(N-q)$ and $SSH/\sigma^2 \sim \chi^2(a,\omega)$. The properties lead to the transformation

$$\begin{aligned}F &= \frac{SSH/\text{rank}(\boldsymbol{C})}{SSE/(N-q)} \\ &= \frac{N-q}{a}(\gamma^{-2/N} - 1) \\ &\sim F(a, N-q, \omega)\,,\end{aligned} \tag{15.52}$$

with $\omega = (\boldsymbol{\theta} - \boldsymbol{\theta}_0)'[\boldsymbol{C}(\boldsymbol{X}'\boldsymbol{X})^{-1}\boldsymbol{C}']^{-1}(\boldsymbol{\theta} - \boldsymbol{\theta}_0)/\sigma^2$. Testing $H_0 : \boldsymbol{\theta} = \boldsymbol{\theta}_0$ versus $H_A : \boldsymbol{\theta} \neq \boldsymbol{\theta}_0$ is equivalent to testing $H_0 : \omega = 0$ versus $H_A : \omega \neq 0$ because $\omega = 0$ iff $\boldsymbol{\theta} = \boldsymbol{\theta}_0$.

Here F is a monotone decreasing function of γ and $\gamma \in (0, \gamma_m)$ iff $F \in [F_m, \infty)$ with $F_m = (\gamma_m^{-2/N} - 1)(N - q)/a$. Therefore a test based on F can be made equivalent to the test based on $\gamma : \phi_{\text{LRT}}(\boldsymbol{y}) = \mathbb{B}[\gamma(\boldsymbol{y}) < \gamma_\alpha] = \mathbb{B}[F(\boldsymbol{y}) > f_{\text{crit}}]$. Hypothesis $\boldsymbol{\theta} = \boldsymbol{\theta}_0$ or equivalently $\omega = 0$ is rejected for improbably small values of the likelihood ratio, or equivalently, for improbably large values of the F statistic. □

Corollary 15.6.1 For $\text{GLM}_{N,q}\text{LTFR}(\boldsymbol{y}; \boldsymbol{X\beta}, \sigma^2)$ with Gaussian errors and $\text{rank}(\boldsymbol{X}) = r < q$, $\boldsymbol{\theta} = \boldsymbol{C\beta}$ $(a \times 1)$ with known constant $\boldsymbol{C}$ of rank $a \le r$. If $H_0 = \mathbb{B}(\boldsymbol{\theta} = \boldsymbol{\theta}_0)$ *versus* $H_A = \mathbb{B}(\boldsymbol{\theta} \ne \boldsymbol{\theta}_0)$ is a testable hypothesis, then the likelihood ratio test may be implemented exactly by rejecting H_0 for improbably large values of the statistic

$$F = \left\{(\widehat{\boldsymbol{\theta}} - \boldsymbol{\theta}_0)'[\boldsymbol{C}(\boldsymbol{X}'\boldsymbol{X})^{-}\boldsymbol{C}']^{-1}(\widehat{\boldsymbol{\theta}} - \boldsymbol{\theta}_0)/a\right\}/\widehat{\sigma}^2 . \tag{15.53}$$

Here $\widehat{\boldsymbol{\theta}} = \boldsymbol{C}\widetilde{\boldsymbol{\beta}}$, and $\widetilde{\boldsymbol{\beta}}$ is any least squares estimator of $\boldsymbol{\beta}$ based on $(\boldsymbol{X}'\boldsymbol{X})^{-}$, any generalized inverse of $(\boldsymbol{X}'\boldsymbol{X})$, and $\widehat{\sigma}^2 = (\boldsymbol{y} - \boldsymbol{X}\widetilde{\boldsymbol{\beta}})'(\boldsymbol{y} - \boldsymbol{X}\widetilde{\boldsymbol{\beta}})/(N - r)$.

Proof. Left as an exercise. One possible approach centers on finding a linearly equivalent FR model.

Corollary 15.6.2 For $\text{GGLM}_{N,q}\text{LTFR}(\boldsymbol{y}; \boldsymbol{X\beta} | \boldsymbol{R\beta} = \boldsymbol{r}, \sigma^2\boldsymbol{D})$ with Gaussian errors, $\boldsymbol{D} = \boldsymbol{D}'$ $(N \times N)$ is known and positive definite, $\boldsymbol{R\beta} = \boldsymbol{r}$ is a consistent system of r equations, with $\text{rank}(\boldsymbol{R}) = r \le q$. If $\boldsymbol{\theta} = \boldsymbol{C\beta}$ is an estimable parameter and $H_0 = \mathbb{B}(\boldsymbol{\theta} = \boldsymbol{\theta}_0)$ *versus* $H_A = \mathbb{B}(\boldsymbol{\theta} \ne \boldsymbol{\theta}_0)$ is testable, then the LRT consists of rejecting H_0 for improbably large values of the F statistic ______________________________ (to be completed by the reader).

Proof. Left as an exercise.

Corollary 15.6.3 The $\text{GLM}_{N,q}(\boldsymbol{y}; \boldsymbol{X\beta}, \sigma^2)$ with Gaussian errors has estimable scalar secondary parameter $\theta = \boldsymbol{C\beta}$ with $1 \times q$ $\boldsymbol{C} \ne \boldsymbol{0}$ and 1×1 $m = \boldsymbol{C}(\boldsymbol{X}'\boldsymbol{X})^{-}\boldsymbol{C}' \ne 0$. If $t = \widehat{\theta}/(m\widehat{\sigma}^2)^{1/2}$, then the following hold.

(a) The *likelihood ratio test* of size α for testing $H_0 = \mathbb{B}(\theta = \theta_0)$ versus $H_A = \mathbb{B}(\theta \ne \theta_0)$ is $\phi(\boldsymbol{y}) = \mathbb{B}(|t| > t_{\text{crit}})$ in which t_{crit} is the $100(1 - \alpha/2)$ percentile of the $t(N - q)$ distribution.

(b) The *likelihood ratio test* of size α for testing $H = \mathbb{B}(\theta > \theta_0)$ versus $H_A = \mathbb{B}(\theta \le \theta_0)$ is $\phi(\boldsymbol{y}) = \mathbb{B}(t > t_{\text{crit}})$, in which t_{crit} is the $100(1 - \alpha)$ percentile of the $t(N - q)$ distribution.

For multiparameter hypotheses such as $H = \mathbb{B}(\boldsymbol{\theta} = \boldsymbol{\theta}_0)$ we have a "generalized two-tail" alternative hypothesis $H_A = \mathbb{B}(\boldsymbol{\theta} \ne \boldsymbol{\theta}_0)$. For a one-parameter hypothesis $H = \mathbb{B}(\theta = \theta_0)$, either a two-tail or one-tail test may be chosen.

The notation of Boolean algebra (Definition 2.5) includes three operators, namely $\vee$ (OR), $\wedge$ (AND), as well as $\neg$ (NOT). As throughout, $\mathbb{B}() = 1$ results from a "TRUE" argument, while $\mathbb{B}() = 0$ results from a "FALSE" argument.

Definition 15.3 (a) With $\boldsymbol{a}$ an arbitrary real vector, $\boldsymbol{a} \in \Re^a$, and unknown parameter vector $\boldsymbol{\theta}$, the *composite hypothesis* is $H = \mathbb{B}(\boldsymbol{\theta} = \mathbf{0})$, with $H(\boldsymbol{a}) = \mathbb{B}(\boldsymbol{a}'\boldsymbol{\theta} = 0)$ a *component hypothesis*.
(b) Similarly, the *composite alternative* is $H_A = \mathbb{B}(\boldsymbol{\theta} \neq \mathbf{0}) = \neg H$, and $H_A(\boldsymbol{a}) = \mathbb{B}(\boldsymbol{a}'\boldsymbol{\theta} \neq 0) = \neg H(\boldsymbol{a})$ is the *component alternative*.

Lemma 15.3 With the notation of the preceding definition,

$$H = \bigwedge_{\boldsymbol{a} \neq \mathbf{0}} H(\boldsymbol{a}) \tag{15.54}$$

$$H_A = \bigvee_{\boldsymbol{a} \neq \mathbf{0}} H_A(\boldsymbol{a}) \,. \tag{15.55}$$

Proof. Here $\boldsymbol{\theta} = \mathbf{0}$ iff $\boldsymbol{a}'\boldsymbol{\theta} = 0 \; \forall \boldsymbol{a} \in \Re^a$. Consequently

$$\mathbb{B}(\boldsymbol{\theta} = \mathbf{0}) = \bigwedge_{\boldsymbol{a} \neq \mathbf{0}} \mathbb{B}(\boldsymbol{a}'\boldsymbol{\theta} = 0) \,. \tag{15.56}$$

The results follow immediately. □

Definition 15.4 (a) Hypothesis H can be decomposed as an intersection of component hypotheses,

$$H = \bigwedge_{\boldsymbol{a} \neq \mathbf{0}} H(\boldsymbol{a}) \,, \tag{15.57}$$

and $\phi(\boldsymbol{y}; \boldsymbol{a})$ is a test of component hypothesis $H(\boldsymbol{a})$ with *rejection region* (*critical region*) $RR(\boldsymbol{a}) = \{\boldsymbol{y} : \phi(\boldsymbol{y}; \boldsymbol{a}) = 1\}$ and *acceptance region* $AR(\boldsymbol{a}) = \{\boldsymbol{y} : \phi(y; \boldsymbol{a}) = 0\}$.
(b) The *union-intersection test* of hypothesis H is the test specified by the acceptance region

$$RR = \bigcup_{\boldsymbol{a} \neq \mathbf{0}} RR(\boldsymbol{a}) \tag{15.58}$$

$$AR = \bigcap_{\boldsymbol{a} \neq \mathbf{0}} AR(\boldsymbol{a}) \,. \tag{15.59}$$

Theorem 15.7 Union-Intersection Test (Roy, 1957) For $\text{GLM}_{N,q}(y_i; \boldsymbol{X}_i\boldsymbol{\beta}, \sigma^2)$ with Gaussian errors, $\text{rank}(\boldsymbol{X}) = r \leq q$, and $a \times 1$ $\boldsymbol{\theta} = \boldsymbol{C}\boldsymbol{\beta}$ is an a priori testable secondary parameter [$\boldsymbol{\theta}$ is estimable and $\text{rank}(\boldsymbol{C}) = a \leq q$, $\boldsymbol{C}$ and $\boldsymbol{\theta}_0$ are known constants]. For testing $H = \mathbb{B}(\boldsymbol{\theta} = \boldsymbol{\theta}_0)$ versus $H_A = \mathbb{B}(\boldsymbol{\theta} \neq \boldsymbol{\theta}_0)$, the *union-*

intersection test (UIT) of size α is $\phi(\boldsymbol{y}) = \mathbb{B}[F(\boldsymbol{y}) > f_{\text{crit}}]$ in which

$$F(\boldsymbol{y}) = \{(\widehat{\boldsymbol{\theta}} - \boldsymbol{\theta}_0)'[\boldsymbol{C}(\boldsymbol{X}'\boldsymbol{X})^{-}\boldsymbol{C}']^{-1}(\widehat{\boldsymbol{\theta}} - \boldsymbol{\theta}_0)/a\}/\widehat{\sigma}^2 , \tag{15.60}$$

with

$$\widetilde{\boldsymbol{\beta}} = (\boldsymbol{X}'\boldsymbol{X})^{-}\boldsymbol{X}'\boldsymbol{y} \tag{15.61}$$

$$\widehat{\boldsymbol{\theta}} = \boldsymbol{C}\widetilde{\boldsymbol{\beta}} \tag{15.62}$$

$$\widehat{\sigma}^2 = \boldsymbol{y}'[\boldsymbol{I} - \boldsymbol{X}(\boldsymbol{X}'\boldsymbol{X})^{-}\boldsymbol{X}']\boldsymbol{y}/(N - r) , \tag{15.63}$$

and f_{crit} is the $100(1-\alpha)$ percentile of the $F(a, N - q, 0)$ distribution.

Proof. By the union-intersection principle, we define the estimator of the value of $H \equiv \mathbb{B}(\boldsymbol{\theta} = \boldsymbol{\theta}_0)$ as

$$\widehat{H} = \bigwedge_{\boldsymbol{a} \neq \boldsymbol{0}} \widehat{H}(\boldsymbol{a}) . \tag{15.64}$$

An appropriate test procedure for the component sub-hypothesis, $H(\boldsymbol{a}) = \mathbb{B}(\boldsymbol{a}'\boldsymbol{\theta} = \boldsymbol{a}'\boldsymbol{\theta}_0)$, is a two-sided t test, $\phi(\boldsymbol{y}; \boldsymbol{a}) = \mathbb{B}(t^2(\boldsymbol{a}) > t^2_{\text{crit}})$. Here t_{crit} is the $100(1 - \alpha/2)$ percentile of the $t(N - r, 0)$ distribution and $t^2_{\text{crit}} = f_{\text{crit}}$ is the $100(1-\alpha)$ percentile of the $F(1, N - r, 0)$ distribution. In turn,

$$t(\boldsymbol{a}) = \frac{\boldsymbol{a}'\widehat{\boldsymbol{\theta}} - \boldsymbol{a}'\boldsymbol{\theta}_0}{\left(\widehat{\sigma}^2\boldsymbol{a}'\boldsymbol{M}\boldsymbol{a}\right)^{1/2}} , \tag{15.65}$$

with $\widehat{\mathcal{V}}(\widehat{\boldsymbol{\theta}}) = \widehat{\sigma}^2\boldsymbol{M}$ and $\boldsymbol{M} = \boldsymbol{C}(\boldsymbol{X}'\boldsymbol{X})^{-}\boldsymbol{C}'$. Since $\boldsymbol{\theta}$ is estimable, $\boldsymbol{M}$ is nonsingular. The acceptance region for the test is

$$\begin{aligned} AR(\boldsymbol{a}) &= \{\boldsymbol{y} : \phi(\boldsymbol{y}; \boldsymbol{a}) = 0\} \\ &= \left\{\boldsymbol{y} : t^2(\boldsymbol{a}) \le f_{\text{crit}}\right\} , \end{aligned} \tag{15.66}$$

and the decision rule is

$$\widehat{H}(\boldsymbol{a}) = \mathbb{B}\left[t^2(\boldsymbol{a}) \le f_{\text{crit}}\right] . \tag{15.67}$$

The union-intersection decision rule for the composite test is

$$\begin{aligned} \widehat{H} &= \bigwedge_{\boldsymbol{a} \neq \boldsymbol{0}} \widehat{H}(\boldsymbol{a}) \\ &= \bigwedge_{\boldsymbol{a} \neq \boldsymbol{0}} \mathbb{B}\left[t^2(\boldsymbol{a}) \le f_{\text{crit}}\right] \\ &= \mathbb{B}\{\sup_{\boldsymbol{a} \neq \boldsymbol{0}}\left[t^2(\boldsymbol{a})\right] \le f_{\text{crit}}\} . \end{aligned} \tag{15.68}$$

Thus

$$\gamma(\boldsymbol{y}) = \sup_{\boldsymbol{a}\neq\boldsymbol{0}}\left[t^2(\boldsymbol{a})\right] \tag{15.69}$$

could be used as a UIT statistic. However, it is convenient to simplify the form as

$$\begin{aligned}
\gamma(\boldsymbol{y}) &= \sup_{\boldsymbol{a}\neq\boldsymbol{0}}\left[\frac{\left(\boldsymbol{a}'\widehat{\boldsymbol{\theta}} - \boldsymbol{a}'\boldsymbol{\theta}_0\right)^2}{\widehat{\sigma}^2\boldsymbol{a}'\boldsymbol{M}\boldsymbol{a}}\right] \\
&= \sup_{\boldsymbol{a}\neq\boldsymbol{0}}\left[\left(\boldsymbol{a}'\widehat{\boldsymbol{\theta}} - \boldsymbol{a}'\boldsymbol{\theta}_0\right)(\boldsymbol{a}'\boldsymbol{M}\boldsymbol{a})^{-1}\left(\boldsymbol{a}'\widehat{\boldsymbol{\theta}} - \boldsymbol{a}'\boldsymbol{\theta}_0\right)/\widehat{\sigma}^2\right] \\
&= \frac{1}{\widehat{\sigma}^2}\sup_{\boldsymbol{a}\neq\boldsymbol{0}}\left[\left(\boldsymbol{a}'\widehat{\boldsymbol{\theta}} - \boldsymbol{a}'\boldsymbol{\theta}_0\right)(\boldsymbol{a}'\boldsymbol{M}\boldsymbol{a})^{-1}\left(\boldsymbol{a}'\widehat{\boldsymbol{\theta}} - \boldsymbol{a}'\boldsymbol{\theta}_0\right)\right] \\
&= (\widehat{\boldsymbol{\theta}} - \boldsymbol{\theta}_0)'\boldsymbol{M}^{-1}(\widehat{\boldsymbol{\theta}} - \boldsymbol{\theta}_0)/\widehat{\sigma}^2 \\
&= (\boldsymbol{C}\widehat{\boldsymbol{\beta}} - \boldsymbol{\theta}_0)'[\boldsymbol{C}(\boldsymbol{X}'\boldsymbol{X})^{-}\boldsymbol{C}']^{-1}(\boldsymbol{C}\widehat{\boldsymbol{\beta}} - \boldsymbol{\theta}_0)/\widehat{\sigma}^2\,.
\end{aligned} \tag{15.70}$$

The second from the last step uses a matrix theory result about quadratic forms (Schott, 2005, problem 9.42). We know $F(\boldsymbol{y}) = \gamma(\boldsymbol{y})/a \sim F(a, N-r, \omega)$, with

$$\omega = (\boldsymbol{C}\boldsymbol{\beta} - \boldsymbol{\theta}_0)'[\boldsymbol{C}(\boldsymbol{X}'\boldsymbol{X})^{-}\boldsymbol{C}']^{-1}(\boldsymbol{C}\boldsymbol{\beta} - \boldsymbol{\theta}_0)/\sigma^2. \tag{15.71}$$

Thus $F(\boldsymbol{y})$ can be used as the UIT statistic.

In summary, the union-intersection decision rule for the composite test is

$$\begin{aligned}
\widehat{H} &= \bigwedge_{\boldsymbol{a}\neq\boldsymbol{0}}\widehat{H}(\boldsymbol{a}) \\
&= \mathbb{B}\left\{\sup_{\boldsymbol{a}\neq\boldsymbol{0}}\left[t^2(\boldsymbol{a})\right] \le f_{\text{crit}}\right\} \\
&= \mathbb{B}\left[F(\boldsymbol{y}) \le \frac{1}{a}f_{\text{crit}}\right].
\end{aligned} \tag{15.72}$$

However, one problem remains, because the size of the test is too large. In fact,

$$\begin{aligned}
\Pr\{\text{type I error}\} &= \Pr\left\{\widehat{H} = \text{False} \mid H = \text{True}\right\} \\
&= \Pr\{F(\boldsymbol{y}) > a^{-1}f_{\text{crit}}(1, N-r)\} \\
&> \Pr\{F(\boldsymbol{y}) > f_{\text{crit}}(a, N-r)\} = \alpha
\end{aligned} \tag{15.73}$$

because $a^{-1}f_{\text{crit}}(1, N-r) < f_{\text{crit}}(a, N-r)$. Since the individual tests for the component subhypotheses can be of any size, we need only specify that the critical value to be used throughout the proof should be $af_{\text{crit}}(a, N-r)$ rather than $f_{\text{crit}}(1, N-r)$. □

15.4 RELATED DISTRIBUTIONS

With Gaussian errors, all components of test statistics for the univariate linear model are quadratic forms in Gaussian variables that reduce to scaled chi squares. Here we give explicit forms for the distributions.

The formulation of the likelihood ratio test makes it clear that testing a general linear hypothesis *always* corresponds to comparing two models, with the smaller nested inside the larger. The simplicity of univariate theory allows computing tests in terms of results from fitting a single model. Hence the following theorem applies to a wide variety of tests. Muller and Fetterman (2002, Chapter 5) included explicit expressions for sums of squares components of added-in-order tests and added-last tests, among others.

Theorem 15.8 With Gaussian errors and $\text{rank}(\boldsymbol{X}) = r \leq q$, a $\text{GLM}_{N,q}(y_i; \boldsymbol{X}_i\boldsymbol{\beta}, \sigma^2)$ has

$$\begin{aligned} SSE/\sigma^2 &= \widehat{\sigma}^2(N-r)/\sigma^2 \\ &= \boldsymbol{y}'[\boldsymbol{I}_N - \boldsymbol{X}(\boldsymbol{X}'\boldsymbol{X})^-\boldsymbol{X}']\boldsymbol{y}/\sigma^2 \\ &\sim \chi^2(N-r). \end{aligned} \tag{15.74}$$

Estimable $\boldsymbol{\theta}$ and full rank $\boldsymbol{M} = \boldsymbol{C}(\boldsymbol{X}'\boldsymbol{X})^{-1}\boldsymbol{C}'$ ensure

$$\begin{aligned} SSH/\sigma^2 &= (\widehat{\boldsymbol{\theta}} - \boldsymbol{\theta}_0)'\boldsymbol{M}^{-1}(\widehat{\boldsymbol{\theta}} - \boldsymbol{\theta}_0)/\sigma^2 \\ &\sim \chi^2\left[a, (\boldsymbol{\theta} - \boldsymbol{\theta}_0)'\boldsymbol{M}^{-1}(\boldsymbol{\theta} - \boldsymbol{\theta}_0)/\sigma^2\right]. \end{aligned} \tag{15.75}$$

Proof. Corollary 1.15 gives that $[\boldsymbol{I}_N - \boldsymbol{X}(\boldsymbol{X}'\boldsymbol{X})^-\boldsymbol{X}'] = \boldsymbol{L}_0\boldsymbol{L}_0'$ is idempotent of rank $N - r$ with $\boldsymbol{L}_0'\boldsymbol{L}_0 = \boldsymbol{I}_{N-r}$ and $\boldsymbol{L}_0'\boldsymbol{X} = \boldsymbol{0}$. With $\boldsymbol{y} \sim \mathcal{N}_N(\boldsymbol{X\beta}, \sigma^2\boldsymbol{I}_N)$ and $\boldsymbol{\beta}'\boldsymbol{X}'\boldsymbol{L}_0\boldsymbol{L}_0'\boldsymbol{X\beta} = 0$, Theorem 9.4 gives $SSE/\sigma^2 \sim \chi^2(N-r, 0)$. If

$$\boldsymbol{y}_t = \boldsymbol{y} - \boldsymbol{X}\boldsymbol{C}'(\boldsymbol{C}\boldsymbol{C}')^{-1}\boldsymbol{\theta}_0, \tag{15.76}$$

having $\boldsymbol{y} \sim \mathcal{N}_N(\boldsymbol{X\beta}, \boldsymbol{I}_N\sigma^2)$ implies $\boldsymbol{y}_t \sim \mathcal{N}_N[\text{E}(\boldsymbol{y}_t), \boldsymbol{I}_N\sigma^2]$, with

$$\text{E}(\boldsymbol{y}_t) = \boldsymbol{X\beta} - \boldsymbol{X}\boldsymbol{C}'(\boldsymbol{C}\boldsymbol{C}')^{-1}\boldsymbol{\theta}_0. \tag{15.77}$$

If $\boldsymbol{T} = \boldsymbol{C}(\boldsymbol{X}'\boldsymbol{X})^-\boldsymbol{X}'$ then $\widehat{\boldsymbol{\theta}} = \boldsymbol{Ty}$. Testable $\boldsymbol{\theta}$ ensures $\boldsymbol{C}(\boldsymbol{X}'\boldsymbol{X})^-(\boldsymbol{X}'\boldsymbol{X}) = \boldsymbol{C}$. Hence $\boldsymbol{TXC}'(\boldsymbol{CC}')^{-1} = \boldsymbol{I}_a$ and

$$\begin{aligned} \widehat{\boldsymbol{\theta}} - \boldsymbol{\theta}_0 &= \boldsymbol{Ty} - \boldsymbol{T}[\boldsymbol{XC}'(\boldsymbol{CC}')^{-1}]\boldsymbol{\theta}_0 \\ &= \boldsymbol{Ty}_t. \end{aligned} \tag{15.78}$$

In turn

$$\begin{aligned} \widehat{\delta} &= (\widehat{\boldsymbol{\theta}} - \boldsymbol{\theta}_0)'\boldsymbol{M}^{-1}(\widehat{\boldsymbol{\theta}} - \boldsymbol{\theta}_0) \\ &= \boldsymbol{y}_t'\boldsymbol{T}'\boldsymbol{M}^{-1}\boldsymbol{Ty}_t. \end{aligned} \tag{15.79}$$

If $\boldsymbol{A} = \boldsymbol{T}'\boldsymbol{M}^{-1}\boldsymbol{T}$, then $\widehat{\delta} = \boldsymbol{y}_t'\boldsymbol{Ay}_t$ and

$$\begin{aligned} A^2 &= (T'M^{-1}T)(T'M^{-1}T) \\ &= T'M^{-1}C(X'X)^{-}\{X'X[(X'X)^{-}]'C'\}M^{-1}T \\ &= T'M^{-1}C(X'X)^{-}\{C'\}M^{-1}T \\ &= A. \end{aligned} \tag{15.80}$$

Furthermore, with $\theta_{c0} = C'(CC')^{-1}\theta_0$,

$$\begin{aligned} [\mathrm{E}(y_t)]'A\mathrm{E}(y_t) &= [X\beta - X\theta_{c0}]'T'M^{-1}T[X\beta - X\theta_{c0}] \\ &= [TX\beta - TX\theta_{c0}]'M^{-1}[TX\beta - TX\theta_{c0}] \\ &= (\theta - \theta_0)'M^{-1}(\theta - \theta_0). \end{aligned} \tag{15.81}$$

The idempotency of A combines with $y_t \sim \mathcal{N}_N[\mathrm{E}(y_t), I_N\sigma^2]$ and Theorem 9.4 to give $\widehat{\delta}/\sigma^2 \sim \chi^2\left[a, (\theta - \theta_0)'M^{-1}(\theta - \theta_0)/\sigma^2\right]$. □

15.5 TRANSFORMATIONS AND INVARIANCE PROPERTIES

A test which does not vary when the data have been transformed in an unimportant way will usually be preferred to any test without the same invariance property. Any one of a large number of often mutually incompatible invariance properties may seem ideal, depending upon the application. However, (1) the scale of the data (nominal, ordinal, interval, ratio), (2) the mathematical constraints imposed among parameters by the hypothesis (inequalities, orderings, differences, ratios), and (3) the statistical distribution assumed combine to greatly narrow the range of sensible choices.

Our focus centers nearly all of the time on linear models of interval-scale data with Gaussian errors. Furthermore, most tests of interest involve linear relationships among parameters. Consequently, and not surprisingly, statisticians have focused on invariance under linear transformations in such settings. Transformation of the data (or functions of the data, such as parameter estimators), contrast matrices, and parameters all hold some interest.

Definition 15.5 (a) A test ϕ has *scale invariance* iff its value does not vary with changes in the units of the original observations, i.e., $\phi(\{y_{ij}\}) = \phi(\{b_j y_{ij}\})$.
(b) A test ϕ has *location invariance* iff its value its value does not vary with changes in the origin of the original observations, i.e., $\phi(\{y_{ij}\}) = \phi(\{a_j + y_{ij}\})$.
(c) A test ϕ is *linearly invariant*, which is often abbreviated as invariant, iff its value is scale invariant. Many, but not all, such applications also have location invariance.

For random vector x, with A and b conforming constants, statisticians often describe the transformation of $y = Ax + b$ as a linear transformation. However, a precise use of mathematical definitions of transformations would classify $\boldsymbol{y} = \boldsymbol{Ax}$ as a linear transformation and $\boldsymbol{y} = \boldsymbol{Ax} + \boldsymbol{b}$ as a linear transformation plus a translation.

The great majority of linear models used in practice either have $\mathbf{1}_N$ in $\boldsymbol{X}$ or have $\mathbf{1}_N$ spanned by the columns of $\boldsymbol{X}$ (i.e., $\mathbf{1}_N = \boldsymbol{X t}_0$ for some $q \times 1$ constant $\boldsymbol{t}_0$). In either case the columns of $\boldsymbol{X}$ span an intercept (Definition 2.7). At the same time, most hypotheses tested in practice exclude the intercept ($\boldsymbol{C t}_0 = \mathbf{0}$). The parameter deserves the name because it equals the y-axis intercept in plotting the regression function. For the univariate GLM, the usual F test is always invariant to a scale transformation of the response and predictors. In addition, if the model spans an intercept and the hypothesis excludes the intercept, the test has location invariance for the response. All other tests lack location invariance. Section 16.8 contains details and proofs for the more general multivariate case.

Multiplying all y_i by a nonzero constant k has simple consequences. Both $\boldsymbol{\beta}$ and $\boldsymbol{\theta}$ are multiplied by k, while *SSH* and *SSE* are multiplied by k^2. Most importantly, the F statistic, p value, and R^2 values do not change (are invariant).

Theorem 15.9 (a) Both *SSH* and *SSE* are invariant to a square, full-rank transformation of the rows of $\boldsymbol{C}$ and $\boldsymbol{\theta}_0$, with $\boldsymbol{C}_T = \boldsymbol{TC}$ and $\boldsymbol{\theta}_{0T} = \boldsymbol{T\theta}_0$.
(a) The F statistic, p value, and R^2 are invariant to the same transformation.

Proof. *SSE* is not a function of $\boldsymbol{C}$. If $\boldsymbol{T}$ $(a \times a)$ is full rank, with $\boldsymbol{C}_T = \boldsymbol{TC}$,

$$\begin{aligned}
SSH &= (\widehat{\boldsymbol{\theta}} - \boldsymbol{\theta}_0)'\boldsymbol{M}^{-1}(\widehat{\boldsymbol{\theta}} - \boldsymbol{\theta}_0)\\
&= (\boldsymbol{C}\widetilde{\boldsymbol{\beta}} - \boldsymbol{\theta}_0)'[\boldsymbol{C}(\boldsymbol{X}'\boldsymbol{X})^{-}\boldsymbol{C}']^{-1}(\boldsymbol{C}\widetilde{\boldsymbol{\beta}} - \boldsymbol{\theta}_0)\\
&= (\boldsymbol{C}\widetilde{\boldsymbol{\beta}} - \boldsymbol{\theta}_0)'(\boldsymbol{T}^{-1}\boldsymbol{T})'[\boldsymbol{C}(\boldsymbol{X}'\boldsymbol{X})^{-}\boldsymbol{C}']^{-1}\boldsymbol{T}^{-1}\boldsymbol{T}(\boldsymbol{C}\widetilde{\boldsymbol{\beta}} - \boldsymbol{\theta}_0)\\
&= (\boldsymbol{T}'\boldsymbol{C}\widetilde{\boldsymbol{\beta}} - \boldsymbol{T\theta}_0)'[\boldsymbol{TC}(\boldsymbol{X}'\boldsymbol{X})^{-}\boldsymbol{C}'\boldsymbol{T}']^{-1}(\boldsymbol{TC}\widetilde{\boldsymbol{\beta}} - \boldsymbol{T\theta}_0)\\
&= (\boldsymbol{C}_T\widetilde{\boldsymbol{\beta}} - \boldsymbol{\theta}_{0T})'[\boldsymbol{C}_T(\boldsymbol{X}'\boldsymbol{X})^{-}\boldsymbol{C}_T']^{-1}(\boldsymbol{C}_T\widetilde{\boldsymbol{\beta}} - \boldsymbol{\theta}_{0T}), \qquad (15.82)
\end{aligned}$$

which suffices to prove the invariance of *SSH*. □

Theorem 15.10 If the model spans an intercept ($\boldsymbol{X t}_0 = \mathbf{1}_N$ for constant $\boldsymbol{t}_0$) and the hypothesis excludes the intercept ($\boldsymbol{C t}_0 = \mathbf{0}$), the following hold.
(a) Both *SSH* and *SSE* are invariant to a location shift of the form $\boldsymbol{y} + \mathbf{1}_N t_1$ for constant t_1.
(b) The F statistic, p value, and R^2 are similarly invariant.

Proof. The proof of Theorem 16.13 provides the result.

15.6 CONFIDENCE REGIONS FOR θ

Confidence regions (defined in Section 2.10) can be obtained by inverting hypothesis tests, and a confidence region can be inverted to yield a hypothesis test.

We now prove the main results which describe how to create confidence intervals. We also prove that confidence regions exist only for parameters that are testable.

Theorem 15.11 If for any $\boldsymbol{\theta}_0 \in S$ there exists a size-α test, $\phi(\boldsymbol{y})$, of hypothesis $H(\boldsymbol{\theta}_0) = \mathbb{B}(\boldsymbol{\theta} = \boldsymbol{\theta}_0)$, then there exists a corresponding confidence region for $\boldsymbol{\theta}$ with confidence coefficient $c(\alpha) = 1 - \alpha$. Furthermore, if the acceptance region of $\phi(\boldsymbol{y})$ is $AR(\boldsymbol{\theta}_0) = [\boldsymbol{y}_* : \phi(\boldsymbol{y}_*) = 0]$, then $R(\boldsymbol{y}) = [\boldsymbol{\theta}_0 : \boldsymbol{y} \in AR(\boldsymbol{\theta}_0)]$ is the corresponding confidence region.

Proof. Acceptance region $AR(\boldsymbol{\theta}_0)$ is a fixed, constant region within the sample space with boundaries defined by the choice of $\boldsymbol{\theta}_0$ and α. The proof strategy is to define $R(\boldsymbol{y}) = [\boldsymbol{\theta}_0 : \boldsymbol{y} \in AR(\boldsymbol{\theta}_0)]$ and then show that $R(\boldsymbol{y})$ is a confidence region. The first step is to note that $\boldsymbol{\theta}_0 \in R(\boldsymbol{y})$ if and only if $\boldsymbol{y} \in AR(\boldsymbol{\theta}_0)$. It follows that $\Pr\{\boldsymbol{\theta}_0 \in R(\boldsymbol{y})\} = \Pr\{\boldsymbol{y} \in AR(\boldsymbol{\theta}_0)\}$. Evaluating the probability at $\boldsymbol{\theta}_0 = \boldsymbol{\theta}$ reveals $\Pr\{\boldsymbol{\theta} \in R(\boldsymbol{y})\} = \Pr\{\boldsymbol{y} \in \mathrm{AR}(\boldsymbol{\theta}_0) | \boldsymbol{\theta}_0 = \boldsymbol{\theta}\} = 1 - \Pr\{\text{type I error}\} = 1 - \alpha$. □

Theorem 15.12 If an exact $100(1 - \alpha)$ percent confidence region exists for $\boldsymbol{\theta}$, then **(a)** a corresponding test procedure of size α for testing $H(\boldsymbol{\theta}_0) = \mathbb{B}(\boldsymbol{\theta} = \boldsymbol{\theta}_0)$ exists for any $\boldsymbol{\theta}_0 \in S$ and
(b) $H(\boldsymbol{\theta}_0)$ is a *testable* hypothesis.
(c) If the confidence region is $R(\boldsymbol{y})$, then the acceptance region of the corresponding test is $AR(\boldsymbol{\theta}_0) = [\boldsymbol{y}_* : \boldsymbol{\theta}_0 \in R(\boldsymbol{y}_*)]$.

Proof. For $R(y)$ a confidence region for θ with confidence coefficient $1 - \alpha$, $\Pr\{\boldsymbol{\theta} \in R(\boldsymbol{y})\} = 1 - \alpha$. A particular fixed value $\boldsymbol{\theta}_0 \in S$ gives $AR(\boldsymbol{\theta}_0) = [\boldsymbol{y}_* : \boldsymbol{\theta}_0 \in R(\boldsymbol{y}_*)]$, the set of all values in the sample space for which the confidence region would contain $\boldsymbol{\theta}_0$. Having $AR(\boldsymbol{\theta}_0)$ defined allows defining a Boolean function $\phi(\boldsymbol{y})$ and proving that $\phi(\boldsymbol{y})$ is a size-α test. For $RR(\boldsymbol{\theta}_0)$ the complement of $AR(\boldsymbol{\theta}_0)$, *AR* and *RR* define a mutually exclusive and together exhaustive partition of the sample space. The Boolean function $\phi(\boldsymbol{y})$ indicates whether y is in *RR*, with $\phi(\boldsymbol{y}) = 0$ if $y \in AR(\boldsymbol{\theta}_0)$ and $\phi(\boldsymbol{y}) = 1$ if $y \in RR(\boldsymbol{\theta}_0)$. Showing that the test is size α begins by noting $\boldsymbol{y} \in AR(\boldsymbol{\theta}_0)$ if and only if $\boldsymbol{\theta}_0 \in R(\boldsymbol{y})$. Now $\Pr\{\boldsymbol{y} \in AR(\boldsymbol{\theta}_0) | \boldsymbol{\theta}_0 = \boldsymbol{\theta}\} = \Pr\{\boldsymbol{\theta} \in R(\boldsymbol{y})\} = 1 - \alpha$. It follows that $\phi(\boldsymbol{y})$ is a size-α test of hypothesis $H(\boldsymbol{\theta}_0) = \mathbb{B}(\boldsymbol{\theta} = \boldsymbol{\theta}_0)$. Tests do not exist for nontestable hypotheses. Therefore we conclude that $H(\boldsymbol{\theta}_0)$ is testable. □

The last two theorems and proofs reveal the relationship between tests and confidence regions. The confidence region $R(\boldsymbol{y}) = [\boldsymbol{\theta}_0 : \boldsymbol{y} \in AR(\boldsymbol{\theta}_0)]$ partitions the parameter space as a function of the data. On the other hand, the acceptance region $AR(\boldsymbol{\theta}_0) = [\boldsymbol{y}_* : \boldsymbol{\theta}_0 \in R(\boldsymbol{y}_*)]$ partitions the data space as a function of the parameter. For a given value of $\boldsymbol{y}$, confidence region $R(\boldsymbol{y})$ is the set of all choices of $\boldsymbol{\theta}_0 \in S$ such that hypothesis $H(\boldsymbol{\theta}_0)$ is not rejected. The fact that $R(\boldsymbol{y})$ depends on the data makes it stochastic with boundaries that depend on the realization of $\boldsymbol{y}$.

Example 15.3 In the context of a $\text{GLM}_{N,q}(y_i; \boldsymbol{X}_i\boldsymbol{\beta}, \sigma^2)$ with Gaussian errors, inverting a test to obtain a confidence region is easy to illustrate when θ $(a \times 1)$ is a scalar $(a = 1)$. The two-sided t test, $\phi(\boldsymbol{y}) = \mathbb{B}(t^2 > f_{\text{crit}})$, of $H(\boldsymbol{\theta}_0) = \mathbb{B}(\boldsymbol{\theta} = \boldsymbol{\theta}_0)$ has scalar $\boldsymbol{m} = \boldsymbol{C}(\boldsymbol{X}'\boldsymbol{X})^{-}\boldsymbol{C}'$ (because $a = 1$) and acceptance region

$$\begin{aligned} AR(\boldsymbol{\theta}_0) &= \left\{\boldsymbol{y} : t^2(y) = (\widehat{\boldsymbol{\theta}} - \boldsymbol{\theta}_0)^2/(\widehat{\sigma}^2\boldsymbol{m}) \le f_{\text{crit}}\right\} \\ &= \left\{\boldsymbol{y} : \widehat{\boldsymbol{\theta}} - (f_{\text{crit}}\widehat{\sigma}^2 m)^{1/2} \le \boldsymbol{\theta}_0 \le \widehat{\boldsymbol{\theta}} + (f_{\text{crit}}\widehat{\sigma}^2 m)^{1/2}\right\} \\ &= \left\{\boldsymbol{y} : \boldsymbol{\theta}_0 \in \left[\widehat{\boldsymbol{\theta}} \pm (f_{\text{crit}}\widehat{\sigma}^2 m)^{1/2}\right]\right\}. \end{aligned} \tag{15.83}$$

It follows that the confidence region is interval

$$R(\boldsymbol{y}) = [\widehat{\boldsymbol{\theta}} \pm (f_{\text{crit}}\widehat{\sigma}^2 m)^{1/2}]. \tag{15.84}$$

Here, f_{crit} is the appropriate critical value defined as the $100(1 - \alpha)$ percentile of the $F(1, N - r, 0)$ distribution and $\widehat{\mathcal{V}}(\widehat{\boldsymbol{\theta}}) = \widehat{\sigma}^2\boldsymbol{m}$. By Theorem 15.11, the confidence coefficient for $R(\boldsymbol{y})$ is $1 - \alpha$.

Example 15.4 In the context of a $\text{GLM}_{N,q}(y_i; \boldsymbol{X}_i\boldsymbol{\beta}, \sigma^2)$ with Gaussian errors, more generality arises by inverting an F test to obtain a confidence region for θ $(a \times 1)$. The test $\phi(\boldsymbol{y}) = \mathbb{B}(F > f_{\text{crit}})$ of $H(\boldsymbol{\theta}_0) = \mathbb{B}(\boldsymbol{\theta} = \boldsymbol{\theta}_0)$ has acceptance region

$$AR(\boldsymbol{\theta}_0) = \left[\boldsymbol{y} : F(\boldsymbol{y}) = (\widehat{\boldsymbol{\theta}} - \boldsymbol{\theta}_0)'m^{-1}(\widehat{\boldsymbol{\theta}} - \boldsymbol{\theta}_0)/(a\widehat{\sigma}^2) \le f_{\text{crit}}\right]. \tag{15.85}$$

It follows that the $100(1 - \alpha)$ percent ellipsoidal confidence region is

$$R(\boldsymbol{y}) = [\boldsymbol{\theta}_0 : (\widehat{\boldsymbol{\theta}} - \boldsymbol{\theta}_0)'m^{-1}(\widehat{\boldsymbol{\theta}} - \boldsymbol{\theta}_0) \le (a\widehat{\sigma}^2)f_{\text{crit}}]. \tag{15.86}$$

The appropriate critical value f_{crit} is defined in Theorem 15.7.

In practice, high dimensional $(a > 2)$ hyperellipsoidal confidence regions are of limited utility. To facilitate graphical representations, a practical approach is to create a set of simultaneous confidence intervals for the elements of $\boldsymbol{\theta}$. Inverting a Bonferroni test, a Tukey test, a Scheffé test, or any of the other multiple-comparison test procedures gives the desired confidence region. The intervals for the individual elements of $\boldsymbol{\theta}$ collectively define a high-dimensional rectangular confidence region for $\boldsymbol{\theta}$ which has confidence coefficient no more than $1-\alpha$.

Example 15.5 A $\text{GLM}_{N,q}(y_i; \boldsymbol{X}_i\boldsymbol{\beta}, \sigma^2)$ with Gaussian errors allows many approaches to formulating simultaneous confidence intervals for the k elements of set $\boldsymbol{S_a} = [\boldsymbol{a}'\boldsymbol{\theta} : \boldsymbol{a} \in \{\boldsymbol{a_1}, \boldsymbol{a_2}, ..., \boldsymbol{a_k}\}]$, with k finite. For each of the k choices of a, the Bonferroni test of $H(\boldsymbol{a}, \boldsymbol{\theta}_0) = \mathbb{B}(\boldsymbol{a}'\boldsymbol{\theta} = \boldsymbol{a}'\boldsymbol{\theta}_0)$ has acceptance region

$$AR(\boldsymbol{a}, \boldsymbol{\theta}_0) = \left[\boldsymbol{y} : F(\boldsymbol{y}) = [\boldsymbol{a}'(\widehat{\boldsymbol{\theta}} - \boldsymbol{\theta}_0)]^2/(\widehat{\sigma}^2\boldsymbol{a}'\boldsymbol{m}\boldsymbol{a}) \le f_{\text{crit}}\right]. \tag{15.87}$$

The critical value f_{crit} depends on k and is the $100[1-(\alpha/k)]$ percentile of $F(1, N-r, 0)$. The corresponding Bonferroni confidence interval for $\boldsymbol{a}'\boldsymbol{\theta}$,

$$R(\boldsymbol{a}, \boldsymbol{y}) = \left[\boldsymbol{a}'\widehat{\boldsymbol{\theta}} \pm (f_{\text{crit}}\, \widehat{\sigma}^2 \boldsymbol{a}'\boldsymbol{M}\boldsymbol{a})^{1/2}\right], \tag{15.88}$$

has a confidence coefficient no greater than $1-\alpha$. Similarly, the Union-Intersection test (also known as the Scheffé test) can be inverted to obtain k simultaneous confidence intervals. The Scheffé test of $H(\boldsymbol{a}, \boldsymbol{\theta}_0) = \mathbb{B}(\boldsymbol{a}'\boldsymbol{\theta} = \boldsymbol{a}'\boldsymbol{\theta}_0)$ has acceptance region

$$AR(\boldsymbol{a}, \boldsymbol{\theta}_0) = \left[\boldsymbol{y} : F(\boldsymbol{y}) = \left[\boldsymbol{a}'(\widehat{\boldsymbol{\theta}} - \boldsymbol{\theta}_0)\right]^2 / (a\widehat{\sigma}^2 \boldsymbol{a}'\boldsymbol{M}\boldsymbol{a}) \leq f_{a,\text{crit}}\right]. \tag{15.89}$$

Here $f_{a,\text{crit}}$ is the $100(1-\alpha)$ percentile of $F(a, N-r, 0)$, which depends only on a and not on k. The Scheffé confidence interval for $\boldsymbol{a}'\boldsymbol{\theta}$,

$$R(\boldsymbol{a}, \boldsymbol{y}) = \left[\boldsymbol{a}'\widehat{\boldsymbol{\theta}} \pm (a f_{a,\text{crit}}\, \widehat{\sigma}^2 \boldsymbol{a}'m\boldsymbol{a})^{1/2}\right], \tag{15.90}$$

has a confidence coefficient no greater than $1-\alpha$. Unless k is roughly on the order of 2^a, the Bonferroni intervals will be narrower than the Scheffé intervals. Comparing $a \cdot f_{a,\text{crit}}$ to f_{crit} illustrates the claim. When set $S_{\boldsymbol{a}}$ comprises k points on a regression curve or surface, then the set of confidence intervals define a confidence band. Depending on the kind of confidence band desired, k can be very small or extremely large. Stewart (1987, 1991) discussed confidence band procedures and compared various methods with respect to their graphical advantages and disadvantages.

EXERCISES

15.1 Prove the following.

Lemma. If $\boldsymbol{X} = \mathbf{1}_n \otimes \boldsymbol{X}_{\text{Es}}$, then $(\boldsymbol{X}'\boldsymbol{X})^- = n^{-1}(\boldsymbol{X}_{\text{Es}}'\boldsymbol{X}_{\text{Es}})^-$.

15.2 Model 1 is

$$\begin{aligned}\boldsymbol{y} &= \left(\mathbf{1}_n \otimes \begin{bmatrix} 1 & 1 & 0 & 0 \\ 1 & 0 & 1 & 0 \\ 1 & 0 & 0 & 1 \end{bmatrix}\right)\boldsymbol{\beta}_1 + \boldsymbol{e} \\ &= \boldsymbol{X}_1\boldsymbol{\beta}_1 + \boldsymbol{e}.\end{aligned}$$

Finding a valid and unambiguous test of the general linear hypothesis $H_0 : \boldsymbol{C\beta} = \boldsymbol{\theta}_0$ requires verifying that (a) $\boldsymbol{\theta} = \boldsymbol{C\beta}$ is estimable and (b) the hypothesis is testable. In the following, with $\boldsymbol{\theta}_j = \boldsymbol{C}_j\boldsymbol{\beta}$, verify (a) holds or does not hold and verify (b) holds or does not hold. It is not sufficient to merely state the correct answer. You must briefly justify each positive or negative answer. For each estimable $\boldsymbol{\theta}$, describe each element very briefly as a function of cell means.

15.2.1 $\boldsymbol{C}_1 = [1 \;\; 0 \;\; 0 \;\; 0]$

15.2.2 $\boldsymbol{C}_2 = [0 \;\; 1 \;\; 0 \;\; 0]$

15.2.3 $\boldsymbol{C}_3 = [1\ 1\ 0\ 0]$
15.2.4 $\boldsymbol{C}_4 = [3\ 1\ 1\ 1]/3$
15.2.5 $\boldsymbol{C}_5 = \begin{bmatrix} 0 & 1 & -1 & 0 \\ 0 & 1 & 0 & -1 \\ 0 & 0 & 1 & -1 \end{bmatrix}$
15.2.6 $\boldsymbol{C}_6 = \begin{bmatrix} 0 & 1 & 0 & -1 \\ 0 & 0 & 1 & -1 \end{bmatrix}$

15.3 A complete, unbalanced, G group one-way ANOVA may be written as

$$\boldsymbol{y} = \begin{bmatrix} \mathbf{1}_{N_1} & \mathbf{0} & \mathbf{0} \\ \mathbf{0} & \ddots & \mathbf{0} \\ \mathbf{0} & \mathbf{0} & \mathbf{1}_{N_G} \end{bmatrix} \begin{bmatrix} \mu_1 \\ \vdots \\ \mu_G \end{bmatrix} + \boldsymbol{e}.$$

Assuming i.i.d. $e_i \sim \mathcal{N}(0, \sigma^2)$, a testable hypothesis of the form $H_0 : \boldsymbol{\theta} = \boldsymbol{\theta}_0$ with $\boldsymbol{\theta} = \boldsymbol{C}\boldsymbol{\beta}$ may be expressed in terms of a possibly noncentral F statistic. The total sample size is $N_+ = \sum_{g=1}^{G} N_g$.

15.3.1 Briefly explain what "complete" requires of a particular sample for the design being considered.

15.3.2 Briefly explain what "unbalanced" requires of a particular sample for the design being considered.

15.3.3 Consider $\boldsymbol{C} = [1\ -1\ 0 \cdots 0]$, with $\boldsymbol{\theta}_0 = \mathbf{0}$, and the corresponding F test in this particular model.

(a) What are the degrees of freedom for the test in terms of $\{N_g, G\}$?

(b) Given the complete unbalanced design, what constraints on $\{N_g, G\}$ must be present for the test to be well defined?

(c) Find a convenient form for the noncentrality parameter in terms of $\{\mu_g, N_g, \sigma^2\}$.

15.3.4 In this part, consider the special case of a completely balanced design.

(a) Simplify the expression for the design matrix given the special case.

(b) If $G = 3$ and $\boldsymbol{C} = [\,\mathbf{1}_2\ -\boldsymbol{I}_2\,]$, with $\boldsymbol{\theta}_0 = \mathbf{0}$, find a convenient form for the noncentrality parameter in terms of $\{\mu_g, N_g, \sigma^2\}$ (with a completely balanced design).

CHAPTER 16

Tests for Multivariate Linear Models

16.1 MOTIVATION

Throughout the chapter, we always assume Gaussian errors in considering a multivariate linear model, a $\text{GLM}_{N,p,q}(\boldsymbol{Y}_i; \boldsymbol{X}_i\boldsymbol{B}, \Sigma)$ with $\text{rank}(\boldsymbol{X}) = r \leq q$. With fixed, known constants $\{\boldsymbol{C}, \boldsymbol{U}, \boldsymbol{\Theta}_0\}$, a secondary parameter is $a \times b$ $\boldsymbol{\Theta} = \boldsymbol{CBU}$. Most of the time, but not always, we will require $\text{rank}(\boldsymbol{C}) = a \leq q$ and $\text{rank}(\boldsymbol{U}) = b \leq p$. The associated general linear hypothesis (GLH) may be stated as

$$H_0 : \boldsymbol{\Theta} = \boldsymbol{\Theta}_0 , \tag{16.1}$$

with corresponding alternative $H_A : \boldsymbol{\Theta} \neq \boldsymbol{\Theta}_0$. In terms of Boolean variables, with $\mathbb{B}() \in \{0, 1\}$, the hypothesis may be written as $H_0 = \mathbb{B}(\boldsymbol{\Theta} = \boldsymbol{\Theta}_0)$ *versus* $H_A = \mathbb{B}(\boldsymbol{\Theta} \neq \boldsymbol{\Theta}_0)$. Results for the univariate GLM will always be included as the special case of $p = 1$ column in $\boldsymbol{Y}$ and $\boldsymbol{B}$, with $\boldsymbol{U} = [\,1\,]$. The first issue addressed is testability. The above hypothesis is *testable* iff the parameter $\boldsymbol{\Theta}$ is estimable and the contrast matrices are full rank. We do not have hypothesis testing procedures for nontestable hypotheses.

Another important issue is whether $\boldsymbol{\Theta}$ is truly a fixed constant. If the model and its parameters are defined prior to collection of the data, then $\boldsymbol{\Theta}$ is an unknown, fixed constant. However, one frequently discovers interesting aspects of the data which are directly suggested by the results at hand but not considered prior to data collection. In such cases $\boldsymbol{\Theta}$ may depend in some way upon the observed $\boldsymbol{Y}$. As in the univariate model (Definition 15.1), we shall refer to parameters defined independently of $\boldsymbol{Y}$ as a priori parameters. Others, including ones obviously suggested by the data, will be called post hoc parameters. As indicated by the status of the parameter, the corresponding hypothesis is either an a priori hypothesis or a post hoc hypothesis. Different statistical tests are required for the two different kinds of hypotheses.

16.2 TESTABILITY OF MULTIVARIATE HYPOTHESES

In the theoretical formulation of any hypothesis test procedure, the hypothesis is always manipulated in terms of its linearly independent (LIN) components. Therefore one requires the rows of $a \times q$ $\boldsymbol{C}$ to be LIN and the columns of $p \times b$ $\boldsymbol{U}$ to be LIN [rank$(\boldsymbol{C}) = a \leq q$ and rank$(\boldsymbol{U}) = b \leq p$]. For a univariate model $(p = 1)$, $\boldsymbol{U}$ is a scalar, $\boldsymbol{U} = 1$. In the following, we first address the issue of testability for such full-rank $\boldsymbol{C}$ and $\boldsymbol{U}$ matrices. Later we consider more general $\boldsymbol{C}$ and $\boldsymbol{U}$. The formal definition of a testable hypothesis implies questions of testability arise only in LTFR models.

Definition 16.1 Under the assumptions of $\text{GLM}_{N,p,q}(\boldsymbol{Y}_i; \boldsymbol{X}_i\boldsymbol{B}, \boldsymbol{\Sigma})$ with Gaussian errors, rank$(\boldsymbol{X}) = r \leq q$, and $a \times b$ $\boldsymbol{\Theta} = \boldsymbol{CBU}$, the hypothesis $H_0 = \mathbb{B}(\boldsymbol{\Theta} = \boldsymbol{\Theta}_0)$ *versus* $H_A = \mathbb{B}(\boldsymbol{\Theta} \neq \boldsymbol{\Theta}_0)$ is *testable* iff (1) rank$(\boldsymbol{C}) = a \leq q$, (2) rank$(\boldsymbol{U}) = b \leq p$, and (3) $\boldsymbol{\Theta}$ is estimable.

In FR models, $\boldsymbol{B}$ and $\boldsymbol{\Theta}$ are always estimable and corresponding tests are always testable. In LTFR models $H_0 = \mathbb{B}(\boldsymbol{B} = \boldsymbol{0})$ *versus* $H_A = \mathbb{B}(\boldsymbol{B} \neq \boldsymbol{0})$ is never testable, while $\boldsymbol{\Theta}$ may or may not be estimable. As in the univariate case, estimability and testability require only the least squares assumptions (Homogeneity, Independence, Linearity, Existence of finite second moments) and do not require a particular distribution.

To motivate the relationship between estimable parameters and testable hypotheses, we highlight the following argument. We seek test statistics which are functions of $\widehat{\boldsymbol{\Theta}} = \boldsymbol{C}\widetilde{\boldsymbol{B}}\boldsymbol{U}$, with $\widetilde{\boldsymbol{B}} = \widehat{\boldsymbol{B}}$ in the FR case and $\boldsymbol{U} = 1$ in univariate models. The quantity $\widehat{\boldsymbol{\Theta}}$ must be invariant to the choice of $\widetilde{\boldsymbol{B}}$ for the test to be invariant. The test result should not depend on the choice of generalized inverse! We have previously proven $\widehat{\boldsymbol{\Theta}}$ is invariant iff $\boldsymbol{\Theta}$ is estimable. A collection of theorems in Chapters 11 and 12 give estimability criteria. In particular, $\boldsymbol{C}(\boldsymbol{X}'\boldsymbol{X})^{-}(\boldsymbol{X}'\boldsymbol{X}) = \boldsymbol{C}$ if and only if $\boldsymbol{CBU}$ is estimable.

Lemma 16.1 A $\text{GLM}_{N,p,q}(\boldsymbol{Y}_i; \boldsymbol{X}_i\boldsymbol{B}, \boldsymbol{\Sigma})$ leads to considering constants $a \times a$ $\boldsymbol{M} = \boldsymbol{C}(\boldsymbol{X}'\boldsymbol{X})^{-}\boldsymbol{C}'$ and $p \times b$ $\boldsymbol{U}$.

(a) If $\boldsymbol{X}$ is full rank, then rank$(\boldsymbol{M}) = a$ and rank$(\boldsymbol{U}) = b$ provide necessary and sufficient conditions for testability of $a \times b$ $\boldsymbol{CBU}$.

(b) If $\boldsymbol{X}$ is less than full rank, then the conditions rank$(\boldsymbol{M}) = a$ and rank$(\boldsymbol{U}) = b$ provide necessary but not sufficient conditions for testability.

(c) If $\boldsymbol{X}$ is less than full rank, then the conditions rank$(\boldsymbol{M}) = a$ and rank$(\boldsymbol{U}) = b$ combined with the requirement $\boldsymbol{C} = \boldsymbol{C}(\boldsymbol{X}'\boldsymbol{X})^{-}(\boldsymbol{X}'\boldsymbol{X})$, or any other condition which guarantees estimability, provide a necessary and sufficient set of conditions for testability.

Proof. Left as an exercise.

Certain related conditions are worth remembering. With full rank $\boldsymbol{X}$, the rank of $\boldsymbol{M}$ is the rank of $\boldsymbol{C}$. No matter the rank of $\boldsymbol{X}$, with $\boldsymbol{C}$ $a \times q$, both rank($\boldsymbol{C}$) and rank($\boldsymbol{M}$) are necessarily bounded above by $\min(a, q)$.

Theorem 16.1 A nontestable hypothesis has no reasonable test. For $\text{GLM}_{N,p,q}(\boldsymbol{Y}_i; \boldsymbol{X}_i\boldsymbol{B}, \Sigma)$ with Gaussian errors and $\text{rank}(\boldsymbol{X}) = r \leq q$, nonestimable $\Theta = \boldsymbol{CBU}$ $(a \times b)$ may have $\text{rank}(\boldsymbol{C}) = a \leq q$ and $\text{rank}(\boldsymbol{U}) = b \leq p$. Even with $\text{rank}(\boldsymbol{C}) = a$ and $\text{rank}(\boldsymbol{U}) = b$ the hypothesis $H_0 = \mathbb{B}(\Theta = \Theta_0)$ *versus* $H_A = \mathbb{B}(\Theta \neq \Theta_0)$ is not testable. No reasonable test can be found; i.e., any test procedure will have size $\alpha = 0$ and power $1 - \beta = 0$. Furthermore the test will not be invariant to changes in irrelevant quantities.

Proof. The univariate model is a special case of the multivariate model. Hence the proof of the nonexistence of a valid test in the univariate case demonstrates a valid test can not be guaranteed in the multivariate case. □

Corollary 16.1 If $\Theta_s = [\,\Theta_1' \; \Theta_2'\,]'$ and Θ_1 is estimable but Θ_2 is not, then tests for Θ_s are indistinguishable from tests for Θ_1.

Proof. The previous chapter has a proof for the univariate case. The $\boldsymbol{U}$ matrix plays no role in estimability, although it does affect testability. □

The multivariate test statistics described in Chapter 3 are all functions of

$$\widehat{\Delta} = (\boldsymbol{C}\widetilde{\boldsymbol{B}}\boldsymbol{U} - \Theta_0)'[\boldsymbol{C}(\boldsymbol{X}'\boldsymbol{X})^{-}\boldsymbol{C}']^{-1}(\boldsymbol{C}\widetilde{\boldsymbol{B}}\boldsymbol{U} - \Theta_0)\,. \tag{16.2}$$

The $b \times b$ matrix generalizes $SSH = \widehat{\delta}$ from the univariate setting, which leads to the alternate notation $\boldsymbol{S}_h = \widehat{\Delta}$. In the FR case $\widetilde{\boldsymbol{B}} = \widehat{\boldsymbol{B}}$ and $\boldsymbol{X}'\boldsymbol{X}$ is nonsingular. The test statistics depend on $\boldsymbol{C}$ and Θ_0 only through $\widehat{\Delta}$. In a univariate model ($p = 1$ or $b = 1$), $\widehat{\Delta}$ is a scalar. The ability to compute $\widehat{\Delta}$ depends on the existence of $[\boldsymbol{C}(\boldsymbol{X}'\boldsymbol{X})^{-}\boldsymbol{C}']^{-1}$. The value of $\widehat{\Delta}$ can be computed iff $\boldsymbol{M} = \boldsymbol{C}(\boldsymbol{X}'\boldsymbol{X})^{-}\boldsymbol{C}' \neq \boldsymbol{0}$ has full rank. Testable Θ guarantees full-rank $\boldsymbol{M}$.

In many cases it is possible to compute $\widehat{\Delta}$ (and therefore the test statistic) even though Θ is not estimable and the hypothesis is not testable. The result is formally stated in the following theorem. Subsequently, we prove a theorem which answers the question "What hypothesis is the test statistic addressing?" in such cases.

Theorem 16.2 The $\text{GLM}_{N,p,q}(\boldsymbol{Y}_i; \boldsymbol{X}_i\boldsymbol{B}, \Sigma)$ with Gaussian errors has $\Theta = \boldsymbol{CBU}$ $a \times b$, $\text{rank}(\boldsymbol{X}) = r \leq q$, *and* $\text{rank}(\boldsymbol{C}) = a \leq q$, *and* $\text{rank}(\boldsymbol{U}) = b \leq p$.
(a) Θ is estimable implies $\boldsymbol{M} = \boldsymbol{C}(\boldsymbol{X}'\boldsymbol{X})^{-}\boldsymbol{C}'$ has full rank, although
(b) $\boldsymbol{M} = \boldsymbol{C}(\boldsymbol{X}'\boldsymbol{X})^{-}\boldsymbol{C}'$ having full rank does not imply Θ is estimable.

The preceding chapter includes a proof in the univariate case. As always, $\boldsymbol{U}$ plays no role in determining estimability.

Theorem 16.3 For a $\text{GLM}_{N,p,q}(\boldsymbol{Y}_i;\boldsymbol{X}_i\boldsymbol{B},\boldsymbol{\Sigma})$ with Gaussian errors, $\text{rank}(\boldsymbol{X}) = r < q$, $\text{rank}(\boldsymbol{C}) = a \le q$, and $\text{rank}(\boldsymbol{U}) = b \le p$, $\boldsymbol{\Theta} = \boldsymbol{CBU}$ $(a \times b)$ may not be estimable even though $\boldsymbol{M}^{-1} = [\boldsymbol{C}(\boldsymbol{X}'\boldsymbol{X})^{-}\boldsymbol{C}']^{-1}$ exists. If so, test statistics based on

$$\widehat{\boldsymbol{\Delta}} = (\boldsymbol{C}\tilde{\boldsymbol{B}}\boldsymbol{U} - \boldsymbol{\Theta}_0)'\boldsymbol{M}^{-1}(\boldsymbol{C}\hat{\boldsymbol{B}}\boldsymbol{U} - \boldsymbol{\Theta}_0) \tag{16.3}$$

can be computed and the testable hypothesis actually tested is $H_0 = \mathbb{B}(\boldsymbol{\Theta}_G = \boldsymbol{0})$ *versus* $H_A = \mathbb{B}(\boldsymbol{\Theta}_G \ne \boldsymbol{0})$, with $\boldsymbol{\Theta}_G = \boldsymbol{C}(\boldsymbol{X}'\boldsymbol{X})^{-}(\boldsymbol{X}'\boldsymbol{X})\boldsymbol{BU} - \boldsymbol{\Theta}_0$.

Proof. Searle (1971, p. 195) provided a proof in the univariate case.

Without estimability, $\boldsymbol{\Theta}_G$ is not necessarily invariant to the choice of $(\boldsymbol{X}'\boldsymbol{X})^{-}$. For each choice of generalized inverse a potentially different testable hypothesis is tested! An example of such a poorly defined test is given in the previous chapter.

We now consider more general definitions of $a \times q$ $\boldsymbol{C}$ and $p \times b$ $\boldsymbol{U}$. Previously we have required $\text{rank}(\boldsymbol{C}) = a \le q$ and $\text{rank}(\boldsymbol{U}) = b \le p$. Can the requirement be relaxed? We are occasionally interested in matrices $\boldsymbol{C}$ and $\boldsymbol{U}$ which are not of full rank. In any case we will need $\boldsymbol{C} = \boldsymbol{AX}$ (for some $\boldsymbol{A}$) to have any hope of formulating a testable hypothesis. Otherwise $\boldsymbol{\Theta}$ will not be estimable.

Estimability of $\boldsymbol{\Theta}$ does not require $\boldsymbol{C}$ to have full rank. If $a \times b$ $\boldsymbol{C}_1\boldsymbol{BU}_1 = \boldsymbol{\Theta}_1$ is estimable, then so is the $2a \times 2b$ parameter

$$\begin{bmatrix}\boldsymbol{C}_1\\ \boldsymbol{C}_1\end{bmatrix}\boldsymbol{B}[\boldsymbol{U}_1\ \boldsymbol{U}_1] = \begin{bmatrix}\boldsymbol{C}_1\boldsymbol{B}\\ \boldsymbol{C}_1\boldsymbol{B}\end{bmatrix}[\boldsymbol{U}_1\ \boldsymbol{U}_1] \tag{16.4}$$

$$\boldsymbol{C}_2\boldsymbol{BU}_2 = \begin{bmatrix}\boldsymbol{\Theta}_1 & \boldsymbol{\Theta}_1\\ \boldsymbol{\Theta}_1 & \boldsymbol{\Theta}_1\end{bmatrix} = \boldsymbol{\Theta}_2\,. \tag{16.5}$$

However, $\boldsymbol{\Theta}_2$ is not testable because $\text{rank}(\boldsymbol{C}_2) = \text{rank}(\boldsymbol{C}_1) = a < 2a$ and $\text{rank}(\boldsymbol{U}_2) = \text{rank}(\boldsymbol{U}_1) = b < 2b$. Also $H_{0,2} = \mathbb{B}(\boldsymbol{\Theta}_2 = \boldsymbol{0}) \Leftrightarrow \mathbb{B}[(\boldsymbol{\Theta}_1 = \boldsymbol{0}) \cap (\boldsymbol{\Theta}_1 = \boldsymbol{0}) \cap (\boldsymbol{\Theta}_1 = \boldsymbol{0}) \cap (\boldsymbol{\Theta}_1 = \boldsymbol{0})]$ is overstated and identical to $H_{0,1} = \mathbb{B}(\boldsymbol{\Theta}_1 = \boldsymbol{0})$. We should test subhypothesis $H_{0,1}$ rather than $H_{0,2}$.

Theorem 16.4 A test with a LTFR $\boldsymbol{C}$ matrix cannot be distinguished from a test with $\boldsymbol{C}$ replaced by a full (row) rank matrix with rows that span all rows of $\boldsymbol{C}$. More specifically, $\text{GLM}_{N,p,q}(\boldsymbol{Y}_i;\boldsymbol{X}_i\boldsymbol{B},\boldsymbol{\Sigma})$ has Gaussian errors, $\text{rank}(\boldsymbol{X}) = r \le q$, estimable $\boldsymbol{\Theta} = \boldsymbol{CBU}$ $(a \times b)$, $\text{rank}(\boldsymbol{C}) = a_1 < \min(a, q)$, $\text{rank}(\boldsymbol{U}) = b \le p$,

$$\begin{aligned}\boldsymbol{\Theta} &= \boldsymbol{CBU}\\ &= \begin{bmatrix}\boldsymbol{C}_1\\ \boldsymbol{C}_2\end{bmatrix}\boldsymbol{BU} = \begin{bmatrix}\boldsymbol{\Theta}_1\\ \boldsymbol{\Theta}_2\end{bmatrix},\end{aligned} \tag{16.6}$$

and $\boldsymbol{\Theta}_0 = [\boldsymbol{\Theta}_{0,1}'\ \boldsymbol{\Theta}_{0,2}']'$. For $j \in \{1, 2\}$, $\boldsymbol{C}_j$, $\boldsymbol{\Theta}_j$, and $\boldsymbol{\Theta}_{0,j}$ have a_j rows and $\text{rank}(\boldsymbol{C}) = \text{rank}(\boldsymbol{C}_1) = a_1 < \min\{a, q\}$.

(a) Tests about $\boldsymbol{\Theta}$ cannot be distinguished from tests about $\boldsymbol{\Theta}_1$; i.e., $\boldsymbol{\Theta} = \boldsymbol{\Theta}_0$ iff

$\Theta_1 = \Theta_{0,1}$, and $H_0 = \mathbb{B}(\Theta = \Theta_0) = \mathbb{B}(\Theta_1 = \Theta_{0,1})$.
(b) Tests about $\Theta_1 = C_1 B U$ are testable; i.e., C_1 has full row rank, U has full column rank, and Θ_1 is estimable.

Proof. The rows of C_2 are linear combinations of the rows of C_1; i.e., $C_2 = AC_1$ for some fixed matrix A. Hence $\Theta_2 = AC_1BU = A\Theta_1$ and $\Theta = \Theta_0$ iff $\Theta_1 = \Theta_{0,1}$. Also

$$\begin{aligned} \Theta = CBU &= \begin{bmatrix} I \\ A \end{bmatrix} C_1 BU \\ &= \begin{bmatrix} \Theta_1 \\ \Theta_2 \end{bmatrix} = \begin{bmatrix} I \\ A \end{bmatrix} \Theta_1 . \end{aligned} \tag{16.7}$$

Estimable Θ implies Θ_1 is estimable; combining estimability with full rank of C_1 implies $\widehat{\Delta}_1$ exists. While $\widehat{\Delta} = (\Theta - \Theta_0)'[C(X'X)^- C']^{-1}(\widehat{\Theta} - \Theta_0)$ does *not* exist and cannot be computed, $\widehat{\Delta}_1 = (\widehat{\Theta}_1 - \Theta_{0,1})'[C_1(X'X)^- C_1']^{-1}(\widehat{\Theta}_1 - \Theta_{0,1})$ does exist and can be computed. The commonly used tests require the existence of $\widehat{\Sigma}_*^{-1} = (U'\widehat{\Sigma}U)^{-1}$. Full column rank of U (and the assumption of full-rank Σ) guarantees the desired property. □

Theorem 16.5 A test with a LTFR U matrix is indistinguishable from a test with U replaced by a full (column) rank matrix with columns that span all columns of U. More specifically, a $\text{GLM}_{N,p,q}(Y_i; X_i B, \Sigma)$ has Gaussian errors, $\text{rank}(X) = r \le q$, estimable $\Theta = CBU$ $(a \times b)$, and $\text{rank}(C) = a \le q$. For $j \in \{1,2\}$ U_j $p \times b_j$, $U = [\,U_1\, U_2\,]$, and Θ_j $a \times b_j$, $\Theta = [\,\Theta_1\, \Theta_2\,]$, $\Theta_0 = [\,\Theta_{0,1}\, \Theta_{0,2}\,]$, and $\text{rank}(U) = \text{rank}(U_1) = b_1 \le \min\{b, p\}$.
(a) Tests about Θ are indistinguishable from tests about Θ_1; i.e., $\Theta = \Theta_0$ iff $\Theta_1 = \Theta_{0,1}$, and $H_0 = \mathbb{B}(\Theta = \Theta_0) = \mathbb{B}(\Theta_1 = \Theta_{0,1})$.
(b) Tests about $\Theta_1 = CBU_1$ are testable; i.e., C has full row rank, U_1 has full column rank, and Θ_1 is estimable.

Proof. **(a)** All columns of U_2 are linear combinations of columns of U_1: $U_2 = U_1 A$ for fixed A. Here $\Theta = \Theta_0$ iff $\Theta_1 = \Theta_{0,1}$, because $\Theta_2 = CBU_2 = CBU_1A = \Theta_1 A$. **(b)** Matrix $\widehat{\Delta}_1 = (\widehat{\Theta}_1 - \Theta_{0,1})'[C(X'X)^- C']^{-1}(\widehat{\Theta}_1 - \Theta_{0,1})$ exists and can be computed. The commonly used tests require the existence of $\widehat{\Sigma}_*^{-1} = (U_1'\widehat{\Sigma}U_1)^{-1}$, which is guaranteed by the full column rank of U_1, even though $(U'\widehat{\Sigma}U)$ is singular and $(U'\widehat{\Sigma}U)^{-1}$ does not exist. □

The following theorem justifies using a simpler model which is linearly equivalent. The approach allows avoiding some of the pitfalls with LTFR models.

Theorem 16.6 Any primary or secondary expected- value parameter testable in $\text{GLM}_{N,p,q}(Y_i; X_i B, \Sigma)$ is also testable in a linearly equivalent model.

Proof. The result follows from the parallel result about estimability.

16.3 TESTS OF A PRIORI HYPOTHESES

The present chapter focuses on the GLH, $H_0: \boldsymbol{\Theta} = \boldsymbol{\Theta}_0$, for a multivariate $\text{GLM}_{N,p,q}(\boldsymbol{Y}_i; \boldsymbol{X}_i\boldsymbol{B}, \boldsymbol{\Sigma})$ with $\text{rank}(\boldsymbol{X}) = r \leq q$ and Gaussian errors. Here we consider only testable $\boldsymbol{\Theta} = \boldsymbol{CBU}$ of dimensions $a \times b$. Testable parameters must be estimable and have $\text{rank}(\boldsymbol{C}) = a \leq q$, and $\text{rank}(\boldsymbol{U}) = b \leq p$, for fixed and known $\boldsymbol{C}, \boldsymbol{U}$, and $\boldsymbol{\Theta}_0$. The primary parameters $\boldsymbol{B}$ and $\boldsymbol{\Sigma}$ have estimators

$$\widetilde{\boldsymbol{B}} = (\boldsymbol{X}'\boldsymbol{X})^{-}\boldsymbol{X}'\boldsymbol{Y} \tag{16.8}$$

$$\widehat{\boldsymbol{\Sigma}} = \boldsymbol{Y}'[\boldsymbol{I} - \boldsymbol{X}(\boldsymbol{X}'\boldsymbol{X})^{-}\boldsymbol{X}']\boldsymbol{Y}/(N - r). \tag{16.9}$$

Next we derive tests of $H_0 = \mathbb{B}(\boldsymbol{\Theta} = \boldsymbol{\Theta}_0)$ *versus* $H_A = \mathbb{B}(\boldsymbol{\Theta} \neq \boldsymbol{\Theta}_0)$ in terms of secondary parameters

$$\boldsymbol{\Theta} = \boldsymbol{CBU} \tag{16.10}$$

$$\boldsymbol{\Sigma}_* = \boldsymbol{U}'\boldsymbol{\Sigma}\boldsymbol{U}, \tag{16.11}$$

which are $a \times b$ and $b \times b$ respectively. With $\boldsymbol{M} = \boldsymbol{C}(\boldsymbol{X}'\boldsymbol{X})^{-}\boldsymbol{C}'$ $(a \times a)$, fully specifying distributions under the alternative involves the $b \times b$ shift matrix

$$\boldsymbol{\Delta} = (\boldsymbol{\Theta} - \boldsymbol{\Theta}_0)'\boldsymbol{M}^{-1}(\boldsymbol{\Theta} - \boldsymbol{\Theta}_0) \tag{16.12}$$

or the noncentrality matrix

$$\boldsymbol{\Omega} = \boldsymbol{\Delta}\boldsymbol{\Sigma}_*^{-1}. \tag{16.13}$$

Corresponding estimators, with $\nu_e = N - \text{rank}(\boldsymbol{X})$, are

$$\widehat{\boldsymbol{\Theta}} = \boldsymbol{C}\widetilde{\boldsymbol{B}}\boldsymbol{U} \tag{16.14}$$

$$\nu_e^{-1}\boldsymbol{S}_e = \widehat{\boldsymbol{\Sigma}}_* = \boldsymbol{U}'\widehat{\boldsymbol{\Sigma}}\boldsymbol{U} \tag{16.15}$$

$$\boldsymbol{S}_h = \widehat{\boldsymbol{\Delta}} = (\widehat{\boldsymbol{\Theta}} - \boldsymbol{\Theta}_0)'\boldsymbol{M}^{-1}(\widehat{\boldsymbol{\Theta}} - \boldsymbol{\Theta}_0). \tag{16.16}$$

Also

$$\widehat{\boldsymbol{\Theta}} - \boldsymbol{\Theta}_0 \sim \mathcal{N}_{a,b}(\boldsymbol{CBU} - \boldsymbol{\Theta}_0, \boldsymbol{M}, \boldsymbol{\Sigma}_*) \tag{16.17}$$

$$\boldsymbol{S}_e \sim \mathcal{W}_b(\nu_e, \boldsymbol{\Sigma}_*, \boldsymbol{0}) \tag{16.18}$$

$$\boldsymbol{S}_h \sim \mathcal{W}_b(a, \boldsymbol{\Sigma}_*, \boldsymbol{\Delta}). \tag{16.19}$$

The parameters determining the distributions of the secondary parameter estimators make it obvious that any reasonable test statistic must be a function of $\widehat{\boldsymbol{\Delta}}$ and $\widehat{\boldsymbol{\Sigma}}_*$, which generalize *SSH* and $\widehat{\sigma}^2$ (equivalent to *SSE*) and reduce to them if $p = 1$. Properties of special cases help motivate the choice of statistics. Most importantly, if $b = 1$ (always true for $p = 1$), then the $b \times b$ matrix $\widehat{\boldsymbol{\Omega}}/a = \widehat{\boldsymbol{\Delta}\boldsymbol{\Sigma}_*^{-1}}/a$ is 1×1, a scalar, and exactly equals the usual F random variable from univariate theory. In the same case, $\boldsymbol{\Omega} = \boldsymbol{\Delta}\boldsymbol{\Sigma}_*^{-1}$ is 1×1 and reduces to $\omega = (\boldsymbol{\theta} - \boldsymbol{\theta}_0)'\boldsymbol{M}^{-1}(\boldsymbol{\theta} - \boldsymbol{\theta}_0)\sigma_*^{-2}$, the F noncentrality. More generally, if

$s = \min(a, b) = 1$, then the $b \times b$ matrix $\widehat{\boldsymbol{\Delta}}\widehat{\boldsymbol{\Sigma}}_*^{-1}/a$ has a single nonnegative eigenvalue which is a constant multiple of an F random variable.

The most general case, with $s = \min(a, b) > 1$, creates more complication. In contrast to the univariate results, which are a special case, the multivariate likelihood ratio and union-intersection tests differ. Furthermore, other criteria, such as the substitution principle, lead to distinct tests. A form of linear invariance and exact test size stand as the properties demanded of all candidate statistics.

16.4 LINEAR INVARIANCE

A multivariate test is said to be linearly invariant if the hypothesis test (interpreted as a decision function) does not vary under full-rank transformation of the response variables being tested. Transformations of the underlying model provide a straightforward way to formalize the idea. For testable $\boldsymbol{\Theta} = \boldsymbol{CBU}$, a test of $H_0 : \boldsymbol{CBU} = \boldsymbol{\Theta}_0$ in $\text{GLM}_{N,p,q}(\boldsymbol{Y}_i; \boldsymbol{X}_i\boldsymbol{B}, \boldsymbol{\Sigma})$ with Gaussian errors may be expressed in terms of a model of $\boldsymbol{Y}_U = \boldsymbol{YU}$ $(N \times b)$ by the transformation

$$\begin{aligned} \boldsymbol{YU} &= \boldsymbol{XBU} + \boldsymbol{EU} \\ \boldsymbol{Y}_U &= \boldsymbol{XB}_U + \boldsymbol{E}_U . \end{aligned} \tag{16.20}$$

With $\boldsymbol{\Sigma}_* = \boldsymbol{U}'\boldsymbol{\Sigma}\boldsymbol{U}$, the transformation implicitly defines the $\text{GLM}_{N,b,q}(\boldsymbol{Y}_{Ui}; \boldsymbol{X}_i\boldsymbol{B}_U, \boldsymbol{\Sigma}_*)$. In turn, testing $H_0 : \boldsymbol{CB}_U = \boldsymbol{\Theta}_0$ is obviously equivalent to testing $H_0 : \boldsymbol{CBU} = \boldsymbol{\Theta}_0$.

Definition 16.2 A $\text{GLM}_{N,p,q}(\boldsymbol{Y}_i; \boldsymbol{X}_i\boldsymbol{B}, \boldsymbol{\Sigma})$ with Gaussian errors and testable $\boldsymbol{\Theta} = \boldsymbol{CBU}$ allows testing $H_0 : \boldsymbol{CBU} = \boldsymbol{\Theta}_0$.
(a) If $\boldsymbol{Y}_U = \boldsymbol{YU}$ and $\boldsymbol{B}_U = \boldsymbol{BU}$, the test is *linearly invariant* whenever the same test occurs for $H_0 : \boldsymbol{\Theta T} = \boldsymbol{\Theta}_0\boldsymbol{T}$ in $\text{GLM}_{N,b,q}(\boldsymbol{Y}_{Ui}\boldsymbol{T}; \boldsymbol{X}_i\boldsymbol{B}_U, \boldsymbol{T}'\boldsymbol{\Sigma}_*\boldsymbol{T})$, with constant and full-rank $\boldsymbol{T}$ $(b \times b)$. As always, $\boldsymbol{\Sigma}_* = \boldsymbol{U}'\boldsymbol{\Sigma}_*\boldsymbol{U}$.
(b) The test has *location invariance* whenever the model spans an intercept $(\boldsymbol{X}\boldsymbol{t}_0 = \boldsymbol{1}_N)$ and the hypothesis excludes it $(\boldsymbol{C}\boldsymbol{t}_0 = \boldsymbol{0})$.

Location invariance also can be expressed in terms of the transformation $\boldsymbol{Y}_{Ui} + \boldsymbol{1}_N\boldsymbol{t}_1'$ for constant $\boldsymbol{t}_1$ $(b \times 1)$. Clearly $\text{E}(\boldsymbol{Y}_{Ui} + \boldsymbol{1}_N\boldsymbol{t}_1') = (\boldsymbol{X}_i\boldsymbol{B}_U + \boldsymbol{1}_N\boldsymbol{t}_1')$. However, the specific impact on $\boldsymbol{B}_U = \boldsymbol{BU}$ varies with the design. In the special case of a full-rank model with $\boldsymbol{1}_N$ as the first column (defining the intercept parameter), $\boldsymbol{t}_1'$ is added to the first row of $\boldsymbol{B}_U$.

Some authors include location invariance as part of linear invariance. In any case, the only functions of $\widehat{\boldsymbol{\Delta}}$ and $\widehat{\boldsymbol{\Sigma}}_*$ which allow achieving the invariance properties are the eigenvalues of $\widehat{\boldsymbol{\Omega}} = \widehat{\boldsymbol{\Delta}}\widehat{\boldsymbol{\Sigma}}_*^{-1}$.

Although many appealing size-α tests exist, for general $\boldsymbol{\Sigma}$, no uniformly most powerful (UMP) test of size α (among scale- invariant tests) can be found for the general linear multivariate hypothesis. Next we consider four commonly used test procedures all of which are invariant, unbiased, and admissible.

16.5 FOUR MULTIVARIATE TEST STATISTICS

The multivariate (MULTIREP) statistics involve $\nu_e = N-\text{rank}(\boldsymbol{X})$, $\boldsymbol{M}$ $(a \times a)$ with $\text{rank}(\boldsymbol{M}) = a$, $\widehat{\boldsymbol{\Delta}}$ $(b \times b)$ with $\text{rank}(\widehat{\boldsymbol{\Delta}}) = s = \min\{a, b\}$, and $\text{rank}(\widehat{\boldsymbol{\Sigma}}_*) = b$. The four multivariate test statistics of interest are defined in terms of the $s = \min\{a, b\}$ nonzero eigenvalues of $\widehat{\boldsymbol{\Delta}}\widehat{\boldsymbol{\Sigma}}_*^{-1}$, which is usually not symmetric. The symmetric and full-rank nature of $\widehat{\boldsymbol{\Sigma}}_*$ ensures full-rank $\boldsymbol{F}$ exists such that $\widehat{\boldsymbol{\Sigma}}_* = \boldsymbol{F}\boldsymbol{F}'$ and $\widehat{\boldsymbol{\Sigma}}_* = \boldsymbol{F}^{-t}\boldsymbol{F}^{-1}$. The eigenvalues of $\widehat{\boldsymbol{\Delta}}\widehat{\boldsymbol{\Sigma}}_*^{-1}$ coincide with the eigenvalues of the symmetric matrix $\boldsymbol{F}^{-1}\widehat{\boldsymbol{\Delta}}\boldsymbol{F}^{-t}$, while the eigenvectors differ by the transformation $\boldsymbol{V}_F = \boldsymbol{F}^{-1}\boldsymbol{V}$. The coincidence may be proven by observing

$$\begin{aligned}(\widehat{\boldsymbol{\Delta}}\widehat{\boldsymbol{\Sigma}}_*^{-1} - \lambda \boldsymbol{I}_b)\boldsymbol{v} &= \boldsymbol{0}\\ \boldsymbol{F}^{-1}(\widehat{\boldsymbol{\Delta}}\boldsymbol{F}^{-t}\boldsymbol{F}^{-1} - \lambda \boldsymbol{I}_b)(\boldsymbol{F}\boldsymbol{F}^{-1})\boldsymbol{v} &= \boldsymbol{F}^{-1}\boldsymbol{0}\\ (\boldsymbol{F}^{-1}\widehat{\boldsymbol{\Delta}}\boldsymbol{F}^{-t} - \lambda \boldsymbol{I}_b)(\boldsymbol{F}^{-1}\boldsymbol{v}) &= \boldsymbol{0}\,. \end{aligned} \tag{16.21}$$

With probability 1, $\widehat{\boldsymbol{\Sigma}}_*^{-1}$ is positive definite and $\widehat{\boldsymbol{\Delta}}$ is either positive definite or positive semidefinite, and both are symmetric. Hence the eigenvalues in the last equation are real and nonnegative ($\boldsymbol{F}^{-1}\widehat{\boldsymbol{\Delta}}\boldsymbol{F}^{-t}$ and $\widehat{\boldsymbol{\Delta}}$ are congruent and therefore have the same number of positive, negative, and zero eigenvalues). The $b \times b$ matrices $\widehat{\boldsymbol{\Delta}}\widehat{\boldsymbol{\Sigma}}_*^{-1}$ and $\boldsymbol{F}^{-1}\widehat{\boldsymbol{\Delta}}\boldsymbol{F}^{-t}$ have rank $s = \min\{a, b\}$ because $\widehat{\boldsymbol{\Sigma}}_*$ and $\boldsymbol{F}$ are rank b, while $\text{rank}(\widehat{\boldsymbol{\Delta}}) = a = \text{rank}(\boldsymbol{M})$. Various authors discuss matrices which differ by a simple multiple (Kuhfeld, 1986). The matrix $\boldsymbol{S}_e = \nu_e\widehat{\boldsymbol{\Sigma}}_*$ $(b \times b)$ is the sum of squares of error and reduces to *SSE* in the univariate case. Considering $\boldsymbol{S}_h\boldsymbol{S}_e^{-1} = \widehat{\boldsymbol{\Delta}}\widehat{\boldsymbol{\Sigma}}_*^{-1}/\nu_e$, the multivariate generalization of *SSH*/*SSE*, is often convenient. The multivariate analog of the total sum of squares is $\boldsymbol{S}_t = \boldsymbol{S}_h + \boldsymbol{S}_e$.

The test statistics can also be defined in terms of the eigenvalues of $\boldsymbol{S}_h\boldsymbol{S}_t^{-1}$, which reduces to *SSH*/*SST*, or $\boldsymbol{S}_e\boldsymbol{S}_t^{-1}$, which reduces to *SSE*/*SST*. Writing $\boldsymbol{S}_h = \widehat{\boldsymbol{\Delta}}$ leads to Table 16.1, which summarizes algebraic relationships among the different sets of eigenvalues. Table 1 in Muller, LaVange, Ramey, and Ramey (1992) contains additional information. Here $0 \leq \widehat{\rho}_k \leq 1$ is one of s generalized canonical correlations. It reduces to the sole multiple correlation for a univariate model which spans an intercept combined with a hypothesis excluding the intercept.

Table 16.1 Eigenvalues of Matrix Labeling Column as a Function of Eigenvalues of Matrix Labeling Row

	$\boldsymbol{S}_h\boldsymbol{S}_e^{-1}$	$\boldsymbol{S}_h\boldsymbol{S}_t^{-1}$	$\boldsymbol{S}_e\boldsymbol{S}_t^{-1}$
$\boldsymbol{S}_h\boldsymbol{S}_e^{-1}$	$\widehat{\phi}_k$	$\widehat{\phi}_k/(1+\widehat{\phi}_k)$	$1/(1+\widehat{\phi}_k)$
$\boldsymbol{S}_h\boldsymbol{S}_t^{-1}$	$\widehat{\rho}_k^2/(1-\widehat{\rho}_k^2)$	$\widehat{\rho}_k^2$	$(1-\widehat{\rho}_k^2)$
$\boldsymbol{S}_e\boldsymbol{S}_t^{-1}$	$(1-\widehat{\lambda}_k)/\widehat{\lambda}_k$	$(1-\widehat{\lambda}_k)$	$\widehat{\lambda}_k$

Definition 16.3 Wilks lambda ($\widehat{\Lambda}$), the likelihood ratio statistic, is

$$\begin{aligned}\widehat{\Lambda} &= |\boldsymbol{S}_h \boldsymbol{S}_e^{-1} + \boldsymbol{I}_b|^{-1} \\ &= |\boldsymbol{S}_e|/|\boldsymbol{S}_h + \boldsymbol{S}_e| = |\boldsymbol{S}_e|/|\boldsymbol{S}_t| \\ &= |\boldsymbol{S}_e \boldsymbol{S}_t^{-1}| \\ &= \prod_{k=1}^{b} (1 - \widehat{\rho}_k^2) . \end{aligned} \tag{16.22}$$

Definition 16.4 Roy's largest root (RLR), the union-intersection principle statistic, is $\text{RLR} = \widehat{\phi}_{\max}(\boldsymbol{S}_h \boldsymbol{S}_e^{-1})$, which is the largest eigenvalue of $\boldsymbol{S}_h \boldsymbol{S}_e^{-1}$, say $\widehat{\phi}_1$. The variable $\widehat{\phi}_1$ is a one-to-one function of and hence equivalent to $\widehat{\rho}_1^2$.

Definition 16.5 The Hotelling-Lawley *trace* (HLT, ANOVA analog) is

$$\text{HLT} = \text{tr}(\boldsymbol{S}_h \boldsymbol{S}_e^{-1}) = \sum_{k=1}^{b} \widehat{\phi}_k = \sum_{k=1}^{b} \widehat{\rho}_k^2 / (1 - \widehat{\rho}_k^2) . \tag{16.23}$$

Definition 16.6 The Pillai-Bartlett trace (PBT), the R^2 analog statistic, is

$$\text{PBT} = \text{tr}(\boldsymbol{S}_h \boldsymbol{S}_t^{-1}) = \sum_{k=1}^{b} \widehat{\rho}_k^2 . \tag{16.24}$$

Pillai (1955) justified the PBT on heuristic grounds. It is linearly invariant, and the distribution depends only on the dimensions of the problem under the null. The other three statistics considered in the chapter share the same properties.

Roy (1957) proposed $\widehat{\phi}_1$ as a multivariate test statistic and derived the central distribution of $\widehat{\rho}_1^2 = \widehat{\phi}_1/(1 + \widehat{\phi}_1)$. Some authors (including Muller and Peterson, 1984) refer to $\widehat{\rho}_1^2$ as Roy's largest root.

Lawley (1938) and Hotelling (1951) proposed T^2 as a test statistic and derived the central distribution of $T^2/\nu_e = \text{tr}(\boldsymbol{S}_h \boldsymbol{S}_e^{-1})$. Some authors refer to $\text{tr}(\boldsymbol{S}_h \boldsymbol{S}_e^{-1})$ as "the trace criterion" or as "Hotelling's trace criterion" (Timm, 1975).

Timm (1975, p. 148–149) noted the noncentral distributions have been very difficult to compute, although recent advances in algorithms have led to published tables. Muller and Peterson (1984) also discussed the noncentral distributions.

For a multivariate model, $\boldsymbol{S}_e \sim \mathcal{W}_b(\nu_e, \boldsymbol{\Sigma}_*)$, independently of

$$\boldsymbol{S}_h = \widehat{\boldsymbol{\Delta}} \sim \mathcal{W}_b(a, \boldsymbol{\Sigma}_*, \boldsymbol{\Delta}) . \tag{16.25}$$

Under H_0, $\boldsymbol{\Delta} = \boldsymbol{0}$ and $\widehat{\boldsymbol{\Delta}}$ has a central Wishart distribution. Aside from certain special cases, addressed later, the distribution of $\widehat{\Lambda}$ is complicated.

Theorem 16.7 Under H_0, the statistic $\widehat{\Lambda}$ equals a product of b independent beta-1 random variables. More specifically, if $\boldsymbol{S}_e \sim \mathcal{W}_b(\nu_e, \boldsymbol{\Sigma})$ independently of $\boldsymbol{S}_h \sim \mathcal{W}_b(a, \boldsymbol{\Sigma})$ and $\nu_e \geq b$, then

$$\widehat{\Lambda} = |\boldsymbol{S}_h \boldsymbol{S}_e^{-1} + \boldsymbol{I}|^{-1} = \prod_{k=1}^{b} x_k \,, \tag{16.26}$$

with $\{x_k\}$ independent beta random variables, $x_k \sim \beta[(\nu_e+k-b)/2, a/2]$. If $a = 1$, $\widehat{\Lambda}$ has the same distribution as a single beta random variable with parameters $\{(\nu_e + 1 - b)/2\,, b/2\}$ and hence is a one-to-one function of an F.

Proof. Rao (1973, Section 8.b.2.xi, Theorem 3.4.3, Corollary 3.4.3) gave a proof.

Corollary 16.7 With $s = \min\{a, b\}$,
(a) if $s = 1$, then

$$\begin{aligned} F &= (\widehat{\Lambda}^{-1} - 1)(\nu_e + 1 - b)/(|a - b| + 1) \\ &\sim F(|a - b| + 1, \nu_e + 1 - b)\,, \end{aligned} \tag{16.27}$$

(b) if $s = 2$, then

$$\begin{aligned} F &= \left(\widehat{\Lambda}^{-1} - 1\right)(\nu_e + 1 - b)/(|a - b| + 2) \\ &\sim F[2|a - b| + 8, 2(\nu_e + 1 - b)]\,. \end{aligned} \tag{16.28}$$

Proof. Morrison (1990) provided a proof.

The results can be used for an exact LRT when $s = 1$ or $s = 2$, which covers a useful proportion of hypotheses tested in practice.

Theorem 16.8 The likelihood ratio test (LRT) statistic is $\widehat{\Lambda}$. More specifically, $\text{GLM}_{N,p,q}(\boldsymbol{Y}_i; \boldsymbol{X}_i\boldsymbol{B}, \boldsymbol{\Sigma})$ with Gaussian errors has estimable $\boldsymbol{\Theta} = \boldsymbol{CBU}$ $a \times b$, $\text{rank}(\boldsymbol{X}) = r \leq q$, $\text{rank}(\boldsymbol{C}) = a \leq q$, and $\text{rank}(\boldsymbol{U}) = b \leq p$, $\boldsymbol{C}$, $\boldsymbol{U}$, and $\boldsymbol{\Theta}_0$ known constants. The LRT of $H_0 = \mathbb{B}(\boldsymbol{\Theta} = \boldsymbol{\Theta}_0)$ versus $H_A = \mathbb{B}(\boldsymbol{\Theta} \neq \boldsymbol{\Theta}_0)$ is $\widehat{H}_A = \mathbb{B}(\widehat{\Lambda} < t_0)$ with $\widehat{\Lambda} = |\boldsymbol{S}_h \boldsymbol{S}_e^{-1} + \boldsymbol{I}|^{-1} = |\boldsymbol{S}_e|/|\boldsymbol{S}_h + \boldsymbol{S}_e|$ and appropriate critical value t_0. Here $\widehat{\boldsymbol{\Theta}} = \boldsymbol{C}\widehat{\boldsymbol{B}}\boldsymbol{U}$, $\boldsymbol{S}_h = (\widehat{\boldsymbol{\Theta}} - \boldsymbol{\Theta}_0)'[\boldsymbol{C}(\boldsymbol{X}'\boldsymbol{X})^{-}\boldsymbol{C}']^{-1}(\widehat{\boldsymbol{\Theta}} - \boldsymbol{\Theta}_0)$, $\widehat{\boldsymbol{\Sigma}} = \boldsymbol{Y}'[\boldsymbol{I} - \boldsymbol{X}(\boldsymbol{X}'\boldsymbol{X})^{-}\boldsymbol{X}']\boldsymbol{Y}/(N - r)$, and $\boldsymbol{S}_e = \boldsymbol{U}'\widehat{\boldsymbol{\Sigma}}\boldsymbol{U}(N - r)$.

Proof. Anderson (2004) included a proof. The result follows from expressing the test in terms of the unconstrained and constrained likelihood expressions in the proofs for MLEs of $\boldsymbol{B}$ and $\boldsymbol{\Theta}$. The proof directly generalizes the univariate proof.

Hypothesis H_0 is rejected for improbably small values of $\widehat{\Lambda}$, relative to an appropriate critical value.

Theorem 16.9 The union-intersection (UI) test statistic is $\widehat{\phi}_{\max}(\boldsymbol{S}_h\boldsymbol{S}_e^{-1})$. The $\text{GLM}_{N,p,q}(\boldsymbol{Y}_i;\boldsymbol{X}_i\boldsymbol{B},\boldsymbol{\Sigma})$ with Gaussian errors has estimable $\boldsymbol{\Theta}=\boldsymbol{CBU}$ $a\times b$, $\text{rank}(\boldsymbol{X})=r\le q$, $\text{rank}(\boldsymbol{C})=a\le q$, and $\text{rank}(\boldsymbol{U})=b\le p$, with $\boldsymbol{C}$, $\boldsymbol{U}$, and $\boldsymbol{\Theta}_0$ known constants. For testing $H_0=\mathbb{B}(\boldsymbol{\Theta}=\boldsymbol{\Theta}_0)$ versus $H_A=\mathbb{B}(\boldsymbol{\Theta}\neq\boldsymbol{\Theta}_0)$ the UI test of H_0 versus H_A is $\widehat{H}_A=\mathbb{B}[\widehat{\phi}_{\max}(\boldsymbol{S}_h\boldsymbol{S}_e^{-1})>l_\alpha]$ and has test statistic $\widehat{\phi}_{\max}(\boldsymbol{S}_h\boldsymbol{S}_e^{-1})$, the largest eigenvalue of $\boldsymbol{S}_h\boldsymbol{S}_e^{-1}$ and sometimes described as Roy's largest root statistic. The appropriate critical value is l_α. Hypothesis H_0 is rejected for improbably large values of the test statistic relative to l_α, the appropriate critical value.

Proof. Any hypothesis involving $\boldsymbol{\Theta}_0\neq\boldsymbol{0}$ can be converted to a test with $\boldsymbol{\Theta}_0=\boldsymbol{0}$. The proof (in the section on invariance below) is based on the fact $\boldsymbol{\Theta}=\boldsymbol{0}$ iff $\boldsymbol{\Theta b}=\boldsymbol{0}\ \forall\boldsymbol{b}\in\Re^b$ ($\boldsymbol{b}$ is $b\times 1$). In terms of Boolean algebra we have the following decomposition of the null hypothesis into infinitely many univariate hypotheses:

$$\begin{aligned} H_0 &= \mathbb{B}(\boldsymbol{\Theta}=\boldsymbol{0}) \\ &= \bigwedge_{\boldsymbol{b}\neq\boldsymbol{0}}\mathbb{B}(\boldsymbol{\Theta b}=\boldsymbol{0}) \\ &= \bigwedge_{\boldsymbol{b}\neq\boldsymbol{0}}H_0(\boldsymbol{b})\,. \end{aligned} \tag{16.29}$$

For a given $\boldsymbol{b}\neq\boldsymbol{0}$, $H_0(\boldsymbol{b})$ is a univariate joint hypothesis and $\boldsymbol{\Theta b}$ is an $(a\times 1)$ vector with $\widehat{\boldsymbol{\Theta b}}\sim\mathcal{N}_a[\boldsymbol{\Theta b},\boldsymbol{b}'\boldsymbol{\Sigma b C}(\boldsymbol{X}'\boldsymbol{X})^{-1}\boldsymbol{C}']$. The result is easily derived from $\boldsymbol{Yb}\sim\mathcal{N}_N(\boldsymbol{XBb},\boldsymbol{b}'\boldsymbol{\Sigma b I}_N)$. The univariate UI test of $H_0(\boldsymbol{b})\equiv\mathbb{B}(\boldsymbol{\Theta b}=\boldsymbol{0})$ is based upon the statistic

$$\begin{aligned} F(\boldsymbol{b};\boldsymbol{Y}) &= \frac{\boldsymbol{b}'\boldsymbol{S}_h\boldsymbol{b}/a}{\boldsymbol{b}'\boldsymbol{S}_e\boldsymbol{b}/\nu_e} \\ &= \frac{(\boldsymbol{b}'\widehat{\boldsymbol{\Theta}}')[\boldsymbol{C}(\boldsymbol{X}'\boldsymbol{X})^-\boldsymbol{C}']^{-1}(\widehat{\boldsymbol{\Theta}}\boldsymbol{b})/a}{\boldsymbol{b}'\boldsymbol{U}'\boldsymbol{Y}'[\boldsymbol{I}_N-\ \boldsymbol{X}(\boldsymbol{X}'\boldsymbol{X})^-\boldsymbol{X}']\boldsymbol{YUb}/\nu_e}\,. \end{aligned} \tag{16.30}$$

The distributions of $\boldsymbol{S}_h$ and $\boldsymbol{S}_e$ are both Wishart. Hence $q_j=\boldsymbol{b}'\boldsymbol{S}_j\boldsymbol{b}$ is distributed as a scaled chi square. The univariate UI test is

$$\begin{aligned} \widehat{H}_A(\boldsymbol{b}) &\equiv \widehat{H}_A(\boldsymbol{b};\boldsymbol{Y}) \\ &= \mathbb{B}[F(\boldsymbol{b};\boldsymbol{Y})>f_{\text{crit}}]\,, \end{aligned} \tag{16.31}$$

with decision function

$$\begin{aligned} \widehat{H}_0(\boldsymbol{b}) &\equiv \widehat{H}_0(\boldsymbol{b};\boldsymbol{Y}) \\ &= \mathbb{B}[F(\boldsymbol{b};\boldsymbol{Y})\le f_{\text{crit}}] \\ &= 1-\widehat{H}_A(\boldsymbol{b})\,. \end{aligned} \tag{16.32}$$

The appropriate critical value, f_{crit}, is the $100(1-\alpha)$ percentile of the $F(a,\nu_e)$ distribution. The appropriate critical value does *not* depend on $\boldsymbol{b}$.

Corresponding acceptance and rejection regions are

$$AR(H_0) = \bigcap_{b \neq 0} AR[H_0(\boldsymbol{b})] \tag{16.33}$$

$$RR(H_0) = \bigcup_{b \neq 0} RR[H_0(\boldsymbol{b})] \,. \tag{16.34}$$

The regions are sets in the sample space, $AR(H_0) = \{\boldsymbol{Y} : \widehat{H}_A = \text{TRUE}\}$ and $RR(H_0) = \{\boldsymbol{Y} : \widehat{H}_0 = \text{FALSE}\}$, while

$$\begin{aligned} AR[H_0(\boldsymbol{b})] &= \{\boldsymbol{Y} : \widehat{H}_0(\boldsymbol{b}) = \text{TRUE}\} \\ &= \{\boldsymbol{Y} : F(\boldsymbol{b}; \boldsymbol{Y}) \leq f_{\text{crit}}\} \end{aligned} \tag{16.35}$$

and

$$\begin{aligned} RR[H_0(\boldsymbol{b})] &= \{\boldsymbol{Y} : \widehat{H}_0(\boldsymbol{b}) = \text{FALSE}\} \\ &= \{\boldsymbol{Y} : F(\boldsymbol{b}; \boldsymbol{Y}) > f_{\text{crit}}\} \,. \end{aligned} \tag{16.36}$$

The decision function is $\widehat{H}_0 = \mathbb{B}[\boldsymbol{Y} \in \text{AR}(H_0)]$ and the test is $\widehat{H}_A = \mathbb{B}[\boldsymbol{Y} \in \text{RR}(H_0)]$. Thus by the union-intersection principle we have

$$\widehat{H}_A = \bigvee_{b \neq 0} \widehat{H}_A(\boldsymbol{b}) \tag{16.37}$$

$$H_0 = \bigwedge_{b \neq 0} H_0(\boldsymbol{b}) \,. \tag{16.38}$$

With the above notation the heart of the proof is simple:

$$\begin{aligned} \widehat{H}_0 &= \mathbb{B}[\boldsymbol{Y} \in AR(H_0)] \\ &= \bigwedge_{b \neq 0} \mathbb{B}\{\boldsymbol{Y} \in AR[H_0(\boldsymbol{b})]\} \\ &= \bigwedge_{b \neq 0} \mathbb{B}[F(\boldsymbol{b}; \boldsymbol{Y}) \leq f_{\text{crit}}] \\ &= \mathbb{B}\left[\sup_{b \neq 0}\{F(\boldsymbol{b}; \boldsymbol{Y})\} \leq f_{\text{crit}}\right] = \mathbb{B}(\lambda \leq f_{\text{crit}}) \,. \end{aligned} \tag{16.39}$$

Similar equalities allow proving $\widehat{H}_A = \mathbb{B}(\lambda > f_{\text{crit}})$. Thus the UI statistic is

$$\begin{aligned} \lambda &= \sup_{b \neq 0}\{F(\boldsymbol{b}; \boldsymbol{Y})\} \\ &= \sup_{b \neq 0}\left(\frac{\boldsymbol{b}'\boldsymbol{S}_h\boldsymbol{b}}{\boldsymbol{b}'\boldsymbol{S}_e\boldsymbol{b}}\right)\left(\frac{\nu_e}{a}\right) . \end{aligned} \tag{16.40}$$

Given $\boldsymbol{S}_e^{-1}$ exists with probability 1, so does the Cholesky decomposition $\boldsymbol{S}_e^{-1} = \boldsymbol{L}'\boldsymbol{L}$ and $\boldsymbol{L}^{-1}$. If $\boldsymbol{w} = (\boldsymbol{L}^{-1})'\boldsymbol{b}$, then $\boldsymbol{b} = \boldsymbol{w}'\boldsymbol{L}$ and

$$\begin{aligned}\widehat{\phi} &= \sup_{\boldsymbol{w}\neq\boldsymbol{0}}\left(\frac{\boldsymbol{w}'\boldsymbol{L}\boldsymbol{S}_h\boldsymbol{L}'\boldsymbol{w}}{\boldsymbol{w}'\boldsymbol{w}}\right)\left(\frac{\nu_e}{a}\right)\\ &= \left(\frac{\nu_e}{a}\right)\widehat{\phi}_{\max}(\boldsymbol{L}\boldsymbol{S}_h\boldsymbol{L}'),\end{aligned} \tag{16.41}$$

with $\widehat{\phi}_{\max}$ the largest root of characteristic polynomial

$$\begin{aligned}|\boldsymbol{L}\boldsymbol{S}_h\boldsymbol{L}' - \widehat{\phi}\boldsymbol{I}| &\equiv |\boldsymbol{S}_h - \widehat{\phi}\boldsymbol{S}_e|\\ &\equiv |\boldsymbol{S}_h\boldsymbol{S}_e^{-1} - \widehat{\phi}\boldsymbol{I}| = 0\,.\end{aligned} \tag{16.42}$$

Thus the U-I test is

$$\begin{aligned}\widehat{H}_a &= \mathbb{B}\left(\widehat{\phi} > f_{\text{crit}}\right)\\ &= \mathbb{B}\left[(\nu_e/a)\widehat{\phi}_{\max}(\boldsymbol{S}_h\boldsymbol{S}_e^{-1}) > f_{\text{crit}}\right]\\ &= \mathbb{B}\left[\widehat{\phi}_{\max}(\boldsymbol{S}_h\boldsymbol{S}_e^{-1}) > f_{\text{crit}}a/\nu_e\right].\end{aligned} \tag{16.43}$$

□

In practice, the statistic $\widehat{\rho}_1^2 = \widehat{\phi}_{\max}/(1+\widehat{\phi}_{\max})$ (which is a monotone function of $\widehat{\phi}_{\max}$) is preferred because $0 \le \widehat{\phi}_{\max} < \infty$ while $0 \le \widehat{\rho}_1^2 \le 1$. Therefore $\widehat{\rho}_1^2$ may be interpreted simply as the fraction of variance controlled. Since $\widehat{\phi}_{\max} = \widehat{\rho}_1^2/(1-\widehat{\rho}_1^2)$, the critical values bear the relationship $l_\alpha = x_\alpha/(1-x_\alpha)$.

The distribution of $\widehat{\rho}_1^2$ has been tabulated and charted by Heck, with the symbol θ_s. Heck's charts are given in the appendix of Morrison's (1990) book. Timm's book gives both tables and charts of the distribution. Also, Pillai (1956) gave an algorithm to approximate the right tail probability. Code for Pillai's approximation is listed in an appendix of Harris (1975). Free software LINMOD also implements the algorithm. The Appendix (Section A.2) of the present book contains a description of LINMOD and where to retrieve it from the Web. A missing value is returned if the p value exceeds 0.10 because the algorithm is guaranteed to work for small p values. Under H_0, the distribution of $\widehat{\rho}_1^2$ has parameters $s = \min(a,b)$, $m = (|a-b|-1)/2$, and $n = (\nu_e-b-1)/2$. Hypothesis H_0 is rejected if $\widehat{\rho}_1^2 > x_\alpha$.

Theorem 16.10 The substitution test (ST) principle leads to the following test for the $\text{GLM}_{N,p,q}(\boldsymbol{Y}_i; \boldsymbol{X}_i\boldsymbol{B}, \boldsymbol{\Sigma})$ with Gaussian errors, estimable $\boldsymbol{\Theta} = \boldsymbol{C}\boldsymbol{B}\boldsymbol{U}$ $(a \times b)$, $\text{rank}(\boldsymbol{X}) = r \le q$, $\text{rank}(\boldsymbol{C}) = a \le q$, $\text{rank}(\boldsymbol{U}) = b \le p$, and known constants $\boldsymbol{C}$, $\boldsymbol{U}$, $\boldsymbol{\Theta}_0$. The ST of $H_0 = \mathbb{B}(\boldsymbol{\Theta} = \boldsymbol{\Theta}_0)$ versus $H_A = \neg H_0$ is $\widehat{H}_A = \mathbb{B}[\text{tr}(\boldsymbol{S}_h\boldsymbol{S}_e^{-1}) > t_0]$ with test statistic $\text{tr}(\boldsymbol{S}_h\boldsymbol{S}_e^{-1})$, the Hotelling-Lawley trace, and appropriate critical value t_0. Hypothesis H_0 is rejected for improbably large values of the test statistic relative to the appropriate critical value.

Proof. Arnold (1981, p. 363–364) included a proof.

The substitution principle may be described as the conditioning principle. If $\mathbf{\Sigma}_*$ were known, then the UMP, invariant, size-α test would reject the hypothesis for improbably large values of $\text{tr}(\boldsymbol{S}_h\mathbf{\Sigma}_*^{-1})$. Substituting $\widehat{\mathbf{\Sigma}}_*$ for $\mathbf{\Sigma}_*$ removes the conditioning and suggests a test which may or may not have desirable properties.

Under the null, the eigenvalues of $\boldsymbol{S}_h\boldsymbol{S}_t^{-1}$ will be close to zero. Data that do not support the null yield large eigenvalues. Hence each test statistic discussed can be thought of as a means of summarizing the information in the $s = \min(a, b)$ nonzero eigenvalues. Roy's criterion uses the largest eigenvalue, the LRT uses a product of nonzero eigenvalues (a one-to-one function of a geometric mean), and the two trace criteria look at weighted sums of the eigenvalues. Each statistic has a reasonable interpretation in terms of the eigenvalues.

The one- and two-sample versions of T^2 generalize the one sample and independent groups t tests to allow simultaneous consideration of p response variables. Both correspond to special cases of the four general statistics, which provide the same p value and conclusion in any situation with $a = 1$.

16.6 WHICH MULTIVARIATE TEST IS BEST?

The existence of four tests leads to an obvious question: Which one is best? Not surprisingly, the answer varies with the definition of "best" and the particular hypothesis being tested. Here we emphasize accuracy of test size, robustness to violation of assumptions, and power. The multipart answer to the question derives from a few dimensions and one set of eigenvalues for data analysis and a corresponding set of eigenvalues for power analysis.

All four test statistics are functions of the data solely through the $s = \min(a, b)$ nonzero eigenvalues of the $b \times b$ matrix $\widehat{\mathbf{\Omega}} = \widehat{\mathbf{\Delta}}\widehat{\mathbf{\Sigma}}_*^{-1} = \boldsymbol{S}_h\boldsymbol{S}_e^{-1}\nu_e$. Equivalently, the s generalized, squared canonical correlations, $\{\widehat{\rho}_k^2\}$, the eigenvalues of $\widehat{\mathbf{\Delta}}(\widehat{\mathbf{\Delta}} + \nu_e\widehat{\mathbf{\Sigma}}_*)^{-1} = \boldsymbol{S}_h(\boldsymbol{S}_h + \boldsymbol{S}_e)^{-1}$, suffice. The only additional values needed to compute any of the tests are the constants $\{a, b, \nu_e\}$ with $\nu_e = N - r$. Under the null, $\{a, b, \nu_e\}$ completely determine the distribution, and all four are exact size α.

Although all four tests are size α, typically they give four distinct p values. Reporting only the smallest is statistically dishonest due to the bias introduced. More precisely, the reported value gives a test with inflated test size.

Specifying the distributions of the four test statistics under the alternative requires knowing only $\{a, b, \nu_e\}$ and the nonzero eigenvalues of the noncentrality matrix $\mathbf{\Omega} = \mathbf{\Delta}\mathbf{\Sigma}_*^{-1}$, or equivalently $\{\rho_k^2\}$, the eigenvalues of $\mathbf{\Delta}(\mathbf{\Delta} + \nu_e\mathbf{\Sigma}_*)^{-1}$. Although $\text{rank}(\widehat{\mathbf{\Omega}}) = s$ with probability 1 and $1 \leq s \leq b$, $\text{rank}(\mathbf{\Omega}) = s_*$ with $0 \leq s_* \leq s$. Here $s_* = 0$ iff $\mathbf{\Omega} = \mathbf{0}$ iff $\mathbf{\Theta} = \mathbf{\Theta}_0$ iff the null hypothesis is true.

If $s = 1$, then the four multivariate test statistics (1) are one to one functions of each other, and (2) provide exactly the same p value and power and, (3) the unique test is optimal in many ways. The test is then exactly size α, invariant, provides the likelihood ratio test and the union-intersection test, and is uniformly most powerful among all size-α and invariant tests.

If $s > 1$, no uniformly most powerful test exists among the class of size-α and invariant tests for finite samples. Which test is most powerful depends on the particular set of eigenvalues of $\boldsymbol{\Omega}$ or equivalently the particular set of $\{\rho_k^2\}$. Some performance differences among the tests are known. Olson (1974, 1976, 1979) provided extensive studies of power and robustness for the multivariate tests. He characterized alternative hypotheses as involving either concentrated noncentrality if $s_* = 1$ or diffuse noncentrality if $s_* > 1$. Concentrated noncentrality ensures RLR is the most powerful test of the four, while diffuse noncentrality ensures RLR is the least powerful test of the four. Olson concluded RLR was the least robust while PBT was the most robust. For the range of cases he considered, he strongly preferred PBT, with LRT in second place.

Example 16.1 The following unusual example occurred in an actual data analysis. For a data set with $s = 5$, $m = 0$, $N = 10$, the following p values were obtained: Wilks lambda $p = 0.1896$, Pillai's trace $p = 0.7799$, Hotelling-Lawley trace $p = 0.0030$, Roy's largest root $p = 0.0001$. Such examples are uncommon. Usually the various test multivariate statistics are in approximate agreement.

16.7 UNIVARIATE APPROACH TO REPEATED MEASURES: UNIREP

Muller and Barton (1989) detailed the origin and history of the UNIREP tests. Although they are not invariant to a full-rank transformation of the transformed response data, they do meet a weaker criterion, namely invariance to a full-rank orthonormal transformation. All linearly invariant tests, including the four "MULTIREP" tests, are also orthonormal invariant.

The sphericity property is met whenever $\boldsymbol{\Sigma}_*$ has all eigenvalues of the same size. With sphericity, $\epsilon = 1$, and the uncorrected test provides an exact size α with uniformly most power, among the class of similarly invariant tests. Without sphericity, the uncorrected test typically has inflated test size, the Geisser-Greenhouse and Huynh-Feldt tests are approximately size α, and the Box test is nearly always very conservative.

John (1972) discussed the likelihood ratio test of sphericity (Mauchly's test) and the locally (near the null) most powerful size-α test of sphericity. The likelihood ratio statistic is a one-to-one function of $|\widehat{\boldsymbol{\Sigma}}_*|/\text{tr}(\widehat{\boldsymbol{\Sigma}}_*)$, while the locally most powerful test is a one-to-one function of $\widehat{\epsilon} = b^{-1}\text{tr}(\widehat{\boldsymbol{\Sigma}}_*)/\text{tr}(\widehat{\boldsymbol{\Sigma}}_*^2)$. In our opinion, neither should be used as a "screening test" to choose a UNIREP test, for the same reasons a test of heterogeneity should not be used to choose between an unadjusted or Satterthwaite t test. The locally most powerful nature of the sphericity test based on $\widehat{\epsilon}$ leads to the speculation that the Geisser-Greenhouse test shares the property.

Coffey and Muller (2003) proved the following lemma, which characterizes the noncentral distribution of the UNIREP statistic. The lemma uses the spectral decomposition $\boldsymbol{\Sigma}_* = \boldsymbol{\Upsilon}\text{Dg}(\boldsymbol{\lambda})\boldsymbol{\Upsilon}'$, with $\boldsymbol{\Upsilon}'\boldsymbol{\Upsilon} = \boldsymbol{I}_b$, which allows defining $\boldsymbol{\Delta}_\Upsilon = \boldsymbol{\Upsilon}'\boldsymbol{\Delta}\boldsymbol{\Upsilon}$ and $\boldsymbol{\Omega}_\Upsilon = \boldsymbol{\Delta}_\Upsilon\text{Dg}(\boldsymbol{\lambda})^{-1}$.

Lemma 16.2 Except for known constants, the distribution of the UNIREP test statistic, $T_u = \left[\text{tr}(\widehat{\boldsymbol{\Delta}})/a\right]/\text{tr}(\widehat{\boldsymbol{\Sigma}}_*)$, is completely and exactly determined by the b eigenvalues of $\boldsymbol{\Sigma}_*$, $\boldsymbol{\lambda}$, and the b noncentralities $\boldsymbol{\omega}_\Upsilon = \{\omega_{\Upsilon kk}\} = \{\boldsymbol{v}_k'(\boldsymbol{\Theta}-\boldsymbol{\Theta}_0)'\boldsymbol{M}^{-1}(\boldsymbol{\Theta}-\boldsymbol{\Theta}_0)\boldsymbol{v}_k/\lambda_k\}$, the diagonal elements of $\boldsymbol{\Omega}_\Upsilon = \boldsymbol{\Delta}_\Upsilon \text{Dg}(\boldsymbol{\lambda})^{-1}$. The same characterization holds for all four UNIREP tests.

Proof. Coffey and Muller (2003) provided a proof.

It is straightforward to prove that the eigenvalues of $\boldsymbol{\Omega}_\Upsilon$ and $\boldsymbol{\Omega}$ coincide. Of course, the eigenvalues of $\boldsymbol{\Omega}_\Upsilon$ coincide with the diagonal values only when $\boldsymbol{\Omega}_\Upsilon$ is diagonal. Hence it is not surprising that Monte Carlo simulations demonstrate the UNIREP tests may be more or less powerful than any of the MULTIREP tests, depending on $\boldsymbol{\lambda}$, $\boldsymbol{\omega}_\Upsilon$, and the eigenvalues of $\boldsymbol{\Omega}_\Upsilon$, which are equivalent to $\{\rho_k^2\}$.

16.8 MORE ON INVARIANCE PROPERTIES

Theorem 16.11 (a) The eigenvalues of $\boldsymbol{S}_h\boldsymbol{S}_e^{-1}$, **(b)** the canonical correlations, **(c)** the usual four MULTIREP test statistics, and **(d)** the associated p values are invariant to a square and full-rank transformation simultaneously applied to $\boldsymbol{Y}_U = \boldsymbol{Y}\boldsymbol{U}$ and $\boldsymbol{\Theta}_0$, with $\boldsymbol{Y}_T = \boldsymbol{Y}_U\boldsymbol{T}$ and $\boldsymbol{\Theta}_{0T} = \boldsymbol{\Theta}_0\boldsymbol{T}$. The same invariance holds under the same type of transformation to $\boldsymbol{U}$ and $\boldsymbol{\Theta}_0$.

Proof. Here $\boldsymbol{S}_e = \boldsymbol{F}\boldsymbol{F}'$ and $\boldsymbol{S}_e^{-1} = \boldsymbol{F}^{-t}\boldsymbol{F}^{-1}$. A similarity transformation allows proving the eigenvalues of $\boldsymbol{S}_h\boldsymbol{S}_e^{-1} = \boldsymbol{S}_h\boldsymbol{F}^{-t}\boldsymbol{F}^{-1}$ coincide with the eigenvalues of $\boldsymbol{F}^{-1}\boldsymbol{S}_h\boldsymbol{F}^{-t}$. If $\boldsymbol{H} = \boldsymbol{X}(\boldsymbol{X}'\boldsymbol{X})^{-}\boldsymbol{X}'$ and $\boldsymbol{M} = \boldsymbol{C}(\boldsymbol{X}'\boldsymbol{X})\boldsymbol{C}'$, then

$$\begin{aligned}\boldsymbol{S}_e &= \boldsymbol{U}'\widehat{\boldsymbol{\Sigma}}\boldsymbol{U}\nu_e \\ &= \boldsymbol{U}'[\boldsymbol{Y}'(\boldsymbol{I}-\boldsymbol{H})\boldsymbol{Y}/\nu_e]\boldsymbol{U}\nu_e \\ &= \boldsymbol{U}'\boldsymbol{Y}'(\boldsymbol{I}-\boldsymbol{H})\boldsymbol{Y}\boldsymbol{U} = \boldsymbol{Y}_U'\boldsymbol{A}_e\boldsymbol{Y}_U \end{aligned} \tag{16.44}$$

and

$$\begin{aligned}\boldsymbol{S}_h &= (\widehat{\boldsymbol{\Theta}}-\boldsymbol{\Theta}_0)'\boldsymbol{M}^{-1}(\widehat{\boldsymbol{\Theta}}-\boldsymbol{\Theta}_0) \\ &= \left(\boldsymbol{C}\widehat{\boldsymbol{B}}\boldsymbol{U}-\boldsymbol{\Theta}_0\right)'\boldsymbol{M}^{-1}\left(\boldsymbol{C}\widehat{\boldsymbol{B}}\boldsymbol{U}-\boldsymbol{\Theta}_0\right) \\ &= [\boldsymbol{C}(\boldsymbol{X}'\boldsymbol{X})^{-}\boldsymbol{X}'\boldsymbol{Y}\boldsymbol{U}-\boldsymbol{\Theta}_0]'\boldsymbol{M}^{-1}[\boldsymbol{C}(\boldsymbol{X}'\boldsymbol{X})^{-}\boldsymbol{X}'\boldsymbol{Y}\boldsymbol{U}-\boldsymbol{\Theta}_0] \\ &= [\boldsymbol{Y}_U'\boldsymbol{X}(\boldsymbol{X}'\boldsymbol{X})^{-}\boldsymbol{C}'-\boldsymbol{\Theta}_0']\boldsymbol{M}^{-1}[\boldsymbol{C}(\boldsymbol{X}'\boldsymbol{X})^{-}\boldsymbol{X}'\boldsymbol{Y}_U-\boldsymbol{\Theta}_0]. \end{aligned} \tag{16.45}$$

If $\boldsymbol{\Theta}_0 = \boldsymbol{0}$, then

$$\begin{aligned}\boldsymbol{S}_h &= \boldsymbol{Y}_U'\boldsymbol{X}(\boldsymbol{X}'\boldsymbol{X})^{-}\boldsymbol{C}'\boldsymbol{M}^{-1}\boldsymbol{C}(\boldsymbol{X}'\boldsymbol{X})^{-}\boldsymbol{X}'\boldsymbol{Y}_U \\ &= \boldsymbol{Y}_U'\boldsymbol{A}_h\boldsymbol{Y}_U\,. \end{aligned} \tag{16.46}$$

In turn,

$$
\begin{aligned}
(\boldsymbol{S}_h\boldsymbol{S}_e^{-1} - \lambda\boldsymbol{I})\boldsymbol{x} &= \boldsymbol{0} \\
\left[(\boldsymbol{Y}_U'\boldsymbol{A}_h\boldsymbol{Y}_U)(\boldsymbol{Y}_U'\boldsymbol{A}_e\boldsymbol{Y}_U)^{-1} - \lambda\boldsymbol{I}\right]\boldsymbol{x} &= \boldsymbol{0} \\
\left[(\boldsymbol{Y}_U'\boldsymbol{A}_h\boldsymbol{Y}_U)\boldsymbol{T}\boldsymbol{T}^{-1}(\boldsymbol{Y}_U'\boldsymbol{A}_e\boldsymbol{Y}_U)^{-1}\boldsymbol{T}^{-t}\boldsymbol{T}' - \lambda\boldsymbol{T}^{-t}\boldsymbol{T}'\right]\boldsymbol{x} &= \boldsymbol{0} \\
\left[(\boldsymbol{Y}_U'\boldsymbol{A}_h\boldsymbol{Y}_U\boldsymbol{T})(\boldsymbol{T}'\boldsymbol{Y}_U'\boldsymbol{A}_e\boldsymbol{Y}_U\boldsymbol{T})^{-1} - \lambda\boldsymbol{T}^{-t}\right]\boldsymbol{T}'\boldsymbol{x} &= \boldsymbol{0} \\
\boldsymbol{T}'\left[(\boldsymbol{Y}_U'\boldsymbol{A}_h\boldsymbol{Y}_U\boldsymbol{T})(\boldsymbol{T}'\boldsymbol{Y}_U'\boldsymbol{A}_e\boldsymbol{Y}_U\boldsymbol{T})^{-1} - \lambda\boldsymbol{T}^{-t}\right]\boldsymbol{T}'\boldsymbol{x} &= \boldsymbol{T}'\boldsymbol{0} \\
\left[(\boldsymbol{T}'\boldsymbol{Y}_U'\boldsymbol{A}_h\boldsymbol{Y}_U\boldsymbol{T})(\boldsymbol{T}'\boldsymbol{Y}_U'\boldsymbol{A}_e\boldsymbol{Y}_U\boldsymbol{T})^{-1} - \lambda\boldsymbol{I}\right]\boldsymbol{x}_T &= \boldsymbol{0} \\
\left[(\boldsymbol{Y}_T'\boldsymbol{A}_h\boldsymbol{Y}_T)(\boldsymbol{Y}_T'\boldsymbol{A}_e\boldsymbol{Y}_T)^{-1} - \lambda\boldsymbol{I}\right]\boldsymbol{x}_T &= \boldsymbol{0}\,. \qquad (16.47)
\end{aligned}
$$

The last form demonstrates the eigenvalues coincide (although eigenvectors differ by the factor of $\boldsymbol{T}'$) if $\boldsymbol{\Theta}_0 = \boldsymbol{0}$. Allowing $\boldsymbol{\Theta}_0 \neq \boldsymbol{0}$ is covered below. □

Corollary 16.11 Without loss of generality, analysis or simulation of the MULTIREP tests for any $\{\boldsymbol{\Sigma}_{*1} = \boldsymbol{\Phi}\boldsymbol{\Phi}', \boldsymbol{\Delta}_1, \boldsymbol{\Omega}_1 = \boldsymbol{\Delta}_1\boldsymbol{\Sigma}_{*1}^{-1}\}$ can be based on the equivalent model with $\boldsymbol{\Sigma}_{*2} = \boldsymbol{I}_b$, $\boldsymbol{\Delta}_2 = \boldsymbol{\Phi}^{-1}\boldsymbol{\Delta}_1\boldsymbol{\Phi}^{-t}$, and $\boldsymbol{\Omega}_2 = \boldsymbol{\Delta}_2\boldsymbol{\Sigma}_{*2}^{-1} = \boldsymbol{\Delta}_2$. Also, $\boldsymbol{\Omega}_2 = \boldsymbol{\Omega}_2'$ has the same eigenvalues as $\boldsymbol{\Omega}_1$ (which need not be symmetric).

Proof. If $\boldsymbol{\Sigma}_{*1} = \boldsymbol{\Phi}\boldsymbol{\Phi}'$ and $\boldsymbol{T} = \boldsymbol{\Phi}^{-t}$, then $[\text{row}_i(\boldsymbol{Y}_T)]' \sim \mathcal{N}_b[(\boldsymbol{X}_i\boldsymbol{B}\boldsymbol{U}\boldsymbol{\Phi}^{-t})', \boldsymbol{I}_b]$. Also, $\widehat{\boldsymbol{\Delta}}_1 \sim \mathcal{W}_b(a, \boldsymbol{\Sigma}_*, \boldsymbol{\Delta}_1)$ and $\boldsymbol{\Phi}^{-1}\widehat{\boldsymbol{\Delta}}_1\boldsymbol{\Phi}^{-t} \sim \mathcal{W}_b(a, \boldsymbol{I}_b, \boldsymbol{\Phi}^{-1}\boldsymbol{\Delta}_1\boldsymbol{\Phi}^{-t})$. Finally, $\boldsymbol{\Omega}_1$ is similar to $\boldsymbol{\Phi}^{-1}(\boldsymbol{\Omega}_1)\boldsymbol{\Phi} = \boldsymbol{\Phi}^{-1}(\boldsymbol{\Delta}_1\boldsymbol{\Phi}^{-t}\boldsymbol{\Phi}^{-1})\boldsymbol{\Phi} = \boldsymbol{\Phi}^{-1}\boldsymbol{\Delta}_1\boldsymbol{\Phi}^{-t} = \boldsymbol{\Delta}_2 = \boldsymbol{\Omega}_2$ and therefore has the same eigenvalues. □

Theorem 16.12 (a) The matrices $\boldsymbol{S}_e$ and $\boldsymbol{S}_h$ are invariant to a square, full-rank, simultaneous transformation of the rows of $\boldsymbol{C}$ and $\boldsymbol{\Theta}_0$, with $\boldsymbol{C}_T = \boldsymbol{T}\boldsymbol{C}$ and $\boldsymbol{\Theta}_{0T} = \boldsymbol{T}\boldsymbol{\Theta}_0$.
(b) The eigenvalues of $\boldsymbol{S}_h\boldsymbol{S}_e^{-1}$, the canonical correlations, the usual four MULTIREP test statistics, and their associated p values are all invariant in the same way.

Proof. (a) $\boldsymbol{S}_e$ is not a function of $\boldsymbol{C}$. If $\boldsymbol{T}$ is $a \times a$ and full rank, with $\boldsymbol{C}_T = \boldsymbol{T}\boldsymbol{C}$,

$$
\begin{aligned}
\boldsymbol{S}_h &= (\boldsymbol{C}\widetilde{\boldsymbol{B}}\boldsymbol{U} - \boldsymbol{\Theta}_0)'[\boldsymbol{C}(\boldsymbol{X}'\boldsymbol{X})^-\boldsymbol{C}']^{-1}(\boldsymbol{C}\widetilde{\boldsymbol{B}}\boldsymbol{U} - \boldsymbol{\Theta}_0) \qquad (16.48) \\
&= (\boldsymbol{C}\widetilde{\boldsymbol{B}}\boldsymbol{U} - \boldsymbol{\Theta}_0)'(\boldsymbol{T}^{-1}\boldsymbol{T})'[\boldsymbol{C}(\boldsymbol{X}'\boldsymbol{X})^-\boldsymbol{C}']^{-1}\boldsymbol{T}^{-1}\boldsymbol{T}(\boldsymbol{C}\widetilde{\boldsymbol{B}}\boldsymbol{U} - \boldsymbol{\Theta}_0) \\
&= (\boldsymbol{T}'\boldsymbol{C}\widetilde{\boldsymbol{B}}\boldsymbol{U} - \boldsymbol{T}\boldsymbol{\Theta}_0)'[\boldsymbol{T}\boldsymbol{C}(\boldsymbol{X}'\boldsymbol{X})^-\boldsymbol{C}'\boldsymbol{T}']^{-1}(\boldsymbol{T}\boldsymbol{C}\widetilde{\boldsymbol{B}}\boldsymbol{U} - \boldsymbol{T}\boldsymbol{\Theta}_0) \\
&= (\boldsymbol{C}_T\widetilde{\boldsymbol{B}}\boldsymbol{U} - \boldsymbol{\Theta}_{0T})'[\boldsymbol{C}_T(\boldsymbol{X}'\boldsymbol{X})^-\boldsymbol{C}_T']^{-1}(\boldsymbol{C}_T\widetilde{\boldsymbol{B}}\boldsymbol{U} - \boldsymbol{\Theta}_{0T})\,,
\end{aligned}
$$

which suffices to prove the invariance of $\boldsymbol{S}_h$. Part **(b)** follows immediately. □

As noted earlier, the parameters determining the distributions of the secondary parameter estimators make it obvious that any reasonable test statistic must be a function of $\widehat{\boldsymbol{\Delta}}$ and $\widehat{\boldsymbol{\Sigma}}_*$. Given that the eigenvalues of $\boldsymbol{S}_h\boldsymbol{S}_e^{-1}$ are invariant, how do we know they are the only invariants for a testable hypothesis? We have proven all four statistics depend on the data *only* through the eigenvalues. Therefore any other feature may change and not affect the test, unless the change carries over into the eigenvalues. They are also obviously the minimal sufficient statistics because changing any one of them changes the test statistics (except perhaps for RLR).

Theorem 16.13 If the model spans an intercept ($\boldsymbol{X t}_0 = \boldsymbol{1}_N$) and the hypothesis excludes the intercept ($\boldsymbol{C t}_0 = \boldsymbol{0}$), then the following hold.
(a) Matrices $\boldsymbol{S}_h$ and $\boldsymbol{S}_e$, the eigenvalues of $\boldsymbol{S}_h \boldsymbol{S}_e^{-1}$, and the canonical correlations, are invariant to a location shift of the form $\boldsymbol{Y}_U + \boldsymbol{1}_N \boldsymbol{t}_1'$ for $b \times 1$ constant vector $\boldsymbol{t}_1$ and $\boldsymbol{Y}_U = \boldsymbol{YU}$.
(b) The MULTIREP test statistics and associated p values are similarly invariant.
(c) The UNIREP test statistics and associated p values are similarly invariant.

Proof. With $\boldsymbol{H} = \boldsymbol{X}(\boldsymbol{X}'\boldsymbol{X})^{-}\boldsymbol{X}'$, the matrices $\boldsymbol{S}_h$ and $\boldsymbol{S}_e$ and the test statistics depend on $\boldsymbol{Y}_U$ only through

$$\widehat{\boldsymbol{\Theta}} = \boldsymbol{C}(\boldsymbol{X}'\boldsymbol{X})^{-}\boldsymbol{X}'\boldsymbol{Y}_U \tag{16.49}$$
$$\widehat{\boldsymbol{E}}_U = (\boldsymbol{I}_N - \boldsymbol{H})\boldsymbol{Y}_U\,, \tag{16.50}$$

with $\widehat{\boldsymbol{\Sigma}}_* = \widehat{\boldsymbol{E}}_U'\widehat{\boldsymbol{E}}_U/\nu_e$. Writing

$$\boldsymbol{C}(\boldsymbol{X}'\boldsymbol{X})^{-}\boldsymbol{X}'(\boldsymbol{Y}_U + \boldsymbol{1}_N\boldsymbol{t}_1') = \boldsymbol{C}(\boldsymbol{X}'\boldsymbol{X})^{-}\boldsymbol{X}'\boldsymbol{Y}_U + \boldsymbol{C}(\boldsymbol{X}'\boldsymbol{X})^{-}\boldsymbol{X}'\boldsymbol{1}_N\boldsymbol{t}_1' \tag{16.51}$$
$$(\boldsymbol{I}_N - \boldsymbol{H})(\boldsymbol{Y}_U + \boldsymbol{1}_N\boldsymbol{t}_1') = (\boldsymbol{I}_N - \boldsymbol{H})\boldsymbol{Y}_U + (\boldsymbol{I}_N - \boldsymbol{H})\boldsymbol{1}_N\boldsymbol{t}_1' \tag{16.52}$$

allows concluding location invariance reduces to proving

$$\boldsymbol{0} = \boldsymbol{C}(\boldsymbol{X}'\boldsymbol{X})^{-}\boldsymbol{X}'\boldsymbol{1}_N\boldsymbol{t}_1' \tag{16.53}$$
$$\boldsymbol{0} = [\boldsymbol{I}_N - \boldsymbol{X}(\boldsymbol{X}'\boldsymbol{X})^{-}\boldsymbol{X}']\boldsymbol{1}_N\boldsymbol{t}_1'\,. \tag{16.54}$$

The zero matrices of the two equations have dimensions $a \times b$ and $N \times b$. In turn, it suffices to prove

$$\boldsymbol{0} = \boldsymbol{C}(\boldsymbol{X}'\boldsymbol{X})^{-}\boldsymbol{X}'\boldsymbol{1}_N \tag{16.55}$$
$$\boldsymbol{0} = [\boldsymbol{I}_N - \boldsymbol{X}(\boldsymbol{X}'\boldsymbol{X})^{-}\boldsymbol{X}']\boldsymbol{1}_N\,. \tag{16.56}$$

The zero matrices have dimensions $a \times 1$ and $n \times 1$.

In the simplest case, $\boldsymbol{X}$ is full rank, the first column is $\boldsymbol{1}_N$, and the remaining columns are centered. Therefore

$$\boldsymbol{X}'\boldsymbol{1}_N = \begin{bmatrix} N \\ \boldsymbol{0}_{(q-1)\times 1} \end{bmatrix} \tag{16.57}$$

and

$$(\boldsymbol{X}'\boldsymbol{X})^{-} = \begin{bmatrix} N & \boldsymbol{0} \\ \boldsymbol{0} & \boldsymbol{X}_d'\boldsymbol{X}_d \end{bmatrix}^{-} = \begin{bmatrix} N^{-1} & \boldsymbol{0} \\ \boldsymbol{0} & (\boldsymbol{X}_d'\boldsymbol{X}_d)^{-1} \end{bmatrix}. \tag{16.58}$$

If the test excludes the intercept, then $\boldsymbol{C} = [\,\boldsymbol{0}_{a\times 1} \quad \boldsymbol{C}_d\,]$ and

$$C(X'X)^{-}X'\mathbf{1}_N = \begin{bmatrix}\mathbf{0}_{a\times 1} & C_d\end{bmatrix}\begin{bmatrix}N^{-1} & \mathbf{0}\\ \mathbf{0} & (X_d'X_d)^{-1}\end{bmatrix}\begin{bmatrix}N\\ \mathbf{0}_{(q-1)\times 1}\end{bmatrix}$$
$$= \begin{bmatrix}\mathbf{0}_{a\times 1} & C_d\end{bmatrix}\begin{bmatrix}1\\ \mathbf{0}_{(q-1)\times 1}\end{bmatrix} = \mathbf{0}_{a\times 1}\,. \tag{16.59}$$

Furthermore

$$[I_N - X(X'X)^{-}X']\mathbf{1}_N = \mathbf{1}_N - X\begin{bmatrix}N^{-1} & \mathbf{0}\\ \mathbf{0} & (X_d'X_d)^{-1}\end{bmatrix}\begin{bmatrix}N\\ \mathbf{0}_{(q-1)\times 1}\end{bmatrix}$$
$$= \mathbf{1}_N - X\begin{bmatrix}1\\ \mathbf{0}_{(q-1)\times 1}\end{bmatrix}$$
$$= \mathbf{1}_N - \mathbf{1}_N = \mathbf{0}_{N\times 1}\,, \tag{16.60}$$

which completes the proof of the special case. The theory of linearly equivalent models allows extending the result to estimable parameters in LTFR cases. □

Theorem 16.14 The UNIREP test statistic, for any population covariance pattern, is invariant to a FR transformation of columns of U and Θ_0 by orthonormal T (but not general FR T). Here $T^{-1} = T'$ and hence $TT' = I_b$.

Proof.

$$F_u = \frac{\text{tr}(S_h)/(ab)}{\text{tr}(S_e)/(b\nu_e)} = \frac{\text{tr}(S_hTT')/(ab)}{\text{tr}(S_eTT')/(b\nu_e)} = \frac{\text{tr}(T'S_hT)/(ab)}{\text{tr}(T'S_eT)/(b\nu_e)}\,. \qquad □$$

Corollary 16.14 In a simulation, choosing $T = \Upsilon$, with $\Sigma_* = \Upsilon\text{Dg}(\lambda)\Upsilon'$ and $Y_T = YUT$, gives $[\text{row}_i(Y_T)]' \sim \mathcal{N}_b[(X_iBU\Upsilon)', \text{Dg}(\lambda)]$.

Theorem 16.15 Without loss of generality, $\Theta_0 = \mathbf{0}$ may be assumed.

Proof. A multivariate $\text{GLM}_{N,p,q}(Y_i; X_iB, \Sigma)$ with Gaussian errors, fixed X, $N \gg r = \text{rank}(X)$, has model equation $Y = XB + E$, with testable secondary parameter $\Theta = CB$ and corresponding hypothesis $H_0 : \Theta = \Theta_0$. Choosing

$$Y_T = Y - XC'(CC')^{-1}\Theta_0 \tag{16.61}$$

defines the model

$$Y_T = XB_T + E_T\,, \tag{16.62}$$

with associated estimable secondary parameters $\Theta_T = CB_T$ and corresponding hypothesis $H_0 : \boldsymbol{\Theta}_T = \mathbf{0}$. Estimability ensures $\boldsymbol{C}(\boldsymbol{X}'\boldsymbol{X})^-\boldsymbol{X}'\boldsymbol{X} = \boldsymbol{C}$. Therefore

$$\begin{aligned}\widehat{\boldsymbol{\Theta}}_T &= \boldsymbol{C}\widetilde{\boldsymbol{B}}_T \\ &= \boldsymbol{C}(\boldsymbol{X}'\boldsymbol{X})^-\boldsymbol{X}'\left[\boldsymbol{Y} - \boldsymbol{X}\boldsymbol{C}'(\boldsymbol{C}\boldsymbol{C}')^{-1}\boldsymbol{\Theta}_0\right] \\ &= \boldsymbol{C}(\boldsymbol{X}'\boldsymbol{X})^-\boldsymbol{X}'\boldsymbol{Y} - [\boldsymbol{C}(\boldsymbol{X}'\boldsymbol{X})^-\boldsymbol{X}'\boldsymbol{X}]\boldsymbol{C}'(\boldsymbol{C}\boldsymbol{C}')^{-1}\boldsymbol{\Theta}_0 \\ &= \boldsymbol{C}(\boldsymbol{X}'\boldsymbol{X})^-\boldsymbol{X}'\boldsymbol{Y} - [\boldsymbol{C}]\boldsymbol{C}'(\boldsymbol{C}\boldsymbol{C}')^{-1}\boldsymbol{\Theta}_0 \\ &= \widehat{\boldsymbol{\Theta}} - \boldsymbol{\Theta}_0 \,. \end{aligned} \tag{16.63}$$

□

The result holds for hypothesis testing, no matter what kind of predictors. It is unclear whether the result applies to power analysis for random predictors.

Theorem 16.16 For any testable general linear hypothesis for a GLM, a linearly equivalent model may be found in which $\boldsymbol{C} = [\,\boldsymbol{I}_a \;\; \mathbf{0}\,]$ provides the original test.

Proof. A GLM (with any sort of predictors), with $N \gg r = \text{rank}(\boldsymbol{X})$, has model equation $\boldsymbol{Y} = \boldsymbol{X}\boldsymbol{B} + \boldsymbol{E}$, with testable secondary parameter $\boldsymbol{\Theta} = \boldsymbol{C}\boldsymbol{B}$ and corresponding hypothesis $H_0 : \boldsymbol{\Theta} = \mathbf{0}$. For any $\boldsymbol{C}$ of dimension $a \times q$, rank a, the singular value decomposition allows writing $\boldsymbol{C} = \boldsymbol{L}[\,\text{Dg}(\boldsymbol{s}_C)\;\mathbf{0}_{a\times(q-a)}\,]\boldsymbol{R}'$ with $\boldsymbol{L}$ and $\text{Dg}(\boldsymbol{s}_C)$ $a \times a$ and $\boldsymbol{R}$ $q \times q$. In turn, $\boldsymbol{C}\boldsymbol{C}' = \boldsymbol{L}\text{Dg}(\boldsymbol{s}_C^2)\boldsymbol{L}'$, and $\boldsymbol{L}\boldsymbol{L}' = \boldsymbol{L}'\boldsymbol{L} = \boldsymbol{I}_a$. With $\boldsymbol{R}_1$ $q \times a$ and $\boldsymbol{R}_0$ $q \times (q-a)$,

$$\boldsymbol{C}'\boldsymbol{C} = [\,\boldsymbol{R}_1 \;\; \boldsymbol{R}_0\,]\begin{bmatrix}\text{Dg}(\boldsymbol{s}_C^2) & \mathbf{0} \\ \mathbf{0} & \mathbf{0}\end{bmatrix}\begin{bmatrix}\boldsymbol{R}_1' \\ \boldsymbol{R}_0'\end{bmatrix}, \tag{16.64}$$

with $\boldsymbol{R}\boldsymbol{R}' = \boldsymbol{R}'\boldsymbol{R} = \boldsymbol{I}_q$. If $\boldsymbol{C}_\perp = \boldsymbol{R}_0'$, which is $(q-a) \times q$, then

$$\begin{aligned}\boldsymbol{T} &= \begin{bmatrix}\boldsymbol{C} \\ \boldsymbol{R}_0'\end{bmatrix} \\ &= \begin{bmatrix}\boldsymbol{L}\text{Dg}(\boldsymbol{s}_C)\boldsymbol{R}_1' \\ \boldsymbol{R}_0'\end{bmatrix} \\ &= \begin{bmatrix}\boldsymbol{L} & \mathbf{0} \\ \mathbf{0} & \boldsymbol{I}\end{bmatrix}\begin{bmatrix}\text{Dg}(\boldsymbol{s}_C) & \mathbf{0} \\ \mathbf{0} & \boldsymbol{I}\end{bmatrix}\boldsymbol{R}' \end{aligned} \tag{16.65}$$

and

$$\begin{aligned}\boldsymbol{T}^{-1} &= [\,\boldsymbol{R}_1 \; \boldsymbol{R}_0\,]\begin{bmatrix}\text{Dg}(\boldsymbol{s}_C)^{-1} & \mathbf{0} \\ \mathbf{0} & \boldsymbol{I}\end{bmatrix}\begin{bmatrix}\boldsymbol{L}' & \mathbf{0} \\ \mathbf{0} & \boldsymbol{I}\end{bmatrix} \\ &= [\,\boldsymbol{R}_1 \; \boldsymbol{R}_0\,]\begin{bmatrix}\text{Dg}(\boldsymbol{s}_C)^{-1}\boldsymbol{L}' & \mathbf{0} \\ \mathbf{0} & \boldsymbol{I}\end{bmatrix} \\ &= [\,\boldsymbol{R}_1\text{Dg}(\boldsymbol{s}_C)^{-1}\boldsymbol{L}' \;\; \boldsymbol{R}_0\,] \\ &= [\,\boldsymbol{C}^+ \;\; \boldsymbol{R}_0\,] \\ &= \left[\,\boldsymbol{C}'(\boldsymbol{C}\boldsymbol{C}')^{-1} \;\; \boldsymbol{R}_0\,\right]. \end{aligned} \tag{16.66}$$

In turn,

$$\begin{aligned} Y &= XT^{-1}TB + E \\ &= \left[XC'(CC')^{-1} \quad XR_0 \right] \begin{bmatrix} \Theta \\ \Theta_\perp \end{bmatrix} + E \\ &= \left[X_{Tc} \; X_{T0} \right] \begin{bmatrix} \Theta \\ \Theta_\perp \end{bmatrix} + E \\ &= X_T \begin{bmatrix} \Theta \\ \Theta_\perp \end{bmatrix} + E \,. \end{aligned} \tag{16.67}$$

□

The proof is a special case of one approach to a "completion" theorem, seen in various contexts in linear models and matrix theory. The basic idea is part of the theory and process behind the concept of linearly equivalent models. A parallel approach is useful for identifying the estimable part of a LTFR model.

A similar result may be proven for $U = I$. The result holds for hypothesis testing, no matter whether predictors are fixed or only conditionally fixed. It is unclear whether the result applies to random predictors.

16.9 TESTS OF HYPOTHESES ABOUT Σ

Multivariate models raise the interesting possibility of estimating constrained structures and testing hypotheses about the structure of $\boldsymbol{\Sigma}$. Many exact and approximate results have been developed for such tasks. Such techniques may be roughly described as falling into one of four types.

One class of tests concerns testing hypotheses about functions of $\boldsymbol{\Sigma}$, such as the trace or determinant. For central Wishart matrices, both have simple forms. Each can be thought of as the generalized variance, with the trace corresponding to the arithmetic average eigenvalue and the determinant to the geometric mean,

$$\text{tr}(\boldsymbol{\Sigma})/p = \sum_{k=1}^{p} \lambda_k / p = \overline{\lambda}_A \tag{16.68}$$

$$|\boldsymbol{\Sigma}|^{1/p} = \left(\prod_{k=1}^{p} \lambda_k \right)^{1/p} = \overline{\lambda}_G \,. \tag{16.69}$$

The interpretation is strengthened by recognizing that the eigenvalues of a covariance matrix equal the variances of the underlying principal components.

A second class of tests involves $H_0 : \boldsymbol{\Sigma} = \boldsymbol{\Sigma}_0$, based on $\widehat{\boldsymbol{\Sigma}} \sim \mathcal{W}_p(\nu, \boldsymbol{\Sigma}_0, \mathbf{0})$ and $\boldsymbol{\Sigma}_0$ a particular structure. Anderson (2004), Morrison (1990), and Timm (2002, Section 3.8) provided useful treatments of some techniques. Some specific tests have been developed for special patterns, including $\boldsymbol{\Sigma}_0 = \sigma^2 \boldsymbol{I}_p$, sphericity, and $\boldsymbol{\Sigma}_0 = \sigma^2 [\rho \mathbf{1}_p \mathbf{1}_p' + (1-\rho)\boldsymbol{I}_p]$, compound symmetry (Lemma 1.33).

Tests for sphericity have an unfortunate history in the "univariate approach" to repeated measures. Violation of sphericity may greatly inflate the test size of the uncorrected UNIREP test. The problem led to using a nonsignificant test of sphericity as justification for using an uncorrected test. The process is the multivariate generalization of testing for homogeneity of variance between groups as part of a univariate ANOVA for independent observations. The process fails to control test size while providing a false sense of security.

In contrast, the following decision path is valid. One should assume compound symmetry if and only if the sampling scheme guarantees it. Most importantly, we have never encountered repeated measures in time which seemed likely to meet the assumption. If the sampling scheme does imply meeting the assumption, then use the uncorrected test. Otherwise, use either the Geisser-Greenhouse corrected test (preferred by the present authors) or the Huynh-Felt corrected test, which is likely somewhat more widely used. Muller, Edwards, Simpson, and Taylor (2006) and Muller and Barton (1989) provided additional discussion. Careful consideration of Kirk's (1995) recommendation for a three-step process allows concluding it always gives the same test result as always using the Geisser-Greenhouse test.

A third class of tests involve techniques known as *factor analysis*, which provide tools for developing models of correlation and covariance matrices. The key idea is to assume each response variable equals a linear combination of a set of unobserved, underlying, *latent variables* plus a component unique to the variable. The name *common factor model* reflects the decomposition in terms of shared, or common, and unique pieces. Advances in computing power allowed focusing on maximum likelihood methods for *covariance structure* models. From the perspective of the multivariate GLM, such a model allows only a simple design matrix, such as one between group factor, and complex specification of covariance structure. McDonald (1985) and Jöreskog (1993) gave thorough presentations.

The fourth and final type of tests involves mixed models. A general linear mixed model requires an explicit choice for a covariance model, commonly described as specifying the random effects. The validity and numerical feasibility of such a model depends on choosing a scientifically defensible covariance model with sufficiently few parameters. Current methods use large-sample theory for comparing covariance structures. The fragility of mixed models with respect to the accuracy of the covariance model makes such statistics, especially in small samples, important in fitting mixed models. Regrettably, even less is known about statistics focused on such random effects than is known about statistics focused on fixed effects.

16.10 CONFIDENCE REGIONS FOR Θ

Section 2.10 contains a definition of confidence regions for $\boldsymbol{\theta}$ in the univariate model. Inverting a hypothesis test creates a confidence region, and inverting a confidence region yields a unique hypothesis test. Confidence regions exist only for parameters that are estimable and testable. Section 15.6 contains proofs of the results as well as a description of the inversion process and examples.

In Section 3.10, replacing $\{\boldsymbol{y}, \boldsymbol{\theta}, \sigma^2\}$ by $\{\text{vec}(\boldsymbol{Y}), \text{vec}(\boldsymbol{\Theta}), \text{vec}(\boldsymbol{\Sigma})\}$ allowed extending the definiton of confidence regions from the univariate model to the multivariate model. Consequently, separate multivariate proofs are not needed for the following two theorems. In multivariate notation, the essence of the relationship between hypothesis tests and confidence regions is that $R(\boldsymbol{Y}) = [\boldsymbol{\Theta}_0 : \boldsymbol{Y} \in AR(\boldsymbol{\Theta}_0)]$ while $AR(\boldsymbol{\Theta}_0) = [\boldsymbol{Y}_* : \boldsymbol{\Theta}_0 \in R(\boldsymbol{Y}_*)]$.

Theorem 16.17 If for any $\boldsymbol{\Theta}_0 \in S$ a size-α test exists, $\phi(\boldsymbol{Y})$, of hypothesis $H(\boldsymbol{\Theta}_0) = \mathbb{B}(\boldsymbol{\Theta} = \boldsymbol{\Theta}_0)$, then there exists a corresponding confidence region for $\boldsymbol{\Theta}$ with confidence coefficient $c(\alpha) = 1 - \alpha$. Furthermore, if the acceptance region of $\phi(\boldsymbol{Y})$ is $AR(\boldsymbol{\Theta}_0) = [\boldsymbol{Y}_* : \phi(\boldsymbol{Y}_*) = 0]$, then the corresponding confidence region is $R(\boldsymbol{Y}) = [\boldsymbol{\Theta}_0 : \boldsymbol{Y} \in AR(\boldsymbol{\Theta}_0)]$.

Theorem 16.18 If there exists an exact $100(1 - \alpha)$ percent confidence region for Θ, then **(a)** a corresponding test procedure of size α exists for testing $H(\Theta_0) = \mathbb{B}(\Theta = \Theta_0)$ for any $\Theta_0 \in \Theta$ and
(b) $H(\Theta_0)$ is a testable hypothesis.
(c) If the confidence region is $R(\boldsymbol{Y})$, then the acceptance region of the corresponding test is $AR(\Theta_0) = [\boldsymbol{Y}_* : \Theta_0 \in R(\boldsymbol{Y}_*)]$.

Example 16.2 A $\text{GLM}_{N,p,q}(\boldsymbol{Y}_i; \boldsymbol{X}_i\boldsymbol{B}, \Sigma)$ with Gaussian errors has $\text{rank}(\boldsymbol{X}) = r \leq q$, while $t(Y; \Theta_0)$ represents any of the multivariate test statistics based on $\boldsymbol{S}_h = (\widehat{\boldsymbol{\Theta}} - \boldsymbol{\Theta}_0)'\boldsymbol{M}^{-1}(\widehat{\boldsymbol{\Theta}} - \boldsymbol{\Theta}_0)$ and $\boldsymbol{S}_e = \nu_e \boldsymbol{U}'\widehat{\boldsymbol{\Sigma}}\boldsymbol{U}$. When inverting the test based on t to obtain a confidence region for Θ $(a \times b)$, the test $\phi(\boldsymbol{Y}) = \mathbb{B}(t > t_{\text{crit}})$ of $H(\boldsymbol{\Theta}_0) = \mathbb{B}(\boldsymbol{\Theta} = \boldsymbol{\Theta}_0)$ has acceptance region

$$AR(\boldsymbol{\Theta}_0) = [\boldsymbol{Y} : t(Y; \Theta_0) \leq t_{\text{crit}}]. \tag{16.70}$$

Here t_{crit} is the appropriate critical value. The corresponding $100(1 - \alpha)$ percent ellipsoidal confidence region is

$$R(\boldsymbol{Y}) = [\boldsymbol{\Theta}_0 : \boldsymbol{Y} \in AR(\boldsymbol{\Theta}_0)]. \tag{16.71}$$

The region is a hyperellipsoidal and has ab dimensions. If $a = 1$, then

$$\begin{aligned} t(Y; \Theta_0) &= \text{tr}(\boldsymbol{S}_h \boldsymbol{S}_e^{-1}) \\ &= \text{tr}\{[\boldsymbol{C}(\boldsymbol{X}'\boldsymbol{X})^- \boldsymbol{C}']^{-1}\}(\widehat{\boldsymbol{\Theta}} - \Theta_0)\boldsymbol{S}_e^{-1}(\widehat{\boldsymbol{\Theta}} - \Theta_0)' \\ &= F(Y; \Theta_0)\frac{p(N - r)}{N(N - p)}, \end{aligned} \tag{16.72}$$

with $F(Y; \Theta_0) \sim F(p, n - p)$. It follows that the $100(1 - \alpha)$ percent ellipsoidal confidence region is

$$R(\boldsymbol{Y}) = [\boldsymbol{\Theta}_0 : (\widehat{\boldsymbol{\Theta}} - \boldsymbol{\Theta}_0)\boldsymbol{S}_e^{-1}(\widehat{\boldsymbol{\Theta}} - \boldsymbol{\Theta}_0)' \leq \lambda] , \tag{16.73}$$

in which $\lambda = t_{\text{crit}}/\mathbf{tr}\{[\boldsymbol{C}(\boldsymbol{X}'\boldsymbol{X})^{-}\boldsymbol{C}']^{-1}\}$.

EXERCISES

16.1 Model 1 is

$$[\boldsymbol{y}_1 \quad \boldsymbol{y}_2 \quad \boldsymbol{y}_3] = \left(\mathbf{1}_n \otimes \begin{bmatrix} 1 & 1 & 0 & 0 \\ 1 & 0 & 1 & 0 \\ 1 & 0 & 0 & 1 \end{bmatrix}\right)\boldsymbol{B}_1 + \boldsymbol{E}$$
$$\boldsymbol{Y} = \boldsymbol{X}_1\boldsymbol{B}_1 + \boldsymbol{E}.$$

Finding a valid test of the general linear hypothesis $H_0 : \boldsymbol{CBU} = \boldsymbol{\Theta}_0$ requires verifying that (a) $\boldsymbol{\Theta} = \boldsymbol{CBU}$ is estimable and (b) the hypothesis is testable. In the following, with $\boldsymbol{\Theta}_j = \boldsymbol{C}_j\boldsymbol{B}\boldsymbol{U}_j$, verify (a) holds or does not hold, and (b) holds or does not hold. It is not sufficient to merely state the correct answer. You must briefly justify each positive or negative answer. For each estimable $\boldsymbol{\Theta}_j$, describe each element very briefly as a function of cell means.

16.1.1 $\boldsymbol{C}_1 = [1 \quad 0 \quad 0 \quad 0]$ $\quad$ $\boldsymbol{U}_1 = \boldsymbol{I}_3$

16.1.2 $\boldsymbol{C}_2 = [0 \quad 1 \quad 0 \quad 0]$ $\quad$ $\boldsymbol{U}_2 = [\mathbf{1}_2 \quad -\boldsymbol{I}_2]'$

16.1.3 $\boldsymbol{C}_3 = [1 \quad 1 \quad 0 \quad 0]$ $\quad$ $\boldsymbol{U}_3 = [\boldsymbol{U}_2 \quad -\boldsymbol{U}_2]$

16.1.4 $\boldsymbol{C}_4 = [3 \quad 1 \quad 1 \quad 1]/3$ $\quad$ $\boldsymbol{U}_4 = \boldsymbol{I}_3$

16.1.5 $\boldsymbol{C}_5 = \begin{bmatrix} 0 & 1 & -1 & 0 \\ 0 & 1 & 0 & -1 \\ 0 & 0 & 1 & -1 \end{bmatrix}$ $\quad$ $\boldsymbol{U}_5 = \begin{bmatrix} 1 & -1 & 1 \\ 1 & 0 & -2 \\ 1 & 1 & 1 \end{bmatrix}$

16.1.6 $\boldsymbol{C}_6 = \begin{bmatrix} 0 & 1 & 0 & -1 \\ 0 & 0 & 1 & -1 \end{bmatrix}$ $\quad$ $\boldsymbol{U}_6 = \boldsymbol{U}_5$

16.2 For this exercise, you may use any results in the book up to immediately before Lemma 16.1.

16.2.1 Prove part (a) of Lemma 16.1.

16.2.2 Prove part (b) of Lemma 16.1.

16.2.3 Prove part (c) of Lemma 16.1.

16.3.1 With $\boldsymbol{S}_h$ as defined in Chapter 3 and Chapter 16, prove directly that $\text{tr}(\boldsymbol{S}_h)$ is invariant to a full-rank and orthonormal transformation of the columns of $\boldsymbol{U}$. Assume that $\boldsymbol{\Theta}_0 = \mathbf{0}$. (A separate proof, which is *not* part of this exercise, allows concluding that this may be done without loss of generality.)

16.3.2 With $\widehat{\boldsymbol{\Sigma}}_*$ as defined in Chapter 3 and Chapter 16, prove directly that $\text{tr}(\widehat{\boldsymbol{\Sigma}}_*)$ is invariant to a full rank and orthonormal transformation of the columns of $\boldsymbol{U}$.

16.3.3 Prove directly that the UNIREP F statistic is invariant to multiplying $\boldsymbol{C}$ (between-subject contrast matrix) by a nonzero constant. Assume that $\boldsymbol{\Theta}_0 = \mathbf{0}$.

16.3.4 Prove directly that the UNIREP F statistic is invariant to multiplying $\boldsymbol{U}$ (within-subject contrast matrix) by a nonzero constant. Assume that $\boldsymbol{\Theta}_0 = \mathbf{0}$.

16.4 A complete, unbalanced, repeated-measures ANOVA, with p repeated measures, may be written using cell mean coding as

$$\boldsymbol{Y} = \begin{bmatrix} \mathbf{1}_{N_1} & \mathbf{0} & \mathbf{0} \\ \mathbf{0} & \ddots & \mathbf{0} \\ \mathbf{0} & \mathbf{0} & \mathbf{1}_{N_G} \end{bmatrix} \begin{bmatrix} \mu_{1,1} & \cdots & \mu_{1,p} \\ \vdots & \ddots & \vdots \\ \mu_{G,1} & \cdots & \mu_{G,p} \end{bmatrix} + \boldsymbol{E}.$$

Assuming i.i.d. $[\text{row}_i(\boldsymbol{E})]' \sim \mathcal{N}_p(\mathbf{0}, \boldsymbol{\Sigma})$, a testable general linear hypothesis has the form $H_0 : \boldsymbol{\Theta} = \boldsymbol{\Theta}_0$, with $\boldsymbol{\Theta} = \boldsymbol{CBU}$. It is convenient to write $\sum_{g=1}^{G} N_g = N_+$.

16.4.1 Express the distribution of $\boldsymbol{Y}$ in terms of a (direct-product) matrix Gaussian.

16.4.2 *Ignoring group*, the set of p averages (grand means), with one for each response variable, can be computed directly as a linear transformation of $\boldsymbol{Y}$. Provide simple expressions for the transformation. Also provide an explicit form for the joint distribution of the set of p averages.

16.4.3 For any full-rank model, $\widehat{\boldsymbol{B}} \sim \mathcal{N}_{q,p}\left[\boldsymbol{B}, (\boldsymbol{X}'\boldsymbol{X})^{-1}, \boldsymbol{\Sigma}\right]$ (here $q = G$).

(a) Provide a simple expression for $(\boldsymbol{X}'\boldsymbol{X})^{-1}$.

(b) With $\boldsymbol{C} = \mathbf{1}_G'/G$ and $\boldsymbol{U} = \boldsymbol{I}_p$, provide an explicit form for the distribution of $\widehat{\boldsymbol{\Theta}}$ (mean of the group means, one per variable).

(c) Testing the hypothesis defined by $\boldsymbol{C} = \mathbf{1}_G'/G$ and $\boldsymbol{U} = \boldsymbol{I}_p$ involves $\boldsymbol{S}_h = \widehat{\boldsymbol{\Theta}}'\boldsymbol{M}^{-1}\widehat{\boldsymbol{\Theta}}$. Provide an explicit form for $\boldsymbol{M}^{-1}$. What relevance, if any, does the term "harmonic mean" have?

(d) Assuming the alternative hypothesis holds, name the distribution and provide explicit forms for the parameters of the distribution of $\boldsymbol{S}_h$.

16.5 Assume $y_{ig} \sim \mathcal{N}_1\left(\mu_g, \sigma_g^2\right)$ for constant $0 < \sigma_g^2 < \infty$, constant finite μ_g, $i \in \{1, \dots, n\}$, and $g \in \{1, 2\}$, with y_{ig} independent of $y_{i'g'}$ unless $i = i'$ and $g = g'$. The exercise concerns the model

$$\boldsymbol{y}_s = \boldsymbol{X\beta} + \boldsymbol{e}_s$$
$$\begin{bmatrix} \boldsymbol{y}_1 \\ \boldsymbol{y}_2 \end{bmatrix} = \begin{bmatrix} \mathbf{1}_n & \mathbf{0} \\ \mathbf{0} & \mathbf{1}_n \end{bmatrix} \begin{bmatrix} \mu_1 \\ \mu_2 \end{bmatrix} + \begin{bmatrix} \boldsymbol{e}_1 \\ \boldsymbol{e}_2 \end{bmatrix},$$

with $\boldsymbol{y}_g' = [\, y_{1g} \cdots y_{ng} \,]$ and testing hypothesis 1, $H_0 : \mu_1 = \mu_2$ versus $H_A : \mu_1 \neq \mu_2$.

16.5.1 Fully specify the distribution of $\boldsymbol{e}_s$.

16.5.2 Specify the likelihood function for $\boldsymbol{y}_s$.

16.5.3 Describe maximum likelihood estimators for $\{\mu_1, \mu_2, \sigma_1^2, \sigma_2^2\}$. You do not need to verify the validity of solution as a maximum (to save time in the exercises).

16.5.4 Briefly indicate why the likelihood ratio test statistic for hypothesis 1 is *not* a one-to-one function of a t (or F) random variable, such as occurs for the usual independent groups t.

16.5.5 Alternately, consider testing hypothesis 2, $H_0 : (\mu_1 = \mu_2) \cap (\sigma_1^2 = \sigma_2^2)$, versus $H_A : (\mu_1 \neq \mu_2) \cup (\sigma_1^2 \neq \sigma_2^2)$. Give an explicit form for the maximum log likelihood under H_0.

Hint: Take advantage of special cases of results covered in the text. Doing so will greatly reduce the work for the exercise.

16.6 (*Optional, noncredit.*) Assume $[\text{row}_i(\boldsymbol{Y}_g)]' = \boldsymbol{Y}_{ig}' \sim \mathcal{N}_p(\boldsymbol{\mu}_g, \boldsymbol{\Sigma}_g)$ for constant $0 < |\boldsymbol{\Sigma}_g| < \infty$, constant finite $\boldsymbol{\mu}_g$, $i \in \{1, \ldots, n\}$, and $g \in \{1, 2\}$, with $\boldsymbol{Y}_{ig}$ independent of $\boldsymbol{Y}_{i'g'}$ unless $i = i'$ and $g = g'$. This question concerns the model

$$\boldsymbol{Y}_s = \boldsymbol{X}\boldsymbol{B} + \boldsymbol{E}_s$$
$$\begin{bmatrix} \boldsymbol{Y}_1 \\ \boldsymbol{Y}_2 \end{bmatrix} = \begin{bmatrix} \mathbf{1}_n & \mathbf{0} \\ \mathbf{0} & \mathbf{1}_n \end{bmatrix} \begin{bmatrix} \boldsymbol{\mu}_1' \\ \boldsymbol{\mu}_2' \end{bmatrix} + \begin{bmatrix} \boldsymbol{E}_1 \\ \boldsymbol{E}_2 \end{bmatrix}$$

and testing hypothesis 1, $H_0 : \boldsymbol{\mu}_1 = \boldsymbol{\mu}_2$, versus $H_A : \boldsymbol{\mu}_1 \neq \boldsymbol{\mu}_2$.
16.6.1 Fully specify the distribution of $\boldsymbol{E}_s$.
16.6.2 Specify the likelihood function for $\boldsymbol{Y}_s$.
16.6.3 Describe maximum likelihood estimators for $\{\boldsymbol{\mu}_1, \boldsymbol{\mu}_2, \boldsymbol{\Sigma}_1, \boldsymbol{\Sigma}_2\}$. You do not need to verify the validity of solution as a maximum (to save time in the exercises).
16.6.4 Briefly indicate why the likelihood ratio test statistic for hypothesis 1 is *not* a one-to-one function of a t (or F) random variable, such as occurs for the usual independent groups Hotelling T^2.
16.6.5 Alternately, consider testing hypothesis 2, $H_0 : (\boldsymbol{\mu}_1 = \boldsymbol{\mu}_2) \cap (\boldsymbol{\Sigma}_1 = \boldsymbol{\Sigma}_2)$, versus $H_A : (\boldsymbol{\mu}_1 \neq \boldsymbol{\mu}_2) \cup (\boldsymbol{\Sigma}_1 \neq \boldsymbol{\Sigma}_2)$. Give an explicit form for the maximum log likelihood under H_0.

CHAPTER 17

Tests for Generalizations of Multivariate Linear Models

17.1 MOTIVATION

Generalizations of the multivariate linear model typically allow straightforward estimation. In contrast, accurate inference in small samples often proves difficult.

Example 17.1 A $\text{GLM}_{N,p,q}(\boldsymbol{Y}_i; \boldsymbol{X}_i\boldsymbol{B}, \Sigma)$ for blood sugar measured once per day at noon on Monday–Friday has $N \times 5$ $\boldsymbol{Y}$. The scientists wish to assess how well the amount of sugars consumed at breakfast (measured once per day, Monday–Friday) predicts blood sugar levels. A model with only linear effects is

$$\boldsymbol{Y} = \boldsymbol{X}\boldsymbol{B}_1 + \boldsymbol{E}_1 \tag{17.1}$$

$$[\,\boldsymbol{y}_{\text{M}}\ \boldsymbol{y}_{\text{Tu}}\ \boldsymbol{y}_{\text{W}}\ \boldsymbol{y}_{\text{Th}}\ \boldsymbol{y}_{\text{F}}\,] = [\,\boldsymbol{1}_N\ \boldsymbol{x}_{\text{M}}\ \boldsymbol{x}_{\text{Tu}}\ \boldsymbol{x}_{\text{W}}\ \boldsymbol{x}_{\text{Th}}\ \boldsymbol{x}_{\text{F}}\,] \begin{bmatrix} \beta_{0,\text{M}} & \beta_{0,\text{Tu}} & \beta_{0,\text{W}} & \beta_{0,\text{Th}} & \beta_{0,\text{F}} \\ \beta_{1,\text{M}} & \beta_{1,\text{Tu}} & \beta_{1,\text{W}} & \beta_{1,\text{Th}} & \beta_{1,\text{F}} \\ \beta_{2,\text{M}} & \beta_{2,\text{Tu}} & \beta_{2,\text{W}} & \beta_{2,\text{Th}} & \beta_{2,\text{F}} \\ \beta_{3,\text{M}} & \beta_{3,\text{Tu}} & \beta_{3,\text{W}} & \beta_{3,\text{Th}} & \beta_{3,\text{F}} \\ \beta_{4,\text{M}} & \beta_{4,\text{Tu}} & \beta_{4,\text{W}} & \beta_{4,\text{Th}} & \beta_{4,\text{F}} \\ \beta_{5,\text{M}} & \beta_{5,\text{Tu}} & \beta_{5,\text{W}} & \beta_{4,\text{Th}} & \beta_{5,\text{F}} \end{bmatrix} + \boldsymbol{E}_1 .$$

Unfortunately, the model uses sugar consumed on Friday to help predict blood sugar on Monday. A more reasonable model constrains all illogical coefficients to be zero, with nonzero coefficients only for predictors on or before the day:

$$\boldsymbol{Y} = \boldsymbol{X}\boldsymbol{B}_2 + \boldsymbol{E}_2 \tag{17.2}$$

$$[\,\boldsymbol{y}_{\text{M}}\ \boldsymbol{y}_{\text{Tu}}\ \boldsymbol{y}_{\text{W}}\ \boldsymbol{y}_{\text{Th}}\ \boldsymbol{y}_{\text{F}}\,] = [\,\boldsymbol{1}_N\ \boldsymbol{x}_{\text{M}}\ \boldsymbol{x}_{\text{Tu}}\ \boldsymbol{x}_{\text{W}}\ \boldsymbol{x}_{\text{Th}}\ \boldsymbol{x}_{\text{F}}\,] \begin{bmatrix} \beta_{0,\text{M}} & \beta_{0,\text{Tu}} & \beta_{0,\text{W}} & \beta_{0,\text{Th}} & \beta_{0,\text{F}} \\ \beta_{1,\text{M}} & \beta_{1,\text{Tu}} & \beta_{1,\text{W}} & \beta_{1,\text{Th}} & \beta_{1,\text{F}} \\ 0 & \beta_{2,\text{Tu}} & \beta_{2,\text{W}} & \beta_{2,\text{Th}} & \beta_{2,\text{F}} \\ 0 & 0 & \beta_{3,\text{W}} & \beta_{3,\text{Th}} & \beta_{3,\text{F}} \\ 0 & 0 & 0 & \beta_{4,\text{Th}} & \beta_{4,\text{F}} \\ 0 & 0 & 0 & 0 & \beta_{5,\text{F}} \end{bmatrix} + \boldsymbol{E}_2 .$$

A simpler model assumes that only the covariate from the same day is a predictor:

$$\boldsymbol{Y} = \boldsymbol{X}\boldsymbol{B}_3 + \boldsymbol{E}_3 \tag{17.3}$$

$$\left[\, \boldsymbol{y}_{\mathrm{M}}\ \boldsymbol{y}_{\mathrm{Tu}}\ \boldsymbol{y}_{\mathrm{W}}\ \boldsymbol{y}_{\mathrm{Th}}\ \boldsymbol{y}_{\mathrm{F}} \,\right] = \left[\, \boldsymbol{1}_N\ \boldsymbol{x}_{\mathrm{M}}\ \boldsymbol{x}_{\mathrm{Tu}}\ \boldsymbol{x}_{\mathrm{W}}\ \boldsymbol{x}_{\mathrm{Th}}\ \boldsymbol{x}_{\mathrm{F}} \,\right] \begin{bmatrix} \beta_{0,\mathrm{M}} & \beta_{0,\mathrm{Tu}} & \beta_{0,\mathrm{W}} & \beta_{0,\mathrm{Th}} & \beta_{0,\mathrm{F}} \\ \beta_{1,\mathrm{M}} & 0 & 0 & 0 & 0 \\ 0 & \beta_{2,\mathrm{Tu}} & 0 & 0 & 0 \\ 0 & 0 & \beta_{3,\mathrm{W}} & 0 & 0 \\ 0 & 0 & 0 & \beta_{4,\mathrm{Th}} & 0 \\ 0 & 0 & 0 & 0 & \beta_{5,\mathrm{F}} \end{bmatrix} + \boldsymbol{E}_2 \,.$$

All three models have a simple Gaussian likelihood. Derivatives give equations that can be solved iteratively for estimators of $\boldsymbol{B}_j$ and Σ. and the likelihood ratio test is simple to compute. However, any pair of $\{\boldsymbol{B}_1, \boldsymbol{B}_2, \boldsymbol{B}_3\}$ differ due to *nonlinear* constraints, which disallows applying standard linear hypothesis theory.

17.2 DOUBLY MULTIVARIATE MODELS

As discussed in Chapter 4, some doubly multivariate settings may be analyzed with combinations of multivariate models. The approach has the advantage of exact properties (often a Bonferroni correction must control overall test size).

Timm (Section 6.7, 2002) reviewed doubly multivariate models based on a direct-product covariance assumption. The work of Boik (1988, 1991) deserves special attention because his methods appear to control test size in small samples.

17.3 MISSING RESPONSES IN MULTIVARIATE LINEAR MODELS

Section 3.12 includes a brief discussion of missing data in the multivariate linear model. For data missing at random, the methods of Catellier and Muller (2000) very nearly control test size, even in small samples ($N = 12$). In contrast, current mixed model tests may dramatically inflate test size. Section A.2 in the Appendix contains a description of free software (which may be downloaded from the Web) to implement the methods.

17.4 EXACT AND APPROXIMATE WEIGHTED LEAST SQUARES

Knowing exact weights allows transforming the linear model to a new model which exactly follows the usual assumptions. Some commercial software incorporates options to simplify the process.

Approximate and iterated approximate weighted least squares typically leads to very optimistic tests and inference, at least in small samples. Some form of corrected or test should be used, if available. Estimation is typically well behaved.

17.5 SEEMINGLY UNRELATED REGRESSIONS

Seemingly unrelated regression models (Srivastava and Giles, 1987) also allow accurate estimation. However, the specific nature of the model has allowed statisticians to develop approximate tests which work reasonably well in small samples (Rocke, 1989). Also, Timm (2002, Section 5.14) gave a useful review of testing, including some newer work.

17.6 GROWTH CURVE MODELS (GMANOVA)

From the perspective of today, early formulations of growth curve models may be transformed exactly into special cases of the general linear multivariate model. Often a GCM expresses repeated measures as a polynomial function of time. A quadratic model uses only the average, linear, and quadratic trends and hence $p_T = 3$ coefficients. The approach corresponds to restricting attention to a reduced model defined by a transformation. For a $\text{GLM}_{N,p,q}(\boldsymbol{Y}_i; \boldsymbol{X}_i\boldsymbol{B}, \Sigma)$ with $p = 4$ equally spaced repeated measures,

$$\boldsymbol{U}_T = \begin{bmatrix} 1 & -3 & 1 \\ 1 & -1 & -1 \\ 1 & 1 & -1 \\ 1 & 3 & 1 \end{bmatrix} \begin{bmatrix} 4 & 0 & 0 \\ 0 & 20 & 0 \\ 0 & 0 & 4 \end{bmatrix}^{-1/2} \tag{17.4}$$

defines an $N \times 3$ matrix of lower order trend scores,

$$\boldsymbol{Y}_T = \boldsymbol{Y}\boldsymbol{U}_T. \tag{17.5}$$

Transforming the original model gives

$$\begin{aligned} \boldsymbol{Y}\boldsymbol{U}_T &= \boldsymbol{X}\boldsymbol{B}\boldsymbol{U}_T + \boldsymbol{E}\boldsymbol{U}_T \\ \boldsymbol{Y}_T &= \boldsymbol{X}\boldsymbol{B}_T + \boldsymbol{E}_T . \end{aligned} \tag{17.6}$$

The theory of the multivariate linear model applies exactly to the reduced model.

The $N \times 1$ set of cubic trend scores has been ignored in the reduced model. Later formulations introduced the use of the ignored high-order trends as covariates to increase precision of the estimators. Unfortunately, the process makes hypothesis tests and confidence intervals optimistic (Berger, 1986).

17.7 TESTING HYPOTHESES IN THE GCM

The GCM is a multivariate GLM with Gaussian errors and $p - q$ linear restrictions on $\boldsymbol{B}$. The unrestricted model is $\text{GLM}_{N,p,q}(\boldsymbol{Y}_i; \boldsymbol{X}_i\boldsymbol{B}_*, \Sigma)$, while the restrictions give $\text{GLM}_{N,p,m}(\boldsymbol{Y}_i;\boldsymbol{X}_i\boldsymbol{B}_*|\boldsymbol{B}_*=\boldsymbol{B}_1\boldsymbol{T}_1,\Sigma)$ with $\boldsymbol{B}_*$ $(q\times p)$. Equivalently

$$\text{E}(\boldsymbol{Y}) = \boldsymbol{X}\boldsymbol{B}_* \tag{17.7}$$

$$\mathcal{V}[\text{vec}(\boldsymbol{Y}')] = \boldsymbol{I} \otimes \Sigma . \tag{17.8}$$

The general GCM is obtained from the GLM by choosing $\boldsymbol{B}_* = \boldsymbol{B}_1\boldsymbol{T}_1$ $(q \times m \times p)$ or, equivalently,

$$\boldsymbol{B}_* = [\,\boldsymbol{B}_1\ \boldsymbol{B}_2\,]\begin{bmatrix}\boldsymbol{T}_1\\ \boldsymbol{T}_2\end{bmatrix}, \tag{17.9}$$

with $\boldsymbol{B}_2 = \boldsymbol{0}$. Here we assume $\boldsymbol{T}$ $(p \times p)$ has full rank and $\boldsymbol{T}^{-1} = \boldsymbol{G} = [\,\boldsymbol{G}_1\ \boldsymbol{G}_2\,]$, with $\boldsymbol{G}_1$ $(p \times m)$ and $\boldsymbol{G}_2$ $[p \times (p-m)]$. The relationship between $\boldsymbol{B}_*$ and $\boldsymbol{B}$ can be written as $\boldsymbol{B}_*[\,\boldsymbol{G}_1\ \boldsymbol{G}_2\,] = [\,\boldsymbol{B}_1\ \boldsymbol{B}_2\,]$, with $\boldsymbol{B}_1$ $(q \times m)$ and $\boldsymbol{B}_2$ $[q \times (p-m)]$. Since $\boldsymbol{B}_2 = \boldsymbol{0}$, the equation makes it clear an equivalent notation is $\mathrm{GLM}_{N,p,q}(\boldsymbol{Y}_i; \boldsymbol{X}_i\boldsymbol{B}_* | \boldsymbol{B}_*\boldsymbol{G}_2 = \boldsymbol{0}, \boldsymbol{\Sigma})$ with $\boldsymbol{G}_2$ of dimension $p \times (p-m)$. The matrix $\boldsymbol{T}$ has specified structure, such as a polynomial structure, and the restriction imposes the structure on $\boldsymbol{B}_*$ $(q \times p)$ in the form of $p-m$ linearly independent constraints. With $\boldsymbol{T}$ based on polynomials, the restriction indicates the elements of $\boldsymbol{B}_*$ are functionally related to one another as specified by the polynomial.

Example 17.2 If the children's weights in Example 13.1 had been recorded monthly from the ages of 5 to 7 years, then $\boldsymbol{B}_*$ would be 1×25. If growth is truly linear during the time of observation, then only two parameters (intercept and slope) are needed to characterize the growth curve. The 25 elements of $\boldsymbol{B}_*$ are the mean weights over 25 months. Since the elements are functionally related (they form a straight line), the 25 elements (means) can be condensed into 2 columns:

$$\boldsymbol{B}_* = [\,\beta_0\ \beta_1\ 0\ \cdots\ 0\,]\boldsymbol{T}\,. \tag{17.10}$$

Therefore it is appropriate to impose the restriction $\boldsymbol{B}_*\boldsymbol{G}_2 = \boldsymbol{0}$, with $\boldsymbol{G}_2$ (25×23).

Theorem 17.1 The $\mathrm{GCM}_{N,p,q,m}(\boldsymbol{Y}_i;\ \boldsymbol{X}_i\boldsymbol{B}_1\boldsymbol{T}_1,\ \boldsymbol{\Sigma})$ with $\boldsymbol{B}$ and $\boldsymbol{T}$ defined in terms of an orthogonal or orthonormal polynomial allows testing $H_0 : \boldsymbol{C}\boldsymbol{B}_1\boldsymbol{U} = \boldsymbol{0}$ $(a \times b)$ against the general alternative. Under $\mathrm{GLM}_{N,p,q}(\boldsymbol{Y}_i; \boldsymbol{X}_i\boldsymbol{B}_* | \boldsymbol{B}_*\boldsymbol{G}_2 = \boldsymbol{0}, \boldsymbol{\Sigma})$ with $\boldsymbol{G}_2$ of dimension $p \times (p-m)$, an equivalent hypothesis is $H_0 : \boldsymbol{C}\boldsymbol{B}_*\boldsymbol{G}_1\boldsymbol{U} = \boldsymbol{0}$.

Proof. With $\boldsymbol{T}^{-1} = \boldsymbol{G} = [\,\boldsymbol{G}_1\ \boldsymbol{G}_2\,]$ we have $\boldsymbol{B}_*\boldsymbol{G}_1 = \boldsymbol{B}_1$. □

A more general version of the linear restrictions on $\boldsymbol{B}_*$ involves an arbitrary constant matrix $\boldsymbol{D}$, with $\boldsymbol{B}_* = \boldsymbol{B}\boldsymbol{T} + \boldsymbol{D}$. No new problems arise. The form leads to an ordinary GCM having expectation of the form $\mathrm{E}(\boldsymbol{Y} - \boldsymbol{X}\boldsymbol{D}) = \boldsymbol{X}\boldsymbol{B}\boldsymbol{T}$.

17.8 CONFIDENCE BANDS FOR GROWTH CURVES

For the polynomial $\mathrm{GCM}_{N,p,q,m}(\boldsymbol{Y}_i; \boldsymbol{X}_i\boldsymbol{B}\boldsymbol{T}, \boldsymbol{\Sigma})$ with $m = 2$, Stewart (1987) developed a finite-interval confidence band in terms of a type III bivariate t distribution, in a manner analogous to Bowden's (1970) univariate results. The procedure generalizes to $m > 2$ in terms of type III multivariate t distributions.

CHAPTER 18

Tests for Linear Mixed Models

18.1 OVERVIEW

We introduced the notation $\text{LMM}_{N,p_i,q,m}[\boldsymbol{y}_i; \boldsymbol{X}_i\boldsymbol{\beta}, \boldsymbol{Z}_i\boldsymbol{\Sigma}_{di}(\boldsymbol{\tau}_d)\boldsymbol{Z}_i' + \boldsymbol{\Sigma}_{ei}(\boldsymbol{\tau}_e)]$ in Chapter 5 to indicate a general linear mixed model, with model equation $\boldsymbol{y}_i = \boldsymbol{X}_i\boldsymbol{\beta} + \boldsymbol{Z}_i\boldsymbol{d}_i + \boldsymbol{e}_i$ for one independent sampling unit (ISU). In the present chapter, we always add the Gaussian assumption, which has three parts: $\boldsymbol{e}_i \sim \mathcal{N}_{p_i}[\boldsymbol{0}, \boldsymbol{\Sigma}_{ei}(\boldsymbol{\tau}_e)]$, $\boldsymbol{d}_i \sim \mathcal{N}_{p_i}[\boldsymbol{0}, \boldsymbol{\Sigma}_{di}(\boldsymbol{\tau}_d)]$, and $\boldsymbol{e}_i \perp\!\!\!\perp \boldsymbol{d}_i$. The model implies $\text{E}(\boldsymbol{y}_i) = \boldsymbol{X}_i\boldsymbol{\beta}$ and $\mathcal{V}(\boldsymbol{y}_i) = \boldsymbol{Z}_i\boldsymbol{\Sigma}_{di}(\boldsymbol{\tau}_d)\boldsymbol{Z}_i' + \boldsymbol{\Sigma}_{ei}(\boldsymbol{\tau}_e)$. Hence $\boldsymbol{\beta}$ determines the expected values, the first moments of the observations. In turn, $\boldsymbol{\tau}_d$ determines the covariance matrix, the second moments, of the unobserved (latent) error components varying within the ISU due to observed characteristics in $\boldsymbol{Z}_i$. Similarly, $\boldsymbol{\tau}_e$ determines the covariance matrix of the unobserved (latent) error components varying within the ISU due to observation. The Gaussian structure ensures parameters $\boldsymbol{\beta}$ and $\boldsymbol{\tau}' = [\boldsymbol{\tau}_d' \, \boldsymbol{\tau}_e']$ fully determine the distributions of all model statistics.

The presence of just mean and covariance parameters leads to three possible types of inferences. Most often, data analysts seek to draw inferences about functions of $\boldsymbol{\beta}$, the parameters determining the means (the fixed effects). Less often, data analysts seek to draw inferences about functions of $\boldsymbol{\tau}$, the covariance parameters, especially $\boldsymbol{\tau}_d$ (the random effects parameters). Even less often, data analysts seek to draw inferences about functions directly involving both $\boldsymbol{\beta}$ and $\boldsymbol{\tau}$.

The univariate and multivariate models allow computing exact or nearly exact likelihood-based inferences as an adjunct to the noniterative calculation of estimates of a single model. In contrast, the mixed model requires iterative calculation of estimates for two or more models. The desire for speed and convenience has led to a heavy emphasis on approximate tests based on single-model fits. The inexorable increase in computer speed has made the computer time needed to fit an additional model much less consequential. However, the control language for contemporary software makes comparing two model fits awkward and hence creates a barrier to conducting tests based on two model fits.

At the present time, confidence intervals and hypothesis tests based on a single model fit dominate practice. Two related reasons reinforce the habit. First, popular contemporary commercial software does not directly implement any inference based on fitting two or more models. Second, invoking software twice to fit two distinct models takes extra time and care for the data analyst to implement correctly.

Vonesh and Chinchilli (1997), Verbeke and Molenberghs (2000), and Demidenko (2004) provided excellent book-length treatments of mixed models. Khuri, Mathew, and Sinha (1998) focused specifically on tests in mixed models.

18.2 ESTIMABILITY OF $\theta = C\beta$

For an estimable secondary parameter $\boldsymbol{\theta} = \boldsymbol{C\beta}$, we consider testing

$$H_0 : \boldsymbol{\theta} = \boldsymbol{\theta}_0 \,. \tag{18.1}$$

In parallel to extending the concept of estimable $\boldsymbol{\theta}$ to mixed models, the concept of a testable hypothesis extends to the mixed model. In particular, we require estimable $\boldsymbol{\theta}$ and full (row) rank of $\boldsymbol{C}$. Results about testability center on guaranteeing existence of a valid test. Such statements involve population parameter and design matrix properties and hence apply to mixed models.

18.3 LIKELIHOOD RATIO TESTS OF $C\beta$

Hypotheses concerning $\boldsymbol{\theta} = \boldsymbol{C\beta}$ can be tested with a likelihood ratio test. However the distribution theory is not well developed. The process involves finding estimates for two models, with one of them false. The iterative calculations for the false model may be unstable. Although theoretically appealing, the method is less used than methods based on fitting a single model.

Example 18.1 The likelihood ratio approach allows testing $H_0 : \boldsymbol{C\beta} = \boldsymbol{\theta}_0$ in the $\text{GLM}_{N,q}(y_i; \boldsymbol{X}_i\boldsymbol{\beta}, \sigma^2)$ with (i.i.d.) Gaussian errors, as a special case of the mixed model. With $r = \text{rank}(\boldsymbol{X})$, the exact distribution of the likelihood ratio test statistic can be expressed in terms of

$$\begin{aligned} F = \frac{(\widehat{\boldsymbol{\theta}} - \boldsymbol{\theta}_0)'\boldsymbol{M}^{-1}(\widehat{\boldsymbol{\theta}} - \boldsymbol{\theta}_0)/a}{\widehat{\sigma}^2} &= \frac{SSH/a}{SSE/(N-r)} \\ &\sim F(a, N-r)\,. \end{aligned} \tag{18.2}$$

The mixed model likelihood ratio test approximation corresponds to saying $(Fa) \stackrel{\cdot}{\sim} \chi^2(a)$, which is correct only asymptotically. With $SSE/\sigma^2 = \widehat{\sigma}^2/\sigma^2 \sim \chi^2(N-r)$, $\widehat{\sigma}^2 \to \sigma^2$ as $N \to \infty$. For any N, under the null $SSH/\sigma^2 \sim \chi^2(a)$. The denominator degrees-of-freedom parameter of the F distribution accounts for having estimated σ^2 in a finite sample.

It is typically straightforward, and always possible, to code a linear model design matrix so that a hypothesis test corresponds to deleting one or more variables from the model. Doing so allows computing the likelihood ratio test statistic as the difference between the log likelihood values for the full and reduced model. The asymptotic approximation is $-2\ln[L(\widehat{\boldsymbol{\theta}}|\boldsymbol{C\beta}=\boldsymbol{\theta}_0)/L(\widehat{\boldsymbol{\theta}})] \dot{\sim} \chi^2(a)$. Unfortunately, more accurate approximations are available only for some special cases. Zucker, Lieberman, and Manor (2000) described a second-order approximation for a particular class of models and also demonstrated the substantial improvement in accuracy that the approach provides.

18.4 LIKELIHOOD RATIO TESTS INVOLVING $\boldsymbol{\tau}$

Many linear and some nonlinear hypotheses concerning $\boldsymbol{\tau}$ alone or $\boldsymbol{\gamma}' = [\boldsymbol{\beta}' \; \boldsymbol{\tau}']$ can be tested with the likelihood ratio approach. As always, the hypothesis of interest must correspond to comparing a full and constrained model with the constrained model nested within the full. Maximum likelihood estimates are used. Some special case tests for some models have known distributions (Demidenko, 2004). Otherwise only large-sample results are available.

18.5 TEST SIZE OF WALD-TYPE TESTS OF $\boldsymbol{\beta}$ USING REML

Among the widely available approximate F methods based on fitting a single model, the Kenward and Roger (1997) technique appears to provide the most accurate test size. However, substantial room for improvement in small-sample performance remains. Simulations of Park, Park, and Davis (2001) as well as simulations of Schaalje, McBride, and Fellingham (2003) support the conclusion.

Table 18.1 Estimated Test Size, *Target = 0.04*, for UNIREP (Std. Err. < 0.0004) and Mixed (Std. Err. < 0.003)

		UNIREP		Mixed		
N	ϵ	GG[1]	HF[2]	Resid[3]	Satter[4]	K/R[5]
10	0.28	0.042	0.045	0.254	0.138	0.114
	0.51	0.039	0.052	0.263	0.137	0.081
	1.00	0.021	0.052	0.263	0.144	0.043
20	0.28	0.041	0.042	0.116	0.077	0.040
	0.51	0.040	0.046	0.115	0.072	0.036
	1.00	0.029	0.038	0.116	0.075	0.038
40	0.28	0.040	0.041	0.075	0.054	0.040
	0.51	0.041	0.043	0.076	0.061	0.045
	1.00	0.034	0.039	0.074	0.056	0.040

[1]Geisser Greenhouse, [2]Huynh-Feldt
[3]Residual, [4]Satterthwaite, [5]Kenward Roger

Muller, Edwards, Simpson and Taylor (2006) illustrated some limitations of the approach with a simulation summarized in Table 18.1. Simulated Gaussian data were generated which met all assumptions of a $\text{GLM}_{N,p,q}(\boldsymbol{Y}_i; \boldsymbol{X}_i\boldsymbol{B}, \boldsymbol{\Sigma})$ with $p = 9$ and $q = 1$, with $\boldsymbol{X} = \boldsymbol{1}_N$, a grand mean model. No missing or mistimed data were allowed. The 9 responses reflected a 3×3 within-subject factorial design, with factor names Clip and Region. The study design used $\alpha = 0.04$, *not* 0.05. For the UNIREP test of Clip $\times$ Region interaction, $H_0 : \boldsymbol{CBU} = \boldsymbol{\Theta}_0$ used $\boldsymbol{C} = 1$, $\boldsymbol{\Theta}_0 = \boldsymbol{0}$, and

$$\begin{aligned} \boldsymbol{U}_{\text{cr}} &= \boldsymbol{T}_{\text{c}} \otimes \boldsymbol{T}_{\text{r}} \\ &= \begin{bmatrix} -4/\sqrt{42} & 2/\sqrt{14} \\ -1/\sqrt{42} & -3/\sqrt{14} \\ 5/\sqrt{42} & 1/\sqrt{14} \end{bmatrix} \otimes \begin{bmatrix} -1/\sqrt{2} & 1/\sqrt{6} \\ 0 & -2/\sqrt{6} \\ 1/\sqrt{2} & 1/\sqrt{6} \end{bmatrix} . \end{aligned} \tag{18.3}$$

Columns of $\boldsymbol{T}_{\text{c}}$ contain the linear and quadratic orthonormal trends for $\log_2(\text{Clip}) \in \{1, 2, 4\}$, and columns of $\boldsymbol{T}_{\text{r}}$ contain the orthonormal contrasts of linear and quadratic trends for $\log_2(\text{Region}) \in \{1, 3, 5\}$. Four eigenvalues sets, $\boldsymbol{\lambda}_1' \approx [0.4796\ 0.0100\ 0.0100\ 0.0100]$, $\boldsymbol{\lambda}_2' \approx [0.3455\ 0.0612\ 0.0556\ 0.0472]$, $\boldsymbol{\lambda}_3' \approx [0.2355\ 0.1712\ 0.0556\ 0.0472]$, and $\boldsymbol{\lambda}_4' \approx [0.1274\ 0.1274\ 0.1274\ 0.1274]$, were used. Corresponding values of ϵ are approximately 0.28, 0.51, 0.72, and 1. Having $\epsilon = 1$ corresponds to having the underlying repeated measures being compound symmetric. The conditions are the same as conditions 5–8 in Table III in Coffey and Muller (2003). Given $\boldsymbol{\Sigma}_* = \text{Dg}(\boldsymbol{\lambda}_j)$ for $j \in \{1, 2, 3, 4\}$, $\boldsymbol{\Sigma} = \boldsymbol{U}_{\text{cr}}\boldsymbol{\Sigma}_*\boldsymbol{U}_{\text{cr}}'$. A total of 500,000 replications were tabulated for each condition for the UNIREP tests (Geisser-Greenhouse and Huynh-Feldt). A total of 5000 replications were tabulated for the mixed tests based on fitting a single model with SAS PROC MIXED® and using an F approximation (residual sum of squares, Satterthwaite, and Kenward-Roger approximations for denominator degrees of freedom). The "unstructured" covariance option on the repeated statement was always used. In addition to the type I error rate inflation, the mixed model approach also failed to converge in a fraction of the cases. A number of adjustments to the inputs controlling the estimation algorithms greatly reduced but did not completely eliminate the problem. Each set of data which led to convergence failure was subsequently analyzed with the UNIREP approach, which always gave well-behaved and reasonable estimates for the data observed. As noted earlier, no missing or mistimed data were present.

In interpreting the results, it helps to keep in mind the distinction between number of independent sampling units (ISUs), with $N \in \{10, 20, 40\}$ in the simulations, and number of observations, with $n = 9N \in \{90, 180, 360\}$ in the simulations. The residual sums-of-squares approximation inflates test size the most. It implicitly uses weighted least squares estimators and makes no adjustment for having estimated the covariance parameters. Consequently, denominator degrees of freedom are based on $n - \text{rank}(\boldsymbol{X}_s)$, which coincides with the degrees of freedom for the uncorrected UNIREP test. The corrected tests use reduced numerator and denominator degrees of freedom.

18.6 USING WALD-TYPE TESTS OF $\boldsymbol{\beta}$ WITH REML

We next describe the steps needed to compute the Kenward and Roger (1997) approximation for inference about elements of $\boldsymbol{\beta}$. The approach assumes $\Sigma_i(\boldsymbol{\tau})$ has been correctly specified and does not change. Except for notation differences, our presentation closely follows the original.

Using notation introduced in Chapter 5, we consider $\text{LMM}_{N,p_i,q,m}[\boldsymbol{y}_i; \boldsymbol{X}_i\boldsymbol{\beta}, \boldsymbol{Z}_i\Sigma_{di}(\boldsymbol{\tau}_d)\boldsymbol{Z}_i' + \Sigma_{ei}(\boldsymbol{\tau}_e)]$ with Gaussian errors. The model for all of the data stacked by participant (ISU) may be written

$$\begin{array}{ccccc} \boldsymbol{y}_s & = & \boldsymbol{X}_s\boldsymbol{\beta} + & \boldsymbol{Z}_s\boldsymbol{d}_s & + \quad \boldsymbol{e}_s \end{array}$$

$$\begin{bmatrix} \boldsymbol{y}_1 \\ \boldsymbol{y}_2 \\ \vdots \\ \boldsymbol{y}_N \end{bmatrix} = \begin{bmatrix} \boldsymbol{X}_1 \\ \boldsymbol{X}_2 \\ \vdots \\ \boldsymbol{X}_N \end{bmatrix} \boldsymbol{\beta} + \begin{bmatrix} \boldsymbol{Z}_1 & 0 & \cdots & 0 \\ 0 & \boldsymbol{Z}_2 & & \vdots \\ \vdots & & \ddots & 0 \\ 0 & \cdots & 0 & \boldsymbol{Z}_N \end{bmatrix} \begin{bmatrix} \boldsymbol{d}_1 \\ \boldsymbol{d}_2 \\ \vdots \\ \boldsymbol{d}_N \end{bmatrix} + \begin{bmatrix} \boldsymbol{e}_1 \\ \boldsymbol{e}_2 \\ \vdots \\ \boldsymbol{e}_N \end{bmatrix} . \tag{18.4}$$

For an $a \times 1$ testable secondary parameter $\boldsymbol{\theta} = \boldsymbol{C}\boldsymbol{\beta}$, necessarily with $\text{rank}(\boldsymbol{C}) = a \leq q$, we consider testing the hypothesis

$$H_0 : \boldsymbol{\theta} = \boldsymbol{\theta}_0 \tag{18.5}$$

against the general alternative. Here $\text{E}(\boldsymbol{y}_i) = \boldsymbol{X}_i\boldsymbol{\beta}$ implies $\text{E}(\boldsymbol{y}_s) = \boldsymbol{X}_s\boldsymbol{\beta}$, while $\Sigma_i(\boldsymbol{\tau}) = \boldsymbol{Z}_i\Sigma_{di}(\boldsymbol{\tau}_d)\boldsymbol{Z}_i' + \Sigma_{ei}(\boldsymbol{\tau}_e)$ implies

$$\mathcal{V}(\boldsymbol{y}_s) = \bigoplus_{i=1}^{N} \Sigma_i(\boldsymbol{\tau}) = \Sigma_s(\boldsymbol{\tau}) . \tag{18.6}$$

The unbiased REML estimator of $\boldsymbol{\beta}$ ($q \times 1$) is

$$\widehat{\boldsymbol{\beta}} = [\boldsymbol{X}_s'\Sigma_s^{-1}(\widehat{\boldsymbol{\tau}})\boldsymbol{X}_s]^{-1}\boldsymbol{X}_s'\Sigma_s^{-1}(\widehat{\boldsymbol{\tau}})\boldsymbol{y}_s . \tag{18.7}$$

The asymptotic covariance matrix of $\widehat{\boldsymbol{\beta}}$ is

$$\mathcal{V}_{\text{a}}(\widehat{\boldsymbol{\beta}}) = \boldsymbol{\Phi}(\boldsymbol{\tau}) = [\boldsymbol{X}_s'\Sigma_s^{-1}(\boldsymbol{\tau})\boldsymbol{X}_s]^{-1} . \tag{18.8}$$

More precisely, with s elements in $\boldsymbol{\tau}$,

$$[\boldsymbol{\Phi}(\boldsymbol{\tau})]^{-1/2}(\widehat{\boldsymbol{\beta}} - \boldsymbol{\beta}) \overset{\text{asy}}{\sim} \mathcal{N}_s(\boldsymbol{0}, \boldsymbol{I}) . \tag{18.9}$$

As always, the distinction between the number of ISUs, N, and the number of observations, $n = \sum_{i=1}^{N} p_i$, must be treated carefully. Fixing N and having $p_i \to \infty$ implies $n \to \infty$ but does not guarantee convergence of $\mathcal{V}(\widehat{\boldsymbol{\beta}})$, except with side conditions to greatly simplify the covariance structure.

The following asymptotically correct estimator has often been used for evaluating the precision of $\widehat{\boldsymbol{\beta}}$ and for testing hypotheses:

$$\begin{aligned}\widetilde{\mathcal{V}}_1(\widehat{\boldsymbol{\beta}}) &= \boldsymbol{\Phi}(\widehat{\boldsymbol{\tau}}) = [\boldsymbol{X}_s'\boldsymbol{\Sigma}_s^{-1}(\widehat{\boldsymbol{\tau}})\boldsymbol{X}_s]^{-1} \\ &= \widetilde{\boldsymbol{\Phi}}_1 . \end{aligned} \tag{18.10}$$

The estimator treats $\widehat{\boldsymbol{\tau}}$ in equation (18.7) as if it were a fixed constant. Doing so corresponds to assuming $\mathcal{V}(\widehat{\boldsymbol{\tau}}) \approx \mathbf{0}$, which is not true for finite sample sizes. In turn, $\boldsymbol{\Phi}(\widehat{\boldsymbol{\tau}}) = \widetilde{\boldsymbol{\Phi}}_1$ is a biased estimator of $\boldsymbol{\Phi}(\boldsymbol{\tau})$, which is an approximation of $\mathcal{V}(\widehat{\boldsymbol{\beta}})$. Consequently, when N is small, $\widetilde{\mathcal{V}}_1(\widehat{\boldsymbol{\beta}})$ underestimates $\mathcal{V}(\widehat{\boldsymbol{\beta}})$ due to a combination of bias and approximation error.

Kackar and Harville (1984) used a first-order Taylor series expansion to better approximate $\mathcal{V}(\widehat{\boldsymbol{\beta}})$ in small sample sizes. The expansion depends on

$$\boldsymbol{P}_j = \boldsymbol{X}_s'\left\{\frac{\partial \boldsymbol{\Sigma}_s^{-1}(\boldsymbol{\tau})}{\partial \tau_j}\right\}\boldsymbol{X}_s \tag{18.11}$$

and

$$\boldsymbol{Q}_{jk} = \boldsymbol{X}_s'\left\{\frac{\partial \boldsymbol{\Sigma}_s^{-1}(\boldsymbol{\tau})}{\partial \tau_j}\right\}\boldsymbol{\Sigma}_s(\boldsymbol{\tau})\left\{\frac{\partial \boldsymbol{\Sigma}_s^{-1}(\boldsymbol{\tau})}{\partial \tau_k}\right\}\boldsymbol{X}_s . \tag{18.12}$$

The expansion leads to the approximation

$$\mathcal{V}(\widehat{\boldsymbol{\beta}}) \approx \boldsymbol{\Phi}(\boldsymbol{\tau}) + \boldsymbol{\Phi}(\boldsymbol{\tau})\left\{\sum_{j=1}^{t}\sum_{k=1}^{t}\langle\mathcal{V}(\widehat{\boldsymbol{\tau}})\rangle_{jk}[\boldsymbol{Q}_{jk} - \boldsymbol{P}_j\boldsymbol{\Phi}(\boldsymbol{\tau})\boldsymbol{P}_k]\right\}\boldsymbol{\Phi}(\boldsymbol{\tau}) . \tag{18.13}$$

An estimator of $\mathcal{V}(\widehat{\boldsymbol{\beta}})$ may be computed in three steps: (1) evaluate $\boldsymbol{P}_j$ and $\boldsymbol{Q}_{jk}$ at $\widehat{\boldsymbol{\tau}}$ to give $\widehat{\boldsymbol{P}}_k$ and $\widehat{\boldsymbol{Q}}_{jk}$; (2) replace $\boldsymbol{\Phi}(\boldsymbol{\tau})$ by $\widetilde{\boldsymbol{\Phi}}_1 = \boldsymbol{\Phi}(\widehat{\boldsymbol{\tau}})$ from equation 18.10; and (3) replace $\mathcal{V}(\widehat{\boldsymbol{\tau}})$ with an estimator, $\widetilde{\mathcal{V}}(\widehat{\boldsymbol{\tau}})$. The value of $\widetilde{\mathcal{V}}(\widehat{\boldsymbol{\tau}})$ may be computed either as the inverse of the $\boldsymbol{\tau}$ submatrix of the expected information matrix or as the $\boldsymbol{\tau}$ submatrix of the inverse of the observed information matrix. In turn,

$$\widetilde{\mathcal{V}}_2(\widehat{\boldsymbol{\beta}}) = \widetilde{\boldsymbol{\Phi}}_1 + \widetilde{\boldsymbol{\Phi}}_1\left[\sum_{j=1}^{t}\sum_{k=1}^{t}\langle\widetilde{\mathcal{V}}(\widehat{\boldsymbol{\tau}})\rangle_{jk}\left(\widehat{\boldsymbol{Q}}_{jk} - \widehat{\boldsymbol{P}}_j\widetilde{\boldsymbol{\Phi}}_1\widehat{\boldsymbol{P}}_k\right)\right]\widetilde{\boldsymbol{\Phi}}_1 . \tag{18.14}$$

Kenward and Roger (1997) used the Kackar and Harville result to devise a further improved estimator of $\mathcal{V}(\widehat{\boldsymbol{\beta}})$. The estimator adds an adjustment in estimating $\boldsymbol{\Phi}(\widehat{\boldsymbol{\tau}})$ to compensate for finite N [and $\mathcal{V}(\widehat{\boldsymbol{\tau}}) \neq \mathbf{0}$]. With

$$\boldsymbol{R}_{jk} = \boldsymbol{X}_s'\boldsymbol{\Sigma}_s^{-1}(\boldsymbol{\tau})\left\{\frac{\partial^{(2)}\boldsymbol{\Sigma}_s^{-1}(\boldsymbol{\tau})}{\partial \tau_j\,\partial \tau_k}\right\}[\boldsymbol{\Sigma}_s^{-1}(\boldsymbol{\tau})]\boldsymbol{X}_s/4 \tag{18.15}$$

and $\widehat{\boldsymbol{R}}_{jk}$ the value of $\boldsymbol{R}_{jk}$ evaluated at $\widehat{\boldsymbol{\tau}}$, the adjusted estimator is

$$\begin{aligned}\widetilde{\mathcal{V}}_3(\widehat{\boldsymbol{\beta}}) &= \widetilde{\boldsymbol{\Phi}}_1 + 2\widetilde{\boldsymbol{\Phi}}_1\left[\sum_{j=1}^{t}\sum_{k=1}^{t}\langle\widetilde{\mathcal{V}}(\widehat{\boldsymbol{\tau}})\rangle_{jk}\left(\widehat{\boldsymbol{Q}}_{jk}-\widehat{\boldsymbol{P}}_j\widetilde{\boldsymbol{\Phi}}_1\widehat{\boldsymbol{P}}_k-\widehat{\boldsymbol{R}}_{jk}\right)\right]\widetilde{\boldsymbol{\Phi}}_1 . \\ &= \widetilde{\boldsymbol{\Phi}}_2 .\end{aligned} \tag{18.16}$$

Kenward and Roger found, in limited simulations, that replacing the expected information matrix by the observed information matrices had no discernible effect on accuracy of tests and standard errors. However, the missing data mechanisms in their examples and simulations were arguably MCAR. More generally, many kinds of MAR mechanisms can require using the observed information matrix to avoid bias (Verbeke and Molenberghs, 2000, Chapter 21).

Kenward and Roger (1997) provided an F approximation for hypothesis tests. Computing the test statistic, F_{KR}, requires a denominator degrees of freedom and a scaling constant λ. Both depend on

$$\boldsymbol{T} = \boldsymbol{C}'(\boldsymbol{C}\boldsymbol{\Phi}^{-1}\boldsymbol{C}')^{-1}\boldsymbol{C}, \tag{18.17}$$

$$d_1 = \sum_{j=1}^{t}\sum_{k=1}^{t}\langle\mathcal{V}(\widehat{\boldsymbol{\tau}})\rangle_{jk}\mathrm{tr}(\boldsymbol{T}\boldsymbol{\Phi}\boldsymbol{P}_j\boldsymbol{\Phi})\mathrm{tr}(\boldsymbol{T}\boldsymbol{\Phi}\boldsymbol{P}_k\boldsymbol{\Phi}), \tag{18.18}$$

$$d_2 = \sum_{j=1}^{t}\sum_{k=1}^{t}\langle\mathcal{V}(\widehat{\boldsymbol{\tau}})\rangle_{jk}\mathrm{tr}(\boldsymbol{T}\boldsymbol{\Phi}\boldsymbol{P}_j\boldsymbol{\Phi}\boldsymbol{T}\boldsymbol{\Phi}\boldsymbol{P}_k\boldsymbol{\Phi}), \tag{18.19}$$

$g=[(a+1)d_1-(a+4)d_2]/[(a+2)d_2]$, $c_1=g/[3a+2(1-g)]$, $c_2=(a-g)/[3a+2(1-g)]$, $c_3=(a-g+2)/[3a+2(1-g)]$, $b=(d_1+6d_1)/(2a)$, $c_4=(1-d_2/a)^{-1}$, $c_5=2(1+c_1b)/\left[a(1-c_2b)^2(1-c_3b)\right]$, and $\rho=c_5/(2c_4)^2$. In turn,

$$\nu_* = 4 + (a+2)/(a\rho - 1) \tag{18.20}$$

and

$$\lambda = \nu_*/[(\nu_* - 2)c_4] . \tag{18.21}$$

The simplest approximation uses $F_m = (\widehat{\boldsymbol{\theta}} - \boldsymbol{\theta}_0)'(\boldsymbol{C}\widetilde{\boldsymbol{\Phi}}_1^{-1}\boldsymbol{C}')^{-1}(\widehat{\boldsymbol{\theta}} - \boldsymbol{\theta}_0)/a$. Replacing $\widetilde{\boldsymbol{\Phi}}_1$ by $\widetilde{\boldsymbol{\Phi}}_2$ in the expression for F_m and using $\widetilde{\boldsymbol{\Phi}}_2$ to compute $\widehat{\nu}_*$ and $\widehat{\lambda}$ gives

$$\begin{aligned}F_{\mathrm{KR}} &= \widehat{\lambda}\cdot(\widehat{\boldsymbol{\theta}} - \boldsymbol{\theta}_0)'(\boldsymbol{C}\widehat{\boldsymbol{\Phi}}_2^{-1}\boldsymbol{C}')^{-1}(\widehat{\boldsymbol{\theta}} - \boldsymbol{\theta}_0)/a \\ &\mathrel{\dot\sim} F(a, \widehat{\nu}_*) .\end{aligned} \tag{18.22}$$

18.7 USING WALD-TYPE TESTS OF $\{\boldsymbol{\beta}, \boldsymbol{\tau}\}$ WITH REML

A simpler approximation for tests of $H_0 : \boldsymbol{\theta} = \boldsymbol{\theta}_0$, with $\boldsymbol{\theta} = \boldsymbol{C}\boldsymbol{\beta}$, involves using

$$F_m = (\widehat{\boldsymbol{\theta}} - \boldsymbol{\theta}_0)'(\boldsymbol{C}\widetilde{\boldsymbol{\Phi}}_1^{-1}\boldsymbol{C}')^{-1}(\widehat{\boldsymbol{\theta}} - \boldsymbol{\theta}_0)/a\,. \tag{18.23}$$

A variation on the test statistic was discussed in Helms (1992). The distribution of F_m is approximated by the $F(a, \nu_2, \omega)$ distribution in which $a = \text{rank}(\boldsymbol{C})$, $\nu_2 = N - \text{rank}([\,\boldsymbol{X}_s\ \boldsymbol{Z}_s\,])$, and $\omega = (\boldsymbol{\theta} - \boldsymbol{\theta}_0)'[\boldsymbol{C}(\boldsymbol{X}_s'\boldsymbol{\Sigma}_s^{-1}\boldsymbol{X}_s)^{-1}\boldsymbol{C}']^{-1}(\boldsymbol{\theta} - \boldsymbol{\theta}_0)$. The appropriate critical value for testing H_0 versus the general alternative is $f_{\text{C}} = F_F^{-1}(1 - \alpha; a, \nu_2)$. The power of the approximately size-α test is approximately Power $= 1 - F(f_{\text{C}}; a, \nu_2, \omega)$. Only limited simulation results are available. As noted earlier, the Kenward-Roger approximation provides greater accuracy for inference about $\boldsymbol{\theta} = \boldsymbol{C}\boldsymbol{\beta}$.

The same form of approximate test can be applied to tests involving both $\{\boldsymbol{\beta}, \boldsymbol{\tau}\}$. It seems likely that the roughness of approximation requires a substantial number of independent sampling units to provide the desired distribution. If, as defined in equation 5.9, $\boldsymbol{\Sigma}_{ds} = \bigoplus_{i=1}^{N}\boldsymbol{\Sigma}_{di}$ and $\boldsymbol{\Sigma}_{es} = \bigoplus_{i=1}^{N}\boldsymbol{\Sigma}_{ei}$, then

$$\mathcal{V}\left(\begin{bmatrix} \widehat{\boldsymbol{\beta}} \\ \widehat{\boldsymbol{d}}_s - \boldsymbol{d}_s \end{bmatrix}\right) = \begin{bmatrix} \boldsymbol{X}_s'\boldsymbol{\Sigma}_{es}^{-1}\boldsymbol{X}_s & \boldsymbol{X}_s'\boldsymbol{\Sigma}_{es}^{-1}\boldsymbol{Z}_s \\ \boldsymbol{Z}_s'\boldsymbol{\Sigma}_{es}^{-1}\boldsymbol{X}_s & \boldsymbol{\Sigma}_{ds}^{-1} + \boldsymbol{Z}_s'\boldsymbol{\Sigma}_{es}^{-1}\boldsymbol{Z}_s \end{bmatrix}^{-} = \boldsymbol{M}^{-}. \tag{18.24}$$

For $\boldsymbol{M}$ positive definite,

$$\begin{aligned} \boldsymbol{M}^{-1} &= \begin{bmatrix} \boldsymbol{M}^{(11)} & \boldsymbol{M}^{(12)} \\ \boldsymbol{M}^{(21)} & \boldsymbol{M}^{(22)} \end{bmatrix} \\ &= \begin{bmatrix} \boldsymbol{M}_{11} & \boldsymbol{M}_{12} \\ \boldsymbol{M}_{21} & \boldsymbol{M}_{22} \end{bmatrix}^{-1}. \end{aligned} \tag{18.25}$$

Tests of hypotheses involving $\boldsymbol{d}_s$, such as

$$H_0 : \boldsymbol{L}'\begin{bmatrix} \boldsymbol{\beta} \\ \boldsymbol{d}_s \end{bmatrix} = \boldsymbol{\theta}_0\,, \tag{18.26}$$

are testable against the general alternative using test statistic

$$F_L = \boldsymbol{L}'\left(\begin{bmatrix} \widehat{\boldsymbol{\beta}} \\ \widehat{\boldsymbol{d}}_s \end{bmatrix} - \boldsymbol{\theta}_0\right)'\left(\boldsymbol{L}\widehat{\boldsymbol{M}}^{-}\boldsymbol{L}'\right)^{-1}\left(\begin{bmatrix} \widehat{\boldsymbol{\beta}} \\ \widehat{\boldsymbol{d}}_s \end{bmatrix} - \boldsymbol{\theta}_0\right)\Big/\nu_1\,. \tag{18.27}$$

Here $\nu_1 = \text{rank}(\boldsymbol{L})$, $\nu_2 = n - \text{rank}([\,\boldsymbol{X}_s\ \boldsymbol{Z}_s\,])$, and

$$\omega = \boldsymbol{L}'\left(\begin{bmatrix} \boldsymbol{\beta} \\ \boldsymbol{d}_s \end{bmatrix} - \boldsymbol{\theta}_0\right)'(\boldsymbol{L}\boldsymbol{M}^{-}\boldsymbol{L}')^{-1}\left(\begin{bmatrix} \boldsymbol{\beta} \\ \boldsymbol{d}_s \end{bmatrix} - \boldsymbol{\theta}_0\right). \tag{18.28}$$

The appropriate critical value for testing H_0 versus the general alternative is $f_{\text{C}} = F_F^{-1}(1 - \alpha, \nu_1, \nu_2)$. The power of the approximate size-α test is approximately Power $= 1 - F(f_{\text{C}}; \nu_1, \nu_2, \omega)$.

CHAPTER 19

A Review of Multivariate and Univariate Linear Models

19.1 MATRIX GAUSSIAN AND WISHART PROPERTIES

The results in the present section give the basic distribution theory needed for parameter estimators in the multivariate and univariate linear models with Gaussian errors. Only the test statistic distributions require separate treatment. Univariate results occur as the special case with $p = 1$.

We begin by reproducing key results about Gaussian and Wishart distribtutions from Chapter 8 and Chapter 10. The results provide the basis for nearly all distributional results in the remainder of the chapter.

Copy of Definition 8.4 The $n \times p$ random matrix Y follows a *direct-product matrix Gaussian* distribution, typically abbreviated *matrix Gaussian* and written $\boldsymbol{Y} \sim \mathcal{N}_{n,p}(\boldsymbol{M}, \Xi, \Sigma)$, if and only if
(a) $\text{vec}(\boldsymbol{Y}) \sim (\mathcal{S})\mathcal{N}_{n,p}[\text{vec}(\boldsymbol{M}), \Sigma \otimes \Xi]$; if and only if
(b) $\text{vec}(\boldsymbol{Y}') \sim (\mathcal{S})\mathcal{N}_{n,p}[\text{vec}(\boldsymbol{M}'), \Xi \otimes \Sigma]$; if and only if
(c) $\boldsymbol{Y} = \boldsymbol{\Psi Z \Phi}' + \boldsymbol{M}$ with $\text{vec}(\boldsymbol{Z}) \sim \mathcal{N}_{n_1 p_1}(\boldsymbol{0}, \boldsymbol{I})$ and
$\boldsymbol{\Psi}$ $(n \times n_1)$ of rank $n_1 \geq 1$, $\Xi = \boldsymbol{\Psi\Psi}'$,
$\boldsymbol{\Phi}'$ $(p_1 \times p)$ of rank $p_1 \geq 1$, $\Sigma = \boldsymbol{\Phi\Phi}'$.

Writing $\boldsymbol{Y} \sim \mathcal{SN}_{n,p}(\boldsymbol{M}, \Xi, \Sigma)$ indicates $n_1 = \text{rank}(\boldsymbol{\Psi}) = \text{rank}(\Xi) < n$, or $p_1 = \text{rank}(\boldsymbol{\Phi}) = \text{rank}(\Sigma) < p$, or both.

Writing $\boldsymbol{Y} \sim (\mathcal{S})\mathcal{N}_{n,p}(\boldsymbol{M}, \Xi, \Sigma)$ emphasizes allowing any combination of $n_1 \leq n$ and $p_1 \leq p$.

Copy of Theorem 8.12 If $\boldsymbol{Y} \sim (\mathcal{S})\mathcal{N}_{n,p}(\boldsymbol{M}, \Xi, \Sigma)$, while $\boldsymbol{A} \neq \boldsymbol{0}$ $(n_1 \times n)$, $\boldsymbol{B} \neq \boldsymbol{0}$ $(p \times p_1)$ and $\boldsymbol{C}$ $(n_1 \times p_1)$ are finite constant matrices, then

$$\boldsymbol{AYB} + \boldsymbol{C} \sim (\mathcal{S})\mathcal{N}_{n_1,p_1}(\boldsymbol{AMB} + \boldsymbol{C}, \boldsymbol{A}\Xi\boldsymbol{A}', \boldsymbol{B}'\Sigma\boldsymbol{B})\,. \qquad (19.1)$$

Copy of Definition 10.2 **(a)** If $\boldsymbol{Y} \sim \mathcal{N}_{\nu,p}(\mathbf{0}, \boldsymbol{I}_\nu, \boldsymbol{\Sigma})$ then $\boldsymbol{Y}'\boldsymbol{Y} \sim \mathcal{W}_p(\nu, \boldsymbol{\Sigma})$ indicates $\boldsymbol{Y}'\boldsymbol{Y}$ follows a *central (integer) Wishart* distribution with (integer) $\nu > 0$ degrees of freedom.
(b) If $\boldsymbol{Y} \sim \mathcal{N}_{\nu,p}(\boldsymbol{M}, \boldsymbol{I}_\nu, \boldsymbol{\Sigma})$, then $\boldsymbol{Y}'\boldsymbol{Y} \sim \mathcal{W}_p(\nu, \boldsymbol{\Sigma}, \boldsymbol{M}'\boldsymbol{M})$ indicates $\boldsymbol{Y}'\boldsymbol{Y}$ follows a *noncentral (integer) Wishart* distribution with (integer) $\nu > 0$ degrees of freedom, *shift* $\boldsymbol{\Delta} = \boldsymbol{M}'\boldsymbol{M}$, and *noncentrality* $\boldsymbol{\Omega} = \boldsymbol{M}'\boldsymbol{M}\boldsymbol{\Sigma}^+$.
(c) Singular $\boldsymbol{\Sigma}$ may be emphasized by writing $\mathcal{SW}_p(\nu, \boldsymbol{\Sigma})$ or $\mathcal{SW}_p(\nu, \boldsymbol{\Sigma}, \boldsymbol{\Delta})$.
(d) Writing $(\mathcal{S})\mathcal{W}_p(\nu, \boldsymbol{\Sigma})$ or $(\mathcal{S})\mathcal{W}_p(\nu, \boldsymbol{\Sigma}, \boldsymbol{\Delta})$ indicates possibly singular $\boldsymbol{\Sigma}$.

Copy of Theorem 10.4 If $\boldsymbol{S} \sim (\mathcal{S})\mathcal{W}_p(\nu, \boldsymbol{\Sigma}, \boldsymbol{\Delta})$ and $\boldsymbol{T}$ is *any* $p \times p_1$ constant matrix, then

$$\boldsymbol{T}'\boldsymbol{S}\boldsymbol{T} \sim (\mathcal{S})\mathcal{W}_{p_1}(\nu, \boldsymbol{T}'\boldsymbol{\Sigma}\boldsymbol{T}, \boldsymbol{T}'\boldsymbol{\Sigma}\boldsymbol{T})\,. \tag{19.2}$$

Copy of Theorem 10.8 If $\boldsymbol{Y} \sim (\mathcal{S})\mathcal{N}_{N,p}(\boldsymbol{M}, \boldsymbol{I}_N, \boldsymbol{\Sigma})$, $\text{rank}(\boldsymbol{\Sigma}) = p_1 \leq p$ with $N \geq p_1$, $N \times N$ $\boldsymbol{A}$ and $\boldsymbol{B}$ are constants, the following hold.
(a) $\boldsymbol{AY}$ and $\boldsymbol{BY}$ are independent if and only if $\boldsymbol{AB}' = \mathbf{0}$.
(b) If $\boldsymbol{A} = \boldsymbol{A}'$ is positive definite or positive semidefinite and $\boldsymbol{AB}' = \mathbf{0}$, then $\boldsymbol{BY}$ and $\boldsymbol{Y}'\boldsymbol{AY}$ are independent.
(c) If $\boldsymbol{A} = \boldsymbol{A}'$ and $\boldsymbol{B} = \boldsymbol{B}'$ are positive definite or positive semidefinite and $\boldsymbol{BA} = \mathbf{0}$, then $\boldsymbol{Y}'\boldsymbol{AY}$ and $\boldsymbol{Y}'\boldsymbol{BY}$ are independent.
(d) If $\boldsymbol{A} = \boldsymbol{A}' = \boldsymbol{A}^2$, then $\nu = \text{rank}(\boldsymbol{A}) = \text{tr}(\boldsymbol{A})$ and

$$\boldsymbol{S} = \boldsymbol{Y}'\boldsymbol{AY} \sim (\mathcal{S})\mathcal{W}_p(\nu, \boldsymbol{\Sigma}, \boldsymbol{M}'\boldsymbol{AM})\,. \tag{19.3}$$

Copy of Corollary 10.8.1 If $p = 1$ and $\sigma^2 > 0$, then $\boldsymbol{y} \sim \mathcal{N}_{n,1}(\boldsymbol{\mu}, \boldsymbol{I}_n, \sigma^2) \Leftrightarrow \boldsymbol{y} \sim \mathcal{N}_n(\boldsymbol{\mu}, \boldsymbol{I}_n\sigma^2)$ and $s = \boldsymbol{y}'\boldsymbol{Ay} \sim \mathcal{W}_1(\nu, \sigma^2, \boldsymbol{\mu}'\boldsymbol{A\mu})$. Equivalently $s/\sigma^2 \sim \chi^2(\nu, \boldsymbol{\mu}'\boldsymbol{A\mu}/\sigma^2)$.

19.2 DESIGN MATRIX PROPERTIES

Section 1.15 includes many useful results on functions of a design matrix $\boldsymbol{X}$ of dimension $N \times q$, with $\text{rank}(\boldsymbol{X}) = r \leq q < N$. With $\boldsymbol{s}_1$ an $r \times 1$ vector of strictly positive values, $\boldsymbol{L}_1$ an $N \times r$ and columnwise orthonormal matrix $\boldsymbol{L}_1'\boldsymbol{L}_1 = \boldsymbol{I}_r$, and $\boldsymbol{R}_1$ a $q \times r$ and orthonormal matrix $\boldsymbol{R}_1'\boldsymbol{R}_1 = \boldsymbol{I}_r$, the SVD gives

$$\boldsymbol{X} = [\boldsymbol{L}_1 \;\; \boldsymbol{L}_0] \begin{bmatrix} \text{Dg}(\boldsymbol{s}_1) & \mathbf{0} \\ \mathbf{0} & \mathbf{0} \end{bmatrix} \begin{bmatrix} \boldsymbol{R}_1' \\ \boldsymbol{R}_0' \end{bmatrix} = \boldsymbol{L}_1\text{Dg}(\boldsymbol{s}_1)\boldsymbol{R}_1'\,. \tag{19.4}$$

Without loss of generality, the decomposition may be chosen such that

$$\boldsymbol{X}'\boldsymbol{X} = \boldsymbol{R}\text{Dg}(\boldsymbol{s}_1, \mathbf{0})^2\boldsymbol{R}' = \boldsymbol{R}_1\text{Dg}(\boldsymbol{s}_1)^2\boldsymbol{R}_1' \tag{19.5}$$

and

$$\boldsymbol{XX}' = \boldsymbol{L}\begin{bmatrix} \mathrm{Dg}(\boldsymbol{s}_1) & \mathbf{0} & \mathbf{0} \\ \mathbf{0} & \mathbf{0} & \mathbf{0} \end{bmatrix}^2 \boldsymbol{L}' = \boldsymbol{L}_1 \mathrm{Dg}(\boldsymbol{s}_1)^2 \boldsymbol{L}_1' . \tag{19.6}$$

Even though $(\boldsymbol{X}'\boldsymbol{X}) = (\boldsymbol{X}'\boldsymbol{X})'$, a one-condition inverse $(\boldsymbol{X}'\boldsymbol{X})^-$ need not be symmetric. However, the special properties of $\boldsymbol{X}(\boldsymbol{X}'\boldsymbol{X})^-\boldsymbol{X}'$, a projection matrix, ensures uniqueness and symmetry of

$$\begin{aligned} \boldsymbol{H} &= \boldsymbol{X}(\boldsymbol{X}'\boldsymbol{X})^-\boldsymbol{X}' \\ &= \boldsymbol{X}(\boldsymbol{X}'\boldsymbol{X})^+\boldsymbol{X}' \\ &= \boldsymbol{L}_1\boldsymbol{L}_1' . \end{aligned} \tag{19.7}$$

The matrix $\boldsymbol{H}$ is rank r and idempotent. In turn, $\boldsymbol{I}_N - \boldsymbol{H}$ is rank $\nu_e = N - r$ and idempotent. Furthermore

$$\begin{aligned} \boldsymbol{I}_N - \boldsymbol{H} &= \boldsymbol{I}_N - \boldsymbol{L}_1\boldsymbol{L}_1' \\ &= [\,\boldsymbol{L}_1\ \boldsymbol{L}_0\,]\begin{bmatrix} \boldsymbol{I}_r & \mathbf{0} \\ \mathbf{0} & \boldsymbol{I}_{N-r} \end{bmatrix}[\,\boldsymbol{L}_1\ \boldsymbol{L}_0\,]' - [\,\boldsymbol{L}_1\ \boldsymbol{L}_0\,]\begin{bmatrix} \boldsymbol{I}_r & \mathbf{0} \\ \mathbf{0} & \mathbf{0} \end{bmatrix}[\,\boldsymbol{L}_1\ \boldsymbol{L}_0\,]' \\ &= [\,\boldsymbol{L}_1\ \boldsymbol{L}_0\,]\begin{bmatrix} \mathbf{0} & \mathbf{0} \\ \mathbf{0} & \boldsymbol{I}_{N-r} \end{bmatrix}[\,\boldsymbol{L}_1\ \boldsymbol{L}_0\,]' \\ &= \boldsymbol{L}_0\boldsymbol{L}_0' . \end{aligned} \tag{19.8}$$

Lemma 19.1 In the multivariate GLM, whenever $\boldsymbol{X}$ has full rank and $\mathbf{1}_N$ is the first column, the matrix $(\boldsymbol{X}'\boldsymbol{X})^{-1}$ has a simple form in terms of the first two moments of the remaining $q-1$ predictors, $\boldsymbol{X}_2$. With $\overline{\boldsymbol{x}} = \boldsymbol{X}_2'\mathbf{1}_N/N$ and $\boldsymbol{S}_X = (\boldsymbol{X}_2'\boldsymbol{X}_2 - N\overline{\boldsymbol{x}}\,\overline{\boldsymbol{x}}')/N$,

$$(\boldsymbol{X}'\boldsymbol{X})^{-1} = N^{-1}\begin{bmatrix} 1 + \overline{\boldsymbol{x}}'\boldsymbol{S}_X^{-1}\overline{\boldsymbol{x}} & \overline{\boldsymbol{x}}'\boldsymbol{S}_X^{-1} \\ \boldsymbol{S}_X^{-1}\overline{\boldsymbol{x}} & \boldsymbol{S}_X^{-1} \end{bmatrix} . \tag{19.9}$$

Proof. Without loss of generality, the intercept may be assumed to be the first column, which allows writing

$$\begin{aligned} \boldsymbol{X}'\boldsymbol{X} &= [\,\mathbf{1}_N\ \ \boldsymbol{X}_2\,]'[\,\mathbf{1}_N\ \ \boldsymbol{X}_2\,] \\ &= \begin{bmatrix} N & N\overline{\boldsymbol{x}}' \\ N\overline{\boldsymbol{x}} & \boldsymbol{X}_2'\boldsymbol{X}_2 \end{bmatrix} \\ &= \begin{bmatrix} a_{11} & \boldsymbol{a}_{21}' \\ \boldsymbol{a}_{21} & \boldsymbol{A}_{22} \end{bmatrix} . \end{aligned} \tag{19.10}$$

Standard results on partitioned matrices (Theorem 1.16) gives

$$(\boldsymbol{X}'\boldsymbol{X})^{-1} = \begin{bmatrix} b_{11} & \boldsymbol{b}_{21}' \\ \boldsymbol{b}_{21} & \boldsymbol{B}_{22} \end{bmatrix} , \tag{19.11}$$

with

$$\begin{aligned}\boldsymbol{B}_{22} &= \left(\boldsymbol{A}_{22} - \boldsymbol{a}_{21}a_{11}^{-1}\boldsymbol{a}_{21}'\right)^{-1} \\ &= \left(\boldsymbol{X}_2'\boldsymbol{X}_2 - N\overline{\boldsymbol{x}}N^{-1}N\overline{\boldsymbol{x}}'\right)^{-1} \\ &= \left(\boldsymbol{X}_2'\boldsymbol{X}_2 - N\overline{\boldsymbol{x}}\,\overline{\boldsymbol{x}}'\right)^{-1} \\ &= N^{-1}\boldsymbol{S}_X^{-1}\,,\end{aligned} \tag{19.12}$$

$$\begin{aligned}b_{11} &= a_{11}^{-1} + a_{11}^{-1}\boldsymbol{a}_{21}'\boldsymbol{B}_{22}\boldsymbol{a}_{21}a_{11}^{-1} \\ &= N^{-1} + N^{-1}N\overline{\boldsymbol{x}}'N^{-1}\boldsymbol{S}_X^{-1}N\overline{\boldsymbol{x}}N^{-1} \\ &= N^{-1}\left(1 + \overline{\boldsymbol{x}}'\boldsymbol{S}_X^{-1}\overline{\boldsymbol{x}}\right),\end{aligned} \tag{19.13}$$

and

$$\begin{aligned}\boldsymbol{b}_{12} &= -a_{11}^{-1}\boldsymbol{a}_{21}'\boldsymbol{B}_{22} \\ &= -N^{-1}N\overline{\boldsymbol{x}}'N^{-1}\boldsymbol{S}_X^{-1} \\ &= -N^{-1}\overline{\boldsymbol{x}}'\boldsymbol{S}_X^{-1}\,.\end{aligned} \tag{19.14}$$

□

19.3 MODEL COMPONENTS

For a $\text{GLM}_{N,p,q}(\boldsymbol{Y}_i; \boldsymbol{X}_i\boldsymbol{B}, \Sigma)$ with Gaussian errors and $\boldsymbol{X}$ fixed at least conditionally,

$$\boldsymbol{E} \sim \mathcal{N}_{N,p}(\boldsymbol{0}, \boldsymbol{I}, \Sigma) \tag{19.15}$$

and

$$\boldsymbol{Y} \sim \mathcal{N}_{N,p}(\boldsymbol{X}\boldsymbol{B}, \boldsymbol{I}, \Sigma)\,. \tag{19.16}$$

In the special case of the univariate model $p = 1$ and

$$\boldsymbol{e} \sim \mathcal{N}_{N,1}\left(\boldsymbol{0}, \boldsymbol{I}, \sigma^2\right) \sim \mathcal{N}_N\left(\boldsymbol{0}, \boldsymbol{I}\sigma^2\right) \tag{19.17}$$

and

$$\boldsymbol{y} \sim \mathcal{N}_{N,1}\left(\boldsymbol{X}\boldsymbol{B}, \boldsymbol{I}, \sigma^2\right) \sim \mathcal{N}_N\left(\boldsymbol{X}\boldsymbol{B}, \boldsymbol{I}\sigma^2\right). \tag{19.18}$$

19.4 PRIMARY PARAMETER AND RELATED ESTIMATORS

With a LTFR design matrix

$$\widetilde{\boldsymbol{B}} = (\boldsymbol{X}'\boldsymbol{X})^{-}\boldsymbol{X}'\boldsymbol{Y}\,, \tag{19.19}$$

which reduces to $\widetilde{\boldsymbol{\beta}} = (\boldsymbol{X}'\boldsymbol{X})^{-}\boldsymbol{X}'\boldsymbol{y}$ if $p = 1$ (the univariate model). Choosing a particular $(\boldsymbol{X}'\boldsymbol{X})^{-}$ which meets the following condition is equivalent to choosing a two-condition generalized inverse:

$$[(\boldsymbol{X}'\boldsymbol{X})^{-}\boldsymbol{X}']\boldsymbol{I}[(\boldsymbol{X}'\boldsymbol{X})^{-}\boldsymbol{X}']' = (\boldsymbol{X}'\boldsymbol{X})^{-}\boldsymbol{X}'\boldsymbol{X}[(\boldsymbol{X}'\boldsymbol{X})^{-}]' = (\boldsymbol{X}'\boldsymbol{X})^{-}. \quad (19.20)$$

With the condition,

$$\widetilde{\boldsymbol{B}} \sim (\mathcal{S})\mathcal{N}_{q,p}[(\boldsymbol{X}'\boldsymbol{X})^{-}\boldsymbol{X}'\boldsymbol{X}\boldsymbol{B}, (\boldsymbol{X}'\boldsymbol{X})^{-}, \boldsymbol{\Sigma}] \quad (19.21)$$

$$\text{vec}(\widetilde{\boldsymbol{B}}) \sim (\mathcal{S})\mathcal{N}_{qp}\{\text{vec}[(\boldsymbol{X}'\boldsymbol{X})^{-}\boldsymbol{X}'\boldsymbol{X}\boldsymbol{B}], \boldsymbol{\Sigma} \otimes (\boldsymbol{X}'\boldsymbol{X})^{-}\}. \quad (19.22)$$

The univariate model has $p = 1$ and

$$\widetilde{\boldsymbol{\beta}} \sim (\mathcal{S})\mathcal{N}_{q,1}[(\boldsymbol{X}'\boldsymbol{X})^{-}\boldsymbol{X}'\boldsymbol{X}\boldsymbol{\beta}, (\boldsymbol{X}'\boldsymbol{X})^{-}, \sigma^2] \quad (19.23)$$

$$\sim (\mathcal{S})\mathcal{N}_{q}[(\boldsymbol{X}'\boldsymbol{X})^{-}\boldsymbol{X}'\boldsymbol{X}\boldsymbol{\beta}, \sigma^2(\boldsymbol{X}'\boldsymbol{X})^{-}]. \quad (19.24)$$

The relationships $\boldsymbol{X}(\boldsymbol{X}'\boldsymbol{X})^{-}\boldsymbol{X}'\boldsymbol{X}\boldsymbol{B} = \boldsymbol{X}\boldsymbol{B}$ and $\boldsymbol{H}\boldsymbol{I}\boldsymbol{H} = \boldsymbol{H}$ give

$$\begin{aligned}\widehat{\boldsymbol{Y}} &= \boldsymbol{X}\widetilde{\boldsymbol{B}} \\ &= \boldsymbol{H}\boldsymbol{Y} \\ &\sim \mathcal{SN}_{N,p}(\boldsymbol{X}\boldsymbol{B}, \boldsymbol{H}, \boldsymbol{\Sigma})\end{aligned} \quad (19.25)$$

$$\begin{aligned}\widehat{\boldsymbol{E}} &= \boldsymbol{Y} - \widehat{\boldsymbol{Y}} \\ &= (\boldsymbol{I}_N - \boldsymbol{H})\boldsymbol{Y} \\ &\sim \mathcal{SN}_{N,p}\{[\boldsymbol{I}_N - \boldsymbol{X}(\boldsymbol{X}'\boldsymbol{X})^{-}\boldsymbol{X}']\boldsymbol{X}\boldsymbol{B}, (\boldsymbol{I}_N - \boldsymbol{H})\boldsymbol{I}(\boldsymbol{I}_N - \boldsymbol{H}), \boldsymbol{\Sigma}\} \\ &\sim \mathcal{SN}_{N,p}[\boldsymbol{0}, (\boldsymbol{I}_N - \boldsymbol{H}), \boldsymbol{\Sigma}].\end{aligned} \quad (19.26)$$

The univariate case has $p = 1$,

$$\begin{aligned}\widehat{\boldsymbol{y}} &= \boldsymbol{X}\widetilde{\boldsymbol{\beta}} \\ &= \boldsymbol{H}\boldsymbol{y} \\ &\sim \mathcal{SN}_{N,1}(\boldsymbol{X}\boldsymbol{\beta}, \boldsymbol{H}, \sigma^2) \\ &\sim \mathcal{SN}_{N}(\boldsymbol{X}\boldsymbol{\beta}, \boldsymbol{H}\sigma^2),\end{aligned} \quad (19.27)$$

and

$$\begin{aligned}\widehat{\boldsymbol{E}} &= \boldsymbol{y} - \widehat{\boldsymbol{y}} \\ &= (\boldsymbol{I}_N - \boldsymbol{H})\boldsymbol{y} \\ &\sim \mathcal{SN}_{N,1}[\boldsymbol{0}, (\boldsymbol{I}_N - \boldsymbol{H}), \sigma^2] \\ &\sim \mathcal{SN}_{N}[\boldsymbol{0}, (\boldsymbol{I}_N - \boldsymbol{H})\sigma^2].\end{aligned} \quad (19.28)$$

With $\nu_e = N - r$ and $r = \text{rank}(\boldsymbol{X})$, in general

$$\begin{aligned}\nu_e\widehat{\boldsymbol{\Sigma}} &= \widehat{\boldsymbol{E}}'\widehat{\boldsymbol{E}} \\ &= \boldsymbol{Y}'(\boldsymbol{I}_N - \boldsymbol{H})(\boldsymbol{I}_N - \boldsymbol{H})\boldsymbol{Y} \\ &= \boldsymbol{Y}'(\boldsymbol{I}_N - \boldsymbol{H})\boldsymbol{Y} \\ &\sim \mathcal{W}_p(\nu_e, \boldsymbol{\Sigma}),\end{aligned} \quad (19.29)$$

and for $p = 1$

$$\begin{aligned}\nu_e\widehat{\sigma}^2 &= \widehat{e}'\widehat{e} \\ &= SSE \\ &= y'(I_N - H)y \\ &\sim \mathcal{W}_1(\nu_e, \sigma^2)\,.\end{aligned} \tag{19.30}$$

Equivalently,

$$\nu_e\widehat{\sigma}^2/\sigma^2 \sim \chi^2(\nu_e)\,. \tag{19.31}$$

19.5 SECONDARY PARAMETER ESTIMATION

As always, $M = C(X'X)^{-}C'$. Having an estimable Θ requires $C(X'X)^{-}X'X = C$. Hence for estimable Θ

$$\begin{aligned}\widehat{\Theta} &= C\widetilde{B}U \\ &= [C(X'X)^{-}X']YU\,,\end{aligned} \tag{19.32}$$

$$C(X'X)^{-}X'XBU = CBU\,, \tag{19.33}$$

and

$$\begin{aligned}[C(X'X)^{-}X'][C(X'X)^{-}X']' &= C(X'X)^{-}X'X[(X'X)^{-}]'C' \\ &= C(X'X)^{-}\{C(X'X)^{-}X'X\}' \\ &= C(X'X)^{-}C'\,.\end{aligned} \tag{19.34}$$

Using the results just described gives

$$\widehat{\Theta} - \Theta_0 \sim \mathcal{N}_{a,b}[CBU - \Theta_0, C(X'X)^{-}C', \Sigma] \tag{19.35}$$

$$\text{vec}(\widehat{\Theta} - \Theta_0) \sim \mathcal{N}_{ab}[\text{vec}(CBU - \Theta_0), \Sigma_* \otimes M] \tag{19.36}$$

and

$$\nu_e\widehat{\Sigma}_* = (N - r)U'\widehat{\Sigma}U \sim \mathcal{W}_b(\nu_e, U'\Sigma U, 0)\,. \tag{19.37}$$

The univariate special case has $p = 1$ and

$$\begin{aligned}\widehat{\theta} &= C\widetilde{\beta} \\ &= [C(X'X)^{-}X']y\,.\end{aligned} \tag{19.38}$$

In turn,

$$\widehat{\theta} - \theta_0 \sim \mathcal{N}_{a,1}[C\beta - \theta_0, C(X'X)^{-}C', \sigma^2] \tag{19.39}$$

$$\sim \mathcal{N}_a(C\beta - \theta_0, \sigma^2 M) \tag{19.40}$$

and $\nu_e\widehat{\Sigma}_*$ reduces to $\nu_e\widehat{\sigma}^2 = \widehat{e}'\widehat{e}$ with $\nu_e\widehat{\sigma}^2/\sigma^2 \sim \chi^2(\nu_e, 0)$.

A *testable* hypothesis requires estimable $\boldsymbol{\Theta}$, full (row) rank of $\boldsymbol{C}$, and full (column) rank of $\boldsymbol{U}$. The requirements ensure full rank of the $a \times a$ symmetric matrix $\boldsymbol{M} = \boldsymbol{C}(\boldsymbol{X}'\boldsymbol{X})^{-}\boldsymbol{C}'$ as well as nonsingular $\boldsymbol{\Sigma}_* = \boldsymbol{U}'\boldsymbol{\Sigma}\boldsymbol{U}$. In turn,

$$\begin{aligned}\boldsymbol{M} &= \boldsymbol{C}(\boldsymbol{X}'\boldsymbol{X})^{-}\boldsymbol{C}' \\ &= \boldsymbol{V}_M \mathrm{Dg}(\boldsymbol{\lambda}_M)\boldsymbol{V}_M' \\ &= [\boldsymbol{V}_M \mathrm{Dg}(\boldsymbol{\lambda}_M)^{1/2}][\mathrm{Dg}(\boldsymbol{\lambda}_M)^{1/2}\boldsymbol{V}_M'] \\ &= \boldsymbol{F}_M\boldsymbol{F}_M' \end{aligned} \tag{19.41}$$

and

$$\begin{aligned}\widehat{\boldsymbol{\Delta}} &= (\widehat{\boldsymbol{\Theta}} - \boldsymbol{\Theta}_0)'\boldsymbol{M}^{-1}(\widehat{\boldsymbol{\Theta}} - \boldsymbol{\Theta}_0) \\ &= (\widehat{\boldsymbol{\Theta}} - \boldsymbol{\Theta}_0)'\boldsymbol{F}_M^{-t}\boldsymbol{F}_M^{-1}(\widehat{\boldsymbol{\Theta}} - \boldsymbol{\Theta}_0) \\ &= [\boldsymbol{F}_M^{-1}(\widehat{\boldsymbol{\Theta}} - \boldsymbol{\Theta}_0)]'[\boldsymbol{F}_M^{-1}(\widehat{\boldsymbol{\Theta}} - \boldsymbol{\Theta}_0)] . \end{aligned} \tag{19.42}$$

Furthermore

$$\begin{aligned}\boldsymbol{F}_M^{-1}(\widehat{\boldsymbol{\Theta}} - \boldsymbol{\Theta}_0) &\sim \mathcal{N}_{a,b}\left[\boldsymbol{F}_M^{-1}(\boldsymbol{\Theta} - \boldsymbol{\Theta}_0), \boldsymbol{F}_M^{-1}\boldsymbol{M}\boldsymbol{F}_M^{-t}, \boldsymbol{\Sigma}_*\right] \\ &\sim \mathcal{N}_{a,b}\left[\boldsymbol{F}_M^{-1}(\boldsymbol{\Theta} - \boldsymbol{\Theta}_0), \boldsymbol{F}_M^{-1}\boldsymbol{F}_M\boldsymbol{F}_M'\boldsymbol{F}_M^{-t}, \boldsymbol{\Sigma}_*\right] \\ &\sim \mathcal{N}_{a,b}\left[\boldsymbol{F}_M^{-1}(\boldsymbol{\Theta} - \boldsymbol{\Theta}_0), \boldsymbol{I}_a, \boldsymbol{\Sigma}_*\right] . \end{aligned} \tag{19.43}$$

The row covariance structure being $\boldsymbol{I}_a$ allows concluding

$$\begin{aligned}\widehat{\boldsymbol{\Delta}} &= \left[\boldsymbol{F}_M^{-1}(\widehat{\boldsymbol{\Theta}} - \boldsymbol{\Theta}_0)\right]'\left[\boldsymbol{F}_M^{-1}(\widehat{\boldsymbol{\Theta}} - \boldsymbol{\Theta}_0)\right] \\ &\sim \mathcal{W}_b\left\{a, \boldsymbol{\Sigma}_*, [\boldsymbol{F}_M^{-1}(\boldsymbol{\Theta} - \boldsymbol{\Theta}_0)]'\boldsymbol{F}_M^{-1}(\boldsymbol{\Theta} - \boldsymbol{\Theta}_0)\right\} \\ &\sim \mathcal{W}_b\left[a, \boldsymbol{\Sigma}_*, (\boldsymbol{\Theta} - \boldsymbol{\Theta}_0)'\boldsymbol{M}^{-1}(\boldsymbol{\Theta} - \boldsymbol{\Theta}_0)\right] . \end{aligned} \tag{19.44}$$

A more traditional derivation of the last result follows from considering

$$\boldsymbol{Y}_U = \boldsymbol{Y}\boldsymbol{U} - \boldsymbol{X}\boldsymbol{C}'(\boldsymbol{C}\boldsymbol{C}')^{-1}\boldsymbol{\Theta}_0 . \tag{19.45}$$

Having $\boldsymbol{Y} \sim \mathcal{N}_{N,p}(\boldsymbol{X}\boldsymbol{B}, \boldsymbol{I}_N, \boldsymbol{\Sigma})$ implies $\boldsymbol{Y}_U \sim \mathcal{N}_{N,b}[\mathrm{E}(\boldsymbol{Y}_U), \boldsymbol{I}_N, \boldsymbol{\Sigma}_*]$ with

$$\mathrm{E}(\boldsymbol{Y}_U) = \boldsymbol{X}\boldsymbol{B}\boldsymbol{U} - \boldsymbol{X}\boldsymbol{C}'(\boldsymbol{C}\boldsymbol{C}')^{-1}\boldsymbol{\Theta}_0 \tag{19.46}$$

$$\boldsymbol{\Sigma}_* = \boldsymbol{U}'\boldsymbol{\Sigma}\boldsymbol{U} . \tag{19.47}$$

If $\boldsymbol{T} = \boldsymbol{C}(\boldsymbol{X}'\boldsymbol{X})^{-}\boldsymbol{X}'$, then $\widehat{\boldsymbol{\Theta}} = \boldsymbol{T}\boldsymbol{Y}\boldsymbol{U}$. *A* testable $\boldsymbol{\Theta}$ ensures $\boldsymbol{T}\boldsymbol{X}\boldsymbol{C}'(\boldsymbol{C}\boldsymbol{C}')^{-1} = \boldsymbol{I}_a$. Therefore

$$\begin{aligned}\widehat{\boldsymbol{\Theta}} - \boldsymbol{\Theta}_0 &= \boldsymbol{T}\boldsymbol{Y}\boldsymbol{U} - \boldsymbol{T}[\boldsymbol{X}\boldsymbol{C}'(\boldsymbol{C}\boldsymbol{C}')^{-1}]\boldsymbol{\Theta}_0 \\ &= \boldsymbol{T}\boldsymbol{Y}_U . \end{aligned} \tag{19.48}$$

Furthermore

$$\begin{aligned}\widehat{\boldsymbol{\Delta}} &= (\widehat{\boldsymbol{\Theta}} - \boldsymbol{\Theta}_0)'\boldsymbol{M}^{-1}(\widehat{\boldsymbol{\Theta}} - \boldsymbol{\Theta}_0) \\ &= \boldsymbol{Y}_U'\boldsymbol{T}'\boldsymbol{M}^{-1}\boldsymbol{T}\boldsymbol{Y}_U . \end{aligned} \tag{19.49}$$

If $\boldsymbol{A} = \boldsymbol{T}'\boldsymbol{M}^{-1}\boldsymbol{T}$ then $\widehat{\boldsymbol{\Delta}} = \boldsymbol{Y}_U'\boldsymbol{A}\boldsymbol{Y}_U$ and

$$\begin{aligned}\boldsymbol{A}^2 &= (\boldsymbol{T}'\boldsymbol{M}^{-1}\boldsymbol{T})(\boldsymbol{T}'\boldsymbol{M}^{-1}\boldsymbol{T}) \\ &= \boldsymbol{T}'\boldsymbol{M}^{-1}\boldsymbol{C}(\boldsymbol{X}'\boldsymbol{X})^{-}\boldsymbol{X}'\{\boldsymbol{X}[(\boldsymbol{X}'\boldsymbol{X})^{-}]'\boldsymbol{C}'\}\boldsymbol{M}^{-1}\boldsymbol{T} \\ &= \boldsymbol{T}'\boldsymbol{M}^{-1}\boldsymbol{C}(\boldsymbol{X}'\boldsymbol{X})^{-}\{\boldsymbol{C}'\}\boldsymbol{M}^{-1}\boldsymbol{T} = \boldsymbol{A}.\end{aligned} \tag{19.50}$$

Furthermore, with $\boldsymbol{\Theta}_{c0} = \boldsymbol{C}'(\boldsymbol{C}\boldsymbol{C}')^{-1}\boldsymbol{\Theta}_0$,

$$\begin{aligned}[\mathrm{E}(\boldsymbol{Y}_U)]'\boldsymbol{A}\mathrm{E}(\boldsymbol{Y}_U) &= [\boldsymbol{X}\boldsymbol{B}\boldsymbol{U} - \boldsymbol{X}\boldsymbol{\Theta}_{c0}]'\boldsymbol{T}'\boldsymbol{M}^{-1}\boldsymbol{T}[\boldsymbol{X}\boldsymbol{B}\boldsymbol{U} - \boldsymbol{X}\boldsymbol{\Theta}_{c0}] \\ &= [\boldsymbol{T}\boldsymbol{X}\boldsymbol{B}\boldsymbol{U} - \boldsymbol{T}\boldsymbol{X}\boldsymbol{\Theta}_{c0}]'\boldsymbol{M}^{-1}[\boldsymbol{T}\boldsymbol{X}\boldsymbol{B}\boldsymbol{U} - \boldsymbol{T}\boldsymbol{X}\boldsymbol{\Theta}_{c0}] \\ &= (\boldsymbol{\Theta} - \boldsymbol{\Theta}_0)'\boldsymbol{M}^{-1}(\boldsymbol{\Theta} - \boldsymbol{\Theta}_0).\end{aligned} \tag{19.51}$$

The idempotency of $\boldsymbol{A}$ combines with $\boldsymbol{Y}_U \sim \mathcal{N}_{N,b}[\mathrm{E}(\boldsymbol{Y}_U), \boldsymbol{I}_N, \boldsymbol{\Sigma}_*]$ to give $\widehat{\boldsymbol{\Delta}} \sim \mathcal{W}_b\left[a, \boldsymbol{\Sigma}_*, (\boldsymbol{\Theta} - \boldsymbol{\Theta}_0)'\boldsymbol{M}^{-1}(\boldsymbol{\Theta} - \boldsymbol{\Theta}_0)\right]$ (Theorem 10.8).

19.6 ADDED-LAST AND ADDED-IN-ORDER TESTS

Muller and Fetterman (2002) gave detailed discussions of the interpretations and uses of added-last and added-in-order tests in univariate models. Here we prove some important properties.

Lemma 19.2 In a $\mathrm{GLM}_{N,p,q}\mathrm{FR}(\boldsymbol{Y}_i; \boldsymbol{X}_i\boldsymbol{B}, \boldsymbol{\Sigma})$ with Gaussian errors, the added-last *SS* for all predictors are independent if and only if $\boldsymbol{X}'\boldsymbol{X}$ is a diagonal matrix.

Proof. Testing $\boldsymbol{\beta} = \boldsymbol{0}$ uses $\boldsymbol{C} = \boldsymbol{I}_q$, while testing a particular slope uses $\boldsymbol{C}_j = \mathrm{row}_j(\boldsymbol{C})$. With $\langle(\boldsymbol{X}'\boldsymbol{X})^{-1}\rangle_{jk}$ the j,k element of $(\boldsymbol{X}'\boldsymbol{X})^{-1}$,

$$m_j = \boldsymbol{C}_j(\boldsymbol{X}'\boldsymbol{X})^{-1}\boldsymbol{C}_j' = [0\,0\cdots 1\cdots 0](\boldsymbol{X}'\boldsymbol{X})^{-1}\boldsymbol{C}_j' = \langle(\boldsymbol{X}'\boldsymbol{X})^{-1}\rangle_{jj} \tag{19.52}$$

and

$$SSH_j = \widehat{\theta}_j'\left[\boldsymbol{C}_j(\boldsymbol{X}'\boldsymbol{X})^{-1}\boldsymbol{C}_j'\right]^{-1}\widehat{\theta}_j = \widehat{\theta}_j^2 m_j^{-1}. \tag{19.53}$$

Furthermore $\widehat{\boldsymbol{\theta}}_j = \boldsymbol{C}_j\widehat{\boldsymbol{\beta}}$ and

$$q_j = \widehat{\boldsymbol{\theta}}_j' m_j^{-1}\widehat{\boldsymbol{\theta}}_j = \widehat{\boldsymbol{\beta}}'\boldsymbol{C}_j' m_j^{-1}\boldsymbol{C}_j\widehat{\boldsymbol{\beta}} = \widehat{\boldsymbol{\beta}}'\boldsymbol{A}_j\widehat{\boldsymbol{\beta}}. \tag{19.54}$$

Here $\widehat{\boldsymbol{\beta}} \sim \mathcal{N}_q[\boldsymbol{\beta}, \sigma^2(\boldsymbol{X}'\boldsymbol{X})^{-1}]$. With $j \neq k$, Theorem 8.13 gives $q_j \perp\!\!\!\perp q_k \Leftrightarrow \boldsymbol{A}_j[\sigma^2(\boldsymbol{X}'\boldsymbol{X})^{-1}]\boldsymbol{A}_k = \boldsymbol{0}$. The scalar $\boldsymbol{C}_j(\boldsymbol{X}'\boldsymbol{X})^{-1}\boldsymbol{C}_k' = \langle(\boldsymbol{X}'\boldsymbol{X})^{-1}\rangle_{jk}$ gives

$$\begin{aligned}\boldsymbol{A}_j\sigma^2(\boldsymbol{X}'\boldsymbol{X})^{-1}\boldsymbol{A}_k &= \sigma^2\boldsymbol{C}_j' m_j^{-1}\left[\boldsymbol{C}_j(\boldsymbol{X}'\boldsymbol{X})^{-1}\boldsymbol{C}_k'\right]m_k^{-1}\boldsymbol{C}_k \\ &= (\sigma^2 m_j^{-1}m_k^{-1})(\boldsymbol{C}_j'\boldsymbol{C}_k)\left[\boldsymbol{C}_j(\boldsymbol{X}'\boldsymbol{X})^{-1}\boldsymbol{C}_k'\right] \\ &= (\sigma^2 m_j^{-1}m_k^{-1})(\boldsymbol{C}_j'\boldsymbol{C}_k)\langle(\boldsymbol{X}'\boldsymbol{X})^{-1}\rangle_{jk}.\end{aligned} \tag{19.55}$$

Given the assumptions, the scalar $(\sigma^2 m_j^{-1} m_k^{-1})$ is never zero. Furthermore $(\boldsymbol{C}_j'\boldsymbol{C}_k)$ is a $q \times q$ matrix with a 1 in location j, k and zeros elsewhere. Hence the last expression is zero if and only if $\langle(\boldsymbol{X}'\boldsymbol{X})^{-1}\rangle_{jk}$ is zero. The proof is completed by noting $(\boldsymbol{X}'\boldsymbol{X})^{-1}$ is diagonal if and only if $(\boldsymbol{X}'\boldsymbol{X})$ is diagonal. □

Lemma 19.3 In a $\text{GLM}_{N,p,q}\text{FR}(\boldsymbol{Y}_i; \boldsymbol{X}_i\boldsymbol{B}, \boldsymbol{\Sigma})$ with Gaussian errors, if $\boldsymbol{X}$ includes an intercept, then added-last *SS* for predictor variables other than the intercept are mutually independent if and only if they are mutually uncorrelated.

Proof. Without loss of generality, we may assume the intercept is the leftmost column in $\boldsymbol{X}$. Testing all slopes equal to zero uses $\boldsymbol{C} = [\,\boldsymbol{0}\;\; \boldsymbol{I}_{q-1}\,]$, while testing a particular slope corresponds to considering $\boldsymbol{C}_j = \text{row}_j(\boldsymbol{C})$. Lemma 19.1 gives

$$(\boldsymbol{X}'\boldsymbol{X})^{-1} = \frac{1}{N}\begin{bmatrix} 1 + \overline{\boldsymbol{x}}'\boldsymbol{C}_X^{-1}\overline{\boldsymbol{x}} & \overline{\boldsymbol{x}}'\boldsymbol{C}_X^{-1} \\ \boldsymbol{C}_X^{-1}\overline{\boldsymbol{x}} & \boldsymbol{C}_X^{-1} \end{bmatrix}. \tag{19.56}$$

With $\langle \boldsymbol{C}_X^{-1}\rangle_{jk}$ indicating element j, k of $\boldsymbol{C}_X^{-1}$,

$$m_j = \boldsymbol{C}_j(\boldsymbol{X}'\boldsymbol{X})^{-1}\boldsymbol{C}_j' = [\,0\,0\cdots 1\cdots 0\,](\boldsymbol{X}'\boldsymbol{X})^{-1}\boldsymbol{C}_j = N^{-1}\langle \boldsymbol{C}_X^{-1}\rangle_{jj} \tag{19.57}$$

and

$$SSH_j = \widehat{\theta}_j'\left[\boldsymbol{C}_j(\boldsymbol{X}'\boldsymbol{X})^{-1}\boldsymbol{C}_j'\right]^{-1}\widehat{\theta}_j = \widehat{\theta}_j^2 m_j^{-1}. \tag{19.58}$$

Furthermore $\widehat{\boldsymbol{\theta}}_j = \boldsymbol{C}_j\widehat{\boldsymbol{\beta}}$,

$$q_j = \widehat{\boldsymbol{\theta}}_j' m_j^{-1}\widehat{\boldsymbol{\theta}}_j = \widehat{\boldsymbol{\beta}}'\boldsymbol{C}_j' m_j^{-1}\boldsymbol{C}_j\widehat{\boldsymbol{\beta}} = \widehat{\boldsymbol{\beta}}'\boldsymbol{A}_j\widehat{\boldsymbol{\beta}}. \tag{19.59}$$

Here $\widehat{\boldsymbol{\beta}} \sim \mathcal{N}_q\left[\boldsymbol{\beta}, \sigma^2(\boldsymbol{X}'\boldsymbol{X})^{-1}\right]$. With $j \neq k$, Theorem 8.13 gives $q_j \perp\!\!\!\perp q_k \Leftrightarrow \boldsymbol{A}_j\left[\sigma^2(\boldsymbol{X}'\boldsymbol{X})^{-1}\right]\boldsymbol{A}_k = \boldsymbol{0}$. The scalar $\boldsymbol{C}_j(\boldsymbol{X}'\boldsymbol{X})^{-1}\boldsymbol{C}_k' = \langle \boldsymbol{C}_X^{-1}\rangle_{jk}$ gives

$$\begin{aligned} \boldsymbol{A}_j\sigma^2(\boldsymbol{X}'\boldsymbol{X})^{-1}\boldsymbol{A}_k &= \sigma^2\boldsymbol{C}_j' m_j^{-1}\left[\boldsymbol{C}_j(\boldsymbol{X}'\boldsymbol{X})^{-1}\boldsymbol{C}_k'\right] m_k^{-1}\boldsymbol{C}_k \\ &= (\sigma^2 m_j^{-1} m_k^{-1})(\boldsymbol{C}_j'\boldsymbol{C}_k)\left[\boldsymbol{C}_j(\boldsymbol{X}'\boldsymbol{X})^{-1}\boldsymbol{C}_k'\right] \\ &= (\sigma^2 m_j^{-1} m_k^{-1})(\boldsymbol{C}_j'\boldsymbol{C}_k)\langle \boldsymbol{C}_X^{-1}\rangle_{jk}. \end{aligned} \tag{19.60}$$

Given the assumptions, the scalar $(\sigma^2 m_j^{-1} m_k^{-1})$ is never zero. Furthermore $(\boldsymbol{C}_j'\boldsymbol{C}_k)$ is $q \times q$ with a 1 in location j, k and zeros elsewhere. Hence the last expression is zero if and only if $\langle \boldsymbol{C}_X^{-1}\rangle_{jk}$ is zero. The proof is completed by noting $\boldsymbol{C}_X^{-1}$ is diagonal if and only if $\boldsymbol{C}_X$ is diagonal if and only if the corresponding correlation matrix is diagonal. □

Lemma 19.4 Added-in-order *SS* are always independent in a $\text{GLM}_{N,q}\text{FR}(y_i; \boldsymbol{X}_i\boldsymbol{\beta}, \sigma^2)$ with Gaussian errors.

Proof. In the following, $\boldsymbol{X}_j$ indicates the first j columns of $\boldsymbol{X}$, and $\boldsymbol{X}_{i,j} = \text{row}_i(\boldsymbol{X}_j)$. In turn, model j is $\text{GLM}_{N,q}\text{FR}(y_i; \boldsymbol{X}_{i,j}\boldsymbol{\beta}_j, \sigma_j^2)$ with Gaussian errors. Furthermore $\boldsymbol{y} = \boldsymbol{X}_j\boldsymbol{\beta}_j + \boldsymbol{e}_j$, with full-rank $\boldsymbol{X}_j$ of dimension $N \times j$. Also

$$\boldsymbol{X}_1\boldsymbol{\beta}_1 = [\,\boldsymbol{x}_1\,][\,\beta_1\,]$$
$$\vdots$$
$$\boldsymbol{X}_j\boldsymbol{\beta}_j = [\,\boldsymbol{x}_1\;\boldsymbol{x}_2\cdots\boldsymbol{x}_j\,]\begin{bmatrix}\beta_1\\ \vdots\\ \beta_j\end{bmatrix}. \tag{19.61}$$

The matrix $\boldsymbol{H}_j = \boldsymbol{X}_j(\boldsymbol{X}_j'\boldsymbol{X}_j)^{-1}\boldsymbol{X}_j' = \boldsymbol{H}_j^2$ is $N \times N$, symmetric, idempotent, and $\text{rank}(\boldsymbol{H}_j) = \text{rank}(\boldsymbol{X}_j) = j$. Similarly, $(\boldsymbol{I}_N - \boldsymbol{H}_j)$ is symmetric, $N \times N$, idempotent, of rank $N - j$, and $\boldsymbol{H}_j(\boldsymbol{I}_N - \boldsymbol{H}_j) = (\boldsymbol{I}_N - \boldsymbol{H}_j)\boldsymbol{H}_j = \boldsymbol{0}$. Also

$$\begin{aligned}\boldsymbol{y}'\boldsymbol{y} = \boldsymbol{y}'\boldsymbol{I}_N\boldsymbol{y} &= \boldsymbol{y}'\boldsymbol{H}_j\boldsymbol{y} + \boldsymbol{y}'(\boldsymbol{I}_N - \boldsymbol{H}_j)\boldsymbol{y}\\ SST_j &= SSH_j + SSE_j\,.\end{aligned} \tag{19.62}$$

The ANOVA theorem may be applied by defining $\boldsymbol{y}_{1,j} = \boldsymbol{y}\sigma_j^{-1}$ and writing

$$\boldsymbol{A} = \boldsymbol{A}_1 + \boldsymbol{A}_2 \tag{19.63}$$
$$\boldsymbol{I}_N = \boldsymbol{H}_j + (\boldsymbol{I}_N - \boldsymbol{H}_j)\,, \tag{19.64}$$

with a corresponding decomposition of ranks, namely $N = j + (N - j)$.

The test of adding variables $\{j+1, j+2, \ldots, j+k\}$ in order compares model j to model $j+k$ (models in the added-in-order pool) and uses the added-in-order sums of squares, $SS_{j+k} = SSE_j - SSE_{j+k} = SSH_{j+k} - SSH_j$. In particular,

$$SS_{j+k} = \boldsymbol{y}'\boldsymbol{H}_{j+k}\boldsymbol{y} - \boldsymbol{y}'\boldsymbol{H}_j\boldsymbol{y} = \boldsymbol{y}'(\boldsymbol{H}_{j+k} - \boldsymbol{H}_j)\boldsymbol{y} = \boldsymbol{y}'\boldsymbol{A}_{j+k}\boldsymbol{y}\,. \tag{19.65}$$

The design matrix for model $j+k$ may be partitioned as $\boldsymbol{X}_{j+k} = [\,\boldsymbol{X}_j\;\boldsymbol{X}_{k(-j)}\,]$, with j columns in $\boldsymbol{X}_j$ and $k-j$ columns in $\boldsymbol{X}_{k(-j)}$, which contains variables $\{j+1, j+2, \ldots, j+k\}$. In turn,

$$\boldsymbol{X}_{j+k}'\boldsymbol{X}_{j+k} = \begin{bmatrix}\boldsymbol{X}_j'\boldsymbol{X}_j & \boldsymbol{X}_j'\boldsymbol{X}_{k(-j)}\\ \boldsymbol{X}_{k(-j)}'\boldsymbol{X}_j & \boldsymbol{X}_{k(-j)}'\boldsymbol{X}_{k(-j)}\end{bmatrix} \tag{19.66}$$

has a similarly partitioned inverse,

$$(\boldsymbol{X}_{j+k}'\boldsymbol{X}_{j+k})^{-1} = \begin{bmatrix}\boldsymbol{B}_{11} & \boldsymbol{B}_{12}\\ \boldsymbol{B}_{12}' & \boldsymbol{B}_{22}\end{bmatrix}, \tag{19.67}$$

with

$$\boldsymbol{B}_{11} = (\boldsymbol{X}_j'\boldsymbol{X}_j)^{-1} + (\boldsymbol{X}_j'\boldsymbol{X}_j)^{-1}\boldsymbol{X}_j'\boldsymbol{X}_{k(-j)}\boldsymbol{B}_{22}\boldsymbol{X}_{k(-j)}'\boldsymbol{X}_j(\boldsymbol{X}_j'\boldsymbol{X}_j)^{-1} \tag{19.68}$$
$$\boldsymbol{B}_{12} = -(\boldsymbol{X}_j'\boldsymbol{X}_j)^{-1}\boldsymbol{X}_j'\boldsymbol{X}_{k(-j)}\boldsymbol{B}_{22}\,. \tag{19.69}$$

Using the partitioned matrix inverse in the hat matrix for the larger model gives

$$\begin{aligned}
\boldsymbol{H}_{j+k} &= \boldsymbol{X}_{j+k}\left(\boldsymbol{X}'_{j+k}\boldsymbol{X}_{j+k}\right)^{-1}\boldsymbol{X}'_{j+k} \\
&= \left[\boldsymbol{X}_j\ \boldsymbol{X}_{k(-j)}\right]\begin{bmatrix}\boldsymbol{B}_{11} & \boldsymbol{B}_{12}\\ \boldsymbol{B}'_{12} & \boldsymbol{B}_{22}\end{bmatrix}\begin{bmatrix}\boldsymbol{X}'_j\\ \boldsymbol{X}'_{k(-j)}\end{bmatrix} \\
&= \left[\boldsymbol{X}_j\ \boldsymbol{X}_{k(-j)}\right]\begin{bmatrix}\boldsymbol{B}_{11} & \boldsymbol{B}_{12}\\ \boldsymbol{B}'_{12} & \boldsymbol{B}_{22}\end{bmatrix}\begin{bmatrix}\boldsymbol{X}'_j\\ \boldsymbol{X}'_{k(-j)}\end{bmatrix} \\
&= \left[\boldsymbol{X}_j\boldsymbol{B}_{11}+\boldsymbol{X}_{k(-j)}\boldsymbol{B}'_{12}\quad \boldsymbol{X}_j\boldsymbol{B}_{12}+\boldsymbol{X}_{k(-j)}\boldsymbol{B}_{22}\right]\begin{bmatrix}\boldsymbol{X}'_j\\ \boldsymbol{X}'_{k(-j)}\end{bmatrix} \\
&= \boldsymbol{X}_j\boldsymbol{B}_{11}\boldsymbol{X}'_j+\boldsymbol{X}_{k(-j)}\boldsymbol{B}'_{12}\boldsymbol{X}_j+\boldsymbol{X}_j\boldsymbol{B}_{12}\boldsymbol{X}'_{k(-j)}+\boldsymbol{X}_{k(-j)}\boldsymbol{B}_{22}\boldsymbol{X}'_{k(-j)}\,. \qquad (19.70)
\end{aligned}$$

The first term in the last line of the preceding equation may be expressed as

$$\begin{aligned}
\boldsymbol{X}_j\boldsymbol{B}_{11}\boldsymbol{X}'_j &= \boldsymbol{X}_j\left(\boldsymbol{X}'_j\boldsymbol{X}_j\right)^{-1}\boldsymbol{X}'_j + \\
&\qquad \boldsymbol{X}_j\left(\boldsymbol{X}'_j\boldsymbol{X}_j\right)^{-1}\boldsymbol{X}'_j\boldsymbol{X}_{k(-j)}\boldsymbol{B}_{22}\boldsymbol{X}'_{k(-j)}\boldsymbol{X}_j\left(\boldsymbol{X}'_j\boldsymbol{X}_j\right)^{-1}\boldsymbol{X}'_j \\
&= \boldsymbol{H}_j + \boldsymbol{H}_j\boldsymbol{X}_{k(-j)}\boldsymbol{B}_{22}\boldsymbol{X}'_{k(-j)}\boldsymbol{H}_j\,, \qquad (19.71)
\end{aligned}$$

while the third term may be expressed as

$$\begin{aligned}
\boldsymbol{X}_j\boldsymbol{B}_{12}\boldsymbol{X}'_{k(-j)} &= \boldsymbol{X}_j\left[-\left(\boldsymbol{X}'_j\boldsymbol{X}_j\right)^{-1}\boldsymbol{X}'_j\boldsymbol{X}_{k(-j)}\boldsymbol{B}_{22}\right]\boldsymbol{X}'_{k(-j)} \\
&= -\boldsymbol{H}_j\boldsymbol{X}_{k(-j)}\boldsymbol{B}_{22}\boldsymbol{X}'_{k(-j)}\,. \qquad (19.72)
\end{aligned}$$

Substituting the alternate forms back into the expression for $\boldsymbol{H}_{j+k}$ gives

$$\begin{aligned}
\boldsymbol{H}_{j+k} &= \boldsymbol{X}_j\boldsymbol{B}_{11}\boldsymbol{X}'_j + \boldsymbol{X}_{k(-j)}\boldsymbol{B}'_{12}\boldsymbol{X}'_j + \boldsymbol{X}_j\boldsymbol{B}_{12}\boldsymbol{X}'_{k(-j)} + \boldsymbol{X}_{k(-j)}\boldsymbol{B}_{22}\boldsymbol{X}'_{k(-j)} \\
&= \boldsymbol{H}_j + \boldsymbol{H}_j\boldsymbol{X}_{k(-j)}\boldsymbol{B}_{22}\boldsymbol{X}'_{k(-j)}\boldsymbol{H}_j - \\
&\qquad \left(\boldsymbol{X}_{k(-j)}\boldsymbol{B}_{22}\boldsymbol{X}'_{k(-j)}\boldsymbol{H}_j+\boldsymbol{H}_j\boldsymbol{X}_{k(-j)}\boldsymbol{B}_{22}\boldsymbol{X}'_{k(-j)}\right) + \\
&\qquad \boldsymbol{X}_{k(-j)}\boldsymbol{B}_{22}\boldsymbol{X}'_{k(-j)}\,. \qquad (19.73)
\end{aligned}$$

In turn,

$$\begin{aligned}
\boldsymbol{H}_j\boldsymbol{H}_{j+k} &= \boldsymbol{H}_j + \boldsymbol{H}_j\boldsymbol{X}_{k(-j)}\boldsymbol{B}_{22}\boldsymbol{X}'_{k(-j)}\boldsymbol{H}_j - \\
&\qquad \left(\boldsymbol{H}_j\boldsymbol{X}_{k(-j)}\boldsymbol{B}_{22}\boldsymbol{X}'_{k(-j)}\boldsymbol{H}_j + \boldsymbol{H}_j\boldsymbol{X}_{k(-j)}\boldsymbol{B}_{22}\boldsymbol{X}'_{k(-j)}\right) + \\
&\qquad \boldsymbol{H}_j\boldsymbol{X}_{k(-j)}\boldsymbol{B}_{22}\boldsymbol{X}'_{k(-j)} \\
&= \boldsymbol{H}_j + \boldsymbol{H}_j\boldsymbol{X}_{k(-j)}\boldsymbol{B}_{22}\boldsymbol{X}'_{k(-j)}\boldsymbol{H}_j - \\
&\qquad \left(\boldsymbol{H}_j\boldsymbol{X}_{k(-j)}\boldsymbol{B}_{22}\boldsymbol{X}'_{k(-j)}\boldsymbol{H}_j + \boldsymbol{H}_j\boldsymbol{X}_{k(-j)}\boldsymbol{B}_{22}\boldsymbol{X}'_{k(-j)}\right) + \\
&\qquad \boldsymbol{H}_j\boldsymbol{X}_{k(-j)}\boldsymbol{B}_{22}\boldsymbol{X}_{k(-j)} = \boldsymbol{H}_j\,. \qquad (19.74)
\end{aligned}$$

Also $\boldsymbol{H}_{j+k}=\boldsymbol{H}_j+(\boldsymbol{H}_{j+k}-\boldsymbol{H}_j) = \boldsymbol{H}_j+\boldsymbol{A}_{j+k}$, $\boldsymbol{H}_j\boldsymbol{A}_{j+k}= \boldsymbol{H}_j(\boldsymbol{H}_{j+k}-\boldsymbol{H}_j) = \boldsymbol{0}$, and $\text{rank}(\boldsymbol{H}_{j+k}) = \text{rank}(\boldsymbol{H}_j) + \text{rank}(\boldsymbol{A}_{j+k})$. Finally

$$\begin{aligned}
\boldsymbol{A}_{j+k_1}\boldsymbol{A}_{(j+k_1)+k_2} &= (\boldsymbol{H}_{j+k_1} - \boldsymbol{H}_j)(\boldsymbol{H}_{j+k_1+k_2} - \boldsymbol{H}_{j+k_1}) \\
&= \boldsymbol{H}_{j+k_1}\boldsymbol{H}_{j+k_1+k_2} - \boldsymbol{H}_j\boldsymbol{H}_{j+k_1+k_2} - \boldsymbol{H}_{j+k_1}\boldsymbol{H}_{j+k_1} + \boldsymbol{H}_j\boldsymbol{H}_{j+k_1} \\
&= \boldsymbol{H}_{j+k_1} - \boldsymbol{H}_j - \boldsymbol{H}_{j+k_1} + \boldsymbol{H}_j \\
&= \boldsymbol{0}
\end{aligned} \tag{19.75}$$

and

$$\begin{aligned}
\boldsymbol{A}_{j+k_1}\boldsymbol{A}_{(j+k_1+m)+k_2} &= (\boldsymbol{H}_{j+k_1} - \boldsymbol{H}_j)(\boldsymbol{H}_{j+k_1+m+k_2} - \boldsymbol{H}_{j+k_1+m}) \\
&= \boldsymbol{H}_{j+k_1}\boldsymbol{H}_{j+k_1+m+k_2} - \boldsymbol{H}_j\boldsymbol{H}_{j+k_1+m+k_2} - \\
&\qquad \boldsymbol{H}_{j+k_1}\boldsymbol{H}_{j+k_1+m} + \boldsymbol{H}_j\boldsymbol{H}_{j+k_1+m} \\
&= \boldsymbol{H}_{j+k_1} - \boldsymbol{H}_j - \boldsymbol{H}_{j+k_1} + \boldsymbol{H}_j \\
&= \boldsymbol{0}.
\end{aligned} \tag{19.76}$$

□

CHAPTER 20

Sample Size for Univariate Linear Models

20.1 SAMPLE SIZE CONSULTING: BEFORE YOU BEGIN

A scientist describes, in three sentences, a study requiring six months and costing tens of thousands of dollars. The scientist then asks "How many subjects do I need?" Finding a good answer requires intensive and iterative collaboration between the scientist and statistician to clearly detail the (1) ethical, monetary, and time constraints, (2) scientific goals, both long term (vague) and short term (concrete), (3) study design, and (4) data analysis plan. We believe some form of sample size analysis, such as power analysis for test procedures and analysis of precision for estimators, should play a key role in the planning of most studies. Muller, Barton, and Benignus (1984), Muller and Benignus (1992), O'Brien and Muller (1993), and Catellier and Muller (2002) provided introductory and tutorial presentations.

In practice, scientists typically seek to achieve more than one goal in each study. Finding a satisfactory design and analysis plan requires eliciting the complete set of goals and the relative importance of each. Although estimation essentially always has importance, the importance of statistical hypothesis testing can range from negligible (or not applicable) to critically essential. An important question for the statistician to ask the scientist is, "Will the study be a success if none of the planned hypothesis tests turn out to be statistically significant?" Jiroutek, Muller, Kupper, and Stewart (2003) discussed the question from the perspective of choosing a criterion probability to control when choosing a sample size.

In addition to choosing a sample size, the collaborative process should include comparing a variety of designs and associated analyses. Exploring and describing the variation in performance as a function of design features, assumptions about the population, and choice of analysis inform and improve the decision process.

Tables and plots for a range of different scenarios help greatly. Figure 20.1 illustrates the power tradeoffs between sample size and mean difference for an independent groups t-test. As with all univariate linear models with Gaussian errors and fixed predictors, given α, only 1) sample size, 2) mean differences and 3) error variance affect power. Example E01 in the manual for the free software

POWERLIB203.IML described in the Appendix (Section A.2) gives the code used to produce Figure 20.1.

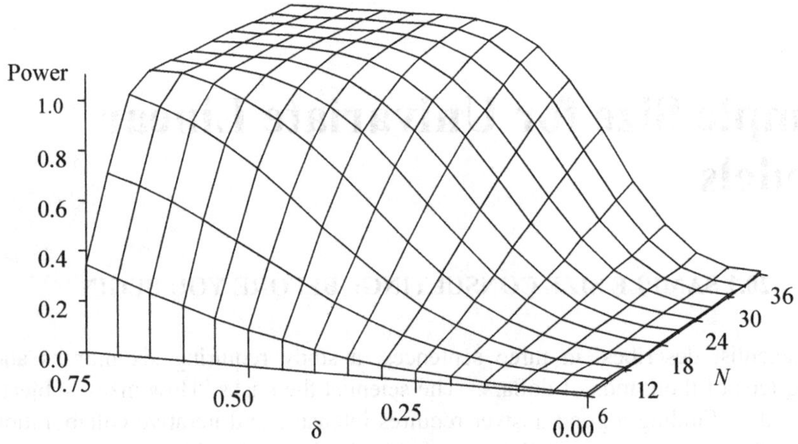

Figure 20.1 Power as a function of sample size (N) and mean difference (δ) for a balanced independent groups t test with $\sigma^2 = 0.068$ and $\alpha = 0.01$.

20.2 THE MACHINERY OF A POWER ANALYSIS

The discussion assumes Gaussian errors and fixed predictors. We restrict attention to testable hypotheses, which requires full-rank $\boldsymbol{C} = \boldsymbol{C}(\boldsymbol{X}'\boldsymbol{X})^{-}(\boldsymbol{X}'\boldsymbol{X})$. A univariate GLM power calculation with fixed predictors is fully specified by α, σ^2, $\boldsymbol{X}$, $\boldsymbol{\beta}$, $\boldsymbol{C}$, and $\boldsymbol{\theta}_0$. The size of the test, α, as well as the test statistic must be chosen a priori. The dimensions of the model fix the degrees of freedom. Although specifying $\boldsymbol{\beta}$ and σ^2 suffices to complete the power analysis, specifying the (usually smaller and simpler) matrices $\boldsymbol{\theta} = \boldsymbol{C}\boldsymbol{\beta}$, $\boldsymbol{\theta}_0$, and σ^2 also suffices and usually proves easier. A further simplification occurs because the noncentrality $\omega = (\boldsymbol{\theta} - \boldsymbol{\theta}_0)'\boldsymbol{M}^{-1}(\boldsymbol{\theta} - \boldsymbol{\theta}_0)/\sigma^2$ suffices. With $\omega = N\rho^2/(1-\rho^2)$, ρ^2 is a (generalized) squared correlation. Also $0 \leq \omega < \infty$ and $0 \leq \rho^2 \leq 1$, while $\omega \equiv 0$ $\Leftrightarrow \rho^2 \equiv 0 \Leftrightarrow \boldsymbol{\theta} = \boldsymbol{\theta}_0 \Leftrightarrow H_0$ holds $\Leftrightarrow$ power $= \alpha$.

Deleting any duplicate rows from the design matrix creates the essence matrix (Definition 11.5), Es($\boldsymbol{X}$). It allows easily determining essential properties of a design, such as rank. Comparing essence matrices allows determining relationships between alternate parameter definitions. The concept allows conveniently separating total sample size from the coding scheme. The separation simplifies computing and interpreting power in linear models.

20.3 INDEPENDENT *t* EXAMPLE

For cell mean coding, $\text{Es}(\boldsymbol{X}) = \boldsymbol{I}_2$, $\boldsymbol{C} = [1\ \ -1]$, and $\boldsymbol{U} = 1$, while for reference cell coding $\text{Es}(\boldsymbol{X}) = \begin{bmatrix} 1 & 0 \\ 1 & 1 \end{bmatrix}$ and $\boldsymbol{C} = [0\ \ 1]$. Here n indicates the number of replicates (number of observations per unique row of $\boldsymbol{X}$, which corresponds to a cell for any factorial design). In turn,

$$\begin{aligned} \boldsymbol{y} &= (\mathbf{1}_n \otimes \boldsymbol{I}_2)\begin{bmatrix} \mu_1 \\ \mu_2 \end{bmatrix} + \boldsymbol{e} \\ &= \left(\mathbf{1}_n \otimes \begin{bmatrix} 1 & 0 \\ 1 & 1 \end{bmatrix}\right)\begin{bmatrix} \mu \\ \delta \end{bmatrix} + \boldsymbol{e}. \end{aligned} \tag{20.1}$$

Here $[\mu\ \delta]'$ is equivalent to $\delta \cdot [0\ 1]'$ in terms of power for the test of interest.

Example 20.1 The following code computes power for the model just described.

```
TITLE1 "P0802.SAS--simple independent t test";
PROC IML SYMSIZE=4000 WORKSIZE=4000;
%INCLUDE "..\IML\POWERLIB.IML";
C={1 -1};    U={1};     THETA0=0;
ALPHA=.01;
SIGMA={2.1};*Variance if univariate, as here;  SIGSCAL={1};
RHOSCAL={1};
ESSENCEX=I(2); REPN=5;   or REPN={5,10};
BETA={7.0,7.0 };  BETASCAL={1}; RUN POWER;
BETA={7.0,7.3 };  BETASCAL={1}; RUN POWER;
BETA={7.0,7.15}; BETASCAL={1}; RUN POWER;
*last 3 lines together equivalent to either of next 2 lines;
*   BETA={0,1}; BETASCAL={0,.15,.30}; RUN POWER;
*   BETA={0,1}; BETASCAL=DO(0,.30,.15); RUN POWER;
```

Here $\theta = (\mu_1 - \mu_2) = \delta$. Hence one may choose BETA={0,1} and BETASCAL=δ or BETASCAL=DO(δlow, δhigh, δincrement). Doing so avoids specifying the grand mean, $(\mu_1+\mu_2)/2$, which does not affect the power of interest.

_HOLDPOW CASE	ALPHA	SIGSCAL	RHOSCAL	BETASCAL	TOTAL_N	POWER
1	0.01	1	1	0	10	0.01
2	0.01	1	1	0	20	0.01
3	0.01	1	1	0.15	10	0.011
4	0.01	1	1	0.15	20	0.012
5	0.01	1	1	0.3	10	0.013
6	0.01	1	1	0.3	20	0.017

What if an unbalanced design is required? Two methods are available.

Method 1. REPN=1.

Example with $N_1 = 11$ participants in one group and $N_2 = 7$ in the second group.

REPN=1;

N1=11; N2=7; ESSENCEX=BLOCK(J(N2,1,1) , J(N1,1,1)); $\boldsymbol{X} = \begin{bmatrix} \mathbf{1}_7 & \mathbf{0} \\ \mathbf{0} & \mathbf{1}_{11} \end{bmatrix}$;

Method 2. REPN $\geq$ 1.

Example proportional design with two controls for every treatment participant.

ESSENCEX=BLOCK(J(2,1,1) , J(1,1,1)); $\boldsymbol{X} = \begin{bmatrix} 1 & 0 \\ 1 & 0 \\ 0 & 1 \end{bmatrix}$;

Allows using REPN={3,6}; to consider N=9 and N=18.

20.4 PAIRED t EXAMPLE

A paired data t test can be conducted as a one-sample t test with the model

$$\boldsymbol{y} = \mathbf{1}_n \delta + \boldsymbol{e}\,. \tag{20.2}$$

Example 20.2

```
TITLE1 "P0804.SAS--paired t, one sample/difference version";
PROC IML SYMSIZE=4000 WORKSIZE=4000;
%INCLUDE "..\IML\POWERLIB.IML";
C={1}; U={1}; THETA0=0;
ALPHA=.01;
SIGMA={ 1.1 };*Variance of difference; SIGSCAL={1}; RHOSCAL={1};
ESSENCEX=I(1);
REPN={5,10};
BETA={1}; BETASCAL=DO(0,.3,.15);
RUN POWER;
```

20.5 THE IMPACT OF USING $\widehat{\sigma}^2$ OR $\widehat{\boldsymbol{\beta}}$ IN POWER ANALYSIS

Data analysts often use $\widehat{\sigma}^2$, an estimator of σ^2, in a power analysis. In turn, the power value inherits the randomness of the estimator. Taylor and Muller (1995) derived exact methods to account for the randomness in the context of the $\text{GLM}_{N,q}(y_i; \boldsymbol{X}_i\boldsymbol{\beta}, \sigma^2)$ with fixed $\boldsymbol{X}$ and Gaussian errors. Accounting for the randomness leads to creating a confidence region around the power curve, as illustrated in Figure 20.2. The particular range of values used in the figure reflect the following scientific context. In humans, substantially elevated creatinine levels in the blood typically indicate severe kidney disease. With normal values falling below 1 mg/dL, values of 2 or higher are important and a bad sign. Values in the range of 10–20 usually reflect kidneys near complete failure (leading to dialysis, kidney transplant, or death). Experience with the measure has led to the realization that 1/creatinine (in units of dL/mg) is approximately Gaussian. The vertical reference line at 0.5 dL/mg in Figure 20.2 indicates the change in reciprocal creatinine deemed of clinical significance by the nephrologist (kidney specialist) who asked for guidance on sample size.

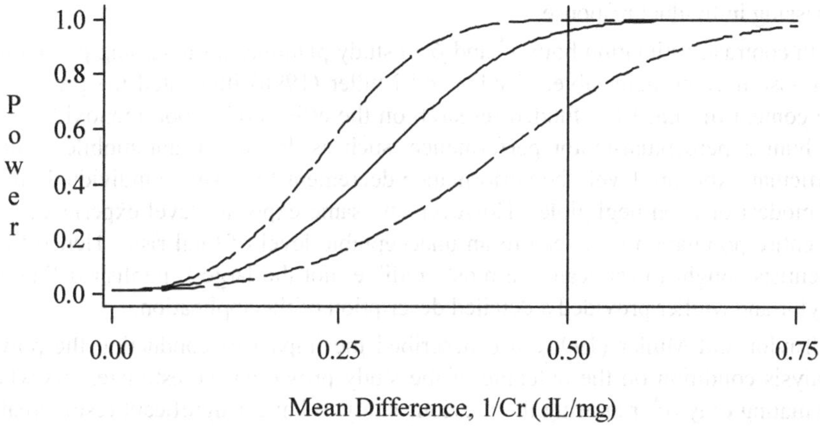

Figure 20.2 Power curve with two-sided 95% confidence region due to $\widehat{\sigma}^2$ for independent groups t test power with $N = 12 + 12$.

In some cases, data analysts conduct power analysis conditional on the outcome of a study. At the end of a trial in the drug discovery process, planning for future studies, including power analysis, centers on the conditions and variables with the smallest p values. In other settings, a strong intellectual commitment to a particular hypothesis may lead a scientist to conduct a power analysis following a nonsignificant result. The first scenario implies $\widehat{\sigma}^2$ arose from a density truncated on the right (excluding large values), while the second scenario implies $\widehat{\sigma}^2$ arose from a density truncated on the left (excluding small values). Muller and Pasour (1997) extended the results of Taylor and Muller (1995) to allow for such truncation. The first scenario creates optimistic bias (estimated power too large), while the second scenario creates pessimistic bias (estimated power too small).

Taylor and Muller (1996) described the impact of using estimates of *both* σ^2 and $\boldsymbol{\beta}$ in estimating power of a univariate linear model. We have seen the approach used in three settings: data analysis, study planning for assessing individual response, and study planning for population response. Only the last application seems defensible.

In the context of data analysis, Lenth (2001) appropriately criticized the computation of such an estimate ("retrospective power," $\widehat{P}$). Taylor and Muller's (1996) results make it clear that $\widehat{P}$ is a one-to-one function of the p value. Hence it adds no information or value to any data analysis. Finding the largest $\widehat{P}$ is equivalent to finding the smallest p value, which is often very misleading.

Lenth (2001) also appropriately criticized estimating both σ^2 and $\boldsymbol{\beta}$ when planning a future study ("prospective power"). Allowing an observed difference to drive the sample size makes no reference to the concept of scientific importance. In Figure 20.2, the clinically important difference of 0.5 dL/mg (1/Cr) drove the power analysis. In the context of evaluating a drug intended to improve kidney function, with response variable 1/Cr, it does not seem defensible to use an estimate of $\boldsymbol{\beta}$ to drive the power analysis. The setting involves study planning for assessing individual response.

In contrast, estimating both σ^2 and $\boldsymbol{\beta}$ in study planning for assessing population response may be defensible. Taylor and Muller (1996) illustrated the process in the context of U.S. EPA funded research on the effects of carbon monoxide (CO) on human perceptual-motor performance, such as driving an automobile. For a particular exposure level, the performance decrement for a single individual might be modest or even negligible. However, the same exposure level experienced by an entire population may lead to an unacceptable level of total risk. Hence EPA scientists sought to replicate the most credible, not the largest, published finding. Taylor and Muller provided a detailed description of the application.

Taylor and Muller (1996) also described the impact of conducting the power analysis condition on the outcome of the study providing the estimate. As when estimating only σ^2, requiring the previous study to have a significant result creates optimistic bias (estimated power too large). Similarly, requiring the previous study to have a nonsignificant result creates pessimistic bias (estimated power too small).

20.6 RANDOM PREDICTORS

Although errors in measurement could introduce additional randomness in both fixed and random predictors, we assume that the scientist measures all predictors without appreciable error. Given the assumption, in practice, the distinction between random and fixed predictors does not affect the distribution theory. However, in power analysis the distinction between random and fixed predictors changes and complicates the distribution theory. Sampson (1974) detailed many of the basic issues and known results for both the univariate and multivariate model with Gaussian predictors.

Jayakar (1970), Soller and Genizi (1978), Genizi and Soller (1979), and Gatsonis and Sampson (1989) developed methods for models involving random dichotomous and Gaussian predictors in the univariate GLM. To our knowledge, power with any other predictor distribution has not been studied.

Many questions remain, especially with combinations of fixed and random predictors. Most importantly, do the simple approximations often used in practice lead to poor approximations of power?

As with any probability, power can be interpreted as the expected value of an indicator variable. Power computed with random predictors can be thought of as *expected power* due to expectation with respect to the choice of predictor values. Glueck and Muller (2003) recommended considering *quantile power*, such as median power, in lieu of expected power. A more conservative approach would use a lower quantile to reduce risk of study failure.

20.7 INTERNAL PILOT DESIGNS

Having chosen a $\boldsymbol{\beta}$ of scientific importance (such as a pattern of mean differences), a valid choice of σ^2 usually stands as the biggest barrier to an appropriate choice of sample size in the univariate linear model. An *internal pilot design* (Wittes and Brittain, 1990) solves the problem as follows. First, a traditional power analysis is conducted based on a planning value σ_0^2 and a target power which together imply a total sample size of n_0. Second, the first $n_1 < n_0$ observations are collected, and $\widehat{\sigma}_1^2$ is computed from residuals from the appropriate linear model. However, no interim data analysis is conducted, and the data analysts and scientists remain masked with respect to treatment assignment. Third, a new total sample size, $N_+ = n_1 + N_2$, is computed, based on $\widehat{\sigma}_1^2$. Fourth, an additional N_2 observations are collected. Fifth, and finally, the analysis is conducted for the complete set of N_+ observations. The interim power analysis typically increases sample size when needed to compensate for σ_0^2 being too small and decreases sample size to compensate for σ_0^2 being too large. Hence expected sample size and power are improved.

Unfortunately, an internal pilot design can inflate test size, at least in small samples. Hence Coffey and Muller (1999, 2000a, 2000b, 2001) described many small-sample results which allow using an internal pilot with any univariate linear model with Gaussian errors and fixed predictors. The methods control test size while still providing the desired advantages in power and expected sample size.

20.8 OTHER CRITERIA FOR CHOOSING A SAMPLE SIZE

In the context of the Neyman-Pearson approach to testing a hypothesis, test size, indicated α, equals the probability of rejecting the null hypothesis given the null holds: $\alpha = \Pr\{\text{reject } H_0 | H_0 = \text{TRUE}\} = \Pr\{\text{reject } H_0 | H_A = \text{FALSE}\}$. While α equals the probability of a type I error, a false positive, $\beta = \Pr\{\text{fail to reject } H_0 | H_A = \text{TRUE}\}$ gives the probability of a type II error, a false negative. In turn, the conventional definition of power is $1 - \beta = \Pr\{\text{reject } H_0 | H_A = \text{TRUE}\}$. It is mathematically convenient to define the power function, $\Pr\{\boldsymbol{y} \in RR | H\}$, as a function of $\boldsymbol{\theta}$ by adding a single point so that power under the null is at most α.

A goal other than rejecting a hypothesis leads to using design criteria other than power. Historically, interest in controlling confidence interval width has been the next most popular criterion. For scalar hypotheses in the general linear model, Jiroutek, Muller, Kupper and Stewart (2003) reviewed criteria in terms of three basic events. The event *width* (W) occurs when the observed confidence interval is less than a fixed constant chosen a priori. The event *validity* (V) occurs if the confidence interval contains the parameter of interest. The event *rejection* (R) occurs if the confidence interval excludes the null value of the hypothesis. For a two-sided test in the general linear model, the probability of rejection is slightly greater than power due to the slight chance that the hypothesis is rejected in the "wrong" direction. For a t test with positive difference in population means, rejection in the wrong direction involves rejecting the null with a negative difference observed. For a one-sided test, the probability of rejection and the (unconditional) definition of power coincide.

Jiroutek et al. (2003) observed that various authors have considered controlling only power, essentially $\Pr\{R\}$, or $\Pr\{R|V\}$, or $\Pr\{W\}$, or $\Pr\{W|V\}$. They argued that scientists often desire not only a valid hypothesis test with good power but also a valid confidence interval of a reasonable size. If so, then $\Pr\{(W \cap R)|V\}$ captures the combined goals, with all previously studied criteria occurring as special cases. Jiroutek et al. described convenient single-integral formulas for computing $\Pr\{(W \cap R)|V\}$. Not surprisingly, controlling $\Pr\{W\}$ or $\Pr\{W|V\}$ when study goals include rejection can lead to collecting far too few or far too many observations. Either problem may also occur when controlling power and ignoring confidence interval goals. Free SAS/IML® software to compute $\Pr\{(W \cap R)|V\}$, $\Pr\{R\}$, $\Pr\{W\}$, and $\Pr\{W|V\}$ for any scalar hypothesis in a univariate or multivariate GLM with fixed predictors can be found at the website documented in the Appendix (Section A.2).

EXERCISES

20.1 Use the POWERLIB software described in the Appendix (Section A.2) to reproduce the results in Example 20.1.

20.2 A $\text{GLM}_{N,q}(y_i; \boldsymbol{X}_i\boldsymbol{\beta}, \sigma^2)$ with Gaussian errors and $N = 10$ observations has

$$\boldsymbol{X\beta} = [\,\mathbf{1}_{10} \quad \boldsymbol{x}\,]\begin{bmatrix}\alpha \\ \gamma\end{bmatrix}.$$

With $\boldsymbol{C} = [\,0\;1\,]$ and $\boldsymbol{M} = [\boldsymbol{C}(\boldsymbol{X}'\boldsymbol{X})^{-1}\boldsymbol{C}']$, a two-sided test of $H_0 : \gamma = 0$ uses

$$F_{\text{obs}} = (\boldsymbol{C}\widehat{\boldsymbol{\beta}})'\boldsymbol{M}^{-1}(\boldsymbol{C}\widehat{\boldsymbol{\beta}})/\widehat{\sigma}^2 = \frac{\widehat{\gamma}^2}{\widehat{\sigma}^2}\sum_{i=1}^{N}(x_i - \overline{x})^2 \sim F(1, N-2, \omega),$$

and noncentrality $\omega = (\boldsymbol{C\beta})'\boldsymbol{M}^{-1}(\boldsymbol{C\beta})/\sigma^2 = (\gamma^2/\sigma^2)\sum_{i=1}^{N}(x_i - \overline{x})^2$.
Consider testing the null hypothesis $H_0 : \gamma = 0$ with a significance level of 0.05.
20.2.1 What is the probability of rejecting the null hypothesis if $\alpha = 1$, $\gamma = 1/4$, $\sigma^2 = 1$, and $\boldsymbol{x} = [0.3\;1.0\;1.5\;1.7\;1.8\;2.5\;2.6\;4.0\;7.4\;14.5]'$?
20.2.2 Suppose that instead of being fixed, $\boldsymbol{x}$ is the realization of a vector-valued random variable. Specifically, let $\boldsymbol{x}$ be a random sample of size 10 from an exponential distribution with mean $\mu_x = 4$. Use computer simulations to estimate the *unconditional* probability of rejecting the null hypothesis.
Hints. (a) $\Pr\{\text{REJECT NULL}\} = \text{E}_{\boldsymbol{x}}(\Pr\{\text{REJECT NULL}|\boldsymbol{x}\}) = \text{E}_{\boldsymbol{x}}[g(\boldsymbol{x})]$ (say).
(b) Let $\{x_1, \ldots, x_N\}$ denote a random sample and let $y_i = g(x_i)$. The statistic $\overline{y} = \sum_{i=1}^{N} y_i/N$ is an unbiased estimator of the unconditional probability.

(c) The simulations can be conducted in SAS/IML or another matrix-based language. The following is a skeleton program.

```
SEED=nnnnn;  *you must assign a value to start
              pseudo-random number generator;
*Ideally, choose a prime number as large as possible;
*In practice, choose 5-7 digit # not divisible by 2, 3, or 5;
*Number nmopq is divisible by 3 <=> n+m+o+p+q divisible by 3;
NREPS = 1000; * THE NUMBER OF REPLICATIONS;
power = J(NREPS,1, . );   * AN EMPTY VECTOR TO STORE RESULTS;
  do i = 1 to NREPS;
  x=4*ranexp(J(10,1,SEED)); * GENERATE RANDOM SAMPLE, SIZE 10;
  NCP = ... ;   * A FUNCTION OF γ, σ², and x.
  CRIT =  ...;   * CALCULATE THE CRITICAL VALUE;
  power[i] = ...;   * A FUNCTION of NCP, CRIT df1, df2;
  end;
estimate = sum(power)/NREPS;   * ESTIMATED UNCONDITONAL POWER;
```

The approach suggested is an extremely inefficient programming strategy. The data should be "buffered" (written to disk after, perhaps, every 100 replications). We accept such inefficiency to help keep the exercise manageable.

20.3 A researcher plans a randomized trial to compare three drugs (H_2 receptor antagonists) used to treat stomach ulcers. An equal number of patients will be randomized to three treatment groups (Cimetidine, Ranitidine, Famotidine). The main outcome is gastric acidity measured in pH. The goal is to test whether the three drugs increase pH the same amount. There are no covariates. The investigator is not sure whether to obtain baseline measurements. She knows that baseline measurements can improve power but is not sure whether it is worth the cost. With a fixed budget, money spent on baseline measurements reduces the number of participants. She has asked help in choosing the most cost-effective and powerful design. All tests will be based on a type I error rate of $\alpha = 0.05$.

Design 1: one measurement per participant

Measure gastric acidity after a 6 month treatment regimen. Use a univariate one-way ANOVA model with 3 groups. The outcome variable is gastric acidity (call it y_1). The null hypothesis is that $\mathrm{E}(y_1)$ is the same for all participants in all treatments. The test statistic is the ANOVA test of overall regression, which has 2 numerator degrees of freedom.

Design 2: difference scores, two measurements per participant (baseline and follow-up)

Obtain a "baseline" measurement prior to initiating therapy (call it y_0) and a follow-up measurement (y_1) after the 6-month treatment. Due to randomization, $\mathrm{E}(y_0)$ is the same for all patients regardless of treatment assignment. Let $z = y_1 - y_0$. The null hypothesis is that $\mathrm{E}(z)$ is the same for all participants regardless of treatment assignment. The data analysis is the same as for design option 1, but the outcome is z rather than y_1. The approach is equivalent to fitting a general linear multivariate model with two outcome variables, y_0 and y_1, and using $\boldsymbol{U} = [1 \;\; -1]'$.

Participants are statistically independent with data assumed to be Gaussian. Which design yields the better power depends on unknown parameters. The anticipated difference between follow-up and baseline, $\mathrm{E}(y_1 - y_0)$, is 0.75 for Cimetidine, 1.00 for Ranitidine, and 1.25 for Famotidine.
Previous studies suggest that $\mathcal{V}(y_0) \approx \mathcal{V}(y_1) \approx 1$ and $\rho(y_0, y_1) \equiv \rho \approx 0.6$.
20.3.1 Under design 1, what is the probability of rejecting the null hypothesis when there are $n = 50$ participants per group ($N = 150$ participants total)?
20.3.2 What is the smallest sample size needed to achieve 0.80 power for design 1?
20.3.3 Under design 2, what is the probability of rejecting the null hypothesis when there are $n = 50$ participants per group ($N = 150$ participants total)?
20.3.4 What is the smallest sample size needed to achieve 0.80 power for design 2?
20.3.5 Repeat 20.3.3 using several different choices of ρ between 0 and 1. Plot a graph illustrating the relationship between ρ and power (an approximate, hand-drawn graph is acceptable).
20.3.6 The physician wants to enroll as many participants as possible, but can only spend a fixed amount of money. You're asked to help her determine which design is more cost effective. Under design option 1, the study costs \$75 per participant. Under design option 2, the study costs \$100 per participant. Which design do you recommend and why?

CHAPTER 21

Sample Size for Multivariate Linear Models

21.1 THE MACHINERY OF A POWER ANALYSIS

We focus here mostly on the theory of power analysis and ignore the many important practical issues surrounding the task. The reader may wish to consult O'Brien and Muller (1993) for a tutorial and many related references.

The discussion assumes Gaussian errors and fixed predictors. For data analysis, the presence of random predictors causes no additional complexity. In contrast, allowing random predictors introduces an additional layer of distributional complexity to power analysis. Glueck and Muller (2003) reviewed the limited range of cases that have been solved. The same authors considered fixed predictors in combination with Gaussian covariates. An important open question remains concerning how to compute power for the interaction of fixed and Gaussian predictors. More generally, power analysis for random but not Gaussian predictors have received little attention.

We restrict attention to testable hypotheses, with full-rank $\boldsymbol{C}, \boldsymbol{U}, \Sigma_*$, and $\boldsymbol{M}$, while $\boldsymbol{C} = \boldsymbol{C}(\boldsymbol{X}'\boldsymbol{X})^{-}(\boldsymbol{X}'\boldsymbol{X})$. As discussed in Section 2.5 of Muller, LaVange, Ramey, and Ramey (1992), α, Σ, $\boldsymbol{X}$, $\boldsymbol{B}$, $\boldsymbol{C}$, $\boldsymbol{U}$, and Θ_0 fully specifes a multivariate general linear model power calculation. Both α, the size of the test, as well as the test statistic must be chosen a priori. The dimensions of the model fix the degrees of freedom. Although specifying $\boldsymbol{B}$ and Σ suffices to complete the power analysis, specifying the (usually smaller and simpler) matrices $\Theta = \boldsymbol{CBU}$, Θ_0, and $\Sigma_* = \boldsymbol{U}'\Sigma\boldsymbol{U}$ also suffices and often proves easier. Further simplification occurs because the $b \times b$ matrix $\boldsymbol{\Omega} = (\Theta - \Theta_0)'\boldsymbol{M}^{-1}(\Theta - \Theta_0)\Sigma_*^{-1}$ suffices. As a final simplification, the eigenvalues of $\boldsymbol{\Omega}$ provide the minimally sufficient (additional) information required to complete the power analysis. Eigenvalue k of $\boldsymbol{\Omega}$ is $\omega_k = N\rho_k^2/(1-\rho_k^2)$, with ρ_k^2 a (generalized) squared canonical correlation. Also $0 \leq \omega_k < \infty$ and $0 \leq \rho_k^2 \leq 1$. Writing $\Sigma_* = \boldsymbol{\Phi}\boldsymbol{\Phi}'$ gives $\Sigma_*^{-1} = \boldsymbol{\Phi}^{-t}\boldsymbol{\Phi}^{-1}$ and allows defining the symmetric matrix $\boldsymbol{\Omega}_\Phi = \boldsymbol{\Phi}^{-1}(\Theta - \Theta_0)'\boldsymbol{M}^{-1}(\Theta - \Theta_0)\boldsymbol{\Phi}^{-t}$, which is a quadratic form in the symmetric matrix $\boldsymbol{M}^{-1}$. The eigenvalues of $\boldsymbol{\Omega}$ equal the eigenvalues of $\boldsymbol{\Omega}_\Phi$. For $(a \times b)$ Θ, $s = \min(a, b)$ and $\text{rank}(\Theta) = s_*$, with $0 \leq s_* \leq s$. In turn, rank$(\boldsymbol{\Omega}) = s_*$, the number of nonzero ω_k, which equals the number of nonzero ρ_k^2. Also $s_* = 0 \; \Leftrightarrow \omega_k \equiv 0 \; \Leftrightarrow \rho_k^2 \equiv 0 \; \Leftrightarrow \boldsymbol{\Theta} = \boldsymbol{\Theta}_0$ $\Leftrightarrow H_0$ holds $\Leftrightarrow$ power $= \alpha$.

The sizes of s and s_* control many features of power analysis in the multivariate GLM. If $s = 1$, then all tests coincide and $s_* = 1$. Having $s > 1$ allows increasing s_*, which always increases power (with all other properties held constant). The theoretical property has a very important and practical implication. A study design with sufficient power to detect a linear dose effect will have even more power if an additional quadratic effect occurs (above and beyond the linear). Hence power analysis with $s_* = 1$ often may be taken as a conservative approach.

Deleting any duplicate rows from the design matrix ($\boldsymbol{X}$) creates the essence matrix (Definition 11.5). It simplifies determining properties of a design, such as rank. Comparing essence matrices allows determining relationships between alternate parameter sets. The concept provides convenient separation of total sample size from the coding scheme. The separation helps simplify interpretation and computation of power.

21.2 PAIRED t TEST EXAMPLE

A paired data t test can be conducted with $\text{Es}(\boldsymbol{X}) = \boldsymbol{I}_1 = 1$, $\boldsymbol{C} = 1$, and $\boldsymbol{U} = [-1 \;\; 1]'$. The power for $H_0 : \mu_2 = \mu_1$ will be the same for either of two different models:

$$\boldsymbol{Y} = \mathbf{1}_n[\,\mu_1 \; \mu_2\,] + \boldsymbol{E} \tag{21.1}$$

$$\boldsymbol{Y} = \mathbf{1}_n[0 \quad 1]\delta + \boldsymbol{E}\,. \tag{21.2}$$

Here $\boldsymbol{CBU} = \theta = (\mu_2 - \mu_1) = \delta$. The second model avoids the need to specify the grand mean, $(\mu_1 + \mu_1)/2$, which has no effect on the power analysis.

Example 21.1 The code can use BETA={0 1}=({0,1})` and BETASCAL=δ or BETASCAL=DO(δlow, δhigh, δincrement);.

```
TITLE1 "P0803.SAS--simple paired t example";
PROC IML SYMSIZE=4000 WORKSIZE=4000;
%INCLUDE "..\IML\POWERLIB.IML";
U=({1 -1})`;
C={1}; THETA0=0;
ALPHA=.01;
SIGSCAL={1};
RHOSCAL={1};
SIGMA={ 2.1 3.2 ,
        3.2 2.4   };
*or;    P=2; VARIANCE=2.1;  RHO=.4;
        SIGMA=VARIANCE#( I(P)#(1-RHO) +J(P,P,RHO) );
ESSENCEX=I(1);
REPN={5,10};
```

```
BETA={0 1};
BETASCAL=DO(0,.30,.15);
RUN POWER;

_HOLDPOW CASE   ALPHA SIGSCAL RHOSCAL BETASCAL TOTAL_N  POWER
          1     0.01        1       1        0       5  0.01
          2     0.01        1       1        0      10  0.01
          3     0.01        1       1     0.15       5  0.011
          4     0.01        1       1     0.15      10  0.012
          5     0.01        1       1     0.3        5  0.013
          6     0.01        1       1     0.3       10  0.02
```

21.3 TIME BY TREATMENT EXAMPLE

For $\text{GLM}_{N,p,q}(\boldsymbol{Y}_i; \boldsymbol{X}_i\boldsymbol{B}, \boldsymbol{\Sigma})$ with Gaussian errors, the hypothesis of time-by-treatment interaction often generates particular interest. All participants are assumed to have been measured at the same times, $\{t_1, \ldots, t_p\}$, with $\text{Es}(\boldsymbol{X}) = \boldsymbol{I}_q$ and $\boldsymbol{X} = \mathbf{1}_n \otimes \text{Es}(\boldsymbol{X})$. In the simplest case $s_* = 1$.

The hypothesis of treatment-by-time interaction may be expressed in terms of differences of differences of means, with

$$\boldsymbol{C}_1 = [\,\mathbf{1}_{q-1} \quad -\boldsymbol{I}_{q-1}\,] \tag{21.3}$$
$$\boldsymbol{U}_1 = [\,\mathbf{1}_{p-1} \quad -\boldsymbol{I}_{p-1}\,]' \tag{21.4}$$
$$\boldsymbol{\Theta}_0 = \mathbf{0}\,. \tag{21.5}$$

Useful corresponding canonical forms are, with $q \times p$ $\boldsymbol{B}_1$ and $a \times b$ $\boldsymbol{\Theta}_1$,

$$\begin{aligned}\boldsymbol{B}_1 &= \begin{bmatrix} \theta & \mathbf{0}_{1\times(p-1)} \\ \mathbf{0}_{(q-1)\times 1} & \mathbf{0}_{(q-1)\times(p-1)} \end{bmatrix} \\ &= \begin{bmatrix} 1 & \mathbf{0}_{1\times(p-1)} \\ \mathbf{0}_{(q-1)\times 1} & \mathbf{0}_{(q-1)\times(p-1)} \end{bmatrix} \delta\,,\end{aligned} \tag{21.6}$$

$$\begin{aligned}\boldsymbol{\Theta}_1 &= \begin{bmatrix} \theta & \mathbf{0}_{1\times(b-1)} \\ \mathbf{0}_{(a-1)\times 1} & \mathbf{0}_{(a-1)\times(b-1)} \end{bmatrix} \\ &= \begin{bmatrix} 1 & \mathbf{0}_{1\times(b-1)} \\ \mathbf{0}_{(a-1)\times 1} & \mathbf{0}_{(a-1)\times(b-1)} \end{bmatrix} \delta\,.\end{aligned} \tag{21.7}$$

The alternative hypothesis of interest may be a linear-by-linear (treatment-by-time) interaction. If so, $\boldsymbol{C}_2$ and $\boldsymbol{U}_2$ give the orthonormal polynomial trends for q and p values, respectively. If $p = 3$ and $q = 4$ (assuming equal spacing for both between and within factors), $\boldsymbol{\Theta}_0 = \mathbf{0}$ and

$$C_2 = \text{Dg}(\{20,4,4\})^{-1/2}\begin{bmatrix} -3 & -1 & 1 & 3 \\ 1 & -1 & -1 & 1 \\ 1 & -1 & 1 & -1 \end{bmatrix} = \begin{bmatrix} C_{2,1} \\ \vdots \\ C_{2,q-1} \end{bmatrix} \tag{21.8}$$

$$U_2 = \begin{bmatrix} -1 & 1 \\ 0 & -2 \\ 1 & 1 \end{bmatrix}\text{Dg}(2,6)^{-1/2} = [\, \boldsymbol{u}_{2,1}\ \boldsymbol{u}_{2,2} \cdots \boldsymbol{u}_{2,p-1} \,]. \tag{21.9}$$

Also, with $q \times p\ \boldsymbol{B}_2$, $q \times 1\ \boldsymbol{C}'_{2,1}$, $1 \times p\ \boldsymbol{u}'_{2,1}$, $a \times b\ \boldsymbol{\Theta}_2$,

$$\boldsymbol{B}_2 = (\boldsymbol{C}'_{2,1}\boldsymbol{u}'_{2,1})\delta \tag{21.10}$$

and

$$\begin{aligned}
\boldsymbol{\Theta}_2 &= \boldsymbol{C}_2(\boldsymbol{C}'_{2,1}\boldsymbol{u}'_{2,1}\delta)[\, \boldsymbol{u}_1\ \boldsymbol{u}_1 \cdots \boldsymbol{u}_{p-1} \,]\delta \\
&= \begin{bmatrix} \boldsymbol{C}_{2,1} \\ \vdots \\ \boldsymbol{C}_{2,q-1} \end{bmatrix} \boldsymbol{C}'_{2,1}\boldsymbol{u}'_{2,1}[\, \boldsymbol{u}_1\ \boldsymbol{u}_1 \cdots \boldsymbol{u}_{p-1} \,]\delta \\
&= \begin{bmatrix} 1 \\ 0 \\ \vdots \\ 0 \end{bmatrix} [\, 1\, 0 \cdots 0 \,]\delta \\
&= \begin{bmatrix} 1 & \mathbf{0}_{1\times(b-1)} \\ \mathbf{0}_{(a-1)\times 1} & \mathbf{0}_{(a-1)\times(b-1)} \end{bmatrix}\delta = \boldsymbol{\Theta}_1 .
\end{aligned} \tag{21.11}$$

More generally, if $\boldsymbol{\delta} = [\, \delta_1\ \delta_2 \cdots \delta_{s_*} \,]'$, $\boldsymbol{C}_{2*}$ indicates the first s_* rows of $\boldsymbol{C}_2$, and $\boldsymbol{U}_{2*}$ indicates the first s_* columns of $\boldsymbol{U}_2$, then $\boldsymbol{B}_{2*} = \boldsymbol{C}'_{2*}\text{Dg}(\boldsymbol{\delta})\boldsymbol{U}'_{2*}$ has rank s_* and

$$\begin{aligned}
\boldsymbol{\Theta}_{2*} &= \boldsymbol{C}_2\boldsymbol{B}_{2*}\boldsymbol{U}_2 \\
&= \underset{[(a\times q)(q\times s_*)][(s_*\times s_*)][(s_*\times p)(p\times b)]}{(\boldsymbol{C}_2\boldsymbol{C}'_{2*})\text{Dg}(\boldsymbol{\delta})(\boldsymbol{U}'_{2*}\boldsymbol{U}_2)} \\
&= \begin{bmatrix} \boldsymbol{I}_{s_*} \\ \mathbf{0}_{(a-s_*)\times s_*} \end{bmatrix}\text{Dg}(\boldsymbol{\delta})[\, \boldsymbol{I}_{s_*} \quad \mathbf{0}_{s_*\times(b-s_*)} \,] \\
&= \begin{bmatrix} \text{Dg}(\boldsymbol{\delta}) & \mathbf{0} \\ \mathbf{0} & \mathbf{0}_{(a-s_*)\times(b-s_*)} \end{bmatrix}.
\end{aligned} \tag{21.12}$$

Choosing

$$\boldsymbol{B}_3 = \boldsymbol{\Theta}_{2*} \tag{21.13}$$

$$\boldsymbol{C}_3 = \begin{bmatrix} \boldsymbol{I}_{s_*} & \mathbf{0} \\ \mathbf{0}_{(a-s_*)\times s_*} & \mathbf{0} \end{bmatrix} \tag{21.14}$$

$$\boldsymbol{U}_3 = \begin{bmatrix} \boldsymbol{I}_{s_*} \\ \mathbf{0}_{(b-s_*)\times s_*} \end{bmatrix} \tag{21.15}$$

implies $\boldsymbol{C}_3\boldsymbol{B}_3\boldsymbol{U}_3 = \boldsymbol{\Theta}_{2*}$.

In some cases, the alternative hypothesis of interest may concern a main effect. For a between-groups main effect, examples are

$$B = \frac{1}{6}\begin{bmatrix} 0 & 0 & 0 \\ 1 & 1 & 1 \\ 2 & 2 & 2 \\ 3 & 3 & 3 \end{bmatrix}\delta \tag{21.16}$$

and

$$B = C_{2,1}\mathbf{1}_p'\delta \tag{21.17}$$

while a within-subject main effect example is

$$B = \mathbf{1}_q'\boldsymbol{u}_{2,1}'\delta\,. \tag{21.18}$$

Example 21.2 The following code provides an example calculation of C_2C_{2*}'.

```
PROC IML; RESET PRINT;
Q=4; SSTAR=2;
POLY=ORPOL(1:Q);
C=(POLY[*,2:Q])`;
CSTAR=C[1:SSTAR,*];
CCSTARP=C*CSTAR`;
```

The following program computes power for a time by treatment interaction.

```
PROC IML SYMSIZE=4000 WORKSIZE=4000;
%INCLUDE "..\IML\POWERLIB.IML";
P=3;   U=( J(P-1,1,1)||(-I(P-1)) )`;
Q=4;   C=J(Q-1,1,1)||(-I(Q-1));
ALPHA=.01;

VARIANCE=2.1;
RHO=.4;
SIGMA=VARIANCE#( I(P)#(1-RHO) +J(P,P,RHO) );
SIGSCAL={1, 2};
RHOSCAL={1};

ESSENCEX=I(Q);
REPN={5,10};
BETA=J(Q,P,0);
BETA[1,1]=1;
BETASCAL=DO(0,.30,.15);

RUN POWER;
```

21.4 COMPARING BETWEEN AND WITHIN DESIGNS

A multivariate GLM with Gaussian errors for a paired data t test, with $\boldsymbol{y}_j$ dimension $N \times 1$, may be written

$$[\,\boldsymbol{y}_1\;\boldsymbol{y}_2\,] = \mathbf{1}_N[\,\mu_1\;\mu_2\,] + \boldsymbol{E}\,. \tag{21.19}$$

For simplicity of comparison to an independent t test analysis, we assume homogeneity of variance holds across repeated measures (within subject):

$$\boldsymbol{\Sigma} = \sigma^2\begin{bmatrix} 1 & \rho \\ \rho & 1 \end{bmatrix}. \tag{21.20}$$

Using $\boldsymbol{C} = [\,1\,]$ and $\boldsymbol{U} = [\,1\;\; -1\,]'$ tests $H_0 : \mu_1 = \mu_2$. In turn, $\boldsymbol{\Theta} = (\mu_1 - \mu_2)$,

$$\begin{aligned}\boldsymbol{\Sigma}_* &= \boldsymbol{U}'\boldsymbol{\Sigma}\boldsymbol{U} \\ &= \sigma^2[1\; -1]\begin{bmatrix} 1 & \rho \\ \rho & 1 \end{bmatrix}\begin{bmatrix} 1 \\ -1 \end{bmatrix} \\ &= \sigma^2 2(1-\rho)\,, \end{aligned} \tag{21.21}$$

$$\boldsymbol{M} = \boldsymbol{C}(\boldsymbol{X}'\boldsymbol{X})^{-}\boldsymbol{C}' = N^{-1}\,, \tag{21.22}$$

and

$$\begin{aligned}\boldsymbol{\Omega} &= (\boldsymbol{\Theta} - \boldsymbol{\Theta}_0)'\boldsymbol{M}^{-1}(\boldsymbol{\Theta} - \boldsymbol{\Theta}_0)\boldsymbol{\Sigma}_*^{-1} \\ &= \frac{(\mu_1 - \mu_2)^2}{\sigma^2}\left[\frac{N}{2(1-\rho)}\right]. \end{aligned} \tag{21.23}$$

For $\nu_e = N - \text{rank}(\boldsymbol{X})$, the degrees of freedom are $ab = 1$ and $s(\nu_e - b + s) = N - 1$.

An independent groups t test with a balanced design corresponds to

$$\begin{bmatrix} \boldsymbol{y}_1 \\ \boldsymbol{y}_2 \end{bmatrix} = \boldsymbol{I}_2 \otimes \mathbf{1}_{N/2}\begin{bmatrix} \mu_1 \\ \mu_2 \end{bmatrix} + \boldsymbol{e}\,. \tag{21.24}$$

Assuming $\boldsymbol{e} \sim \mathcal{N}(\mathbf{0}, \sigma^2\boldsymbol{I}_N)$ implies homogeneity of variance (between groups). Choosing $\boldsymbol{C} = [\,1\;\; -1\,]$ and $\boldsymbol{U} = [\,1\,]$ allows testing $H_0 : \mu_1 = \mu_2$. In turn, $\boldsymbol{\Theta} = (\mu_1 - \mu_2)$, $\boldsymbol{\Sigma}_* = \sigma^2$,

$$\boldsymbol{M} = [\,1\;\; -1\,][(N/2)\boldsymbol{I}_2]^{-}\begin{bmatrix} 1 \\ -1 \end{bmatrix} = 4/N\,, \tag{21.25}$$

and

$$\boldsymbol{\Omega} = \frac{(\mu_1 - \mu_2)^2}{\sigma^2}\left(\frac{N}{4}\right). \tag{21.26}$$

The degrees of freedom are $ab = 1$ and $s(\nu_e - b + s) = N - 2$.

Two possible comparisons seem interesting, based on having either a constant number of independent sampling units (ISUs) or a constant number of observations. With constant ISUs, N has the same value in the two settings, and the error degrees of freedom for paired minus independent equals $(N-1)-(N-2)=1$. In turn,

$$\begin{aligned}\frac{\omega_{\text{paired}}}{\omega_{\text{indep}}} &= \frac{N/[2(1-\rho)]}{N/4} \\ &= \frac{2}{(1-\rho)} > 1\,. \end{aligned} \tag{21.27}$$

Both discrepancies favor the paired design (with constant ISUs, which requires collecting more observations).

With a constant number of observations, the error degrees of freedom for paired minus independent equals $(N/2-1)-(N-2)=-N/2+1$. In turn,

$$\begin{aligned}\frac{\omega_{\text{paired}}}{\omega_{\text{indep}}} &= \frac{(N/2)/[2(1-\rho)]}{N/4} \\ &= \frac{1}{(1-\rho)}\,. \end{aligned} \tag{21.28}$$

If $\rho \le 0$, then $(1-\rho)^{-1} \le 1$ and both discrepancies favor the independent design. If $\rho \ge 0$, then $(1-\rho)^{-1} \ge 1$ and for each N there exists $\rho_0(N)$ such that $\rho \ge \rho_0$ implies the paired design is superior, and otherwise the independent design is superior. Equivalently, for each ρ there exists $N_0(\rho)$ such that $N \ge N_0(\rho)$ and the paired design is superior, while otherwise the independent design is superior. Some regions may be undefined. A three-dimensional plot, with power difference vertically, N and ρ as the floor plane, can be especially informative. Critical values are $f_{\text{p}} = F_F^{-1}(1-\alpha; 1, N/2-1)$ and $f_{\text{I}} = F_F^{-1}(1-\alpha, 1, N-2)$. In turn, for method m, Power$(m) = 1 - F_F(f_m; 1, \nu_m; \omega_m)$.

Other comparisons are also interesting. For paired data

$$\begin{aligned}\text{vec}(\widehat{\boldsymbol{\Theta}}) &\sim \mathcal{N}_{ab}[\text{vec}(\boldsymbol{\Theta}), \boldsymbol{\Sigma}_* \otimes \boldsymbol{M}] \\ &\sim \mathcal{N}\left[(\mu_1-\mu_2), 2(1-\rho)\sigma^2/N\right], \end{aligned} \tag{21.29}$$

while for independent data

$$\begin{aligned}\text{vec}(\widehat{\boldsymbol{\Theta}}) &\sim \mathcal{N}_{ab}[\text{vec}(\boldsymbol{\Theta}), \boldsymbol{\Sigma}_* \otimes \boldsymbol{M}] \\ &\sim \mathcal{N}\left[(\mu_1-\mu_2), 4\sigma^2/N\right]. \end{aligned} \tag{21.30}$$

With equal sample sizes

$$\begin{aligned}\gamma &= \frac{2(1-\rho)\sigma^2/N}{4\sigma^2/N} \\ &= (1-\rho)/2\,, \end{aligned} \tag{21.31}$$

while for equal observations

$$\begin{aligned}\gamma &= \frac{2(1-\rho)\sigma^2/(N/2)}{4\sigma^2/N} \\ &= (1-\rho)/4 .\end{aligned} \tag{21.32}$$

The difference in error degrees of freedom affects the confidence interval. Hence one should compare quantiles of confidence interval widths, which are of the form

$$w = 2\sigma_m \sqrt{F_F^{-1}(1-\alpha; 1, \nu_m)} . \tag{21.33}$$

21.5 SOME INVARIANCE PROPERTIES

Any testable secondary parameter $\boldsymbol{\Theta} = \boldsymbol{CBU}$ ($a \times b$), with $H_0 : \boldsymbol{\Theta} = \boldsymbol{\Theta}_0$, has $\text{rank}(\boldsymbol{C}) = a \leq q$ and $\text{rank}(\boldsymbol{U}) = b \leq p$. Eigenvalues of $\boldsymbol{S}_h \boldsymbol{S}_e^{-1}$, and hence all multivariate test statistics and associate p values, are invariant to full-rank transformation of rows of $\boldsymbol{C}$ and columns of $\boldsymbol{U}$ (Section 16.8). Specifying $\boldsymbol{B}$ and $\boldsymbol{\Sigma}$ usually requires the most thought. Specifying $\boldsymbol{\Theta} = \boldsymbol{CBU}$, $\boldsymbol{\Theta}_0$, and $\boldsymbol{\Sigma}_* = \boldsymbol{U}'\boldsymbol{\Sigma}\boldsymbol{U}$ suffices. In turn, $\boldsymbol{B}$ and $\boldsymbol{\Sigma}$ reduce to canonical forms. For $H_0 : \mu_1 = \mu_2$ either $\boldsymbol{\beta} = [\, \mu_1 \; \mu_2 \,]'$ or $\boldsymbol{\beta} = [\, 0 \; \delta \,]'$ (with $\delta = \mu_2 - \mu_1$) leads to the same power. In practice, most tests involve hypotheses which exclude the intercept, while the model does span an intercept. Such situations allow assuming a grand mean of zero because the mean has no effect on power.

For fixed α, $\boldsymbol{\Sigma}$, $\boldsymbol{X}$, $\boldsymbol{C}$, $\boldsymbol{U}$, and $\boldsymbol{\Theta}_0$, power for any $s_* = 1$ alternative may be expressed and plotted as a function of a scalar parameter, such as a mean difference or a squared canonical correlation. Choosing such a representation can be extremely helful in defining a range of alternate hypotheses of interest. Plotting power in terms of the scalar parameter is especially enlightening.

21.6 RANDOM PREDICTORS

As discussed in the previous chapter, the presence of random predictors greatly complicates noncentral distribution theory. Sampson (1974) detailed many of the issues for the univariate and multivariate models with Gaussian predictors.

Glueck and Muller (1998) reviewed the limited work on random predictor power in multivariate models. They also described methods for accurately approximating power for a $\text{GLM}_{N,p,q}(\boldsymbol{Y}_i; \boldsymbol{X}_i\boldsymbol{B}, \boldsymbol{\Sigma})$ with combinations of fixed and Gaussian predictors corresponding to a baseline covariate design. Approximating power for the more general model allowing unequal slopes for each group remains an open and important question.

21.7 INTERNAL PILOT DESIGNS

The principles and advantages of internal pilot designs for univariate models were introduced in Section 20.7. An interim residual analysis (without any interim data analysis) provides a variance estimate $\widehat{\sigma}_1^2$. In turn, a power analysis based on $\widehat{\sigma}_1^2$ leads to increasing or decreasing total sample size. Avoiding the potential for inflated test size due to using an internal pilot requires special testing procedures.

Coffey and Muller (2003) considered using internal pilot designs with the UNIREP approach to repeated measures. They (incorrectly) speculated that the inherent conservatism of the Geisser-Greenhouse test in small samples might compensate for the test size inflation induced by an internal pilot. The interim analysis produces $\widehat{\boldsymbol{\Sigma}}_{*1} = \boldsymbol{U}'\widehat{\boldsymbol{\Sigma}}_1\boldsymbol{U}$ and an updated sample size choice. Simulations demonstrated that test size was inflated above the target level in small samples. Hence work on developing other strategies for controlling test size was begun.

Coffey and Muller drew an additional conclusion from their simulations. For $\epsilon = \text{tr}^2(\boldsymbol{\Sigma}_*)/[b\text{tr}(\boldsymbol{\Sigma}_*^2)]$ near the lower boundary of $1/b$ and small N, the UNIREP power approximations of Muller and Barton (1989) lack sufficient accuracy for internal pilot use. In the worst case, $b = 1$, $\epsilon \approx 0.29$, and $N = 20$, a particular pattern of means gave a predicted power of 0.65, while a power of 0.87 was observed in a simulation with 100,000 replications (standard error ≈ 0.0015). Muller, Edwards, Simpson, and Taylor (2006) demonstrated approximations for power of the UNIREP tests which almost entirely eliminate the inaccuracy. In the worst-case condition just described, the new method predicted power of 0.84.

EXERCISES

21.1 Use the POWERLIB software described in the Appendix A (Section A.2) to reproduce the results in Example 21.1.

21.2 A clinical trial is planned to compare a new drug with an existing one. Treatment starts on a Monday morning. The outcome is measured that afternoon and the four following afternoons. A multivariate $\text{GLM}_{N,p,q}(\boldsymbol{Y}_i; \boldsymbol{X}_i\boldsymbol{B}, \boldsymbol{\Sigma})$ with Gaussian errors has $\boldsymbol{Y}\,(20 \times 5)$, and

$$\boldsymbol{XB} = \begin{bmatrix} \mathbf{1}_{10} & \mathbf{0}_{10} \\ \mathbf{0}_{10} & \mathbf{1}_{10} \end{bmatrix} \begin{bmatrix} \mu_{11} & \mu_{12} & \mu_{13} & \mu_{14} & \mu_{15} \\ \mu_{21} & \mu_{22} & \mu_{23} & \mu_{24} & \mu_{25} \end{bmatrix}.$$

Suppose that $\boldsymbol{\Sigma} = \sigma^2[(1-\rho)\boldsymbol{I} + \mathbf{1}\mathbf{1}'\rho]$, $\sigma^2 = 7$, $\rho = 0.5$, and $\mu_{ij} = (i-1)(j-1)$ for $i \in \{1,2\}$, $j \in \{1,\ldots,5\}$. The Wilks statistic has been chosen for all tests. Define

$$\boldsymbol{\Theta}_1 = [\mu_{11} - \mu_{21} \quad \mu_{12} - \mu_{22} \quad \mu_{13} - \mu_{23} \quad \mu_{14} - \mu_{24} \quad \mu_{15} - \mu_{25}]$$

$$\theta_2 = [(\mu_{11} + \mu_{12} + \mu_{13} + \mu_{14} + \mu_{15})/5 - (\mu_{21} + \mu_{22} + \mu_{23} + \mu_{24} + \mu_{25})/5]$$

$$\boldsymbol{\Theta}_3 = \begin{bmatrix} \mu_{11} - \mu_{12} & \mu_{11} - \mu_{13} & \mu_{11} - \mu_{14} & \mu_{11} - \mu_{15} \\ \mu_{21} - \mu_{22} & \mu_{21} - \mu_{23} & \mu_{21} - \mu_{24} & \mu_{21} - \mu_{25} \end{bmatrix}$$

$$\boldsymbol{\Theta}_4 = (\mu_{11}+\mu_{21}) \cdot \mathbf{1}_4' - [(\mu_{12}+\mu_{22}) \ (\mu_{13}+\mu_{23}) \ (\mu_{14}+\mu_{24}) \ (\mu_{15}+\mu_{25})].$$

21.2.1 Write a one-sentence scientific interpretation, aimed at the scientists, of $H_0 : \boldsymbol{\Theta}_1 = \mathbf{0}$.
21.2.2 Specify the $\boldsymbol{C}$ and $\boldsymbol{U}$ matrices needed.
21.2.3 What is the probability of rejecting $H_0 : \boldsymbol{\Theta}_1 = \mathbf{0}$ using a 0.05 level test?
21.2.4 Write a one-sentence scientific interpretation, aimed at the scientists, of $H_0 : \theta_2 = 0$.
21.2.5 Specify the $\boldsymbol{C}$ and $\boldsymbol{U}$ matrices needed.
21.2.6 What is the probability of rejecting the null hypothesis $\theta_2 = 0$ using a 0.05 level test?
21.2.7 Write a one sentence scientific interpretation, aimed at the scientists, of $H_0 : \boldsymbol{\Theta}_3 = \mathbf{0}$.
21.2.8 Specify the $\boldsymbol{C}$ and $\boldsymbol{U}$ matrices needed.
21.2.9 What is the probability of rejecting the null hypothesis $\boldsymbol{\Theta}_3 = \mathbf{0}$ using a 0.05 level test?
21.2.10 Write a one-sentence scientific interpretation, aimed at the scientists, of $H_0 : \boldsymbol{\Theta}_4 = \mathbf{0}$.
21.2.11 Specify the C and U matrices needed.
21.2.12 What is the probability of rejecting the null hypothesis $\boldsymbol{\Theta}_4 = \mathbf{0}$ using a 0.05 level test?
21.2.13 The various tests arise from either a MANOVA or repeated-measures approach. A third alternative is a set of Bonferroni corrected tests of drug difference, one for each day.
21.2.14 Specify the $\boldsymbol{C}$ and $\boldsymbol{U}$ matrices needed to conduct the five tests and the appropriate nominal level of α.
21.2.15 What is the probability of rejecting the null hypothesis $H_0 : \Theta_{5,1} = 0$?
21.2.16 What is the probability of rejecting the null hypothesis $H_0 : \Theta_{5,5} = 0$?
21.2.17 In practice, balancing control of test size and maximizing power leads to using only one of the approaches (MANOVA, REPM, Bonferroni univariate). Which seems preferable? Which particular tests are most logically consistent (not necessarily the most powerful) with the speculation that $\mu_{ij} = (i-1)(j-1)$? Without doing any additional calculations, do the results suggest a different sample size, assuming a target power of 0.90 or better?

21.3 Consider Table 6, p. 553, in Muller and Barton (1989). Row 9 provides data about condition 113, which is detailed in their Table 4 and associated text.
21.3.1 The canonical form of the $\boldsymbol{B}$ matrix is described in the right-hand column of p. 552, line 9. Think of the form as $\boldsymbol{B} = c\boldsymbol{B}_c$. For the conditions of the simulations associated with their Tables 4–6, give a numerical value for $\boldsymbol{B}_c$. Write

this as a matrix of numbers.
21.3.2 For the conditions of the simulations associated with their Tables 4–6, use their Table 4 and associated text to give a numerical value of the covariance matrix used for condition 113. Write the results as a matrix of numbers.
21.3.3 Use power software to find the value of c which was used to produce the predicted power value of 0.80 for the GG test for the ninth line of Table 6 (p. 553), for condition 113.

Hint: In SAS/IML®, use the DO function to create a list of candidate c values and assign the result to BETASCAL, as in the examples.
21.3.4 Produce a high-resolution plot of power as a function of c.
Hint: You will likely wish to use the DS option in POWERLIB.

21.4 Assume the notation and setting of Muller, LaVange, Ramey, and Ramey, 1992). As throughout, $\mathbf{1}_j$ indicates a $j \times 1$ vector of 1's. With $n = 9$, define $\boldsymbol{X} = \boldsymbol{I}_3 \otimes \mathbf{1}_n$,

$$\boldsymbol{B} = \mu \mathbf{1}_3 \mathbf{1}_5' + \delta \frac{1}{16} \begin{bmatrix} 0 & 1 & 4 & 9 & 16 \\ 0 & 1 & 2 & 3 & 4 \\ 0 & 0 & 0 & 0 & 0 \end{bmatrix},$$

$$\boldsymbol{\Sigma} = \sigma^2 \begin{bmatrix} 1 & 0 & 0 & 0 & 0 \\ 0 & 2 & 0 & 0 & 0 \\ 0 & 0 & 3 & 0 & 0 \\ 0 & 0 & 0 & 4 & 0 \\ 0 & 0 & 0 & 0 & 5 \end{bmatrix}^{1/2} \begin{bmatrix} 1 & \rho & \rho^2 & \rho^3 & \rho^4 \\ \rho & 1 & \rho & \rho^2 & \rho^3 \\ \rho^2 & \rho & 1 & \rho & \rho^2 \\ \rho^3 & \rho^2 & \rho & 1 & \rho \\ \rho^4 & \rho^3 & \rho^2 & \rho & 1 \end{bmatrix} \begin{bmatrix} 1 & 0 & 0 & 0 & 0 \\ 0 & 2 & 0 & 0 & 0 \\ 0 & 0 & 3 & 0 & 0 \\ 0 & 0 & 0 & 4 & 0 \\ 0 & 0 & 0 & 0 & 5 \end{bmatrix}^{1/2},$$

$$\boldsymbol{C} = \begin{bmatrix} 1 & -1 & 0 \\ 1 & 0 & -1 \end{bmatrix},$$

$$\boldsymbol{U}' = \begin{bmatrix} 1 & -1 & 0 & 0 & 0 \\ 1 & 0 & -1 & 0 & 0 \\ 1 & 0 & 0 & -1 & 0 \\ 1 & 0 & 0 & 0 & -1 \end{bmatrix},$$

and $\boldsymbol{\Theta}_0 = \mathbf{0}$.
Assume $\delta = 2$, $\sigma^2 = 1$, and $\rho = 0.6$. Use $\alpha = 0.025$.
21.4.1 Prove that $\boldsymbol{\Theta}$ is invariant to the value of μ for this hypothesis. You may use IML, or any other matrix language, for the purely numerical calculations. (Hence, without loss of generality, assume $\mu = 0$.)
21.4.2 Compute $\boldsymbol{\Omega}$.
21.4.3 Compute the eigenvalues of $\boldsymbol{\Omega}$.
21.4.4 Compute the squared canonical correlations.
21.4.5 Compute the (population values) of the measures of multivariate association for the four MULTIREP (invariant) tests.

21.5 Write your own matrix language code to directly implement the steps in Section 2.5 of Muller, LaVange, Ramey, and Ramey (1992, p. 1214) to compute

the approximate power of the Pillai-Bartlett test *only*. In particular, compute
21.5.1 Step 1 in Section 2.5 for the Pillai-Bartlett test only.
21.5.2 Step 2 in Section 2.5 for the Pillai-Bartlett test only.
21.5.3 Step 3 in Section 2.5 for the Pillai-Bartlett test only.
21.5.4 Step 4 in Section 2.5 for the Pillai-Bartlett test only.

21.6. Use POWERLIB to compute the same power value as in the last part of the exercise.

CHAPTER 22

Sample Size for Generalizations of Multivariate Models

22.1 MOTIVATION

In developing new statistical methods, statisticians have historically focused first on estimation and then on inference. Methods for choosing a sample size typically come last. Practical and mathematical reasons stimulate the order. Distribution theory for sample size involves greater complexity than estimation or inference under the null. Furthermore, the enthusiasm that practicing data analysts have for sample size methods has never been shared by more theoretical statisticians. Hence few results are available for power and sample size analysis of generalizations of multivariate linear models. In keeping with our discussions of estimation and testing, we sketch some results for growth curve models.

22.2 SAMPLE SIZE METHODS FOR GROWTH CURVE MODELS

Some noncentral theory has been developed for growth curves with higher order trends used as covariates, especially for large samples. However, the methods have not been studied carefully in small samples. Berger (1986) used simulations to study both test size and power for a variety of methods for analyzing growth curves. His results support avoiding the use of high-order polynomials as covariates due to the likelihood of inflating test size.

Without the use of high-order polynomials as covariates, growth curve analysis reduces to a special case of a multivariate GLM. Hence methods in the previous chapter apply. Particular care must be taken to choose $\boldsymbol{U}$ matrices to correctly reflect the specified model.

An example will illustrate the issue. If the data involve five time points while the desired model includes only linear and quadratic trends (along with zero order), then two within-subject contrasts would be used, $\boldsymbol{U}_0' = 5^{-1/2} \cdot [\,1\;1\;1\;1\;1\,]$ and

$$\boldsymbol{U}_{1-2}' = \begin{bmatrix} 10 & 0 \\ 0 & 14 \end{bmatrix}^{-1/2} \begin{bmatrix} -2 & -1 & 0 & 1 & 2 \\ 2 & -1 & -2 & -1 & 2 \end{bmatrix} . \tag{22.1}$$

The omitted cubic and quartic trends are spanned by

$$\boldsymbol{U}_{3-4}' = \begin{bmatrix} 10 & 0 \\ 0 & 70 \end{bmatrix}^{-1/2} \begin{bmatrix} -1 & -2 & 0 & -2 & 1 \\ 1 & -4 & 6 & -4 & 1 \end{bmatrix} . \tag{22.2}$$

The default choices for a multivariate linear model (the multivariate approach to repeated measures, MULTIREP) would use $\boldsymbol{U}_0$ and

$$\boldsymbol{U}_{\mathrm{W}}' = \begin{bmatrix} -1 & 1 & 0 & 0 & 0 \\ -1 & 0 & 1 & 0 & 0 \\ -1 & 0 & 0 & 1 & 0 \\ -1 & 0 & 0 & 0 & 1 \end{bmatrix} . \tag{22.3}$$

However, the columns of $\boldsymbol{U}_{\mathrm{W}}$ span $\boldsymbol{U}_{1-2}$ *and* $\boldsymbol{U}_{3-4}$. More precisely, if $\boldsymbol{U}_{\mathrm{T}} = [\,\boldsymbol{U}_{1-2}\ \boldsymbol{U}_{3-4}\,]$ then $\boldsymbol{U}_{\mathrm{T}} = \boldsymbol{U}_{\mathrm{W}}\boldsymbol{P}$ for 4×4 and full rank. Invariance properties of the multivariate model imply that any test with $\boldsymbol{U}_{\mathrm{W}}$ gives the same result as a test with $\boldsymbol{U}_{\mathrm{W}} \neq \boldsymbol{U}_{1-2}$.

Using $\boldsymbol{U}_{\mathrm{W}}$ rather than $\boldsymbol{U}_{1-2}$ provides another example of an alignment error, as discussed in the previous chapter. Checking the degrees of freedom associated with the hypothesis and parameters leads to recognizing that $4 \neq 2$. Making sure dimensions and degrees of freedom correspond to the desired inference can help avoid alignment errors.

CHAPTER 23

Sample Size for Linear Mixed Models

23.1 MOTIVATION

Mixed models have become one of the most widely used methods for data analysis. Unfortunately, as Verbeke and Molenberghs (2000, Section 23.2) noted, very little is known about nonnull distributions in mixed models. The wide variety of test statistics used in mixed models adds a substantial complexity to the task. Additional complexity arises from interest in random predictors. As always with questions of power, Monte Carlo simulations provide a completely defensible, although onerous, method of approximating power.

In linear models, full specification of noncentral distributions requires knowing the (population) distributions of any random predictors. For mixed models, the comment applies to both $\boldsymbol{X}_i$ and $\boldsymbol{Z}_i$.

23.2 METHODS

Two approaches have been suggested for approximating power in general mixed models. Both are based on a supposition that reflects a property of many univariate and multivariate linear models. Only limited simulations are available to support the methods which use F approximations for Wald-type tests. However, the basic ideas are promising. In the particular case of a $\text{GLM}_{N,q}(y_i; \boldsymbol{X}_i\boldsymbol{\beta}, \sigma^2)$ with fixed predictors and Gaussian errors, the (testable) general linear hypothesis is $H_0 : \boldsymbol{C\beta} = \boldsymbol{\theta}_0$. With $\omega = (\boldsymbol{\theta} - \boldsymbol{\theta}_0)'\boldsymbol{M}^{-1}(\boldsymbol{\theta} - \boldsymbol{\theta}_0)/\sigma^2$ and $\boldsymbol{M} = \boldsymbol{C}(\boldsymbol{X}'\boldsymbol{X})\boldsymbol{C}'$, the usual F statistic (for a *univariate* model) is given in equation 2.32 as $F = [(\widehat{\boldsymbol{\theta}} - \boldsymbol{\theta}_0)'\boldsymbol{M}^{-1}(\widehat{\boldsymbol{\theta}} - \boldsymbol{\theta}_0)/a]/\widehat{\sigma}^2 = \widehat{\omega}/a \sim F(a, N-r, \omega)$. Hence the noncentrality parameter may be characterized as $\omega = f_A a$, with f_A the F statistic occurring in the very special case with $\widehat{\boldsymbol{\theta}} = \boldsymbol{\theta}$ and $\widehat{\sigma}^2 = \sigma^2$. O'Brien and Muller (1993) described the concept as the *exemplary data* approach to power calculation and credited Graybill (1976) with earlier promotion of the idea. The exemplary data approach leads to noncentral F approximations for power of the mixed model.

Helms (1992) described a noncentral F power approximation for an F statistic (F_H) defined in terms of a novel modification of the usual REML estimators (equations 14–15 in his paper). Verbeke and Molenberghs (2000, Section 23.2) provided some related discussion. However, the latter authors implicitly allow using standard REML or ML variance estimators (in the definition of the F statistic, their equation 6.6) while emphasizing the importance of specifying the degrees of freedom parameter. As mentioned earlier, additional simulations seem to be necessary to assess the accuracy of the approximations discussed.

Stroup (2003) described a strategy for approximating mixed model power which also uses the exemplary data concept. A simulation with 500 replications for one design gave power estimates consistent with the approximation. For computing convenience, Stroup used the Satterthwaite approach to determine approximate error degrees of freedom in power approximation. His simulations (for a narrow range of models) used the Kenward and Roger (1997) null case approximation (which currently provides the best control of test size).

23.3 INTERNAL PILOT DESIGNS

The principles and advantages of internal pilot designs for univariate models were introduced in Section 20.7. Zucker and Denne (2002) examined internal pilot designs for a two group clinical trial with at least $N = 40$ participants. In most cases, they disallowed a decrease in sample size. They used a likelihood ratio test with a second-order (Bartlett) approximation specifically derived for the covariance structure chosen, as derived by Zucker, Lieberman, and Manor (2000). Applying the method to any other design requires design-specific derivatives. An adjustment for using an internal pilot was required to avoid test size inflation.

APPENDIX

Computing Resources

A.1 EXAMPLE DATA: DETECTING BREAST CANCER IN MAGNETIC RESONANCE IMAGING WITH A CONTRAST AGENT

Research data used by permission of Dr. M. P. Braeuning and Dr. E. D. Pisano.

Approximately 1 in 9 American women develop breast cancer, based on full life expectancy. Breast cancer kills more American women than any other cancer. Regular self-examination for all women and screening mammography (x-rays of the breasts) for older women provide the main lines of defense. Having detected a suspicious region, the physician must discriminate between malignant and benign tissue. The life-and-death consequences of diagnosis make expensive procedures worthwhile. More information can be found at www.cancer.org.

The goal was to compare image brightness over time in different types of breast tissue on MRIs that followed injection of the contrast agent intravenous gadolinium-DTPA. For each woman, average brightness was recorded on each of four images, taken at 0, 45, 90, or 135 sec (+/- a few seconds) after injection. For each image, measurements were recorded from one region of fat, one of parenchyma, and one or two regions of interest (ROI). Each ROI was classified as benign or malignant (cancer), based on a subsequent pathologist's reading of a biopsy. Ten women requiring diagnosis were imaged. The following diagram shows the data available for each patient. Multicenter trials typically pay for a second pathology reading to increase reliability and validity of the (high quality but not perfect) single reading.

Patient	Cancer	Benign	Fat	Parenchyma
2	✓	✓	✓	✓
6	✓	✓	✓	✓
7	✓	✓	✓	✓
1	✓		✓	✓
3		✓	✓	✓
4	✓		✓	✓
5	✓		✓	✓
8		✓	✓	✓
9		✓	✓	✓
10	✓		✓	✓

Each ✓ indicates the presence of four observations (baseline=0, 45, 90, and 135 sec). Except where indicated in the diagram, for each patient, tissue type and time, the mean recorded MRI signal, the standard deviation, and the area (mm²) were available. Missing data indicates that no such tissue was found in the images.

Conversations with physicists led us to conclude that the logarithm of (signal at time j/signal at time 0) is proportional to concentration, which was chosen as the response variable. Exploratory analysis of Box-Cox power transformations of the response, coupled with evaluation of jackknife residuals, supported the choice.

All files can be downloaded from the Web site http://ehpr.ufl.edu/muller/. Raw data are in P0101.DAT (a text file). Programs P0101.SAS–P0105.SAS created SAS files P0101.SD2–P0105.SD2 (version 6.12, created on a PC). If you have trouble transporting the files to another platform, and feel compelled to start from raw data, you must use the programs provided to ensure that the same variable names, labels, and data are being analyzed.

The exercises use P0104.SD2, produced by P0104.SAS, which has the following statement:

```
LABEL DLOGROI1="Dif log ROI 45sec"
      DLOGROI2="Dif log ROI 90sec"
      DLOGROI3="Dif log ROI 135sec"
      DLOG_F_1="Dif log Fat 45sec"
      DLOG_F_2="Dif log Fat 90sec"
      DLOG_F_3="Dif log Fat 135sec"
      DLOG_P_1="Dif log Parenchyma 45sec"
      DLOG_P_2="Dif log Parenchyma 90sec"
      DLOG_P_3="Dif log Parenchyma 135sec"
      BENIGN  ="1=> Benign, else 0"
      MALIGN  ="1=> Malign, else 0";
```

The program P0104.SAS produced the following:

```
P0104.SAS-Create file using only one benign or malig per case
All variables in file
        D     D     D     D     D     D      D      D      D C
        L     L     L     L     L     L      L      L      L O
        O     O     O     O     O     O      O      O      O N B M
        G     G     G     G     G     G      G      G      G S E A
                                             R      R      R T N L
        _     _     _     _     _     _
O       F     F     F     P     P     P      O      O      O A I I
b I     _     _     _     _     _     _      I      I      I N G G
s D     1     2     3     1     2     3      1      2      3 T N N
1 2 0.002 0.008 0.041 0.210 0.259 0.278  0.944  1.127  1.195 1 1 0
2 6 0.083 0.071 0.137 0.054 0.001 0.142  0.084  0.037  0.088 1 1 0
3 3 0.118 0.271 0.234 0.001 0.055 0.056  0.081  0.184  0.260 1 1 0
4 8 -.178 -.243 0.056 0.007 0.056 0.066  0.087  0.209  0.193 1 1 0
5 9 0.137 0.080 0.191 0.174 0.162 0.127  0.062  0.197  0.184 1 1 0
6 7 0.016 0.165 0.135 0.038 0.149 0.168  0.121  0.162  0.261 1 0 1
```

```
 7  1 0.054 0.057 0.017 0.037 -.031 0.019  0.099  0.003  0.082 1 0 1
 8  4 0.623 0.518 0.604 -.096 0.180 0.154 -0.578 -0.124  0.092 1 0 1
 9  5 -.085 0.231 0.153 -.033 0.026 0.071  0.071  0.136  0.293 1 0 1
10 10 -.122 -.285 -.084 0.029 0.070 0.064  0.232  0.225  0.253 1 0 1
```

Before you begin the exercises, please read programs P0101.SAS through P0104.SAS, as well as the associated log and list files (contained in the same directory) to understand both the programming and scientific decisions behind creating the file P0104.SD2. Most importantly, note that P0104.SD2 and P0105.SD2 have no missing data, while P0101.SD2–P0103.SD2 do.

A.2 FREE SOFTWARE

A.2.1 Overview

The software described here is available at no cost on the Web at http://ehpr.ufl.edu. All of the software is written in SAS/IML®. Hence the source code is included, which allows embedding the software in other programs, as well as translating modules to other languages. Each must be downloaded separately. The files include user manuals and many examples.

A.2.2 LINMOD: Multivariate Linear Models Analysis

LINMOD (LINear MODels) performs a wide variety of computations in SAS/IML for a general linear multivariate model with Gaussian errors. LINMOD allows the analyst familiar with matrix algebra notation and IML syntax to efficiently compute tests, estimates, and all associated statistics. Muller, LaVange, Ramey, and Ramey (1992) presented a succinct statement of the model, the general linear hypothesis, and associated statistics. Chapters 3 and 6 provide a more detailed overview. Chapters 4, 12, 13, 16, 17, 19, and 21 contain more detail.

All of the results can be computed with some combination of PROC GLM and REG in SAS. The primary advantage of LINMOD lies in the efficiency which results from the direct relationship between the syntax of the program and the matrix algebra formulation of models and tests. The primary disadvantage of LINMOD lies in the statistical sophistication required to use the program to produce valid results. However, anyone who has spent time trying to deduce exactly how an estimate, test, confidence interval, or correlation was computed by a particular option in a particular program will welcome the ability to explicitly control the calculations in a formula-based syntax. The additional overhead of LINMOD will usually only be worthwhile for complex designs or tests. The less simple and traditional the design and/or hypothesis, the more likely that LINMOD will appeal to someone able to define and analyze linear models in matrix notation. A further advantage lies in the fact that all of the computed values remain available to the user in matrices. This allows storing all results in a permanent SAS database (as distinct from the listing file) and exercising complete control of formatting.

Having results in matrix format makes subsequent custom processing very easy. Using ODS with GLM or REG provides an alternative approach.

Comparing LINMOD to PROC REG (or GLM) demonstrates a classic tradeoff in program design: increased flexibility and control for the sophisticated user versus friendly interface for the naive user. For example, LINMOD does not add or in any way recognize an intercept term in any model. The user must code one if desired and may choose to test it if present. Using LINMOD requires the ability to define and analyze linear models in matrix notation.

Starting from the assumption of the user having matrix knowledge, great effort was expended to give LINMOD an interface with consistent design, extensive error checking, and informative messages. The wide assortment of matrix operators and functions in PROC IML greatly enhances power and flexibility. Furthermore, the extensive editing, printing, plotting, and data management facilities of SAS are available for pre- and post-processing.

A.2.3 MISSMOD: Multivariate Linear Models with Missing Data

MISSMOD provides accurate test size in small samples for many kinds of Gaussian repeated measures and multivariate data with missing values. The software computes approximate tests described by Catellier and Muller (2000) for general linear multivariate models. Simulations support the conclusion that, in contrast to current mixed model competitors, the methods control test size even with as few as 12 observations for 6 repeated measures and 5% missing data. Assuming data missing at random (MAR), the EM algorithm provides maximum likelihood estimates using all of the available data. The tests generalize standard "multivariate" and "univariate" approaches to repeated-measures tests by reducing the error degrees of freedom by replacing the number of independent sampling units by various functions of the numbers of nonmissing pairs of responses. The program is based closely on LINMOD. Source code, an extensive user's guide, and example programs are included in the free download.

A.2.4 POWERLIB: Multivariate and Repeated-Measures Power

POWERLIB provides convenient power calculations for a wide range of multivariate linear models with Gaussian errors. The multivariate and univariate approaches to repeated measures as well as MANOVA tests are covered. F approximations are used throughout and are reduced to exact forms whenever possible. Approximate or exact power for the Wilks, Pillai-Bartlett, and Hotelling-Lawley tests is available. Approximate or exact power for the Box conservative test, Geisser-Greenhouse, Huynh-Feldt, and uncorrected test is also available. Confidence limits may be requested for most power values to reflect the uncertainty due to using estimated variances and, when appropriate, means and variances. A simple option causes SAS data files to be produced automatically, which simplifies producing plots and tables for manuscripts. Documentation includes a range of designs and examples for both UNIX and Windows systems. Source code is included in the free download.

A.2.5 CISIZE: Sample Size Involving Confidence Intervals

CISIZE generalizes ideas about sample size to achieve confidence intervals and power properties. It computes $\Pr\{(W \cap R)|V\}$, $\Pr\{R\}$, $\Pr\{W\}$ and $\Pr\{W|V\}$ for any scalar hypothesis in a univariate or multivariate GLM with fixed predictors. The software implements the techniques discussed in Jiroutek et al. (2003).

A.3 ORTHOGONAL POLYNOMIAL COEFFICIENTS

d = Number of Distinct Points, p = Power of Trend

d	p	c_1	c_2	c_3	c_4	c_5	c_6	c_7	c_8	c_9	$\sum_{j=1}^{d} c_j^2$
2	1	-1	1	.	.	.	.	.	.	.	2
3	1	-1	0	1	.	.	.	.	.	.	2
	2	1	-2	1	.	.	.	.	.	.	6
4	1	-3	-1	1	3	.	.	.	.	.	20
	2	1	-1	-1	1	.	.	.	.	.	4
	3	-1	3	-3	1	.	.	.	.	.	20
5	1	-2	-1	0	1	2	.	.	.	.	10
	2	2	-1	-2	-1	2	.	.	.	.	14
	3	-1	2	0	-2	1	.	.	.	.	10
	4	1	-4	6	-4	1	.	.	.	.	70
6	1	-5	-3	-1	1	3	5	.	.	.	70
	2	5	-1	-4	-4	-1	5	.	.	.	84
	3	-5	7	4	-4	-7	5	.	.	.	180
	4	1	-3	2	2	-3	1	.	.	.	28
	5	-1	5	-10	10	-5	1	.	.	.	252
7	1	-3	-2	-1	0	1	2	3	.	.	28
	2	5	0	-3	-4	-3	0	5	.	.	84
	3	-1	1	1	0	-1	-1	1	.	.	6
	4	3	-7	1	6	1	-7	3	.	.	154
	5	-1	4	-5	0	5	-4	1	.	.	84
	6	1	-6	15	-20	15	-6	1	.	.	924
8	1	-7	-5	-3	-1	1	3	5	7	.	168
	2	7	1	-3	-5	-5	-3	1	7	.	168
	3	-7	5	7	3	-3	-7	-5	7	.	264
	4	7	-13	-3	9	9	-3	-13	7	.	616
	5	-7	23	-17	-15	15	17	-23	7	.	2184
	6	1	-5	9	-5	-5	9	-5	1	.	264
	7	-1	7	-21	35	-35	21	-7	1	.	3432
9	1	-4	-3	-2	-1	0	1	2	3	4	60
	2	28	7	-8	-17	-20	-17	-8	7	28	2772
	3	-14	7	13	9	0	-9	-13	-7	14	990
	4	14	-21	-11	9	18	9	-11	-21	14	2002
	5	-4	11	-4	-9	0	9	4	-11	4	468
	6	4	-17	22	1	-20	1	22	-17	4	1980
	7	-1	6	-14	14	0	-14	14	-6	1	858
	8	1	-8	28	-56	70	-56	28	-8	1	12870

References

Anderson, T. W. (2004) *An Introduction to Multivariate Statistical Analysis*, 3rd ed., New York: Wiley.

Andrade, D. F. and Helms, R. W. (1986) ML estimation and LR tests for the multivariate normal distribution with general linear model mean and linear-structure covariance matrix, *Communications in Statistics A*, **15**, 89–107.

Andrews, D. F., Gnanadesikan, R. and Warner, J. L. (1971) Transformations of multivariate data, *Biometrics*, **27**, 825–840.

Arnold, S. F. (1981) *The Theory of Linear Models and Multivariate Analysis*, New York: Wiley.

Banerjee, K. S. (1964) A note on idempotent matrices, *Annals of Mathematical Statistics*, **35**, 880–882.

Barton, C. N. and Cramer, E. C. (1989) Hypothesis testing in multivariate linear models with randomly missing data, *Communications in Statistics B,* **18**, 875–895.

Basilevsky, A. (1994) *Statistical Factor Analysis and Related Methods*, New York: Wiley.

Benignus, V. A., Muller, K. E., Smith, M. V., Pieper, K. S. and Prah, J. D. (1990) Compensatory tracking in humans with elevated carboxyhemoglobin, *Neurotoxicology and Teratology*, **12**, 105–110.

Berger, M. P. F. (1986) A Monte Carlo study of the power of alternative tests under the generalized MANOVA model, *Communications in Statistics A*, **15**, 1251–1283.

Billio, M. and Monfort, A. (1998) Switching state-space models: likelihood function, filtering and smoothing, *Statistical Planning and Inference*, **68**, 65–103.

Boik, R. J. (1988) The mixed model for multivariate repeated measures: validity conditions and an approximate test, *Psychometrika*, **53**, 469–486.

Boik, R. J. (1991) Scheffé mixed model for repeated measures a relative efficiency evaluation, *Communications in Statistics - Theory and Methods*, **A20**, 1233–1255.

Bowden, D. C. (1970) Simultaneous confidence bands for linear regression models, *Journal of the American Statistical Association*, **65**, 413–421.

Box, G. E. P. (1954a) Some theorems on quadratic forms applied in the study of analysis of variance problems: I. effects of inequality of variance in the one-way classification, *Annals of Mathematical Statistics*, **25**, 290–302.

Box, G. E. P. (1954b) Some theorems on quadratic forms applied in the study of analysis of variance problems: II. effects of inequality of variance and of correlation between errors in the two-way classification, *Annals of Mathematical Statistics*, **25**, 484–498.

Box, G. E. P. and Cox, D. R. (1964) An analysis of transformations, *Journal of the Royal Statistical Society B*, **26**, 211–43 (with discussion, 244–252).

Box, G. E. P. and Cox, D. R. (1984) An analysis of transformations revisited, rebuttal, *Journal of the American Statistical Association*, **77**, 209–210.

Box, G. E. P., Jenkins, G. M., and Reinsel, G. C. (1994) *Time Series Analysis: Forecasting and Control,* 3rd ed., Englewood Cliffs, NJ: Prentice-Hall.

Carroll, R. J. and Rupert, D. (1981) On prediction and the power transformation family, *Biometrika*, **68**, 609–615.

Carroll R. J. and Rupert, D. (1985) Transformations in regression: a robust analysis, *Technometrics*, **27**, 1–12.

Carroll, R. J. and Rupert, D. (1988) *Transformations and Weighting in Regression*, London: Chapman and Hall.

Casella, G. (1983) Leverage and regression through the origin, *The American Statistician*, **37**, 147–1512.

Catellier, D. J. and Muller, K. E. (2000) Tests for Gaussian repeated measures with missing data in small samples, *Statistics in Medicine*, **19**, 1101–1114.

Catellier, D. J. and Muller, K. E. (2002) Sample size and power considerations in physical activity research. In *Physical Activity Assessments for Health-Related Research*, Chapter 6, G. J. Welk, ed., 93-104. Champaign, IL: Human Kinetics.

Cochran, W. G. (1934) The distribution of quadratic forms in a normal system with applications to the analysis of variance, *Proceedings of the Cambridge Philosophical Society*, **30**, 178–191.

Coffey, C. S. and Muller, K. E. (1999) Exact test size and power of a Gaussian error linear model for an internal pilot study, *Statistics in Medicine* **18**, 1199–1214.

Coffey, C. S. and Muller, K. E. (2000a) Properties of doubly-truncated Gamma variables, *Communications in Statistics A*, **29**, 851–857.

Coffey, C. S. and Muller, K. E. (2000b) Some distributions and their implications for an internal pilot study with a univariate linear model, *Communications in Statistics A*, **29**, 2677–2691.

Coffey, C. S. and Muller, K. E. (2001) Controlling test size while gaining the benefits of an internal pilot design, *Biometrics,* **57**, 625–631.

Coffey, C. S. and Muller, K. E. (2003) Properties of internal pilots with the univariate approach to repeated measures, *Statistics in Medicine*, **22**, 2469–2485.

Cox, D. R. and Small, N. J. H. (1978) Testing multivariate normality, *Biometrika*, **65**, 263–272.

Cramer, E. M. and Nicewander, W. A. (1979) Some symmetric, invariant measures of multivariate association, *Pyschometrika*, **44**, 43–54.

Cramér, H. (1946) *Mathematical Methods of Statistics*, Princeton, NJ: University Press.

Cressie, N. (1991) *Statistics for Spatial Data*, New York: Wiley InterScience.

D'Agostino, R. B. and Stephens, M. A. (1986) *Goodness-of-Fit-Techniques*, New York: Marcel Dekker.

Daintith, J. and Nelson, R. D. (1989) *The Penguin Dictionary of Mathematics*, London, UK: Penguin.

Davies, R. B. (1980) Algorithm AS 155: the distribution of a linear combination of χ^2 random variables, *Applied Statistics*, **29**, 323–333.

Demidenko, E. (2004) *Mixed Models Theory and Applications*, New York: Wiley.

Diggle, P. and Kenward, M. G. (1994) Informative drop-out in longitudinal data analysis, *Applied Statistics*, **43**, 49–93.

Edwards, L. J., Stewart, P. W., Muller, K. E. and Helms, R. W. (2001) Linear equality constraints in the general mixed model, *Biometrics*, **57**, 1185–1190.

Epps, T. W. (1993) Characteristic functions and their empirical counterparts: geometrical interpretations and applications to statistical inference, *American Statistician*, **47**, 33–38.

Fairclough, D. L. and Helms, R. H. (1986) A mixed linear model with linear covariance structure: a sensitivity analysis of maximum likelihood estimators, *Journal of Statistical Computation and Simulation*, **25**, 205–236.

Feller, W. (1968) *An Introduction to Probability Theory and Its Applications*, 3rd ed., Vol. I, New York: Wiley.

Gallant, A. R. and Fuller, W. A. (1973) Fitting segmented polynomial regression models whose join points have to be estimated, *Journal of the American Statistical Association*, **68**, 144–147.

Gatsonis, C. and Sampson, A. R. (1989) Multiple correlation: exact power and sample size calculations, *Psychological Bulletin*, **106**, 516–524.

Geisser, S. and Greenhouse, S. W. (1958) An extension of Box's results on the use of the F distribution in multivariate analysis, *Annals of Mathematical Statistics*, **29**, 885–891.

Genizi, A. and Soller, M. (1979) Power derivation in an ANOVA model which is intermediate between the "fixed-effects" and the "random-effects" models, *Journal of Statistical Planning and Inference*, **3**, 127–134.

Gil-Pelaez, J. (1951) Note on the inversion theorem, *Biometrika*, **38**, 481–482.

Glueck, D. H. and Muller, K. E. (1998) On the trace of a Wishart, *Communications in Statistics A*, **27**, 2137–2141. Corrigenda, 2002, **31**, 159–160.

Glueck, D. H. and Muller, K. E. (2003) Adjusting power for a baseline covariate in a linear model, *Statistics in Medicine*, **22**, 2535–2551.

Gnanadesikan, R. and Kettenring, J. R. (1972) Robust estimates, residuals, and outlier detection with multiresponse data, *Biometrics*, **28**, 81–124. Corrigenda, 1972, **28**, 1142.

Good, I. J. (1963) On the independence of quadratic expressions, *Journal of the Royal Statistical Society B*, **25**, 377–382. Corrigenda, 1966, **28**, 584.

Good, I. J. (1969) Conditions for a quadratic form to have a chi-squared distribution, *Biometrika*, 1969, **54**, 215–216. Corrigenda, 1970, **57**, 225.

Goodnight, J. H. (1979) A tutorial on the SWEEP operator, *The American Statistician*, **33**, 149–158.

Graybill, F. A. (1969) *Introduction to Matrices with Applications in Statistics.* Belmont, CA: Wadsworth.

Graybill, F. A. (1976) *Theory and Applications of the Linear Model*, North Scituate, MA: Duxbury Press.

Greenhouse, S. W. and Geisser, S. (1959) On methods in the analysis of profile data, *Psychometrika*, **24**, 95–112.

Grizzle, J. E. and Allen, D. M. (1969) Analysis of growth and dose response curves, *Biometrics*, **25**, 357–381.

Gupta, A. K. and Nagar, D. K. (2000) *Matrix Variate Distributions*, Boca Raton, FL: Chapman Hall/CRC.

Harris, R. J. (1975) *Primer of Multivariate Statistics*, New York: Academic.

Harvey, J. R. (1972) On expressing moments in terms of cumulants and vice versa, *American Statistician*, **26**, 38–39. Corrigenda, 1973, **27**, 44.

Harville, D. A. (1974) Bayesian inference for variance components using only error contrasts, *Biometrika*, **61**, 383–385.

Harville, D. A. (1977) Maximum likelihood approaches to variance component estimation and to related problems, *Journal of the American Statistical Association*, **72**, 320–338. Comments and responses 338–340.

Harville, D. A. (1990) BLUP (best linear unbiased prediction) and beyond, *Advances in Statistical Methods for Genetic Improvement of Livestock*, pp. 239–276, D. Gianola and K. Hammond Eds., New York: Springer.

Harville, D. A. (1996) The posterior distribution of the fixed and random effects in a mixed-effects linear model, *Journal of Statistical Computation and Simulation*, **54**, 211–229.

Helms, R. W. (1988a) Definitions of linear model parameters and hypotheses as functions of E(Y), *Communications in Statistics A*, **17**, 2715–2723.

Helms, R. W. (1988b) Comparisons of parameter and hypothesis definitions in a general linear model. *Communications in Statistics A*, **17**, 2725–2753.

Helms, R. W. (1988c) Manipulating statistical hypotheses and tests as Boolean functions, *The American Statistician*, **42**, 253–256.

Helms, R. W. (1992) Intentionally incomplete longitudinal designs: I. methodology and comparison of some full span designs, *Statistics in Medicine*, **11**, 1889–1913.

Heitjan, D. F. (1994) Ignorability in general incomplete-data models, *Biometrika*, **81**, 701-708.

Henderson, C. R. (1953) Estimation of variance and variance components, *Biometrics*, **9**, 226-252.

Henderson, C. R. (1963) Selection index and expected genetic advance, *Statistical Genetics in Plant Breeding*, National Academy of Sciences, National Research Council Publication #982.

Hinkley, D. V. and Runger, G. (1984) The analysis of transformed data, *Journal of the American Statistical Association*, **79**, 302–320.

Hocking, R. R. (1985) *The Analysis of Linear Models*, Monterey, CA: Brooks Cole.

Hotelling, H. (1951) A generalized *t* test and measure of multivariate dispersion, *Proceedings of the Second Berkeley Symposium on Mathematical Statistics and Probability*, Berkeley, CA: University of California Press, pp. 23–41.

Huynh, H. and Feldt, L. S. (1976) Estimation of the Box correction for degrees of freedom from sample data in randomized block and split-plot designs, *Journal of Educational Statistics,* **1**, 69–82.

Imhof, J. P. (1961) Computing the distribution of quadratic forms in normal variables, *Biometrika*, **48**, 419–426.

Jackson, G. M. (1991) *A User's Guide to Principal Components*, Chichester, UK: Wiley.

Jayakar, A. D. (1970) On the detection and estimation of linkage between a locus influencing a quantitative character and a marker locus, *Biometrics*, **26**, 451–464.

Jennrich, R. I. and Schluchter, M. D. (1985) Unbalanced repeated-measures models with structured covariance matrices, *Biometrics*, **42**, 805–820.

Jiroutek, M. R., Muller, K. E., Kupper, L. L. and Stewart, P. W. (2003) A new method for choosing sample size for confidence interval-based inferences, *Biometrics*, **59**, 580–590.

John, S. (1972) The distribution of a statistic used for testing sphericity of normal distributions, *Biometrika*, **59**, 169–173.

Johnson, N. L. and Kotz, S. (1970) *Continuous Univariate Distributions,* Vol. 2, Boston: Houghton Mifflin.

Johnson, N. L. and Kotz, S. (1972) *Distributions in Statistics: Continuous Multivariate Distributions*, New York: Wiley.

Johnson, N. L., Kotz, S. and Balakrishnan, N. (1994) *Continuous Univariate Distributions*, Vol. 1, 2nd ed., New York: Wiley.

Johnson, N. L., Kotz, S. and Balakrishnan, N. (1995) *Continuous Univariate Distributions*, Vol. 2, 2nd ed., New York: Wiley.

Johnson, N. L., Kotz, S. and Balakrishnan, N. (1997) *Discrete Multivariate Distributions*, New York: Wiley.

Johnson, N. L., Kotz, S. and Balakrishnan, N. (1995) *Continuous Univariate Distributions*, Vol. 1, 2nd ed., New York: Wiley.

Jolliffe, I. T. (2002) *Principal Components Analysis*, 2nd ed., New York: Springer-Verlag.

Jöreskog, K. G. (1993) *Applied Factor Analysis in the Natural Sciences*, Cambridge: Cambridge University Press.

Kackar, R. N. and Harville, D. A. (1984) Approximations for standard errors of estimators of fixed and random effects in mixed linear models, *Journal of the American Statistical Association*, **79**, 853–862.

Kendall, M. G. and Stuart, A. (1977) *The Advanced Theory of Statistics*, Vol. 1, 4th ed., London: Charles Griffin and Co.

Kenward, M. G. and Roger, J. H. (1997) Small sample inference for fixed effects from restricted maximum likelihood, *Biometrics*, **53**, 983–997.

Khatri, C. G. (1966) A note on a MANOVA model applied to problems in growth curves, *Annals of the Institute of Statistical Mathematics*, **18**, 75-86.

Khuri, A. I., Mathew, T. and Sinha B. K. (1998) *Statistical Tests for Mixed Linear Models*, New York: Wiley.

Kim, H., Gribbin, M. J., Muller, K. E. and Taylor, D. J. (2006) Analytic, computational and approximate forms for ratios of noncentral and central Gaussian quadratic forms, *Journal of Computational and Graphical Statistics*, **15**, 443-459.

Kirk, R. E. (1995) *Experimental Design: Procedures for the Behavioral Sciences*, , 3rd ed., Monterey, CA: Brooks Cole.

Kleinbaum, D. G., Kupper, L. L., Muller, K. E., and Nizam, A. (1998) *Applied Regression Analysis and Other Multivariable Methods,* 3rd ed., Boston: Duxbury Press.

Kotz, S., Balakrishnan, N. and Johnson, N. L., (2000) *Continuous Multivariate Distributions*, Vol. 1, 2nd ed., New York: Wiley.

Kotz, S. and Nadarajah, S. (2004) *Multivariate t Distributions and Their Applications*, Cambridge: Cambridge University Press.

Kshirsagar, A. M., and Smith, W. B. (1995) *Growth Curves*, New York: Marcel Dekker.

Kuhfeld, W. F. (1986) A note on Roy's largest root, *Psychometrika*, **51**, 479–481.

Kvålseth, T. O. (1985) Cautionary note about R^2, *The American Statistician*, **39**, 279–285.

Laird, N. M. and Ware, J. (1982) Random-effects models for longitudinal data, *Biometrics*, **38**, 963–974.

Lancaster, P. (1969) *Theory of Matrices*, New York: Academic Press.

Lawley, D. N. (1938) A generalization of Fisher's z test, *Biometrika*, **30**, 180–187.

Lenth, R. V. (2001) Some practical guidelines for effective sample size determination, *American Statistician*, **55**, 187–193.

Lindgren, B. W. (1976) *Statistical Theory*, 3rd ed., New York: Macmillan.

Lindstrom, M. J. and Bates, D. F. (1988) Newton-Raphson and EM algorithms for linear mixed effects models for repeated measures data, *Journal of the American Statistical Association*, **83**, 1014–1022. Corrigendum, 1994, **89**, 1572.

Lipsitz, S. R., Garrett I., Fitzmaurice, M., Ibrahim, J. G., Gelber, R. and Lipshultz, S. (2002) Parameter estimation in longitudinal studies with outcome-dependent follow-up, *Biometrics*, **58**, 621–630.

Littel, R. (2003) Analysis of unbalanced mixed models: a case study comparison of ANOVA vs REML/GLS, *Journal of Agricultural, Biological, and Environmental Statistics*, **7**, 472–490.

Little, R. J. A. (1992), Regression with missing X's: a review, *Journal of the American Statistical Association*, **87**, 1227–1237.

Little, R. J. A. and Rubin, D. B. (2002) *Statistical Analysis with Missing Data*, 2nd ed., New York: Wiley.

Loperfido, N. (1999) A conditional normality test based on the conditional probability integral transformation, *Statistica Applicata*, **11**, 281–291.

Loynes, R. M. (1966) On idempotent matrices, *Annals of Mathematical Statistics*, **37**, 491–494.

Lukacs, E. (1983) *Developments in Characteristic Function Theory*, London: Charles Griffin and Co.

Mecklin, C. J. and Mundfrom, D. J. (2005) A Monte Carlo comparison of the type I and type II error rates of test of multivariate normality, *Journal of Statistical Computation and Simulation*, **75**, 93–107.

Magnus, J. R. (1978) Maximum likelihood estimation of the GLS model with unknown parameters in the disturbance covariance matrix, *Journal of Econometrics*, **7**, 281–312.

Magnus, J. R. and Neudecker, H. (1988) *Matrix Differential Calculus With Applications in Statistics and Econometrics*, New York: Wiley.

Malkovich, J. F. and Afifi, A. A. (1973) On tests for multivariate normality, *Journal of the American Statistical Association*, **68**, 176–179.

Mardia, K. V. (1970) Measures of multivariate skewness and kurtosis with applications, *Biometrika*, **57**, 519–530.

Mardia, K. V. (1975) [Algorithm AS 84] Measures of multivariate skewness and kurtosis, *Applied Statistics*, **24**, 262–265.

Mardia, K. V., Kent, J. T. and Bibby, J. M. (1979) *Multivariate Analysis*, London: Academic Press.

McCullagh, P. and Nelder, J. A. (1989) *Generalized Linear Models*, 2nd ed., London: Chapman & Hall.

McDonald, R. P. (1985) *Factor Analysis and Related Methods*, NJ: Lawrence Erlbaum.

McKeon, J. J. (1974) F approximations to the distribution of Hotelling's T_0^2, *Biometrika*, **61**, 381–383.

Mecklin, C. J. and Mundfrom, D. J. (2005) A Monte Carlo comparison of the type I and type II error rates of tests of multivariate normality, *Journal of Statistical Computation and Simulation*, **75**, 93-107.

Meng, X. and Rubin, D. B. (1991) Using EM to obtain asymptotic variance-covariance matrices: the SEM algorithm, *Journal of the American Statistical Association*, **86**, 899–909.

Mitra, S. K. (1970) A density-free approach to the matrix variate beta distribution, *Sankhya A*, **32**, 81–88.

Morrison, D. F. (1990) *Multivariate Statistical Methods*, 3rd ed., New York: McGraw-Hill.

Muirhead, R. J. (1984) *Aspects of Multivariate Statistical Theory*, New York: Wiley.

Muller, K. E. (1982) Understanding canonical correlation through the general linear model and principal components, *American Statistician*, **36**, 342-354.

Muller, K. E. (1998) A new F approximation for the Pillai-Bartlett trace under H_0, *Journal of Computational and Graphical Statistics*, **7**, 131–137.

Muller, K. E. and Barton, C. N. (1989) Approximate power for repeated-measures ANOVA lacking sphericity, *Journal of the American Statistical Association*, **84**, 549–555. Corrigenda, 1991, **86**, 255–256.

Muller, K. E., Barton, C. N., and Benignus, V.A. (1984) Recommendations for appropriate statistical practice in toxicologic experiments, *Neurotoxicology*, **5**, 113-126.

Muller, K. E. and Benignus, V. A. (1992) Increasing scientific power with statistical power, *Neurotoxicology and Teratology*, **14**, 211–219.

Muller, K. E. and Chi, Y. (2006) Extending the central Wishart to high dimension, low sample size, with implications for simulations, *manuscript in review*.

Muller, K. E., Edwards, L. J., Simpson, S. and Taylor D. T. (2006) Accurate test size and power in small samples of repeated measures without missing data, *manuscript in review*.

Muller, K. E. and Fetterman, B. A. (2002) *Regression and ANOVA: An Integrated Approach Using SAS® Software*, Cary, NC: SAS Institute.

Muller, K. E., LaVange, L. M., Ramey, S. L. and Ramey, C. T. (1992) Power calculations for general linear multivariate models including repeated measures applications, *Journal of the American Statistical Association*, **87**, 1209–1226.

Muller, K. E., and Pasour, V. B. (1997) Bias in linear model power and sample size due to estimating variance, *Communications in Statistics A*, **26**, 839–851.

Muller, K. E. and Peterson, B. L. (1984) Practical methods for computing power in testing the multivariate general linear hypothesis, *Computational Statistics and Data Analysis*, **2**, 143–158.

O'Brien, R. G. (1979) A general ANOVA method for robust tests of additive models for variances, *Journal of the American Statistical Association*, **74**, 877–880.

O'Brien, R. G. and Muller, K. E. (1993) A unified approach to statistical power for *t*-tests to multivariate models. In *Applied Analysis of Variance in Behavioral Sciences*, Chapter 8, pp. 297–344, L. K. Edwards Ed., New York: Marcel Dekker.

Odell, P. L. and Feiveson, A. H. (1966) A numerical procedure to generate a sample covariance matrix, *Journal of the American Statistical Association*, **61**, 199–203. Corrigendum, 1249–1250, with acknowledgment of priority to Wijsman, 1959.

Olson, C. L. (1974) Comparative robustness of six tests in multivariate analysis of variance, *Journal of the American Statistical Association*, **69**, 894–908.

Olson, C. L. (1976) On choosing a test statistic in multivariate analysis, *Psychological Bulletin*, **83**, 579–586.

Olson, C. L. (1979) Practical considerations in choosing a MANOVA test statistic: a rejoinder to Stevens, *Psychological Bulletin*, **86**, 1350–1352.

Pan, W. K. (2003) *Multilevel Spatial and Statistical Analyses to Examine the Relationship between Population and Environment: A Case Study of the Ecuadorian Amazon*, Dissertation, Department of Biostatistics, University of North Carolina, Chapel Hill.

Park, T., Park, J. K. and Davis, C. S. (2001) Effects of covariance model assumptions on hypothesis tests for repeated measurements: analysis of ovarian hormone data and pituitary-pteryomaxillary distance data, *Statistics in Medicine*, **20**, 2441–2453.

Patterson, H. D. and Thompson, R. (1971) Recovery of inter-block information when block sizes are unequal, *Biometrika*, **58**, 545–554.

Pillai, K. C. S. (1955) Some new test criteria in multivariate analysis, *Annals of Mathematical Statistics*, **26**, 117–121.

Pillai, K. C. S. (1956) On the distribution of the largest or the smallest root of a matrix in multivariate analysis, *Biometrika*, **43**, 122–127.

Pizer, S. M., Chen, J. Z., Fletcher, P. T., Friedman, Y., Fritsch, D. S., Gash, A. G., Glotzer, J. M., Jiroutek, M. R., Joshi, S., Lu. C., Muller, K. E., Thall, A., Tracton, G., Yushkevich, P. and Chaney, E. L. (2003) Deformable m-reps for 3D medical image segmentation, *International Journal of Computer Vision*, **55**, 85–106.

Potthoff, R. F. and Roy, S. N. (1964) A generalized multivariate analysis of variance model useful especially for growth curve problems, *Biometrics*, **51**, 313–326.

Rao, C. R. (1951) An asymptotic expansion of the distribution of Wilks' criterion, *Bulletin of the Institute of International Statistics*, **XXXIII**, 177–180.

Rao, C. R. (1965) The theory of least squares when the parameters are stochastic and its application to the analysis of growth curves, *Biometrika*, **52**, 447–468.

Rao, C. R. (1973) *Linear Statistical Inference and Its Applications*, 2nd ed., New York: Wiley.

Raudenbush, S. W. and Bryk, A. S. (2002) *Hierarchical Linear Models: Applications and Data Analysis Methods*, Newbury Park, CA: Sage Publications.

Rocke, D. M. (1989) Bootstrap Bartlett adjustment in seemingly unrelated regression, *Journal of the American Statistical Association*, **84**, 598–601.

Rousseeuw, P. J. and Van Zomeren, B. C. (1990) Unmasking multivariate outliers and leverage points, *Journal of the American Statistical Association*, **85**, 633–639. Also see comments and rejoinder, 640–651.

Roy, S. N. (1957) *Some Aspects of Multivariate Analysis*, New York: Wiley.

Rubin, D. R. (1976) Inference and missing data, *Biometrika*, **63**, 581-590.

Sampson, A. R. (1974) A tale of two regressions, *Journal of the American Statistical Association*, **69**, 682–689.

SAS Institute (1999) *SAS/IML User's Guide, Version 8,* Cary, NC: SAS Institute, Inc.

Satterthwaite, F. E. (1946) An approximate distribution of estimates of variance components, *Biometrics Bulletin*, **2**, 110–114.

Schaalje, G. B., McBride J. B. and Fellingham, G. W. (2003) Adequacy of approximations to distributions of test statistics in complex mixed linear models, *Journal of Agricultural, Biological, and Environmental Statistics*, **7**, 512–524.

Schott, J. R. (2005) *Matrix Analysis for Statistics*, 2nd ed. New York: Wiley.

Scott, A. and Wild, C. (1991) Transformations and R^2, *The American Statistician*, **45**, 127–129.

Searle, S. R. (1971) *Linear Models*, New York: Wiley.

Searle, S. R. (1982) *Matrix Algebra Useful for Statistics*, New York: Wiley.

Searle, S. R., Casella, G. and Mcculloch, C. M. (1992) *Variance Components*, New York: Wiley.

Shanbhag, D. N. (1966) On the independence of quadratic forms, *Journal of the Royal Statistical Society B*, **28**, 582–583.

Siegel, A. F. (1979) The noncentral chi-squared distribution with zero degrees of freedom and testing for uniformity, *Biometrika*, **66**, 381–386.

Small, N. J. H. (1980) Marginal skewness and kurtosis in testing multivariate normality, *Applied Statistics*, **29**, 85–87.

Small, N. J. H. (1988) In *The Encyclopedia of Statistical Sciences*, Vol. 9, pp. 95–100, N. L. Johnson and S. Kotz, Eds., New York: Wiley.

Smith, P. L. (1979) Splines as a useful and convenient statistical tool, *The American Statistician*, **33**, 57–62.

Soller, M. and Genizi, A. (1978) The efficiency of experimental designs for the detection of linkage between a marker locus and a locus affecting a quantitative trait in segregating populations, *Biometrics*, **34**, 37–55.

Srivastava, V. K., and Giles, D. E. A. (1987) *Seemingly Unrelated Regression Equation Models: Estimation and Inference*, New York: Marcel Dekker.

Stevens, S. S. (1946) On the theory of scales of measurement, *Science*, **103**, 677–680.

Stevens, S. S. (1951) Mathematics, measurement, and psychophysics. In *Handbook of Experimental Psychology*, S. S. Stevens, ed., New York: Wiley.

Stewart, P. W. (1987) Line segment confidence limits, *Biometrics*, **43**, 629–640.

Stewart, P. W. (1991) The graphical advantages of finite interval confidence band procedures, *Communications in Statistics A*, **20**, 3975–3994.

Stewart, P. W. (2000) Estimation and comparison of growth and dose-response curves in the presence of purposeful censoring. In *Handbook of Statistics, Vol. 18:*

Bioenvironmental and Public Health Statistics, P. K. Sen, P. K. and C. R. Rao, Eds. Ch.10, pp. 325–354, Amsterdam and New York: North-Holland/Elsevier.

Stroup, W. W. (2003) Power analysis based on spatial effects mixed models: a tool for comparing design and analysis strategies in the presence of spatial variability, *Journal of Agricultural, Biological, and Environmental Statistics*, **7**, 491–511.

Taylor, D. J. and Muller, K. E. (1995) Computing confidence bounds for power and sample size of the general linear univariate model, *American Statistician*, **49**, 43–47.

Taylor, D. J. and Muller, K. E. (1996) Bias in linear model power and sample size calculations due to estimating noncentrality, *Communications in Statistics A*, **25**, 1595–1610.

Thode, H. C. Jr. (2002) *Testing for Normality*, New York: Marcel Dekker.

Timm, N. H. (1975) *Multivariate Analysis with Applications in Education and Psychology*, Monterey, CA: Brooks/Cole.

Timm, N. H. (2002) *Applied Multivariate Analysis*, New York: Springer-Verlag.

Velleman, P. F. and Wilkinson, L. (1993) Nominal, ordinal, interval, and ratio typologies are misleading, *American Statistician*, **47**, 65–72.

Verbeke, G. and Molenberghs, G. (2000) *Linear Mixed Models for Longitudinal Data*, New York: Springer.

Vonesh, E. F, and Chinchilli, V. M. (1997) *Linear and Nonlinear Models for the Analysis of Repeated Measurements*, New York: Marcel Dekker.

Wackerly, D. D., Mendenhall, W., and Scheaffer, R. L. (1996) *Mathematical Statistics with Applications*, Belmont, CA: Duxbury Press, 5th ed.

Weisstein, E. W. (2003) *CRC Concise Encyclopedia of Mathematics*, 2nd ed., Boca Raton, FL: Chapman & Hall/CRC.

Widaman, K. F. (1993) Common factor analysis versus principal component analysis: differential bias in representing model parameters?, *Multivariate Behavioral Research*, **28**, 263–311.

Widaman, K. F. (2004) Common factors vs. components: principals and principles, errors and misconceptions, presentation at *Factor Analysis at 100: Historical Developments and Future Directions*, conference at the University of North Carolina, Chapel Hill, May 2004.

Wijsman, R. A. (1959) Applications of a certain representation of the Wishart matrix, *Annals of Mathematical Statistics*, **30**, 597–601.

Willet, J. B. and Singer, J. D. (1988) Another cautionary note about R^2: its use in weighted least-squares regression analysis, *The American Statistician*, **42**, 236–238.

Wishart, J. (1928) The generalized product moment distribution in samples from a normal multivariate population, *Biometrika*, **20A**, 32–52.

Wittes, J. and Brittain, E. (1990) The role of internal pilot studies in increasing the efficiency of clinical trials, *Statistics in Medicine*, **9**, 65–72.

Wolfinger, R., Tobias, R. and Sall, J. (1994) Computing Gaussian likelihoods and their derivatives for general linear mixed models, *SIAM Journal on Scientific and Statistical Computing*, **15**, 1294–1310.

Zucker, D. M. and Denne, J. (2002) Sample size redetermination for repeated measures studies, *Biometrics*, **58**, 548–559.

Zucker, D. M., Lieberman, O. and Manor, O. (2000) Improved small sample inference in the mixed linear model: Bartlett correction and adjusted likelihood, *Journal of the Royal Statistical Society B*, **62**, 827–838.

Index

A
Acceptance region, 51
Added last tests, 356–360
Added-in-order tests, 356–360
Algebraic multiplicity, 20
Approximate weighted least squares (AWLS), 82
Associative laws for matrix algebra, 9
Autoregressive covariance matrix, 38

B
Balanced design, 105
Balanced random coefficient model, 283
Basis, 11
Best linear unbiased estimator (BLUE), 222, 227
Best linear unbiased predictor (BLUP), 287
Bickel-Doksum transformation, 128
Bilinear form, 10
Biorthogonal matrices, 10
Block diagonal matrix, 3
Boolean algebra, 44
Bonferroni correction, 67, 77, 107–108
Box-Cox power transformation, 128
Breakdown point, 162

C
Canonical correlation, 64
Casewise missing, 74-75
Cauchy-Schwartz inequality, 135
Characteristic function (CF)
  Vector, 119
  Matrix, 138
Chi-square distribution, 169-170
Cholesky factor matrix, 25, 72
Coefficient of determination, 47
Column, extracting from matrix, 4
Commensurate, 101
Common factor model, 332
Commutative laws for matrix algebra, 9
Complete design, 105
Composite acceptance region, 301
Composite alternative, 301
Component hypothesis, 301
Composite hypothesis, 300
Compound symmetric covariance matrix, 37–38
Concatenating matrices, 4
Conditional distribution, 132
Confidence
  Band, finite interval, 340
  Coefficient, 52
  Intervals, 52
  Regions, 52, 73, 306, 332
Conform, conformation of matrices, 5
Congruent matrices, 21
Consistent system of equations, 17

Constituent matrix decomposition, 23
Continuous
  Data, 103
  Random variable, including absolutely, 117
Correlation matrix, 33, 135
Covariance design matrix, 92
Covariance (dispersion) matrix, 8, 25, 33, 134
  sample value, 34
  For Gaussian data, also see *Wishart*
Critical region (rejection region), 51
Critical value, 51
Cross product, 6
Cumulant, generating function (CGF), 124, 125
Cumulative distribution function (CDF), 116

D
Data matrix, 34
Determinant, 14–15
Diagonal of matrix, 2
Diagonal matrix, 2, 14, 22
Direct product (matrix multiplication), 7
Direct-product (matrix)
  Gaussian distribution, 156, 157, 349
Direct sum, 7
Discrete Fourier transform, 273
Discrete random variable, 117
Distributions, 116, 117
Distributive laws for matrix algebra, 9
Dot product, 6
Doubly Multivariate outcomes, 101

E
Eigenanalysis, 18, 19, 20, 21
Eigenvalue, 19
Eigenvector, 20
Error sum of squares, 49, 65
Essence matrix, 218
Estimable, 212, 249
Exact weighted least squares, 238
Exchangeable observations, 105
Excludes an intercept, 46
Expected value, 132

F
Factor matrix, 24
Factor analysis, 332
Fisher Scoring algorithm, 284
Full rank, 12–15

G
Gaussian (nonsingular or singular, Standard or not, central or noncentral)
  Scalar, 142
  Vector (multivariate), 143–144
  General matrix, 156
  Direct-product matrix, 156
General linear (null) hypothesis, 60
General linear mixed model, 92
  With Gaussian errors, 96
General linear multivariate model, 56
  With Gaussian errors, 59
Generalized inverse matrix, 14, 16–17, 23, 25
Geometric multiplicity, 20
Generalized general linear model (GGLM), 80–81
  linearly equivalent, 238
General linear model (GLM), univariate
  Analysis of Variance (ANOVA) models, 42
  Definition, 40
  Coefficient of determination, 47, 48
  Concepts, 39
  Estimated errors, 65
  Estimation, 39
  Estimators, 209, 246

Explicit and implicit restrictions, 40, 231
Full rank, 40, 56
Inference, 39
Least squares assumptions, 41
Less than full rank, 40, 224
Noncentrality parameter, 44
Numerical methods, 39
One-to-one linear transformations, 267
Predicted values, 65
Primary parameters, 40, 209
Secondary parameters, 43, 209
With Gaussian errors, 41
General linear model (GLM), multivariate
Estimators, 246
Primary parameters, 59, 246
Secondary parameters, 59, 246
Growth curve model (GCM), 84–86, 89, 266–277

H

Hat matrix, 49, 65
Horizontal direct product, 6
Horizontally concatenated, 4
Hotelling-Lawley trace statistic, 61, 67, 319
Hypothesis
A priori parameter, 289
Full and constrained model, 45
General linear hypothesis (GLH), 43–44, 46, 48, 59, 289
Post hoc parameters, 289
Sum of squares, 49, 65
Testability, 290–295
Tests, multivariate, 66-73, Chapter 16
Tests, univariate, 50-51, Chapter 15

I

Idempotent, 10
Identity matrix, 3
Ignorable (missing data), 76
Independence
Mutual (total), 130
Observation, 101
Pairwise, 130
Statistical, 130
Independent sampling unit (ISU), 101
Inner product, 6, 10
Interval scale, 103
Inverse (matrix operator), 15
Inverse Wishart, 205
Iterated approximate weighted least squares (ITAWLS), 82, 263

J

Joint cumulative distribution function, 117

K

Knots, see *Spline*
Kurtosis, 164
Kronecker covariance, 245
Kronecker design, 245
Kronecker (direct) product, 7

L

Less than full rank, 12, 15, 48, 95
Least squares estimator, 224
Linear covariance structure, 282
Linear mixed models tests, 341–343
Linear transformation, including full rank (nonsingular) and less than full rank (singular), 126
Linearly equivalent, 210, 246, 269
Linearly dependent, 11
Linearly independent, 11
Linear (scale) invariance, 305, 317
Location invariance, 305, 317
Lower triangular matrix, 3

M
Mahalanobis distance, 62, 162
Marginal distribution, 128–129
Masking, 162
Matrix addition and subtraction, 5
Matrix, defined, 1
Mean
  Population, 32, 132
  Sample, 34
  Also see *moments*
Missing
  Data, 75
  At random (MAR), 76
  Completely at random (MCAR), 75
Model notation, summary, 111
Moment generating function (MGF)
  Vector, 123
  Matrix, 138
Moments, 133
Moore-Penrose generalized inverse, 16
Multiplication
  Direct product of two matrices, 7
  Elementwise for two matrices, 5
  Matrix with matrix, 2, 6
  Horizontal direct product, 6
  Scalar with matrix, 5
Multivariate analysis of variance (MANOVA), 107
Multivariate association, 70
Multivariate outcomes, 101
Multivariate quadratic form, 193
Multivariate tests, including multivariate approach to repeated measures (MULTIREP), 59, 61, 70, 318

N
Negative definite, 22
Negative semidefinite, 22
Nominal scale, 103
Noncentral chi-square distribution, 170
Noncentral Wishart distribution, 350
Noncentrality parameter, 44, 59
Nonnegative definite, 22
Nonpositive definite, 22
Nonsingular matrix, 15
Nonsingular Gaussian vector, 144, 146
Nonsymmetric square matrices, 12
Nonzero eigenvalues, 21
Null hypothesis, 44

O
Observational unit, 101
Operations in matrix algebra, rules for, 8–9
Ordinal scale, 103
Ordinary least squares (OLS), 263
Orthogonal matrix, 10
Orthonormal matrix, 10
Outer product, 6, 10

P
Partitioned matrix, 3, 4, 30, 33
Permutation matrix, 30
Pillai-Bartlett trace, ANOVA analog statistic, 67, 319
Polynomial growth curve, 269
Positive definite, 22
Positive semidefinite, 22
Power of a test, 51
Power function, 51
Predicted values, 49, 65
Principal component, 36
Projection, projection matrix, 27–28
Probability density function (PDF), 116, 118, 129
Projection matrix, 28

Q
Quadratic form
  Matrix expression, 10
  Random, 174

univariate Gaussian, 174
multivariate Gaussian, 193

R

Random deviation, 94
Random vector, 143
Rank of a matrix, defined, 12
Rank of a vector space, 12
Ratio scale, 103
Rejection region (critical region), 51
Repeated measures, 101
Residuals, 49, 65
Restricted maximum likelihood (REML), 285
Restricted linear model (explicit or implicit), 231
Row, extracting from matrix, 4
Roy's largest root, the union-intersection principle statistic, 66, 319
Robust distance, 162

S

Sample mean (vector), 34
Sample SSCP matrix, 34
Sample covariance matrix, 34
Scalar multiplication of a matrix, 5
Scalar, defined, 1
Scale (linear) invariance, 305, 317
Secondary parameter, 43, 209, 246
Semidefinite, 22
Seemingly unrelated regressions, 83
Shift parameter, 44, 59
Similar matrices, 21
Simple matrix, 20
Singular matrix, 15
Singular value decomposition, 13, 26
Singular Gaussian vector, 144
Size of a test, 51
Skewness, 164
Spans an intercept, 46
Spectral decomposition, 13, 36
Spectrum (of a matrix), 19
Sphericity, spherical distribution, 37
Spline, 235
Square matrix, 2
Squared error loss function, 223
Stacked by column, 4
Standard Gaussian: see Gaussian
Stieltjes integral, 119
Subpopulation mean, 94
Sums of squares and cross products (SSCP) matrix, 34
Sylvester's Law of Inertia, 21
Symmetric matrix, 2

T

Testable, 290, 312
Testability of hypothesis, 312, 316
Test statistic, 51
Trace, matrix, 5
Transpose, 2
Translation, 126
Triangular matrix (upper, lower), 3
Type I error, 51
Type II error, 51

U

Uncorrelated, 136
Uniformly minimum variance estimator, (UMVUE), 222, 227
Union-intersection test, 301
Univariate approach to repeated measures (UNIREP) tests, 61, 67
Unstructured covariance matrix, 37
Upper triangular matrix, 3

V

Variance, 134
Vector space, defined, 1, 11

Vertically concatenated, 4

W

Weighted least squares (WLS), 82
Wilks lambda, likelihood test, 67, 319
Wishart distribution, 193, 350

Z

Zero matrix, 3